Control System Design

Control System Design

Graham C. Goodwin
Centre for Integrated Dynamics and Control
University of Newcastle
Newcastle, Australia

Stefan F. Graebe
OMV Aktiengesellschaft
Department of Optimization/Automation
Vienna, Austria

Mario E. Salgado
Departamento de Electrónica
Universidad Técnica Federico Santa María
Valparaíso, Chile

Prentice Hall
Upper Saddle River, New Jersey 07458

Library of Congress Cataloging-in-Publication Data

Goodwin, Graham C. (Graham Clifford)
     Control system design / Graham C. Goodwin, Stefan F. Graebe, Mario E. Salgado.
      p.cm.
     Includes bibliographical references and index.
     ISBN 0-13-958653-9
     1. Automatic control. 2. System design. I. Graebe, Stefan F. II. Salgado, Mario E. III. Title.

     629.8--dc21   00-042744

Vice president and editorial director of ECS: MARCIA HORTON
Acquistions editor: ERIC FRANK
Publisher: TOM ROBBINS
Production editor: IRWIN ZUCKER
Executive managing editor: VINCE O'BRIEN
Managing editor: DAVID A. GEORGE
Manufacturing buyer: PAT BROWN
Manufacturing manager: TRUDY PISCIOTTI
Vice president and director of production and manufacturing, ESM: DAVID W. RICCARDI
Director of creative services: PAUL BELFANTI
Cover designer: BRUCE KENSALAAR
Editorial assistant: JENNIFER DIBLASI

© 2001 by Prentice Hall
Prentice-Hall, Inc.
Upper Saddle River, New Jersey 07458

Printed in the United States of America

10 9 8 7 6 5 4 3 2 1

ISBN 0-13-958653-9

Prentice-Hall International (UK) Limited, London
Prentice-Hall of Australia Pty. Limited, Sydney
Prentice-Hall Canada Inc., Toronto
Prentice-Hall Hispanoamericana, S.A., Mexico
Prentice-Hall of India Private Limited, New Delhi
Prentice-Hall of Japan, Inc., Tokyo
Pearson Education Asia Pte. Ltd.
Editora Prentice-Hall do Brasil, Ltda., Rio de Janeiro

Dedicated, in thankful appreciation
for support and understanding, to

Rosslyn

Alice

Mariví

CONTENTS OVERVIEW

II   SISO CONTROL ESSENTIALS   **117**

APPENDICES (*These can be viewed on the accompanying CD-ROM or at: http://www.prenhall.com/goodwin.*)

A NOTATION, SYMBOLS, AND ACRONYMS

B SMITH–MCMILLAN FORMS

C RESULTS FROM ANALYTIC FUNCTION THEORY

ACKNOWLEDGEMENTS

The authors wish to thank the large number of colleagues and friends who have worked with us in the area of control over the years. This book is really a synthesis of ideas that they helped us to formulate. All three authors spent time together in the Centre for Industrial Control Science at the University of Newcastle, Australia. This was a fertile breeding ground for many discussions on the principles of control. Financial support from the Australian Government for this centre, under the Commonwealth Special Centres program, is gratefully acknowledged. Also, financial and other support was provided by the Universidad Técnica Federico Santa María, covering, amongst other things, several visits to Chile by the first author during the writing of this book. Many students and colleagues read drafts of the book over a five-year period. The authors accept full responsibility for the views expressed in the book (and all remaining errors). Nonetheless, they wish particularly to acknowledge suggestions from Thomas Brinsmead, Arthur Conley, Sam Crisafulli, Jose De Doná, Arie Feuer, Jaime Glaría, William Heath, Kazuo Komatsu, David Mayne, Tristan Perez, María Seron, Gustavo Vergara, Liuping Wang, and Steve Weller. The book was composed and typed by many people, including the authors; however, in the final stages of producing the book, Jayne Disney gave considerable help. Also, Tim Wylie and Adrian Bastiani kindly produced the Engineering Drawings shown in the text. The authors also wish to thank the staff of Prentice-Hall, especially Eric Frank, for their advice, guidance, and support during this project. The authors also gratefully acknowledge very valuable, and initially anonymous, feedback received from the following reviewers of the first draft: Kemin Zhou (Louisiana State University), Rick Johnson (Cornell University), J.B. Pearson (Rice University), Chaooki Abdallah (University of New Mexico), Steven Chin (The Catholic University of America), Andy Grace (The Mathworks, Inc.), Jim Freudenberg (University of Michigan), Bill Perkins (University of Illinois at Urbana-Champaign), and Hassan Khalil (Michigan State University). We trust that the final manuscript adequately reflects their very helpful and insightful suggestions.

PREFACE

Introduction to Control Engineering

Control Engineering plays a fundamental role in modern technological systems. The benefits of improved control in industry can be immense. They include improved product quality, reduced energy consumption, minimization of waste material, increased safety levels, and reduction of pollution. A difficulty with the subject, however, is that some of the more advanced aspects depend on a sophisticated mathematical background. Arguably, mathematical systems theory is one of the most significant achievements of twentieth-century science, but its practical impact is only as important as the benefits it can bring. Thus, we include in this book a strong emphasis on design, ultimately striking a balance between theory and practice.

It was the authors' involvement in several industrial control-system design projects that provided part of the motivation to write this book. In a typical industrial problem, we found ourselves investigating fluid and thermal dynamics, experiencing the detrimental effects of nonconstant PLC scan rates, dealing with system integration and network communication protocols, building trust with plant operators, and investigating safe bumpless transfer schemes for testing tentative control designs on potentially dangerous plants. In short, we experienced the day-to-day excitement, frustration, set-backs, and progress in getting advanced control to contribute to a commercial company's bottom line. This is not an easy task. Success in this type of venture typically depends on the application of a wide range of multidisciplinary skills; however, it is rewarding and exciting work for those who do it.

One of the main aims of this book is to share this excitement with our readers. We hope to contribute to the development of skills and attitudes within readers and students that will better equip them to face the challenges of real-world design problems. The book is thus intended to contribute to the ongoing reform of the Control Engineering curriculum. This topic continues to receive considerable international attention as educators strive to convey the excitement and importance of control engineering. Indeed, entire issues of the IEEE Control Systems Magazine have been devoted to this theme.

Reforming the curriculum will not, however, be done by books alone. It will be done by people: students, teachers, researchers, practitioners, publication and grant reviewers, and by market pressures. Moreover, for these efforts to be efficient and sustainable, the control engineering community will need to communicate their experiences via a host of new books, laboratories, simulations, and web-based resources. Thus, there will be a need for several different and complementary approaches. In this context, the authors believe that this book will have been successful if it contributes, in some way, to the revitalization of interest by students in the exciting discipline of control engineering.

We stress that this is not a *how-to book*. On the contrary, we provide a comprehensive, yet condensed, presentation of rigorous control engineering. We employ, and thus require, mathematics as a means to *model* the process, *analyze* its properties under feedback, *synthesize* a controller with particular properties, and arrive at a *design* addressing the inherent trade-offs and constraints applicable to the problem.

In particular, we believe that success in control projects depends on two key ingredients: (i) having a comprehensive understanding of the process itself, gained by studying the relevant physics, chemistry, and so on; and (ii) by having mastery of the fundamental concepts of signals, systems, and feedback. The first ingredient typically occupies more than fifty per cent of the effort. It is an inescapable component of the complete design cycle; however, it is impractical for us to give full details of the processes to which control might be applied, because they cover chemical plants, electromechanical systems, robots, power generators, and so on. We thus emphasize the fundamental control engineering aspects that are common to all applications and we leave readers to complement this emphasis with process knowledge relevant to their particular problem. Thus, the book is principally aimed at the second ingredient of control engineering. Of course, we do give details of several real-world examples, so as to put the methods into a proper context.

The central theme of this book is continuous-time control; however, we also treat digital control in detail, because most modern control systems will usually be implemented on some form of computer hardware. This approach inevitably led to a book of larger volume than originally intended, but one with the advantage of providing a comprehensive treatment within an integrated framework. Naturally, there remain specialized topics that are not covered in the book; however, we trust that we provide a sufficiently strong foundation so that the reader can comfortably turn to the study of appropriate complementary literature.

Goals

Thus, in writing this book we chose as our principal goals the following:

- providing accessible treatment of rigorous material selected with applicability in mind;

- giving early emphasis to design, including methods for dealing with fundamental trade-offs and constraints;

- providing additional motivation through substantial interactive web-based support; and

- demonstrating the relevance of the material through numerous industrial case studies.

Indeed, the material in the book is illustrated by numerous industrial case studies with which the authors have had direct involvement. Most of these case studies were carried out, in collaboration with industry, by the *Centre for Integrated Dynamics and Control* (CIDAC) (a Commonwealth Special Research Centre) at the University of Newcastle.

The projects that we have chosen to describe include the following:

- satellite tracking

- pH control

- control of a continuous casting machine

- sugar mill control

- distillation column control

- ammonia-synthesis plant control

- zinc coating-mass estimation in a continuous-galvanizing line

- BISRA gauge for thickness control in rolling mills

- roll-eccentricity compensation in rolling mills

- hold-up effect in reversing rolling mills

- flatness control in steel rolling

- vibration control

Design is a complex process, one that requires judgment and iteration. The design problem normally is incompletely specified, sometimes is ill-defined, and many times is without solution. A key element in design is an understanding of those factors that limit the achievable performance. This naturally leads to a viewpoint of control design that takes account of these fundamental limitations. This viewpoint is a recurring theme throughout the book.

Our objective is not to explore the full depth of mathematical completeness but instead to give enough detail so that a reader can begin applying the ideas as soon as possible. This approach is connected to our assumption that readers will have ready access to modern computational facilities, including the software package MATLAB–SIMULINK. This assumption allows us to put the emphasis on fundamental ideas rather than on the tools. Every chapter includes worked examples and problems for the reader.

Overview of the Book

The book is divided into eight parts. A brief summary of each of the parts is given here.

Part I: The Elements

This part covers basic continuous-time signals and systems and would be suitable for an introductory course on this topic. Alternatively, it could be used to provide review material before starting the study of control in earnest.

Part II: SISO Control Essentials

This part deals with basic *single-input single-output* (SISO) control, including classical *proportional, integral and derivative* (PID) tuning. This section, together with Part I, covers the content of many of the existing curricula for basic control courses.

Part III: SISO Control Design

This part covers design issues in SISO Control. We consider many of these ideas to be crucial to achieving success in practical control problems. In particular, we believe that the chapter dealing with constraints should be mentioned, if at all possible, in all introductory courses. Also, feedforward and cascade structures, which are covered in this part, are very frequently employed in practice.

Part IV: Digital Computer Control

This part covers material essential to the understanding of digital control. We go beyond traditional treatments of this topic by studying inter-sample issues.

Part V: Advanced SISO Control

This part could be the basis of a second course on control at an undergraduate level. It is aimed at the introduction of ideas that flow through to *multi-input multi-output* (MIMO) systems later in the book.

Part VI: MIMO Control Essentials

This part gives the basics required for a junior-level graduate course on MIMO control. In particular, this part covers basic MIMO system theory. It also shows how one can exploit SISO methods in some MIMO design problems.

Part VII: MIMO Control Design

This part describes tools and ideas that can be used in industrial MIMO design. In particular, it includes *linear quadratic optimal control theory* and *optimal filtering*. These two topics have major significance in applications. We also include a chapter on Model Predictive Control. We believe this to be important material, because of the widespread use of this technique in industrial applications.

Part VIII: Advanced MIMO Control

This final part of the book could be left for private study. It is intended to test the reader's understanding of the other material by examining advanced issues. Alternatively, instructors could use this part to extend parts VI and VII in a more senior graduate course on MIMO Control.

Using this Book

This is a comprehensive book on control system design that can be used in many different course patterns. If one adopts the book for an early course on control, then the unused material is excellent reference material for later use in practice or for review. If one uses the book for a later course, then the early material gives an excellent summary of the basic building blocks on which the subject rests.

The book can be used for many different course patterns. Some suggested patterns are outlined as follows:

(i) Signals and Systems

This would be taught from Part I of the book.

(ii) Basic Control Theory

This would typically be taught for Part II of the book, together with some material for Part I (depending on the student's prior exposure to signals and systems) and some material from Part III. In particular, the chapter on design limitations (Chapter 8) requires only elementary knowledge of Laplace Transforms and gives students an understanding of those issues which limit achievable performance. This is an extremely important ingredient in all real-world control design problems. Also, Chapter 11 which deals with constraints is very important in practice. Finally, the ideas of feedforward and cascade architectures that are covered in Chapter 10 are central to solving real-world design problems.

(iii) Digital Control

This can be taught from Part IV. Indeed, we feel our treatment here is better focused on applications than many of the traditional treatments because of the emphasis we place on inter-sample behavior. In the various courses taught by the authors of this book some of the material on digital control is typically included in the Basic Control Theory Course. This is possible because the students are well prepared having taken a Signals and System course prior to the control course.

(iv) Second Course on Control

A second course on control typically includes an introduction to state space design, observers, and state-variable feedback. This material can be taught from Parts V to VII of the book. Part V is relatively straightforward and is intended to bridge the gap from single-input single-output systems (which are principally the focus of

Parts I to IV) and multi-input multi-output systems (which are principally covered in Parts VI, VII, and VIII). We consider Chapter 22 on optimal control and filtering to be very important and have included in this chapter many real world design case studies. Also, Chapter 23 on Model Predictive Control is important as this technique is widely used in industrial control.

Two of the authors (Goodwin and Salgado) have taught undergraduate and postgraduate courses of the type mentioned above, using draft versions of this book, in Australia and South America.

Website

We have created a comprehensive website to support the book. This website contains the following:

- Full Appendices (So that this material can be read at the same time as the printed text in the book.)

- Full Matlab Support (This can be downloaded and used to reproduce all of the designs in the book.)

- Interactive Java Laboratories (These illustrate the material in the book but can also be used for fun interaction.)

- Selected Solutions for Problems (This allows students to see how certain key problems can be solved. Of course instructors adopting the book will be sent a copy of the comprehensive solutions manual that covers *every* problem set in the book.)

- On-Line Forum (So that topics of general interest to control-system design can be raised and discussed.)

- An Errata Section (This is used to give details of any errors occurring in the book.)

- Extensive PowerPoint Slides (Approximately 2,500 slides are available for use with the book.)

We see the use of this material as follows:

For the Instructor
We believe that the Matlab support and PowerPoint slides should be particularly helpful to an instructor. For example, it would be possible to teach the course entirely using the resources provided. Also, we have found that students enjoy using the Virtual Laboratories. These can be displayed in the classroom as part of a lecture or given to students to enhance their understanding of the material.

For the Student

We believe that the PowerPoint slides are an excellent and easily understood summary of the book which by-passes all unnecessary technicalities. Even if your instructor does not use these slides in his/her presentations, we consider that they are an excellent summary for study purposes. If you print them out and annotate them, then remembering the material should be easy. Also, students should enjoy the Java Applets. If you can understand the case studies covered by these applets then you will be well on the way to understanding this exciting subject.

The website can be accessed at either of the following URLs:

`http://www.prenhall.com/goodwin`

`http://csd.newcastle.edu.au/control/`

Alternatively, see the authors' home pages for a link.

Also note that the website is under continuous development, so the resources provided will continue to grow and evolve as time proceeds.

Newcastle, Australia
Valparaíso, Chile
Vienna, Austria

Part I

THE ELEMENTS

PREVIEW

Designing and operating an automated process so that it maintains specifications on, for instances, profitability, quality, safety, and environmental impact requires a close interaction between experts from different disciplines. These include, for example, computer-, process-, mechanical-, instrumentation- and control-engineering experts.

Each of these disciplines views the process and its control from a different perspective, so each has adopted different categories, or elements, in terms of which they think about the automated system. The computer engineer, for example, would think in terms of computer hardware, network infrastructure, operating system and application software. The mechanical engineer would emphasize the mechanical components from which the process is assembled; the instrumentation engineer would think in terms of actuators, sensors, and their electrical wiring.

The control engineer, in turn, thinks of the elements of a control system in terms of such abstract elements as signals, systems, and dynamic responses. These elements can be further specified by their physical realization, the associated model, or their properties. (See Table 1.)

	Tangible examples	*Examples of mathematical approximation*	*Examples of properties*
Signals	set-point, control input, disturbances, measurements, ...	continuous function, sample-sequence, random process, ...	analytic, stochastic, sinusoidal, standard deviations
Systems	process, controller, sensors, actuators, ...	differential equations, difference equations, transfer functions, state space models, ...	continuous-time, sampled, linear, nonlinear, ...

Table 1. Systems and signals in a control loop

This book emphasizes the control engineer's perspective, one of process automation; however, the reader should bear the other perspectives in mind, because they form essential elements in a holistic view of the subject.

This first part of the book is the first stage of our journey into control engineering. It gives an introduction to the core elements of continuous-time signals and systems and describes the pivotal role of feedback in control-system design. These are the basic building blocks on which the remainder of the development rests.

Chapter 1

THE EXCITEMENT OF CONTROL ENGINEERING

1.1 Preview

This chapter is intended to provide motivation for studying control engineering. In particular, it covers

- an overview of the scope of control,

- historical periods in the development of control theory,

- types of control problems,

- introduction to system integration, and

- economic benefits analysis.

1.2 Motivation for Control Engineering

Feedback control has a long history, which began with the early desire of humans to harness the materials and forces of nature to their advantage. Early examples of control devices include clock-regulating systems and mechanisms for keeping windmills pointed into the wind.

A key step forward in the development of control occurred during the industrial revolution. At that time, machines were developed that greatly enhanced the capacity to turn raw materials into products of benefit to society. The associated machines, specifically steam engines, involved large amounts of power, and it was soon realized that this power needed to be *controlled* in an organized fashion if the systems were to operate safely and efficiently. A major development at this time was Watt's fly-ball governor. This device regulated the speed of a steam engine by throttling the flow of steam; see Figure 1.1. These devices remain in service to this day.

5

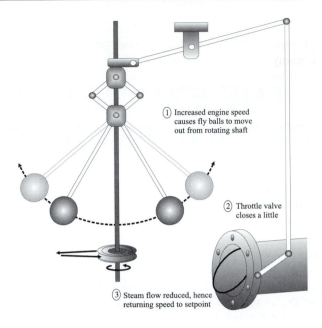

① Increased engine speed
causes fly balls to move
out from rotating shaft

② Throttle valve
closes a little

③ Steam flow reduced, hence
returning speed to setpoint

Figure 1.1. Watt's fly-ball governor

The World Wars also led to many developments in control engineering. Some of these were associated with guidance systems whilst others were connected with the enhanced manufacturing requirements necessitated by the war effort.

The push into space in the 1960's and 1970's also depended on control developments. These developments then flowed back into consumer goods, as well as into commercial, environmental, and medical applications. These applications of advanced control have continued at a rapid pace. To quote just one example from the author's direct experience, centre-line thickness control in rolling mills has been a major success story for the application of advanced control ideas. Indeed, the accuracy of centre-line thickness control has improved by two orders of magnitude over the past 50 years, thanks, in part, to enhanced control. For many companies, these developments were central not merely to increased profitability but even to remaining in business.

By the end of the twentieth century, control has become a ubiquitous (but largely unseen) element of modern society. Virtually every system we come in contact with is underpinned by sophisticated control systems. Examples range from simple household products (temperature regulation in air-conditioners, thermostats in hot-water heaters, etc.), to more sophisticated systems, such as the family car (which has hundreds of control loops), to large-scale systems (such as chemical plants, aircraft, and manufacturing processes). For example, Figure 1.2 on page 8 shows the process schematic of a Kellogg ammonia plant. There are about 400 of these plants around the world. An integrated chemical plant of the type shown in Figure 1.2 will typically have many hundreds of control loops. Indeed, for simplicity, we have not shown many of the utilities in Figure 1.2, yet these also have substantial numbers of control loops associated with them.

Many of these industrial controllers involve *cutting edge* technologies. For example, in the case of rolling mills (illustrated in Figure 1.3 on page 13), the control system involves forces on the order of 2,000 tonnes, speeds up to 120 km/hour, and (in the aluminum industry) tolerances of 5 micrometers or 1/500th of the thickness of a human hair! All of this is achieved with precision hardware, advanced computational tools, and sophisticated control algorithms.

Beyond these *industrial* examples, feedback regulatory mechanisms are central to the operation of biological systems, communication networks, national economies, and even human interactions. Indeed, if one thinks carefully, control in one form or another can be found in every aspect of life.

In this context, control engineering is concerned with designing, implementing, and maintaining these systems. As we shall see later, this is one of the most challenging and interesting areas of modern engineering. Indeed, to carry out control successfully, one needs to combine many disciplines, including modeling (to capture the underlying physics and chemistry of the process), sensor technology (to measure the status of the system), actuators (to apply corrective action to the system), communications (to transmit data), computing (to perform the complex task of changing measured data into appropriate actuator actions), and interfacing (to allow the multitude of different components in a control system to *talk* to each other in a seamless fashion).

Thus, control engineering is an exciting multidisciplinary subject with an enormously large range of practical applications. Moreover, interest in control is unlikely to diminish in the foreseeable future. On the contrary, it is likely to become ever more important, because of the increasing globalization of markets and environmental concerns.

1.2.1   Market Globalization Issues

Market globalization is increasingly occurring, and this situation means that, to stay in business, manufacturing industries are necessarily placing increasing emphasis on issues of quality and efficiency. Indeed, in today's society, few if any companies can afford to be second best. In turn, this focuses attention on the development of improved control systems, so that processes operate in the best possible way. In particular, improved control is a key enabling technology underpinning

- enhanced product quality,

- waste minimization,

- environmental protection,

- greater throughput for a given installed capacity,

- greater yield,

- deferring of costly plant upgrades, and

- higher safety margins.

All of these issues are relevant to the control of an integrated plant such as that shown in Figure 1.2.

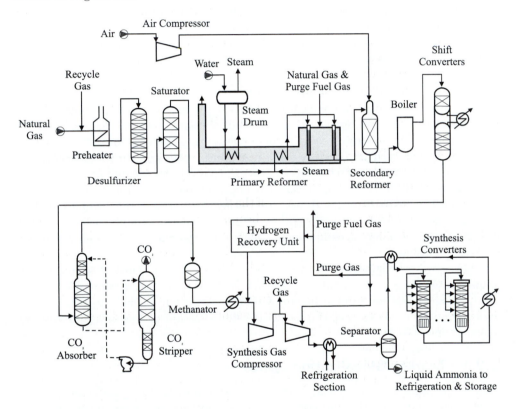

Figure 1.2. Process schematic of a Kellogg ammonia plant

1.2.2 Environmental Issues

All companies and governments are becoming increasingly aware of the need to achieve the benefits outlined above, whilst still respecting finite natural resources and preserving our fragile environment. Again, control engineering is a core enabling technology in reaching these goals. To quote one well-known example, the changes in legislation covering emissions from automobiles in California have led car manufacturers to significant changes in technology, including enhanced control strategies for internal combustion engines.

Thus, we see that control engineering is driven by major economic, political, and environmental forces. The rewards for those who can get all the factors right can be enormous.

1.3   Historical Periods of Control Theory

We have seen above that control engineering has taken several major steps forward at crucial events in history (e.g., the industrial revolution, the Second World War, the push into space, economic globalization, and shareholder-value thinking). Each of these steps has been matched by a corresponding burst of development in the underlying theory of control.

Early on, when the compelling concept of feedback was applied, engineers sometimes encountered unexpected results. These then became catalysts for rigorous analysis. For example, if we go back to Watt's fly-ball governor, it was found that, under certain circumstances, these systems could produce self-sustaining oscillations. Toward the end of the 19th century, several researchers (including Maxwell) showed how these oscillations could be described via the properties of ordinary differential equations.

The developments around the period of the Second World War were also matched by significant developments in Control Theory. For example, the pioneering work of Bode, Nyquist, Nichols, Evans, and others appeared at this time. This resulted in simple graphical means for analyzing single-input single-output feedback control problems. These methods are now generally known by the generic term *Classical Control Theory*.

The 1960's saw the development of an alternative *state space* approach to control. This followed the publication of work by Wiener, Kalman (and others) on optimal estimation and control. This work allowed multivariable problems to be treated in a unified fashion. This had been difficult, if not impossible, in the classical framework. This set of developments is loosely termed *Modern Control Theory*.

By the 1980's, these various approaches to control had reached a sophisticated level, and emphasis then shifted to other related issues, including the effect of model error on the performance of feedback controllers. This can be classified as the period of *Robust Control Theory*.

In parallel, there has been substantial work on nonlinear control problems. This has been motivated by the fact that many real-world control problems involve nonlinear effects.

There have been numerous other developments, including adaptive control, autotuning, and intelligent control. These are too numerous to detail here. Anyway, our purpose is not to give a comprehensive history but simply to give a flavor for the evolution of the field.

At the time of writing this book, control has become a mature discipline. It is thus possible to give a treatment of control which takes account of many different viewpoints and to unify these in a common framework. This is the approach we will adopt here.

1.4   Types of Control-System Design

Control-system design in practice requires cyclic effort, in which one iterates through modeling, design, simulation, testing, and implementation.

Control-system design also takes several different forms, and each requires a slightly different approach.

One factor that affects the form that the effort takes is whether the system is part of a predominantly commercial mission. Examples where this is not the case include research, education and missions such as landing the first man on the moon. Although cost is always a consideration, these types of control design are mainly dictated by technical, pedagogical, reliability, and safety concerns.

On the other hand, if the control design is motivated commercially, one again gets different situations depending on whether the controller is a small subcomponent of a larger commercial product (such as the cruise controller or ABS in a car) or whether it is part of a manufacturing process (such as the motion controller in the robots assembling a car). In the first case, one must also consider the cost of including the controller in every product, which usually means that there is a major premium on cost and hence one is forced to use rather simple microcontrollers. In the second case, one can usually afford significantly more complex controllers, provided that they improve the manufacturing process in a way that significantly enhances the value of the manufactured product.

In all of these situations, the control engineer is further affected by where the control system is in its lifecycle:

- initial *grass roots* design;

- commissioning and tuning;

- refinement and upgrades;

- forensic studies.

1.4.1   Initial *Grass Roots* Design

In this phase, the control engineer is faced by a *green-field* or so-called *grass roots* project, and thus the designer can steer the development of a system from the beginning. This includes ensuring that the design of the overall system takes account of the subsequent control issues. All too often, systems and plants are designed on the basis of steady-state considerations alone. It is, then, small wonder that operational difficulties can appear down the track. It is our belief that control engineers should be an integral part of all design teams. The control engineer needs to interact with the design specifications and to ensure that dynamic as well as steady-state issues are considered.

1.4.2 Commissioning and Tuning

Once the basic architecture of a control system is in place, then the control engineer's job becomes one of tuning the control system to meet the required performance specifications as closely as possible. This phase requires a deep understanding of feedback principles to ensure that the tuning of the control system is carried out in an expedient, safe, and satisfactory fashion.

1.4.3 Refinement and Upgrades

Once a system is up and running, then the control engineer's job turns into one of maintenance and refinement. The motivation for refinement can come from many directions. They include the following:

- internal forces–e.g., the availability of new sensors or actuators may open the door for improved performance;

- external forces–e.g., market pressures or new environmental legislation may necessitate improved control performance.

1.4.4 "Forensic" Studies

Forensic investigations are often the role of control engineering consultants. Here, the aim is to suggest remedial actions that will rectify an observed control problem. In these studies, it is important that the control engineer take a holistic view, because successful control performance usually depends on satisfactory operation of many interconnected components. In our experience, poor control performance is as likely to be associated with basic plant design flaws, poor actuators, inadequate sensors, or computer problems as it is to be the result of poor control tuning. However, all of these issues can, and should be, part of the control engineer's domain. Indeed, it is often only the control engineer who has the necessary overview to resolve these complex issues successfully.

1.5 System Integration

As is evident from the above discussion, success in control engineering depends on taking a holistic viewpoint. The issues that are embodied in a typical control design include the following:

- plant–i.e., the process to be controlled;

- objectives;

- sensors;

- actuators;

- communications;

- computing;

- architectures and interfacing;

- algorithms;

- accounting for disturbances and uncertainty.

These issues are briefly discussed below.

1.5.1   Plant

As mentioned in subsection 1.4.1, the physical layout of a plant is an intrinsic part of control problems. Thus, a control engineer needs to be familiar with the *physics* of the process under study. This includes a rudimentary knowledge of the basic energy balance, mass balance, and material flows in the system. The physical dimensions of equipment and how they relate to performance specifications must also be understood. In particular, we recommend the production of *back of the envelope* physical models as a first step in designing and maintaining control systems. These models will typically be refined as one progresses.

1.5.2   Objectives

Before designing sensors, actuators, or control architectures, it is important to know the goal, that is, to formulate the control objectives. This knowledge includes the following:

- what does one want to achieve (energy reduction, yield increase, . . .)?

- what variables need to be controlled to achieve these objectives?

- what level of performance is necessary (accuracy, speed, . . .)?

1.5.3   Sensors

Sensors are the *eyes* of control enabling one to *see* what is going on. Indeed, one statement that is sometimes made about control is: *If you can measure it, you can control it.* This is obviously oversimplified and not meant literally. Nonetheless, it is a catchy phrase highlighting the fact that being able to make appropriate measurements is an intrinsic part of the overall control problem. Moreover, new sensor technologies often open the door to improved control performance.

Alternatively, in those cases where particularly important measurements are not readily available, then one can often infer these vital pieces of information from other observations. This leads to the idea of a *soft* or *virtual* sensor. We will see that this is one of the most powerful techniques in the control engineer's *bag of tools*.

1.5.4   Actuators

Once sensors are in place to report on the *state* of a process, then the next issue is the ability to affect or actuate the system, in order to move the process from the current state to a desired state. Thus, we see that actuation is another intrinsic element in control problems. The availability of new or improved actuators also often opens the door to significant improvements in performance. Conversely, inadequate or poor actuators often lie at the heart of control difficulties. A typical industrial control problem will usually involve many different actuators–see, for example, the flatness-control set-up shown in Figure 1.3.

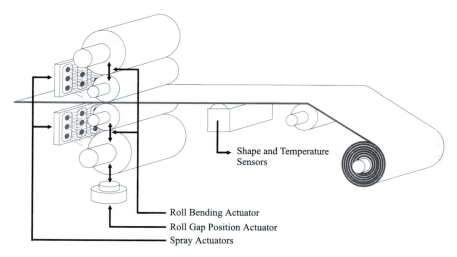

Shape and Temperature
Sensors

Roll Bending Actuator
Roll Gap Position Actuator
Spray Actuators

Figure 1.3. Typical flatness-control set-up for rolling mill

1.5.5   Communications

Interconnecting sensors to actuators involves the use of communication systems. A typical plant can have many thousands of separate signals to be sent over long distances. Thus, the design of communication systems and their associated protocols is an increasingly important aspect of modern control engineering.

There are special issues and requirements for communication systems with real-time data. For example, in voice communication, small delays and imperfections in transmission are often unimportant by virtue of being transparent to the recipient. However, in high-speed real-time control systems, these issues could be of major importance. For example, there is an increasing tendency to use Ethernet-type connections for data transmission in control. However, as is well known by those familiar with this technology, if a delay occurs on the transmission line, then the transmitter simply *tries again* at some later random time. This obviously introduces a nondeterministic delay into the transmission of the data. All control systems depend upon precise knowledge of, not only *what* has happened, but *when* it happened, so attention to such delays is very important for the performance of the overall system.

1.5.6   Computing

In modern control systems, the connection between sensors and actuators is invariably made via a computer of some sort. Thus, computer issues are necessarily part of the overall design. Current control systems use a variety of computational devices including DCS's (Distributed Control Systems), PLC's (Programmable Logic Controllers), and PC's (Personal Computers). In some cases, these computer elements are rather limited with respect to the facilities they offer. As with communication delays, computational delays can be crucial to success or failure in the operation of control systems. Determinism in timing is important, so a multi-tasking real-time operating system may be required.

Another aspect of computing is that of numerical precision. We know of several control systems that failed to meet the desired performance specifications simply because of inadequate attention to numerical issues. For this reason, we will devote some attention to this issue in the sequel.

A final computer-based question in control concerns the ease of design and implementation. Modern computer-aided tools for rapid prototyping of control systems provide integrated environments for control-system modeling, design, simulation and implementation. These *pictures to real-time code* facilities have allowed development times for advanced control algorithms to be reduced from many months to periods on the order of days or, in some cases, hours.

1.5.7   Architectures and Interfacing

The issue of what to connect to what is a nontrivial one in control-system design. One might feel that the best solution would always be to bring all signals to a central point, so that each control action would be based on complete information (leading to so-called centralized control). However, this is rarely (if ever) the best solution in practice. Indeed, there are very good reasons why one might not wish to bring all signals to a common point. Obvious objections to this include complexity, cost, time constraints in computation, maintainability, and reliability.

Thus, one usually partitions the control problem into manageable subsystems. How one does this is part of the control engineer's domain. Indeed, we will see, in the case studies presented in the text, that these architectural issues can be crucial to the final success, or otherwise, of a control system.

Indeed, one of the principal tools that a control-system designer can use to improve performance is to exercise lateral thinking relative to the architecture of the control problem. As an illustration, we will present a real example later in the text (see Chapter 8) where thickness-control performance in a reversing rolling mill is irrevocably constrained by a particular architecture. It is shown that no improvement in actuators, sensors, or algorithms (within this architecture) can remedy the problem. However, by simply changing the architecture so as to include extra actuators (namely the currents into coiler and uncoiler motors) then the difficulty is resolved. (See Chapter 10.) As a simpler illustration, the reader is invited to compare the difference in trying to balance a broom on one's finger with one's eyes open or shut. Again there is an architectural difference here–this time it is a function of available sensors. A full analysis of the reasons behind the observed differences in the difficulty of these types of control problems will be explained in Chapters 8 and 9 of the book.

We thus see that architectural issues are of paramount importance in control-design problems. A further architectural issue revolves around the need to *divide and conquer* complex problems. This leads to a hierarchical view of control as illustrated in Table 1.1.

Level	Description	Goal	Time frame	Typical design tool
4	Plant-wide optimization	Meeting customer orders and scheduling supply of materials	Every day (say)	Static optimization
3	Steady-state optimization at unit operational level	Efficient operation of a single unit (e.g., distillation column)	Every hour (say)	Static optimization
2	Dynamic control at unit operation level	Achieving set-points specified at level 3 and achieving rapid recovery from disturbances	Every minute (say)	Multivariable control (e.g., Model Predictive Control)
1	Dynamic control at single-actuator level	Achieving liquid flow rates as specified at level 2 by manipulation of available actuators (e.g., valves)	Every second (say)	Single variable control (e.g., PID)

Table 1.1. Typical control hierarchy

Having decided what connections need to be made, there is the issue of interfacing the various subcomponents. This is frequently a nontrivial job, because it is often true that special interfaces are needed between different equipment. Fortunately vendors of control equipment are aware of this difficulty and increasing attention is being paid to standardization of interfaces.

1.5.8   Algorithms

Finally, we come to the real *heart* of control engineering–the algorithms that connect the sensors to the actuators. It is all to easy to underestimate this final aspect of the problem.

As a simple example from the reader's everyday experience, consider the problem of playing tennis at top international level. One can readily accept that one needs good eyesight (sensors) and strong muscles (actuators) to play tennis at this level, but these attributes are not sufficient. Indeed, eye–hand coordination (i.e., control) is also crucial to success.

Thus beyond sensors and actuators, the control engineer has to be concerned with the science of dynamics and feedback control. These topics will actually be the central theme of the remainder of this book. As one of our colleagues put it: *Sensors provide the eyes and actuators the muscle, but control science provides the finesse.*

1.5.9   Disturbances and Uncertainty

One of the things that makes control science interesting is that all real-life systems are acted on by noise and external disturbances. These factors can have a significant impact on the performance of the system. As a simple example, aircraft are subject to disturbances in the form of wind gusts, and cruise controllers in cars have to cope with different road gradients and different car loadings. However, we will find that, by appropriate design of the control system, quite remarkable insensitivity to external disturbances can be achieved.

Another related issue is that of model uncertainty. All real-world systems have very complex models, but an important property of feedback control is that one can often achieve the desired level of performance by using relatively simple models. Of course, it is incumbent on designers to appreciate the effect of model uncertainty on control performance and to decide whether attention to better modeling would enable better performance to be achieved.

Both of the issues raised above are addressed, in part, by the remarkable properties of feedback. This concept will underpin much of our development in the book.

1.5.10   Homogeneity

A final point is that all interconnected systems, including control systems, are only as good as their weakest element. The implications of this in control-system design are that one should aim to have all components (plant, sensors, actuators, communications, computing, interfaces, algorithms, etc,) be of roughly comparable accuracy and performance. If this is not possible, then one should focus on the weakest component to get the best return for a given level of investment. For example, there is no point placing all one's attention on improving linear models (as has become fashionable in parts of modern control theory) if the performance-limiting factor is that one needs to replace a sticking valve or to develop a virtual sensor for a key missing measurement. Thus, a holistic viewpoint is required with an accurate assessment of error budgets associated with each subcomponent.

1.5.11   Cost-Benefit Analysis

While we are on the subject of ensuring best return for a given amount of effort, it is important to raise the issue of benefits analysis. Control engineering, in common with all other forms of engineering, depends on being able to convince management that there is an attractive cost-benefit trade-off in a given project. Payback periods in modern industries are often as short as 6 months, and thus this aspect requires careful and detailed attention. Typical steps include the following:

- assessing the range of control opportunities;

- developing a short list for closer examination;

- deciding on a project with high economic or environmental impact;

- consulting appropriate personnel (management, operators, production staff, maintenance staff, etc.);

- identifying the key action points;

- collecting base-case data for later comparison;

- deciding on revised performance specifications;

- updating actuators, sensors, etc.;

- developing algorithms;

- testing the algorithms via simulation;

- testing the algorithms on the plant by using a rapid prototyping system;

- collecting preliminary performance data for comparison with the base case;

- final implementation;

- collecting final performance data;

- final reporting on project.

1.6   Summary

- Control Engineering is present in virtually all modern engineering systems.

- Control is often the hidden technology, as its very success removes it from view.

- Control is a key enabling technology with respect to

    o enhanced product quality,

    o waste and emission minimization,

    o environmental protection,

    o greater throughput for a given installed capacity,

    o greater yield,

    o deferring costly plant upgrades, and

    o higher safety margins.

- Examples of controlled systems include the following:

System	Controlled outputs include	Controller	Desired performance includes
Aircraft	Course, pitch, roll, yaw	Autopilot	Maintain flight path on a safe and smooth trajectory
Furnace	Temperature	Temperature controller	Follow warm-up temperature profile, then maintain temperature
Wastewater treatment	pH value of effluent	pH controller	Neutralize effluent to specified accuracy
Automobile	Speed	Cruise controller	Attain, then maintain selected speed without undue fuel consumption

- Control is a multidisciplinary subject that includes

    o sensors,

    o actuators,

    o communications,

    o computing,

    o architectures and interfacing, and

    o algorithms.

- Control design aims to achieve a desired level of performance in the face of disturbances and uncertainty.

- Examples of disturbances and uncertainty include the following:

System	Actuators	Sensors	Disturbances	Uncertainties
Aircraft	Throttle servo, rudder and flap actuators, etc.	Navigation instruments	Wind, air pockets, etc.	Weight, exact aerodynamics, etc.
Furnace	Burner valve actuator	Thermocouples, heat sensors	Temperature of incoming objects, etc.	Exact thermodynamics, temperature distribution
Wastewater treatment	Control acid valve servo	pH sensor	Inflow concentration	pH gain curve, measurement errors
Automobile	Throttle positioning	Tachometer	Hills	Weight, exact dynamics

1.7 Further Reading

Historical notes

The IEEE History Centre and its resources are an excellent source for the historically interested reader.

Other useful sources are listed here:

Black, H.W. (1934). Stabilized Feedback Amplifiers. *Bell Systems Tech. Journal*, 13:1-18.

Bode, H. (1969). Feedback: the history of an idea. In *Selected Papers on Mathematical Trends in Control Theory*, pages 106-123. Dover, New York.

Fuller, A. (1976). The early development of control theory. *Trans. ASME, J. Dyn. Sys. Meas. Contr.*, 98:109-118, 224-235.

James, H.M., Nichols, N.B., and Phillips, R.S., editors (1947). *Theory of Servomechanisms*. McGraw-Hill, New York.

Maxwell, T. (1868). On governors. *Proc. Royal Soc. London*, 16:270-283.

Mayr, O. (1970). *The Origins of Feedback Control*. MIT Press, Cambridge, Mass.

Chapter 2

INTRODUCTION TO THE
PRINCIPLES OF FEEDBACK

2.1 Preview

This chapter delineates the path we have chosen for our control engineering journey. In particular, this chapter contains

- an industrial motivational example,

- a statement of the fundamental nature of the control problem,

- the idea of inversion as the central ingredient in solving control problems, and

- evolution from open-loop inversion to closed-loop feedback solutions.

2.2 The Principal Goal of Control

As we have seen in Chapter 1, examples of dynamic systems with automatic controllers abound: advanced process controllers are operating in virtually every industrial domain; micro-controllers pervade an immense array of household and entertainment electronics; thermostats regulate temperatures in domestic- to industrial-sized ovens, and autopilots control aircraft.

Designing any one of these systems requires the close cooperation of experts from various disciplines.

To particularize the principal goal of control engineering within this team effort, it is helpful to distinguish between a system's tangible realization and its behavior. The aircraft's physical realization, for example, includes fuselage, wings, and ailerons. Its behavior, on the other hand, refers to the aircraft's dynamic response to a change in throttle, aileron, or flap position.

To control such a system automatically, one needs to interface the system to a controller, which will also have a physical realization and behavior. Depending on the application, the controller could be realized in a chip, analog electronics, a PLC, or a computer. There also needs to be a channel by which the controller and system can interact via sensors and actuators: sensors to report the state of the system, actuators as a means for the controller to act on the system.

With this process and control infrastructure in place, the key remaining question pertains to the controller behavior. In the aircraft application, for example, if the controller (here called an autopilot) detects a deviation in speed, height or heading via the sensors, just how should it command throttle and ailerons to get back on target?

This is the control engineer's key concern, or, stated in general terms, the fundamental goal of control engineering is to find technically, environmentally, and economically feasible ways of acting on systems to control their outputs to desired values, thus ensuring a desired level of performance. As discussed earlier, finding a good solution to this question frequently requires an involvement in process design, actuator and sensor selection, mathematical analysis, and modeling.

The control engineer's perspective on the aircraft navigation example described above includes a cyclical dependency: autopilot commands affect the aircraft, whose changed speed, height, and heading in turn affect the further actions of the autopilot.

Such a cyclically dependent interaction between system behaviors is called *feedback*.

Feedback phenomena exist both in nature and technology. The periodic population growth and reduction in the famous predator-and-prey interactions are an example of feedback occurring in nature. The high-pitch whistling sound occurring as a result of interaction between microphones and loudspeakers in a concert hall is a technical example of feedback.

In both of these cases, neither of the two interacting systems can be clearly designated as controller or process–they are simply two systems interacting in feedback. Nevertheless, the feedback interaction is seen to have a profound impact on the behavior of the engaged systems.

This behavior-altering effect of feedback is a key mechanism that control engineers exploit deliberately to achieve the objective of *acting on a system to ensure that the desired performance specifications are achieved.*

2.3   A Motivating Industrial Example

To make the above general discussion more concrete, we next present a simplified, yet essentially authentic, example of an industrial control problem. The example, taken from the steel industry, is of a particular nature; however, the principal elements of specifying a desired behavior, modeling and the necessity for trade-off decisions are generic. Some details of the example might not be quite clear at this early stage, but they will set the scene for future work.

One of the products of the steel industry is a so-called bloom, which is a rectangular slab of steel. Blooms are produced in a process called a *continuous caster*. A diagram of an industrial bloom caster is given in Figure 2.1. The principal components of such a system relevant to our discussion here are shown in Figure 2.2. A photograph of a bloom caster can be found on the web page.

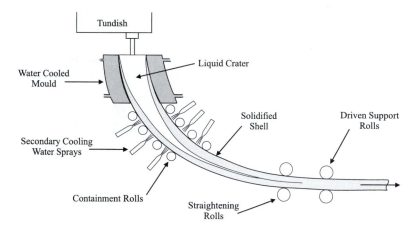

Figure 2.1. Process schematic of an industrial bloom caster

The *tundish* can be thought of as a large container that acts as a reservoir for molten steel. A control valve regulates the rate of flow of steel that enters the mould mounted under the tundish. The mould, whose cross-sectional area equals the cross-sectional area of the desired bloom, is open from above and below. By intense cooling, steel in the mould is cooled to a semi-solid state. In this state, it is sufficiently firm so that the strand can be withdrawn continuously from the mould by rolls. The resulting continuous strand is then subjected to further cooling and finally cut into blooms.

2.3.1   Performance Specifications

There are two fundamental specifications for the continuous caster: safety and profitability.

These ultimate requirements can be further broken down to derive a control-system specification that quantifies the target to be met by controller-design.

- *Safety*: In this example, the safety requirement translates rather straightforwardly into a constraint on the level of molten steel in the mould (called the
  mould level). Clearly, the mould level must never be in danger of overflowing
  or emptying, because either case would result in molten metal spilling, with
  disastrous consequences.

- *Profitability*: Obviously, the system needs to be operated in a cost-effective
  fashion. Aspects which contribute to this requirement include the following:

  - *Product quality*: It turns out that high-frequency changes in the mould level
    reduce bloom quality, because molten metal has a viscosity similar to that

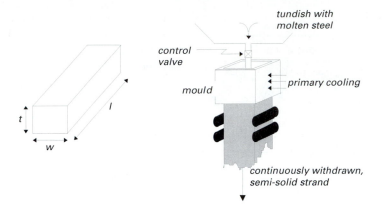

Figure 2.2. Continuous caster: typical bloom (left) and simplified diagram (right)

of water, and turbulence mixes impurities into the steel. Furthermore, because of the cooling applied to the mould, an oscillating mould level results in uneven cooling of the steel, which also reduces its quality. Thus, product quality demands steady control of the mould level.

○ *Maintenance*: Through an oxidizing reaction, the nozzle through which the metal flows into the mould is subject to intense wear at the mould level. The nozzle's lifetime is therefore maximized if the mould level is controlled so that it slowly ramps across the wear band of the nozzle. Maintenance costs are further reduced if the lifetime of the control valve is maximized by having relatively nonaggressive control actions. Thus, keeping the cost of maintenance low requires steady regulation around a set-point that slowly ramps across the nozzle wear band while avoiding aggressive valve movements.

○ *Throughput*: Throughput is a direct function of casting speed. The casting speed, however, is also a function of numerous up-stream process factors (availability of the necessary grades of raw material, furnace capacity, etc.) as well as down-stream factors (product demand, shipping capacity, etc.). Thus, the mould-level control system should maintain performance at all casting speeds to avoid being a limiting factor in the overall operation of the process.

2.3.2   Modeling

To make progress on the control-system design problem as set out above, it is first necessary to gain an understanding of how the process operates. This understanding is typically expressed in the form of a mathematical model that describes the steady-state and the dynamic behavior of the process. To construct such a model, we first define relevant process variables. Thus, we introduce the following:

$$
\begin{array}{rcl}
h^* & : & \text{commanded level of steel in mould} \\
h(t) & : & \text{actual level of steel in mould} \\
v(t) & : & \text{valve position} \\
\sigma(t) & : & \text{casting speed} \\
q_{in}(t) & : & \text{inflow of matter into the mould} \\
q_{out}(t) & : & \text{outflow of matter from the mould.}
\end{array}
$$

Physics suggests that the mould level will be proportional to the integral of the difference between in- and outflow:

$$
h(t) = \int_{-\infty}^{t} \left(q_{in}(\tau) - q_{out}(\tau) \right) d\tau \tag{2.3.1}
$$

where we have assumed a unit cross-section of the mould for simplicity. We also assume, again for simplicity, that the measurements of valve position, $v(t)$, and casting speed, $\sigma(t)$, are calibrated such that they actually indicate the corresponding in- and outflows:

$$
v(t) = q_{in}(t) \tag{2.3.2}
$$
$$
\sigma(t) = q_{out}(t) \tag{2.3.3}
$$

Hence, the process model becomes

$$
h(t) = \int_{-\infty}^{t} \left(v(\tau) - \sigma(\tau) \right) d\tau \tag{2.3.4}
$$

The casting speed can be measured fairly accurately, but mould-level sensors are typically prone to high-frequency measurement noise, which we take into account by introducing an additive spurious signal $n(t)$:

$$
h_m(t) = h(t) + n(t) \tag{2.3.5}
$$

where $h_m(t)$ is the measurement of $h(t)$ corrupted by noise. A block diagram of the overall process model and the measurements is shown in Figure 2.3.

This is a very simple model, but it captures the essence of the problem.

2.3.3   Feedback and Feedforward

We will find later that the core idea in control is that of inversion. Moreover, inversion can be achieved conveniently by the use of two key mechanisms (namely,

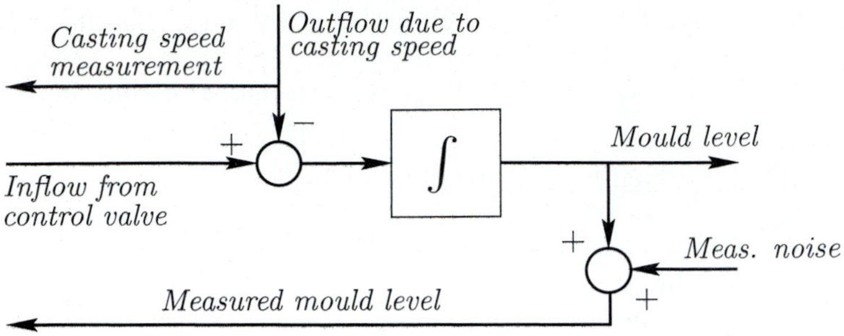

Figure 2.3. Block diagram of the simplified mould-level dynamics, sensors, and actuator

feedback and feedforward). These tools give an elegant and robust solution to many control-design problems. In the context of mould-level control, the simplest *feedback* controller is a constant gain, K, driving the valve proportionally to the error between the commanded mould level, h^*, and the measurement of the actual mould level, $h_m(t)$:

$$v(t) = K(h^* - h_m(t)) \qquad (2.3.6)$$

To anticipate how a controller of this form might perform, we observe that a deviation between set-point and measurement must first occur before the controller can react. We know, however, that a change in casting speed requires a modified operating point for the valve. Thus, rather than letting a change in casting speed occur, which then leads to an error in the mould level to which the feedback controller reacts, we can improve the strategy by changing the valve-position proactively. This is called *feedforward*. This leads to a final controller of the form:

$$v(t) = K\left([h^* - h_m(t)] + \left[\frac{1}{K}\sigma(t) \right] \right) \qquad (2.3.7)$$

Note that this controller features joint feedback and a preemptive action (feedforward). In particular, the second term gives the predictable action necessary to compensate for the casting speed changes, while the first term reacts to the remaining error.

A block diagram for the final control system is shown in Figure 2.4.

Further discussion of this problem, together with the capability of the reader interacting with the design, is contained in the book's web page.

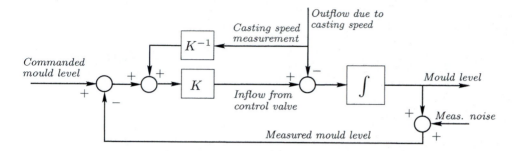

Figure 2.4. Model of the simplified mould-level control with feedforward compensation for casting speed

2.3.4   A First Indication of Trade-offs

On simulating the performance of the above control loop for $K = 1$ and $K = 5$ (see Figure 2.5), we find that the smaller controller gain ($K = 1$) results in a slower response to a change in the mould-level set-point. On the other hand, the larger controller gain ($K = 5$) results in a faster response but also increases the effects of measurement noise as seen by the less steady level control and by the significantly more aggressive valve movements.

Thus, the performance requirements derived in subsection §2.3.1 appear to be in conflict with each other, at least to some degree.

At this point, a control engineer who does not have a systematic background in control-system design would have a difficult time in assessing whether this conflict is merely a consequence of having such a simple controller or whether it is fundamental. How much effort should be spent in finding a good value for K? Should one choose a more complex controller? Should one spend more effort in modeling the mould-level process?

The remainder of the book is devoted to developing systematic answers to these and other related questions.

2.4   Definition of the Problem

The example presented in section §2.3 motivates the following more formal statement of the nature of the control problem:

Definition 2.1. *The fundamental control problem*

The central problem in control is to find a technically feasible way to act on a given process so that the process adheres, as closely as possible to some desired behavior. Furthermore, this approximate behavior should be achieved in the face of uncertainty of the process and in the presence of uncontrollable external disturbances acting on the process.

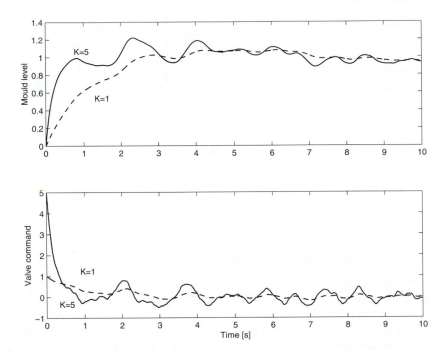

Figure 2.5. A first indication of trade-offs: Increased responsiveness to set-point changes also increases sensitivity to measurement noise and actuator wear.

□□□

The above definition introduces several ideas:

- **Desired behavior**. *This needs to be specified as part of the design problem.*

- **Feasibility**. *This means that the solution must satisfy various constraints, which can be of technical, environmental, economic, or other nature.*

- **Uncertainty**. *The available knowledge about a system will usually be limited and of limited accuracy.*

- **Action**. *The solution requires that action be somehow applied to the process, typically via one or more manipulated variables which command the actuators.*

- **Disturbances**. *The process to be controlled will typically have inputs other than those that are manipulated by the controller. These other inputs are called disturbances.*

- **Approximate behavior**. *A feasible solution will rarely be perfect. There will invariably be a degree of approximation in achieving the specified goal.*

- **Measurements**. *These are crucial to let the controller know what the system is actually doing and how the unavoidable disturbances are affecting it.*

In the sequel, we will refer to the process to be controlled as the *plant*, and we will say that the plant is under *automatic control* when the control objectives are achieved with infrequent human intervention.

2.5   Prototype Solution to the Control Problem via Inversion

One particularly simple, yet insightful way of thinking about control problems is via inversion. To describe this idea we argue as follows:

- say that we know what effect an action at the input of a system produces at the output, and

- say that we have a desired behavior for the system output;
  then one simply needs to invert the relationship between input and output to determine what input action is necessary to achieve the desired output behavior.

In spite of the apparent naivety of this argument, its embellished ramifications play a profound role in control-system design. In particular, most of the real-world difficulties in control relate to the search for a strategy that captures the intent of the above inversion idea, while respecting a myriad of other considerations, such as insensitivity to model errors, disturbances, and measurement noise.

To be more specific, let us assume that the required behavior is specified by a scalar target signal or *reference*, $r(t)$, for a particular process variable, $y(t)$, that has an additive disturbance $d(t)$. Say we also have available a single manipulated variable, $u(t)$. We denote by y a function of time: $y = \{y(t) : t \in \mathbb{R}\}$.

In describing the prototype solution to the control problem below, we will make a rather general development that, in principle, can apply to general nonlinear dynamical systems. In particular, we will use a function, $f\langle \circ \rangle$, to denote an operator mapping one function space to another. So as to allow this general interpretation, we introduce the following notation:

The symbol y (without brackets) will denote an element of a function space: $y \overset{\triangle}{=} \{y(t) : \mathbb{R} \to \mathbb{R}\}$. An operator, $f\langle \circ \rangle$, will then represent a mapping from a function space, say χ, onto χ.

What we suggest is that the reader, on a first reading, simply interpret f as a *static linear gain* linking one real number, the input u, to another real number, the output y. On a subsequent reading, the more general interpretation, using nonlinear dynamic operators, can be used.

Let us also assume (for the sake of argument) that the output is related to the input by a known functional relationship of the form

$$y = f\langle u\rangle + d \qquad\qquad (2.5.1)$$

where f is a transformation or mapping (possibly dynamic) that describes the input-output relations in the plant.[1] We call a relationship of the type given in (2.5.1) a *model*.

The control problem then requires us to find a way to generate u in such a way that $y = r$. In the spirit of inversion, a direct, although somewhat naive, approach to obtain a solution would thus be to set

$$y = r = f\langle u\rangle + d \qquad\qquad (2.5.2)$$

from which we could derive a *control law*, by solving for u. This leads to

$$u = f^{-1}\langle r - d\rangle \qquad\qquad (2.5.3)$$

This idea is illustrated in Figure 2.6.

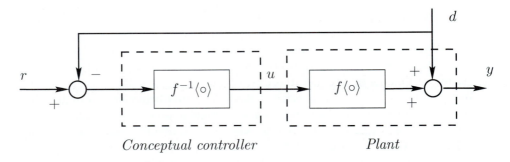

Conceptual controller   *Plant*

Figure 2.6. Conceptual controller

This is a conceptual solution to the problem. However, a little thought indicates that the answer given in (2.5.3) presupposes certain stringent requirements for its success. For example, inspection of equations (2.5.1) and (2.5.3) suggests the following requirements:

R1 The transformation f clearly needs to describe the plant exactly.

---

[1] We introduce this term here loosely. A more rigorous treatment will be deferred until Chapter 19.

R2 The transformation f should be well-formulated in the sense that a bounded output is produced when u is bounded–we then say that the transformation is *stable*.

R3 The inverse f^{-1} should also be well-formulated in the sense used in *R2*.

R4 The disturbance needs to be measurable, so that u is computable.

R5 The resulting action u should be realizable and should not violate any constraint.

Of course, these are very demanding requirements. Thus, a significant part of Automatic Control theory deals with the issue of how to change the control architecture so that inversion is achieved but in a more robust fashion and so that the stringent requirements set out above can be relaxed.

To illustrate the meaning of these requirements in practice, we briefly review a number of situations.

Example 2.1 (Heat exchanger). *Consider the problem of a heat exchanger in which water is to be heated by steam having a fixed temperature. The plant output is the water temperature at the exchanger output and the manipulated variable is the air pressure (3 to 15 [psig]) driving a pneumatic valve that regulates the amount of steam feeding the exchanger.*

In the solution of the associated control problem, the following issues should be considered:

- *Pure time delays might be a significant factor, because this plant involves mass and energy transportation. However a little thought indicates that a pure time delay does not have a realizable inverse (otherwise we could predict the future), and hence R3 will not be met.*

- *It can easily happen that, for a given reference input, the control law (2.5.3) leads to a manipulated variable outside the allowable input range (3 to 15 [psig] in this example). This will lead to saturation in the plant input. Condition R5 will then not be met.*

□□□

Example 2.2 (Flotation in mineral processing). *In copper processing, one crucial stage is the flotation process. In this process, the mineral pulp (water and ground mineral) is continuously fed to a set of agitated containers where chemicals are added to separate (by flotation) the particles with high copper concentration. From a control point of view, the goal is to determine the appropriate addition of chemicals and the level of agitation to achieve maximal separation.*

Characteristics of this problem are as follows:

- *The process is complex (physically distributed, time varying, highly nonlinear, multivariable, and so on) and hence it is difficult to obtain an accurate model for it. Thus, R1 is hard to satisfy.*

- *One of the most significant disturbances in this process is the size of the mineral particles in the pulp. This disturbance is actually the output of a previous stage (grinding). To apply a control law derived from (2.5.3), one would need to measure the size of all these particles or (at least) to obtain some average measure of this. Thus, condition R4 is hard to satisfy.*

- *Pure time delays are also present in this process, and thus condition R3 cannot be satisfied.*

□□□

One could imagine various other practical cases where one or more of the requirements listed above cannot be satisfied. Thus, the only sensible way to proceed is to accept that there will inevitably be intrinsic limitations and to pursue the solution within those limitations. With this in mind, we will impose constraints that will allow us to solve the problem subject to the limitations that the physical set-up imposes. The most commonly used constraints are as follows:

L1 to *restrict* attention to those problems where the prescribed behavior (reference signals) belong to restricted classes and where the desired behavior is achieved only asymptotically;

L2 To seek *approximate* inverses.

In summary, we can conclude the following:

In principle, all controllers implicitly generate an inverse of the process, in so far as this is feasible. Controllers differ with respect to the mechanism used to generate the required approximate inverse.

2.6   High-Gain Feedback and Inversion

As we will see later, typical models used to describe real plants cannot be inverted exactly. We will next show, however, that there is a rather intriguing property of feedback that implicitly generates an approximate inverse of dynamic transformations, without the inversions having to be carried out explicitly.

To develop this idea, let us replace the conceptual controller shown in Figure 2.6 by the realization shown in Figure 2.7. As before, f represents the process model. The transformation h will be described further below.

As in section §2.5, r, u, y can be interpreted as real numbers, and $h\langle \circ \rangle$, $f \langle \circ \rangle$ are scalar linear gains, on a first reading. On a second reading, these can be given the general nonlinear interpretation introduced in section §2.5.

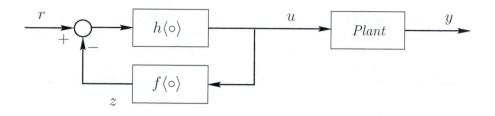

Figure 2.7. Realization of conceptual controller

From Figure 2.7, we see that

$$u = h\langle r - z\rangle = h\langle r - f\langle u\rangle\rangle \tag{2.6.1}$$

Thus,

$$h^{-1}\langle u\rangle = r - f\langle u\rangle \tag{2.6.2}$$

from which we finally obtain

$$u = f^{-1}\langle r - h^{-1}\langle u\rangle\rangle \tag{2.6.3}$$

Equation (2.6.3) suggests that the loop in Figure 2.7 implements an approximate inverse of $f\langle \circ\rangle$, that is, $u = f\langle r\rangle$, if

$$r - h^{-1}\langle u\rangle \approx r \tag{2.6.4}$$

We see that this is achieved if h^{-1} is small, that is, if h is a *high-gain* transformation.

Hence, if f characterizes our knowledge of the plant and if h is a high-gain transformation, then the architecture illustrated in Figure 2.7 effectively builds an approximate inverse for the plant model without requiring that the model of the plant, f, be explicitly inverted. We illustrate this idea by an example.

Example 2.3. *Assume that a plant can be described by the model*

$$\frac{dy(t)}{dt} + 2\sqrt{y(t)} = u(t) \tag{2.6.5}$$

and that a control law is required to ensure that $y(t)$ follows a slowly varying reference.

One way to solve this problem is to construct an inverse for the model that is valid in the low-frequency region. Using the architecture in Figure 2.7 on the preceding page, we obtain an approximate inverse, provided that $h\langle \circ \rangle$ has large gain in the low-frequency region. A simple solution is to choose $h\langle \circ \rangle$ to be an integrator that has infinite gain at zero frequency. The output of the controller is then fed to the plant. The result is illustrated in Figure 2.8, which shows the reference and the plant outputs. The reader might wish to explore this example further by using the SIMULINK file **tank1.mdl** *on the accompanying CD.*

□□□

2.7   From Open- to Closed-Loop Architectures

A particular scheme has been suggested in Figure 2.7 for realizing an approximate inverse of a plant model. Although the controller in this scheme is implemented as a feedback system, the control is actually applied to the plant in *open loop*. In particular, we see that the control signal $u(t)$ is *independent of what is actually happening in the plant.* This is a serious drawback, because the methodology will not lead to a satisfactory solution to the control problem unless

- the model on which the design of the controller has been based is a very good representation of the plant,

- the model and its inverse are stable, and

- disturbances and initial conditions are negligible.

We are thus motivated to find an alternative solution to the problem, one that retains the key features but does not suffer from the above drawback. This is indeed possible by changing the scheme slightly so that feedback is placed around the *plant* itself rather than around the *model*.

To develop this idea, we begin with the basic feedback structure as illustrated in Figure 2.9. We proceed as follows.

If we assume, for the moment, that the model in Figure 2.9 is perfect, then we can rearrange the diagram to yield the alternative scheme shown in Figure 2.10

This scheme, which has been derived from an open-loop architecture, is the basis of feedback control. The key feature of this scheme is that the controller output depends not only on the a-priori data provided by the model but also on what is actually happening at the plant output at every instant. It has other interesting features, which are discussed in detail below. However, at this point, it will be worthwhile to carry out an initial discussion of the similarities and differences between open- and closed-loop architectures of the types shown in Figure 2.9 and Figure 2.10.

- The first thing to note is that, provided that the model represents the plant exactly, and that all signals are bounded (i.e., the loop is stable), then the

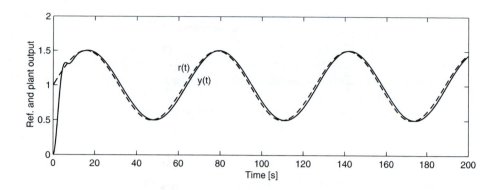

Figure 2.8. Tank level control by using approximate inversion

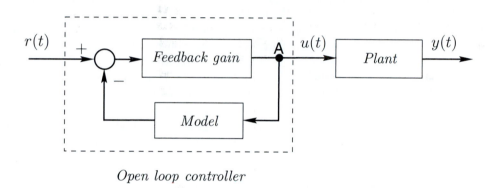

Open loop controller

Figure 2.9. Open-loop control with built-in inverse

schemes are equivalent *regarding the relation between $r(t)$ and $y(t)$*. The key differences are due to disturbances and different initial conditions.

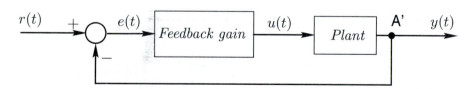

Figure 2.10. Closed-loop control

- In the open-loop control scheme the controller incorporates feedback internally: a signal at point A is fed back. In closed-loop control, the signal at A' is fed back. The fundamental difference is that, in the first case, everything happens *inside the controller*, either in a computer or in some external hardware connection. In the second case, the signal fed back is a process variable: *measuring devices are used* to determine what is actually happening. The heuristic advantages of the latter alternative are undoubtedly clear to the reader. We will develop the formal background to these advantages as we proceed.

2.8   Trade-offs Involved in Choosing the Feedback Gain

The preliminary insights of the previous two sections could seem to imply that all that is needed to generate a controller is to place high-gain feedback around the plant. This is true insofar that it goes. However, nothing in life is cost-free, and this also applies to the use of high-gain feedback.

For example, if a plant disturbance leads to a nonzero error, $e(t)$, in Figure 2.10, then high-gain feedback will result in a very large control action, $u(t)$. This might lie outside the available input range and thus invalidate the solution.

Another potential problem with high-gain feedback is that it is often accompanied by the very substantial risk of instability. Instability is characterized by self-sustaining (or growing) oscillations. As an illustration, the reader will probably have witnessed the high-pitch whistling sound that is heard when a loudspeaker is placed too close to a microphone. This is a manifestation of instability resulting from excessive feedback gain. Tragic manifestations of instability include aircraft crashes and the Chernobyl, disaster in which a runaway condition occurred.

Yet another potential disadvantage of high loop gain was hinted at in subsection §2.3.4. There, we saw that increasing the controller gain leads to increased sensitivity to measurement noise.

In summary, high loop gain is desirable from many perspectives, but it is undesirable when viewed from other perspectives. Thus, when choosing the feedback gain, one needs to make a conscious trade-off between competing issues.

The previous discussion can be summarized in the following statement.

High loop gain gives approximate inversion, which is the essence of control. However, in practice, the choice of feedback gain is part of a complex web of design trade-offs. Understanding and balancing these trade-offs is the essence of control-system design.

2.9   Measurements

We have seen that one of the key issues in feedback control is that there must exist suitable measurements to feed back. Indeed, if one can measure a variable, then there is a good chance that one can design a controller to bring it to a desired reference value.

A more accurate description of the feedback control loop, including sensors, is shown in Figure 2.11. From this figure, it can be seen that what we actually control is the measured value rather than the true output. These can be quite different.

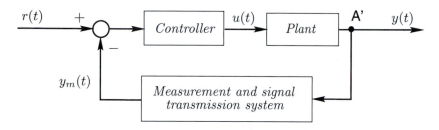

Figure 2.11. Closed-loop control with sensors

Hence the measurement system should ideally satisfy requirements such as the following:

- *Reliability.* It should operate within the necessary range.

- *Accuracy.* For a variable with a constant value, the measurement should settle to the correct value.

- *Responsiveness.* If the variable changes, the measurement should be able to follow the changes. Slow responding measurements can not only affect the quality of control but can actually make the feedback loop unstable. Loop instability can arise even though the loop has been designed to be stable for exact measurement of the process variable.

- *Noise immunity.* The measurement system, including the transmission path, should not be significantly affected by exogenous signals, such as measurement noise.

- *Linearity.* If the measurement system is not linear, then at least the nonlinearity should be known, so that it can be compensated for.

- *Nonintrusive measurement.* The measuring device should not significantly affect the behavior of the plant.

In most of the sequel, we will assume that the measurement system is sufficiently good, so that only measurement noise needs to be accounted for. This ideal measurement loop will be known as a unity feedback loop .

2.10 Summary

- Control is concerned with finding technically, environmentally, and commercially feasible ways of acting on a technological system to control its outputs to desired values while ensuring a desired level of performance.

- Fundamental to control engineering is the concept of inversion.

- Inversion can be achieved by a feedback architecture.

- Feedback refers to an iterative cycle of

    o quantifying the desired behavior,

    o measuring the actual values of relevant system variables by sensors,

    o inferring the actual system state from the measurements,

    o comparing the inferred state to the desired state,

    o computing a corrective action to bring the actual system to the desired state,

    o applying the corrective action to the system via actuators, and then

    o repeating the above steps.

- The principal components in a feedback loop are shown in Figure 2.12.

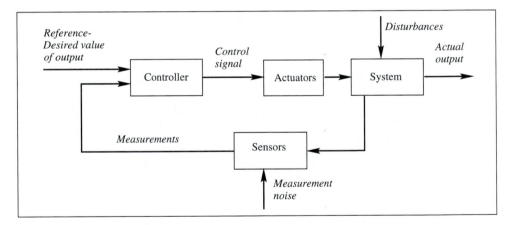

Figure 2.12. Typical feedback loop

- The desired performance is typically quantified on the following bases:

    o accuracy with which the outputs should attain their desired values;

    o required tolerance level to inaccurate assumptions, disturbances and plant changes;

   o specification of transients;

   o constraints on acceleration, overshoot, energy consumption, and so forth.

  • Control system objectives usually include the following:

   o maximization of throughput, speed, yield safety, and more;

   o minimization of energy consumption, waste production, emissions, and more;

   o decreasing the impact of disturbances, noise uncertainty, time variations, and so forth.

  • The chapter gives a first indication that the desired performance objectives are usually in conflict with each other and therefore form a network of trade-offs.

  • By *control-system design* we mean the process of

   o understanding the trade-off network,

   o making deliberate design decisions consistent with these trade-offs, and

   o being able to systematically translate the chosen goal into a controller.

2.11 Further Reading

Introduction to feedback control

Åström, K. and Wittenmark, B. (1990). *Computer Controlled Systems. Theory and Design.* Prentice-Hall, Englewood Cliffs, N.J., 2nd edition.

D'Azzo, J. and Houpis, C. (1988). *Feedback Control Systems Analysis and Synthesis.* McGraw-Hill, New York.

Doeblin, E.O. (1985). *Control System Principles and Design.* Wiley, New York.

Dorf, R. (1989). *Modern Control Systems.* Addison-Wesley, Reading, Mass., 5th edition.

Franklin, G.F., Powell, J.D., and Emami-Naeini, A. (1991). *Feedback Control of Dynamics Systems.* Addison-Wesley, Reading, Mass., 2nd edition.

Kuo, B.C. (1995). *Automatic Control Systems.* Prentice-Hall, Englewood Cliffs, N.J., 7th edition.

Levine, W.S., editor (1996). *The Control Handbook.* CRC Press, Boca Raton, FL.

Ogata, K. (1997). *Modern Control Engineering.* Prentice-Hall, Upper Saddle River, N.J., 3rd edition.

Truxal, J.G. (1955). *Control Systems Synthesis.* McGraw-Hill, New York.

Continuous casting

Graebe, S.F., Goodwin, G.C., and Elsley, G. (1995). Rapid prototyping and implementation of control in continuous steel casting. *IEEE Control Systems Magazine*, 15(4):64-71.

Chapter 3

MODELING

3.1 Preview

The design of a control system typically requires a delicate interplay between fundamental constraints and trade-offs. To accomplish this, a designer must have a comprehensive understanding of how the process operates. This understanding is usually captured in the form of a mathematical model. With the model in hand, the designer can proceed to use the model to predict the impact of various design choices. The aim of this chapter is to give a brief introduction to modeling. Specific topics to be covered include the following:

- how to select the appropriate model complexity;

- how to build models for a given plant;

- how to describe model errors;

- how to linearize nonlinear models.

It also provides a brief introduction to certain commonly used models, including

- state space models and

- high-order differential and high-order difference-equation models.

3.2 The *Raison d'être* for Models

The basic idea of feedback is tremendously compelling. Recall the mould-level control problem from Chapter 2. Actually, there are only three ways that a controller could manipulate the valve: open, close, or leave as is. Nevertheless, we have seen already that the precise way this is done involves subtle trade-offs between conflicting objectives, such as speed of response and sensitivity to measurement noise.

For many problems, it is both possible and feasible to find these precise controller settings by simple trial-and-error. However, many problems preclude this approach, on account of, complexity, efficiency, cost, or even danger. Also, a trial-and-error approach cannot answer, *before* trial, questions such as the following:

- Given a physical plant and objective, what controller can achieve the given objective? Can it be achieved at all?

- Given controller and plant, how will they perform in closed loop?

- Why is a particular loop behaving the way it is? Can it be done better? If so, by which controller?

- How would the loop performance change if the system parameters were to change, or disturbances were larger, or a sensor were to fail?

To answer these questions systematically, we need a means of capturing the behavior of the system in such a way that it can be manipulated outside the constraints of physical reality.

An appropriate means for achieving this goal is to express the impact that initial conditions, control inputs, and disturbances have on internal variables and on the output, by a set of mathematical equations. This set of equations is called a *model*.

The power of a mathematical model lies in the fact that it can be simulated in hypothetical situations, be subject to states that would be dangerous in reality, and can be used as a basis for synthesizing controllers.

To tap into this power however, it is first necessary to come up with the model– an effort that can range from trivial to near impossible.

Therefore, just as in control-system design itself, there is also an art and a science in the building of models. The following sections briefly raise some of the issues involved.

3.3   Model Complexity

In building a model, it is important to bear in mind that all real processes are complex, and hence any attempt to build an exact description of the plant is usually an impossible goal. Fortunately, feedback is usually very forgiving and hence, in the context of control-system design, one can usually get away with rather simple models, provided that they capture the essential features of the problem.

Actually both *art* and *science* are involved in deriving a model which, on the one hand, captures the plant features which are relevant to designing the controller, yet, on the other hand, is not so complex as to mask the essence of the problem. This is a nontrivial task, and it often involves iteration and refinement as the solution proceeds. It is usually best to start simple, then add features as the solution evolves.

It is also important to note that models for control purposes usually differ from those intended for such other purposes as process design. Control-relevant models describe the dynamic quantitative relationships between plant inputs and outputs.

Fine internal details of a plant are relevant only to the extent that they are necessary to achieve the desired level of performance.

All real systems are arbitrarily complex, so all models must be approximate descriptions of the process. We introduce several terms to make this clear:

- **Nominal model.** This is an approximate description of the plant used for control-system design.

- **Calibration model.** This is a more comprehensive description of the plant. It includes other features not used for control-system design but having a direct bearing on the achieved performance.

- **Model error.** This is the difference between the nominal model and the calibration model. Details of this error might be unknown, but various bounds might be available for it.

In the sequel, we will often refer to the calibration model as the *true plant*. However, the reader should note that the calibration model will also generally not be an exact description of the true plant, and hence one needs to exercise caution when interpreting the results.

To decide what is a sensible nominal model is usually not easy. For the moment, let us just say that it should capture the control-relevant features of the plant dynamics and plant nonlinearities. The authors have seen several examples in industry where extremely intricate physical models were developed but they ultimately turned out to be of limited value for control-system design, either because they contained so many free parameters that they were incapable of being calibrated on the real process or because they failed to describe certain key features (such as backlash in a valve) that were ultimately found to dominate controller performance. This will be illustrated by two examples.

Example 3.1 (Control of temperature in an industrial furnace). *Consider a very large industrial furnace that is heated by oil. We want to control the temperature inside the furnace by throttling the valves that regulate the oil flow to the burners. A nominal model in this case should include the dynamics of the furnace heating process. Typically, the speed at which the valves open or close will not be significant in comparison with the relatively slower dynamics of the heating, and hence the inclusion of this effect will usually not add relevant accuracy to the description for the purpose of control. However, careful consideration should be given to the fact that the valves might saturate or stick, and this may point to difficulties in deriving a suitable control law. We should thus include these issues as part of the nominal model if we suspect they are important to the operation of the control system.*

□□□

Example 3.2 (Control of water flow in a pipe). *Consider a pipe where water flow has to be controlled. To that end, a valve is installed, and the control input*

*u(t) is the signal driving the valve positioner. Assuming that the pipe is always
full of water, it is then clear that the model of this plant is dominated by the valve,
because it is the only source of dynamics in the problem. Stickiness of the valve and
saturation might well be significant issues for this control problem.*

□□□

The examples sketched above suggest that modeling errors and model complexity
are relative concepts.

A final point in this discussion relates to the idea of *robustness*. Control design is
typically based on the nominal plant model. However, the controller will be used to
control the real plant. One of the challenges for the designer is to obtain a controller
which, when in command of the true plant, continues to perform essentially as
predicted by the model, without significant degradation. When this is the case,
we will say that we have designed a *robust controller*. To achieve robustness, it is
usually necessary to have a measure of the modeling error, in the form of a bound
of some sort, so that appropriate precautions can be taken in the design phase.

3.4   Building Models

A first possible approach to building a plant model is to postulate a specific model
structure and to use what is known as a *black-box* approach to modeling. In this
approach, one varies the model parameters either by trial-and-error or by an algorithm, until the dynamic behavior of model and plant match sufficiently well.

An alternative approach for dealing with the modeling problem is to use physical laws (such as conservation of mass, energy and momentum) to construct the
model. In this approach, one uses the fact that, in any real system, there are *basic
phenomenological laws* which determine the relationships between all the signals in
the system. These laws relate to the nature of the system and can include physics,
chemistry, and economic theory. One can use these principles to derive a model, as
was done in subsection §2.3.2.

In practice, it is common to combine both black-box and phenomenological ideas
to build a model. Phenomenological insights are often crucial to understanding the
key dynamics (including dominant features), nonlinearities, and significant time
variations in a given system. It will thus help in making an initial choice regarding
the complexity of the model. On the other hand, the black-box approach often
allows one to fit models to sections of the plant where the underlying physics is too
complex to derive a suitable phenomenological model.

To illustrate the point, consider a simple cylindrical tank containing water and
having cross-sectional area A. This tank discharges through an orifice in the bottom
of the tank. Physical principles indicate that the discharge flow $q(t)$ ought to
be reasonably modeled as $q(t) = K\sqrt{h(t)}$, where $h(t)$ is the water level in the
tank and K is a constant to be determined. This constant could, alternatively,
be determined by using physical principles, but it would be a substantial effort.
A simpler method would be to measure $h(t)$ every T seconds, where T is chosen

such that $|h(t) - h(t - T)|$ is small. Then a good estimate of the flow is $\hat{q}(t) = |h(t) - h(t - T)|A/T$. We could then estimate the value of K by using linear regression of $\hat{q}(t)$ on $\sqrt{h(t)}$ for different values of t. We can see that, in this example, the final model combines physical knowledge with experimental observations. This situation occurs very frequently in modeling for control purposes.

Another issue of practical relevance is the inclusion of the actuator in the modeling process. Actuators are, in many cases, highly nonlinear. They usually also have their own dynamic behavior. Indeed, in some cases, the actuator dynamics may actually dominate other plant features. This is, for instance, the situation arising with valves, hydraulic actuators, and controlled rectifiers. Thus, in the sequel, when we refer to the *plant model*, it should be understood that this model also includes the actuators, if necessary.

To summarize

> Control-relevant models are often quite simple compared to the true process and usually combine physical reasoning with experimental data.

3.5   Model Structures

Given the dynamic nature of real processes, the standard mathematical description of process models includes, apart from algebraic relations, the following:

- dependencies on the accumulated (or integrated) effect of process variables, and

- dependencies on the rate of change (or differential) of variables.

These two features determine what is generally called the *plant dynamics* and point to the fact that the behavior of a real process cannot be described satisfactorily without including its past history and the way it deals with changes.

Models can usually be reduced to the form of differential equations (continuous-time), difference equations (discrete-time), or a combination of them (hybrid or sampled-data systems). These models relate plant inputs to selected plant outputs, and they deal with a limited description of the system under study.

In the next two sections, we describe two possible ways that are commonly used to describe models.

3.6   State Space Models

A very valuable and frequently used tool for plant modeling is a *state-variable* description. State variables form a set of inner variables which is a complete set, in the sense that, if these variables are known at some time, then any plant output, $y(t)$, can be computed, at all future times, as a function of the state variables and the present and future values of the inputs.

3.6.1   The General Case

If we denote by x the vector corresponding to a particular choice of state variables, then the general form of a state-variable model is as follows:

for continuous-time systems,

$$\frac{dx}{dt} = f(x(t), u(t), t) \tag{3.6.1}$$

$$y(t) = g(x(t), u(t), t) \tag{3.6.2}$$

for discrete-time systems,

$$x[k+1] = f_d(x[k], u[k], k) \tag{3.6.3}$$

$$y[k] = g_d(x[k], u[k], k) \tag{3.6.4}$$

The fact that a state space description is structured around a first-order vector differential equation (see (3.6.1)) frequently facilitates numerical solutions to various control problems. This is particularly true in the linear case, where very substantial effort has been devoted to numerically robust ways for solving the associated control problems. We will devote Chapter 17 to a more in-depth treatment of linear state space models. We give a brief overview here.

3.6.2   Linear State Space Models

We say that a system is linear if the principle of superposition holds. By this we mean that if initial conditions x_{01} and x_{02} produce responses $h_{01}(t)$ and $h_{02}(t)$, respectively, with zero input, and inputs $u_1(t)$ and $u_2(t)$ produce responses $h_{11}(t)$ and $h_{12}(t)$, respectively, with zero initial conditions, then the response to input $u_1(t) + u_2(t)$ with initial conditions $x_{01} + x_{02}$ is $h_{01}(t) + h_{01}(t) + h_{11}(t) + h_{12}(t)$.

A system is said to be time-invariant if the response to a time-translated input is simply a time translation of the original response–that is, if the input $u_1(t)$ $(\forall t \in \mathbb{R})$ produces a response $g_1(t)$ $(\forall t \in \mathbb{R})$, then the input $u_2(t) = u_1(t+\tau)$ $(\forall t \in \mathbb{R})$, produces the response $g_2(t) = g_1(t+\tau)$ $(\forall t \in \mathbb{R})$.

In the linear, time-invariant case, equations (3.6.1) and (3.6.2) become

$$\frac{dx(t)}{dt} = \mathbf{A}x(t) + \mathbf{B}u(t) \tag{3.6.5}$$

$$y(t) = \mathbf{C}x(t) + \mathbf{D}u(t) \tag{3.6.6}$$

where $\mathbf{A}, \mathbf{B}, \mathbf{C}$, and $\mathbf{D}$ are matrices of appropriate dimensions.

We illustrate by building a state space model for an electrical network.

Example 3.3. *Consider the simple electrical network shown in Figure 3.1. Assume that we want to model the voltage $v(t)$.*

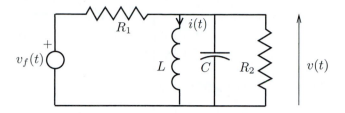

Figure 3.1. Electrical network–state space model

By applying fundamental network laws, we obtain the following equations:

$$v(t) = L\frac{di(t)}{dt} \tag{3.6.7}$$

$$\frac{v_f(t) - v(t)}{R_1} = i(t) + C\frac{dv(t)}{dt} + \frac{v(t)}{R_2} \tag{3.6.8}$$

These equations can be rearranged as follows:

$$\frac{di(t)}{dt} = \frac{1}{L}v(t) \tag{3.6.9}$$

$$\frac{dv(t)}{dt} = -\frac{1}{C}i(t) - \left(\frac{1}{R_1 C} + \frac{1}{R_2 C}\right)v(t) + \frac{1}{R_1 C}v_f(t) \tag{3.6.10}$$

Equations (3.6.9) and (3.6.10) have the form of vector equation (3.6.1) if the state variables are chosen to be $x_1(t) = i(t)$ and $x_2(t) = v(t)$–that is, the state vector becomes $x(t) = [x_1(t) \quad x_2(t)]^T$. The equation corresponding to (3.6.2) is generated by noting that $y(t) = v(t) = x_2(t)$.

Also, equations (3.6.9) and (3.6.10) represent a linear *state space model, one of the form in (3.6.5)–(3.6.6), with*

$$\mathbf{A} = \begin{bmatrix} 0 & \frac{1}{L} \\ -\frac{1}{C} & -\left(\frac{1}{R_1 C} + \frac{1}{R_2 C}\right) \end{bmatrix}; \quad \mathbf{B} = \begin{bmatrix} 0 \\ \frac{1}{R_1 C} \end{bmatrix}; \quad \mathbf{C} = \begin{bmatrix} 0 & 1 \end{bmatrix}; \quad \mathbf{D} = \mathbf{0} \tag{3.6.11}$$

□□□

A further example is provided by a motor.

Example 3.4. *Consider a separately excited d.c. motor. Let $v_a(t)$ denote the armature voltage, $\theta(t)$ the output angle. A simplified schematic diagram of this system is shown in Figure 3.2.*
    Let

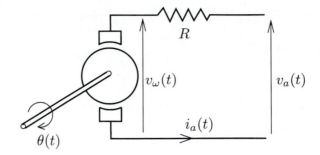

Figure 3.2. Simplified model of a d.c. motor

J -   *be the inertia of the shaft*
$\tau_e(t)$ -   *the electrical torque*
$i_a(t)$ -   *the armature current*
k_1, k_2 -   *constants*
R -   *the armature resistance*

Application of well-known principles of physics tells us that these variables are related as follows:

$$J\ddot{\theta}(t) = \tau_e(t) = k_1 i_a(t) \tag{3.6.12}$$

$$v_\omega(t) = k_2 \dot{\theta}(t) \tag{3.6.13}$$

$$i_a(t) = \frac{v_a(t) - k_2\dot{\theta}(t)}{R} \tag{3.6.14}$$

Combining these equations, we obtain the following second-order differential equation model:

$$J\ddot{\theta}(t) = k_1 \left[\frac{v_a(t) - k_2\dot{\theta}(t)}{R} \right] \tag{3.6.15}$$

We can easily convert this model to state space form by introducing

$$x_1(t) = \theta(t) \tag{3.6.16}$$

$$x_2(t) = \dot{\theta}(t) \tag{3.6.17}$$

The model (3.6.15) can be rewritten as

$$\frac{d}{dt}\begin{pmatrix} x_1(t) \\ x_2(t) \end{pmatrix} = \begin{bmatrix} 0 & 1 \\ 0 & \frac{-k_1 k_2}{R} \end{bmatrix} \begin{bmatrix} x_1(t) \\ x_2(t) \end{bmatrix} + \begin{bmatrix} 0 \\ \frac{k_1}{R} \end{bmatrix} v_a(t) \tag{3.6.18}$$

This seems very simple, but the reader may be surprised to find how often simple servos of the type shown in Figure 3.2 arise in practice. They are the basis of many positioning systems and robots.

□□□

3.7   Solution of Continuous-Time State Space Models

Because state space models are described by (a set of) first-order differential equations, it is quite easy to solve them.

A key quantity in determining solutions to state equations is the *matrix exponential*, defined as

$$e^{\mathbf{A}t} = \mathbf{I} + \sum_{i=1}^{\infty} \frac{1}{i!} \mathbf{A}^i t^i \qquad (3.7.1)$$

The explicit solution to the linear state equation is then given by

$$x(t) = e^{\mathbf{A}(t-t_o)} x_o + \int_{t_o}^{t} e^{\mathbf{A}(t-\tau)} \mathbf{B} u(\tau) d\tau \qquad (3.7.2)$$

This claim can be verified by direct substitution of (3.7.2) into (3.6.5). To perform the necessary differentiation, note that

$$\frac{de^{\mathbf{A}t}}{dt} = \mathbf{A} e^{\mathbf{A}t} = e^{\mathbf{A}t} \mathbf{A} \qquad (3.7.3)$$

and also recall Leibnitz's rule :

$$\frac{d}{dt} \int_{f(t)}^{g(t)} H(t,\tau) d\tau = H(t, g(t)) \dot{g}(t) - H(t, f(t)) \dot{f}(t) + \int_{f(t)}^{g(t)} \frac{\partial}{\partial t} H(t,\tau) d\tau \quad (3.7.4)$$

Applying (3.7.4) and (3.7.3) to (3.7.2) yields

$$\dot{x}(t) = \mathbf{A} e^{\mathbf{A}(t-t_o)} x_o + \mathbf{B} u(t) + \mathbf{A} \int_{t_o}^{t} e^{\mathbf{A}(t-\tau)} \mathbf{B} u(\tau) d\tau \qquad (3.7.5)$$

$$= \mathbf{A} x(t) + \mathbf{B} u(t) \qquad (3.7.6)$$

as claimed.

Notice that, if $u(t) = 0 \quad \forall t \geq t_o$, then the matrix $e^{\mathbf{A}(t-\tau)}$ alone determines the transition from $x(\tau)$ to $x(t) \quad \forall t \geq \tau$. This observation justifies the name *transition matrix* that is usually given to the matrix $e^{\mathbf{A}t}$. The model output $y(t)$ is found from (3.6.6) and (3.7.5) to be

$$y(t) = \mathbf{C}e^{\mathbf{A}(t-t_o)}x_o + \mathbf{C}\int_{t_o}^{t} e^{\mathbf{A}(t-\tau)}\mathbf{B}u(\tau)d\tau + \mathbf{D}u(t) \qquad (3.7.7)$$

Notice that the state and the output solutions (3.7.2) and (3.7.7) consist of two terms each, namely, an initial-condition response due to x_o and a forced response that depends on the input $u(t)$ over the interval $[t_o, t]$. The reader is invited to check that the principle of superposition holds for the response given in (3.7.7).

We shall use state space models sparingly in Parts I to III of this book. However, whenever convenient, we will make side comments to show how the various results relate to these models. In parts V to VIII, where more advanced material is presented, we will use state space models frequently, because they simplify much of the presentation.

3.8 High-Order Differential and Difference-Equation Models

An alternative model format that is frequently used is a high-order differential equation that directly relates inputs to outputs. These models are normally known as *input–output models*.

In the continuous-time case, these models have the form

$$l\left(\frac{d^n y(t)}{dt^n}, \cdots, y(t), \frac{d^{n-1}u(t)}{dt^{n-1}}, \cdots, u(t)\right) = 0 \qquad (3.8.1)$$

where l is some nonlinear function.

A simple example of such a model was given in equation (3.6.15).

Similarly, for the discrete case, we can write

$$m\left(y[k+n], y[k+n-1], \cdots, y[k], u[k+n-1], \cdots, u[k]\right) = 0 \qquad (3.8.2)$$

where m is a nonlinear function and where we have used the notation $\{y[k]\}$ to denote the sequence $\{y[k] : k = 0, 1, \dots\}$.

We will describe this kind of model in more detail in Chapter 4.

3.9 Modeling Errors

We argued earlier that models for real processes will invariably involve some level of approximation. It is desirable, if possible, to include knowledge of the degree of approximation into the design procedure.

Say that the true plant and its nominal model are described, respectively, by

$$y = g\langle u \rangle \tag{3.9.1}$$
$$y_o = g_o\langle u \rangle \tag{3.9.2}$$

where g and g_o are general transformations (see section §2.5).

The so-called *additive modeling error* (AME) is then defined by a transformation g_ϵ:

$$y = y_o + g_\epsilon\langle u \rangle \tag{3.9.3}$$

A difficulty with the AME is that it is not scaled relative to the *size* of the nominal model. This is the advantage of the so-called *multiplicative modeling error* (MME), g_Δ, defined by

$$y = g_o\langle u + g_\Delta\langle u \rangle \rangle \tag{3.9.4}$$

Example 3.5. *The output of a plant is assumed to be exactly described by*

$$y = f\langle \mathrm{sat}_\alpha\langle u \rangle \rangle \tag{3.9.5}$$

where $f\langle \circ \rangle$ is a linear transformation and sat_α denotes the α–saturation operator,

$$\mathrm{sat}_\alpha\langle x \rangle = \begin{cases} \alpha & |x(t)| > |\alpha| \\ x & |x(t)| \le |\alpha| \end{cases} \tag{3.9.6}$$

If the nominal model is chosen as $g_o\langle \circ \rangle = f\langle \circ \rangle$ (i.e., if the saturation is ignored), determine the additive and the multiplicative modeling errors.

Solution

Because f is a linear transformation, it is then distributive and additive. Hence, the plant output is

$$\begin{aligned}
y &= f\langle u + \mathrm{sat}_\alpha\langle u \rangle - u \rangle \\
&= f\langle u \rangle + f\langle \mathrm{sat}_\alpha\langle u \rangle - u \rangle \\
&= f\langle u \rangle + g_\epsilon\langle u \rangle \\
&= f\langle u + \mathrm{sat}_\alpha\langle u \rangle - u \rangle \\
&= f\langle u + g_\Delta\langle u \rangle \rangle
\end{aligned} \tag{3.9.7}$$

The AME and the MME are thus described by the block diagrams shown in Figure 3.3.

□□□

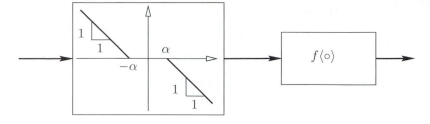

Additive modeling error

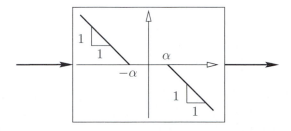

Multiplicative modeling error

Figure 3.3. AME and MME due to saturation

Of course, the exact model errors are rarely known, because the true plant is not precisely known. However, certain information about the size of errors might still be available. This will typically be expressed in terms of a bound on the AME or MME between the nominal model and some other (more complex) calibration model. For example, we might have a description of the form: $||g_\Delta|| < \epsilon$, where $|| \circ ||$ is a suitable norm.

3.10   Linearization

Although almost every real system includes nonlinear features, many systems can be reasonably described, at least within certain operating ranges, by linear models. The incentive to try to approximate a nonlinear system by a linear model is that the science and art of linear control are vastly more complete and simpler than they are for the nonlinear case. A useful way to obtain these linear models is to begin with a nonlinear model and then to build a linear approximation in the neighborhood of a chosen operating point. This approach is not particular to control-system analysis, synthesis, and design but is also a key modeling tool in other fields, for example in analog electronics.

    The linearization strategy can be applied equally well to models incorporating continuous or discrete time, and to state space and input-output models (high-order differential and difference equations). For simplicity, we sketch below the linearization for state space models. Thus, consider

$$\dot{x}(t) = f(x(t), u(t)) \tag{3.10.1}$$
$$y(t) = g(x(t), u(t)) \tag{3.10.2}$$

Say that $\{x_Q(t), u_Q(t), y_Q(t); t \in \mathbb{R}\}$ is a given set of trajectories that satisfy the above equations:

$$\dot{x}_Q(t) = f(x_Q(t), u_Q(t)); \qquad x_Q(t_o) \text{ given} \tag{3.10.3}$$
$$y_Q(t) = g(x_Q(t), u_Q(t)) \tag{3.10.4}$$

The trajectory $\{x_Q(t), u_Q(t), y_Q(t); t \in \mathbb{R}\}$ might correspond to an equilibrium point of the state space model. In this case, x_Q, u_Q, y_Q will not depend on time, and (x_Q, y_Q) will satisfy $\dot{x}_Q = 0$, i.e.

$$f(x_Q, u_Q) = 0 \tag{3.10.5}$$

Say, now, that we want to describe a trajectory $\{x(t), u(t), y(t); t \in \mathbb{R}\}$, where $x(t)$, $u(t)$, and $y(t)$ are *close to* $\{x_Q(t), u_Q(t), y_Q(t); t \in \mathbb{R}\}$. We can then use a first-order Taylor series to approximate the model. This approach leads to

$$\dot{x}(t) \approx f(x_Q, u_Q) + \left.\frac{\partial f}{\partial x}\right|_{\substack{x=x_Q \\ u=u_Q}} (x(t) - x_Q) + \left.\frac{\partial f}{\partial u}\right|_{\substack{x=x_Q \\ u=u_Q}} (u(t) - u_Q) \tag{3.10.6}$$

$$y(t) \approx g(x_Q, u_Q) + \left.\frac{\partial g}{\partial x}\right|_{\substack{x=x_Q \\ u=u_Q}} (x(t) - x_Q) + \left.\frac{\partial g}{\partial u}\right|_{\substack{x=x_Q \\ u=u_Q}} (u(t) - u_Q) \tag{3.10.7}$$

In the above , we have used the notation $\frac{\partial f}{\partial x}$ to denote the matrix having the $ijth$ element $\frac{\partial f_i}{\partial x_j}$. Note that the derivatives are evaluated at the nominal trajectory. In the case of a fixed equilibrium point, these derivative matrices will be constant matrices.

Equations (3.10.7) and (3.10.6) have the following form:

$$\dot{x}(t) = \mathbf{A}x(t) + \mathbf{B}u(t) + \mathbf{E} \tag{3.10.8}$$
$$y(t) = \mathbf{C}x(t) + \mathbf{D}u(t) + \mathbf{F} \tag{3.10.9}$$

where

$$\mathbf{A} = \left.\frac{\partial f}{\partial x}\right|_{\substack{x=x_Q \\ u=u_Q}} ; \qquad \mathbf{B} = \left.\frac{\partial f}{\partial u}\right|_{\substack{x=x_Q \\ u=u_Q}} \tag{3.10.10}$$

$$\mathbf{C} = \left.\frac{\partial g}{\partial x}\right|_{\substack{x=x_Q \\ u=u_Q}} ; \qquad \mathbf{D} = \left.\frac{\partial g}{\partial u}\right|_{\substack{x=x_Q \\ u=u_Q}} \tag{3.10.11}$$

$$\mathbf{E} = f(x_Q, u_Q) - \left.\frac{\partial f}{\partial x}\right|_{\substack{x=x_Q \\ u=u_Q}} x_Q - \left.\frac{\partial f}{\partial u}\right|_{\substack{x=x_Q \\ u=u_Q}} u_Q \tag{3.10.12}$$

$$\mathbf{F} = g(x_Q, u_Q) - \left.\frac{\partial g}{\partial x}\right|_{\substack{x=x_Q \\ u=u_Q}} x_Q - \left.\frac{\partial g}{\partial u}\right|_{\substack{x=x_Q \\ u=u_Q}} u_Q \tag{3.10.13}$$

In general, $\mathbf{A}$, $\mathbf{B}$, $\mathbf{C}$, $\mathbf{D}$, $\mathbf{E}$, and $\mathbf{F}$ will depend on time. However, in the case in which we linearize about an equilibrium point, they will be time-invariant.

It is also possible to write approximate equations in terms of the increments $\Delta x(t) = x(t) - x_Q(t)$, $\Delta u(t) = u(t) - u_Q(t)$. From (3.10.7) and (3.10.6), upon using (3.10.3) and (3.10.4), we have

$$\frac{d\Delta x(t)}{dt} = \mathbf{A}\Delta x(t) + \mathbf{B}\Delta u(t) \tag{3.10.14}$$

$$\Delta y(t) = \mathbf{C}\Delta x(t) + \mathbf{D}\Delta u(t) \tag{3.10.15}$$

Remark 3.1. *The linearization procedure presented above produces a model that is linear in the **incremental** components of inputs and outputs around a chosen operating point (i.e., small-signal model).*

We illustrate by two examples.

Example 3.6. *Consider a continuous-time system having a true model given by*

$$\frac{dx(t)}{dt} = f(x(t), u(t)) = -\sqrt{x(t)} + \frac{(u(t))^2}{3} \tag{3.10.16}$$

Assume that the input $u(t)$ fluctuates around $u = 2$. Find an operating point with $u_Q = 2$ and a linearized model around it.

Solution

(i) *The operating point is computed from (3.10.16) with $u_Q = 2$ and by taking $\frac{dx(t)}{dt} = 0$. This leads to*

$$\sqrt{x_Q} - \frac{(u_Q)^2}{3} = 0 \quad \implies \quad x_Q = \frac{16}{9} \tag{3.10.17}$$

(ii) *The following linearized model is then obtained by expanding (3.10.16) in a Taylor series.*

$$\frac{d\Delta x(t)}{dt} = -\frac{1}{2\sqrt{x_Q}}\Delta x(t) + \frac{2u_Q}{3}\Delta u(t) \qquad (3.10.18)$$

When the numerical values for the operating point are used, we obtain the following linearized model:

$$\frac{d\Delta x(t)}{dt} = -\frac{3}{8}\Delta x(t) + \frac{4}{3}\Delta u(t) \qquad (3.10.19)$$

To appreciate the quality of the approximation, we consider the original system and its linearized model and run a simulation where the input to the original system is a constant equal to 2 plus an increasing-amplitude pulse sequence. The results are shown in Figure 3.4. We can see there that the linearization error increases as the system is driven away from the operating point around which the linearized model was computed.

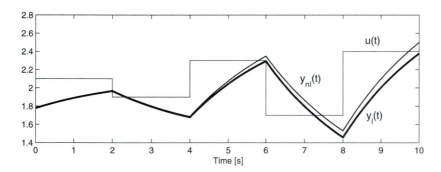

Figure 3.4. Nonlinear system output, $y_{nl}(t)$, and linearized system output, $y_l(t)$, for a square-wave input of increasing amplitude, $u(t)$

□□□

As a slightly more complex example, we present the following:

Example 3.7 (Inverted pendulum). *Many readers will be familiar with the objective of balancing a broom (or rod) on the tip of one's finger. Common experience indicates that this is a difficult control task. Many universities around the world have built inverted-pendulum systems to demonstrate control issues. A photograph*

*of one built at the University of Newcastle, Australia is shown on the book's web-
site. The reason that the problem is interesting from a control perspective is that
it illustrates many of the difficulties associated with real-world control problems.
For example, the model is very similar to that used in rudder roll stabilization of
ships. The latter problem will be addressed in Chapter 23. A sketch of a typical
inverted-pendulum system is shown in Figure 3.5.*

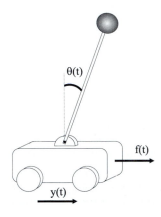

Figure 3.5. Inverted pendulum

In Figure 3.5, we have used the following notation:

$y(t)$ -   *distance from some reference point*
$\theta(t)$ -   *angle of pendulum*
M -   *mass of cart*
m -   *mass of pendulum (assumed concentrated at tip)*
ℓ -   *length of pendulum*
$f(t)$ -   *forces applied to pendulum*

Application of Newtonian physics to this system leads to the following model:

$$\ddot{y} = \frac{1}{\lambda_m + \sin^2\theta(t)}\left[\frac{f(t)}{m} + \dot{\theta}^2(t)\ell\sin\theta(t) - g\cos\theta(t)\sin\theta(t)\right] \qquad (3.10.20)$$

$$\ddot{\theta} = \frac{1}{\ell\lambda_m + \sin^2\theta(t)}\left[-\frac{f(t)}{m}\cos\theta(t) + \dot{\theta}^2(t)\ell\sin\theta(t)\cos\theta(t) + (1-\lambda_m)g\sin\theta(t)\right]$$
$$(3.10.21)$$

where $\lambda_m = (M/m)$.

*These equations are nonlinear. However, for small departures of θ from the
vertical position we can linearize about $\theta_o = 0$, $\dot{\theta}_o = 0$. Using the methods outlined
above, we obtain:*

$$\ddot{y} = \frac{1}{\lambda_m}\left[\frac{f(t)}{m} - g\theta(t)\right] \tag{3.10.22}$$

$$\ddot{\theta} = \frac{1}{\ell\lambda_m}\left[-\frac{f(t)}{m} + (1+\lambda_m)g\theta(t)\right] \tag{3.10.23}$$

We can now convert this to state space form, with input $u(t) = f(t)$ *and output* $y(t)$, *by introducing*

$$
\begin{aligned}
x_1(t) &= y(t)\\
x_2(t) &= \dot{y}(t)\\
x_3(t) &= \theta(t)\\
x_4(t) &= \dot{\theta}(t)
\end{aligned}
$$

This leads to a linear state space model as in (3.6.5), (3.6.6) *where*

$$
\mathbf{A} = \begin{bmatrix} 0 & 1 & 0 & 0\\ 0 & 0 & \frac{-mg}{M} & 0\\ 0 & 0 & 0 & 1\\ 0 & 0 & \frac{(M+m)g}{M\ell} & 0 \end{bmatrix}; \quad \mathbf{B} = \begin{bmatrix} 0\\ \frac{1}{M}\\ 0\\ -\frac{1}{M\ell} \end{bmatrix}; \quad \mathbf{C} = \begin{bmatrix} 1 & 0 & 0 & 0 \end{bmatrix} \tag{3.10.24}
$$

More will be said later about control issues associated with this system.

□□□

Remark 3.2. *Modern computational packages include special commands to compute linearized models around a user-defined (precomputed) operating point. In the case of MATLAB–SIMULINK, the appropriate commands are* **linmod** *(for continuous-time systems) and* **dlinmod** *(for discrete-time and hybrid systems).*

Remark 3.3. *It is obvious that linearized models are only approximate models. Thus, these models should be used with appropriate caution (as indeed should all models). In the case of linearized models, the next term in the Taylor series expansion can often be employed to tell us something about the size of the associated modeling error.*

Linear models often give deep insights and lead to simple control strategies. Linear models can be obtained by linearizing a nonlinear model at an operating point. Caution is needed to deal with unavoidable modeling errors.

3.11   Case Studies

Space precludes us from giving a more detailed treatment of modeling. Indeed, this usually falls into the realm of other courses that deal specifically with this topic.

Nonetheless, simple models for all of the case studies will be presented when these problems are discussed. The reader is referred to the following:

- Satellite Tracking (Chapter 22)

- pH Control (Chapter 19 and the website)

- Continuous Casting Machine (Chapters 2 and 8, and the website)

- Sugar Mill (Chapter 24 and the website)

- Distillation Column (Chapter 6 and the website)

- Ammonia Synthesis (Chapter 20)

- Zinc Coating-Mass Estimation (Chapter 22)

- BISRA Gauge (Chapter 8 and the website)

- Roll Eccentricity (Chapters 10 and 22, and the website)

- Hold-up Effect in Rolling Mills (Chapters 8 and 10, and the website)

- Flatness Control in Steel Rolling (Chapter 21 and the website)

- Vibration Control (Chapter 22)

- Servomechanism (Chapter 3)

- Tank Level Estimation (Chapter 18 and the website)

- Four Coupled Tanks (Chapters 21 and 24, and the website)

- Ball and Plate Mechanism (the website)

- Heat Exchanger (Chapter 4)

- Inverted Pendulum (Chapters 3, 9, and 24, and the website)

- Ship Rudder Roll Stabilization (Chapter 23)

3.12 Summary

- In order to design a controller systematically for a particular system, one needs a formal–though possibly simple–description of the system. Such a description is called a model.

- A model is a *set of mathematical equations* that are *intended* to *capture* the effect of *certain system variables* on certain other system variables.

- The italicized expressions above should be understood as follows:

○ *certain system variables*: It is usually neither possible nor necessary to model the effect of every variable on every other variable; one therefore limits oneself to certain subsets. Typical examples include the effect of input on output, the effect of disturbances on output, the effect of a reference-signal change on the control signal, or the effect of various unmeasured internal system variables on each other.

○ *capture*: A model is never perfect, and it is therefore always associated with a modeling error. The word *capture* highlights the existence of errors, but it does not yet concern itself with the precise definition of their type and effect.

○ *intended*: This word is a reminder that one does not always succeed in finding a model with the desired accuracy, and hence some iterative refinement may be needed.

○ *set of mathematical equations*: There are numerous ways of describing the system behavior, such as linear or nonlinear differential or difference equations.

• Models are classified according to properties of the equation they are based on. Examples of classification include the following:

Model attribute	*Contrasting attribute*	*Asserts whether . . .*
Single-input single-output	Multiple-input multiple-output	. . . the model equations have one input and one output only.
Linear	Nonlinear	. . . the model equations are linear in the system variables.
Time-varying	Time-invariant	. . . the model parameters are constant.
Continuous	Sampled	. . . model equations describe the behavior at every instant of time or only in discrete *samples* of time.
Input-output	state space	. . . the model equations rely on functions of input and output variables only or also include the so-called *state variables*.
Lumped parameter	Distributed parameter	. . . the model equations are ordinary or partial differential equations.

• In many situations, nonlinear models can be linearized around a user-defined operating point.

3.13   Further Reading

Modeling

Campbell, D.P. (1958). *Process Dynamics*. Wiley, New York.

Cannon, R. (1967). *Dynamics of Physical Systems*. McGraw-Hill.

Ogata, K. (1998). *System Dynamics*. Prentice-Hall, Upper Saddle River, N.J., 3rd edition.

Stephanopoulos, G. (1984). *Chemical Process Control: An Introduction to Theory and Practice*. Prentice-Hall, Englewood Cliffs, N.J.

Identification

Bohlin, T. and Graebe, S.F. (1995). Issues in nonlinear stochastic grey box identification. *International Journal of Adaptive Control and Signal Processing*, 9(6):465-490.

Goodwin, G.C. and Payne, R.L. (1977). *Dynamic System Identification*. Academic Press, New York.

Ljung, L. (1999). *System Identification. Theory for the User*. Prentice-Hall, Englewood Cliffs, N.J., 2nd edition.

Ljung, L. and Söderström, T. (1983). *Theory and Practice of Recursive Identification*. MIT Press, Cambridge, Mass.

3.14   Problems for the Reader

Problem   3.1. *Consider an electronic amplifier with input voltage $v_i(t)$ and output voltage $v_o(t)$. Assume that*

$$v_o(t) = 8v_i(t) + 2 \tag{3.14.1}$$

3.1.1 *Show that the amplifier does not strictly satisfy the principle of superposition. Thus, this system is not strictly linear. (A better term for this system would be* affine..)

3.1.2 *Note that the system can also be written as follows:*

$$v_o(t) = 8v_i(t) + 2d_i(t) \tag{3.14.2}$$

where $d_i(t)$ is a constant offset (equal to 1).

Show that the principle of superposition does hold for the input vector $[v_i(t) \quad d_i(t)]^T$.

3.1.3 *Obtain an incremental model for $\Delta v_o(t) = v_o(t) - v_{oQ}$, $\Delta v_i(t) = v_i(t) - v_{iQ}$, where (v_{iQ}, v_{oQ}) is any point satisfying the model (3.14.1). Show that this incremental model is the same for all choices of the pair (v_{iQ}, v_{oQ}).*

Problem   3.2. *Consider an electronic amplifier, similar to that in Problem 3.1, but having input voltage $v_i(t)$ and output voltage $v_o(t)$ related by*

$$v_o(t) = 8(v_i(t))^2 \tag{3.14.3}$$

3.2.1 *Show that this amplifier is nonlinear, by showing that if we compose $v_i(t)$ from two subcomponents, for example, $v_i(t) = v_{i1}(t) + v_{i2}(t)$, then the response to $v_i(t)$ is not the sum of the responses to $v_{i1}(t)$ and $v_{i2}(t)$.*

3.2.2 *Assume that $v_i(t) = 5 + \cos(100t)$. What is the output?*

3.2.3 *Obtain an incremental model for $\Delta v_o(t) = v_o(t) - v_{oQ}$, $\Delta v_i(t) = v_i(t) - v_{iQ}$, where (v_{iQ}, v_{oQ}) is any point satisfying the model (3.14.3). Show that this incremental model depends on the choice of the pair (v_{iQ}, v_{oQ}). Compare this with the result in Problem 3.1, and discuss.*

Problem   3.3. *A system has an input-output model given by*

$$y(t) = 2|u(t)| \tag{3.14.4}$$

3.3.1 *Build a small-signal model around $u_Q = 5$.*

3.3.2 *Discuss why this cannot be done at $u_Q = 0$.*

Problem 3.4. *A discrete-time system with input $u[k]$ and output $y[k]$ is described by the difference equation*

$$y[k] - 0.8y[k-1] + 0.15y[k-2] = 0.2u[k-i] \qquad (3.14.5)$$

3.4.1 *Build a state space model for $i = 0$.*

3.4.2 *Repeat for $i = 1$.*

Problem 3.5. *Consider a savings account yielding a 5% interest p.a. Assume that the opening deposit is US\$200 and that a yearly deposit of $d[i]$ is made at the end of the ith year. Build a model that describes the evolution of the balance of this account at the end of the kth year, for $k = 1, 2, 3, \ldots$.*

Problem 3.6. *Consider the mechanical system in Figure 3.6, where $u(t)$ is an externally applied force, $v(t)$ is the speed of the mass w.r.t. the wall inertial system, and $y(t)$ is the displacement from the wall.*

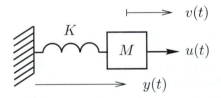

Figure 3.6. Dynamic mechanical system

Find the differential equation describing the relationship between the input $u(t)$ and the variables $v(t)$ and $y(t)$.

Problem 3.7. *Consider a system, with input $u(t)$ and output $y(t)$, having a (non-linear) calibration model given by*

$$\frac{dy(t)}{dt} + \left(2 + 0.1\left(y(t)\right)^2\right) y(t) = 2u(t) \qquad (3.14.6)$$

Assume that we associate with this system a (linear) nominal model given by

$$\frac{dy(t)}{dt} + 2y(t) = 2u(t) \tag{3.14.7}$$

Simulate the two systems, and plot the model error for $u(t) = A\cos(0.5t)$, *with* $A = 0.1$, 1.0, *and* 10.

Discuss why the modeling error grows when A is made larger.

Problem 3.8. *Consider the following nonlinear state space model*

$$\dot{x}_1(t) = -2x_1(t) + 0.1x_1(t)x_2(t) + u(t) \tag{3.14.8}$$
$$\dot{x}_2(t) = -x_1(t) - 2x_2(t)\left(x_1(t)\right)^2 \tag{3.14.9}$$
$$y(t) = x_1(t) + (1 + x_2(t))^2 \tag{3.14.10}$$

Build a linearized model around the operating point given by $u_Q = 1$.

Problem 3.9. *Consider a nonlinear plant having a model given by*

$$\frac{d^2y(t)}{dt^2} + \left[1 + 0.2\sin\left(y(t)\right)\right]\frac{dy(t)}{dt} + 0.5y(t) = 3u(t) - sign\left(u(t)\right) \tag{3.14.11}$$

3.9.1 *Find an approximate inverse for this plant by using the architecture shown in Figure 2.7 on page 33, with* $h\langle\circ\rangle$ *a linear nondynamic gain.*

3.9.2 *Set up a simulation, as shown in Figure 3.7,*

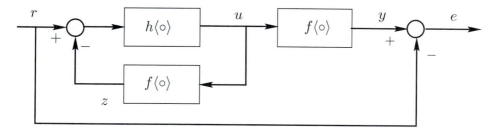

Figure 3.7. Scheme to evaluate the quality of the inversion

where $f\langle\circ\rangle$ *denotes the nonlinear dynamic system described in* (3.14.11).

3.9.3 *Evaluate the quality of your inverse, by examining e, with the SIMULINK schematic in Figure 3.7. Use sine waves of frequency in the range 0 to 0.5[rad/s].*

Problem 3.10. *Consider a nonlinear system having a model given by*

$$\frac{d^2y(t)}{dt^2} + 3\big[y(t) + 0.2\big(y(t)\big)^3\big]\frac{dy(t)}{dt} + 2y(t) = 2\frac{du(t)}{dt} + u(t) \qquad (3.14.12)$$

Build a linear (small-signal) model around the operating point defined by a constant input u = 2.

Problem 3.11. *Consider the mechanical system shown in Figure 3.8. An external force, $f(t)$, is applied at one end of the lever, and it is counterbalanced by a spring connected to the other end. The lever rotates around its center, where the friction torque is proportional to the rotational speed.*

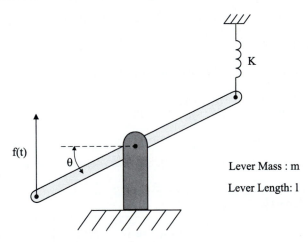

Figure 3.8. Lever system

3.11.1 *Without building any model, tell how many states the system has.*

3.11.2 *Build an input-output linear model for the system, with input $\Delta f(t)$ and output $\Delta\theta(t)$. Take $\theta_q = 0$.*

3.11.3 *Build a linear state space model.*

Chapter 4

CONTINUOUS-TIME SIGNALS AND SYSTEMS

4.1 Preview

The advantage of being able to cast a modeling problem into the form of a linear approximation is that subsequent analysis, as well as controller design, can draw on the wealth of information that is available about the operation of linear systems. In this chapter, we will cover the fundamentals of this theory for linear models of continuous-time processes. Specific topics to be covered include the following:

- linear high-order differential equation models;

- Laplace transforms, which convert linear differential equations to algebraic equations, thus greatly simplifying their study;

- methods for assessing the stability of linear dynamic systems; and

- frequency response.

4.2 Linear Continuous-Time Models

A fundamentally important linear model is the linear version of the general high-order differential model briefly introduced in section §3.8 of Chapter 3. The linear form of this model is:

$$\frac{d^n y(t)}{dt^n} + a_{n-1}\frac{d^{n-1}y(t)}{dt^{n-1}} + \ldots + a_0 y(t) = b_{n-1}\frac{d^{n-1}}{dt^{n-1}}u(t) + \ldots + b_0 u(t) \quad (4.2.1)$$

The reader is invited to recall the model given in equation (3.6.15) for a simple servo. This model was in the form of (4.2.1).

Sometimes it is convenient to use operator notation to denote the operation of differentiation. Consequently, we introduce the Heaviside, or differential, operator, $\rho\langle\circ\rangle$, defined by

$$\rho\langle f(t)\rangle = \rho f(t) \triangleq \frac{df(t)}{dt} \tag{4.2.2}$$

$$\rho^n \langle f(t)\rangle = \rho^n f(t) = \rho\left\langle \rho^{n-1}\langle f(t)\rangle\right\rangle = \frac{df^n(t)}{dt^n} \tag{4.2.3}$$

In terms of this operator, the model (4.2.1) can be written

$$\rho^n y(t) + a_{n-1}\rho^{n-1} y(t) + \ldots + a_0 y(t) = b_{n-1}\rho^{n-1} u(t) + \ldots + b_0 u(t) \tag{4.2.4}$$

Of major importance for linear systems is that the *Principle of Superposition* holds. As described in subsection §3.6.2, this implies that, if two inputs are applied simultaneously, then the response due to these inputs is simply the sum of the responses resulting from applying the inputs separately. This has wide ranging implications. For example, one can understand the response to complex inputs by decomposing them into elementary subcomponents. Regrettably, the same principle fails to hold for nonlinear systems, and this implies that one cannot analyze responses in a piecemeal fashion, but, instead, one is forced to consider the combined inputs in total. Thus, one can gain much more insight into linear systems (for example, by considering benchmark inputs) than one, in general, can for nonlinear systems.

4.3   Laplace Transforms

The study of differential equations of the type described above is a rich and interesting subject. Of all the methods available for studying linear differential equations, one particularly useful tool is provided by Laplace Transforms. A powerful feature of this transform is that it converts linear differential equations into algebraic equations. This is very helpful for analysis purposes.

Definition of the Transform

Consider a continuous-time signal $y(t), 0 \leq t < \infty$. The Laplace-transform pair associated with $y(t)$ is defined as follows:

$$\mathcal{L}[y(t)] = Y(s) = \int_{0^-}^{\infty} e^{-st} y(t)dt \tag{4.3.1}$$

$$\mathcal{L}^{-1}[y(s)] = y(t) = \frac{1}{2\pi j} \int_{\sigma - j\infty}^{\sigma + j\infty} e^{st} Y(s)ds \tag{4.3.2}$$

$Y(s)$ is referred to as the Laplace transform of $y(t)$. The transform pair is well-defined if there exist $\sigma \in \mathbb{R}$ and a positive constant $k < \infty$ such that

$$|y(t)| < ke^{\sigma t}\,;\; \forall t \geq 0 \qquad\qquad (4.3.3)$$

The region $\Re\{s\} \geq \sigma$ is known as the *region of convergence* of the transform.

The preceding transform pair can be used to derive a table of transform pairs. Examples of transforms frequently used in control applications are given in Table 4.1. The reader is encouraged to derive some of the results from first principles.

There are also many interesting properties that flow from the definition of the transforms. Some of these are listed in Table 4.2. Again, the reader is encouraged to revise these properties or to derive them from first principles.

4.4   Laplace Transform. Properties and Examples

Equations (4.3.1) and (4.3.2) provide a way to take signal and system descriptions to the s-domain and back to the time domain. However, equation (4.3.2) is rarely used to obtain the inverse Laplace transform, because the Laplace transforms of most signals of interest are rational fractions in s. Thus, a partial-fraction approach is normally used to evaluate the inverse transform by comparing the transform with standard results.

The reader is asked to verify the following key result (see Table 4.2) regarding the transform of the derivative of a function

$$\mathcal{L}\left[\frac{dy(t)}{dt}\right] = sY(s) - y(0^-) \qquad\qquad (4.4.1)$$

where $\mathcal{L}\left[y(t)\right] = Y(s)$. This result can be used to convert differential equations into algebraic equations in the variable s.

We illustrate by an example.

Example 4.1. *Consider the d.c. motor problem described in Example 3.4. Say, by way of illustration, that $a_1 = 2$, $b_0 = 1$. (Note that, in this particular example, $a_0 = 0$.) Then, applying (4.4.1), and taking Laplace transforms of the model, we obtain*

$$s^2\Theta(s) + 2s\Theta(s) - (s+2)\theta(0^-) - \dot{\theta}(0^-) = V_a(s) \qquad\qquad (4.4.2)$$

Say that the initial conditions are specified as $\theta(0^-) = 0$, $\dot{\theta}(0^-) = 0$, and say that the input is a unit step applied at $t = 0$. Then

$f(t) \qquad (t \geq 0)$	$\mathcal{L}\left[f(t)\right]$	Region of Convergence
1	$\dfrac{1}{s}$	$\sigma > 0$
$\delta_D(t)$	1	$\lvert \sigma \rvert < \infty$
t	$\dfrac{1}{s^2}$	$\sigma > 0$
$t^n \qquad n \in \mathbb{Z}^+$	$\dfrac{n!}{s^{n+1}}$	$\sigma > 0$
$e^{\alpha t} \qquad \alpha \in \mathbb{C}$	$\dfrac{1}{s - \alpha}$	$\sigma > \Re\{\alpha\}$
$te^{\alpha t} \qquad \alpha \in \mathbb{C}$	$\dfrac{1}{(s - \alpha)^2}$	$\sigma > \Re\{\alpha\}$
$\cos(\omega_o t)$	$\dfrac{s}{s^2 + \omega_o^2}$	$\sigma > 0$
$\sin(\omega_o t)$	$\dfrac{\omega_o}{s^2 + \omega_o^2}$	$\sigma > 0$
$e^{\alpha t}\sin(\omega_o t + \beta)$	$\dfrac{(\sin\beta)s + \omega_o^2\cos\beta - \alpha\sin\beta}{(s - \alpha)^2 + \omega_o^2}$	$\sigma > \Re\{\alpha\}$
$t\sin(\omega_o t)$	$\dfrac{2\omega_o s}{(s^2 + \omega_o^2)^2}$	$\sigma > 0$
$t\cos(\omega_o t)$	$\dfrac{s^2 - \omega_o^2}{(s^2 + \omega_o^2)^2}$	$\sigma > 0$
$\mu(t) - \mu(t - \tau)$	$\dfrac{1 - e^{-s\tau}}{s}$	$\lvert \sigma \rvert < \infty$

Table 4.1. Laplace-transform table

$f(t)$	$\mathcal{L}\left[f(t)\right]$	Names
$\displaystyle\sum_{i=1}^{l} a_i f_i(t)$	$\displaystyle\sum_{i=1}^{l} a_i F_i(s)$	Linear combination
$\dfrac{dy(t)}{dt}$	$sY(s) - y(0^-)$	Derivative Law
$\dfrac{d^k y(t)}{dt^k}$	$s^k Y(s) - \sum_{i=1}^{k} s^{k-i} \left.\dfrac{d^{i-1}y(t)}{dt^{i-1}}\right|_{t=0^-}$	High-order derivative
$\displaystyle\int_{0^-}^{t} y(\tau)d\tau$	$\dfrac{1}{s}Y(s)$	Integral Law
$y(t-\tau)\mu(t-\tau)$	$e^{-s\tau}Y(s)$	Delay
$ty(t)$	$-\dfrac{dY(s)}{ds}$	
$t^k y(t)$	$(-1)^k \dfrac{d^k Y(s)}{ds^k}$	
$\displaystyle\int_{0^-}^{t} f_1(\tau)f_2(t-\tau)d\tau$	$F_1(s)F_2(s)$	Convolution
$\displaystyle\lim_{t\to\infty} y(t)$	$\displaystyle\lim_{s\to 0} sY(s)$	Final-Value Theorem
$\displaystyle\lim_{t\to 0+} y(t)$	$\displaystyle\lim_{s\to\infty} sY(s)$	Initial Value Theorem
$f_1(t)f_2(t)$	$\dfrac{1}{2\pi j}\displaystyle\int_{\sigma-j\infty}^{\sigma+j\infty} F_1(\zeta)F_2(s-\zeta)d\zeta$	Time-domain product
$e^{at}f_1(t)$	$F_1(s-a)$	Frequency Shift

Table 4.2. Laplace-transform properties–note that $F_i(s) = \mathcal{L}\left[f_i(t)\right]$, $Y(s) = \mathcal{L}\left[y(t)\right]$, $k \in \{1,2,3,\dots\}$, and $f_1(t) = f_2(t) = 0 \quad \forall t < 0$.

$$\Theta(s) = \left(\frac{1}{s^2 + 2s}\right) V_a(s) \tag{4.4.3}$$

$$= \left(\frac{1}{s^2 + 2s}\right) \frac{1}{s} \tag{4.4.4}$$

Expanding in partial fractions [1] *gives*

$$\Theta(s) = \frac{1}{4(s+2)} + \frac{1}{2s^2} - \frac{1}{4s} \tag{4.4.5}$$

Hence, applying the results from Table 4.1, the output response for $t \geq 0$ is

$$\theta(t) = \frac{1}{4}e^{-2t} + \frac{1}{2}t - \frac{1}{4} \tag{4.4.6}$$

□□□

> Laplace transforms are useful in the study of linear differential systems, because
> they convert differential equations into algebraic equations.

4.5   Transfer Functions

4.5.1   High-Order Differential Equation Models

Consider again the linear high-order differential equation model (4.2.1). Taking
Laplace Transforms converts the differential equation into the following algebraic
equation:

$$s^n Y(s) + a_{n-1} s^{n-1} Y(s) + \ldots + a_0 Y(s)$$
$$= b_{n-1} s^{n-1} U(s) + \ldots + b_0 U(s) + f(s, x_o) \tag{4.5.1}$$

where $f(s, x_o)$ stands for a function of the initial conditions. In the case of *zero
initial conditions*, we have

$$Y(s) = G(s)U(s) \tag{4.5.2}$$

where

---
[1] Use the MATLAB command **residue**.

$$G(s) = \frac{B(s)}{A(s)} \tag{4.5.3}$$

and

$$A(s) = s^n + a_{n-1}s^{n-1} + \ldots + a_0 \tag{4.5.4}$$
$$B(s) = b_{n-1}s^{n-1} + b_{n-2}s^{n-2} + \ldots + b_0 \tag{4.5.5}$$

$G(s)$ is called the *transfer function*. The representation (4.5.3) is a very useful one in gaining insights into various control-design questions.

Example 4.2. *The transfer function for the system in example 4.1 is*

$$G(s) = \frac{1}{s^2 + 2s} \tag{4.5.6}$$

4.5.2   Transfer Functions for Continuous-Time State Space Models

As usual, we can use the Laplace transform to *solve* the differential equations arising from a state space model. Taking the Laplace transform in (3.6.5), (3.6.6) yields

$$sX(s) - x(0) = \mathbf{A}X(s) + \mathbf{B}U(s) \tag{4.5.7}$$
$$Y(s) = \mathbf{C}X(s) + \mathbf{D}U(s) \tag{4.5.8}$$

and, hence,

$$X(s) = (s\mathbf{I} - \mathbf{A})^{-1}x(0) + (s\mathbf{I} - \mathbf{A})^{-1}\mathbf{B}U(s) \tag{4.5.9}$$
$$Y(s) = [\mathbf{C}(s\mathbf{I} - \mathbf{A})^{-1}\mathbf{B} + \mathbf{D}]U(s) + \mathbf{C}(s\mathbf{I} - \mathbf{A})^{-1}x(0) \tag{4.5.10}$$

Actually, these equations can be used to rederive the formula for the solution to the state space model. In particular, equation (3.7.7) can be obtained by applying the inverse Laplace transform to (4.5.10) and by noting that

$$\mathcal{L}\left[e^{\mathbf{A}t}\right] = (s\mathbf{I} - \mathbf{A})^{-1} \tag{4.5.11}$$

We see that, with zero initial conditions, the Laplace Transform of the output, $Y(s)$, is related to the transform of the input $U(s)$ as follows:

$$Y(s) = \mathbf{G}(s)U(s) \tag{4.5.12}$$
$$\mathbf{G}(s) = \mathbf{C}(s\mathbf{I} - \mathbf{A})^{-1}\mathbf{B} + \mathbf{D} \tag{4.5.13}$$

Thus $\mathbf{G}(s)$ is the system transfer function.

Example 4.3. *Consider again the inverted pendulum from Chapter 3. If we derive the transfer function from U to Y, then we obtain*

$$\mathbf{G}(s) = \mathbf{C}(s\mathbf{I} - \mathbf{A})^{-1}\mathbf{B} \tag{4.5.14}$$

$$= \frac{1}{M} \frac{(s^2 - b^2)}{s^2(s^2 - a^2)} \tag{4.5.15}$$

where $a = \sqrt{\frac{(M+m)g}{M\ell}}$; $b = \sqrt{\frac{g}{\ell}}$

□□□

We next define some terms used in connection with transfer functions.

Consider the transfer function given in equations (4.5.3) to (4.5.5).

For simplicity, we assume for the moment that $B(s)$ and $A(s)$ are not simultaneously zero for the same value of s. Also, let the degree of $A(s)$ and $B(s)$ be n and m, respectively. (This assumption will be further clarified in connection with controllability and observability in Chapter 17.) We then define the following terms.

   (i) The roots of $B(s) = 0$ are called the system *zeros*.

  (ii) The roots of $A(s) = 0$ are called the system *poles*.

 (iii) If $A(s) = 0$ has n_k roots at $s = \lambda_k$, the pole λ_k is said to have *multiplicity* n_k.

  (iv) The difference in degrees between $A(s)$ and $B(s)$ is called the *relative degree*.

   (v) If $m < n$ we say that the model is *strictly proper*. This means positive relative degree.

  (vi) If $m = n$ we say that the model is *biproper*. This means zero relative degree.

 (vii) If $m \leq n$ we say that the model is *proper*.

(viii) If $m > n$ we say that the model is *improper* (or has *negative relative degree*).

□□□

Remark 4.1. *Real systems are almost always strictly proper. However, some controller-design methods lead to biproper or even to improper transfer functions. To be implemented, these controllers are usually made proper–for example, by augmenting $A(s)$ with factors of the type $(\alpha_i s + 1)$, where $\alpha_i \in \mathbb{R}^+$.*

Remark 4.2. *Often, practical systems have a time delay between input and output. This is usually associated with the transport of material from one point to another. For example, if there is a conveyor belt or pipe connecting different parts of a plant, then it will invariably introduce a delay.*

The transfer function of a pure delay is of the following form (see Table 4.2):

$$H(s) = e^{-sT_d} \qquad (4.5.16)$$

where T_d is the delay (in seconds). T_d will typically vary with the transportation speed.

Example 4.4 (Heating system). *As a simple example of a system having a pure time delay, consider the heating system shown in Figure 4.1.*

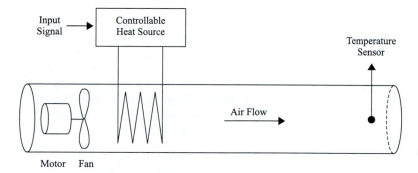

Figure 4.1. Heat-transfer system

The transfer function from the input (the voltage applied to the heating element) to the output (the temperature as seen by the thermocouple) is approximately of the form:

$$H(s) = \frac{Ke^{-sT_d}}{(\tau s + 1)} \qquad (4.5.17)$$

Note that K, T_d, and τ all depend on the speed of the fan, which changes the transport lag from the heater to the measured output as well as various heat-transfer coefficients.

Although this is a very simple example, the kind of model given in (4.5.17) is extremely prevalent in process-control applications.

□□□

Our conclusion regarding transfer functions can be summarized as follows:

Transfer functions describe the input-output properties of linear systems in algebraic form.

4.6   Stability of Transfer Functions

We have seen that the response of a system having transfer function $G(s)$ is of the form:

$$Y(s) = G(s)U(s) + \sum_{k=1}^{p} \sum_{i=1}^{n_k} \frac{\beta_{ki}}{(s - \lambda_k)^i} \qquad (4.6.1)$$

where every β_{ki} is a function of the initial conditions and where, for completeness, we have assumed that every pole, at $s = \lambda_k$, has multiplicity n_k. This latter assumption implies that $n_1 + n_2 + \ldots + n_p = n$.

We say that the system is *stable* if any bounded input produces a bounded output for all bounded initial conditions. In particular, we can use a partial-fraction expansion to decompose the total response into the response of each pole taken separately. For continuous-time systems, we then see that stability requires that the poles have strictly negative real parts: They need to be in the open left-half plane (OLHP) of the complex plane $\boxed{s}$. This also implies that, for continuous-time systems, the stability boundary is the imaginary axis.

Example 4.5. *Consider the system of example 4.1. The poles of this transfer function are -2 and 0. These do not lie in the open left-half plane, OLHP (0 is in the closed left-half plane, CLHP). Thus the system is not stable. Indeed, the reader should verify that a nonzero constant input will produce an output that increases without bound.*

4.7   Impulse and Step Responses of Continuous-Time Linear Systems

Of special interest in the study of linear systems is the response to a dirac-delta function. This can be considered as the limit as $\triangle \to 0$ of the pulse shown in Figure 4.2. The Laplace Transorm of the Dirac Delta is 1 (see Table 4.1). Hence, if we were to apply such an input to a system with zero initial conditions, then the output response would simply be $Y(s) = G(s)U(s) = G(s)$. We can summarize this observation as follows:

> The transfer function of a continuous-time system is the Laplace transform of its response to an impulse (Dirac's delta) with zero initial conditions.

Because of the idealization implicit in the definition of an impulse, it is more common to study a system's dynamic behavior under the step response–that is, when $U(s) = 1/s$. This leads to the so-called *step response*.

$$Y(s) = G(s)\frac{1}{s} \qquad (4.7.1)$$

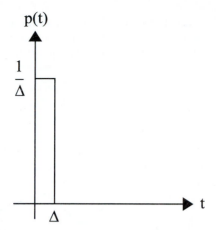

Figure 4.2. Discrete pulse

Application of the final-value theorem (see Table 4.2) shows that the steady-state response (provided that it exists) for a unit step is given by

$$\lim_{t \to \infty} y(t) = y_\infty = \lim_{s \to 0} sG(s)\frac{1}{s} = G(0) \tag{4.7.2}$$

If the system is stable, then the transient part of the step response will decay exponentially to zero, and hence y_∞ will exist. Note that, if $G(s)$ has one or more zeros at $s = 0$, then $y_\infty = 0$.

It is also useful to define a set of parameters that succinctly describe certain relevant properties of the system dynamics. To introduce these definitions, we consider a stable transfer function having the step response shown in Figure 4.3.

We then define the following indicators.

Steady-state value, y_∞: the final value of the step response. (This is meaningless if the system has poles in the CRHP.)

Rise time, t_r : the time elapsed up to the instant at which the step response reaches, for the first time, the value $k_r y_\infty$. The constant k_r varies from author to author, being usually either 0.9 or 1.

Overshoot, M_p : the maximum instantaneous amount by which the step response exceeds its final value. (It is usually expressed as a percentage of y_∞.)

Undershoot, M_u : the (absolute value of the) maximum instantaneous amount by which the step response falls below zero.

Settling time, t_s : the time elapsed until the step response enters (without leaving it afterwards) a specified deviation band, $\pm\delta$, around the final value. This deviation, δ, is usually defined as a percentage of y_∞, say 2% to 5%.

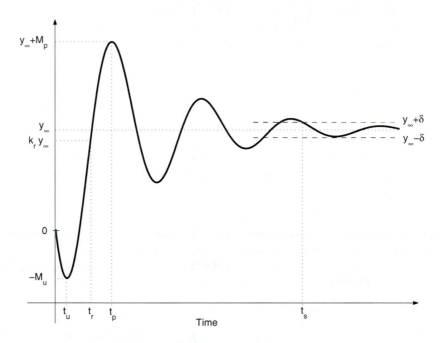

Figure 4.3. Step-response indicators

4.8   Poles, Zeros, and Time Responses

We next examine a number of fundamental properties of poles and zeros of transfer functions. At this point, we are not interested in how these transfer functions arise. Later, we will relate these results to specific transfer functions that arise in feedback systems.

For the moment, we will consider a general transfer function of the form

$$H(s) = K\frac{\prod_{i=1}^{m}(s - \beta_i)}{\prod_{l=1}^{n}(s - \alpha_l)} \tag{4.8.1}$$

where $\alpha_l \in \mathbb{C}$ and $\beta_i \in \mathbb{C}$. If we presume again that there are no values of l and i such that $\alpha_l = \beta_i$ then $\beta_1, \beta_2, \dots, \beta_m$ and $\alpha_1, \alpha_2, \dots, \alpha_n$ are the zeros and poles of the transfer function, respectively. The relative degree is $n_r \stackrel{\triangle}{=} n - m$.

We will be particularly interested in those zeros which lie on, or in the neighborhood of, the imaginary axis, and in those poles which lie in the RHP. Poles and zeros in these locations play a fundamental role on the dynamic behavior of systems.

A special class of transfer functions arises when all poles and zeros lie in the left half of the complex plane $\boxed{s}$. Traditionally, these transfer functions have been called *minimum-phase transfer functions* . In the sequel, however, we will use this name as referring simply to transfer functions with no RHP zeros, irrespective of whether they have RHP poles. We will say that a zero is *nonminimum phase* (NMP) if it lies in the closed RHP, otherwise it will be known as a minimum-phase zero. If a transfer function is referred to as a stable transfer function, the implication is that all its poles are in the open LHP; if it is said to be unstable, it has at least one pole in the closed RHP. The poles themselves are also called *stable* or *unstable poles*, depending on whether they lie in the open LHP or closed RHP. [2]

We next investigate the transient behavior arising from poles and zeros.

4.8.1   Poles

The reader will recall that any scalar rational transfer function can be expanded into partial fractions, each term of which contains either a single real pole, a complex-conjugate pair, or multiple combinations with repeated poles. Thus, an understanding of the effect of poles on transient performance reduces to an understanding of the transients due to first- and second-order poles and their interactions.

A general first-order pole contributes

$$H_1(s) = \frac{K}{\tau s + 1} \tag{4.8.2}$$

The response of this system to a unit step can be computed as

$$y(t) = \mathcal{L}^{-1}\left[\frac{K}{s(\tau s + 1)}\right] = \mathcal{L}^{-1}\left[\frac{K}{s} - \frac{K\tau}{\tau s + 1}\right] = K(1 - e^{-\frac{t}{\tau}}) \tag{4.8.3}$$

The signal $y(t)$ in equation (4.8.3) can be depicted as in Figure 4.4.

From Figure 4.4, we see that the parameters $K \, (= y_\infty)$ and τ (the time constant) can be computed graphically.

For the case of a pair of complex-conjugate poles, it is customary to study a *canonical second-order system* with the transfer function

$$H(s) = \frac{\omega_n^2}{s^2 + 2\psi\omega_n s + \omega_n^2} \tag{4.8.4}$$

---

[2]Sometimes, with abuse of language, nonminimum-phase zeros are also called *unstable* zeros, because they lie in the s-plane region where poles are unstable.

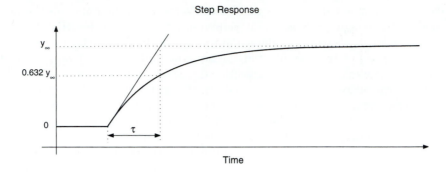

Figure 4.4. Step response of a first-order system

where ψ $(0 < \psi < 1)$ is known as the *damping factor* and ω_n as the natural or *undamped natural frequency.* We also define for future use, *the damped natural frequency, ω_d,* as

$$\omega_d = \omega_n \sqrt{1 - \psi^2} \tag{4.8.5}$$

This system has two complex-conjugate poles, s_1 and s_2, which are given by

$$s_{1,2} = -\psi\omega_n \pm j\omega_d = \omega_n e^{\pm j(\pi - \beta)} \tag{4.8.6}$$

where β is the angle such that $\cos\beta = \psi$.

For this system, the Laplace transform of its unit step response is given by

$$Y(s) = \frac{\omega_n^2}{(s^2 + 2\psi\omega_n s + \omega_n^2)s} = \frac{\omega_n^2}{[(s + \psi\omega_n)^2 + \omega_d^2]\, s} \tag{4.8.7}$$

Expanding in partial fractions, we obtain

$$Y(s) = \frac{1}{s} - \frac{s + \psi\omega_n}{(s + \psi\omega_n)^2 + \omega_d^2} - \frac{\psi\omega_n}{(s + \psi\omega_n)^2 + \omega_d^2} \tag{4.8.8}$$

$$= \frac{1}{s} - \frac{1}{\sqrt{1 - \psi^2}}\left[\sqrt{1 - \psi^2}\, \frac{s + \psi\omega_n}{(s + \psi\omega_n)^2 + \omega_d^2} - \psi\frac{\omega_d}{(s + \psi\omega_n)^2 + \omega_d^2}\right] \tag{4.8.9}$$

On applying the inverse Laplace transform, we finally obtain

$$y(t) = 1 - \frac{e^{-\psi\omega_n t}}{\sqrt{1 - \psi^2}}\sin(\omega_d t + \beta) \tag{4.8.10}$$

The main characteristics of this response are shown in Figure 4.5. In that figure, $y_\infty = 1$ and $T_d = 2\pi/\omega_d$.

We can also compute some of the indicators described in Figure 4.3 on page 76.

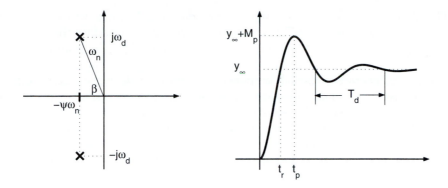

Figure 4.5. Pole location and unit step response of a canonical second-order system.

Rise time

For this case, we use $k_r = 1$ (see Figure 4.3 on page 76), and we thus have that

$$\frac{e^{-\psi\omega_n t_r}}{\sqrt{1-\psi^2}}\sin(\omega_d t_r + \beta) = 0 \tag{4.8.11}$$

From this, we obtain

$$t_r = \frac{\pi - \beta}{\omega_d} \tag{4.8.12}$$

Overshoot

The maximum overshoot, M_p, and the instant at which it occurs, t_p, can be computed by differentiating $y(t)$ and then equating this derivative to zero.

$$\frac{dy(t)}{dt} = -\frac{e^{-\psi\omega_n t}}{\sqrt{1-\psi^2}}\left[-\psi\omega_n \sin(\omega_d t + \beta) + \omega_d \cos(\omega_d t_r + \beta)\right] \tag{4.8.13}$$

Thus, setting this derivative equal to zero, we have that $\omega_d t_p = \pi$, and the overshoot time t_p turns out to be

$$t_p = \frac{\pi}{\omega_d} = \frac{T_d}{2} \tag{4.8.14}$$

In turn, the overshoot is given by

$$M_p = y(t_p) - 1 = -\frac{e^{-\frac{\pi\psi\omega_n}{\omega_d}}}{\sqrt{1-\psi^2}}\sin(\pi + \beta) = e^{-\frac{\pi\psi}{\sqrt{1-\psi^2}}} \tag{4.8.15}$$

The above expressions suggest that a small damping factor ψ will lead to small rise time, at the expense of a high overshoot. We can also appreciate that the decay speed and, consequently, the settling time, are determined by the product $\psi\omega_n$.

□□□

Every pole generates a special component or *natural mode* in the system response to an impulsive input. These modes are present in the system response to any given input (except in very special cases when poles coincide with zeros).

We will usually refer to *fast poles* as those poles that are much farther away from the stability boundary than the other system poles. This is also equivalent to saying that the transients associated with fast poles extinguish faster than those associated with other poles. On the other side, we will use the expression *dominant* or *slow pole(s)* to denote OLHP system pole(s) that are closer to the stability boundary than the rest of the system poles. This is equivalent to saying that the transients associated with the dominant poles decay more slowly than the rest.

If, for example, the system poles are $(-1; -2 \pm j6; -4; -5 \pm j3)$, we can say that the dominant pole is -1, and the fast poles are $-5 \pm j3$.

4.8.2   Zeros

The effect that zeros have on the response of a transfer function is a little more subtle than that due to poles. One reason for this is that while poles are associated with the states in isolation, zeros arise from additive interactions among the states associated with different poles. Moreover, the zeros of a transfer function depend on where the input is applied and how the output is formed as a function of the states.

Although the location of the poles determines the nature of system modes, it is the location of the zeros which determines the proportion in which these modes are combined. These combinations may look completely different from the individual modes.

In a way that parallels the definitions for the system poles, we define *fast* and *slow zeros*. Fast zeros are those which are much farther away from the stability boundary than the dominant poles. On the other hand, slow zeros are those which are much closer to the stability boundary than the system-dominant poles.

To illustrate some of the issues discussed so far, we consider the following example.

Example 4.6. *Consider a system with transfer function given by*

$$H(s) = \frac{-s + c}{c(s + 1)(0.5s + 1)} \tag{4.8.16}$$

This structure allows us to study the effects of a variable zero location, without affecting the location of the poles and the d.c. gain.

In this system, we observe that there are two natural modes, e^{-t} and e^{-2t}, arising from the two poles at -1 and -2 respectively. The first of these modes will be almost *absent in the response as c approaches -1. The situation is the same for the second mode when c approaches -2.*

The more general situation can be appreciated from Figure 4.6 on page 83. In that figure, the corresponding value for c appears besides each response. We can see that a fast zero, for example, $|c| \gg 1$, has no significant impact on the transient response. When the zero is slow and stable, then one obtains significant overshoot, whereas when the zero is slow and unstable, then significant undershoot is obtained. Indeed the effect of different zero locations can be seen to be quite dramatic in this example, where 400% overshoot and 500% undershoot are observed depending on the zero location. The reader might like to verify that even more dramatic results occur if the zeros are shifted even closer to the origin. These properties actually hold quite generally, as we show below.

□□□

Indeed, the issue of undershoot is readily understood with the use of the following lemma, which is an elementary consequence of the definition of the transform.

Lemma 4.1. *Let $H(s)$ be a strictly proper function of the Laplace variable s with region of convergence $\Re\{s\} > -\alpha$. Denote the corresponding time function by $h(t)$:*

$$H(s) = \mathcal{L}\left[h(t)\right] \tag{4.8.17}$$

Then, for any z_0 such that $\Re\{z_0\} > -\alpha$, we have

$$\int_0^\infty h(t)e^{-z_0 t}dt = \lim_{s \to z_0} H(s) \tag{4.8.18}$$

Proof

From the definition of the Laplace transform, we have that, for all s in the region of convergence of the transform, i.e., for $\Re\{s\} > -\alpha$,

$$H(s) = \int_0^\infty h(t)e^{-st}dt \tag{4.8.19}$$

The result follows, because z_0 is in the region of convergence of the transform.

□□□

We illustrate this result by a simple example.

Example 4.7. *Consider the signal*

$$y(t) = e^{+2t} \qquad t \geq 0 \tag{4.8.20}$$

Then

$$Y(s) = \frac{1}{s-2} \qquad for \quad \Re\{s\} > 2 \qquad (4.8.21)$$

Now consider

$$I(z_0) = \int_0^\infty e^{-z_0 t} y(t) dt \qquad (4.8.22)$$

Clearly, for $z_0 = 3$ we have

$$I(3) = \int_0^\infty e^{-3t} e^{2t} dt = \int_0^\infty e^{-t} dt = 1 \qquad (4.8.23)$$

Notice that this is correctly predicted by Lemma 4.1, because

$$Y(3) = \frac{1}{3-2} = 1 \qquad (4.8.24)$$

However, if we take $z_0 = 1$, then $Y(1) = -1$. Yet, clearly, $I(z_0)$ is ∞. This is in accord with Lemma 4.1, because $z_0 = 1$ does not lie in the region of convergence of the transform.

□□□

The above lemma is used below to quantify the relationships between zeros and certain key indicators of the system dynamics.

Lemma 4.2 (Nonminimum-phase zeros and undershoot). *Assume a linear, stable system with transfer function $H(s)$ having unity d.c. gain and a zero at $s = c$, where $c \in \mathbb{R}^+$. Further assume that the unit step response, $y(t)$, has a settling time t_s (see Figure 4.3 on page 76), i.e. $1 + \delta \geq |y(t)| \geq 1 - \delta$ ($\delta \ll 1$), $\forall t \geq t_s$. Then $y(t)$ exhibits an undershoot M_u which satisfies*

$$M_u \geq \frac{1 - \delta}{e^{ct_s} - 1} \qquad (4.8.25)$$

Proof

Define $v(t) = 1 - y(t)$; then

$$V(s) = (1 - H(s)) \frac{1}{s} \qquad (4.8.26)$$

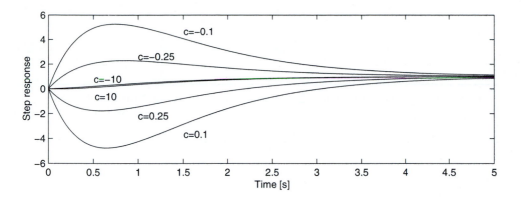

Figure 4.6. Effect of different zero locations on the step response

We notice that the region of convergence for $V(s)$ is given by $\Re\{s\} > 0$. Thus, c is inside that region, and we can apply Lemma 4.1 on page 81 to obtain

$$V(c) = \frac{1 - H(c)}{c} = \frac{1}{c} = \int_0^\infty v(t)e^{-ct}dt \qquad (4.8.27)$$

We then split the integration interval into $[0, t_s] \cup (t_s, \infty)$. Consequently, (4.8.27) leads to

$$\int_0^{t_s} v(t)e^{-ct}dt + \int_{t_s}^\infty |v(t)|e^{-ct}dt \geq \frac{1}{c} \qquad (4.8.28)$$

Also, using the definition for t_s we notice that $|v(t)| \leq \delta \ll 1 \; \forall t \geq t_s$. If we additionally note that

$$\max_{t \geq 0}\{v(t)\} = V_{max} = 1 + M_u > 0 \qquad (4.8.29)$$

then, from (4.8.28), we obtain

$$\frac{1}{c} \leq \int_0^{t_s} V_{max}e^{-ct}dt + \int_{t_s}^\infty \delta e^{-ct}dt = V_{max}\frac{1 - e^{-ct_s}}{c} + \frac{\delta e^{-ct_s}}{c} \qquad (4.8.30)$$

Finally, by using, (4.8.29) the result follows.

□□□

We further observe that, if $ct_s \ll 1$ (and recalling that $\delta \ll 1$), then (4.8.25) becomes

$$M_u > \frac{1}{ct_s} \tag{4.8.31}$$

The lemma above establishes that, when a system has nonminimum-phase zeros, there is a trade-off between having a fast step response and having small
undershoot.

A similar result can be established for a real LHP zero, when the system has a
real LHP zero with magnitude much smaller than the (real part of the) dominant
system pole. This result is established in the following lemma.

Lemma 4.3 (Slow zeros and overshoot). *Assume a linear, stable system with
transfer function $H(s)$ having unity d.c. gain and a zero at $s = c$, $c < 0$. Define
$v(t) = 1 - y(t)$, where $y(t)$ is the unit step response. Further, assume the following.*

A-1 *The system has dominant pole(s) with real part equal to $-p$, $p > 0$.*

A-2 *The zero and the dominant pole are related by*

$$\eta \triangleq \left| \frac{c}{p} \right| \ll 1 \tag{4.8.32}$$

A-3 *The value of δ defining the settling time (see Figure 4.3 on page 76) is chosen
such that there exists $0 < K$ which yields* [3]

$$|v(t)| < Ke^{-pt} \qquad \forall t \ge t_s \tag{4.8.33}$$

*Conclusion: The step response has an overshoot that is bounded below according
to*

$$M_p \ge \frac{1}{e^{-ct_s} - 1} \left(1 - \frac{K\eta}{1 - \eta} \right) \tag{4.8.34}$$

Proof

We first have that

$$V(s) = \mathcal{L}\left[v(t) \right] = (1 - H(s)) \frac{1}{s} = \int_0^\infty v(t) e^{-st} dt \tag{4.8.35}$$

---

[3]Note this value of K is highly dependent on the dominant pole: it is tightly linked to p.

We notice that the region of convergence for $V(s)$ is given by $\Re\{s\} > -p$. Thus, c is inside that region (see (4.8.32)), and so we can apply Lemma 4.1 on page 81 to obtain

$$\frac{1}{c} = \int_0^\infty v(t)e^{-ct}dt = \int_0^{t_s} v(t)e^{-ct}dt + \int_{t_s}^\infty v(t)e^{-ct}dt \qquad (4.8.36)$$

Assume now that the minimum value of $v(t)$ in $[0;\ t_s]$ is $-M_p$, and note that $v(t) > -Ke^{-pt}$, $\forall t > t_s$. Then both integrals on the right-hand side in equation (4.8.36) can be replaced by their minimum values, to yield

$$\frac{1}{c} \geq -M_p \int_0^{t_s} e^{-ct}dt - K \int_{t_s}^\infty e^{-(p+c)t}dt \iff$$

$$-\frac{1}{c} \leq M_p \int_0^{t_s} e^{-ct}dt + K \int_{t_s}^\infty e^{-(p+c)t}dt \quad (4.8.37)$$

Then the result (4.8.34) follows upon solving the above integrals and using (4.8.32).
□□□

Note that if $ct_s \ll 1$ and $K\eta \ll 1$, then an approximate lower bound for M_p is given by

$$M_p \geq \frac{1}{-ct_s} \qquad (4.8.38)$$

The lemma above establishes that, when a stable system has slow zeros, there is a trade-off between having a fast step response and having small overshoot.

We will employ a line of argument similar to that preceding in Chapter 8, when we study the effect of open-loop poles and zeros on the transient response of feedback control systems.

4.9   Frequency Response

We next study the system response to a rather special input, namely, a sine wave. The reason for doing so is that the response to sine waves also contains rich information about the response to other signals. This can be appreciated from Fourier analysis, which tells us that any signal defined on an interval $[t_o, t_f]$ can be represented as a linear combination of sine waves of frequencies $0, \omega_{of}, 2\omega_{of}, 3\omega_{of}, \ldots$, where $\omega_{of} = 2\pi/(t_f - t_o)$ is known as the fundamental frequency. The Principle of Superposition then allows us to combine the response to the individual sine waves to determine the response to the composite waveform.

Consider a linear stable system described in transfer-function form,

$$H(s) = K \frac{\sum_{i=0}^{m} b_i s^i}{s^n + \sum_{k=1}^{n-1} a_k s^k} \tag{4.9.1}$$

and let the system input be an exponential, $e^{s_o t}$. For simplicity, we assume that the poles are distinct and that none of them is equal to s_o. Then the Laplace transform of the system response can be computed by using a partial-fraction expansion to yield the decomposition:

$$Y(s) = \frac{H(s_o)}{s - s_o} + Y_h(s) \tag{4.9.2}$$

where the first term is the response forced by the input (the forced response) and the second term is the decaying response due to initial conditions (the natural response). The corresponding time-domain response is

$$y(t) = H(s_o)e^{s_o t} + \sum_{k=1}^{n} C_k e^{\lambda_k t} \tag{4.9.3}$$

where $\lambda_k, k = 1, 2, \dots, n$ are the system natural frequencies–that is, the poles of $H(s)$, and the $C_k's$ depend on the initial conditions.

Clearly, the second term on the right-hand side of (4.9.3) decays to zero if the system is stable.

Next, note that

$$\sin(\omega t) = \frac{1}{2j}(e^{j\omega t} - e^{-j\omega t}) \tag{4.9.4}$$

Hence, the system response to a sine-wave input can be computed by combining the response to $e^{s_o t}$ with $s_o = j\omega$ and that to $e^{s_o t}$ with $s_o = -j\omega$.

Note that evaluating $H(s)$ at $s = j\omega$ yields a complex number, which can be represented conveniently by its amplitude and phase in polar coordinates, as

$$H(j\omega) = |H(j\omega)|e^{j\phi(\omega)} \tag{4.9.5}$$

Then, the steady-state response to a sine-wave input is obtained from (4.9.3) to (4.9.5).

$$y(t) = \frac{1}{2j}\left[H(j\omega)e^{j\omega t} - H(-j\omega)e^{-j\omega t}\right] \tag{4.9.6}$$

$$= \frac{1}{2j}\left[|H(j\omega)|e^{j(\omega t+\phi(\omega))} - |H(j\omega)|e^{-j(\omega t+\phi(\omega))}\right] \tag{4.9.7}$$

$$= |H(j\omega)|\sin(\omega t + \phi(\omega)) \tag{4.9.8}$$

We thus conclude the following:

A sine-wave input forces a sine wave at the output with the same frequency. Moreover, the amplitude of the output sine wave is modified by a factor equal to the magnitude of $H(j\omega)$ and the phase is shifted by a quantity equal to the phase of $H(j\omega)$.

An interesting observation is that if $H(j\omega)$ is known at least for q nonzero different frequencies, where $q = 1 +$ integer part of $[(m+n)/2]$, then $H(s)$ is uniquely determined at all other frequencies as well.

Remark 4.3. *If the system contains a time delay–that is, if the transfer function is changed to*

$$H(s) = e^{-s\tau} \frac{\sum_{i=0}^{m} b_i s^i}{s^n + \sum_{k=1}^{n-1} a_k s^k} \qquad (4.9.9)$$

–then it can be proven that equation (4.9.8) becomes

$$y(t) = \begin{cases} 0 & \text{if } t < \tau \\ |H(j\omega)| \sin(\omega t + \phi(\omega)) + \sum_{k=1}^{n} C_k e^{\lambda_k(t-\tau)} & \text{if } t \geq \tau \end{cases} \qquad (4.9.10)$$

where $\phi(\omega)$ now includes an additional term $-\tau\omega$. The last term in (4.9.10) is formed by the natural modes, and the constants C_k depend on the initial conditions.
□□□

Frequency responses are a very useful tool for all aspects of analysis, synthesis, and design of controllers and filters. Because of their importance, special plots are used to graph them. They are typically depicted graphically in either magnitude and phase form (commonly called a Bode diagram) or in the form of a polar plot (commonly called a Nyquist plot). Much more will be said about Nyquist diagrams in Chapter 5. We therefore leave that topic until later. We give here a brief treatment of Bode diagrams.

4.9.1   Bode Diagrams

Bode diagrams consist of a pair of plots. One of these plots depicts the magnitude of the frequency response as a function of the angular frequency; the other depicts the angle of the frequency response, also as a function of the angular frequency.

Usually, Bode diagrams are drawn with special axes:

- The abscissa axis is linear in $\log(\omega)$, where the log is to the base 10. This allows a compact representation of the frequency response along a wide range of frequencies. The unit on this axis is the *decade*, where a decade is the distance between ω_1 and $10\omega_1$ for any value of ω_1.

- The magnitude of the frequency response is measured in *decibels* [dB]–that is, in units of $20 \log |H(j\omega)|$. This has several advantages, including good accuracy for small and large values of $|H(j\omega)|$, facility to build simple approximations for $20 \log |H(j\omega)|$, and the fact that the frequency response of cascade systems can be obtained by adding the individual frequency responses.

- The angle is measured on a linear scale in radians or degrees.

Software packages such as MATLAB provide special commands to compute the frequency response and to plot Bode diagrams. However, some simple rules allow one to sketch an approximation for the magnitude and phase diagrams. Consider a transfer function given by

$$H(s) = K \frac{\prod_{i=1}^{m}(\beta_i s + 1)}{s^k \prod_{i=1}^{n}(\alpha_i s + 1)} \tag{4.9.11}$$

Then,

$$20 \log |H(j\omega)| = 20 \log(|K|) - 20k \log |\omega| + \sum_{i=1}^{m} 20 \log |\beta_i j\omega + 1| -$$

$$\sum_{i=1}^{n} 20 \log |\alpha_i j\omega + 1| \tag{4.9.12}$$

$$\angle (H(j\omega)) = \angle(K) - k\frac{\pi}{2} + \sum_{i=1}^{m} \angle(\beta_i j\omega + 1) - \sum_{i=1}^{n} \angle(\alpha_i j\omega + 1) \tag{4.9.13}$$

We thus see that the Bode diagram of any transfer function can be obtained by either adding or subtracting magnitudes (in [dB]) and phases of simple factors. We observe the following:

- A simple gain K has constant magnitude and phase Bode diagram. The magnitude diagram is a horizontal line at $20 \log |K|[dB]$ and the phase diagram is a horizontal line either at $0[rad]$ (when $K \in \mathbb{R}^+$) or at $\pi[rad]$ (when $K \in \mathbb{R}^-$).

- The factor s^k has a magnitude diagram that is a straight line with slope equal to $20k[dB/decade]$ and has constant phase equal to $k\pi/2$. This line crosses the horizontal axis ($0[dB]$) at $\omega = 1$.

- The factor $as + 1$ has a magnitude Bode diagram that can be asymptotically approximated as follows:

o For $|a\omega| \ll 1$, $20 \log |aj\omega + 1| \approx 20 \log(1) = 0[dB]$–that is to say, for low frequencies, this magnitude is a horizontal line. This is known as the *low-frequency asymptote*.

o For $|a\omega| \gg 1$, $20 \log |aj\omega + 1| \approx 20 \log(|a\omega|)$–that is to say, for high frequencies, this magnitude is a straight line with a slope of $20[dB/decade]$ which crosses the horizontal axis $(0[dB])$ at $\omega = |a|^{-1}$. This is known as the *high-frequency asymptote*.

o The phase response is more complex. It roughly changes over two decades. One decade below $|a|^{-1}$ the phase is approximately zero. One decade above $|a|^{-1}$ the phase is approximately $\text{sign}(a)0.5\pi[rad]$. Connecting the points $(0.1|a|^{-1}, 0)$ and $(10|a|^{-1}, 0)$ by a straight line gives $\text{sign}(a)0.25\pi$ for the phase at $\omega = |a|^{-1}$. This is a very rough approximation.

• For $a = a_1 + ja_2$, the phase Bode diagram of the factor $as + 1$ corresponds to the angle of the complex number with real part $1 - \omega a_2$ and imaginary part $a_1\omega$.

Example 4.8. *Consider a transfer function given by*

$$H(s) = 640\frac{(s + 1)}{(s + 4)(s + 8)(s + 10)} \tag{4.9.14}$$

To draw the asymptotic behavior of the gain diagram, we first arrange $H(s)$ into the form shown in equation (4.9.11); this leads to

$$H(s) = 2\frac{(s + 1)}{(0.25s + 1)(0.125s + 1)(0.1s + 1)} \tag{4.9.15}$$

We thus have one constant factor ($K = 2$), and four factors of the form $as + 1$. Using the rules outlined above, we obtain the asymptotic magnitude diagram and the asymptotic phase diagram. Both of these are shown in Figure 4.7, together with the exact Bode plots.

Remark 4.4. *Although, some time ago, substantial effort was devoted to building asymptotic approximations of the type outlined above, the appearance of powerful software packages has made this effort no longer particularly important. Nevertheless, a basic understanding of the impact of poles and zeros on Bode diagrams frequently provides valuable insight for the control engineer.*

4.9.2   Filtering

In an ideal amplifier, the frequency response would be $H(j\omega) = K$, constant $\forall\omega$– that is, every frequency component would pass through the system with equal gain and no phase shift. However, all physical systems and devices have a finite speed with which they can react, and this first implies that $H(j\omega)$ cannot be constant for

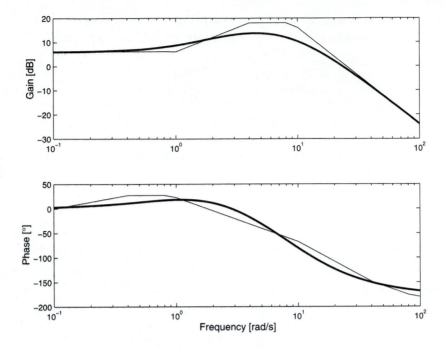

Figure 4.7. Exact (thick line) and asymptotic (thin line) Bode plots

all ω. One way to interpret the fact that $H(j\omega)$ is not equal to a constant for all ω is that the system *filters* inputs of different frequencies to produce the output–that is, the system deals with different sine-wave components selectively, according to their individual frequencies. In this context, it is common to distinguish three frequency sets:

- the *pass band*, in which all frequency components *pass* through the system with approximately the same amplification (or attenuation) and with a phase shift which is approximately proportional to ω;

- the *stop band*, in which all frequency components are *stopped*–in this band, $|H(j\omega)|$ is small compared to the value of $|H(j\omega)|$ in the pass band;

- the *transition band(s)*, which are intermediate between a pass band and a stop band.

Note that a system can have multiple pass or stop bands. These definitions are the origin of the traditional filtering terms: *low-pass, band-pass, high-pass and band-reject filters*.

To further refine these definitions, it is customary to refer to certain quantities:

- **Cut-off frequency, ω_c**–this is a value of ω such that $|H(j\omega_c)| = \hat{H}/\sqrt{2}$, where $\hat{H}$ is, defined as follows:

  ○ $|H(0)|$, for low-pass filters and band-reject filters;
  ○ $|H(\infty)|$, for high-pass filters;
  ○ the maximum value of $|H(j\omega)|$ in the pass band, for band-pass filters.

- **Bandwidth, B_w**–this is a measure of the frequency width of the pass band (or the reject band). It is defined as $B_w = \omega_{c2} - \omega_{c1}$, where $\omega_{c2} > \omega_{c1} \geq 0$. In this definition, ω_{c1} and ω_{c2} are cut-off frequencies on either side of the pass band or reject band (for low-pass filters, $\omega_{c1} = 0$).

The above definitions are illustrated on the frequency response of a band-pass filter shown in Figure 4.8. Here, the lower cut-off frequency is $\omega_{c1} = a \approx 50\ [rad/s]$ and the upper cut-off frequency is $\omega_{c2} = b \approx 200\ [rad/s]$. Thus, the bandwidth can be computed as $B_W = \omega_{c2} - \omega_{c1} = 150\ [rad/s]$.

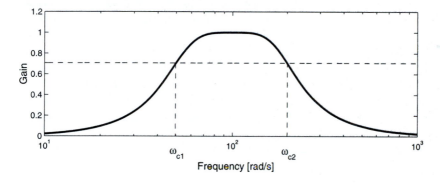

Figure 4.8. Frequency response of a band-pass filter

A system that has constant frequency-response amplitude is known as an *all-pass* filter. The best-known all-pass filter is a pure time delay. Note that the corresponding output amplitude response is equal to the input amplitude regardless of the signal frequency. Rational stable all-pass filters have the general form

$$H_{ap}(s) = K_{ap}\frac{p(-s)}{p(s)} \qquad (4.9.16)$$

where K_{ap} is a constant and $p(s)$ is any stable polynomial.

4.9.3   Distortion and Fidelity

When a system has a nonideal frequency response, we say it introduces *distortion*. To describe the different kinds of distortion that we commonly meet in practice, consider a signal $f(t)$ given by

$$f(t) = \sum_{i=1}^{nf} A_i \sin(\omega_i t + \alpha_i) \qquad\qquad (4.9.17)$$

Say that this signal is the input to a linear stable system (e.g., an audio-power amplifier). Then we say that the system processes this signal with *fidelity* if the amplitude of all sine-wave components are amplified (or attenuated) by approximately the same factor and the individual responses have the same *time delay* through the system. This requires that the frequency response satisfy

$$|H(j\omega_i)| = H_o \qquad \text{constant for} \quad i = 1, 2, \dots, nf \qquad\qquad (4.9.18)$$

$$\phi(\omega_i) = k_o\omega_i \qquad \text{where } k_o \text{ is constant for} \quad i = 1, 2, \dots, nf \qquad (4.9.19)$$

Under these conditions, the waveform at the system output is identical to the one at the input but delayed by k_o. When one or both conditions do not hold, the wave-form at the system output will be different from that of $f(t)$, and the system is then said to *distort* the signal. There can be either amplitude distortion (when (4.9.18) does not hold), phase distortion (when (4.9.19) does not hold), or both phase and amplitude distortion (when either (4.9.18) nor (4.9.19) hold).

The interpretation of this property in terms of filtering is that negligible distortion occurs if the set of frequencies of interest $\{\omega_1, \omega_2, \dots, \omega_{nf}\}$ is well inside the pass band of the system. Note that pure delays do not introduce distortion. However, rational all-pass filters do introduce phase distortion, except at very low frequencies.

4.10   Fourier Transform

A generalization of the idea of frequency response is encapsulated in Fourier transforms. These give a way of representing broad classes of signals in the frequency domain. Fourier transforms are also closely linked to Laplace transforms. However, whereas Laplace transforms are one-sided (i.e., defined for $t \in [0, \infty)$), Fourier transforms are two-sided (i.e., defined for $t \in (-\infty, \infty)$). This difference leads to certain changes in interpretation. For example, Laplace transforms include initial-condition effects, and RHP poles correspond to exponentially increasing signals; however, in Fourier transforms, initial conditions are usually not treated, and RHP poles correspond to noncausal signals (i.e., signals defined for $t \in (-\infty, 0)$). For completeness, we briefly review the key aspects of Fourier transforms below.

4.10.1   Definition of the Fourier Transform

Consider a continuous-time signal $f(t)$, defined for $-\infty \le t < \infty$. Then the Fourier transform pair associated with $f(t)$ is defined as follows:

$$\mathcal{F}\left[f(t)\right] = F(j\omega) = \int_{-\infty}^{\infty} e^{-j\omega t} f(t) dt \qquad (4.10.1)$$

$$\mathcal{F}^{-1}\left[F(j\omega)\right] = f(t) = \frac{1}{2\pi} \int_{-\infty}^{\infty} e^{j\omega t} F(j\omega) d\omega \qquad (4.10.2)$$

$F(jw)$ is referred to as the Fourier transform of $f(t)$. The transform pair is well-defined if the signal $f(t)$ is absolutely integrable–that is, if

$$\int_{-\infty}^{\infty} |f(t)| \, dt < \infty \qquad (4.10.3)$$

It is also possible to use a limiting argument to extend the Fourier transform to certain signals that do not satisfy (4.10.3). This is, for instance, the case for bounded periodic signals. The Fourier transforms for some common signals are shown in Table 4.3, and properties of the Transfer are given in Table 4.4.

4.10.2   Fourier Transform Applications

In addition to the key property of linearity, further useful properties of this transform are shown in Table 4.4 on page 95. When all these properties are put together, they can be applied to solve a wide range of system and signal-analysis problems. To illustrate this, consider the following example.

Example 4.9. *Consider a linear system having its input $u(t)$ and its output $y(t)$ related by the model*

$$\frac{dy(t)}{dt} + 3y(t) = 2u(t) \qquad (4.10.4)$$

It is also known that $u(t) = -0.5 \operatorname{sign}(t)$ [4] and that $y(0) = 0$. Compute $y(t)$ $\forall t$.

Solution

If we apply Fourier transform to (4.10.4), we obtain

$$j\omega Y(j\omega) + 3Y(j\omega) = 2U(j\omega) \Leftrightarrow Y(j\omega) = \frac{2}{j\omega + 3} U(j\omega) \qquad (4.10.5)$$

where $U(j\omega) = -0.5\mathcal{F}\left[\operatorname{sign}(t)\right] = -0.5\mathcal{F}\left[2\mu(t) - 1\right] = \frac{-1}{j\omega}$. Then

---

[4]Recall that the sign function is defined as -1 for a negative argument and +1 for a positive argument.

$f(t) \qquad \forall t \in \mathbb{R}$	$\mathcal{F}\left[f(t)\right]$
1	$2\pi\delta(\omega)$
$\delta_D(t)$	1
$\mu(t)$	$\pi\delta(\omega) + \dfrac{1}{j\omega}$
$\mu(t) - \mu(t - t_o)$	$\dfrac{1 - e^{-j\omega t_o}}{j\omega}$
$e^{\alpha t}\mu(t) \qquad \Re\{\alpha\} < 0$	$\dfrac{1}{j\omega - \alpha}$
$te^{\alpha t}\mu(t) \qquad \Re\{\alpha\} < 0$	$\dfrac{1}{(j\omega - \alpha)^2}$
$e^{-\alpha|t|} \qquad \alpha \in \mathbb{R}^+$	$\dfrac{2\alpha}{\omega^2 + \alpha^2}$
$\cos(\omega_o t)$	$\pi\left(\delta(\omega - \omega_o) + \delta(\omega - \omega_o)\right)$
$\sin(\omega_o t)$	$j\pi\left(\delta(\omega + \omega_o) - \delta(\omega - \omega_o)\right)$
$\cos(\omega_o t)\mu(t)$	$\pi\left(\delta(\omega - \omega_o) + \delta(\omega - \omega_o)\right) + \dfrac{j\omega}{-\omega^2 + \omega_o^2}$
$\sin(\omega_o t)\mu(t)$	$j\pi\left(\delta(\omega + \omega_o) - \delta(\omega - \omega_o)\right) + \dfrac{\omega_o}{-\omega^2 + \omega_o^2}$
$e^{-\alpha t}\cos(\omega_o t)\mu(t) \qquad \alpha \in \mathbb{R}^+$	$\dfrac{j\omega + \alpha}{(j\omega + \alpha)^2 + \omega_o^2}$
$e^{-\alpha t}\sin(\omega_o t)\mu(t) \qquad \alpha \in \mathbb{R}^+$	$\dfrac{\omega_o}{(j\omega + \alpha)^2 + \omega_o^2}$

Table 4.3. Fourier transform table

$f(t)$	$\mathcal{F}\left[f(t)\right]$	Description
$\displaystyle\sum_{i=1}^{l} a_i f_i(t)$	$\displaystyle\sum_{i=1}^{l} a_i F_i(j\omega)$	Linearity
$\dfrac{dy(t)}{dt}$	$j\omega Y(j\omega)$	Derivative law
$\dfrac{d^k y(t)}{dt^k}$	$(j\omega)^k Y(j\omega)$	High-order derivative
$\displaystyle\int_{-\infty}^{t} y(\tau)d\tau$	$\dfrac{1}{j\omega}Y(j\omega) + \pi Y(0)\delta(\omega)$	Integral law
$y(t-\tau)$	$e^{-j\omega\tau}Y(j\omega)$	Delay
$y(at)$	$\dfrac{1}{|a|}Y\left(j\dfrac{\omega}{a}\right)$	Time scaling
$y(-t)$	$Y(-j\omega)$	Time reversal
$\displaystyle\int_{-\infty}^{\infty} f_1(\tau)f_2(t-\tau)d\tau$	$F_1(j\omega)F_2(j\omega)$	Convolution
$y(t)\cos(\omega_o t)$	$\dfrac{1}{2}\left\{Y(j\omega - j\omega_o) + Y(j\omega + j\omega_o)\right\}$	Modulation (cosine)
$y(t)\sin(\omega_o t)$	$\dfrac{1}{j2}\left\{Y(j\omega - j\omega_o) - Y(j\omega + j\omega_o)\right\}$	Modulation (sine)
$F(t)$	$2\pi f(-j\omega)$	Symmetry
$f_1(t)f_2(t)$	$\dfrac{1}{2\pi j}\displaystyle\int_{\sigma-j\infty}^{\sigma+j\infty} F_1(\zeta)F_2(s-\zeta)d\zeta$	Time-domain product
$e^{at}f_1(t)$	$F_1(j\omega - a)$	Frequency shift

Table 4.4. Fourier transform properties; note that $F_i(j\omega) = \mathcal{F}\left[f_i(t)\right]$ and $Y(j\omega) = \mathcal{F}\left[y(t)\right]$

$$Y(j\omega) = -\frac{2}{(j\omega + 3)j\omega} = -\frac{2}{3j\omega} + \frac{2}{3(j\omega + 3)} = -\frac{2}{3}\left(\frac{1}{j\omega} - \frac{1}{(j\omega + 3)}\right) \qquad (4.10.6)$$

from which, upon applying the inverse Fourier transform, we finally obtain

$$y(t) = -\frac{2}{3}\operatorname{sign}(t) + \frac{2}{3}e^{-3t}\mu(t) \qquad (4.10.7)$$

□□□

One of the more attractive features of the Fourier transform is its connection to frequency response: (4.10.2) describes $f(t)$ as a linear combination of exponentials of the form $e^{j\omega t}$, where ω ranges continuously from $-\infty$ to ∞. This connection allows us to interpret the transform as describing the relative frequency content of a given signal. For example, $F(j\omega_e)$ corresponds to the *density* of the component of frequency ω_e. Hence, when an input $u(t)$ is applied to a linear system with model (4.2.1), the output $y(t)$ has a Fourier transform given by

$$Y(j\omega) = H(j\omega)U(j\omega) \qquad (4.10.8)$$

where $H(j\omega)$ has the same interpretation as in section §4.9, namely, the system complex gain for a sine wave of angular frequency ω.

The reader should resist the temptation arising from the similarity of equations (4.5.2) and (4.10.8), to conclude that we can find the Fourier transform from the Laplace transform via a simple variable substitution, $s \Rightarrow j\omega$. This interpretation yields correct results only for the case when the input is absolutely integrable–that is, if it satisfies (4.10.3) and the system is stable. (See, for example, the Laplace and the Fourier transforms for a unit step.) This warning can be further understood by considering an example system with (Laplace) transfer function given by

$$H(s) = \frac{2}{s - 3} \qquad (4.10.9)$$

This implies that the system response to $u(t) = \delta_D(t)$ (with zero initial conditions) is $y(t) = 2e^{3t}\mu(t)$. If we simply replace s by $j\omega$ in (4.10.9), and then apply the inverse Fourier transform, the system response to $u(t) = \delta_D(t)$ would seem to be $y(t) = 2e^{3t}\mu(-t)$–an *anticipatory* reaction. This error is induced by the fact that the Fourier transform for $2e^{3t}\mu(t)$ does not exist, because the integral (4.10.1) does not converge. Actually, within the region of convergence, the corresponding transform corresponds to that of a noncausal signal.

A commonly used result in Fourier analysis is Parseval's Theorem. For completeness, we present it here.

Theorem 4.1. *Let $F(j\omega)$ and $G(j\omega)$ denote the Fourier transform of $f(t)$ and $g(t)$, respectively. Then*

$$\int_{-\infty}^{\infty} f(t)g(t)\, dt = \frac{1}{2\pi} \int_{-\infty}^{\infty} F(j\omega)G(-j\omega)\, d\omega \qquad (4.10.10)$$

Proof

Use the inverse transform formula

$$\int_{-\infty}^{\infty} f(t)g(t)\, dt = \int_{-\infty}^{\infty} \frac{1}{2\pi} \left[\int_{-\infty}^{\infty} F(j\omega)e^{j\omega t} d\omega \right] g(t)\, dt \qquad (4.10.11)$$

Interchange the order of integration

$$\int_{-\infty}^{\infty} f(t)g(t)dt = \frac{1}{2\pi} \int_{-\infty}^{\infty} F(j\omega) \left[\int_{-\infty}^{\infty} e^{j\omega t} g(t)dt \right] d\omega \qquad (4.10.12)$$

$$= \frac{1}{2\pi} \int_{-\infty}^{\infty} F(j\omega)G(-j\omega)\, d\omega \qquad (4.10.13)$$

□□□

A special case arises when $f(t) = g(t)$; then,

$$\int_{-\infty}^{\infty} \{f(t)\}^2\, dt = \frac{1}{2\pi} \int_{-\infty}^{\infty} |F(j\omega)|^2\, d\omega \qquad (4.10.14)$$

Remark 4.5. *The reader will easily verify that Parseval's theorem also applies when $f(t)$ and $g(t)$ are matrices (vectors) of appropriate dimensions.*

4.11   Models Frequently Encountered

Many systems that are met in practice can be modeled by relatively simple first- and second-order linear systems. It is important to be able to recognize these systems when they occur.

Table 4.5 gives the time and frequency responses of simple linear models. The reader is encouraged to evaluate some of these responses to verify the results, because familiarity with these responses can be very helpful in diagnosing a control problem or for estimating models from which a controller can be designed.

For each system, the table presents the step response and the Bode diagrams. Only one parameter at a time is varied, and all of them are assumed to be *positive*. The effect of the variation is shown on each graph by an arrow which points towards the direction along which the parameter *increases* its value.

Some qualitative observations from Table 4.5 are as follows:

System	Parameter	Step response	Bode (gain)	Bode (phase)
$\dfrac{K}{\tau s + 1}$	K			
	τ			
$\dfrac{\omega_n^2}{s^2 + 2\psi\omega_n s + \omega^2}$	ψ			
	ω_n			
$\dfrac{as + 1}{(s + 1)^2}$	a			
$\dfrac{-as + 1}{(s + 1)^2}$	a			

Table 4.5. System models and the influence of parameter variations

$$\frac{K}{\tau s + 1}$$

- The step response is a simple exponential rise.

- The parameter K is the *d.c. gain*. Increasing K increases the final value of the step response.

- The parameter τ is the *time constant*. Increasing τ increases the rise time.

$$\frac{\omega_n^2}{s^2 + 2\psi\omega_n s + \omega^2}$$

- The step response is oscillatory (for $\psi \ll 1$).

- The parameter ψ is the *damping*. Increasing ψ causes the oscillations to decay faster.

- The parameter ω_n is the *undamped natural frequency*. Increasing ω_n leads to oscillations with a shorter period.

$$\frac{as + 1}{(s + 1)^2}$$

- The step response exhibits overshoot that does not oscillate (for $a^{-1} < 1$).

- The parameter $-a^{-1}$ is a minimum-phase zero. Increasing a increases the amount of overshoot.

$$\frac{-as + 1}{(s + 1)^2}$$

- The step response exhibits undershoot which does not oscillate.

- The value a^{-1} is known as a nonminimum-phase zero (NMP). Increasing a increases the amount of undershoot.

Of course, use of these simple models will usually involve some level of approximation. This fact needs to be borne in mind when one is utilizing the models for control-system design. The broader issue of describing modeling errors associated with approximate linear models is taken up in the next section.

4.12   Modeling Errors for Linear Systems

Section §3.9 introduced the idea of errors between the nominal and calibration models. If a linear model is used to approximate a linear system, then modeling errors due to errors in parameters and/or complexity can be expressed in transfer-function form as

$$Y(s) = G(s)U(s) = (G_o(s) + G_\epsilon(s))U(s) = G_o(s)(1 + G_\Delta(s))U(s) \qquad (4.12.1)$$

where $G_\epsilon(s)$ denotes the AME and $G_\Delta(s)$ denotes the MME, introduced in section §3.9.

AME and MME are two different ways of capturing the same modeling error. The advantage of the MME is that it is a relative quantity, whereas the AME is an absolute quantity. This can be seen by noting that

$$G_\Delta(s) = \frac{G_\epsilon(s)}{G_o(s)} = \frac{G(s) - G_o(s)}{G_o(s)} \tag{4.12.2}$$

Usually, linear models will accurately capture the behavior of a plant in the low-frequency region–the behavior of the plant when its inputs are constant or slowly time-varying. This is illustrated in the following example.

Example 4.10. *Time delays do not yield rational functions in the Laplace domain. Thus, a common strategy is to approximate the delay by a suitable rational expression. One possible approximation is*

$$e^{-\tau s} \approx \left(\frac{-\tau s + 2k}{\tau s + 2k} \right)^k \qquad k \in \langle 1, 2, \dots \rangle \tag{4.12.3}$$

where k determines the accuracy of the approximation.

For this approximation, determine the magnitude of the frequency response of the MME.

Solution

We have that

$$G(s) = e^{-\tau s} F(s) \qquad\qquad G_o(s) = \left(\frac{-\tau s + 2k}{\tau s + 2k} \right)^k F(s) \tag{4.12.4}$$

Hence,

$$G_\Delta(s) = e^{-\tau s} \left(\frac{-\tau s + 2k}{\tau s + 2k} \right)^{-k} - 1 \tag{4.12.5}$$

with frequency-response magnitude given by

$$|G_\Delta(j\omega)| = \left| e^{-j\tau\omega} - e^{-j2k\phi} \right| \qquad\qquad \phi = \arctan\frac{\omega\tau}{2k} \tag{4.12.6}$$

The result is depicted in Figure 4.9, where $\omega\tau$ is the normalized frequency.

□□□

Some typical linear-modeling errors include the following cases.

Numerical inaccuracies in the description of poles and zeros

Consider, for instance, a plant described by

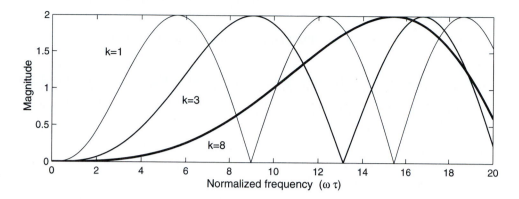

Figure 4.9. MME for all-pass rational approximation of time delays

$$G(s) = \frac{1}{as+1}F(s) \qquad \text{and} \qquad G_o(s) = \frac{1}{(a+\delta)s+1}F(s) \qquad (4.12.7)$$

Here, there is no error in complexity and no d.c. gain error, but the pole is in error. Then

$$G_\epsilon(s) = \frac{\delta s}{(as+1)((a+\delta)s+1)}F(s) \qquad \text{and} \qquad G_\Delta(s) = \frac{\delta s}{as+1} \qquad (4.12.8)$$

Note that the effect of the AME vanishes at low frequencies and at high frequencies. The magnitude of the frequency response of the MME is also small at low frequencies, but it grows up to a maximum value equal to $\frac{\delta}{a}$ for very high frequencies.

Numerical inaccuracies give rise to very special structures for G_ϵ and G_Δ. One of these situations occurs when the real plant has unstable poles. If the unstable pole is not exactly known, then both *the AME and the MME will be unstable*.

Example 4.11. *Consider a plant with $G(s)$ and $G_o(s)$ as in (4.12.7), with $a = -1$, $\delta = 0.2$; then the AME and MME are given by*

$$G_\epsilon(s) = \frac{0.2s}{(-s+1)(-0.8s+1)}F(s) \qquad \text{and} \qquad G_\Delta(s) = \frac{0.2s}{-s+1} \qquad (4.12.9)$$

The instability of the AME and MME is evident.

□□□

Missing pole

Say that the true plant and its nominal model are given, respectively, by

$$G(s) = \frac{1}{as+1} F(s) \qquad \text{and} \qquad G_o(s) = F(s) \qquad (4.12.10)$$

where $F(s)$ is a given transfer function.

Again, there is no d.c. gain error, but

$$G_\epsilon(s) = \frac{-as}{as+1} F(s) \qquad \text{and} \qquad G_\Delta(s) = \frac{-as}{as+1} \qquad (4.12.11)$$

If, as normally happens, $|G_o(j\omega)|$ vanishes at high frequencies, then the AME again has a band-pass-type characteristic in the frequency domain. The magnitude of the frequency response of the MME will again be of high-pass-type, and for very large frequency it will tend to 1.

Error in time delay

Say that the true plant and its nominal model are given, respectively, by

$$G(s) = e^{-\tau s} F(s) \qquad \text{and} \qquad G_o(s) = e^{-\tau_o s} F(s) \qquad (4.12.12)$$

There is no d.c. gain error, and the modeling errors in the frequency domain are given by

$$G_\epsilon(j\omega) = 2je^{-\lambda j\omega} \sin(\kappa\omega) F(j\omega) \qquad G_\Delta(j\omega) = 2je^{-(\lambda-\tau_o)j\omega} \sin(\kappa\omega) \quad (4.12.13)$$

where

$$\kappa = \frac{\tau_o - \tau}{2} \qquad \text{and} \qquad \lambda = \frac{\tau_o + \tau}{2} \qquad (4.12.14)$$

The corresponding magnitudes are

$$|G_\epsilon(j\omega)| = 2 \left| \sin\left(\frac{\omega\tau_o}{2} \left(1 - \frac{\tau}{\tau_o} \right) \right) \right| |F(j\omega)| \qquad (4.12.15)$$

$$|G_\Delta(j\omega)| = 2 \left| \sin\left(\frac{\omega\tau_o}{2} \left(1 - \frac{\tau}{\tau_o} \right) \right) \right| \qquad (4.12.16)$$

These expressions, together with the assumption that $|F(j\omega)|$ vanishes for very high frequencies (as is the case for most real plants), indicate that the AME also vanishes when $\omega \to \infty$. The situation is different for the MME; this error is very small for low frequencies, but, as ω increases, it oscillates between 0 and 2. The above expressions also show that the bigger the absolute error is (as measured by $|\tau - \tau_o|$), the sooner (in frequency) the MME becomes significant.

Missing resonance effect

The omission of resonant modes is very common when one is modeling certain classes of systems, such as robot arms, antennas, and other large flexible structures. This situation may be described by

$$G(s) = \frac{\omega_n^2}{s^2 + 2\psi\omega_n s + \omega_n^2} F(s) \qquad\qquad G_o(s) = F(s) \qquad 0 < \psi < 1$$

$$(4.12.17)$$

The modeling errors are now given by

$$G_\epsilon(s) = \frac{-s(s + 2\psi\omega_n)}{s^2 + 2\psi\omega_n s + \omega_n^2} F(s) \qquad\qquad G_\Delta(s) = \frac{-s(s + 2\psi\omega_n)}{s^2 + 2\psi\omega_n s + \omega_n^2} \quad (4.12.18)$$

Under the assumption that $|G_o(j\omega)|$ vanishes for very high frequencies, the AME effect on the frequency response is of band-pass-type. By contrast, the MME is high-pass, with a resonant peak that can be very large, depending on the value of the damping factor ψ. In Figure 4.10, the magnitudes of the frequency responses for different cases of ψ are shown.

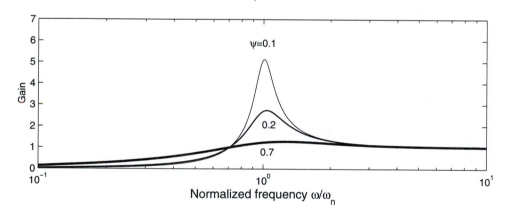

Figure 4.10. MME frequency response for omitted resonance, for different values of the damping factor ψ

4.13   Bounds for Modeling Errors

In control-system design, it often is desirable to account for model errors in some way. Unfortunately, the model errors are rarely known precisely. However, inspection of the examples in section §4.12 indicates that a common situation is for the MME to have a magnitude that is small at low frequencies and grows to 1 or 2 at high frequencies, with missing resonant modes producing peaks and other kinds of behavior in the transition band. It is thus common to assume some knowledge about the MME in the form of a bound. A typical specification might be

$$|G_\Delta(j\omega)| < \epsilon(\omega) \tag{4.13.1}$$

where $\epsilon(\omega)$ is some given positive function of ω.

4.14 Summary

- Ultimately, all physical systems exhibit some form of nonlinearity.

- Nevertheless, there is a strong incentive to find a model that is

    ○ linear, yet
    ○ sufficiently accurate for the purpose.

  Incentives include the following properties of linear models:

    ○ Theory and techniques are significantly simpler and more tractable.
    ○ There are more closed-form solutions and easier-to-use software tools.
    ○ Superposition holds; for example, the response to simultaneously acting set-point changes and disturbances is equal to the sum of these signals acting individually.
    ○ A powerful body of frequency-domain properties and results holds: poles, zeros, transfer functions, and Bode and Nyquist plots with their associated properties and results.
    ○ Relative degree, inversion, stability, and inverse stability are easier to define and check.

- Means to approximate physical systems by linear models include the following:

    ○ *transformations*, such as change of variable;
    ○ *approximations*, such as Taylor series in a neighborhood and black-box identification.

- These points motivate the basic assumption of linearity used in the next few chapters; later chapters introduce techniques for systems that are inherently nonlinear.

- There are two key approaches to linear dynamic models:

    ○ the so-called *time domain*, and

- o the so-called *frequency domain.*

- Although these two approaches are largely equivalent, each has its own particular advantages, and it is therefore important to have a good grasp of each.

- In the time domain,

    - o systems are modeled by differential equations,

    - o systems are characterized by the evolution of their variables (output, etc.) in time, and

    - o the evolution of variables in time is computed by solving differential equations.

- Stable time responses are typically characterized in terms of the following:

Characteristic	*Measure of*
Steady-state gain	How the system, after transients, amplifies or attenuates a constant signal
Rise time	How fast the system reacts to a change in its input
Settling time	How fast the system's transient decays
Overshoot	How far the response grows beyond its final value during transients
Undershoot	How far initial transients grow into the opposite direction relative to the final value

- In the frequency domain, the following situations obtain:

    - o Modeling exploits the key linear-system property that the steady-state response to a sinusoid is again a sinusoid of the same frequency; the system changes amplitude and phase of the input only in a fashion uniquely determined by the system at that frequency.

    - o Systems are modeled by transfer functions, which capture this impact as a function of frequency.

- Terms used to characterize systems in the frequency domain include the following:

Characteristic	*Measure of*
Pass band	Frequency range where the system has minimal impact on the amplitude of a sinusoidal input
Stop band	Frequency range where the system essentially annihilates sinusoidal inputs
Transition band	Frequency range between a system's pass and stop bands
Bandwidth	The frequency range of a system's pass band
Cut-off frequency	A frequency signifying a (somewhat arbitrary) border between a system's pass and transition bands

- In the frequency domain, systems are typically characterized by the following:

Characteristic	*Significance*
Frequency-response plots	Graphical representation of a systems impact on amplitude and phase of a sinusoidal input as a function of frequency.
Poles	The roots of the transfer-function denominator polynomial, they determine stability and, together with the zeros, the transient characteristics.
Zeros	The roots of the transfer-function numerator polynomial, they do not affect stability, but they determine inverse stability and undershoot, and, together with the poles, they have a profound impact on the system's transient characteristics.
Relative degree	Number of poles minus number of zeros; determines whether a system is strictly proper, biproper, or improper.
Strictly proper	The system has more poles than zeros; it is causal and, therefore, implementable; it has an improper inverse and zero high-frequency gain.
Biproper	The system has equal number of poles and zeros; it is implementable, has a biproper inverse, and has a feed-through term, i.e., a nonzero and finite high-frequency gain.
Improper	The system has more zeros than poles; it is not causal, cannot be implemented, has a strictly proper inverse, and has infinite high-frequency gain.

- Particularly important linear models include the following:

  - gain

  - first-order model

  - second-order model

  - integrator

  - a pure time delay (irrational) and its rational approximation

- The importance of these models is due to the following facts:
  - They are frequently observed in practice.
  - More complex systems are decomposable into them by partial-fraction expansion.

- Evaluating a transfer function at any one frequency yields a characteristic complex number.
  - Its magnitude indicates the system's gain at that frequency.
  - Its phase indicates the system's phase shift at that frequency.

- With respect to the important characteristic of stability, a continuous-time system is
  - stable if and only if the real parts of all poles are strictly negative,
  - marginally stable if at least one pole is strictly imaginary and no pole has strictly positive real part,
  - unstable if the real part of at least one pole is strictly positive, and
  - nonminimum phase if the real part of at least one zero is strictly positive.

- The response of linear systems to an arbitrary driving input can be decomposed into the sum of two components:
  - the natural response, which is a function of initial conditions, but independent of the driving input–if the system is stable, the natural response decays to zero;
  - the forced response, which is a function of the driving input, but independent of initial conditions.

- Equivalent ways of viewing transfer-function models include the following:
  - the Laplace transform of a system's differential equation model;
  - the Laplace transform of the system's forced response to an impulse;
  - a model derived directly from experimental observation.

- In principle, the time response of a transfer function can be obtained by taking the inverse Laplace transform of the output; however, in practice, one almost always prefers to transform the transfer function to the time domain and to solve the differential equations numerically.

- Key strengths of time-domain models include the following:
  - they are particularly suitable for solution and simulation on a digital computer;

       ◦ they are extendable to more general classes of models, such as nonlinear systems;

       ◦ they play a fundamental role in state space theory, covered in later chapters.

- Key strengths of frequency-domain models (transfer functions) include the following:

       ◦ they can be manipulated by simple algebraic rules–thus, transfer functions of parallel, series, or feedback architectures can be simply computed;

       ◦ properties such as inversion, stability, inverse stability, and even a qualitative understanding of transients are easily inferred from knowledge of the poles and zeros.

- Time-domain and frequency-domain models can be converted from one to the other.

- All models contain modeling errors.

- Modeling errors can be described as an additive (AME) or multiplicative (MME) quantity.

- Modeling errors are necessarily unknown and frequently are described by upper bounds.

- Certain types of commonly occurring modeling errors, such as numerical inaccuracy, missing poles, inaccurate resonant peaks, or time delays, have certain *finger prints*.

- One can generally assume that modeling errors increase with frequency; the MME typically possesses a high-pass character.

4.15   Further Reading

Laplace transform

Doetsch, G. (1971). *Guide to the applications of Laplace and Z-Transform.* D. van Nostrand. Van Nostrand-Reinhold, London, New York, 2nd English edition.

Lathi, B. (1965). *Signals, Systems and Communication.* Wiley, New York.

Oppenheim, A.V., Wilsky, A.S., and Hamid, N.S. (1997). *Signals and Systems.* Prentice-Hall, Upper Saddle River, N.J., 2nd edition.

Effects of zeros on transient response

Middleton, R.H. (1991). Trade-offs in linear control systems design. *Automatica*, 27(2):281-292.

Middleton, R.H. and Graebe, S.F. (1999). Slow stable open-loop poles: to cancel or not to cancel. *Automatica*, 35:877-886.

Frequency response and Fourier transform

Distefano, J., Stubberud, A., and Williams, I. (1976). *Feedback and Control Systems*. McGraw-Hill, New York.

Papoulis, A. (1977). *Signal Analysis*. McGraw-Hill, New York.

Willems, J.C. (1970). *Stability theory of dynamical systems*. Nelson, London.

4.16   Problems for the Reader

Problem   4.1. *A system transfer function is given by*

$$H(s) = \frac{2}{s+a} \qquad (4.16.1)$$

4.1.1 *Determine conditions under which the step response settles faster than the signal e^{-4t}.*

4.1.2 *Compute the system bandwidth.*

Problem   4.2. *A system transfer function is given by*

$$H(s) = \frac{-s+1}{(s+1)^2} \qquad (4.16.2)$$

Compute the time instant, t_u, at which the step response exhibits maximum undershoot.

Problem   4.3. *The unit step response of a system with zero initial conditions is given by*

$$y(t) = 3 - 2e^{-2t} - e^{-3t} \qquad \forall t \geq 0 \qquad (4.16.3)$$

4.3.1 *Compute the system transfer function.*

4.3.2 *Compute the system response to a unit impulse.*

Problem   4.4. *A nonlinear system has an input-output model given by*

$$\frac{dy(t)}{dt} + y(t)\left(1 - 0.2\,(y(t))^2\right) = 2u(t) \qquad (4.16.4)$$

4.4.1 *Determine the transfer function for the linearized model, as a function of the operating point.*

4.4.2 *Find an operating point for which the above linear model is unstable.*

Problem   4.5. *A system transfer function has its poles at -2, -2, and $-1 \pm j1$ and its zeros at 1 and -3.*

4.5.1 *Determine the dominant pole(s).*

4.5.2 *If the system d.c. gain is equal to 5, build the system transfer function.*

Problem 4.6. *A system transfer function is given by*

$$H(s) = \frac{s + 1 + \epsilon}{(s+1)(s+2)} \tag{4.16.5}$$

4.6.1 *Compute the unit step response.*

4.6.2 *Analyze your result for $\epsilon \in [-1, 1]$.*

Problem 4.7. *The input-output model for a system is given by*

$$\frac{d^2 y(t)}{dt} + 7\frac{dy(t)}{dt} + 12y(t) = 3u(t) \tag{4.16.6}$$

4.7.1 *Determine the system transfer function.*

4.7.2 *Compute the unit step response with zero initial conditions.*

4.7.3 *Repeat with initial conditions $y(0) = -1$ and $\dot{y}(0) = 2$.*

Problem 4.8. *Assume that the Fourier transform of a signal $f(t)$ is given by*

$$\mathcal{F}[f(t)] = F(j\omega) = \begin{cases} 0 & \forall |\omega| > \omega_c \\ 1 & \forall |\omega| \le \omega_c \end{cases} \tag{4.16.7}$$

Show that, if $f(t)$ is the unit impulse response of a linear system, then the system is noncausal, i.e., the system responds   before the input is applied.

Problem 4.9. *The transfer function of a linear stable sytem is given by*

$$H(s) = \frac{-s + 4}{s^2 + 5s + 6} \tag{4.16.8}$$

If the system input is $u(t) = 2\cos(0.5t)$, find the system output in steady state.

Problem 4.10. *Determine the signal $f(t)$ that has a Fourier transform given by*

$$F(j\omega) = \frac{1}{\omega^2 + a^2} \qquad \text{where } a \in \mathbb{R} \tag{4.16.9}$$

Problem 4.11. *Find, if they exist, the Fourier transform of the following signals:*

$$f_1(t) = 2 + \cos(2t) \qquad f_2(t) = (2 + \cos(2t))\mu(t) \qquad f_3(t) = \mu(t) - \mu(t - T)$$
$$f_4(t) = e^{-3t}\cos(0.5t)\mu(t) \qquad f_5(t) = te^{-t} \qquad\qquad f_6(t) = sign(t)$$

Problem 4.12. *Consider the function*

$$F_n(s) = \frac{1}{1 + (-js)^{2n}} \tag{4.16.10}$$

4.12.1 *Find the poles of $F_n(s)$.*

4.12.2 *Find a stable $H_n(s)$ such that $H_n(s)H_n(-s) = F_n(s)$.*

Problem 4.13. *Analyze, for $\beta \in \mathbb{R}$, the frequency response of the AME and the MME when the true and the nominal models are given by*

$$G(s) = \frac{\beta s + 2}{(s + 1)(s + 2)} \qquad and \qquad G_o(s) = \frac{2}{(s + 1)(s + 2)} \tag{4.16.11}$$

respectively.

Problem 4.14. *Consider a linear system with true model given by*

$$G(s) = F(s)\frac{\omega_n^2}{s^2 + 2\psi\omega_n s + \omega_n^2} \tag{4.16.12}$$

where $0 < \psi < 1$. Determine the MME for the nominal model

$$G_o(s) = F(s)\frac{\omega_{no}^2}{s^2 + 2\psi_o\omega_{no}s + \omega_{no}^2} \tag{4.16.13}$$

in the following situations:

(i) $\omega_{no} = \omega_n$ but $\psi \neq \psi_o$

(ii) $\psi = \psi_o$ but $\omega_{no} \neq \omega_n$

Problem 4.15. *Consider the structure shown in Figure 2.7 on page 33.*

4.15.1 *Find linear transformations* $h\langle\circ\rangle$ *and* $f\langle\circ\rangle$ *such that the structure imple-ments a stable approximate inverse for a system having a model given by*

$$G(s) = \frac{-s+4}{(s+4)} \qquad (4.16.14)$$

4.15.2 *Compute the modeling error with respect to the exact inverse.*

Problem 4.16. *Consider the following two transfer functions:*

$$H_1(s) = \frac{0.250}{s^2 + 0.707s + 0.250} \qquad (4.16.15)$$

$$H_2(s) = \frac{0.0625}{s^4 + 1.3066s^3 + 0.8536s^2 + 0.3266s + 0.0625} \qquad (4.16.16)$$

4.16.1 *Draw their Bode diagrams, and verify that each exhibits a low-pass filtering behavior. Compute the bandwidth for each filter.*

4.16.2 *Compute the unit step response for each case and the indicators defined in Figure 4.3 on page 76. Compare and discuss.*

Problem 4.17. *Consider the following transfer functions*

$$H_1(s) = \frac{e^{-s}}{s+e^{-s}} \qquad H_2(s) = \frac{e^{-s}}{s+1+e^{-s}} \qquad (4.16.17)$$

4.17.1 *For each case, compute the system poles.*

4.17.2 *For each case, obtain a plot of the magnitude of the frequency response versus the frequency. Determine the filtering characteristics.*

4.17.3 *Use SIMULINK to obtain each system unit step response.*

Problem 4.18. *Find the impluse and step responses of the following linear trans-fer functions:*

$$G(s) = \frac{1}{s^2 + 2s + 1} \qquad (4.16.18)$$

$$G(s) = \frac{10 - s}{s^2 + 2s + 1} \qquad (4.16.19)$$

$$G(s) = \frac{0.1 - s}{s^2 + 2s + 1} \qquad (4.16.20)$$

$$G(s) = \frac{1}{s^2 + 0.2s + 1} \qquad (4.16.21)$$

Comment on the differences observed.

Problem 4.19. *Calculate the steady-state responses when a unit step is applied to the following systems:*

$$G(s) = \frac{1}{s^3 + 3s^2 + 3s + 1} \tag{4.16.22}$$

$$G(s) = \frac{s^2 + 2s}{s^3 + 3s^2 + 3s + 1} \tag{4.16.23}$$

Comment on the differences observed.

Problem 4.20. *The step response of a system (initially at rest) is measured to be*

$$y(t) = 1 - 0.5e^{-t} - 0.5e^{-2t} \tag{4.16.24}$$

What is the transfer function of the system?

Problem 4.21. *(This is based on a question from an industrial colleague.) The step response of a system is as sketched below*

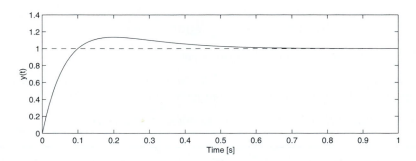

Figure 4.11. Step response

How is it that we can have overshoot without any ringing in the step response? What form do you think the model has?

Problem 4.22. *A parallel connection of 2 systems is shown in Figure 4.12.*

4.22.1 *What is the transfer function from u to y?*

4.22.2 *What are the system poles?*

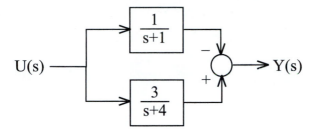

Figure 4.12. Parallel connection of two systems

4.22.3 *What are the system zeros (if any)?*

4.22.4 *Calculate the system step response and comment.*

Part II

SISO CONTROL ESSENTIALS

PREVIEW

The previous part of the book dealt with models for systems. These are the elements in terms of which the control engineer conceptualizes a control loop.

We next turn our attention to the properties of the control loop itself. In particular, we address issues of sensitivity, stability, and loop synthesis. These are the building blocks of design and will be the central topic of the next part of the book.

Chapter 5

ANALYSIS OF SISO CONTROL LOOPS

5.1 Preview

Control-system design makes use of two key enabling techniques: analysis and synthesis. Analysis concerns itself with the impact that a given controller has on a given system when they interact in feedback; synthesis asks how to construct controllers with certain properties. This chapter covers analysis. For a given controller and plant connected in feedback, it asks and answers the following questions:

- Is the loop stable?

- What are the sensitivities to various disturbances?

- What is the impact of linear modeling errors?

- How do small nonlinearities impact on the loop?

We also introduce several analysis tools, specifically,

- Root locus

- Nyquist stability analysis

5.2 Feedback Structures

Recall the preliminary introduction to feedback control given in Chapter 2. We will now focus this topic specifically on linear single-input single-output (SISO) control systems. We will see that feedback can have many desirable properties, such as the capacity to reduce the effect of disturbances, to decrease sensitivity to model errors, or to stabilize an unstable system. We will also see, however, that ill-applied feedback can make a previously stable system unstable, add oscillatory behavior into a previously smooth response, or result in high sensitivity to measurement noise.

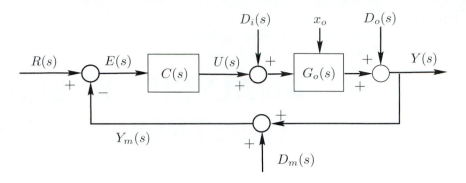

Figure 5.1. Simple feedback control system

We begin our analysis of feedback systems with the linear SISO configuration shown in Figure 5.1. We initially analyze the so-called *nominal loop*, that is, the effect of the controller interacting with the *nominal model* in feedback; later, in section §5.9, we will turn to the impact of modeling errors that arise when the controller is applied to the true system, rather than the model.

In the loop shown in Figure 5.1, we use transfer functions and Laplace transforms to describe the relationships between signals in the loop. In particular, $C(s)$ and $G_o(s)$ denote the transfer functions of the controller and the nominal plant model respectively, which can be represented in fractional form as follows:

$$C(s) = \frac{P(s)}{L(s)} \qquad (5.2.1)$$

$$G_o(s) = \frac{B_o(s)}{A_o(s)} \qquad (5.2.2)$$

where $P(s)$, $L(s)$, $B_o(s)$, and $A_o(s)$ are polynomials in s. $R(s)$, $U(s)$, and $Y(s)$ denote the Laplace transforms of set-point, control signal, and plant output, respectively; $D_i(s)$, $D_o(s)$, and $D_m(s)$ denote the Laplace transforms of input disturbance, output disturbance, and measurement noise, respectively. We also use x_o to denote the initial conditions of the model.

The following relations hold between the variables in Figure 5.1.

$$Y(s) = G_o(s)U(s) + D_o(s) + G_o(s)D_i(s) + \frac{f(s, x_o)}{A_o(s)} \qquad (5.2.3)$$

$$U(s) = C(s)R(s) - C(s)Y(s) - C(s)D_m(s) \qquad (5.2.4)$$

$$= C(s)\left(R(s) - D_m(s) - G_o(s)U(s) - D_o(s) - G_o(s)D_i(s) - \frac{f(s, x_o)}{A_o(s)} \right)$$
$$(5.2.5)$$

where $f(s, x_o)$ is a linear function of the initial state. The above equations can be solved to yield:

$$U(s) = \frac{C(s)}{1 + G_o(s)C(s)}\left(R(s) - D_m(s) - D_o(s) - G_o(s)D_i(s) - \frac{f(s, x_o)}{A(s)} \right) \qquad (5.2.6)$$

and

$$Y(s) = \frac{1}{1 + G_o(s)C(s)}\left[G_o(s)C(s)(R(s) - D_m(s)) + D_o(s) + G_o(s)D_i(s) + \frac{f(s, x_o)}{A(s)} \right]$$
$$(5.2.7)$$

The closed-loop control configuration shown in Figure 5.1, is called a *one-degree-of-freedom* (one-d.o.f.) architecture. This term reflects the fact that there is only one-degree-of-freedom available to shape the two transfer functions from $R(s)$ and $D_m(s)$ to $Y(s)$ and from $D_o(s)$ and $D_i(s)$ to $Y(s)$. Hence, if the controller transfer function $C(s)$ is designed to give a particular reference-signal response–for example,

$$\frac{Y(s)}{R(s)} = \frac{G_o(s)C(s)}{1 + G_o(s)C(s)} \qquad (5.2.8)$$

then this induces a unique output-disturbance response,

$$\frac{Y(s)}{D_o(s)} = \frac{1}{1 + G_o(s)C(s)} \qquad (5.2.9)$$

without any further design freedom.

Frequently, however, it is desirable to be able to shape the reference and disturbance responses separately. This can be achieved by a *two-degree-of-freedom* (two-d.o.f.) architecture, as shown in Figure 5.2. The first-degree-of-freedom is the feedback controller $C(s)$, and $H(s)$; the second-degree-of-freedom is a stable transfer function that is sometimes called a set-point or reference filter.

As with (5.2.7), we see that the two-d.o.f. loop is governed by

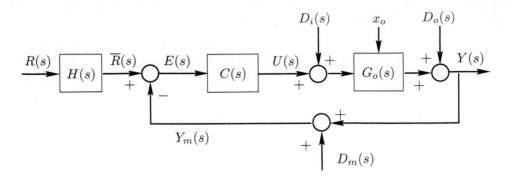

Figure 5.2. Two-degree-of-freedom closed loop

$$Y(s) = \frac{G_o(s)C(s)H(s)}{1 + G_o(s)C(s)} R(s) + \frac{1}{1 + G_o(s)C(s)} \left(D_o(s) + \frac{f(s, x_o)}{A_o(s)} \right)$$
$$+ \frac{G_o(s)}{1 + G_o(s)C(s)} D_i(s) - \frac{G_o(s)C(s)}{1 + G_o(s)C(s)} D_m(s) \tag{5.2.10}$$

The transfer function $C(s)$ can be designed to shape the disturbance response (unchanged from (5.2.7)) and $H(s)$ can be used to shape the reference response independently, where

$$\frac{Y(s)}{R(s)} = \frac{G_o(s)C(s)H(s)}{1 + G_o(s)C(s)} \tag{5.2.11}$$

Note, however, that even in a two-d.o.f. control loop, there remain transfer functions whose dynamics cannot be shaped independently.

Thus, the controller C can be used to shape the response to *one* of the disturbances D_i, D_o, or D_m but once this has been achieved, the remainder are determined.

Also the corresponding controller output response is given by

$$U(s) = \frac{C(s)H(s)}{1 + G_o(s)C(s)} R(s) - \frac{C(s)}{1 + G_o(s)C(s)} \left(D_o(s) + \frac{f(s, x_o)}{A_o(s)} \right)$$
$$- \frac{G_o(s)C(s)}{1 + G_o(s)C(s)} D_i(s) - \frac{C(s)}{1 + G_o(s)C(s)} D_m(s) \tag{5.2.12}$$

which we again see is fixed once $H(s)$ and $C(s)$ have been specified.

5.3   Nominal Sensitivity Functions

We see from equations (5.2.10) and (5.2.12) that the closed-loop response is governed by four transfer functions, which are collectively known as the *sensitivity functions*.

Using equations (5.2.1) and (5.2.2), these sensitivity functions are seen to be given by:

$$T_o(s) \triangleq \frac{G_o(s)C(s)}{1 + G_o(s)C(s)} = \frac{B_o(s)P(s)}{A_o(s)L(s) + B_o(s)P(s)} \tag{5.3.1}$$

$$S_o(s) \triangleq \frac{1}{1 + G_o(s)C(s)} = \frac{A_o(s)L(s)}{A_o(s)L(s) + B_o(s)P(s)} \tag{5.3.2}$$

$$S_{io}(s) \triangleq \frac{G_o(s)}{1 + G_o(s)C(s)} = \frac{B_o(s)L(s)}{A_o(s)L(s) + B_o(s)P(s)} \tag{5.3.3}$$

$$S_{uo}(s) \triangleq \frac{C(s)}{1 + G_o(s)C(s)} = \frac{A_o(s)P(s)}{A_o(s)L(s) + B_o(s)P(s)} \tag{5.3.4}$$

These functions are given specific names as follows:

$$
\begin{aligned}
&T_o(s): &&\text{Nominal complementary sensitivity} \\
&S_o(s): &&\text{Nominal sensitivity} \\
&S_{io}(s): &&\text{Nominal input-disturbance sensitivity} \\
&S_{uo}(s): &&\text{Nominal control sensitivity}
\end{aligned}
$$

The polynomial $A_{cl} \triangleq A_o(s)L(s) + B_o(s)P(s)$ is called the nominal closed-loop characteristic polynomial.

Sensitivity functions are algebraically related. These relations are one of the key manifestations of trade-offs inherent to a feedback loop, and they can be derived from the definitions (5.3.1) to (5.3.4). In particular, we have

$$S_o(s) + T_o(s) = 1 \tag{5.3.5}$$

$$S_{io}(s) = S_o(s)G_o(s) = \frac{T_o(s)}{C(s)} \tag{5.3.6}$$

$$S_{uo}(s) = S_o(s)C(s) = \frac{T_o(s)}{G_o(s)} \tag{5.3.7}$$

With these sensitivities, the closed loop in Figure 5.2 on page 124 is described by

$$Y(s) = T_o(s)\big(H(s)R(s) - D_m(s)\big) + S_o(s)\left(D_o(s) + \frac{f(s, x_o)}{A_o(s)}\right) + S_{io}(s)D_i(s)$$

$$(5.3.8)$$

$$U(s) = S_{uo}(s)\left(H(s)R(s) - D_m(s) - D_o(s) - G_o(s)D_i(s) - \frac{f(s, x_o)}{A_o(s)}\right) \quad (5.3.9)$$

We observe that the effects of initial conditions on plant output and controller output are, respectively, given by

$$S_o(s)\frac{f(s, x_o)}{A_o(s)} = \frac{f(s, x_o)L(s)}{A_o(s)L(s) + B_o(s)P(s)} \qquad (5.3.10)$$

$$-S_{uo}(s)\frac{f(s, x_o)}{A_o(s)} = -\frac{f(s, x_o)P(s)}{A_o(s)L(s) + B_o(s)P(s)} \qquad (5.3.11)$$

From these equations and from (5.3.1)–(5.3.4), one can now see the fundamental role played by the polynomial A_{cl}. It determines both the stability and, together with the individual zeros due to $L(s)$, $P(s)$, $B_o(s)$, and $A_o(s)$, the transient characteristics of the nominal control loop

Equations (5.3.1) to (5.3.4) can be expressed in more compact form as

$$\begin{bmatrix} Y_o(s) \\ U_o(s) \end{bmatrix} = \frac{\begin{bmatrix} G_o(s)C(s) & G_o(s) & 1 & -G_o(s)C(s) \\ C(s) & -G_o(s)C(s) & -C(s) & -C(s) \end{bmatrix}}{1 + G_o(s)C(s)} \begin{bmatrix} H(s)R(s) \\ D_i(s) \\ D_o(s) \\ D_m(s) \end{bmatrix}$$

$$(5.3.12)$$

Example 5.1. *A plant is nominally modeled by $G_o(s) = \frac{4}{(s+1)(s+2)^2}$. The plant has an output disturbance given by $d_o(t) = k + d_v(t)$, where $d_v(t)$ is a zero mean signal with energy in the band $B_d : (0, 4)[rad/s]$.*

A feedback controller $C(s)$ is designed so that

$$T_o(s) = \frac{\alpha}{(s^2 + 1.2\omega_n s + \omega_n^2)(\tau s + 1)^2} \qquad (5.3.13)$$

The factor $(\tau s + 1)^2$ has been introduced to make the controller strictly proper (check that this choice is sufficient to ensure properness!). Discuss the choice of α and ω_n from the point of view of output-disturbance compensation and the magnitude of the required control effort.

Solution

We need $T_o(j\omega) \approx 1$ (which, due to (5.3.5), implies $S_o(j\omega) \approx 0$) at $\omega = 0$ and in B_d. To achieve this, we propose the following choices:

- $\alpha = \omega_n^2$

- ω_n *larger than* $4[rad/s]$. *Say* $\omega_n = 10[rad/s]$

- $\tau = 0.01$ *(which is much smaller than* ω_n^{-1} *).*

This leads to:

$$T_o(s) = \frac{100}{(s^2 + 12s + 100)(0.01s^2 + 1)} \tag{5.3.14}$$

We then check the nominal control sensitivity $S_{uo}(s)$. For the above choice of T_o, we have

$$S_{uo}(s) = \frac{T_o(s)}{G_o(s)} = 25\frac{(s+1)(s+2)^2}{(s^2 + 12s + 100)(0.01s + 1)^2} \tag{5.3.15}$$

At $3[rad/s]$ we have that $|S_{uo}(j\omega)| \approx 20[dB]$. This means that if $d_v(t)$ has a sinusoidal component at $3[rad/s]$, then the controller output would have a component of the same frequency with an amplitude ten times that of the disturbance.

   This could lead to saturation of the input. On reflection we can see that this problem arises because the output disturbance has a bandwidth that is much larger than that of the open-loop plant.

   To reduce control sensitivity in B_d, we see that the only choice available to the designer is to increase the sensitivity to disturbances in B_d, so a design trade-off is called for.

□□□

5.4   Closed-Loop Stability Based on the Characteristic Polynomial

This and subsequent sections cover various tools for analyzing both nominal stability and robust stability.[1]

   We introduce the following definition:

Definition 5.1 (Internal stability). *We say that the nominal loop is internally stable if and only if all eight transfer functions in equation (5.3.12) are stable.*

---

[1] A nominal control loop results from connecting a controller to the nominal model.

The above definition is equivalent to demanding that all signals in the loop be bounded for every set of bounded inputs $r(t), d_i(t), d_o(t)$, and $d_m(t)$. Comparison of (5.3.12) with (5.3.1) through (5.3.4) shows that the nominal closed loop is internally stable if and only if the polynomial $A_o(s)L(s) + B_o(s)P(s)$ has stable factors. We summarize this as follows:

Lemma 5.1. *Nominal internal stability*
Consider the nominal closed loop depicted in Figure 5.2 on page 124 with the model and the controller defined by the representation given by equations (5.2.2) and (5.2.1) respectively. Then the nominal closed loop is internally stable if and only if the roots of the nominal closed-loop characteristic equation

$$A_o(s)L(s) + B_o(s)P(s) = 0 \tag{5.4.1}$$

all lie in the open left-half plane.

Proof

Straightforward from equation (5.3.12) and the definition of internal stability.
□□□

Note that the idea of internal stability implies more than the stability of the reference to output transfer function. In addition, one requires that there be no cancellation of unstable poles between controller and the plant. This is illustrated by the following example.

Example 5.2. *Assume that*

$$G_o(s) = \frac{3}{(s+4)(-s+2)} \qquad\qquad C(s) = \frac{-s+2}{s} \tag{5.4.2}$$

It can be seen that $T_o(s)$ is stable; however, the nominal input-disturbance sensitivity is unstable, because

$$S_{io}(s) = \frac{3s}{(-s+2)(s^2+4s+3)} \tag{5.4.3}$$

Thus, this system is not internally stable and does not satisfy the conditions of Lemma 5.1, because $A_o(s)L(s) + B_o(s)P(s) = (-s+2)(s^2+4s+3)$.

5.5   Stability and Polynomial Analysis

5.5.1   Problem Definition

Consider the polynomial $p(s)$ defined by

$$p(s) = s^n + a_{n-1}s^{n-1} + \ldots + a_1s + a_0 \tag{5.5.1}$$

where $a_i \in \mathbb{R}$.

The problem to be studied deals with the question of whether that polynomial has any root with nonnegative real part. Obviously, this question can be answered by computing the n roots of $p(s)$; however, in many applications, it is of special interest to study the interplay between the location of the roots and certain polynomial coefficients.

Polynomials having all their roots in the closed left plane (i.e., with nonpositive real parts) are known as Hurwitz polynomials. If we restrict the roots to have negative real parts, we then say that the polynomial is strictly Hurwitz.

5.5.2   Some Polynomial Properties of Special Interest

From (5.5.1) we have the following special properties.

Property 1 The coefficient a_{n-1} satisfies

$$a_{n-1} = -\sum_{i=1}^{n} \lambda_i \tag{5.5.2}$$

where λ_1, λ_2, ... λ_n are the roots of $p(s)$.

To prove this property, note that $p(s)$ can be written as

$$p(s) = \prod_{i=1}^{n} (s - \lambda_i) \tag{5.5.3}$$

Then, upon expanding the product in (5.5.3) and upon observing the coefficient accompanying the $(n-1)^{th}$ power of s, (5.5.2) is obtained.

Property 2 The coefficient a_0 satisfies

$$a_0 = (-1)^n \prod_{i=1}^{n} \lambda_i \tag{5.5.4}$$

This property can also be proved expanding (5.5.3) and examining the constant term.

Property 3 If all roots of $p(s)$ have negative real parts, it is *necessary* that $a_i > 0$, $i \in \{0, 1, \ldots, n-1\}$.

To prove this property, we proceed as follows:

a) The roots of $p(s)$ are either real or complex and, if some of them are complex, then these complex roots must appear in conjugate pairs (this is a consequence of the fact that $p(s)$ is a polynomial with real coefficients).

b) Thus, without loss of generality, we can assume that there exist n_1 real roots and n_2 pairs of complex roots. This means that $n_1 + 2n_2 = n$.

c) If all those roots have negative real parts, they can be described as

$$\lambda_i = -|\alpha_i| \qquad i = 1, 2, \ldots, n_1 \tag{5.5.5}$$

$$\lambda_{n_1+i} = \lambda^*_{n_1+n_2+i} = -|\sigma_i| + j\omega_i \qquad i = 1, 2, \ldots, n_2 \tag{5.5.6}$$

d) Hence,

$$p(s) = \prod_{i=1}^{n_1}(s + |\alpha_i|) \prod_{l=1}^{n_2}((s + |\sigma_l|)^2 + \omega_l^2) \tag{5.5.7}$$

where, in the second product, we have grouped the complex pairs in quadratic factors.

e) From (5.5.7), we observe that $p(s)$ corresponds to the product of first- and second-order polynomials, all of which have real and positive coefficients. The coefficients of $p(s)$ are sums of products of the coefficients of these first- and second-order polynomials, so the property is proved.

Note that this property is necessary in a strictly Hurwitz polynomial, but it is not sufficient, *except for the case $n = 1$ and $n = 2$*, when this condition is necessary and sufficient.

Property 4 If any of the polynomial coefficients is nonpositive (negative or zero), then one or more of the roots has nonnegative real part.

This property is a direct consequence of the previous property.

5.5.3   Routh's Algorithm

One of the most popular algorithms to determine whether a polynomial is strictly Hurwitz is Routh's algorithm. We will present it here without proof.

We again consider a polynomial $p(s)$ of degree n, defined as

$$p(s) = \sum_{i=0}^{n} a_i s^i \tag{5.5.8}$$

Routh's algorithm is based on the following numerical array:

s^n	$\gamma_{0,1}$	$\gamma_{0,2}$	$\gamma_{0,3}$	$\gamma_{0,4}$	$\cdots$
s^{n-1}	$\gamma_{1,1}$	$\gamma_{1,2}$	$\gamma_{1,3}$	$\gamma_{1,4}$	$\cdots$
s^{n-2}	$\gamma_{2,1}$	$\gamma_{2,2}$	$\gamma_{2,3}$	$\gamma_{2,4}$	$\cdots$
s^{n-3}	$\gamma_{3,1}$	$\gamma_{3,2}$	$\gamma_{3,3}$	$\gamma_{3,4}$	$\cdots$
s^{n-4}	$\gamma_{4,1}$	$\gamma_{4,2}$	$\gamma_{4,,3}$	$\gamma_{4,4}$	$\cdots$
$\vdots$	$\vdots$	$\vdots$	$\vdots$	$\vdots$	
s^2	$\gamma_{n-2,1}$	$\gamma_{n-2,2}$			
s^1	$\gamma_{n-1,1}$				
s^0	$\gamma_{n,1}$				

Table 5.1. Routh's array

where

$$\gamma_{0,i} = a_{n+2-2i}; \quad i = 1, 2, \ldots, m_0 \qquad \text{and} \quad \gamma_{1,i} = a_{n+1-2i}; \quad i = 1, 2, \ldots, m_1 \tag{5.5.9}$$

with $m_0 = (n+2)/2$ and $m_1 = m_0 - 1$ for n even and $m_1 = m_0$ for n odd. Note that the elements $\gamma_{0,i}$ and $\gamma_{1,i}$ are the coefficients of the polynomials arranged in alternated form.

Furthermore,

$$\gamma_{k,j} = \frac{\gamma_{k-1,1}\,\gamma_{k-2,j+1} - \gamma_{k-2,1}\,\gamma_{k-1,j+1}}{\gamma_{k-1,1}} ; \qquad k = 2, \ldots, n \qquad j = 1, 2, \ldots, m_j \tag{5.5.10}$$

where $m_j = \max\{m_{j-1}, m_{j-2}\} - 1$ and where we must assign a zero value to the coefficient $\gamma_{k-1,j+1}$ when it is not defined in the Routh's array given in Table 5.1.

Note that the definitions of coefficients in (5.5.10) can be expressed by using determinants:

$$\gamma_{k,j} = -\frac{1}{\gamma_{k-1,1}} \begin{vmatrix} \gamma_{k-2,1} & \gamma_{k-2,j+1} \\ \gamma_{k-1,1} & \gamma_{k-1,j+1} \end{vmatrix} \tag{5.5.11}$$

We then have the main result:

Consider a polynomial $p(s)$ given by (5.5.8) and its associated array as in Table 5.1. Then the number of roots with real part greater than zero is equal to the number of sign changes in the first column of the array.

We will next consider some examples.

Example 5.3. *Let $p(s) = s^4 + s^3 + 3s^2 + 2s + 1$. Inspection of the polynomial does not allow us to say whether it is a Hurwitz polynomial. We proceed to use Routh's array.*

s^4	1	3	1
s^3	1	2	
s^3	1	1	
s^1	1		
s^0	1		

From the array, we note that there are no sign changes in the first column. According to Routh's criterion, this means that $p(s)$ is a strictly Hurwitz polynomial.

□□□

Example 5.4. *Let $p(s) = s^5 + 5s^4 + 12s^3 + 13s^2 + 3s + 6$. We first note that all the coefficients of this polynomial are greater than zero; hence, we cannnot discard the possibility that this polynomial is a Hurwitz polynomial. To test this, we build Routh's array.*

s^5	1	12	3
s^4	5	13	6
s^3	$\dfrac{47}{5}$	$\dfrac{9}{5}$	
s^3	$\dfrac{566}{47}$	6	
s^1	$-\dfrac{8160}{235}$		
s^0	6		

From this array, we note that there are two sign changes in the first column. According to Routh's criterion, this means that $p(s)$ has two roots with positive real parts. Hence, $p(s)$ is not a Hurwitz polynomial.

□□□

In using Routh's array there are sometimes special cases that require extra steps to be taken. For example, we see that, in building the array, we cannot proceed when one of the elements in the first column is zero. This can be dealt with in the following way.

Case 1

We first consider the case when the first element in the row associated with s^{n-k} is zero, but there is at least one other term which is nonzero.

In this case, we replace the term $\gamma_{k,1}$ by ϵ, where ϵ is a very small-magnitude number with the same sign as that of $\gamma_{k-1,1}$, i.e., we use either $|\epsilon|$ or $-|\epsilon|$. Then the array is completed and the Routh's criterion is applied to find out whether the polynomial is strictly Hurwitz, for $|\epsilon| \to 0^+$.

Example 5.5. *Consider the polynomial $p(s) = s^5 + 3s^4 + 2s^3 + 6s^2 + 3s + 3$. We then have that the Routh's array is*

s^5	1	2	3
s^4	3	6	3
s^3	0	2	
s^3			
s^1			
s^0			

$$\Longrightarrow$$

s^5	1	2	3
s^4	3	6	3
s^3	$|\epsilon|$	2	
s^3	$6 - \frac{6}{|\epsilon|}$	3	
s^1	$2 + \frac{|\epsilon|^2}{1-2|\epsilon|}$		
s^0	3		

We then observe that, upon $|\epsilon| \to 0^+$, there are two sign changes, that is, $p(s)$ is not Hurwitz because it has two roots with positive real parts.

□□□

Case 2

We next consider the case in which all the elements in the row associated with s^{n-k} are zero, i.e., $\gamma_{k,1} = \gamma_{k,2} = \ldots = 0$.

In this case, the original polynomial can be factorized by the polynomial $p_a(s) = \gamma_{k-1,1}s^{n-k+1} + \gamma_{k-1,2}s^{n-k-1} + \gamma_{k-1,3}s^{n-k-3} + \ldots$. Note that this is a polynomial of only even or only odd powers of s, where the coefficients correspond to the terms in the row just above the zero row. Thus, $p_a(s)$ and, consequently, $p(s)$ are not strictly Hurwitz polynomials.

Example 5.6. *Consider the polynomial $p(s) = s^6 + 5s^5 + 2s^4 + 5s^3 + 4s^2 + 15s + 3$. The associated Routh's array is*

s^6	1	2	4	3
s^5	5	5	15	
s^4	1	1	3	
s^3	0	0		
s^2				
s^1				
s^0				

Thus, $p_a(s) = s^4 + s^2 + 3$. Then, by polynomial division, we obtain $p(s) = p_a(s)(s^2 + 5s + 1)$. Note that the roots of $p_a(s)$ are located at $\pm 0.7849 \pm j1.0564$.

□□□

Routh's algorithm can be applied to the denominator of a transfer function to determine whether the system is stable; however, it can also be used to study the effects of parameter variation on system stability. To illustrate the idea, consider the following example.

Example 5.7. *In a feedback control loop, say that we have* $G_o(s)C(s) = \frac{K}{s(s+1)^2}$. *We want to know which values of K render the closed-loop stable.*

We first compute the closed-loop characteristic polynomial, which is given by $p(s) = s^3 + 2s^2 + s + K$, *and we then build the Routh's array.*

s^3	1	1
s^2	2	K
s^1	$1 - 0.5K$	
s^0	K	

We can now see that there are no closed-loop unstable poles if and only if $1-0.5K > 0$ and $K > 0$. Combining these two requirements, we have that the closed loop is stable if and only if $0 < K < 2$.

□□□

Routh's criterion can also be used to study how fast the system modes decay. To appreciate this possibility, consider the following example.

Example 5.8. *A feedback control loop has an open-loop transfer function given by* $G_o(s)C(s) = \frac{K}{s(s+4)^2}$. *We want to know whether there exists a range of values for K such that all closed-loop modes decay faster than e^{-t}. In other words, we require that all closed-loop poles have real parts smaller than -1.*

The strategy for solving this problem is to consider the vertical line $s = -1$ as a new stability limit in a shifted complex plane. If we call this shifted plane the w complex plane, then, in the above example, we simply make the substitution $w = s + 1$. We then apply Routh's test in this new complex plane.

For the above example, we compute the closed-loop characteristic polynomial as $p(s) = s^3 + 8s^2 + 16s + K$. *In the new complex variable, this becomes* $p_w(w) = p(w - 1) = w^3 + 5w^2 + 3w + K - 9$. *Then, Routh's array for $p_w(w)$ is built, leading to*

w^3	1	3
w^2	5	$K - 9$
w^1	$4.8 - 0.2K$	
w^0	$K - 9$	

We see that, to obtain a strictly Hurwitz $p_w(w)$ polynomial, we need that $9 < K < 24$. Hence, we conclude that this range of K gives closed-loop poles having real parts smaller than -1, as required.

□□□

5.6   Root Locus (RL)

Another classical tool used to study stability of equations of the type given in (5.4.1) is *root locus*. The root-locus approach can be used to examine the location of the roots of the characteristic polynomial as one parameter is varied. For instance, assume that the nominal plant model is given by $G_o(s)$ and that the controller has a transfer function described as $C(s) = KC_a(s)$, where $C_a(s)$ is a known quotient of two monic polynomials in s, and where K is a positive, but unknown, constant. Then the closed-loop poles are the roots of

$$1 + KC_a(s)G_o(s) = 0 \qquad (5.6.1)$$

The set of all points in the complex plane that satisfy (5.6.1) for some positive value of K is known as the *root locus*. This particular problem can be embedded in the following more general problem.

Consider the following equation

$$1 + \lambda F(s) = 0 \qquad \text{where} \qquad F(s) = \frac{M(s)}{D(s)} \qquad (5.6.2)$$

with $\lambda \geq 0$ and

$$M(s) = s^m + b_{m-1}s^{m-1} + \ldots + b_1 s + b_0 = \prod_{i=1}^{m}(s - c_i) \qquad (5.6.3)$$

$$D(s) = s^n + a_{n-1}s^{n-1} + \ldots + a_1 s + a_0 = \prod_{i=1}^{n}(s - p_i) \qquad (5.6.4)$$

where the coefficients of the polynomials $M(s)$ and $D(s)$ are real numbers. Then the solution to the root-locus problem requires us to find the set of all points in the complex plane that are solutions for (5.6.2) for all nonnegative values of λ.

We observe that the solution of equation (5.6.2) is also the solution to the equation

$$D(s) + \lambda M(s) = \prod_{i=1}^{n}(s - p_i) + \lambda \prod_{i=1}^{m}(s - c_i) = 0 \qquad (5.6.5)$$

Prior to modern computers, root-locus methods were an important technique that told the experienced user the impact that one parameter (usually the controller gain) has on the stability and dynamic behavior of the closed loop. Today, the root locus of any particular problem is easily solved by user-friendly software, such as MATLAB. Nevertheless, understanding the underlying patterns of root-locus behavior still provides valuable insights. Root-locus building rules include the following:

R1 The number of roots in equation (5.6.5) is equal to $\max\{m, n\}$. Thus, the root locus has $\max\{m, n\}$ *branches*.

R2 From equation (5.6.2), we observe that s_0 belongs to the root locus (for $\lambda \geq 0$) if and only if

$$\arg F(s_0) = (2k+1)\pi \qquad \text{for} \quad k \in \mathbb{Z}. \tag{5.6.6}$$

R3 From equation (5.6.2), we observe that if s_0 belongs to the root locus, the corresponding value of λ is λ_0, where

$$\lambda_0 = \frac{-1}{F(s_0)} \tag{5.6.7}$$

R4 A point s_0 on the real axis, $s_0 \in \mathbb{R}$, is part of the root locus (for $\lambda \geq 0$) if and only if it is located to the left of an odd number of poles and zeros (so that (5.6.6) is satisfied).

R5 When λ is close to zero, then n of the roots are located at the poles of $F(s)$, that is, at $p_1, p_2, \dots, p_n$ and, if $n < m$, the other $m - n$ roots are located at ∞. (We will be more precise on this issue below.)

R6 When λ is close to ∞, then m of these roots are located at the zeros of $F(s)$, that is, at $c_1, c_2, \dots, c_m$ and, if $n > m$, the other $n - m$ roots are located at ∞. (We will be more precise on this issue below.)

R7 If $n > m$ and λ tends to ∞, then, $n - m$ roots asymptotically tend to ∞, following asymptotes that intersect at $(\sigma, 0)$, where

$$\sigma = \frac{\sum_{i=1}^{n} p_i - \sum_{i=1}^{m} c_i}{n - m} \tag{5.6.8}$$

The angles of these asymptotes are $\eta_1, \eta_2, \dots \eta_{n-m}$, where

$$\eta_k = \frac{(2k-1)\pi}{n-m}; \qquad k = 1, 2, \dots, n - m \tag{5.6.9}$$

R8 If $n < m$ and λ tends to zero, then $m - n$ roots asymptotically tend to ∞, following asymptotes that intersect at $(\sigma, 0)$, where

$$\sigma = \frac{\sum_{i=1}^{n} p_i - \sum_{i=1}^{m} c_i}{m - n} \tag{5.6.10}$$

The angles of these asymptotes are $\eta_1, \eta_2, \dots \eta_{m-n}$, where

$$\eta_k = \frac{(2k-1)\pi}{n-m}; \qquad k = 1, 2, \dots, m - n \tag{5.6.11}$$

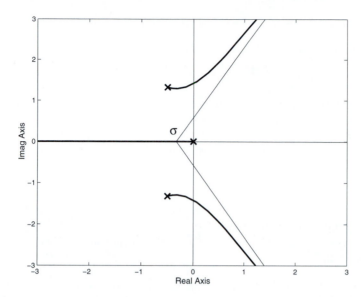

Figure 5.3. Locus for the closed-loop poles when the controller zero varies

R9 When the root locus crosses the imaginary axis, say at $s = \pm j\omega_c$, then ω_c can be computed either by using the Routh Hurwitz algorithm or by using the fact that $s^2 + \omega_c^2$ divides exactly the polynomial $D(s) + \lambda M(s)$, for some positive real value of λ.

Example 5.9. *Consider a plant with transfer function $G_o(s)$ and a feedback controller with transfer function $C(s)$, where*

$$G_o(s) = \frac{1}{(s-1)(s+2)} \qquad and \qquad C(s) = 4\frac{s+\alpha}{s} \qquad (5.6.12)$$

We want to know how the location, of the closed-loop poles change for α moving in $\mathbb{R}^+$.

We first notice that the closed-loop poles are the roots of

$$1 + 4\frac{s+\alpha}{s(s^2+s-2)} = \frac{s(s^2+s-2)+4s+4\alpha}{s(s^2+s-2)} = 0 \Longrightarrow s(s^2+s+2s)+4\alpha = 0$$

$$(5.6.13)$$

After dividing by $s(s^2+s+2s)$, we find that equation (5.6.2) applies, with

$$\lambda = 4\alpha \qquad and \qquad F(s) = \frac{1}{s(s^2+s+2)} \qquad (5.6.14)$$

If we use the rules described, above we obtain the following:

R1 The root locus has three branches ($m = 0$ and $n = 3$).

R4 The negative real axis belongs to the locus.

R5 For α close to zero, that is, λ close to zero, the roots are at the poles of $F(s)$, that is, at $0, -0.5 \pm j0.5\sqrt{7}$.

R7 When α tends to ∞, that is, when λ tends to ∞, the three roots tend to ∞, following asymptotes that intersect the real axis at $(\sigma, 0)$ with $\sigma = -\frac{1}{3}$. The angle of these asymptotes are π, $\pi/3$, and $2\pi/3$. This means that two branches go to the right-half plane.

R9 The characteristic polynomial is $s^3 + s^2 + 2s + \lambda$. When two of the branches cross the imaginary axis, then this polynomial can be exactly divided by $s^2 + \omega_c^2$, leading to a quotient equal to $s + 1$ with a polynomial residue equal to $(2 - \omega_c^2)s + \lambda - \omega_c^2$. If we make the residue equal to zero, we obtain $\omega_c = \sqrt{2}$ and $\lambda_c = 2$.

*The above rules allow one to sketch a root-locus diagram. This can also be obtained by using the MATLAB command **rlocus**; the result is shown in Figure 5.3. A powerful MATLAB environment, **rltool**, allows a versatile root-locus analysis, including shifting and adding of poles and zeros.*

□□□

5.7   Nominal Stability using Frequency Response

A classical and lasting tool that can be used to assess the stability of a feedback loop is Nyquist stability theory. In this approach, stability of the closed loop is predicted by using the open-loop frequency response of the system. This is achieved by plotting a polar diagram of the product $G_o(s)C(s)$ and then counting the number of encirclements of the $(-1, 0)$ point. We show how this works below.

We shall first consider an arbitrary transfer function $F(s)$ (not necessarily related to closed-loop control).

Nyquist stability theory depends upon mappings between two complex planes:

- The independent variable s

- The dependent variable F

The basic idea of Nyquist stability analysis is as follows:

Assume you have a closed oriented curve $\mathcal{C}_s$ in $\boxed{s}$ [2] that encircles Z zeros and P poles of the function $F(s)$. We assume that there are no poles *on* $\mathcal{C}_s$.

If we move along the curve $\mathcal{C}_s$ in a defined direction, then the function $F(s)$ maps $\mathcal{C}_s$ into another oriented closed curve, $\mathcal{C}_F$ in $\boxed{F}$.

---

[2] $\boxed{s}$ and $\boxed{F}$ denote the s- and the F- planes, respectively.

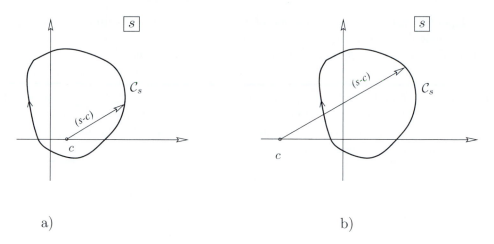

a)   b)

Figure 5.4. Single-zero function and Nyquist analysis

We will show below that the number of times that C_F encircles the origin in the complex plane $\boxed{F}$ is determined by the difference between P and Z. If $G_o(s)C(s)$ is modified, we can observe how the encirclement scenery changes (or how close it is to changing).

In the sequel, it will be helpful to recall that every clockwise (*counterclockwise*) encirclement of the origin by a variable in the complex plane implies that the angle of the variable has changed by $-2\pi[rad]$ ($2\pi[rad]$).

We first analyze the case of a simple function $F(s) = s - c$ with c lying *inside* the region enclosed by C_s. This is illustrated in Figure 5.4, part a).

We see that, as s moves clockwise along C_s, the angle of $F(s)$ changes by $-2\pi[rad]$: the curve C_F will enclose the origin in $\boxed{F}$ once in the clockwise direction.

In Figure 5.4, part b), we illustrate the case when c lies *outside* the region enclosed by C_s. In this case, the angle of $F(s)$ does not change as s moves around the curve C_s and thus, no encirclement of the origin in $\boxed{F}$ occurs.

By using similar reasoning, we see that, for the function $F(s) = (s - p)^{-1}$ (where the pole p lies *inside* the region determined by C_s), there is an angle change of $+2\pi[rad]$ when s moves clockwise along C_s. This is equivalent to saying that the curve C_F encloses the origin in $\boxed{F}$ once in the counterclockwise direction. The phase change is zero if p is *outside* the region, leading again to no encirclements of the origin in $\boxed{F}$.

Consider now the case in which $F(s)$ takes the following more general form:

$$F(s) = K \frac{\prod_{i=1}^{m}(s - c_i)}{\prod_{k=1}^{n}(s - p_k)} \tag{5.7.1}$$

Then any net change in the angle of $F(s)$ results from the sum of the changes of the angle due to the factors $(s - c_i)$ minus the sum of the changes in the angle due to the factors $(s - p_k)$. This leads to the following result.

> Consider a function $F(s)$ as in (5.7.1) and a closed curve $\mathcal{C}_s$ in $\boxed{s}$. Assume that $F(s)$ has Z zeros and P poles inside the region enclosed by $\mathcal{C}_s$. Then, as s moves clockwise along $\mathcal{C}_s$, the resulting curve $\mathcal{C}_F$ encircles the origin in $\boxed{F}$ Z-P times in a clockwise direction.

So far, this result seems rather abstract; however, as hinted earlier, it has a direct connection to the issue of closed-loop stability. Specifically, to apply this result, we consider a special function $F(s)$, related to the open-loop transfer function of plant, G_o, and a controller $C(s)$ by the simple relationship:

$$F(s) = 1 + G_o(s)C(s) \tag{5.7.2}$$

We note that the zeros of $F(s)$ are the closed-loop poles in a unity feedback control system. Also, the poles of $F(s)$ are the open-loop poles of plant and controller. We assume that $G_o(s)C(s)$ is strictly proper, so that $F(s)$ in (5.7.2) satisfies

$$\lim_{|s|\to\infty} F(s) = 1 \tag{5.7.3}$$

In the context of stability assessment, we are particularly interested in the number of closed-loop poles (if any) that lie in the right-half plane, so we introduce a particular curve $\mathcal{C}_s$ which completely encircles the right-half plane (RHP) in $\boxed{s}$ in the *clockwise* direction.

This curve combines the imaginary axis $\mathcal{C}_i$ and the return curve $\mathcal{C}_r$ (a semicircle of infinite radius), as shown in Figure 5.5. This choice of $\mathcal{C}_s$ is known as the Nyquist path.

We now determine the closed curve $\mathcal{C}_F$ in the complex plane $\boxed{F}$ that results from evaluating $F(s)$ for every $s \in \mathcal{C}_s$. $F(s)$ satisfies equation (5.7.3), so the whole mapping of $\mathcal{C}_r$ collapses to the point $(1,0)$ in $\boxed{F}$. Thus, only the mapping of $\mathcal{C}_i$ has to be computed, i.e., we need only to plot the *frequency response* $F(j\omega)$, $\forall \omega \in (-\infty, \infty)$ in the complex plane $\boxed{F}$. This is a polar plot of the frequency response, known as the Nyquist plot.

We have chosen $F(s) = 1 + G_o(s)C(s)$, so the zeros of $F(s)$ correspond to the closed-loop poles. Moreover, we see that $F(s)$ and $G_o(s)C(s)$ share exactly the same pole, (the open-loop poles). We also see that the origin of the complex plane $\boxed{F}$ corresponds to the point $(-1,0)$ in the complex plane $\boxed{G_oC}$. Thus, the Nyquist plot for F can be replaced by that of G_oC by simply counting encirclements about the -1 point. The basic Nyquist theorem, which derives from the previous analysis, is then given by the following:

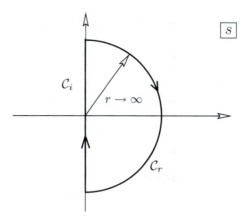

Figure 5.5. Nyquist path

Theorem 5.1. *If a proper open-loop transfer function $G_o(s)C(s)$ has P poles in the open RHP and none on the imaginary axis, then the closed loop has Z poles in the open RHP if and only if the polar plot $G_o(j\omega)C(j\omega)$ encircles the point $(-1, 0)$ clockwise $N=Z-P$ times.*

From this theorem, we conclude that the following:

- If the system is open-loop stable, then, for the closed loop to be internally stable, it is necessary and sufficient that no unstable cancellations occur and that the Nyquist plot of $G_o(s)C(s)$ *not encircle the point $(-1, 0)$.*

- If the system is open-loop unstable, with P poles in the open RHP, then, for the closed loop to be internally stable, it is necessary and sufficient that no unstable cancellations occur and that the Nyquist plot of $G_o(s)C(s)$ *encircle the point $(-1, 0)$ P times counterclockwise.*

- If the Nyquist plot of $G_o(s)C(s)$ passes through the point $(-1, 0)$, there exists an $\omega_o \in \mathbb{R}$ such that $F(j\omega_o) = 0$, i.e., the closed loop has poles located exactly on the imaginary axis. This situation is known as a *critical stability condition.*

There is one important remaining issue, namely, how to apply Nyquist theory when there are open-loop poles exactly on the imaginary axis. The main difficulty in this case can be seen from Figure 5.4 on page 139, part a). If c is located exactly on the curve C_s, then the change in the angle of the vector $s - c$ is impossible to determine. To deal with this problem, a *modified Nyquist path* is employed, as shown in Figure 5.6. The modification is illustrated for the simple case when there is one open-loop pole at the origin.

The Nyquist path $\mathcal{C}_s$ is now composed of three curves: $\mathcal{C}_r$, $\mathcal{C}_a$ and $\mathcal{C}_b$. The resulting closed curve $\mathcal{C}_F$ will differ from the previous case only when s moves along $\mathcal{C}_b$. This is a semicircle of radius ϵ, where ϵ is an infinitesimal quantity. (In that way the encircled region is still the whole RHP, except for an infinitesimal area.)

Using the analysis developed above, we see that we must compute the change in the angle of the vector $(s - p)$ when s travels along $\mathcal{C}_b$ with $p = 0$. This turns out to be $+\pi[rad]$, i.e., the factor s^{-1} maps the curve $\mathcal{C}_b$ into a semicircle with *infinite radius and clockwise direction*.

To compute the number and direction of encirclements, these infinite radius semi-circles must be considered, because they are effectively part of the Nyquist diagram.

This analysis can be extended to include any finite collection of poles of $G_o(s)C(s)$ on $\mathcal{C}_s$.

The modified form of the Nyquist theorem to accommodate the above changes is shown in Figure 5.6.

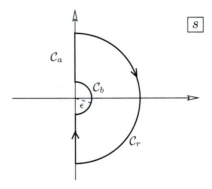

Figure 5.6. Modified Nyquist path

The modified form of Theorem 5.1 is as follows:

Theorem 5.2 (Nyquist theorem). *Given a proper open-loop transfer function $G_o(s)C(s)$ with P poles in the open RHP, then the closed loop has Z poles in the open RHP if and only if the plot of $G_o(s)C(s)$ encircles the point $(-1, 0)$ clockwise N=Z-P times when s travels along the modified Nyquist path.*

Remark 5.1. *To use the Nyquist theorem to evaluate internal stability, we need to have the additional knowledge that no cancellation of unstable poles occurs between*

C(s) and $G_o(s)$. This follows from the fact that the Nyquist theorem applies only to the product $G_o(s)C(s)$, whereas internal stability depends also on the fact that no unstable pole-zero cancellation occurs. (See Lemma 5.1 on page 128.)

5.8   Relative Stability: Stability Margins and Sensitivity Peaks

In control-system design, one often needs to go beyond the issue of closed-loop stability. In particular, it is usually desirable to obtain some quantitative measures of how far from instability the nominal loop is, i.e., to quantify *relative stability*. This is achieved by introducing measures that describe the *distance* from the nominal open-loop frequency response to the critical stability point $(-1; 0)$.

Figure 5.7 shows two sets of relative stability indicators for the case when the open loop has no poles in the open RHP. Part (a) in the figure describes the *gain margin, M_g,* and *the phase margin, M_f.* They are defined as follows:

$$M_g \overset{\triangle}{=} -20\log_{10}(|a|) \tag{5.8.1}$$

$$M_f \overset{\triangle}{=} \phi \tag{5.8.2}$$

Thus, the gain margin indicates the additional gain that would take the closed loop to the critical stability condition. The phase margin quantifies the pure phase delay that should be added to achieve the same critical condition.

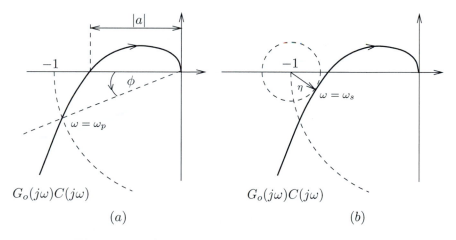

$G_o(j\omega)C(j\omega)$ $G_o(j\omega)C(j\omega)$

(a)   (b)

Figure 5.7. Stability margins and sensitivity peak

Part (b) in Figure 5.7 defines an alternative indicator for relative stability. We first recall that a vector from $(-1; 0)$ to $G_o(j\omega_1)C(j\omega_1)$, for $\omega = \omega_1$, corresponds to $1 + G_o(j\omega_1)C(j\omega_1)$, i.e. to $|S_o(j\omega_1)|^{-1}$. Thus, the radius, η, of the circle tangent to the polar plot of $G_o(j\omega)C(j\omega)$ is the reciprocal of the *nominal sensitivity peak*. The larger the sensitivity peak is, the closer the loop will be to instability.

The sensitivity peak is a more compact indicator of relative stability than the gain and phase margins. The reader is invited to find an example where the gain and phase margins are good, yet a very large sensitivity peak warns of a very delicate stability condition. The converse is not true: sensitivity performance implies minimal values for the gain and phase margins. This is precisely described by the following relationship:

$$M_f \geq 2arcsin\left(\frac{\eta}{2}\right) \tag{5.8.3}$$

Stability margins can also be described and quantified by using Bode diagrams.

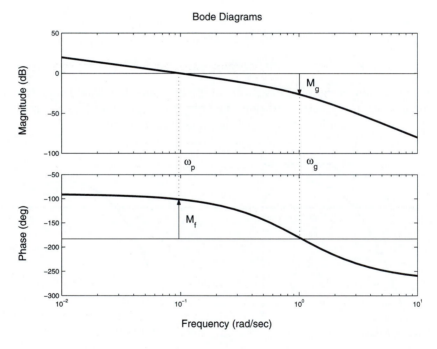

Figure 5.8. Stability margins in Bode diagrams

Consider the Bode diagrams in Figure 5.8; then the frequency ω_p corresponds to the frequency at which the magnitude is 0 $[dB]$. This allows one to compute the phase margin, M_f, as shown in Figure 5.8. The second frequency, ω_g, corresponds to the frequency at which the phase is $-180°$. This allows one to compute the gain margin, M_g, as shown in Figure 5.8.

5.9   Robustness

So far, we have only considered only the effect that the controller has on the nominal closed loop formed with the nominal model for the plant. In practice, however, we are usually interested in not only this nominal performance but also the true performance achieved when the controller is applied to the true plant. This is the so-called "Robustness" issue. We will show below that the nominal sensitivities do indeed tell us something about the true or achieved sensitivities.

5.9.1   Achieved Sensitivities

We contrast the nominal sensitivities derived previously with the achieved (or true) sensitivities when the controller $C(s)$ is applied to some calibration models, $G(s)$. This contrast leads to the following calibration sensitivities:

$$T(s) \triangleq \frac{G(s)C(s)}{1 + G(s)C(s)} = \frac{B(s)P(s)}{A(s)L(s) + B(s)P(s)} \qquad (5.9.1)$$

$$S(s) \triangleq \frac{1}{1 + G(s)C(s)} = \frac{A(s)L(s)}{A(s)L(s) + B(s)P(s)} \qquad (5.9.2)$$

$$S_i(s) \triangleq \frac{G(s)}{1 + G(s)C(s)} = \frac{B(s)L(s)}{A(s)L(s) + B(s)P(s)} \qquad (5.9.3)$$

$$S_u(s) \triangleq \frac{C(s)}{1 + G(s)C(s)} = \frac{A(s)P(s)}{A(s)L(s) + B(s)P(s)} \qquad (5.9.4)$$

where the transfer function of the calibration model is given by

$$G(s) = \frac{B(s)}{A(s)} \qquad (5.9.5)$$

In the sequel, we will again use the words *true plant* to refer to the calibration model, but the reader should recall the comments made in section §3.9 regarding possible reasons why the calibration model may not describe the true plant response.

Note that, in general, $G(s) \neq G_o(s)$, so $T_o \neq T$, etc. A difficulty that we will study later is that $G(s)$ is usually not perfectly known. We thus need to say something about achieved sensitivities, on the basis of only our knowledge of the nominal sensitivities, and the information about the likely model errors. This is addressed below.

5.9.2   Robust Stability

We are concerned with the case in which the nominal model and the true plant differ. It is then necessary that, in addition to nominal stability, we check that stability is retained when the true plant is controlled by the same controller. We call this property *robust stability*.

Sufficient conditions for a feedback loop to be robustly stable are stated in the following:

Theorem 5.3 (Robust stability theorem). *Consider a plant with nominal transfer function $G_o(s)$ and true transfer function given by $G(s)$. Assume that $C(s)$ is the transfer function of a controller that achieves nominal internal stability. Also assume that $G(s)C(s)$ and $G_o(s)C(s)$ have the same number of unstable poles. Then, a sufficient condition for stability of the true feedback loop obtained by applying the controller to the true plant is that*

$$|T_o(j\omega)||G_\Delta(j\omega)| = \left| \frac{G_o(j\omega)C(j\omega)}{1 + G_o(j\omega)C(j\omega)} \right| |G_\Delta(j\omega)| < 1 \qquad \forall \omega \qquad (5.9.6)$$

where $G_\Delta(j\omega)$ is the frequency response of the multiplicative modeling error (MME).

Proof

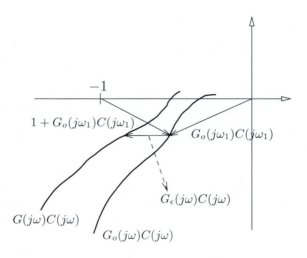

Figure 5.9. Nyquist plot for the nominal and the true loop

We first recall that, by assumption, the nominal loop is stable and that $G(s)C(s)$ and $G_o(s)C(s)$ have the same number of unstable poles. This means that the real loop will be stable if and only if the Nyquist plot of $G(j\omega)C(j\omega)$ encircles the point

$(-1, 0)$ *the same number of times (and in the same direction) that* $G_o(j\omega)C(j\omega)$
does.

   We also have that

$$G(s)C(s) = G_o(s)C(s) + G_\epsilon(s)C(s) \qquad\qquad (5.9.7)$$

that is, the change in the open-loop transfer function is $G_\epsilon(s)C(s)$, *where* $G_\epsilon(j\omega)$ *is
the frequency response of the additive modeling error (AME).*
   Consider now Figure 5.9.
   From that figure, we see that the same number of encirclements occur, if

$$|G_\epsilon(j\omega)C(j\omega)| < |1 + G_o(j\omega)C(j\omega)| \qquad\qquad \forall\omega \qquad\qquad (5.9.8)$$

Recalling that $G_\epsilon(s) = G_o(s)G_\Delta(s)$, *we see that (5.9.8) is equivalent to*

$$\frac{|G_\Delta(j\omega)G_o(j\omega)C(j\omega)|}{|1 + G_o(j\omega)C(j\omega)|} < 1 \qquad\qquad (5.9.9)$$

   *This is equivalent to (5.9.6), upon using the definition of the nominal comple-
mentary sensitivity.*

□□□

Remark 5.2. *Theorem 5.3 gives only a* sufficient *condition for robust stability.
This is illustrated in Example 5.10.*

Remark 5.3. *Theorem 5.3 also applies to discrete-time and sampled-data systems,
provided that the appropriate frequency-response function is used (depending on
whether shift operators or delta operators are used).*

Remark 5.4. *It is also possible to extend Theorem 5.3 on the facing page to the
case when* $G_\Delta(s)$ *is unstable. All that is required is that one keep track of the
appropriate number of encirclements to guarantee stability of the true system.*

Remark 5.5. *When applying the robust stability result, it is usual that* $|G_\Delta(j\omega)|$
be replaced by some known upper bound, say $\epsilon(\omega)$. *The sufficient condition can then
be replaced by* $|T_o(j\omega)\epsilon(\omega)| < 1$, $\forall\omega$.

□□□

Example 5.10. *In a feedback control loop, the open-loop transfer function is given
by*

$$G_o(s)C(s) = \frac{0.5}{s(s+1)^2} \qquad\qquad (5.9.10)$$

and the true plant transfer function is

$$G(s) = e^{-s\tau}G_o(s) \qquad\qquad (5.9.11)$$

5.10.1 *Find the exact value of τ that leads the closed loop to the verge of instability.*

5.10.2 *Use the robust stability theorem 5.3 on page 146 to obtain an estimate for that critical value of τ.*

5.10.3 *Discuss why the result in part 5.10.2 differs from that in part 5.10.1.*

Solution

5.10.1 *The delay introduces a phase change equal to $-\omega\tau$, but it does not affect the magnitude of the frequency response. Thus, the critical stability condition arises when this lag equals the phase margin, M_f, i.e., when the delay is given by*

$$\tau = \frac{M_f}{\omega_p} \tag{5.9.12}$$

where ω_p is defined in Figure 5.7 on page 143 and is such that $|G_o(j\omega_p)| = 1$. This yields $\omega_p = 0.424[rad/s]$ and $M_f = 0.77[rad]$. Hence, the critical value for the delay is $\tau = 1.816[s]$.

5.10.2 *The nominal complementary sensitivity is given by*

$$T_o(s) = \frac{G_o(s)C(s)}{1 + G_o(s)C(s)} = \frac{0.5}{s^3 + 2s^2 + s + 0.5} \tag{5.9.13}$$

and the multiplicative modeling error is

$$G_\Delta(s) = e^{-s\tau} - 1 \implies |G_\Delta(j\omega)| = 2\left|\sin\left(\frac{\omega\tau}{2}\right)\right| \tag{5.9.14}$$

The robust stability theorem states that a sufficient condition for robust stability is that $|T_o(j\omega)G_\Delta(j\omega)| < 1$, $\forall\omega$. Several values of τ were tried. Some of these results are shown in Figure 5.10 on the next page.

Figure 5.10 on the facing page shows that $|T_o(j\omega)G_\Delta(j\omega)| < 1$, $\forall\omega$ for $\tau \leq 1.5$.

5.10.3 *It can be observed that a conservative value of the delay is obtained when the robust stability theorem is used. This is due to the fact that this theorem sets a sufficient condition for robust stability, i.e., it is a worst-case requirement.*

□□□

Further insights into robustness issues can be obtained by relating the nominal and achieved sensitivity functions. In particular, we have the following:

Lemma 5.2. *Consider the nominal sensitivities $S_o(s)$ and $T_o(s)$ and a plant with MME $G_\Delta(s)$. Then the achieved sensitivity functions S and T are given by*

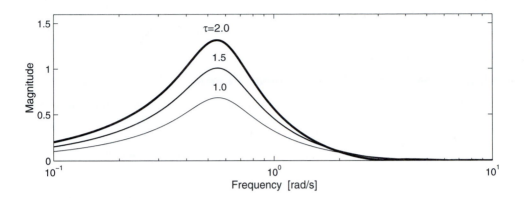

Figure 5.10. Magnitude of the frequency response of $T_o(s)G_\Delta(s)$ for different
values of τ

$$S(s) = S_o(s)S_\Delta(s) \tag{5.9.15}$$
$$T(s) = T_o(s)(1 + G_\Delta(s))S_\Delta(s) \tag{5.9.16}$$
$$S_i(s) = S_{io}(s)(1 + G_\Delta(s))S_\Delta(s) \tag{5.9.17}$$
$$S_u(s) = S_{uo}(s)S_\Delta(s) \tag{5.9.18}$$
$$S_\Delta(s) = \frac{1}{1 + T_o(s)G_\Delta(s)} \tag{5.9.19}$$

$S_\Delta(s)$ *is called the* error sensitivity.

Proof

By direct substitution.

□□□

We see how the performance of the real loop differs from that of the nominal loop
due to modeling errors. This is often called the problem of *performance robustness*.
From equations (5.9.15) to (5.9.19), it can be seen that the nominal performance
will not be too different from the achieved performance if $S_\Delta(j\omega)$ is close to $1 + j0$
for all frequencies. We see from equation (5.9.19) that this will hold if the frequency
response $|T_o(j\omega)|$ rolls off before the MME $|G_\Delta(jw)|$ becomes significant; then, this
ensures $|T_o(jw)G_\Delta(jw)| \ll 1$.

We notice that stability robustness is a less stringent requirement than robust
performance: For the former, it suffices to have $|T_o(jw)G_\Delta(jw)| < 1$, whereas the

latter requires $|T_o(j\omega)G_\Delta(j\omega)| \ll 1$.

Robust stability and performance introduce additional design trade-offs, as we shall show later.

5.9.3   Linear Control of Nonlinear Plants

The robustness analysis presented above has considered only *linear* modeling errors. Of course, in practice, one often has a nonlinear plant, and thus the modeling errors should, more properly, be described by nonlinear operators. As might be expected, small nonlinear errors can be shown to have a correspondingly small effect on closed-loop performance. Also, it is possible to quantify the *size* of the nonlinear model error consistent with maintaining closed-loop stability. These issues will be taken up in Chapter 19, when we address nonlinear control in detail. The more advanced reader is invited to preview section §19.2 at this stage.

5.10   Summary

- This chapter introduced the fundamentals of SISO feedback control-loop analysis.

- Feedback introduces a cyclical dependence between controller and system:

    ○ the controller action affects the systems outputs,

    ○ and the system outputs affect the controller action.

- For better or worse, this situation has a remarkably complex effect on the emergent closed loop.

- Well-designed, feedback can

    ○ make an unstable system stable;

    ○ increase the response speed;

    ○ decrease the effects of disturbances;

    ○ decrease the effects of system parameter uncertainties; and more.

- Poorly designed, feedback can

    ○ introduce instabilities into a previously stable system;

    ○ add oscillations into a previously smooth response;

    ○ result in high sensitivity to measurement noise;

    ○ result in sensitivity to structural modeling errors; and more.

- Individual aspects of the overall behavior of a dynamic system include

- ○ *time domain*: stability, rise time, overshoot, settling time, steady-state errors, among others

- ○ *frequency domain*: bandwidth, cut-off frequencies, gain and phase margins, among others

- Some of these properties have rigorous definitions; others tend to be qualitative.

- Any property or analysis can further be prefixed with the term *nominal* or *robust*:

  - ○ *nominal* indicates a temporarily idealized assumption that the model is perfect;

  - ○ *robust* indicates an explicit investigation of the effect of modeling errors.

- The effect of the controller $C(s) = \frac{P(s)}{L(s)}$ on the nominal model $G_o(s) = \frac{B_o(s)}{A_o(s)}$ in the feedback loop shown in Figure 5.1 on page 122 is

$$
\begin{aligned}
Y &= \frac{G_o C}{1 + G_o C}(R - D_m) + \frac{1}{1 + G_o C}D_o + \frac{G_o}{1 + G_o C}D_i \\
&= \frac{B_o P}{A_o L + B_o P}(R - D_m) + \frac{A_o L}{A_o L + B_o P}D_o + \frac{B_o L}{A_o L + B_o P}D_i \\
U &= \frac{C}{1 + G_o C}(R - D_o - G_o D_i - D_m) = \frac{A_o P}{A_o L + B_o P}(R - D_o - D_m) - \frac{B_o P}{A_o L + B_o P}D_i
\end{aligned}
$$

- Interpretation, definitions, and remarks:

  - ○ The nominal response is determined by four transfer functions.

  - ○ Because of their fundamental importance, they have individual names and symbols:

$$S_o := \frac{1}{1 + G_o C} = \frac{A_o L}{A_o L + B_o P},$$ the nominal sensitivity function;

$$T_o := \frac{G_o C}{1 + G_o C} = \frac{B_o P}{A_o L + B_o P},$$ the nominal complementary sensitivity;

$$S_{io} := \frac{G_o}{1 + G_o C} = \frac{B_o L}{A_o L + B_o P},$$ the nominal input sensitivity;

$$S_{uo} := \frac{C}{1 + G_o C} = \frac{A_o P}{A_o L + B_o P},$$ the nominal control sensitivity.

They are collectively called the *nominal sensitivities*.

  - ○ All four sensitivity functions have the same poles, the roots of $A_o L + B_o P$.

  - ○ The polynomial $A_o L + B_o P$ is also called the nominal *characteristic polynomial*.

- Recall that stability of a transfer function is determined by the roots only.

- Hence, the nominal loop is stable if and only if the real parts of the roots of $A_oL + B_oP$ are all strictly negative. These roots have an intricate relation to the controller and system poles and zeros.

- The properties of the four sensitivities, and therefore the properties of the nominal closed loop, depend on the interlacing of the poles of the characteristic polynomial (the common denominator) and the zeros of A_oL, B_oP, B_oL, and A_oP, respectively.

- Linear modeling errors:

  - If the same controller is applied to a linear system, $G(s)$, that differs from the model by $G(s) = G_o(s)G_\Delta(s)$, then the resulting loop remains stable provided that $|T_o(j\omega)| \, |G_\Delta(j\omega)| < 1, \, \forall \omega$.

  - This condition is also known as the *small-gain theorem*.

  - Obviously, it cannot be easily tested, because the multiplicative modeling error, G_Δ, is typically unknown. Usually, bounds on $G_\Delta(j\omega)$ are used instead.

  - Nevertheless, it gives valuable insight. For example, we see that the closed-loop bandwidth must be tuned to be less than the frequencies where one expects significant modeling errors.

- Nonlinear modeling errors:

  If the same controller is applied to a system, $G_{nl}\langle\circ\rangle$, that not only differs from the model linearly but that is nonlinear (as real systems will always be to some extent), then rigorous analysis becomes very hard in general, but qualitative insights can be obtained into the operation of the system by considering the impact of model errors.

5.11   Further Reading

Root locus

Evans, W. (1950). Control systems synthesis by root locus methods. *Trans. AIEE*, 69:66-69.

Nyquist criterion

Brockett, R. and Willems, J.C. (1965a). Frequency domain stability criteria-Part I. *IEEE Transactions on Automatic Control*, 10(7):255-261.

Brockett, R. and Willems, J.C. (1965b). Frequency domain stability criteria-Part II. *IEEE Transactions on Automatic Control*, 10(10):407-413.

Nyquist, H. (1932). Regeneration theory. *Bell Sys. Tech J.*, 11:126-147.

Poles and zeros

Maddock, R. (1982). *Poles and zeros in Electrical and Control Engineering.* Holt Rinehart and Winston.

5.12   Problems for the Reader

Problem 5.1. *Consider a feedback control loop of a plant with nominal model* $G_o(s) = \frac{1}{(s+1)^2}$. *Assume that the controller $C(s)$ is such that the complementary sensitivity is*

$$T_o(s) = \frac{4}{(s+2)^2} \tag{5.12.1}$$

5.1.1 *Show that the control loop is internally stable.*

5.1.2 *Compute the controller transfer function, $C(s)$.*

5.1.3 *If the reference is a unit step, compute the plant input.*

Problem 5.2. *Consider the same control loop as in Problem 5.1. Compute the maximum instantaneous error.*

Problem 5.3. *Consider the following candidates for the open-loop transfer function, $G_o(s)C(s)$:*

$$(a)\quad \frac{(s+2)}{(s+1)s} \qquad (b)\quad \frac{e^{-0.5s}}{(s+1)s} \qquad (c)\quad \frac{1}{s^2+4}$$

$$(d)\quad \frac{4}{s(s^2+s+1)} \qquad (e)\quad \frac{s^2+4}{(s+1)^3} \qquad (f)\quad \frac{8}{(s-1)(s+2)(s-3)}$$

5.3.1 *For each case, build the Nyquist polar plot, and, hence, by applying the Nyquist theorem, determine the stability of the associated closed loop.*

5.3.2 *For all the cases for which the associated loop is stable, compute the stability margins and the sensitivity peak.*

5.3.3 *Repeat 5.3.1 and 5.3.2, using Bode diagrams.*

Problem 5.4. *In a nominal control loop, the sensitivity is given by*

$$S_o(s) = \frac{s(s+4.2)}{s^2+4.2s+9} \tag{5.12.2}$$

Assume that the reference is a unit step and that the output disturbance is given by $d_o(t) = 0.5\sin(0.2t)$.
Find an expression for the plant output in steady state.

Problem 5.5. *Consider a control loop where* $C(s) = K$.

5.5.1 *Build the root locus to describe the evolution of the closed-loop poles as K goes from 0 to ∞ for the following nominal plants:*

$$
(a) \quad \frac{1}{(s+1)(s-2)} \qquad (b) \quad \frac{1}{(s+1)^4} \qquad (c) \quad \frac{1}{s^2(s+2)}
$$

$$
(d) \quad \frac{(-s+1)(-s+4)}{s(s+2)(s+10)} \qquad (e) \quad \frac{s^2+2s+4}{(s+1)(s+6)(s+8)} \qquad (f) \quad \frac{1}{s^2-4s+8}
$$

5.5.2 *For each case, find the range of values of K, if it exists, such that the nominal loop is stable.*

Problem 5.6. *In a nominal control loop*

$$
G_o(s) = \frac{1}{(s+4)(s-1)} \qquad and \quad C(s) = 8\frac{s+\alpha}{s} \tag{5.12.3}
$$

Use the root-locus technique to find out the evolution of the closed-loop poles for $\alpha \in [0, \infty)$.

Problem 5.7. *Consider a system having the following calibration and nominal models:*

$$
G(s) = F(s)\frac{1}{s-1} \qquad and \quad G_o(s) = F(s)\frac{2}{s-2} \tag{5.12.4}
$$

where $F(s)$ is a proper, stable, and minimum-phase transfer function. Prove the following:

5.7.1 $G_\Delta(2) = -1$

5.7.2 $S_\Delta(s) \triangleq \dfrac{1}{1+T_o(s)G_\Delta(s)}$ *is unstable, having a pole at $s = 2$, where $T_o(s)$ is the complementary sensitivity of an internally stable control loop.*

5.7.3 *The achieved sensitivity $S(s) = S_\Delta(s)S_o(s)$ can be stable even though $S_\Delta(s)$ is unstable.*

Problem 5.8. *Consider a feedback control loop with nominal complementary sensitivity $T_o(s)$. Assume that, in the true feedback control loop, the measurement system is not perfect and the measured output, $Y_m(s)$ is given by*

$$Y_m(s) = Y(s) + G_m(s)Y(s) \tag{5.12.5}$$

where $G_m(s)$ is a stable transfer function.

5.8.1 *Determine an expression for the true complementary sensitivity, $T(s)$, as a function of $T_o(s)$ and $G_m(s)$.*

5.8.2 *Find $G_m(s)$ for the particular case in which the measurement is perfect except for the presence of a pure time delay $\tau > 0$.*

5.8.3 *Analyze the connection between this type of error and those types originating in plant modeling errors.*

Problem 5.9. *Consider a plant with input $u(t)$, output $y(t)$, and transfer function*

$$G(s) = \frac{16}{s^2 + 4.8s + 16} \tag{5.12.6}$$

5.9.1 *Compute the plant output for $u(t) = 0$, $\forall t \geq 0$, $y(0) = 1$ and $\dot{y}(0) = 0$.*

5.9.2 *The same plant, with the same initial conditions, is placed under feedback control, with $r(t) = 0$, $\forall t \geq 0$. Compute the plant output if the controller is chosen in such a way that the complementary sensitivity is given by*

$$T(s) = \frac{\alpha^2}{s^2 + 1.3\alpha s + \alpha^2} \tag{5.12.7}$$

Try different values of $\alpha \in \mathbb{R}^+$–in particular, $\alpha \ll 1$ and $\alpha \gg 1$.

5.9.3 *From the results obtained above, discuss the role of feedback in ameliorating the effects of initial conditions.*

Problem 5.10. *Consider a feedback control loop having $G(s) = \frac{2}{(s+1)(s+2)}$, with reference $r(t) = \mu(t)$, input disturbance $d_i(t) = A\sin(t+\alpha)$, and a controller having the transfer function*

$$C(s) = 600\frac{(s+1)(s+2)(s+3)(s+5)}{s(s^2+1)(s+100)} \tag{5.12.8}$$

5.10.1 *Show that the nominal feedback loop is stable.*

5.10.2 *Explain,* using the inversion principle and sensitivity functions, *why the controller in equation (5.12.8) is a sensible choice.*

5.10.3 *Assume that the true plant transfer function differs from the nominal model by a MME which satisfies the following bound:*

$$|G_\Delta(j\omega)| \le \frac{\omega}{\sqrt{\omega^2 + 400}} \tag{5.12.9}$$

Analyze the loop from the point of view of robust stability and robust performance.

Problem 5.11. *Consider a feedback control loop in which a controller is synthesized to achieve a complementary sensitivity $T_o(s)$. The synthesized controller has a transfer function given by $C_o(s)$; however, the implemented controller has a transfer function given by $C_1(s) \neq C_o(s)$. If*

$$T_o(s) = \frac{100}{s^2 + 13s + 100}; \qquad C_o(s) = \frac{1}{s}F(s); \qquad C_1(s) = \frac{1}{s+\alpha}F(s) \tag{5.12.10}$$

where $F(s)$ is a rational function with $F(0) \neq 0$ and $\alpha > 0$, investigate the stability of the implemented control loop.

Chapter 6

CLASSICAL PID CONTROL

6.1 Preview

In this chapter we review a particular control structure that has become almost universally used in industrial control. It is based on a particular fixed-structure controller family, the so-called PID controller family. The letters 'PID' stand for *Proportional, Integral and Derivative Control.* They have proven to be robust in the control of many important applications.

The simplicity of these controllers is also their weakness: it limits the range of plants that they can control satisfactorily. Indeed, there exists a set of unstable plants which cannot be stabilized with *any* member of the PID family. Nevertheless, the surprising versatility of PID control (really, PID control simply means: control with an up-to-second-order controller) ensures continued relevance and popularity for this controller. It is also important to view this second-order setting as a special case of modern design methods, as presented, for example, in Chapters 7 and 15. This chapter covers the classical approaches to PID design, on account of the historical and practical significance of the methods and their continued use in industry.

6.2 PID Structure

Consider the simple SISO control loop shown in Figure 6.1.

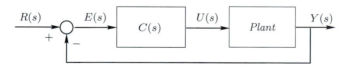

Figure 6.1. Basic feedback control loop

The traditional expressions for PI and PID controllers can be described by their transfer functions, relating error $E(s) = R(s) - Y(s)$ and controller output $U(s)$ as follows:

$$C_P(s) = K_p \tag{6.2.1}$$

$$C_{PI}(s) = K_p \left(1 + \frac{1}{T_r s} \right) \tag{6.2.2}$$

$$C_{PD}(s) = K_p \left(1 + \frac{T_d s}{\tau_D s + 1} \right) \tag{6.2.3}$$

$$C_{PID}(s) = K_p \left(1 + \frac{1}{T_r s} + \frac{T_d s}{\tau_D s + 1} \right) \tag{6.2.4}$$

where T_r and T_d are known as the *reset time* and *derivative time*, respectively.

As seen from (6.2.1) to (6.2.4), the members of this family include, in different combinations, three control modes or actions: *proportional (P), integral (I)* and *derivative (D)*.

Caution must be exercised when applying PID tuning rules, as there are a number of other parameterizations. Equation (6.2.4) is known as the *standard form*. An alternative, *series form*, is as follows:

$$C_{series}(s) = K_s \left(1 + \frac{I_s}{s} \right) \left(1 + \frac{D_s s}{\gamma_s D_s s + 1} \right) \tag{6.2.5}$$

The parallel form is as follows:

$$C_{parallel}(s) = K_p + \frac{I_p}{s} + \frac{D_p s}{\gamma_p D_p s + 1} \tag{6.2.6}$$

Terminology, such as *P-gain* are not uniquely defined and can refer to either K_s in (6.2.5), K_p in (6.2.6) or K_p in (6.2.4). It is therefore important to know which of these parameterizations any one particular technique refers to and, if implemented in a different form, to transform the parameters appropriately. Before the PID was recognized as simply a second-order controller, PID tuning was viewed in terms of the P, I and D parameters. Although their impact on the closed loop is far from independent of each other, their effect was thought as follows:

- **Proportional action** provides a contribution which depends on the instantaneous value of the control error. A proportional controller can control any stable plant, but it provides limited performance and nonzero steady-state errors. This latter limitation is due to the fact that its frequency response is bounded for all frequencies.

  It has also been traditional to use the expression *proportional band* (PB) to describe the proportional action. The equivalence is

$$PB[\%] = \frac{100[\%]}{K_p} \tag{6.2.7}$$

The proportional band is defined as the error required (as a percentage of full scale) to yield a 100% change in the controller output.

- **Integral action**, on the other hand, gives a controller output that is proportional to the accumulated error, which implies that it is a *slow reaction* control mode. This characteristic is also evident in its low-pass frequency response. The integral mode plays a fundamental role in achieving perfect plant inversion at $\omega = 0$. This forces the steady-state error to zero in the presence of a step reference and disturbance. The integral mode, viewed in isolation, has two major shortcomings: its pole at the origin is detrimental to loop stability and it also gives rise to the undesirable effect (in the presence of actuator saturation) known as *wind-up*, which we will discuss in detail in Chapter 11.

- **Derivative action** acts on the rate of change of the control error. Consequently, it is a *fast mode* which ultimately disappears in the presence of constant errors. It is sometimes referred to as a *predictive mode*, because of its dependence on the error trend. The main limitation of the derivative mode, viewed in isolation, is its tendency to yield large control signals in response to high-frequency control errors, such as errors induced by set-point changes or measurement noise. Its implementation requires properness of the transfer functions, so a pole is typically added to the derivative, as is evident in equations (6.2.3) and (6.2.4). In the absence of other constraints, the additional time constant τ_D is normally chosen such that $0.1 T_d \le \tau_D \le 0.2 T_d$. This constant is called the derivative time constant; the smaller it is, the larger the frequency range over which the filtered derivative approximates the exact derivative, with equality in the limit:

$$\lim_{\tau_D \to 0} u_{PID}(t) = K_p e(t) + \frac{K_p}{T_r} \int_{t_o}^{t} e(\tau) d\tau + K_p T_d \frac{de(t)}{dt} + i.c. \qquad (6.2.8)$$

The classical argument to choose $\tau_D \ne 0$ was, apart from ensuring that the controller be proper, to attenuate high-frequency noise. The latter point is illustrated in Figure 6.2, which shows that the filtered derivative approximates the exact derivative well at frequencies up to $\frac{1}{\tau_D}[rad/s]$, but that it has finite gain at high frequencies, whereas an exact derivative has infinite gain.

Because $\tau_D \ne 0$ was largely seen as a necessary evil, i.e., as a necessary departure from a pure proportional, integral, and derivative action, almost all industrial PID controllers once set τ_D as a fixed fraction of T_D, rather than viewing it as an independent design parameter in its own right. Since then it has become clear, however, that the time constant of the derivative is an important degree of freedom available to the designer.

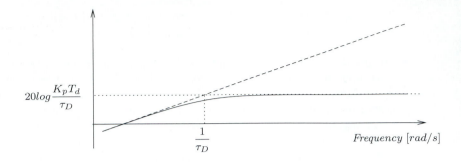

Figure 6.2. Bode magnitude plot of exact derivative (dashed) and filtered
derivative (solid)

Remark 6.1. *Because of frequent misunderstandings, we reiterate that there are
PID parameterizations other than (6.2.4). Some have arisen from the physical implementation of the controller. The series structure of equation (6.2.5), for example,
is essentially due to early physical realization with pneumatic devices.*

*By comparing (6.2.4), (6.2.5), and (6.2.6) one can derive exact (in some cases
approximate) conversion formulas. The important point is to be aware of the existence of different parameterizations and, consequently, different definitions of the
P, I, and D gains.*

6.3   Empirical Tuning

One of the traditional ways to design a PID controller was to use empirical tuning
rules based on measurements made on the real plant. Today, we suggest that
it is preferable for the PID designer to employ model-based techniques, such as
the ones described in sections §7.3, §15.4, and §15.6.2. If necessary, they can be
packaged as simple recipe procedures. The classical techniques are still referred to
by practitioners, so the following sections review the best-known of the classical
tuning methods.

6.4   Ziegler-Nichols (Z-N) Oscillation Method

This procedure is valid only for open-loop stable plants, and it is carried out by
means of the following steps.

- Set the true plant under proportional control, with a very small gain.

- Increase the gain until the loop starts oscillating. *Note that linear oscillation
  is required and that it should be detected at the controller output.*

	K_p	T_r	T_d
P	$0.50K_c$		
PI	$0.45K_c$	$\dfrac{P_c}{1.2}$	
PID	$0.60K_c$	$0.5P_c$	$\dfrac{P_c}{8}$

Table 6.1. Ziegler-Nichols tuning, using the oscillation method

- Record the controller critical gain $K_p = K_c$ and the oscillation period of the controller output, P_c.

- Adjust the controller parameters according to Table 6.1; there is some controversy regarding the PID parameterization for which the Z-N method was developed, but the version described here is, to the best knowledge of the authors, applicable to the parameterization of (6.2.4).

We first observe that the underlying model being obtained in the experiment is only *one point on the frequency response*, namely the one corresponding to a phase equal to $-\pi[rad]$ and a magnitude equal to K_c^{-1}, because the Nyquist plot for $K_pG(j\omega)$ crosses the point $(-1, 0)$ when $K_p = K_c$.

The settings in Table 6.1 were obtained by Ziegler and Nichols, who aimed to achieve an underdamped response to a step for those plants for which a satisfactory model has the form

$$G_o(s) = \frac{K_o e^{-s\tau_o}}{\nu_o s + 1} \qquad \text{where} \qquad \nu_o > 0 \qquad (6.4.1)$$

Figure 6.3 shows the step responses of a PI Z-N tuned control loop. Time, in this figure, has been measured in units of the delay τ_o, and different ratios $x \triangleq \frac{\tau_o}{\nu_o}$ are considered.

Example 6.1. *Consider a plant with a model given by*

$$G_o(s) = \frac{1}{(s+1)^3} \qquad (6.4.2)$$

Find the parameters of a PID controller by using the Z-N oscillation method. Obtain a graph of the response to a unit step input reference and to a unit step input disturbance.

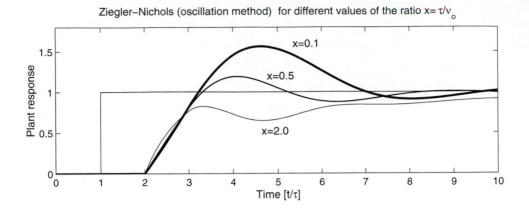

Figure 6.3. PI Z-N tuned (oscillation method) control loop for different values of the ratio $x \triangleq \dfrac{\tau_o}{\nu_o}$.

Solution

We first compute the critical gain K_c and critical frequency ω_c. These values must satisfy

$$K_c G_o(j\omega_c) = -1 \iff K_c = -(j\omega_c + 1)^3 \tag{6.4.3}$$

From this equation we obtain $K_c = 8$ and $\omega_c = \sqrt{3}$. Hence, the critical period is $P_c \approx 3.63$. If we now use the settings in Table 6.1, we obtain the following:

$$K_p = 0.6 * K_c = 4.8; \qquad T_r = 0.5 * P_c \approx 1.81; \qquad T_d = 0.125 * P_c \approx 0.45 \tag{6.4.4}$$

The derivative mode will be attenuated with a fast pole with time constant $\tau_D = 0.1 * T_d = 0.045$. Thus, the final loop-transfer function becomes

$$G_o(s)C(s) = K_p \frac{(T_d + \tau_D)s^2 + (1 + \frac{\tau_D}{T_r})s + \frac{1}{T_r}}{s(\tau_D s + 1)(s + 1)^3} = \frac{52.8s^2 + 109.32s + 58.93}{s(s + 22.2)(s + 1)^3} \tag{6.4.5}$$

With SIMULINK, the loop was simulated with a unit step reference at $t = 0$ and a unit step input disturbance at $t = 10$. The results are shown in Figure 6.4.

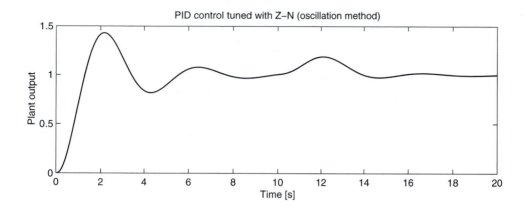

Figure 6.4. Response to step reference and step input disturbance

*The response shown in Figure 6.4 exhibits significant overshoot that might be unacceptable in some applications; however, Z-N tuning provides a starting point for finer tuning. The reader is invited to check, using SIMULINK file **pid1.mdl**, that $T_d = 1$ provides better performance. Of course, if significant measurement noise is present, then one should check whether increasing the derivative action still yields good performance. Similar caution applies regarding saturation in the controller output. The heuristic nature of these arguments highlights the limitations of tuning a PID controller in terms of its classical P, I, and D tuning parameters.*

Remark 6.2. *Figure 6.3 shows that Z-N tuning is very sensitive to the ratio between the delay and the time constant. Another shortcoming of this method is that it requires that the plant be forced to oscillate; this can be dangerous and expensive. Other tuning strategies that do not require this operating condition have been developed. Some of them are discussed below.*

Remark 6.3. *As mentioned earlier, a key issue when applying PID tuning rules (such as Ziegler-Nichols settings) is that of which PID structure these settings are applied to. Some authors claim that the experiments of Ziegler and Nichols and their conclusions were derived by using a controller with series structure of the form described in (6.2.5).*

To obtain an appreciation of these differences, we evaluate the PID control loop for the same plant in Example 6.1, but with the Z-N settings applied to the series structure; in the notation used in (6.2.5), we have that

$$K_s = 4.8 \qquad I_s = 1.81 \qquad D_s = 0.45 \qquad \gamma_s = 0.1 \qquad (6.4.6)$$

If we next compare equations (6.2.4) and (6.2.5), we have that

$$\tau_D = 0.045 \qquad K_p = 5.99 \qquad T_r = 2.26 \qquad T_d = 0.35 \qquad (6.4.7)$$

We finally simulate the same reference and disturbance conditions as those in Example 6.1 on page 163. The results are shown in Figure 6.5.

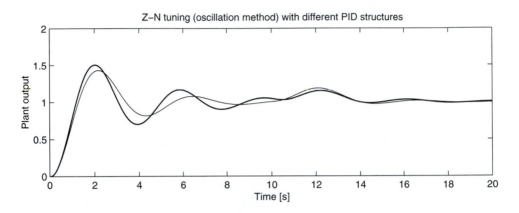

Figure 6.5. PID Z-N settings applied to series structure (thick line) and conventional structure (thin line)

Figure 6.5 shows that, in this particular case, no significant difference arises; however, this conclusion can vary for other systems to be controlled. Thus, it is always advisable to check the parameterization of the PID controller if settings from any given synthesis are going to be implemented.

6.5   Reaction Curve Based Methods

Many plants, particularly the ones arising in the process industries, can be satisfactorily described by the model in (6.4.1). A linearized quantitative version of this model can be obtained with an open-loop experiment, by using the following procedure.

1 With the plant in open loop, take the plant manually to a normal operating point. Say that the plant output settles at $y(t) = y_o$ for a constant plant input $u(t) = u_o$.

2 At an initial time t_o, apply a step change to the plant input, from u_o to u_∞ (this should be in the range of 10 to 20% of full scale).

3 Record the plant output until it settles to the new operating point. Assume that you obtain the curve shown in Figure 6.6. This curve is known as the *process reaction curve*.

In Figure 6.6, *m.s.t.* stands for maximum slope tangent.

4 Compute the parameter model as follows:

$$K_o = \frac{y_\infty - y_o}{u_\infty - u_o}; \qquad \tau_o = t_1 - t_o; \qquad \nu_o = t_2 - t_1 \qquad (6.5.1)$$

The model obtained can be used to derive various tuning methods for PID controllers. One of these methods was also proposed by Ziegler and Nichols. In their proposal, the design objective is to achieve a particular damping in the loop response to a step reference. More specifically, the aim is to obtain a ratio of 4:1 for the first and second peaks in that response. The suggested parameters are shown in Table 6.2.

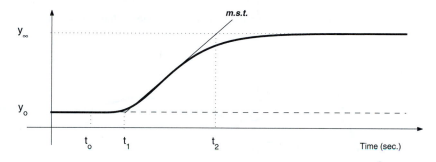

Figure 6.6. Plant step response

The parameter-setting rules proposed in Table 6.2 are applied to the model (6.4.1), where we have again normalized time in delay units. Loop responses for a unit step reference are shown in Figure 6.7 on the following page for different values of the ratio $x = \frac{\tau_o}{\nu_o}$.

Figure 6.7 on the next page reveals the extreme sensitivity of the performance to different values of the ratio x. To improve this limitation, Cohen and Coon carried out further studies to find controller settings that, when based on the same model (6.4.1), lead to a weaker dependence on the ratio of delay to time constant. Their findings are shown in Table 6.3.

To compare the settings proposed in Table 6.3 with the settings suggested by Ziegler and Nichols, the results corresponding to those in Figure 6.7 are shown in Figure 6.8.

A comparison of Tables 6.2 and 6.3 explains why the controller settings proposed by Ziegler-Nichols and Cohen-Coon for small values of x are similar. This similarity

	K_p	T_r	T_d
P	$\dfrac{\nu_o}{K_o\tau_o}$		
PI	$\dfrac{0.9\nu_o}{K_o\tau_o}$	$3\tau_o$	
PID	$\dfrac{1.2\nu_o}{K_o\tau_o}$	$2\tau_o$	$0.5\tau_o$

Table 6.2. Ziegler-Nichols tuning by using the reaction curve

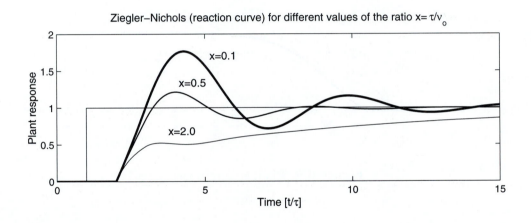

Figure 6.7. PI Z-N tuned (reaction curve method) control loop

is also apparent in the corresponding curves in Figures 6.7 and 6.8 on the next page; however, it is also evident from these figures that Cohen-Coon offers a more homogeneous response for the same range of values for x.

	K_p	T_r	T_d
P	$\dfrac{\nu_o}{K_o\tau_o}\left[1+\dfrac{\tau_o}{3\nu_o}\right]$		
PI	$\dfrac{\nu_o}{K_o\tau_o}\left[0.9+\dfrac{\tau_o}{12\nu_o}\right]$	$\dfrac{\tau_o[30\nu_o+3\tau_o]}{9\nu_o+20\tau_o}$	
PID	$\dfrac{\nu_o}{K_o\tau_o}\left[\dfrac{4}{3}+\dfrac{\tau_o}{4\nu_o}\right]$	$\dfrac{\tau_o[32\nu_o+6\tau_o]}{13\nu_o+8\tau_o}$	$\dfrac{4\tau_o\nu_o}{11\nu_o+2\tau_o}$

Table 6.3. Cohen-Coon tuning by using the reaction curve

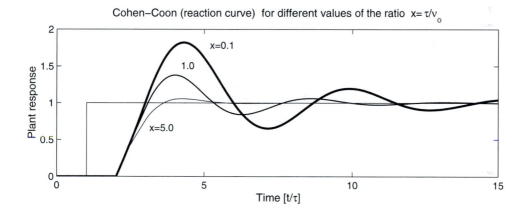

Figure 6.8. PI Cohen-Coon tuned (reaction curve method) control loop

In any case, the reader must be aware that, if one decides to perform PID empirical tuning, the proposed settings delivered by the strategies above are only starting points in the process of obtaining the right controller. In particular, we urge reading sections §7.3, §15.4, and §15.6.2, which give a modern view of PID design.

6.6   Lead-Lag Compensators

Closely related to PID control is the idea of lead-lag compensation. These ideas are frequently used in practice, specially when compensators are built with electronic hardware components. The transfer function of these compensators is of the form

$$C(s) = \frac{\tau_1 s + 1}{\tau_2 s + 1} \qquad (6.6.1)$$

When $\tau_1 > \tau_2$, this is a *lead network*, when $\tau_1 < \tau_2$, this is a *lag network*. The straight-line approximation to the Bode diagrams for these networks are given in Figure 6.9 and Figure 6.10 (where τ_1 and τ_2 differ by a factor of $10 : 1$).

The lead compensator acts like an approximate derivation. In particular, we see from Figure 6.9 that this circuit produces approximately $45°$ of phase advance at $\omega = 1/\tau_1$ without a significant change in gain. Thus, if we have a simple feedback loop which passes through the -1 point at, say, frequency ω_1, then inserting a lead compensator such that $\omega_1 \tau_1 = 1$ will give a $45°$ phase margin. Of course, the disadvantage is an increase in the high-frequency gain, which can amplify high-frequency noise.

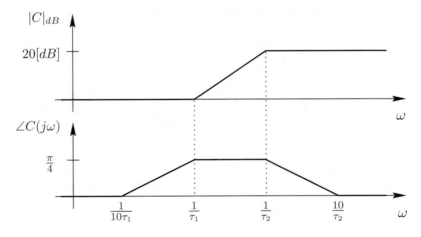

Figure 6.9. Approximate Bode diagrams for lead networks ($\tau_1 = 10\tau_2$)

An alternative interpretation of a lead network is obtained by considering its pole-zero structure. From (6.6.1) and the fact that $\tau_1 > \tau_2$, we see that it introduces a pole-zero pair, where the zero (at $s = -1/\tau_1$) is significantly closer to the imaginary axis than the pole (located at $s = -1/\tau_2$).

On the other hand, a lag compensator acts like an approximate integrator. In particular, we see from Figure 6.10 that the low-frequency gain is $20 \ [dB]$ higher than the gain at $\omega = 1/\tau_1$. Thus, this circuit, when used in a feedback loop, gives better low-frequency tracking and disturbance rejection. A disadvantage of this circuit is the additional phase lag experienced between $1/10\tau_2$ and $10/\tau_1$.

Hence, $1/\tau_2$ is typically chosen to be smaller than other significant dynamics in the plant.

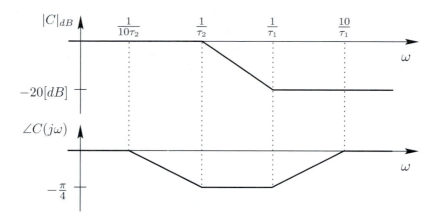

Figure 6.10. Approximate Bode diagrams for lag networks ($\tau_2 = 10\tau_1$)

From the point of view of its pole-zero configuration, the lag network introduces a pole (located at $s = -1/\tau_2$) that is significantly closer to the imaginary axis than the zero (located at $s = -1/\tau_1$).

In summary, the lead network feature used in design is its phase-advance characteristic. The lag network instead is useful due to its gain characteristics at low frequency.

When both effects are required, a lead-lag effect can be obtained by cascading a lead and a lag compensator.

6.7 Distillation Column

PID control is very widely used in industry. Indeed, one would be hard-pressed to find loops that do not use some variant of this form of control.

Here, we illustrate how PID controllers can be utilized in a practical setting by briefly examining the problem of controlling a distillation column.

Distillation columns are an extremely common unit found in a host of chemical processes/plants. Their purpose is to separate mixtures of liquids by using the relative volatilities of the components.

For a detailed account of the theory and operation of distillation columns, together with an interactive simulation, the reader is referred to the website associated with this book. The example we have chosen here is the pilot plant described on the website.

In this example, the distillation column is used to separate ethanol from water. The system is illustrated in Figure 6.11 on the following page.

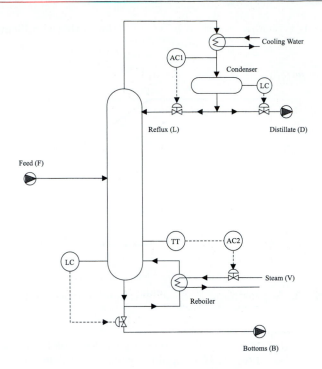

Figure 6.11. Ethanol-water distillation column

All the other key variables are controlled, leaving two inputs and two outputs. We choose the reflux flow (u_1) to control the mole fraction of ethanol in the top product (y_1) and we choose the reboiler steam flow (u_2) to control the composition of the bottom product. (This is inferred from y_2, the bottom plate temperature.)

A locally linearized model for this system is as follows:

$$\begin{bmatrix} Y_1(s) \\ Y_2(s) \end{bmatrix} = \begin{bmatrix} G_{11}(s) & G_{12}(s) \\ G_{21}(s) & G_{22}(s) \end{bmatrix} \begin{bmatrix} U_1(s) \\ U_2(s) \end{bmatrix} \tag{6.7.1}$$

where

$$G_{11}(s) = \frac{0.66e^{-2.6s}}{6.7s + 1} \tag{6.7.2}$$

$$G_{12}(s) = \frac{-0.0049e^{-s}}{9.06s + 1} \tag{6.7.3}$$

$$G_{21}(s) = \frac{-34.7e^{-9.2s}}{8.15s + 1} \tag{6.7.4}$$

$$G_{22}(s) = \frac{0.87(11.6s + 1)e^{-s}}{(3.89s + 1)(18.8s + 1)} \tag{6.7.5}$$

Note that the units of time here are minutes.

Also, note that u_1 affects not only y_1 (via the transfer function G_{11}) but also y_2 (via the transfer function G_{21}). Similarly u_2 effects y_2 (via the transfer function G_{22}) and y_1 (via the transfer function G_{12}).

In designing the two PID controllers, we will initially ignore the two transfer functions G_{12} and G_{21}. This approach leads to two separate (and noninteracting) SISO systems. The resultant controllers are as follows:

$$C_1(s) = 1 + \frac{0.25}{s} \tag{6.7.6}$$

$$C_s(s) = 1 + \frac{0.15}{s} \tag{6.7.7}$$

We see that these are of PI type.

To test the controllers under more realistic conditions, we apply the controllers to the calibration model given in (6.7.1) to (6.7.5). The results are shown in Figure 6.12.

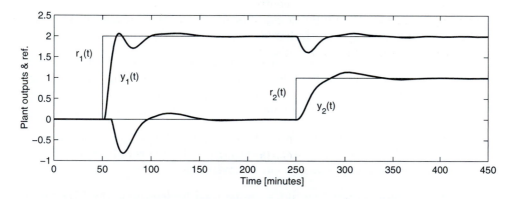

Figure 6.12. Simulation results for PI control of distillation column

In Figure 6.12, a step in the reference (r_1) for output y_1 is applied at $t = 50$ and a step in the reference (r_2) for output y_2 is applied at $t = 250$.

It can be seen from the figure that the PID controllers give quite acceptable performance on this problem; however, the figure also shows something that is very common in practical applications–namely, the two loops interact: a change in reference r_1 not only causes a change in y_1 (as required) but also induces a transient in y_2. Similarly, a change in the reference r_2 causes a change in y_2 (as required) and induces a change in y_1. In this particular example, these interactions are probably sufficiently small to be acceptable. Thus, in common with the majority of industrial problems, we have found that two simple PID (actually PI in this case) controllers give quite acceptable performance for this problem. The reader can interactively try other controllers for this problem on the website.

Finally, the reader might wonder whether the interactions noticed in the above example could become a problem if one demanded more (e.g., faster responses) from the controllers. Under these conditions, the interactions can become an important factor. The reader is thus encouraged to read on to Parts VI, VII, and VIII where we will show how these kinds of interactions can be dealt with rigorously in control-system design.

□□□

6.8 Summary

- PI and PID controllers are widely used in industrial control.

- From a modern perspective, a PID controller is simply a controller (of up-to-second order) containing an integrator. Historically, however, PID controllers were tuned in terms of their **P**, **I**, and **D** terms.

- It has been empirically found that the PID structure often has sufficient flexibility to yield excellent results in many applications.

- The basic term is the proportional term, **P**, which causes a corrective control actuation proportional to the error.

- The integral term, **I**, gives a correction proportional to the integral of the error. This has the positive feature of ultimately ensuring that sufficient control effort is applied to reduce the tracking error to zero; however, integral action tends to have a destabilizing effect due to increased phase shift.

- The derivative term, **D**, gives a predictive capability yielding a control action proportional to the rate of change of the error. This tends to have a stabilizing effect but often leads to large control movements.

- Various empirical tuning methods can be used to determine the PID parameters for a given application. They should be considered as a first guess in a search procedure.

- Attention should also be paid to the PID structure.

- Systematic model-based procedures for PID controllers will be covered in later chapters.

6.9   Further Reading

Åström, K. and Hägglund, T. (1984). A frequency domain method for automatic tuning of simple feedback loops. In *Proceedings of the 23rd IEEE Conference on Decision and Control*, Las Vegas, NV, (1):299-304.

Åström, K. and Hägglund, T. (1995). *PID controllers: theory, design and tuning.* Instrument Society of America, 2nd edition.

Mellquist, M., Crisafulli, S., and MacLeod, I.M. (1997).   Derivative filtering issues for industrial PID controllers.   In *Proceedings of the I.E. Aust. Control '97 Conference*, Sydney, pages 486-491.

Mellquist, M., Crisafulli, S., and MacLeod, I.M. (1997). An investigation of derivative filtering for PID controllers. *Technical Report, Centre for Control Engineering*, University of the Witwatersrand, Johannesburg, South Africa.

Rivera, D., Morari, M., and Skogestad, S. (1986).   Internal model control. PID controller design. *Ind. Eng. Chem. Process Des. Dev.*, 25:252-265.

Voda, A. and Landau, I.D. (1995).   A method for the auto-calibration of PID controllers. *Automatica*, 31(1):41-53.

Ziegler, J. and Nichols, N.B. (1942). Optimum settings for automatic controllers. *Transactions of the ASME*, 64:759-768.

6.10   Problems for the Reader

Problem 6.1. *Consider the following nominal plant models, G_o:*

(a) $\dfrac{10}{(s+1)(s+10)}$

(b) $\dfrac{1}{(s+1)(s-2)}$

(c) $\dfrac{1}{(s^2+s+1)}$

(d) $\dfrac{1}{(s+1)s}$

Find suitable values for the parameters of a member of the PID family to control each of these models.

Problem 6.2. *Consider the model*

$$G_o(s) = \frac{-\alpha s + 1}{(s+1)(s+2)} \qquad (6.10.1)$$

Use the Cohen-Coon tuning strategy to synthesize a PID controller for different values of α in the interval $[0.1;\ 20]$

Problem 6.3. *Discuss whether there exist linear plants that cannot be stabilized by any PID controller. Illustrate your conclusions with some examples.*

Problem 6.4. *Consider a plant, open-loop stable, with complex dynamics and input saturation. Assume that you have the experimental set-up simulated in file **pidemp.mdl** and that you want to tune a PID controller.*
Without knowing the model for the plant,

6.4.1 *Find the critical gain and the oscillation period. Discuss the use of Osc-u and Osc-y.*

6.4.2 *Determine the settings for a PID controller.*

6.4.3 *Simulate the control loop with the tuned PID controller. Discuss.*

Problem 6.5. *Consider a plant with nominal model*

$$G_o(s) = \frac{-s+2}{(s+2)^2} \qquad (6.10.2)$$

Determine the settings that you would obtain on applying the Ziegler-Nichols ultimate-gain method.

Problem 6.6. *Consider a plant with nominal model*

$$G_o(s) = \frac{Ke^{-s\tau}}{Ts+1} \tag{6.10.3}$$

Assume that you apply the Cohen-Coon tuning method.

6.6.1 *Find the gain and phase margins for different ratios τ/T.*

6.6.2 *Find the sensitivity peak for different ratios τ/T.*

Problem 6.7. *Synthesize a PI controller for a plant having a nominal model given by*

$$G_o(s) = \frac{1}{s^2 + 6s + 9} \tag{6.10.4}$$

in such a way that $M_g \geq 10[dB]$ and $M_f \geq \pi/4$.

Problem 6.8. *Consider a plant with nominal model*

$$G_o(s) = \frac{-s+3}{(s+3)(s+5)} \tag{6.10.5}$$

6.8.1 *Synthesize a PID controller to achieve a complementary sensitivity with bandwidth equal to 2 [rad/s].*

6.8.2 *Compare your result with that obtained by using the Cohen-Coon strategy. Use time-domain and frequency-domain tests.*

Chapter 7

SYNTHESIS OF SISO CONTROLLERS

7.1 Preview

In Chapter 5, it was shown how feedback control-loop performance and stability can be characterized by using a set of four sensitivity functions. A key feature of these functions is that all of them have poles belonging to the same set, the set of *closed-loop poles*. Hence, this set determines the stability and the natural modes of the closed loop. A key synthesis question is therefore: given a model, *can one systematically synthesize a controller such that the closed-loop poles are in predefined locations?* This chapter will show that this is indeed possible. We call this *pole assignment*, which is a fundamental idea in control synthesis.

In this chapter, we will use a polynomial description. This evolves naturally from the analysis in Chapter 5. In a later chapter, we will use a state space description. This will provide a natural transition from SISO to MIMO control systems in later chapters.

It is also shown how both approaches can accommodate other design requirements such as zero steady-state tracking errors and disturbance rejection.

Finally, PID controllers are placed into this general framework, and it is shown how they can be synthesized by using pole-assignment methods.

The question of how to choose a set of values for the closed-loop poles to meet the required performance specifications is, of course, a crucial issue. This is actually a nontrivial question, which needs to be considered as part of the intricate web of design trade-offs associated with all feedback loops. The latter will be taken up in the next chapter, where the design problem will be analyzed.

7.2 Polynomial Approach

We recall here the polynomial description of a plant as given in Chapter 4. We will denote the nominal plant transfer function by

179

$$G_o(s) = \frac{B_o(s)}{A_o(s)} \tag{7.2.1}$$

7.2.1   General Formulation

Consider the nominal loop shown in Figure 5.1 on page 122. We recall that, in the nominal control loop, the controller and the nominal model transfer functions are given by

$$C(s) = \frac{P(s)}{L(s)} \qquad\qquad G_o(s) = \frac{B_o(s)}{A_o(s)} \tag{7.2.2}$$

respectively, with

$$P(s) = p_{n_p} s^{n_p} + p_{n_p-1} s^{n_p-1} + \ldots + p_0 \tag{7.2.3}$$

$$L(s) = l_{n_l} s^{n_l} + l_{n_l-1} s^{n_l-1} + \ldots + l_0 \tag{7.2.4}$$

$$B_o(s) = b_{n-1} s^{n-1} + b_{n-2} s^{n-2} + \ldots + b_0 \tag{7.2.5}$$

$$A_o(s) = a_n s^n + a_{n-1} s^{n-1} + \ldots + a_0 \tag{7.2.6}$$

where we have assumed that the nominal model is strictly proper.

Consider now a desired closed-loop polynomial given by

$$A_{cl}(s) = a_{n_c}^c s^{n_c} + a_{n_c-1}^c s^{n_c-1} + \ldots + a_0^c \tag{7.2.7}$$

We will examine the following question: *with the polynomial $A_{cl}(s)$ arbitrarily specified, does there exist a proper $C(s)$ that results in $A_{cl}(s)$ as the closed-loop characteristic polynomial?* To motivate the idea, consider the following example.

Example 7.1. *Let $G_o(s) = \frac{B_o(s)}{A_o(s)}$ be the nominal model of a plant with $A_o(s) = s^2 + 3s + 2$, $B_o(s) = 1$; then, consider a controller of the form*

$$C(s) = \frac{P(s)}{L(s)}; \qquad P(s) = p_1 s + p_0; \qquad L(s) = l_1 s + l_0 \tag{7.2.8}$$

We see that $A_o(s)L(s) + B_o(s)P(s) = (s^2 + 3s + 2)(l_1 s + l_0) + (p_1 s + p_0)$. Say that we would like this to be equal to a polynomial $s^3 + 3s^2 + 3s + 1$; then equating coefficients gives

$$\begin{bmatrix} 1 & 0 & 0 & 0 \\ 3 & 1 & 0 & 0 \\ 2 & 3 & 1 & 0 \\ 0 & 2 & 0 & 1 \end{bmatrix} \begin{bmatrix} l_1 \\ l_0 \\ p_1 \\ p_0 \end{bmatrix} = \begin{bmatrix} 1 \\ 3 \\ 3 \\ 1 \end{bmatrix} \tag{7.2.9}$$

It is readily verified that the 4×4 matrix above is nonsingular, meaning that we can solve for l_1, l_0, p_1, and p_0 leading to $l_1 = 1$, $l_0 = 0$, $p_1 = 1$, and $p_0 = 1$. Hence, the desired characteristic polynomial is achieved by the controller $C(s) = (s+1)/s$.

□□□

In the above example, we saw that the capacity to assign closed-loop poles depending on the nonsingularity of a particular matrix. In the sequel, we will generalize the result; however, to do this, we will need the following mathematical result regarding coprime polynomials.

Theorem 7.1 (Sylvester's theorem). *Consider two polynomials*

$$A(s) = a_n s^n + a_{n-1} s^{n-1} + \ldots + a_0 \tag{7.2.10}$$
$$B(s) = b_n s^n + b_{n-1} s^{n-1} + \ldots + b_0 \tag{7.2.11}$$

together with the following eliminant matrix:

$$
\mathbf{M_e} =
\begin{bmatrix}
a_n & 0 & \cdots & 0 & b_n & 0 & \cdots & 0 \\
a_{n-1} & a_n & \cdots & 0 & b_{n-1} & b_n & \cdots & 0 \\
\vdots & \vdots & \ddots & \vdots & \vdots & \vdots & \ddots & \vdots \\
a_0 & a_1 & \cdots & a_n & b_0 & b_1 & \cdots & b_n \\
0 & a_0 & \cdots & a_{n-1} & 0 & b_0 & \cdots & b_{n-1} \\
\vdots & \vdots & \ddots & \vdots & \vdots & \vdots & \ddots & \vdots \\
0 & 0 & \cdots & a_0 & 0 & 0 & \cdots & a_0
\end{bmatrix}
\tag{7.2.12}
$$

The $A(s)$ and $B(s)$ are relatively prime (coprime) if and only if $\det(\mathbf{M_e}) \neq 0$.

Proof

If: *Assume $A(s)$, $B(s)$ not relatively prime. Then there exists a common root r. Write*

$$A(s) = (s - r)(a'_{n-1} s^{n-1} + \ldots + a'_0) \tag{7.2.13}$$
$$B(s) = (s - r)(b'_{n-1} s^{n-1} + \ldots + b'_0) \tag{7.2.14}$$
$$\tag{7.2.15}$$

Eliminating $(s - r)$ gives

$$A(s)(b'_{n-1} s^{n-1} + \ldots + b'_0) - B(s)(a'_{n-1} s^{n-1} + \ldots + a'_0) = 0 \tag{7.2.16}$$

Equating coefficients on both sides gives

$$\mathbf{M_e}\theta = 0 \qquad\qquad (7.2.17)$$

where

$$\theta^T = \begin{bmatrix} b'_{n-1} & \cdots & b'_0 & -a'_{n-1} & \cdots & -a'_0 \end{bmatrix} \qquad (7.2.18)$$

However, (7.2.17) has a nontrivial solution if and only if $\det(\mathbf{M_e}) = 0$.

Only if: *By reversing the above argument*

□□□

Armed with the above result, we can now generalize Example 7.1 to show that pole assignment is generally possible, provided that some minimal requirements are satisfied.

Lemma 7.1 (SISO pole placement–polynomial approach). *Consider a one-d.o.f. feedback loop with controller and plant nominal model given by (7.2.2) to (7.2.6). Assume that $B_o(s)$ and $A_o(s)$ are relatively prime (coprime)–i.e., they have no common factors. Let $A_{cl}(s)$ be an arbitrary polynomial of degree $n_c = 2n - 1$. Then there exist polynomials $P(s)$ and $L(s)$, with degrees $n_p = n_l = n - 1$, such that*

$$A_o(s)L(s) + B_o(s)P(s) = A_{cl}(s) \qquad (7.2.19)$$

Proof

Equating coefficients in (7.2.19) yields

$$\mathcal{S}\begin{bmatrix} l_{n-1} \\ \vdots \\ l_0 \\ p_{n-1} \\ \vdots \\ p_0 \end{bmatrix} = \begin{bmatrix} a^c_{2n-1} \\ \vdots \\ \vdots \\ \vdots \\ \vdots \\ a^c_0 \end{bmatrix} \quad with \quad \mathcal{S} \triangleq \begin{bmatrix} a_n & & & & 0 & & \\ a_{n-1} & \ddots & & & b_{n-1} & \ddots & \\ \vdots & & \ddots & & \vdots & & \ddots & \ddots \\ a_0 & & & a_n & b_0 & & \ddots & 0 \\ & \ddots & & a_{n-1} & & \ddots & & b_{n-1} \\ & & \ddots & \vdots & & & \ddots & \vdots \\ & & & a_0 & & & & b_0 \end{bmatrix}$$

$$(7.2.20)$$

In view of Sylvester's theorem (see Theorem 7.1), the matrix S is nonsingular if and only if the polynomials B_o and A_0 are relatively prime. The result follows.

□□□

Two additional cases will now be considered:

Case 1, $n_c = 2n - 1 + \kappa$, with κ a positive integer

It is easy to see that, in this case, one feasible solution is obtained if we choose $n_l = n - 1 + \kappa$. Then

$$L(s) = L_{ad}(s) + \bar{L}(s) \tag{7.2.21}$$

$$L_{ad}(s) \triangleq s^n (l_{n-1+\kappa} s^{\kappa-1} + \ldots + l_n) \tag{7.2.22}$$

$$\bar{L}(s) \triangleq l_{n-1} s^{n-1} + \ldots + l_0 \tag{7.2.23}$$

First, the coefficients of $L_{ad}(s)$ can be computed by equating the coefficients corresponding to the highest κ powers of s in equation (7.2.19). Then Lemma 7.1 on the facing page can be applied by replacing $L(s)$ by $\bar{L}(s)$ and $A_{cl}(s)$ by $A_{cl}(s) - A_o(s)L_{ad}(s)$. Thus, a solution always exists, provided that $A_o(s)$ and $B_o(s)$ are coprime.

□□□

Case 2, $n_c < 2n - 1$

In this case, there is no solution, except for very special choices of the closed-loop polynomial $A_{cl}(s)$. This can be seen by arguing as follows:

- The nominal model is strictly proper and the controller should be at least biproper[1], as we have that $n_l = n_c - n$.

- Thus, in the limit, we have that $n_p = n_l = n_c - n$. This means that we have at most $m = 2n_c - 2n + 2$ controller coefficients to achieve the desired closed-loop polynomial. Note that $m = n_c + 1 + (n_c - 2n + 1)$, which, given the assumption for this case $(n_c - 2n + 1 < 0)$, leads to $m < n_c + 1$.

- Equating coefficients in (7.2.19) leads to $n_c + 1$ equations; however, we have already shown that the number of unknowns is less than $n_c + 1$. Thus, the set of equations is, in general, inconsistent.

□□□

Remark 7.1. *There is no loss in generality from assuming that the polynomials $A_o(s)$, $L(s)$, and $A_{cl}(s)$ are monic. In the sequel, we will work under that assumption.*

---

[1]Recall that a rational transfer function is *biproper* if the numerator and denominator are polynomials of equal degree.

□□□

Remark 7.2. *Lemma 7.1 establishes the condition under which a solution exists for the pole-assignment problem, assuming a biproper controller. However, when a strictly proper controller is required, then the minimum degree of $P(s)$ and $L(s)$ have to be $n_p = n - 1$ and $n_l = n$ respectively. Thus, to be able to arbitrarily choose the closed-loop polynomial $A_{cl}(s)$, its degree must be equal to $2n$.*

□□□

Remark 7.3. *No unstable pole-zero cancellations[2] are allowed. This can be argued as follows. Any cancellation between controller and plant model will appear as a factor in $A_o(s)L(s)$ and also in $B_o(s)P(s)$. However, for equation (7.2.19) to be satisfied, it is also necessary that the same factor be present in $A_{cl}(s)$. Since $A_{cl}(s)$ has to be chosen as a stable polynomial, then this common factor must be stable. Only in this case, the nominal closed loop is guaranteed to be internally stable, i.e. the four sensitivity functions will be stable.*

□□□

7.2.2   Constraining the Solution

In subsequent chapters we will frequently find that performance specifications can require the controller to satisfy certain additional constraints. Some of these requirements can be satisfied in the polynomial pole-assignment approach by ensuring that extra poles or zeros are introduced. We illustrate this by several examples below.

Forcing integration in the loop

A standard requirement is that, in steady state, the nominal control loop should yield zero control error due to d.c. components in either the reference, input disturbance or output disturbance. For this to be achieved, a necessary and sufficient condition is that the nominal loop be internally stable and that the controller have, at least, one pole at the origin. This will render the appropriate sensitivity functions zero at zero frequency.

To achieve this we choose

$$L(s) = s\bar{L}(s) \tag{7.2.24}$$

The closed-loop equation (7.2.19) can then be rewritten as

$$\bar{A}_o(s)\bar{L}(s) + B_o(s)P(s) = A_{cl}(s) \qquad \text{with} \qquad \bar{A}_o(s) \triangleq sA_o(s) \tag{7.2.25}$$

---

[2]In the sequel, unless we say something to the contrary, the expression *pole-zero cancellation* will denote the cancellation of a plant pole by a controller zero, or vice versa.

The solution to the pole-assignment problem can now be considered as an equivalent problem having a model of degree $\bar{n} \triangleq n + 1$. Previous results might initially suggest that $A_{cl}(s)$ can be arbitrarily specified if and only if its degree is at least $2\bar{n} - 1 = 2n + 1$; however, a simplification results in this case, because we can allow the degree of $P(s)$ to be higher by one than that of $\bar{L}(s)$, because the controller will still be proper. This means that the minimum degree of $A_{cl}(s)$ is $2n$.

Forcing pole/zero cancellations

Sometimes it is desirable to force the controller to cancel a subset of stable poles and/or zeros of the plant model. To illustrate how this can be included in the pole-assignment strategy, we consider the special case when only one pole in the nominal model, say at $s = -p$, has to be canceled.

To cancel the factor $(s + p)$ in $A_o(s)$, this factor must also be present in $P(s)$. Equation (7.2.19) then has a solution only if the same factor is present in $A_{cl}(s)$. To solve equation (7.2.19), the factor $(s + p)$ can thus be removed from both sides of the equation.

Extension to multiple cancellations is straightforward.

Remark 7.4. *Canceled nominal model poles and zeros will still appear, as poles or zeros, in some closed-loop transfer functions. The discussion above shows that any cancellation will force the canceled factor to be present in the closed-loop polynomial $A_{cl}(s)$.*

To be more precise, any canceled plant-model pole will be a pole in the nominal input sensitivity, $S_{io}(s)$, and a zero in $S_{uo}(s)$. Also, any canceled plant model zero will be a pole in the nominal control sensitivity, $S_{uo}(s)$, and a zero in $S_{io}(s)$.

□□□

The following examples illustrate the polynomial pole-assignment technique.

Example 7.2. *Consider a nominal model given by*

$$G_o(s) = \frac{3}{(s+1)(s+3)} \tag{7.2.26}$$

The initial unconstrained design requires the degree of $A_{cl}(s)$ to be chosen to be equal to at least 3. For example, we could choose $A_{cl}(s) = (s^2 + 5s + 16)(s + 40)$. The polynomials $P(s)$ and $L(s)$ are then of degree 1. The polynomial equation to be solved is

$$(s+1)(s+3)(s+l_0) + 3(p_1 s + p_0) = (s^2 + 5s + 16)(s + 40) \tag{7.2.27}$$

which, upon equating coefficients, leads to

$$l_0 = 41 \qquad p_1 = \frac{49}{3} \qquad p_0 = \frac{517}{3} \tag{7.2.28}$$

Alternatively, if one desires to constrain the design so that the controller features integral action, then $A_{cl}(s)$ must be chosen to have degree of at least 4. We also assume, by way of illustration, that the model pole at $s = -1$ should be canceled. The polynomial equation to be solved now becomes

$$s(s+1)(s+3)(s+l_0) + 3(s+1)(p_1 s + p_0) = (s+1)(s^2 + 5s + 16)(s + 40)$$
$$(7.2.29)$$

After canceling the common factor $(s+1)$, and upon equating coefficients, we finally obtain

$$l_0 = 42 \qquad p_1 = 30 \qquad p_0 = \frac{640}{3} \qquad \Longrightarrow C(s) = \frac{(s+1)(90s + 640)}{3s(s+42)}$$
$$(7.2.30)$$

□□□

Example 7.3. *A plant has a nominal model given by $G_o(s) = \frac{1}{s-1}$. The control aim is to track a sinusoidal reference of $2[rad/s]$ of unknown amplitude and phase.*

Design a controller that stabilizes the plant and that provides zero steady-state control error.

Solution

Polynomial pole assignment will be used. We first consider the order of the polynomial $A_{cl}(s)$ to be specified. The requirement of zero steady error for the given reference implies that

$$S_o(\pm j2) = 0 \iff G_o(\pm j2)C(\pm j2) = \infty \qquad (7.2.31)$$

The constraint in equation (7.2.31) can be satisfied for the given plant if and only if the controller $C(s)$ has poles at $\pm j2$. Thus

$$C(s) = \frac{P(s)}{L(s)} = \frac{P(s)}{(s^2 + 4)L_{ad}(s)} \qquad (7.2.32)$$

This means that $A_{cl}(s)$ can be arbitrarily specified if its degree is equal to at least 3. We choose $A_{cl}(s) = (s^2 + 4s + 9)(s + 10)$. This leads to $L_{ad}(s) = 1$ and $P(s) = p_2 s^2 + p_1 s + p_0$.

The pole-assignment equation is then

$$(s-1)(s^2 + 4) + p_2 s^2 + p_1 s + p_0 = (s^2 + 4s + 9)(s + 10) \qquad (7.2.33)$$

Upon expanding the polynomial products and equating coefficients, we finally obtain

$$P(s) = 15s^2 + 45s + 94 \Longrightarrow C(s) = \frac{15s^2 + 45s + 94}{s^2 + 4} \qquad (7.2.34)$$

In a more numerically demanding problem, the reader may wish to use the MAT-LAB program **paq.m** *on the accompanying CD-ROM.*

□□□

The idea of making $C(\pm j2) = \infty$ in the above example is actually a special case of the Internal Model Principle . This will be discussed in detail in Chapter 10. The above approach to pole assignment introduces additional dynamics in the loop. The basic concept behind this approach is that, by using feedback, one is able to shift the natural system frequencies to arbitrary locations, provided that certain conditions hold.

To solve the polynomial equation, we require that the polynomials A_o and B_o be coprime. We have given an algebraic reason for that requirement. It is also possible to give a state space interpretation to this result. In fact, coprimeness of A_o and B_o turns out to be equivalent to complete reachability and observability of the associated state space model. We will explore these state space issues in Chapter 18.

7.3   PI and PID Synthesis Revisited by using Pole Assignment

The reader will recall that PI and PID controller synthesis by using classical methods was reviewed in Chapter 6. In this section, we place these results in a more modern setting by discussing the synthesis of PI and PID controllers via pole-assignment techniques.

Throughout this section, we will consider one-d.o.f. control loops with PI controllers of the form

$$C_{PI}(s) = K_p + \frac{K_I}{s} \qquad (7.3.1)$$

and proper PID controllers of the form

$$C_{PID}(s) = K_p + \frac{K_I}{s} + \frac{K_D s}{\tau_D s + 1} \qquad (7.3.2)$$

For future reference, we note the following alternative representation of a PID controller.

Lemma 7.2. *Any controller of the form*

$$C(s) = \frac{n_2 s^2 + n_1 s + n_o}{d_2 s^2 + d_1 s} \tag{7.3.3}$$

is identical to the PID controller (7.3.2), where

$$K_p = \frac{n_1 d_1 - n_o d_2}{d_1^2} \tag{7.3.4}$$

$$K_I = \frac{n_o}{d_1} \tag{7.3.5}$$

$$K_D = \frac{n_2 d_1^2 - n_1 d_1 d_2 + n_o d_2^2}{d_1^3} \tag{7.3.6}$$

$$\tau_D = \frac{d_2}{d_1} \tag{7.3.7}$$

Proof

Straightforward upon the equating of (7.3.2) to (7.3.3).

□□□

As indicated by (7.3.1), designing a PI controller requires the tuning of two constants, K_p and K_I, whereas a PID controller requires the additional two constants K_D and τ_D. Rather than tuning them directly, however, one can automatically synthesize them by using pole assignment.

If we assume that the plant can be (at least, approximately) modeled by a second-order model, then we can immediately use pole assignment to synthesize a PID controller. Referring to subsection §7.2.2, we simply choose

degree $A_o(s)$ $= 2$
degree $B_o(s)$ ≤ 1
degree $\bar{L}(s)$ $= 1$
degree $\bar{P}(s)$ $= 2$
degree $A_{cl}(s)$ $= 4$

If the plant contains a time delay, as in (6.4.1), then we need to obtain an approximate second-order model before we can proceed as suggested above. One way to do this is to approximate the time delay by a first-order all-pass system. In this case, (6.4.1) is approximately modeled as

$$G(s) = \frac{Ke^{-s\tau}}{\nu_o s + 1} \approx \left[\frac{K}{\nu_o s + 1}\right]\left[\frac{-\frac{s\tau}{2} + 1}{\frac{s\tau}{2} + 1}\right] = G_o(s) \tag{7.3.8}$$

Example 7.4. *A plant has a nominal model given by*

$$G_o(s) = \frac{2}{(s+1)(s+2)} \tag{7.3.9}$$

Synthesize a PID controller that yields a closed loop with dynamics dominated by the factor $s^2 + 4s + 9$.

Solution

The controller is synthesized by solving the pole-assignment equation, with the following quantities

$$A_{cl}(s) = (s^2 + 4s + 9)(s + 4)^2; \quad B_o(s) = 2; \quad A_o(s) = s^2 + 3s + 2 \quad (7.3.10)$$

where the factor $(s+4)^2$ has been added to ensure that the pole-assignment equation has a solution. Note that this factor generates modes that are faster than those originated in $s^2 + 4s + 9$.

Solving the pole-assignment equation gives

$$C(s) = \frac{P(s)}{s\bar{L}(s)} = \frac{14s^2 + 59s + 72}{s(s + 9)} \quad (7.3.11)$$

We observe, using Lemma 7.2 on page 187, that $C(s)$ is a PID controller with

$$K_p = 5.67; \quad K_I = 8; \quad K_D = 0.93; \quad \tau_D = 0.11 \quad (7.3.12)$$

An important observation is that the solution to this problem has the structure of a PID controller for the given model $G_o(s)$. For a higher-order $G_o(s)$, the resulting controller will not, in general, be a PID controller.

□□□

7.4 Smith Predictor

Time delays are very common in real-world control problems, so it is important to examine whether one can improve on the performance achievable from a simple PID controller. This is especially important when the delay dominates the response.

For the case of *stable* open-loop plants, a useful strategy is provided by the Smith predictor. The basic idea here is to build a parallel model that cancels the delay. (See Figure 7.1.) In this figure, we assume that the plant model takes the form

$$G_o(s) = e^{-s\tau}\bar{G}_o(s) \quad (7.4.1)$$

It can be seen from Figure 7.1 that the design of the controller, $C(s)$, can be carried out by using the *undelayed* part of the model, $\bar{G}_o(s)$, because the delay has been canceled by the parallel model. Thus, we can design the controller with a

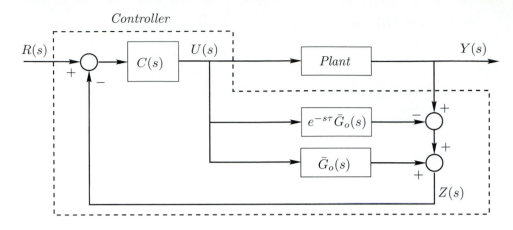

Figure 7.1. Smith-predictor structure

pseudo-complementary sensitivity function, $T_{zr}(s)$, between r and z, one that has no delay in the loop. This would be achieved, for example, via a standard PID block, leading to

$$T_{zr}(s) = \frac{\bar{G}_o(s)C(s)}{1 + \bar{G}_o(s)C(s)} \tag{7.4.2}$$

In turn, this leads to a nominal complementary sensitivity, between r and y, of the form

$$T_o(s) = e^{-s\tau}T_{zr}(s) \tag{7.4.3}$$

Four observations are in order regarding this result.

(i) Although the scheme appears somewhat ad-hoc, it will be shown in Chapter 15 that the architecture is inescapable insofar that it is a member of the set of *all possible stabilizing controllers* for the nominal system (7.4.1).

(ii) Provided $\bar{G}_o(s)$ is simple (e.g., having no nonminimum-phase zero), then $C(s)$ can be designed to yield $T_{zr}(s) \approx 1$; however, from (7.4.3), we see that this leads to the ideal result $T_o(s) = e^{-s\tau}$.

(iii) There are significant robustness issues associated with this architecture. These will be discussed in subsection §8.6.2.

(iv) One cannot use the above architecture when the open-loop plant is unstable.

7.5 Summary

- This chapter addresses the question of synthesis and asks the following questions:

  Given the model $G_0(s) = \frac{B_o(s)}{A_o(s)}$, how can one synthesize a controller, $C(s) = \frac{P(s)}{L(s)}$, such that the closed loop has a particular property?

- Recall:

  - The poles have a profound impact on the dynamics of a transfer function.

  - The poles of the four sensitivities governing the closed loop belong to the same set, namely the roots of the characteristic equation $A_o(s)L(s) + B_o(s)P(s) = 0$.

- Therefore, a key synthesis question is:

  Given a model, can one synthesize a controller such that the closed-loop poles (i.e., sensitivity poles) are in predefined locations?

- Stated mathematically:

  Given polynomials $A_o(s), B_o(s)$ (defining the model) and given a polynomial $A_{cl}(s)$ (defining the desired location of closed-loop poles), is it possible to find polynomials $P(s)$ and $L(s)$ such that $A_o(s)L(s) + B_o(s)P(s) = A_{cl}(s)$? This chapter shows that this is indeed possible.

- The equation $A_o(s)L(s) + B_o(s)P(s) = A_{cl}(s)$ is known as a Diophantine equation.

- Controller synthesis by solving the Diophantine equation is known as *pole placement*. There are several efficient algorithms as well as commercial software to do so.

- Synthesis ensures that the emergent closed loop has particular constructed properties.

  - However, the overall system performance is determined by a number of further properties that are *consequences* of the constructed property.

  - The coupling of constructed and consequential properties generates trade-offs.

- Design is concerned with the following issues:

  - efficiently detecting whether there is no solution that meets the design specifications adequately, and what the inhibiting factors are;

    o choosing the constructed properties such that, whenever possible, the overall behavior emerging from the interacting constructed and the consequential properties meets the design specifications adequately.

- This is the topic of the next chapter.

7.6   Further Reading

General

Horowitz, I. (1991). Survey of quantitative feedback theory (qft). *International Journal of Control*, 53(2):255-291.

Stephanopoulos, G. (1984). *Chemical Process Control: An Introduction to Theory and Practice.* Prentice-Hall, Englewood Cliffs, N.J.

PID design

Åström, K. and Hägglund, T. (1995). *PID controllers: theory, design and tuning.* Instrument Society of America, 2nd edition.

Model-based PID design

Graebe, S.F. and Goodwin, G.C. (1992). Adaptive PID design exploiting partial prior information. *Preprints of the 4th IFAC Symp. on Adaptive Systems in Control and Signal Processing, ACASP '92,* Grenoble, France, 395-400.

Isaksson, A.J. and Graebe, S.F. (1993). Model reduction for PID design. In *Proceedings of the 12th IFAC World Congress*, Sydney, Australia, 5:467-472.

Isaksson, A.J. and Graebe, S.F. (1999). Analytical PID parameter expressions for higher order systems. *Automatica*, to appear.

Morari, M. and Zafiriou, E. (1989). *Robust process control.* Prentice-Hall, Englewood Cliffs, N.J.

Rivera, D., Morari, M., and Skogestad, S. (1986). Internal model control. PID controller design. *Ind. Eng. Chem. Process Des. Dev.*, 25:252-265.

Smith predictor

Smith, O. (1958). *Feedback control systems.* McGraw-Hill, New York

7.7   Problems for the Reader

Problem 7.1. *Consider a plant having nominal model $G_o(s)$. Assume a one-degree-of-freedom control loop with controller $C(s)$, where*

$$G_o(s) = \frac{1}{(s+1)(s+2)}; \qquad C(s) = \frac{as+b}{s} \tag{7.7.1}$$

Find the conditions for a and b under which the nominal feedback loop is stable.

Problem 7.2. *The same nominal plant as in Problem 7.1 has to be controlled to achieve zero steady-state error for step disturbances and lead to a closed loop dominated by three poles at $s = -1$.*

7.2.1 *Find a controller transfer function $C(s)$ that satisfies these requirements.*

7.2.2 *Why is your result special?*

Problem 7.3. *Find a controller $C(s)$ for the following sets of data.*

$B_o(s)$	$A_o(s)$	$A_{cl}(s)$
1	$(s+1)(s+2)$	$(s^2+4s+9)(s+8)$
$-s+6$	$(s+4)(s-1)$	$(s+2)^3$
$s+1$	$(s+7)s^2$	$(s+1)(s+7)(s+4)^3$

Problem 7.4. *A plant has a nominal model given by $G_o(s) = \frac{3}{(s+4)(s+2)}$. Use polynomial pole-placement techniques to synthesize a controller to achieve*

- *zero steady-state errors for constant disturbances*

- *closed-loop modes decaying faster than e^{-3t}*

Problem 7.5. *Closed-loop control has to be synthesized for a plant having nominal model $G_o(s) = \frac{-s+4}{(s+1)(s+4)}$, to achieve the following goals:*

- *zero steady-state errors to a constant reference input;*

- *zero steady-state errors for a sine-wave disturbance of frequency 0.25 [rad/s];*

- *a biproper controller transfer function, $C(s)$.*

Use the pole-placement method to obtain a suitable controller $C(s)$.

Problem 7.6. *Consider a plant having a nominal model*

$$G_o(s) = \frac{8}{(s+2)(s+4)} \tag{7.7.2}$$

7.6.1 *Synthesize a controller $C(s)$ such that the closed-loop polynomial is $A_{cl}(s) = (s+a)^2(s+5)^2$, for $a = 0.1$ and $a = 10$.*

7.6.2 *Discuss your results regarding the pole-zero structure of $C(s)$.*

Problem 7.7. *The nominal model for a plant is given by*

$$G_o(s) = \frac{1}{(s+1)^2} \tag{7.7.3}$$

This plant has to be controlled in a feedback loop with one-degree-of-freedom.

Using the polynomial approach, design a strictly proper *controller $C(s)$ that allocates the closed-loop poles to the roots of $A_{cl}(s) = (s^2 + 4s + 9)(s+2)^k$, $k \in \mathbb{N}$. (Choose an adequate value for k.)*

Problem 7.8. *The nominal model for a plant is given by*

$$G_o(s) = \frac{1}{(s+1)(-s+2)} \tag{7.7.4}$$

Assume that this plant has to be controlled in a feedback loop with one-degree-of-freedom, in such a way that the closed-loop polynomial is dominated by the factor $s^2 + 7s + 25$.

Using the polynomial approach, choose an appropriate minimum degree $A_{cl}(s)$, and synthesize a biproper *controller $C(s)$.*

Problem 7.9. *Consider a linear plant, with input $u(t)$, output $y(t)$, and input disturbance $d(t)$. Assume that the plant model is given by*

$$Y(s) = \frac{1}{s-3}(U(s) + D(s)) \tag{7.7.5}$$

Further assume that the disturbance is a sine wave of frequency 2[rad/s] having unknown magnitude and phase.

Use the pole-assignment approach to synthesize a controller $C(s)$ that allows zero steady-state errors for the given disturbance and a constant set-point.

*Use the polynomial approach, choosing an appropriate closed-loop polynomial $A_{cl}(s)$. (Hint: use MATLAB routine **paq.m** on the accompanying CD-ROM.)*

Problem 7.10. *Consider a plant with nominal model*

$$G_o(s) = \frac{e^{-0.5s}(s+5)}{(s+1)(s+3)} \tag{7.7.6}$$

Build a Smith predictor such that the dominant closed-loop poles are located at $s = -2 \pm j0.5$.

Part III

SISO CONTROL DESIGN

PREVIEW

In the previous part of the book, we introduced the basic techniques of control-system synthesis. These are the elements used to compute a controller to achieve given specifications.

It turns out, however, that the desired performance properties cannot be addressed separately, because they form an interwoven network of trade-offs and constraints. Fast compensation for disturbances, for example, is not a separate degree of freedom from the common requirement of achieving insensitivity to modeling errors or conservation of control energy and utilities. Thus, the control engineer needs to ease a feasible solution into this complex web of trade-offs, compromises, and constraints. Doing this in a systematic and deliberate fashion is what we call control-system design.

Design, which is the key task of the control engineer, is based on a thorough understanding of analysis, synthesis, and design limitations. The first two chapters of this part of the book cover fundamental design limitations in both the time and frequency domains. The third chapter introduces ideas that are very commonly employed in practice, including feedforward and cascade structures. The final chapter discusses ways of dealing with input saturations and slew-rate limits. These are ubiquitous problems in real-world control-system design.

Chapter 8

FUNDAMENTAL LIMITATIONS IN SISO CONTROL

8.1 Preview

The results in the previous chapters have allowed us to determine relationships between the variables in a control loop. We have also defined some key transfer functions (sensitivity functions) that can be used to quantify the control-loop performance and have shown that, under reasonable conditions, the closed-loop poles can be assigned arbitrarily; however, these procedures should all be classified as synthesis rather than design. The related design question is: *where should I assign the closed-loop poles?* These are the kinds of issues we now proceed to study. It turns out that the question of where to assign closed-loop poles is part of a much larger question regarding the fundamental laws of trade-off in feedback design. These fundamental laws, or limitations, govern what is achievable and, conversely, what is not achievable in feedback control systems. Clearly, this lies at the very heart of control engineering. We thus strongly encourage students to gain some feel for these issues. The fundamental laws are related, on the one hand, to the nature of the feedback loop (e.g., whether integration is included) and, on the other hand, to structural features of the plant itself.

The limitations that we examine here include the following:

- sensors

- actuators

    - maximal movements

    - minimal movements

- model deficiencies

- structural issues, including the following:

- ○ poles in the ORHP
- ○ zeros in the ORHP
- ○ zeros that are stable but close to the origin
- ○ poles on the imaginary axis
- ○ zeros on the imaginary axis

We also briefly address possible remedies to these limitations.

8.2   Sensors

Sensors are a crucial part of any control-system design, because they provide the necessary information upon which the controller action is based. They are the *eyes* of the controller. Hence, any error or significant defect in the measurement system will have a significant impact on performance. We recall that, for a nominal plant $G_o(s)$ and a given unity feedback controller $C(s)$, the sensitivity functions take the form:

$$T_o(s) = \frac{G_o(s)C(s)}{1 + G_o(s)C(s)} = \frac{B_o(s)P(s)}{A_o(s)L(s) + B_o(s)P(s)} \tag{8.2.1}$$

$$S_o(s) = \frac{1}{1 + G_o(s)C(s)} = \frac{A_o(s)L(s)}{A_o(s)L(s) + B_o(s)P(s)} \tag{8.2.2}$$

$$S_{io}(s) = \frac{G_o(s)}{1 + G_o(s)C(s)} = \frac{B_o(s)L(s)}{A_o(s)L(s) + B_o(s)P(s)} \tag{8.2.3}$$

$$S_{uo}(s) = \frac{C(s)}{1 + G_o(s)C(s)} = \frac{A_o(s)P(s)}{A_o(s)L(s) + B_o(s)P(s)} \tag{8.2.4}$$

where $G_o(s) = \frac{B_o(s)}{A_o(s)}$ and $C(s) = \frac{P(s)}{L(s)}$.

We will examine two aspects of the sensor limitation problem—noise, and sensor response.

8.2.1   Noise

One of the most common sensor problems is measurement noise. In Chapter 5, it was shown that the effect of measurement noise in the nominal loop is given by

$$Y_m(s) = -T_o(s)D_m(s) \tag{8.2.5}$$

$$U_m(s) = -S_{uo}(s)D_m(s) \tag{8.2.6}$$

where $y_m(t)$ and $u_m(t)$ are respectively the plant output component and the controller output component due to the measurement noise.

From equation (8.2.5), we can see that the deleterious effect of noise can be attenuated if $|T_o(j\omega)|$ is small in the region where $|D_m(j\omega)|$ is significant. Thus, we may conclude the following:

Given the fact that noise is typically dominated by high frequencies, measurement noise usually sets an upper limit on the bandwidth of the loop .

8.2.2 Sensor Response

Another common limitation introduced by measurement systems arises from the fact that sensors themselves often have dynamics. As a simple example, most thermocouples are embedded in a *thermocouple well*, to protect them; however, this *well* introduces an extra lag into the response and can indeed dominate the system response. This can be modeled by assuming that the measured output, $y_m(t)$, is related to the true output, $y(t)$, as follows:

$$Y_m(s) = \frac{1}{\tau_1 s + 1} Y(s) \tag{8.2.7}$$

It might be thought that this limitation can effectively be eliminated by passing $y_m(t)$ through a suitable high-pass filter. For instance, we might use

$$Y_{mf} = \frac{\tau_1 s + 1}{\tau_2 s + 1} Y_m(s) \qquad \tau_1 > \tau_2 \tag{8.2.8}$$

However, if we turn again to the issue of measurement noise, we see that the high-pass filter in (8.2.8) will significantly amplify high-frequency noise and thus the upper limit in the bandwidth again appears.

8.3 Actuators

If sensors provide the *eyes* of control, then actuators provide the *muscle*; however, actuators are also a source of limitations in control performance. We will examine two aspects of actuator limitations. These are maximal movement and minimal movement.

8.3.1 Maximal Actuator Movement

In practice, all actuators have maximal movement constraints in the form of saturation limits on amplitude or slew-rate limits. To see how maximal movement constraints affect achievable performance, we note that peaks in the control action usually occur as a result of large fast changes in either the reference or the output disturbance. (Input disturbance changes are usually filtered by the plant.) Recall that, in a one-d.o.f. loop, the controller output is given by

$$U(s) = S_{uo}(s)(R(s) - D_o(s)) \tag{8.3.1}$$

where

$$S_{uo}(s) \triangleq \frac{T_o(s)}{G_o(s)} \tag{8.3.2}$$

If the loop bandwidth is much larger than that of the open-loop model $G_o(s)$, then the transfer function $S_{uo}(s)$ will significantly enhance the high-frequency components in $R(s)$ and $D_o(s)$. This is illustrated in the following example.

Example 8.1. *Consider a plant and closed loop given by*

$$G_o(s) = \frac{10}{(s+10)(s+1)} \qquad and \qquad T_o(s) = \frac{100}{s^2 + 12s + 100} \tag{8.3.3}$$

Note that the plant and the closed-loop bandwidths have a ratio of approximately 10 : 1. This will be reflected in large control sensitivity, $|S_{uo}(j\omega)|$, at high frequencies, which, in turn, will yield large initial control response in the presence of high-frequency reference signals or disturbances. This is illustrated in Figure 8.1, where the magnitude of the control sensitivity and the control output for a unit-step output disturbance are shown.

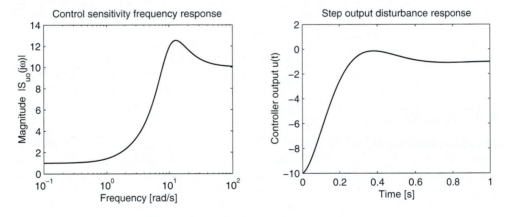

Figure 8.1. Effects of a large ratio of closed-loop bandwidth to plant bandwidth

□□□

We see in the above example that having a closed-loop bandwidth ten times the open-loop bandwidth calls for an initial input signal ten times that of the steady-state value in response to a step demand. This connection between bandwidth and input size is a very rough rule of thumb. The reader is encouraged to look at the various case studies on the website and to change the bandwidth by changing the controller gain to examine the effect it has on maximum input signal.

Actuators also frequently exhibit a limit on the maximum speed with which they can change position. This is usually termed a *slew-rate limit*. If present, this will cause departures from the performance prescribed by a purely linear design.

We can gain a qualitative understanding of the effect of slew-rate limiting by noting that the input is given by

$$U(s) = S_{uo}(s)[R(s) - D_o(s)] \qquad (8.3.4)$$

Hence, the rate of change of the input is given by

$$sU(s) = S_{uo}(s)[sR(s) - sD_o(s)] = \frac{T_o(s)}{G_o(s)}[sR(s) - sD_o(s)] \qquad (8.3.5)$$

Thus, if the bandwidth of the closed loop is much larger than that of the plant dynamics, then the rate of change of the input signal will necessarily be large for fast changes in $r(t)$ and $d_o(t)$.

Hence, we conclude the following:

> To avoid actuator saturation or slew-rate problems, it will generally be necessary to place an upper limit on the closed-loop bandwidth.

8.3.2   Minimal Actuator Movement

In subsection 8.3.1, we learned that control-loop performance is limited by the maximal movement available from actuators. This is heuristically reasonable. What is perhaps less obvious is that control systems are often also limited by minimal actuator movements. Indeed, in the experience of the authors, many real-world control problems are affected by this issue.

Minimal actuator movements are frequently associated with frictional effects: the actuator *sticks*. When the actuator is in this mode, integrators (both in the plant and controller) will *wind-up* until sufficient force is generated to overcome the static-friction component. The manifestations of the problem are usually a self-sustaining oscillation produced as the actuator goes through a cycle of sticking, moving, sticking, and so on. The oscillation frequency is typically at or near the frequency where the loop phase shift is 180°.

Example 8.2 (Continuous casting). *Consider again the mould-level controller described in section §2.3. It is known that many mould-level controllers in industry exhibit poor performances in the form of self-sustaining oscillations. See, for example, the real data shown in Figure 8.2. Many explanations have been proposed for this problem; however, at least on the system with which the authors are familiar, the difficulty was directly traceable to minimal-movement issues associated with the actuator (the slide-gate valve).*

□□□

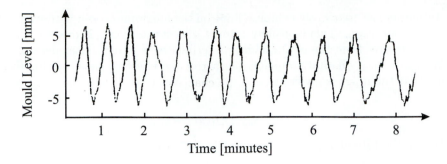

Figure 8.2. Chart recording showing oscillations in conventional mould-level control system

8.4   Disturbances

Another source of performance limitation in real control systems is that arising from disturbances. This effect, too, can be evaluated by using the appropriate loop-sensitivity functions. To be more precise, consider the expression already derived in Chapter 5 for the effect of disturbances on the plant output:

$$Y(s) = S_{io}(s)D_i(s) + S_o(s)D_o(s) \tag{8.4.1}$$

Assume that the input and output disturbances have significant energy only in the frequency bands B_{wi} and B_{wo}, respectively. Then it is clearly desirable to have small values for $|S_o(j\omega)|$ and $|S_{io}(j\omega)|$ in B_{wi} and B_{wo}, respectively. Because $G(s)$ is fixed, this can only be achieved provided that $S_o(j\omega) \approx 0$, and hence $T_o(j\omega) \approx 1$ in the frequency band encompassing the union of B_{wi} and B_{wo}.

Hence,

> To achieve acceptable performance in the presence of disturbances, it will generally be necessary to place a lower bound on the closed-loop bandwidth.

8.5   Model-Error Limitations

Another key source of performance limitation is due to inadequate fidelity in the model used as the basis of control-system design. To analyze this problem, we need to distinguish between nominal performance and the true, or achieved, performance. One usually bases design on the nominal model and then adds the requirement that the resultant performance should be insensitive to the difference between the true and the nominal model. This property, defined in Chapter 5, is usually referred to as robustness.

We have seen in Chapter 5 that these differences can be expressed as differences between the nominal and the corresponding true sensitivity. A key function to quantify these differences defined in that chapter was the error sensitivity $S_\Delta(s)$, given by

$$S_\Delta(s) = \frac{1}{1 + T_o(s)G_\Delta(s)} \tag{8.5.1}$$

where $G_\Delta(s)$ is the multiplicative modeling error.

Modeling is normally good at low frequencies and deteriorates as the frequency increases, because then dynamic features neglected in the nominal model become significant. This implies that $|G_\Delta(j\omega)|$ will typically become increasingly significant with rising frequency.

Hence,

To achieve acceptable performance in the presence of model errors, it will generally be desirable to place an upper limit on the closed-loop bandwidth.

8.6   Structural Limitations

8.6.1   General Ideas

In the previous sections, limitations arising from sensors, actuators, disturbances, and modeling errors have been discussed. All of these aspects have to be considered when carrying out the nominal control design; however, performance *in the nominal linear control loop* is also subject to unavoidable constraints that derive from the particular structure of the nominal model itself.

The aim of this section is to analyze these structural constraints and their effect on overall performance.

8.6.2   Delays

Undoubtedly, the most common source of structural limitation in process-control applications is due to process delays. These delays are usually associated with the transportation of materials from one point to another.

Clearly, by the time one *sees* the effect of a disturbance on the output, it takes a full delay period before it can be cancelled. Thus, the output sensitivity function can, at best, be of the form

$$S_o^*(s) = 1 - e^{-s\tau} \tag{8.6.1}$$

where τ is the delay.

We have actually seen, in section §7.4, that one can approach the above idealized result by appropriate control-system design, using a Smith-predictor-type structure.

Another difficulty is that of robustness against model errors. If we were to achieve the idealized result in (8.6.1), then the corresponding nominal complementary sensitivity would be

$$T_o^*(s) = e^{-s\tau} \tag{8.6.2}$$

However, this has a gain of unity for all frequencies. To examine the consequences of this, we recall the error sensitivity given in (8.5.1). We then see that, if we have any model error (especially in the delay itself), then $G_\triangle(s)$ is likely to grow to near 1 at relatively low frequencies. Indeed, it is readily seen, via the ideas presented in section §4.12, that the relative modeling error will have a magnitude that approaches 1 at approximately a bandwidth of $\frac{1}{\eta\tau}$, where η is the fractional error in the delay. Thus, in practice, one is usually limited by the combined effects of delays and modeling errors to closed-loop bandwidths of the order of $\frac{2\pi}{\tau}$.

In summary, we see that process delays have two effects

(i) Delays limit disturbance rejection by requiring that a delay occur before the disturbance can be cancelled. This is reflected in the ideal sensitivity $S_o^*(s)$ in (8.6.1).

(ii) Delays further limit the achievable bandwidth through the impact of model errors.

Example 8.3 (Thickness control in rolling mills). *We recall the example of thickness control in rolling mills, mentioned in Chapter 1. A schematic diagram for one stand of a rolling mill is given in Figure 8.3.*

In Figure 8.3, we have used the following symbols: F (roll force), σ (unloaded roll gap), H (input thickness), V (input velocity), h (exit thickness), v (exit velocity), h_m (measured exit thickness), d (distance from mill to exit thickness measurement).

The distance from the mill to output thickness measurement introduces a (speed-dependent) time delay of $(\frac{d}{v})$. This introduces a fundamental limit on the controlled performances as described above.

□□□

We have seen above that delays (where the response does not *move* for a given period) represent a very important source of structural limitations in control design. From this, we might conjecture that nonminimum-phase behavior (where the response initially *goes in the wrong direction*) might present even harder challenges to control-system design. This is indeed the case, as we show in the next section.

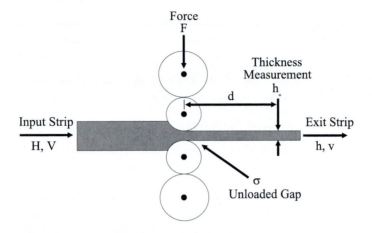

Figure 8.3. Rolling-mill thickness control

8.6.3   Interpolation Constraints

Before reading on, the reader should review the points made in Chapter 4 on the impact of zeros on general transfer functions. In order to relate these general properties of zeros (as described in Chapter 4) to feedback systems, it is first necessary to understand the nature and origin of poles and zeros of sensitivity functions.

We recall that the relevant nominal sensitivity functions for a nominal plant $G_o(s) = \frac{B_o(s)}{A_o(s)}$ and a given unity feedback controller $C(s) = \frac{P(s)}{L(s)}$ are as follows:

$$T_o(s) = \frac{G_o(s)C(s)}{1 + G_o(s)C(s)} = \frac{B_o(s)P(s)}{A_o(s)L(s) + B_o(s)P(s)} \tag{8.6.3}$$

$$S_o(s) = \frac{1}{1 + G_o(s)C(s)} = \frac{A_o(s)L(s)}{A_o(s)L(s) + B_o(s)P(s)} \tag{8.6.4}$$

$$S_{io}(s) = \frac{G_o(s)}{1 + G_o(s)C(s)} = \frac{B_o(s)L(s)}{A_o(s)L(s) + B_o(s)P(s)} \tag{8.6.5}$$

$$S_{uo}(s) = \frac{C(s)}{1 + G_o(s)C(s)} = \frac{A_o(s)P(s)}{A_o(s)L(s) + B_o(s)P(s)} \tag{8.6.6}$$

We note that the poles of all these transfer functions are the set (or, in the case of pole-zero cancellation, a subset) of the closed-loop poles. Although the poles alone determine the stability of the loop, it is the combined pole-zero pattern that determines the precise transient performance.

One of the important implications that derive from the closed-loop sensitivity functions is that all uncancelled plant poles and uncancelled plant zeros impose *algebraic or interpolation constraints* on the sensitivity functions. In particular, we have the following:

(i) The nominal complementary sensitivity $T_o(s)$ has a zero at all uncancelled zeros of $G_o(s)$.

(ii) The nominal sensitivity $S_o(s)$ is equal to 1 at all uncancelled zeros of $G_o(s)$. (This follows from (i) by using the identity $S_o(s) + T_o(s) = 1$.)

(iii) The nominal sensitivity $S_o(s)$ has a zero at all uncancelled poles of $G_o(s)$.

(iv) The nominal complementary sensitivity $T_o(s)$ is equal to 1 at all uncancelled poles of $G_o(s)$. (This follows from (iii) and from the identity $S_o(s) + T_o(s) = 1$.)

It is helpful, at this point, to introduce a typical design constraint. In particular, we will assume that the design achieves an exact inverse of the plant model at zero frequency. This ensures perfect tracking for a fixed set-point and rejection of constant input and output disturbances from the output. To achieve this, we require that $L(0) = 0$, which leads to

$$A_o(0)L(0) = 0 \Leftrightarrow S_o(0) = 0 \Leftrightarrow T_o(0) = 1 \qquad (8.6.7)$$

Note that $A_o(0) = 0$ alone does not lead to full compensation of a constant input disturbance, because the set of zeros of $S_{io}(s)$ does not include the plant poles, hence $A_o(0) = 0$ does not imply that $S_{io}(0)$ is zero.

We will proceed to investigate the relationship between plant poles and zeros and closed-loop time responses. The main result of this section will be that plant open-loop poles and zeros, especially those in the right-half plane, have a dramatic and predictable effect on transient performance.

We will begin by recalling Lemma 4.1 on page 81, which relates (exponentially weighted) integrals of time responses to point values of the corresponding Laplace transform.

The particular transfer functions of interest here will be the various nominal sensitivity functions. In order that we will be able to apply Lemma 4.1 on page 81, we will assume that the closed-loop poles lie to the left of $-\alpha$, for some $\alpha > 0$, which is always true for stable closed-loop systems. We first examine the effects of integrators. Actually, to make this topic more interesting for the reader, we can relate an anecdote from one of our colleagues, whose first job in industry was to design a one-degree-of-freedom controller to reject a ramp disturbance that also had the property that the response to a step reference change did not overshoot. Equation (8.6.12), to be presented below, shows that these design specifications violate fundamental laws of feedback systems. Clearly, a lot of wasted time can be saved by knowing what is not possible (i.e., by appreciating fundamental design constraints).

8.6.4   Effect of Open-Loop Integrators

Lemma 8.1. *We assume that the plant is controlled in a one-degree-of-freedom configuration and that the open-loop plant and controller satisfy:*

$$A_o(s)L(s) = s^i(A_o(s)L(s))' \qquad i \geq 1 \tag{8.6.8}$$

$$\lim_{s \to 0}(A_o(s)L(s))' = c_0 \neq 0 \tag{8.6.9}$$

$$\lim_{s \to 0}(B_o(s)P(s)) = c_1 \neq 0 \tag{8.6.10}$$

that is, that the plant-controller combination has i poles at the origin. Then, for a step-output disturbance or step set-point, the control error, $e(t)$, satisfies

$$\lim_{t \to \infty} e(t) = 0 \qquad \forall i \geq 1 \tag{8.6.11}$$

$$\int_0^\infty e(t)dt = 0 \qquad \forall i \geq 2 \tag{8.6.12}$$

Also, for a negative unit ramp output disturbance or a positive unit ramp reference, the control error, $e(t)$, satisfies

$$\lim_{t \to \infty} e(t) = \frac{c_0}{c_1} \qquad for \ \ i = 1 \tag{8.6.13}$$

$$\lim_{t \to \infty} e(t) = 0 \qquad \forall i \geq 2 \tag{8.6.14}$$

$$\int_0^\infty e(t)dt = 0 \qquad \forall i \geq 3 \tag{8.6.15}$$

Proof

We first recall that the control error satisfies

$$E(s) = S_o(s)(R(s) - D_o(s)) \tag{8.6.16}$$

The expressions for $\lim_{t \to \infty} e(t)$ are then a consequence of the final-value theorem for Laplace transforms. The integral results follow from Lemma 4.1 on page 81, where $h(t)$ has been replaced by $e(t)$ and $z_0 = 0$.

□□□

In the case of input disturbances, a similar result holds. However, caution must be exercised, because the numerator of $S_{io}(s)$ is $B_o(s)L(s)$ rather than $A_o(s)L(s)$, as was the case for $S_o(s)$. This implies that *integration in the plant does not affect the steady-state compensation of input disturbances.* Thus we need to modify Lemma 8.1 to the following:

Lemma 8.2. *Assume that the controller satisfies:*

$$L(s) = s^i (L(s))' \qquad i \geq 1 \tag{8.6.17}$$

$$\lim_{s \to 0} (L(s))' = l_i \neq 0 \tag{8.6.18}$$

$$\lim_{s \to 0} (P(s)) = p_0 \neq 0 \tag{8.6.19}$$

The controller alone has i poles at the origin. Then, for a step input disturbance, the control error, $e(t)$, satisfies

$$\lim_{t \to \infty} e(t) = 0 \qquad \forall i \geq 1 \tag{8.6.20}$$

$$\int_0^\infty e(t)dt = 0 \qquad \forall i \geq 2 \tag{8.6.21}$$

Also, for a negative unit ramp input disturbance, the control error, $e(t)$, satisfies

$$\lim_{t \to \infty} e(t) = \frac{l_i}{p_0} \qquad for \qquad i = 1 \tag{8.6.22}$$

$$\lim_{t \to \infty} e(t) = 0 \qquad \forall i \geq 2 \tag{8.6.23}$$

$$\int_0^\infty e(t)dt = 0 \qquad \forall i \geq 3 \tag{8.6.24}$$

Proof

In this case, the control error satisfies

$$E(s) = -S_{io}(s)D_i(s) \tag{8.6.25}$$

The remainder of the proof parallels that of Lemma 8.1 on the page before.

□□□

Remark 8.1. *Note that, when the integral of the error vanishes, then it follows that the error has equal area above and below zero. Hence, overshoot in the plant output is unavoidable. In particular, it is clearly impossible to design a one-d.o.f. feedback control loop with two integrators in the controller (giving zero steady-state error to a ramp input) that does not overshoot when a step change in the reference is applied. This explains the impossibility of the design specifications in the anecdote referred to just before subsection §8.6.4.*

8.6.5   More General Effects of Open-Loop Poles and Zeros

The results above depend upon the zeros of the various sensitivity functions at the origin; however, it turns out that zeros in the right-half plane have an even more dramatic effect on achievable transient performances of feedback loops. We will show this below. As a prelude to this, we note that Lemma 4.1 on page 81 applies to general zeros. This fact is exploited below to develop a series of integral constraints that apply to the transient response of feedback systems having various combinations of open-loop poles and zeros.

Lemma 8.3. *Consider a feedback control loop having stable closed-loop poles located to the left of $-\alpha$ for some $\alpha > 0$. Also assume that the controller has at least one pole at the origin. Then, for an uncancelled plant zero z_0 or an uncancelled plant pole η_0 to the right of the closed-loop poles–i.e., satisfying $\Re\{z_0\} > -\alpha$ or $\Re\{\eta_0\} > -\alpha$, respectively–we have the following:*

(i) *For a positive unit reference step or a negative unit-step output disturbance, we have*

$$\int_0^\infty e(t)e^{-z_0 t}dt = \frac{1}{z_0} \qquad (8.6.26)$$

$$\int_0^\infty e(t)e^{-\eta_0 t}dt = 0 \qquad (8.6.27)$$

(ii) *For a positive unit step reference and for z_0 in the right-half plane, we have*

$$\int_0^\infty y(t)e^{-z_0 t}dt = 0 \qquad (8.6.28)$$

(iii) *For a negative unit step input disturbance, we have*

$$\int_0^\infty e(t)e^{-z_0 t}dt = 0 \qquad (8.6.29)$$

$$\int\limits_0^\infty e(t)e^{-\eta_0 t}dt = \frac{L(\eta_0)}{\eta_0 P(\eta_0)} \tag{8.6.30}$$

Proof

The proof depends on the key fact that the open-loop poles and zeros that we consider are in the region of convergence of the transform–i.e., they lie to the right of all closed-loop poles. This implies that, even if $e^{-z_0 t}$ (or $e^{-\eta_0 t}$) is a growing exponential, the product $e(t)e^{-z_0 t}$ decays asymptotically.

(i) *In this case, the control error satisfies*

$$E(s) = S_o(s)(R(s) - D_o(s)) \tag{8.6.31}$$

with either $R(s) = \frac{1}{s}$ or $-D_o(s) = \frac{1}{s}$. We also note that $S_o(z_0) = 1$, $S_o(\eta_0) = 0$. The result then follows from Lemma 4.1 on page 81, where $h(t)$ is replaced by $e(t)$.

(ii) *In this case, the plant output satisfies*

$$Y(s) = T_o(s)R(s) \tag{8.6.32}$$

with $R(s) = \frac{1}{s}$. We also note that $T_o(z_0) = 0$. The result then follows from Lemma 4.1 on page 81, where $h(t)$ is replaced by $y(t)$.

(iii) *In this case, the control error satisfies*

$$E(s) = -S_{io}(s)D_i(s) \tag{8.6.33}$$

with $-D_i(s) = \frac{1}{s}$. We also note that $S_{io}(z_0) = 0$, $S_{io}(\eta_0) = \frac{L(\eta_0)}{P(\eta_0)}$. The result then follows from Lemma 4.1 on page 81, where $h(t)$ is replaced by $e(t)$.

□□□

The above results are important, because they provide a link between open-loop plant characteristics and closed-loop transient performance. The following qualitative observations are a consequence of these results. (The reader may first care to review subsection §8.6.3.)

(i) From equation (8.6.26), we see that, if z_0 is a negative real number (i.e., z_o is actually a *minimum-phase zero*), then the error must change sign, because initially it is positive, but the integral is negative. This implies that the plant response *must exhibit overshoot*. Moreover, the size of the overshoot increases as the size of the zero decreases. The size of the overshoot is of the order of $(t_s|z_0|)^{-1}$, where t_s is the settling time. (See Lemma 4.3 on page 84.)

On the other hand, if z_0 is a positive real number (i.e. z_o is a *nonminimum-phase zero*) then the error need not change sign; however, if z_0 is small, then the integral of error will be large and positive. Moreover, for z_0 in the right-half plane, we see from equation (8.6.28) that, for a step change in the set-point the response must undershoot. This implies that the error will peak at a value higher than the initial value of unity. Indeed, the peak error and undershoot magnitude are of the order of $(t_s|z_0|)^{-1}$, where t_s is the settling time. (See Lemma 4.2 on page 82.)

(ii) We see from equation (8.6.27) that any open-loop pole that lies to the right of all closed-loop poles must produce a change in sign in the error and hence overshoot. Furthermore, for a large positive η_0, relative to the closed-loop pole locations, the exponential weighting decays fast relative to the closed-loop settling time. Hence, to achieve a zero weighted integral for the error, it is necessary that either or both of the following conditions hold: the error changes sign rapidly at the beginning of the transient, or the error has a large negative value to compensate for the initial positive value.

(iii) We see from equation (8.6.29) that input disturbances will lead to an error that changes sign and hence overshoots for any real open-loop zero that lies to the right of all closed-loop poles.

(iv) There are subtle interactions between the various constraints. For example, the overshoot produced by open-loop poles that lie to the right of closed-loop poles can usually be linked to stable zeros that are induced in the controller by the action of the pole assignment. These kinds of issues are automatically incorporated in the constraints, because they use the fact that the sum of the numerators of the sensitivity and complementary sensitivity is equal to the denominator, and hence there is a necessary conjunction of the effect of open-loop poles, open-loop zeros, and closed-loop poles.

The above analysis underlies the following performance trade-offs:

- If the magnitude of the real part of the dominant closed-loop poles is greater than the smallest right-half plane zero, then large undershoot is inevitable. This usually means that, in design,

> The bandwidth should in practice be set less than the smallest nonminimum-phase zero.

- If the magnitude of the real part of the dominant closed-loop poles is greater than the magnitude of the smallest stable open-loop zero, then significant overshoot will occur. An alternative is to cancel these zeros in the closed loop by placing them in the denominator of the controller; however, in that case, they will appear in the numerator of the input sensitivity. That might be acceptable: input disturbances can be significantly diminished by passage through the plant.

- If the magnitude of the real part of the dominant closed-loop poles is less than the magnitude of the largest unstable open-loop pole, then significant overshoot will occur, or the error will change sign quickly. Thus:

> It is advisable to set the closed-loop bandwidth greater than the real part of any unstable pole.

- If the magnitude of the real part of the dominant closed-loop poles is greater than the magnitude of the smallest stable open-loop pole, then overshoot will again occur. This can be avoided by canceling these poles in the closed loop by placing them in the numerator of the controller; however, in that case they will appear in the denominator of the input sensitivity, and this effect is usually undesirable, because it means that the effect of input disturbances will decay slowly.

- Note that, for the set-point response, it is often possible to avoid the deleterious effect of overshoot produced by *stable* plant or controller zeros. This can be achieved by canceling them outside the loop in a two-degree-of-freedom design, which then avoids compromising the disturbance performance.

Some of the issues discussed above are illustrated in the following example.

Example 8.4. *Consider a nominal plant model given by*

$$G_o(s) = \frac{s - z_p}{s(s - p_p)} \qquad (8.6.34)$$

The closed-loop poles were assigned to $\{-1, -1, -1\}$. *Then the general controller structure is given by*

$$C(s) = K_c \frac{s - z_c}{s - p_c} \qquad (8.6.35)$$

Five different cases are considered. They are described in Table 8.1.

The different designs were tested with a unit step reference and, in every case, the plant output was observed. The results are shown in Figure 8.4.

From these results we can make the following observations:

Case 1 (Small stable pole) A small amount of overshoot is evident, as predicted by equations (8.6.26) and (8.6.27).

	Case 1	Case 2	Case 3	Case 4	Case 5
	$p_p = -0.2$	$p_p = -0.5$	$p_p = -0.5$	$p_p = 0.2$	$p_p = 0.5$
	$z_p = -0.5$	$z_p = -0.1$	$z_p = 0.5$	$z_p = 0.5$	$z_p = 0.2$
K_c	1.47	20.63	-3.75	-18.8	32.5
p_c	-1.33	18.13	-6.25	-22.0	29.0
z_c	-1.36	-0.48	-0.53	-0.11	0.15

Table 8.1. Case description

Case 2 (Very small stable zero) Here, we see a very large amount of overshoot, as predicted by equation (8.6.26).

Case 3 (Unstable zero, stable pole) Here, we see a significant amount of undershoot.

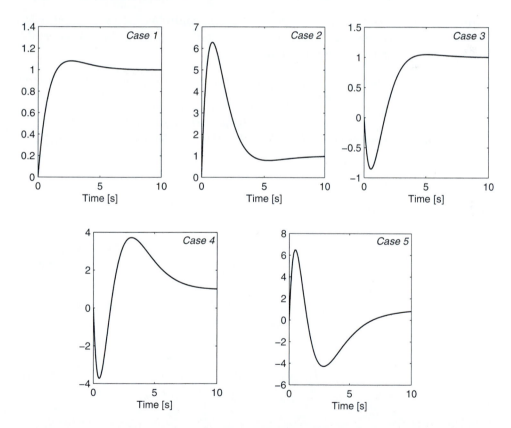

Figure 8.4. Plant output, $y(t)$, for five different pole-zero configurations

This is due to the right-half plane zero. We also observe a small amount of overshoot, that is due to the stable pole at −0.5. This is predicted by equation (8.6.27).

Case 4 (Unstable zero, small unstable pole) We first observe significant undershoot due to the RHP zero and as predicted by equation (8.6.28). We also observe a significant overshoot that is due to the unstable pole, as predicted by equation (8.6.27).

Case 5 (Small unstable zero, large unstable pole) Here, the undershoot is produced by the RHP zero and the overshoot by RHP pole, as predicted by equations (8.6.28) and (8.6.27) respectively. In this case, the overshoot is significantly larger than in Case 4, because the unstable pole is further into the RHP.

□□□

Another illustration of this circle of ideas is presented in the following example.

Example 8.5. *Consider a plant with a one-d.o.f. controller and a nominal model that has a real NMP zero at $s = z_0$ and a real unstable pole at $s = \eta_0$. Assume also that the controller has one pole at the origin.*

8.5.1 Determine time-domain constraints for the controller output $u(t)$, assuming that the reference is a unit step.

8.5.2 If the controller output is constrained to be in the range $[0, U_{max}]$, where U_{max} is a large positive number, then show that the unstable pole will force $u(t)$ into saturation for any step reference.

Solution

8.5.1 We first recall that

$$U(s) = S_{uo}(s)\frac{1}{s} = \int_0^\infty u(t)e^{-st}dt \qquad (8.6.36)$$

We also recall that every open-loop unstable pole is a zero in $S_{uo}(s)$. Then, on applying Lemma 4.1 on page 81, we have that

$$\int_0^\infty u(t)e^{-z_0 t}dt = S_{uo}(z_0)\frac{1}{z_0} \qquad (8.6.37)$$

$$\int_0^\infty u(t)e^{-\eta_0 t}dt = S_{uo}(\eta_0)\frac{1}{\eta_0} = 0 \qquad (8.6.38)$$

8.5.2 *We note that the vanishing integral in (8.6.38) implies that $u(t)$ must be negative during nonzero time intervals. This means that the lower saturation limit will be hit. An additional observation is that this will happen for any bounded reference.*

8.6.6   Effect of Imaginary-Axis Poles and Zeros

An interesting special case of Lemma 8.3 on page 213 occurs when the plant has poles or zeros on the imaginary axis. Under these conditions, we have the following.

Corollary 8.1. *Consider a closed-loop system, as in Lemma 8.3 on page 213. Then, for a unit step reference input,*

(a) *if the plant $G(s)$ has a pair of zeros at $\pm j\omega_0$, then*

$$\int_0^\infty e(t)\cos\omega_0 t\, dt = 0 \tag{8.6.39}$$

$$\int_0^\infty e(t)\sin\omega_0 t\, dt = \frac{1}{\omega_0} \tag{8.6.40}$$

(b) *if the plant $G(s)$ has a pair of poles at $\pm j\omega_0$, then*

$$\int_0^\infty e(t)\cos\omega_0 t\, dt = 0 \tag{8.6.41}$$

$$\int_0^\infty e(t)\sin\omega_0 t\, dt = 0 \tag{8.6.42}$$

where $e(t)$ is the control error, i.e.,

$$e(t) = 1 - y(t) \tag{8.6.43}$$

Proof

(a) *From (8.6.26) we have*

$$\int_0^\infty e(t)e^{\pm j\omega_0 t}\, dt = \frac{1}{\pm j\omega_0} \tag{8.6.44}$$

The result follows upon noting that

$$\cos\omega_0 t = \frac{1}{2}\left(e^{j\omega_0 t} + e^{-j\omega_0 t}\right) \tag{8.6.45}$$

$$\sin\omega_0 t = \frac{1}{2j}\left(e^{j\omega_0 t} + e^{-j\omega_0 t}\right) \tag{8.6.46}$$

(b) Immediate from (8.6.27)

□□□

The above constraints are particularly restrictive in the case of open-loop imaginary-axis zeros near the origin (relative to the closed-loop bandwidth). This is illustrated in the following example.

Example 8.6. *Consider the set-up of Corollary 8.1 on the page before part a). Let us define the exact settling time (assuming it exists) of a system to be*

$$t_s = \inf_T \left\{ |e(t)| = 0 \,; \; \forall t \geq T \right\} \tag{8.6.47}$$

Then, from (8.6.40), we have

$$\int_0^{t_s} e(t) \sin \omega_0 t \, dt = \frac{1}{\omega_0} \tag{8.6.48}$$

Now, assuming that $t_s \leq \dfrac{\pi}{\omega_0}$, we have

$$\int_0^{t_s} e_{\max} \left(\sin \omega_0 \right) dt \geq \int_0^{t_s} e(t) \sin \omega_0 t \, dt = \frac{1}{\omega_0} \tag{8.6.49}$$

where $e_{\max}$ is the maximum value of $|e(t)|$ on the interval $(0, t_s)$. From (8.6.49)

$$\frac{1}{\omega_0} \left[1 - \cos \omega_0 t_s \right] e_{\max} \geq \frac{1}{\omega_0} \tag{8.6.50}$$

or

$$e_{\max} \geq \frac{1}{1 - \cos \omega_0 t_s} \tag{8.6.51}$$

We thus see that, as $\omega_0 t_s \to 0$, so $e_{\max}$ diverges to ∞ !

□□□

Example 8.7. *As a simple numerical example, consider a feedback control loop having its complementary sensitivity transfer function given by*

$$T(s) = \frac{100 s^2 + 1}{s^3 + 3 s^2 + 3 s + 1} \tag{8.6.52}$$

Note that the closed-loop poles are all at -1, while the zeros are at $\pm j 0.1$. The simulation response of $e(t)$ for a unit step input is shown in Figure 8.5 on the facing page.

We can see that the maximum error corresponds to more than 2000% undershoot and 800% overshoot! .

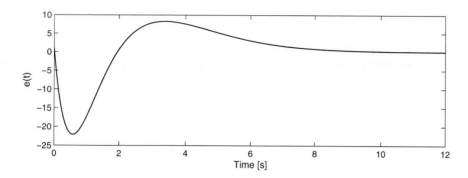

Figure 8.5. Control error for a feedback loop with unit step reference and imaginary zeros

▢▢▢

In this chapter, the derivation of all design constraints has been based on the linear nature of the control system, because we have used sensitivity and complementary sensitivity transfer functions. Thus, a sensible question is whether these constraints will still apply when other control approaches (nonlinear, adaptive, neural, or fuzzy, for instance) are applied. The answer to that question is not straightforward, except in some cases. One of them is that of NMP zeros, as shown below.

Consider a linear plant, with a nominal model characterized by $G_o(s)$, whose output has to be regulated to a constant value. Assume that $G(z_o) = 0$ for $z_o \in \mathbb{R}^+$. Then, for step disturbances and step references, the control (of no matter what type) must generate a plant input of the form

$$u(t) = u_\infty + \Delta u(t) \qquad (8.6.53)$$

where u_∞ is the input final value and $\Delta u(t)$ is the necessary transient action to steer the plant output to the desired set-point, i.e., $\Delta u(\infty) = 0$. Thus, the Laplace transform of the plant input converges for all $\sigma = \Re\{s\} > 0$. On the other hand, the minimum control requirement is stability, so the Laplace transform of the plant output also converges for all $\sigma = \Re\{s\} > 0$. Hence

$$Y(z_o) = G(z_o)U(z_o) = 0 = \int_0^\infty y(t)e^{-z_o t}dt \qquad (8.6.54)$$

Equation (8.6.54) says that the plant output, $y(t)$, must change sign so that the integral in $[0, \infty)$ vanishes. In other words, (8.6.54) implies that $y(t)$ must evolve along the opposite direction to that of the desired value, during at least one interval of nonzero duration. This leads to the terminology of *inverse response* for the behavior of NMP systems. This result is the same as (8.6.29); however, there is a crucial difference: this time, *no specific assumption (linear classical control or otherwise) has been made regarding the control methodology.*

The conclusion is that *NMP systems will lead to performance limitations regardless of the control approach and control architecture being used.* Thus, it can be seen that it is rather important that designers understand these constraints and not aim to achieve any performance specifications that are in violation of the fundamental laws of feedback.

8.7   An Industrial Application (Hold-Up Effect in Reversing Mill)

Here we study a reversing rolling mill. In this form of rolling mill the strip is successively passed from side to side so that the thickness is reduced on each pass.

For a schematic diagram of a single-strand reversing rolling mill, see Figure 8.6.

For this system we define the following variables:

$$
\begin{array}{rl}
h_e(t) & : \quad \text{exit-strip thickness} \\
\tilde{h}_e & : \quad \text{nominal value of } h_e(t) \\
h_i(t) & : \quad \text{input-strip thickness} \\
i_c(t) & : \quad \text{electric current in the coiler motor} \\
\tilde{i}_c & : \quad \text{nominal value of } i_c(t) \\
i_u(t) & : \quad \text{electric current in the uncoiler motor} \\
J_c(t) & : \quad \text{coiler inertia} \\
\tilde{J}_c & : \quad \text{nominal value of } J_c(t) \\
k & : \quad \text{coiler-motor torque constant} \\
\sigma(t) & : \quad \text{gap between rolls without load} \\
\tau_i(t) & : \quad \text{input-strip tension} \\
\tau_e(t) & : \quad \text{exit-strip tension} \\
\omega_c(t) & : \quad \text{rotational speed of the coiler motor} \\
\tilde{\omega}_c & : \quad \text{nominal value of } \omega_c(t) \\
\omega_u(t) & : \quad \text{rotational speed of the uncoiler motor}
\end{array}
$$

The work rolls squeeze the metal so as to achieve a reduction in thickness. A set of larger-diameter rolls (termed *backup rolls*) are mounted above and below the work rolls. A hydraulic cylinder is used to vary the vertical location of the backup roll (in the unloaded position) relative to the frame in which it is mounted. The vertical location of the backup roll with respect to the frame is termed the *unloaded roll gap*. Traditionally, the exit thickness, h, of the steel has been controlled by varying $\sigma(t)$.

We have seen, in Example 8.3 (presented earlier), that there are fundamental design difficulties arising from the delay in measuring the exit thickness; however, these difficulties can largely be overcome by the use of virtual sensors of the type discussed in subsection §8.8.1.

There are also fundamental issues associated with the speed of response of the actuator, which typically will exhibit a slew-rate limit; however, modern hydraulic actuators allow roll-gap response times of the order of 5 to 10 [*ms*]. When this is

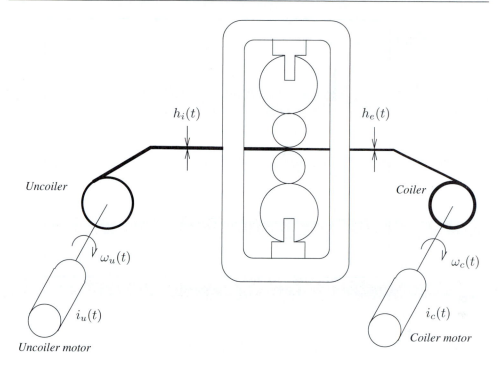

Figure 8.6. Rolling-mill stand

combined with the rapid responses obtained from modern virtual sensors, one is led to the inevitable conclusion that one should be able to design a feedback control system to achieve closed-loop response times in the order of 5 to 10 $[ms]$.

It is with considerable surprise that one finds that, despite great efforts to come up with a suitable design, the closed-loop response of these systems tends to start out fast but then tends to *hold up*. A typical response to a step input disturbance is shown schematically in Figure 8.7.

All manner of hypotheses might be made about the source of the difficulty (nonlinear effects, unmodeled slew-rate limits, etc.). The reader may care to pause and speculate on what might be the source of the problem. Here's a hint: The transfer function from roll gap (σ) to exit thickness (h) turns out to be of the following form (where we have taken a specific real case):

$$G_{h\sigma}(s) = \frac{26.24(s+190)(s+21\pm j11)(s+20)(s+0.5\pm j86)}{(s+143)(s+162\pm j30)(s+30\pm j15)(s+21\pm j6)} \qquad (8.7.1)$$

We see (perhaps unexpectedly) that this transfer function has two zeros located at $s = -0.5 \pm j86$ which are (almost) on the imaginary axis. These zeros arise from the spring action of the strip between the mill, the uncoiler, and the coiler.

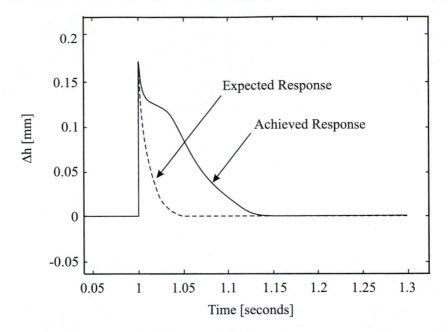

Figure 8.7. Hold-up effect

The resulting poles are turned into zeros by the action of slip. From the analysis carried out in Corollary 8.1 on page 219, it follows that there will be fundamental limitations on the ability of the roll-gap input to control the exit thickness of the steel. In particular, if one aims to have a fast transient response, then significant fluctuations *must* occur that resist the rapid response!

A physical explanation is as follows: If one wishes to reduce the exit thickness, one needs to push down on the roll gap. The exit thickness then drops. However, if the exit mill speed is roughly constant (as it is, through the action of another control loop), then less mass will be transported out of the mill per unit time. Moreover, mass-balance considerations, imply that the entry speed must drop. Now, because of the inertia, the uncoiler cannot change its angular velocity instantaneously. Thus the effect of reduced mill-entry speed will be a drop in input-strip tension. Now, it is also known that strip tension affects the degree of strip-thickness reduction. Hence, reduced tension implies an increase in exit thickness, which impedes the original reduction: the thickness response *holds up*. The associated dynamics depend on the effective strip *spring* between uncoiler and mill. Mathematically, these effects are captured by the near-imaginary-axis zeros in (8.7.1).

The near-imaginary-axis zeros are at 86 $[rad/s]$, so, from (8.6.51), we see that making the effective settling time much less than about 50 $[ms]$ will lead to a large error response. (For $t_s = 50$ $[ms]$, $\omega_0 t_s = 0.68$.) Thus, even though the roll-gap positioning system might be able to be moved very quickly (7 [ms] time constraints are typical with modern hydraulic systems), the effective closed-loop response time is limited to about 50 [ms] by the presence of the imaginary-axis zeros.

To illustrate this effect, consider the plant as given in (8.7.1). Simulations were carried out with the following three PI controllers. (These were somewhat arbitrarily chosen, but the key point here is that the issue of the hold-up effect is fundamental. In particular, *no* controller can improve the situation without at least some radical change!)

$$C_1(s) = \frac{s + 50}{s} \qquad C_2(s) = \frac{s + 100}{s} \qquad C_3(s) = \frac{s + 500}{s} \qquad (8.7.2)$$

For the example given in (8.7.1), the (open-loop) transfer function between the input thickness $H_i(s) = \mathcal{L}[h_i(t)]$ and the output thickness $H_e(s) = \mathcal{L}[h_e(t)]$ (see Figure 8.6) is given by

$$G_{ie}(s) = \frac{0.82(s + 169)(s + 143)(s + 109)(s + 49)(s + 27)(s + 24 \pm j7)}{(s + 143)(s + 162 \pm j30)(s + 30 \pm j15)(s + 21 \pm j6)} \qquad (8.7.3)$$

Then, when the rolling-mill stand is under feedback control, the nominal transfer function between input and output thickness is given by $T_o(s)G_{ie}(s)$, where $T_o(s)$ is the complementary sensitivity and $G_{ie}(s)$ is given in (8.7.3). Note that the zeros of $T_o(s)$ include those located close to the imaginary axis, at $-0.5 \pm j86$.

It is next assumed that a step change of 0.2[mm] occurs, at $t = 1$ in the input thickness. The responses are given in Figure 8.8.

As predicted, the response time is limited to about 50 [ms]. The key point here is that the result given in Corollary 8.1 on page 219 shows that the problem is fundamental and cannot be overcome by *any* control-system design *within the current SISO architecture.*

Remark 8.2. *It is interesting to note that, in this example, the achievable control performance is fundamentally limited by the blocking action of the plant zeros. Moreover, these zeros arise from a physical interaction within the plant.*

Remark 8.3. *The points made in the previous remark hold generally: There is usually a physical reason why certain* bad *zeros arise. Sometimes, it is possible to shift the zeros by changing the physical layout of the plant.*

8.8   Remedies

We next turn to the question of what remedial action one can take to overcome the kinds of limitations discussed above. Because these are fundamental limits, one really only has two options:

  (i) live with the limitation but ensure that the design makes the best of the situation in terms of the desired performance goals, or

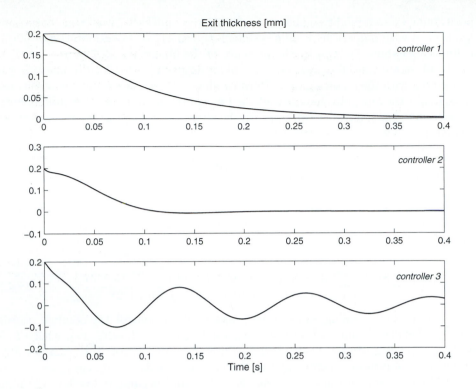

Figure 8.8. Response to a step change in the strip input thickness

(ii) modify the very nature of the problem by changing the system, either through

      ○ new sensors,

      ○ new actuators, or

      ○ alternative architectural arrangements.

We will expand on point (ii) next.

8.8.1 Alternative Sensors

If the sensors are a key stumbling block, then alternative sensors might be needed. One idea that has great potential in control engineering is to use other sensors to replace (or augment) a poor sensor. When other sensors are used together with a model to infer the value of a missing or poor sensor, we say we have used a *virtual* or *soft* sensor.

Example 8.8 (Thickness control in rolling mills, revisited). *We illustrate the use of virtual sensors by returning to Example 8.3. We recall, in that example, that*

the delay between the mill and thickness measuring device was the source of a fundamental limit in rolling-mill thickness performance.

The usual strategy for overcoming this problem is to augment the measurement h_m with a virtual sensor obtained from a model. Two possibilities are the following:

(i) *The force, F, can be related to the thickness, h, and the roll gap, σ, via a simple spring equation of the form*

$$F(t) = M(h(t) - \sigma(t)) \tag{8.8.1}$$

(The spring constant, M, is usually called the Mill Modulus.) $F(t)$ and $\sigma(t)$ can be measured, so an essentially instantaneous estimate of $h(t)$ can be obtained by inverting (8.8.1), leading to

$$\hat{h}(t) = \frac{F(t)}{M} + \sigma(t) \tag{8.8.2}$$

This estimator for existing thickness is called a BISRA gauge and is used extremely commonly in practice.

(ii) *Another possible virtual sensor works as follows:*

It turns out that the strip width is essentially constant in most mills. In this case, conservation of mass across the roll gap leads to the relationship

$$V(t)H(t) \simeq v(t)h(t) \tag{8.8.3}$$

where V, H, v, and h denote the input velocity, input thickness, exit velocity, and exit thickness, respectively.

It is possible to measure V and v by using contact wheels on the strip or by laser techniques. Also, H can be measured upstream of the mill and then delayed appropriately for the incoming strip velocity. Finally, we can estimate the exit thickness from

$$\hat{h}(t) = \frac{V(t)H(t)}{v(t)} \tag{8.8.4}$$

This is called a mass-flow estimator and uses feedforward from the measured upstream input thickness.

Again, this idea is very commonly used in practice.

□□□

8.8.2   Actuator Remedies

If the key limiting factor in control performance is the actuator, then one should think of replacing or resizing the actuator.

Some potential strategies for mitigating the effect of a given poor actuator include the following:

(i) One can sometimes model the saturation effect and apply an appropriate inverse (see Chapter 2), to ensure that appropriate control is executed despite a poor actuator; however, this is usually difficult, because of the complexity of real-world frictional models.

(ii) One can sometimes put a high-gain control loop locally around the offending actuator. This is commonly called *Cascade Control*. As we know, high-gain control tends to reduce the effect of nonlinearities; however, to be able to benefit from this solution, one needs to be able to measure the position of the actuator to a high level of accuracy, and this may not be feasible. (More will be said on the subject of cascade control in Chapter 10.)

(iii) One can sometimes arrange the hardware so that the actuator limitation is removed or, at least, reduced. Examples include the transition from motor-driven screws in rolling mills (which had backlash problems) to modern hydraulic actuators and the use of anti-backlash drives in telescope pointing systems.

8.8.3   Anti-Wind-Up Mechanisms

When an actuator is limited in amplitude or slew rate, then one can often avoid the problem by reducing the performance demands, as discussed in subsection §8.3.1; however, in other applications it is desirable to push the actuator hard up against the limits so as to gain the maximum benefit from the available actuator *authority*. This makes the best of the given situation; however, there is a down-side associated with this strategy.

In particular, one of the costs of driving an actuator into a maximal limit is associated with the problem of integral wind-up. We recall that integrators are very frequently used in control loops to eliminate steady-state errors. However, the performance of integral action depends on the loop remaining in a linear range. If an actuator is forced to a hard limit, then the state of the integrator is likely to build up to a large value. When the input comes out of the limit this *initial condition* will cause a large transient. This effect is called *wind-up*. Much more will be said about ways to avoid this (and other related problems) in Chapter 11. For the moment, it suffices to remark that the core idea used to protect systems against the negative effects of *wind-up* is to turn the integrator off whenever the input reaches a limit. This can be done either by a switch or by implementing the integrator in such a way that it automatically turns off when the input reaches a limit. As a simple illustration, consider the loop given in Figure 8.9.

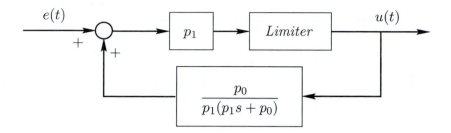

Figure 8.9. Feedback loop with limiter

If we replace the limiter by a unity gain, then elementary block-diagram analysis shows that the transfer function for e to u is

$$\frac{U(s)}{E(s)} = \frac{p_1 s + p_0}{s} \tag{8.8.5}$$

Thus, Figure 8.9 implements a simple PI controller. Note that integral action in this figure has been achieved by placing positive feedback around a *stable* transfer function. If u reaches a limit, then integral action will be temporarily stopped, and anti-wind-up will be achieved. As an illustration of what form the limiter in Figure 8.9 might take, we show a particular limiter in Figure 8.10, which, when used in Figure 8.9, achieves anti-wind-up for an input amplitude limit.

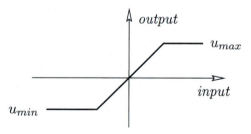

Figure 8.10. Limiter to achieve saturation

An alternative limiter that achieves both slew-rate and amplitude limits is shown in Figure 8.11.

Much more will be said about this topic in Chapter 11.

For the moment, the reader is encouraged to examine the case studies on the website to see the impact that anti-wind-up protection has on practical applications.

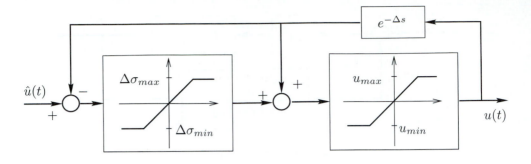

Figure 8.11. Combined saturation and slew-rate limit model

8.8.4   Remedies for Minimal Actuation Movement

Minimal actuator movements are difficult to remedy. In some applications, it is possible to use *dual-range* controllers, wherein a large actuator is used to determine the majority of the control *force* but a smaller actuator is used to give a *fine trim*. An example of the use of this idea is described in connection with pH control on the web.

In other applications, we must *live* with the existing actuator. An example is given below.

Example 8.9 (Continuous caster revisited). *We recall the sustained oscillation problem due to actuator minimal movements described in Example 8.2.*

One cannot use dual-range control in this application, because a small-high-precision valve would immediately clog with solidified steel. A solution we have used to considerable effect in this application is to add a small high-frequency dither signal to the valve. This keeps the valve in motion and hence minimizes "stiction" effects. The high-frequency input dither is filtered out by the dynamics of the process and thus does not have a significant impact on the final product quality. Of course, one does pay the price of having extra wear on the valve due to the presence of the dither signal; however, this cost is off-set by the very substantial improvements in product quality as seen at the output. Some real data is shown in Figure 8.12.

□□□

8.8.5   Architectural Changes

The fundamental limits we have described apply to the given set-up. Clearly, if one changes the physical system in some way, then the situation changes. Indeed, these kinds of change are a very powerful tool in the hands of the control-system designer. Of course, before any change is contemplated, it is desirable to understand

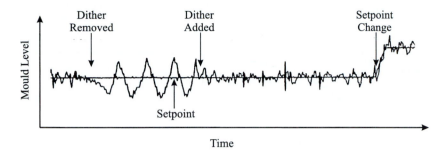

Figure 8.12. Real data, showing effect of adding dither

the source of the limitations–hence, our treatment of fundamental limits as above.

To illustrate how even small structural changes can have an impact, we recall that, in subsection §8.6.5, we showed that open-loop zeros and poles have a profound and predictable effect on the closed-loop performance of a feedback system. In that section, we examined only a one-d.o.f. architecture. Of course, unmeasured disturbances must always be dealt with in this way, because there is no possibility of introducing an extra degree of freedom; however, it is sometimes helpful to exploit a second-d.o.f. when dealing with reference changes. For example, open-loop poles in the CRHP usually induce *slow* stable zeros in the controller, and it is these that lead to the overshoot difficulties in response to a step input. With a two-d.o.f. controller, it is possible to cancel these zeros outside the loop. Note that they remain a difficulty inside the loop and thus contribute to design trade-offs regarding robustness, disturbance rejection, etc. We illustrate by a simple example.

Example 8.10. *Consider the feedback control of plant with nominal model $G_o(s)$ with a PI controller, $C(s)$, where*

$$G_o(s) = \frac{1}{s}; \qquad C(s) = \frac{2s+1}{s} \tag{8.8.6}$$

Then the closed-loop poles are at $(-1; -1)$ and the controller has a zero at $s = -0.5$. Equation (8.6.12) correctly predicts overshoot for the one-d.o.f. design; however, if we first prefilter the reference by $H(s) = \frac{1}{2s+1}$, then no overshoot occurs in response to a unit step change in the reference signal. Figure 8.13 shows the plant output for the one-d.o.f. and the two-d.o.f. architectures. The key difference is that no overshoot appears under the two-d.o.f. design; this improvement is due to the fact that now the transfer function from $R(s)$ to $E(s) = R(s) - Y(s)$ has only one zero at the origin.

□□□

Example 8.11 (Hold-up effect revisited). *We recall the example presented in section §8.7. There, we found that the hold-up effect was inescapable within the*

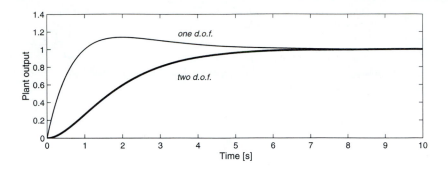

Figure 8.13. Effects of two-d.o.f. architecture

given SISO architecture. We will see in Chapter 10 that it is actually possible to overcome the problem, but this will necessitate that a new architecture for the control system be used.

□□□

8.9   Design Homogeneity, Revisited

We have seen above that limitations arise from different effects. For example, the following factors typically place an upper limit on the usable bandwidth:

- actuator slew-rate and amplitude limits;

- model error;

- delays;

- right-half plane or imaginary-axis zeros.

This leads to the obvious question: *which of these limits, if any, do I need to consider?* The answer is that it is clearly best to focus on that particular issue that has the most impact. In other words, if the presence of a right-half plane zero limits the bandwidth to a point well below the point where model errors become significant, then there is little gain to be achieved from robustifying the controller or improving the model. The reader is invited to compare these comments with those made in subsection §1.5.10, where the issue of homogeneity was first introduced.

8.10   Summary

- This chapter addresses design issues for SISO feedback loops.

- It is shown that the following closed-loop properties cannot be addressed independently by a (linear time-invariant) controller:

  ○ speed of disturbance rejection
  ○ sensitivity to measurement noise
  ○ accumulated control error
  ○ required control amplitude
  ○ required control rate changes
  ○ overshoot, if the system is open-loop unstable
  ○ undershoot, if the system is nonminimum phase
  ○ sensitivity to parametric modeling errors
  ○ sensitivity to structural modeling errors

- Rather, tuning for one of these properties automatically affects the others.

- For example, irrespective of how a controller is synthesized and tuned, if the effect of the *measurement noise* on the output is $T_o(s)$, then the impact of an *output disturbance* is necessarily $1 - T_o(s)$. Thus, any particular frequency cannot be removed from both an output disturbance and the measurement noise, because one would require $T_o(s)$ to be close to 0 at that frequency, whereas the other would require $T_o(s)$ to be close to 1. One can, therefore, only either reject one at the expense of the other or compromise.

- Thus, a faster rejection of disturbance, is generally associated with

  ○ higher sensitivity to measurement noise
  ○ less control error
  ○ larger amplitudes and slew rates in the control action
  ○ higher sensitivity to structural modeling errors
  ○ more undershoot, if the system is nonminimum phase
  ○ less overshoot if the system is unstable

- The trade-offs are made precise by the following fundamental laws of trade-off:

  (1) $S_o(s) = 1 - T_o(s)$
      –that is, an output disturbance is rejected only at frequencies where $|T_o(j\omega)| \approx 1$;

  (2) $Y(s) = -T_o(s)D_m(s)$
      –that is, measurement noise, $d_m(t)$, is rejected only at frequencies where $|T_o(j\omega)| \approx 0$

  (3) $S_{uo}(s) = T_o(s)[G(s)]^{-1}$
      –that is, large control signals arise at frequencies where $|T_o(j\omega)| \approx 1$ but $|G_o(j\omega)| \ll 1$, which occurs when the closed loop is forced to be much more responsive than the open-loop process.

(4) $S_{io}(s) = S_o(s)G_o(s)$

–that is, open-loop poles of the process either necessarily appear as zeros in $S_o(s)$ (resulting in overshoot when rejecting output step disturbances and additional sensitivity) or, if they are stable, can be accepted by the designer as poles in $S_{io}(s)$ instead (where they affect input-disturbance rejection).

(5) $S(s) = S_o(s)S_\Delta(s)$ where $S_\Delta(s) = (1 + T_o(s)G_\Delta(s))^{-1}$

–that is, being responsive to reference changes and against disturbances at frequencies with significant modeling errors jeopardizes stability. Note that the relative (multiplicative) modeling error G_Δ usually accumulates phase and magnitude towards higher frequencies.

(6) Forcing the closed loop faster than unstable zeros necessarily causes substantial undershoot.

- Observing the fundamental laws of trade-off ensures that inadvertently specified, but unachievable, specifications can quickly be identified without wasted tuning effort.

- They also suggest where additional effort would be profitable or wasted:

  - If a design does not fully utilize the actuators, and disturbance rejection is degraded by modeling errors (i.e., the loop is constrained by fundamental trade-off law 5), then additional modeling efforts are warranted.

  - If, on the other hand, loop performance is constrained by nonminimum-phase zeros and a constraint on undershoot (i.e., the loop is constrained by fundamental trade-off law 6), then larger actuators or better models would be wasted.

- It is important to note that the design trade-offs

  - are fundamental to linear time invariant control and

  - are independent of any particular control synthesis methods used.

- However, different synthesis methods

  - choose different closed-loop properties as their constructed property,

  - therefore rendering different properties consequential.

- Some design constraints, such as the inverse response due to NMP zeros, exist not only for linear control systems but also for any other control approach and architecture.

- Remedies for the fundamental limits do exist, but they inevitably require radical changes–for example:

- seeking alternative senses;

- seeking alternative actuators;

- modifying the basic architecture of the plant or controller.

8.11   Further Reading

General

Doyle, J.C., Francis, B.A., and Tannenbaum, A.R. (1992). *Feedback Control Theory*. Macmillan Publishing Company.

Skogestad, S. and Postlethwaite, I. (1996). *Multivariable Feedback Control: Analysis and Design*. Wiley, New York.

Time-domain constraints

Middleton, R.H. (1991). Trade-offs in linear control systems design. *Automatica*, 27(2):281-292.

Slow stable poles

Middleton, R.H. and Graebe, S.F. (1999). Slow stable open-loop poles: To cancel or not to cancel. *Automatica*, 35:877-886.

Rolling-mill control

Bryant, G. (1973). *Automation of tandem mills*. Iron and Steel Institute.

Dendle, D. (1978). Hydraulic position-controled mill and automatic gauge control. Flat rolling–a comparison of rolling-mill types. In *Proceedings of International Conference held by the Metal Society, University College, London*, pages 26-29.

Edwards, W.J. (1978). Design of entry strip thickness control for tandem cold mills. *Automatica*, 14(5):429-441.

Edwards, W.J., Goodwin, G.C., Gómez, G., and Thomas, P. (1995). A review of thickness control on reversing cold rolling mill. *I.E. Aust. Control*, pages 129-134.

King, W. (1973). A new approach to cold mill gauge control. *Iron and Steel Engineering*, pages 40-51.

Hold-up effect in rolling mills

Clark, M. and Mu, Z. (1990). Automatic gauge control for modern high-speed strip mills. In *Proceedings of the 5th International Rolling Conference, Imperial College, London*, pages 63-72.

Goodwin, G.C., Woodyatt, A., Middleton, R.H., and Shim, J. (1999). Fundamental limitations due to $j\omega$-axis zeros in SISO systems. *Automatica*, 35(5):857-863.

Kondo, K., Misaka, Y., Okamato, M., Matsumori, Y., and Miyagi, T. (1988). A new automatic gauge control system for a reversing cold mill. *Trans. ISIJ*, 28:507-513.

8.12   Problems for the Reader

Problem 8.1. *Consider the same conditions as in Lemma 8.2 on page 212.*

8.1.1 *Prove that for a step input disturbance, and $i > 2$,*

$$\int_0^\infty \left\{ \int_0^t e(\tau)d\tau \right\} dt = 0 \tag{8.12.1}$$

$$\int_0^\infty te(t)dt = 0 \tag{8.12.2}$$

8.1.2 *Depict constraints (8.12.1) graphically.*

Problem 8.2. *Consider the same conditions as in Lemma 8.3 on page 213, except for the fact that the pole at $s = \eta_o$ and the zero at $s = z_o$ are* multiple.

8.2.1 *Prove that, for a unit reference step or a unit-step output disturbance, the following holds:*

$$\int_0^\infty te(t)e^{-\eta_o t}dt = 0 \tag{8.12.3}$$

8.2.2 *Prove that, for a unit reference step and z_o in the right-half plane, the following holds:*

$$\int_0^\infty ty(t)e^{-z_o t}dt = 0 \tag{8.12.4}$$

Problem 8.3. *A plant nominal model includes a NMP zero at $s = z_0$. The control specification requires that the control error to a unit step reference satisfy*

$$|e(t)| \le e^{-\alpha(t-t_1)} \qquad \alpha > 0 \qquad t > t_1 > 0 \tag{8.12.5}$$

8.3.1 *Estimate a lower bound for*

$$E_{max} \triangleq \max_{t \in [0, t_1)} e(t) \tag{8.12.6}$$

8.3.2 *Analyze the dependence of the computed bound on the parameters α and t_1.*

Problem 8.4. *The nominal model for a plant is given by*

$$G_o(s) = \frac{5(s-1)}{(s+1)(s-5)} \tag{8.12.7}$$

This plant has to be controlled in a feedback loop with one-degree-of-freedom.

8.4.1 *Determine time-domain constraints for the plant input, the plant output, and the control error in the loop. Assume exact inversion at $\omega = 0$ and step-like reference and disturbances.*

8.4.2 *Why is the control of this nominal plant especially difficult? Discuss.*

Problem 8.5. *Consider a feedback control loop of a stable plant having a (unique) NMP zero at $s = \frac{1}{a}$ $(a \in \mathbb{R}^+)$. Also, assume that a proper controller can be designed such that the complementary sensitivity is given by*

$$T_o(s) = \frac{-as+1}{s^2 + 1.3s + 1} \tag{8.12.8}$$

8.5.1 *Compute the unit step response for different values of a in the range $0 \le a \le 20$.*

8.5.2 *For each case, compute the maximum undershoot M_u, and plot M_u versus a. Discuss.*

Problem 8.6. *Consider a plant having a nominal model given by*

$$G_o(s) = \frac{12(-s+2)}{(s+3)(s+4)^2} \tag{8.12.9}$$

Assume that a control loop has to be designed in such a way that the dominant closed-loop poles are the roots of $s^2 + 1.3\omega_n s + \omega_n^2$.

8.6.1 *Using pole assignment, synthesize a controller for different values of ω_n.*

8.6.2 *Test your designs for step reference and step input disturbances.*

8.6.3 *Compare with the performance of a PID controller tuned by following the Cohen-Coon criterium.*

8.6.4 *Analyze the undershoot observed in the response.*

Problem 8.7. *Consider the feedback control of an unstable plant. Prove that the controller output, $u(t)$, exhibits undershoot for any step reference and for any step-output disturbance.*

Problem 8.8. *Consider two plants having stable nominal models G_a and G_b, given by*

$$G_a(s) = \frac{-s + a}{d(s)}; \qquad G_b(s) = \frac{s + a}{d(s)} \qquad (8.12.10)$$

where $a \in \mathbb{R}^+$ and $d(s)$ is a stable polynomial. Further assume that these plants have to be under feedback control with dominant closed-loop poles having real part smaller than $-a$.

Compare the fundamental design limitations for the two control loops.

Problem 8.9. *Consider the ball-and-beam experiment discussed on the website for the book. This system is known to be nonminimum phase. Explain physically why this is reasonable. (Hint: consider the centripetal force when the beam is rotated.)*

What would you expect to happen to the ball if you rotated the beam suddenly when the ball is near the end of the beam?

Hence, give a physical interpretation to the limitation on response time imposed by the nonminimum-phase zero.

Chapter 9

FREQUENCY-DOMAIN DESIGN LIMITATIONS

9.1 Preview

Chapter 8 showed, via elementary time-domain arguments based on the Laplace transform, that fundamental limitations exist on the achievable performance of linear control systems. For example, it was shown that real right-half plane open-loop plant zeros always lead to undershoot in the reference-to-output step response. Moreover, the extent of this undershoot increases as the response time of the closed-loop system decreases. This implies that right-half plane plant zeros place an inescapable upper limit on the achievable closed-loop bandwidth if excessive undershoot is to be avoided. Similarly, it was shown that real right-half plane open-loop plant poles always lead to overshoot in one-degree-of-freedom feedback systems. The purpose of the current chapter is to develop equivalent frequency-domain constraints and to explore their interpretations.

The results to be presented here have a long and rich history, beginning with the seminal work of Bode published in his 1945 book on network synthesis. Since that time, the results have been embellished in many ways. In this chapter, we summarize the single-input single-output results. In Chapter 24, we will extend the results to the multivariable case.

All of the results follow from the assumption that the sensitivity functions are analytic in the closed right-half plane (i.e., that they are required to be stable). Analytic function theory then implies that the sensitivity functions cannot be specified arbitrarily but, indeed, must satisfy certain precise integral constraints.

The background necessary to appreciate these results is elementary analytic function theory. This theory may not have been met previously by all readers, so we present a brief review in Appendix C, which is available on the book's website. The current chapter can be read in one of two ways. Either the background mathematics can be taken for granted, in which case the emphasis will be on the interpretation of the results, or, alternatively and preferably, the reader can gain a full appreciation by first reviewing the results given in Appendix C. The key result is the Cauchy Integral Formula and its immediate consequences–the Poisson-Jensen Formula. This chapter makes frequent reference to these and other results in Appendix C. It might therefore be useful to have Appendix C displayed on your computer, so that it can be read simultaneously with this chapter.

Most of the results to be discussed below are based on cumulative (i.e., integral or summation) measures of $\ln |S_o(j\omega)|$ and $\ln |T_o(j\omega)|$. Logarithmic functions of sensitivity magnitudes have a sign change when these magnitudes cross the value unity. We recall that 1 is a key value in sensitivity analysis.

9.2 Bode's Integral Constraints on Sensitivity

Bode was the first to establish that the integral of log sensitivity had to be zero (under mild conditions). This meant that, if the sensitivity was pushed down in some frequency range ($|S_o(j\omega)| < 1 \Rightarrow \log |S_o(j\omega)| < 0$) then it had to pop up somewhere else to retain zero for the integral of the logarithm. Gunther Stein, in a memorable Bode lecture at an IEEE Conference on Decision and Control, called this the principle of *Conservation of Sensitivity Dirt*. This image refers to the area under the sensitivity function being analogous to a pile of dirt. If one shovels dirt away from some area (i.e., reduces sensitivity in a frequency band), then it piles up somewhere else (i.e., the sensitivity increases at other frequencies).

The formal statement of the result is as follows:

Lemma 9.1. *Consider a one-d.o.f., stable control loop with the open-loop transfer function*

$$G_o(s)C(s) = e^{-s\tau} H_{ol}(s) \qquad\qquad \tau \geq 0 \qquad\qquad (9.2.1)$$

where $H_{ol}(s)$ is a rational transfer function of relative degree $n_r > 0$, and define

$$\kappa \triangleq \lim_{s \to \infty} s H_{ol}(s) \qquad\qquad (9.2.2)$$

Assume that $H_{ol}(s)$ has no open-loop poles in the open RHP. Then the nominal sensitivity function satisfies

$$\int_0^\infty \ln |S_o(j\omega)| d\omega = \begin{cases} 0 & \text{for } \tau > 0 \\ -\kappa \frac{\pi}{2} & \text{for } \tau = 0 \end{cases} \qquad (9.2.3)$$

Proof

We first treat the case $\tau = 0$.

We make the following changes in notation: $s \to z$, $H_{ol}(s) \to l(z)$, and $g(z) = (1 + l(z))^{-1}$.

We then observe that

$$S_o(z) = (1 + l(z))^{-1} = g(z) \tag{9.2.4}$$

By the assumptions on $H_{ol}(s)$, we observe that $\ln g(z)$ is analytic in the closed RHP; then, by Theorem C.7,

$$\oint_C \ln g(z)dz = 0 \tag{9.2.5}$$

where $C = C_i \cup C_\infty$ is the contour defined in Figure C.4.
    Then

$$\oint_C \ln g(z)dz = j\int_{-\infty}^\infty \ln g(j\omega)d\omega - \int_{C_\infty} \ln(1 + l(z))dz \tag{9.2.6}$$

For the first integral on the right-hand side of equation (9.2.6), we use the conjugate symmetry of $g(z)$ to obtain

$$\int_{-\infty}^\infty \ln g(j\omega)d\omega = 2\int_0^\infty \ln|g(j\omega)|d\omega \tag{9.2.7}$$

For the second integral, we notice that, on C_∞, $l(z)$ can be approximated by

$$\frac{a}{z^{n_r}} \tag{9.2.8}$$

The result follows upon using Example C.7 and upon noticing that $a = \kappa$ for $n_r = 1$. The extension to the case $\tau \neq 0$ is similar and uses the results in Example C.8.

□□□

We conclude from Lemma 9.1 that, independent of the controller design, low sensitivity in certain prescribed frequency bands will result in a sensitivity larger than one in other frequency bands. Graphically, the above statement can be appreciated in Figure 9.1.

In Figure 9.1, when the area A_1 ($\Longrightarrow |S_o(j\omega)| < 1$) is equal to the area A_2 ($\Longrightarrow |S_o(j\omega)| > 1$), then the integral in equation (9.2.3) is equal to zero (or, more generally, $A_1 - A_2 = \kappa\frac{\pi}{2}$).

In theory, this is not a hard constraint, because it is enough to have $|S_o(j\omega)|$ only slightly above one over a large (actually infinite) frequency band. We say *in theory* because robustness and noise immunity, among other factors, will require $|T_o(j\omega)|$ to be very small beyond a certain frequency. Thus, due to the fundamental property of $T_o(j\omega) + S_o(j\omega) = 1$, from that frequency onwards, $|S_o(j\omega)|$ will make

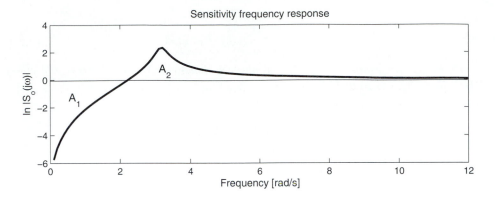

Figure 9.1. Graphical interpretation of the Bode integral

no significant contribution to the balance required in equation (9.2.3). This means that, in practice, the compensation must be achieved within a finite bandwidth. More will be said on this later in the chapter.

The above result assumes that the open-loop system is stable. The extension to *open-loop unstable* systems is as follows:

Lemma 9.2. *Consider a feedback control loop with an open-loop transfer function as in Lemma 9.1 on page 242 and having unstable poles located at $p_1, \ldots, p_N$, pure time delay τ, and relative degree $n_r \geq 1$. Then the nominal sensitivity satisfies the following:*

$$\int_0^\infty \ln |S_o(j\omega)| \, d\omega = \pi \sum_{i=1}^N \Re\{p_i\} \qquad for \quad n_r > 1 \qquad (9.2.9)$$

$$\int_0^\infty \ln |S_o(j\omega)| \, d\omega = -\kappa \frac{\pi}{2} + \pi \sum_{i=1}^N \Re\{p_i\} \qquad for \quad n_r = 1 \qquad (9.2.10)$$

where $\kappa = \lim_{s \to \infty} s H_{ol}(s)$.

Proof

We first treat the case $\tau = 0$.

As in Lemma 9.2, we make the changes in notation $s \to z$, $H_{ol}(s) \to l(z)$, and $g(s) = (1 + l(z))^{-1}$.

We first notice that $\ln g(z)$ is no longer analytic on the RHP (because the open-loop unstable poles, $p_1, \ldots, p_N$, become zeros of $g(z)$) . We then define

$$\tilde{g}(z) \stackrel{\triangle}{=} g(z) \prod_{i=1}^{N} \frac{z + p_i}{z - p_i} \qquad (9.2.11)$$

Thus, $\ln \tilde{g}(z)$ is analytic in the closed RHP. We can then apply Cauchy's integral on the contour C described in Figure C.4 to obtain

$$\oint_C \ln \tilde{g}(z) dz = 0 = \oint_C \ln g(z) dz + \sum_{i=1}^{N} \oint_C \ln \frac{z + p_i}{z - p_i} dz \qquad (9.2.12)$$

The first integral on the right-hand side can be expressed as

$$\oint_C \ln g(z) dz = 2j \int_0^\infty \ln |g(j\omega)| d\omega + \int_{C_\infty} \ln g(z) dz \qquad (9.2.13)$$

where, by using Example C.7

$$\int_{C_\infty} \ln g(z) dz = \begin{cases} 0 & \text{for } n_r > 1 \\ j\kappa\pi & \text{for } n_r = 1 \end{cases} \quad \text{where} \quad \kappa \stackrel{\triangle}{=} \lim_{z \to \infty} z l(z) \qquad (9.2.14)$$

The second integral on the right-hand side of equation (9.2.12) can be computed as follows

$$\oint_C \ln \frac{z + p_i}{z - p_i} dz = j \int_{-\infty}^\infty \ln \frac{j\omega + p_i}{j\omega - p_i} d\omega + \int_{C_\infty} \ln \frac{z + p_i}{z - p_i} dz \qquad (9.2.15)$$

We note that the first integral on the right-hand side is zero, and, by Example C.9, the second integral is equal to $-2j\pi p_i$. Thus, the result follows upon noting that, because $g(z)$ is a real function of z, then

$$\sum_{i=1}^{N} p_i = \sum_{i=1}^{N} \Re\{p_i\} \qquad (9.2.16)$$

When $g(z) = (1 + e^{-z\tau} l(z))^{-1}$ for $\tau > 0$ then the proof follows along the same lines as those above using the result in Example C.8.

□□□

Remark 9.1. *The Bode formulae (and related results) assume that the function under consideration is analytic, not only inside a domain D, but also on its border C. In control design, there may exist singularities on C, such as integrators or purely* *imaginary poles in the controller (aimed at rejecting a particular disturbance–see Chapter 10). These can be dealt with by using an infinitesimal circular indentation in C, constructed so as to leave the singularity outside D. For the functions of interest to us, the integral along the indentation usually vanishes. This is illustrated in Example C.6 for a logarithmic function, when D is the right-half plane and there is a singularity at the origin.*

□□□

9.3 Integral Constraints on Complementary Sensitivity

Equation (9.2.3) refers to the nominal sensitivity. A natural question is whether a similar result exists for the other three nominal sensitivity functions. An obstacle in making this extension is that the logarithms of $|T_o(s)|$ and $|S_{io}(s)|$ are not bounded for large $|s|$, because $T_o(s)$ and $S_{io}(s)$ are strictly proper functions. (Recall that $S_o(s)$ is biproper.) We can circumvent the problem in the case of the complementary sensitivity function by integrating $\omega^{-2}\log|T_o(j\omega)|$.

Again we find that a *conservation* principle applies. The details are given next:

Lemma 9.3. *Consider a one-d.o.f. stable control loop with the open-loop transfer function*

$$G_o(s)C(s) = e^{-s\tau}H_{ol}(s) \qquad\qquad \tau \geq 0 \qquad (9.3.1)$$

where $H_{ol}(s)$ is a rational transfer function of relative degree $n_r > 1$ satisfying

$$H_{ol}(0)^{-1} = 0 \qquad (9.3.2)$$

Furthermore, assume that $H_{ol}(s)$ has no open-loop zeros in the open RHP.

Then the nominal complementary sensitivity function satisfies

$$\int_{0^-}^{\infty} \frac{1}{\omega^2}\ln|T_o(j\omega)|d\omega = \pi\tau - \frac{\pi}{2k_v} \qquad (9.3.3)$$

where k_v is the velocity constant of the open-loop transfer function which satisfies

$$\frac{1}{k_v} = -\lim_{s\to 0}\frac{dT(s)}{ds} \qquad (9.3.4)$$

$$= -\lim_{s\to 0}\frac{1}{sH_{ol}(s)}$$

Proof

Note that

$$T_o(s) = (1 + H_{ol}(s)^{-1}e^{s\tau})^{-1} \tag{9.3.5}$$

Now, consider the function

$$F(s) \triangleq \frac{\ln T_o(s)}{s^2} \tag{9.3.6}$$

We note that $F(s)$ is analytical in the closed RHP, except at the origin. We apply the Cauchy Integral theorem C.7 to this function. We use a contour similar to that shown in Figure C.4 together with an infinitesimal right circular indentation, C_ϵ, at the origin. We then have that

$$\oint_C \frac{\ln T_o(s)}{s^2} ds = 0 = \int_{C_{i-}} \frac{\ln T_o(s)}{s^2} ds + \int_{C_\epsilon} \frac{\ln T_o(s)}{s^2} ds + \int_{C_\infty} \frac{\ln T_o(s)}{s^2} ds \tag{9.3.7}$$

where C_{i-} is the imaginary axis minus the indentation C_ϵ. We can then show, by using (9.3.2), that the integral along C_ϵ is $-\pi \lim_{s \to 0} \frac{dT(s)}{ds}$. It is also straightforward to prove that the integral along C_∞ is equal to $2\pi\tau$. Then the result (9.3.3) follows upon using the conjugate symmetry of the integrand in (9.3.3) and rearranging terms.

□□□

Equation (9.3.3) has the same impact as (9.2.3), because it provides similar insights into design constraints. This can be better seen by noting that

$$\int_{0-}^{\infty} \frac{1}{\omega^2} \ln |T_o(j\omega)| d\omega = \int_0^{\infty} \ln \left| T_o \left(\frac{1}{jv} \right) \right| dv = \pi\tau \tag{9.3.8}$$

where $v = \frac{1}{\omega}$.

Lemma 9.3 has assumed that there are no open-loop *zeros* in the RHP. The result can be extended to the case when there are RHP zeros, as shown in the following lemma, which is analogous to Lemma 9.2 on page 244.

Lemma 9.4. *Consider a one-d.o.f. stable control loop with open-loop transfer function*

$$G_o(s)C(s) = e^{-s\tau} H_{ol}(s) \qquad \text{with} \quad \tau \geq 0 \tag{9.3.9}$$

where $H_{ol}(s)$ is a rational transfer function of relative degree $n_r > 1$ and satisfying

$$H_{ol}(0)^{-1} = 0 \tag{9.3.10}$$

Further assume that $H_{ol}(s)$ has open-loop zeros in the open RHP, located at $c_1, c_2, \ldots, c_M$; then

$$\int_0^\infty \frac{1}{\omega^2} \ln |T_o(j\omega)| d\omega = \pi\tau + \pi \sum_{i=1}^{M} \frac{1}{c_i} - \frac{\pi}{2k_v} \qquad (9.3.11)$$

Proof

We first define

$$\tilde{T}_o(s) \triangleq \prod_{i=1}^{M} \frac{s + c_i}{s - c_i} T_o(s) \qquad (9.3.12)$$

Note that $|\tilde{T}_o(j\omega)| = |T_o(j\omega)|$ and that $s^{-2} \ln \tilde{T}_o(s)$ is analytic in the closed RHP, except at the origin. Thus,

$$\oint_C \frac{\ln \tilde{T}_o(s)}{s^2} ds = 0 = \int_{C_{i-}} \frac{\ln \tilde{T}_o(s)}{s^2} ds + \int_{C_\epsilon} \frac{\ln \tilde{T}_o(s)}{s^2} ds + \int_{C_\infty} \frac{\ln \tilde{T}_o(s)}{s^2} ds \quad (9.3.13)$$

where C_{i-} is the imaginary axis minus the indentation C_ϵ (in the notation of Figure C.4, $C_i = C_\epsilon \cup C_{i-}$). We can then show, by using (9.3.10), that the integral along C_ϵ is zero. Note also that

$$\int_{C_\infty} \frac{\ln \tilde{T}_o(s)}{s^2} ds = \int_{C_\infty} \frac{\ln T_o(s)}{s^2} ds + \sum_{i=1}^{M} \int_{C_\infty} \frac{1}{s^2} \ln \left(\frac{s + c_i}{s - c_i} \right) ds \qquad (9.3.14)$$

The first integral on the right-hand side vanishes, and the second one can be computed from

$$\int_{C_\infty} \frac{1}{s^2} \ln \left(\frac{s + c_i}{s - c_i} \right) ds = -\frac{1}{c_i} \int_{C_\delta} \ln \left(\frac{1 + \phi}{1 - \phi} \right) d\phi \qquad (9.3.15)$$

where $s = \frac{c_i}{\phi}$ and C_δ is an infinitesimal counterclockwise semicircle. The result follows upon using the fact that $\ln(1 + x) \to x$ for $|x| \to 0$.

□□□

We see from (9.3.11) that the presence of (small) right-half plane zeros makes the trade-off in the allocation of complementary sensitivity in the frequency domain more difficult. Also note from (9.3.11) that the zeros combine as the sum of the reciprocals of the zeros (similarly to parallel resistors, for Electrical Engineering readers).

9.4 Poisson Integral Constraint on Sensitivity

The result given in Lemma 9.2 on page 244 shows that the larger is the real part of unstable poles, the more difficult the sensitivity compensation is; however, at least theoretically, the compensation can still be achieved over an infinite frequency band.

When the effect of NMP zeros is included, then the constraint becomes more precise, as shown below from the Poisson-Jensen formula. (See Lemma C.1.)

In the sequel we will need to convert transfer functions, $f(s)$, having right-half plane zeros (at c_k ; $k = 1, \cdots, m$) and/or right-half plane poles (at p_i ; $i = 1, \cdots N$) into functions such that $\ln(f(s))$ is analytic in the right-half plane. A trick we will use is to express $f(s)$ as the product of functions that are nonminimum phase and analytic in the RHP, times the following Blaschke products (or times the inverses of these products):

$$
B_z(s) = \prod_{k=1}^{M} \frac{s - c_k}{s + c_k^*} \qquad B_p(s) = \prod_{i=1}^{N} \frac{s - p_i}{s + p_i^*} \qquad (9.4.1)
$$

Indeed, this technique has already been employed in establishing Lemma 9.2 on page 244 and Lemma 9.4–see equations (9.2.11) and (9.3.12).

We then have the following lemma, which relates a (weighted) integral of $\ln |S_o(j\omega)|$ to the location of both right-half plane zeros and poles of the open-loop transfer function.

Lemma 9.5 (Poisson integral for $S_o(j\omega)$). *Consider a feedback control loop with open-loop NMP zeros located at $c_1, c_2, \ldots, c_M$, where $c_k = \gamma_k + j\delta_k$, and open-loop unstable poles located at $p_1, p_2, \ldots, p_N$. Then the nominal sensitivity satisfies*

$$
\int_{-\infty}^{\infty} \ln |S_o(j\omega)| \frac{\gamma_k}{\gamma_k^2 + (\delta_k - \omega)^2} d\omega = -\pi \ln |B_p(c_k)| \qquad for \quad k = 1, 2, \ldots M
$$

$$(9.4.2)$$

and, for a real unstable zero, i.e., $\delta_k = 0$, equation (9.4.2) simplifies to

$$
\int_{0}^{\infty} \ln |S_o(j\omega)| \frac{2\gamma_k}{\gamma_k^2 + \omega^2} d\omega = -\pi \ln |B_p(c_k)| \qquad for \quad k = 1, 2, \ldots M \qquad (9.4.3)
$$

where $B_p(s)$ is the Blaschke product defined in (9.4.1).

Proof

We first recall that any open-loop NMP zero must be a zero in $T_o(s)$ and thus

$$T_o(c_k) = 0 \iff S_o(c_k) = 1 \iff \ln S_o(c_k) = 0 \tag{9.4.4}$$

Also note that $S_o(s)$ is stable, but nonminimum phase, because its numerator vanishes at all open-loop unstable poles. We remove the unstable poles by using the Blaschke product, by defining $[S_o(s)]_{smp}$ as follows:

$$S_o(s) = [S_o(s)]_{smp} B_p(s) \tag{9.4.5}$$

where $[S_o(s)]_{smp}$ is stable and of minimum phase.

The result (9.4.2) then follows from the application of Lemma C.1, by taking $z = s$, $\tilde{g}(s) = [S_o(s)]_{smp}$, and $g(s) = S_o(s)$.

Equation (9.4.3) results from (9.4.2) upon using (1) the fact that $\delta_k = 0$ and (2) the conjugate symmetry of $S_o(j\omega)$.

□□□

We note that $-\ln|B_p(c_k)| > 0$, and that it grows unbounded as one of the open-loop NMP zeros approaches one of the open-loop unstable poles.

Equation (9.4.2) says that no matter how the controller is designed, a *weighted* compensation of low- and high-sensitivity regions has to be achieved. We note that the weighting function decays with frequency, meaning that this compensation has essentially to be achieved over a finite frequency band. We see from (9.4.2) that the weighting function in these integrals is

$$W(c_k,\omega) \triangleq \frac{\gamma_k}{\gamma_k^2 + (\delta_k - \omega)^2} \tag{9.4.6}$$

We also observe that

$$\int_{-\infty}^{\infty} W(c_k,\omega)d\omega = \pi \tag{9.4.7}$$

Furthermore, we can define a *weighted length of the frequency axis,*

$$2\int_0^{\omega_c} W(c_k,\omega)d\omega = 2\arctan\left(\frac{\omega_c - \delta_k}{\gamma_k}\right) + 2\arctan\left(\frac{\delta_k}{\gamma_k}\right) \triangleq \Omega(c_k,\omega_c) \tag{9.4.8}$$

Note that if $\omega_2 > \omega_1 \geq 0$, then

$$2\int_{\omega_1}^{\omega_2} W(c_k,\omega)d\omega = \Omega(c_k,\omega_2) - \Omega(c_k,\omega_1) \tag{9.4.9}$$

Also, for a *real NMP zero*, i.e., $\delta_k = 0$ we have that

$$\Omega(c_k, \omega_c) = 2 \arctan\left(\frac{\omega_c}{\gamma_k}\right) \qquad (9.4.10)$$

$$\Omega(c_k, \infty) = 2 \lim_{\omega_c \to \infty} \arctan\left(\frac{\omega_c}{\gamma_k}\right) = \pi \qquad (9.4.11)$$

This can be interpreted as giving a length (measure) of π to the frequency axis.

To gain a more complete understanding of the result given in Lemma 9.5 on page 249 , we consider two special cases. For simplicity we assume *real NMP zeros*, i.e., $\delta_k = 0$.

Example 9.1. *Say that a design specification requires that $|S_o(j\omega)| < \epsilon < 1$ over the frequency band $[0, \omega_l]$. We seek a lower bound for the sensitivity peak S_{max}.*

To obtain that bound, we use Lemma 9.5 and the above definitions. Then, upon splitting the integration interval $[0, \infty)$ into $[0, \omega_l] \cup (\omega_l, \infty)$, we obtain

$$2 \int_0^\infty \ln|S_o(j\omega)|W(c_k, \omega)d\omega = 2 \int_0^{\omega_l} \ln|S_o(j\omega)|W(c_k, \omega)d\omega +$$

$$2 \int_{\omega_l}^\infty \ln|S_o(j\omega)|W(c_k, \omega)d\omega \qquad (9.4.12)$$

$$= -\pi \ln|B_p(c_k)|$$

We then substitute the upper bounds for $|S_o(j\omega)|$ in both intervals. By using (9.4.8)-(9.4.11) we obtain

$$2 \int_0^\infty \ln|S_o(j\omega)|W(c_k, \omega)d\omega = -\pi \ln|B_p(c_k)| \qquad (9.4.13)$$

$$< \ln(\epsilon) \int_0^{\omega_l} 2W(c_k, \omega)d\omega + \ln S_{max} \int_{\omega_l}^\infty 2W(c_k, \omega)d\omega$$

$$\qquad (9.4.14)$$

$$= \ln(\epsilon)\Omega(c_k, \omega_l) + (\pi - \Omega(c_k, \omega_l)) \ln S_{max} \qquad (9.4.15)$$

which leads to

$$\ln S_{max} > \frac{1}{\pi - \Omega(c_k, \omega_l)} \left[|\pi \ln|B_p(c_k)|| + |(\ln \epsilon)\Omega(c_k, \omega_l)|\right] \qquad (9.4.16)$$

where we have used the fact that $|B_p(c_k)| < 1 \iff \ln(|B_p(c_k)|) < 0$ for every $c_k > 0$.

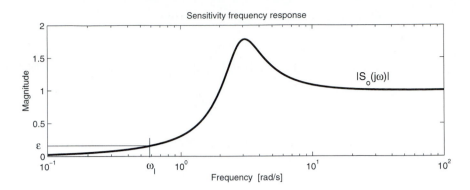

Figure 9.2. Design specification for $|S_o(j\omega)|$

Discussion

(i) Consider the plot of sensitivity versus frequency shown in Figure 9.2. Say that we were to require the closed-loop bandwidth to be greater than the magnitude of a right half plane (real) zero. In terms of the notation used above, this would imply $\omega_l > \gamma_k$. We can then show that there is necessarily a very large sensitivity peak occurring beyond ω_l. To estimate this peak, assume that $\omega_l = 2\gamma_k$ and $\epsilon = 0.3$. Then, *without considering the effect of any possible open-loop unstable pole*, the sensitivity peak will be bounded below by (see equation (9.4.16))

$$\ln S_{max} > \frac{1}{\pi - \Omega(c_k, 2c_k)} \left[|(\ln 0.3)\Omega(c_k, 2c_k)| \right] \qquad (9.4.17)$$

Then, upon using (9.4.10), we have that $\Omega(c_k, 2c_k) = 2\arctan(2) = 2.21$, leading to $S_{max} > 17.7$. Note that this, in turn, implies that the complementary sensitivity peak will be bounded below by $S_{max} - 1 = 16.7$.

(ii) The observation made in (i) is consistent with the analysis carried out in subsection §8.6.5. In both cases, the conclusion is that the closed-loop bandwidth should not exceed the magnitude of the smallest NMP open-loop zero. The penalty for not following this guideline is that a very large sensitivity peak will occur, leading to fragile loops (nonrobust) and large undershoots and overshoots.

(iii) In the presence of unstable open-loop poles, the problem is compounded through the presence of the factor $|\ln|B_p(c_k)||$. This factor grows without bound when one NMP zero approaches an unstable open-loop pole.

□□□

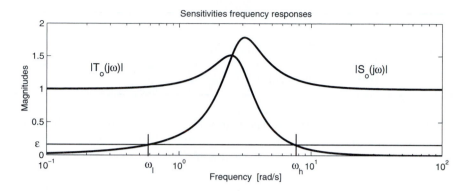

Figure 9.3. Design specifications for $|S_o(j\omega)|$ and $|T_o(j\omega)|$

Example 9.2. *Consider the plots given in Figure 9.3. Say that, in addition to the requirement in the previous example, we ask that $|T_o(j\omega)|$ be less than ϵ for all frequencies larger than ω_h, where $\omega_h > \omega_l$ and $0 < \epsilon \ll 1$. This condition normally originates from the need to ensure noise immunity and robustness against modeling errors. (See Chapter 5.)*

We again seek a lower bound for the sensitivity peak.

We first note that the requirement that $|T_o(j\omega)| < \epsilon$ implies $|S_o(j\omega)| < 1 + \epsilon$, because $S_o(s) + T_o(s) = 1$.

Following the same approach used in the previous example, we split the integration interval $[0, \infty)$ into $[0, \omega_l] \cup (\omega_l, \omega_h] \cup (\omega_h, \infty)$.

$$2 \int_0^\infty \ln|S_o(j\omega)| W(c_k, \omega) d\omega = 2 \int_0^{\omega_l} \ln|S_o(j\omega)| W(c_k, \omega) d\omega +$$

$$2 \int_{\omega_l}^{\omega_h} \ln|S_o(j\omega)| W(c_k, \omega) d\omega + 2 \int_{\omega_h}^\infty \ln|S_o(j\omega)| W(c_k, \omega) d\omega$$

$$= -\pi \ln|B_p(c_k)| \quad (9.4.18)$$

For each interval, we next replace $|S_o(j\omega)|$ by its maximum value, as follows:

$$\max|S_o(j\omega)| = \begin{cases} \epsilon & \omega \in [0, \omega_l] \\ S_{max} & \omega \in [\omega_l, \omega_h] \\ 1 + \epsilon & \omega \in [\omega_h, \infty] \end{cases} \quad (9.4.19)$$

Upon performing these substitutions and using equations (9.4.8) and (9.4.9), we obtain

$$\ln S_{max} > \frac{1}{\Omega(c_k,\omega_h) - \Omega(c_k,\omega_l)} \left[|\pi \ln |B_p(c_k)|| + |(\ln \epsilon)\Omega(c_k,\omega_l)| \right.$$
$$\left. - (\pi - \Omega(c_k,\omega_h)) \ln(1+\epsilon) \right] \qquad (9.4.20)$$

□□□

We can see that, in addition to the trade-offs illustrated in the previous example, a new compromise emerges in this case. Sharp transitions in the sensitivity frequency response, i.e., ω_l close to ω_h, will contribute to large sensitivity peaks.

9.5   Poisson Integral Constraint on Complementary Sensitivity

Analogous to the results given in section §9.4, we can develop constraints for complementary sensitivity.

The result corresponding to Lemma 9.5 is as follows:

Lemma 9.6 (Poisson integral for $T_o(j\omega)$). *Consider a feedback control loop having delay $\tau \geq 0$ and having both open-loop unstable poles located at $p_1, p_2, \ldots, p_N$ (where $p_i = \alpha_i + j\beta_i$) and open-loop zeros in the open RHP, located at $c_1, c_2, \ldots, c_M$.*

Then, the nominal complementary sensitivity satisfies

$$\int_{-\infty}^{\infty} \ln |T_o(j\omega)| \frac{\alpha_i}{\alpha_i^2 + (\beta_i - \omega)^2} d\omega = -\pi \ln |B_z(p_i)| + \pi\tau\alpha_i \qquad for \quad i = 1, 2, \ldots, N$$
$$(9.5.1)$$

Proof

The above result is an almost straightforward application of Lemma C.1. If the delay is nonzero then $\ln |T_o(j\omega)|$ does not satisfy the bounding condition (iv) in Lemma C.1. Thus, we first define

$$\bar{T}_o(s) = T_o(s)e^{s\tau} \implies \ln |\bar{T}_o(j\omega)| = \ln |T_o(j\omega)| \qquad (9.5.2)$$

The result then follows upon applying Lemma C.1 to $\bar{T}_o(s)$ and upon recalling that

$$\ln(T_o(p_i)) = 0 \qquad i = 1, 2, \ldots, N \qquad (9.5.3)$$

□□□

The consequences of this result are illustrated in the following example. For simplicity, we assume that the unstable pole is *real*, i.e., $\beta_i = 0$.

Example 9.3. *Say that the design requirement is that $|T_o(j\omega)| < \epsilon < 1$ for all frequencies above ω_h. See Figure 9.3 again. This requirement defines a closed-loop bandwidth and could originate from robustness and noise-immunity considerations. We seek a lower bound for the sensitivity peak T_{max}. This peak will lie in the frequency region below ω_h.*

Lemma 9.6 can be used to obtain the bound. We split the integration interval $[0, \infty]$ into $[0, \omega_h] \cup (\omega_h, \infty)$. Then, from equation (9.5.1), we have that

$$\int_0^\infty \ln |T_o(j\omega)| \frac{2\alpha_i}{\alpha_i^2 + \omega^2} d\omega = -\pi \ln |B_z(\alpha_i)| + \pi\tau\alpha_i$$

$$= \int_0^{\omega_h} \ln |T_o(j\omega)| \frac{2\alpha_i}{\alpha_i^2 + \omega^2} d\omega + \int_{\omega_h}^\infty \ln |T_o(j\omega)| \frac{2\alpha_i}{\alpha_i^2 + \omega^2} d\omega$$

$$< \Omega(\alpha_i, \omega_h) \ln T_{max} + (\ln \epsilon)(\pi - \Omega(\alpha_i, \omega_h)) \quad (9.5.4)$$

where the functions $\Omega(\circ, \circ)$ were defined in (9.4.6) to (9.4.11).

Rearranging terms in (9.5.4), we obtain

$$\ln T_{max} > \frac{1}{\Omega(\alpha_i, \omega_h)} \left[\pi |\ln |B_z(\alpha_i)|| + \pi\tau\alpha_i + |\ln \epsilon|(\pi - \Omega(\alpha_i, \omega_h))\right] \quad (9.5.5)$$

Discussion

(i) We see that the lower bound on the complementary sensitivity peak is larger for systems with pure delays and that the influence of a delay increases for unstable poles that are far away from the imaginary axis, i.e., large α_i.

(ii) The peak, T_{max}, grows unbounded when a NMP zero approaches an unstable pole, because then $|\ln |B_z(p_i)||$ grows unbounded.

(iii) Say that we ask that the closed-loop bandwidth be much smaller than the magnitude of a right-half plane (real) pole. In terms of the notation used before, we would then have $\omega_h \ll \alpha_i$. Under these conditions, $\Omega(p_i, \omega_h)$ will be very small, leading to a very large complementary sensitivity peak. We note that this is consistent with the results in section §8.6.5, based on time-domain analysis. There, it was shown that, when this condition arises, large overshoot appears.

□□□

The existence of large sensitivity peaks has a time-domain interpretation–namely, that the transient response of the closed loop includes large deviations and slow transients. In the frequency domain, this undesirable feature will usually be reflected in small stability margins.

The results presented above establish design trade-offs originating from the nature of feedback systems in the presence of NMP zeros and unstable open-loop poles. Of course, similar constraints arise from poles and zeros in the controller. If one wishes to have constraints that hold independent of the controller, then equality signs in equations (9.2.9), (9.2.10) , (9.4.2), (9.4.3), and (9.5.1) should be changed to $\geq$ signs, because the terms in these equations should also include the effect of unstable poles or NMP zeros resulting from the controller.

One of the general conclusions that can be drawn from the above analysis is that the design problem becomes more difficult for large unstable open-loop poles and small NMP zeros. The notions of *large* and *small* are relative to the magnitude of the required closed-loop bandwidth.

Remark 9.2. *The accompanying CD-ROM includes two MATLAB routines, **smax.m** and **tmax.m**, to compute lower bounds for the sensitivity and complementary sensitivity under the assumption of certain bandwidth specifications.*

□□□

9.6   Example of Design Trade-offs

To illustrate the application of the above ideas, we will consider the example of the inverted pendulum. This is a classroom example, but similar dynamics apply to rudder roll stabilization of ships–see Chapter 23. We will consider two scenarios–one in which the angle of the pendulum is measured, and one in which this extra measurement is not available.

Example 9.4 (Inverted pendulum without angle measurement). *Consider the system as shown in Figure 9.4.*

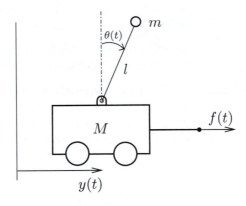

Figure 9.4. Inverted pendulum

Details of the modeling of this system were given in Chapter 3. We recall that, by applying classical mechanics methods, we obtain the following (nonlinear) model:

$$\ddot{y}(t) = \frac{1}{\lambda_m + \sin^2 \theta(t)} \left(\frac{f(t)}{m} + (\dot{\theta}(t))^2 l \sin \theta(t) - g \sin \theta(t) \cos \theta(t) \right) \tag{9.6.1}$$

$$\ddot{\theta}(t) = \frac{1}{l\lambda_m + \sin^2 \theta(t)} \left(\frac{-f(t)}{m} \cos \theta(t) - (\dot{\theta}(t))^2 l \cos \theta(t) \sin \theta(t) + (1 + \lambda_m) g \sin \theta(t) \right) \tag{9.6.2}$$

where $\lambda_m = M/m$ and g is the acceleration due to gravity.

If the above nonlinear model is linearized around the operating point $\theta_Q = 0$, $y_Q = 0$, and $f_Q = 0$, we obtain the following transfer function from input u to output y.

$$\frac{Y(s)}{F(s)} = K \frac{(s-b)(s+b)}{s^2(s-a)(s+a)} \tag{9.6.3}$$

where

$$K = \frac{1}{M}; \quad b = \sqrt{\frac{g}{l}}; \quad a = \sqrt{\frac{(1+\lambda_m)g}{l\lambda_m}} \tag{9.6.4}$$

We observe that there is a NMP zero, located at $s = b$ (which is to the left of the unstable pole located at $s = a$). This induces a control-design conflict because, on the one hand, the closed-loop bandwidth should be less than b, but, on the other hand, it should be larger than the largest unstable pole, i.e., larger than a. To quantify the effects of this unresolvable conflict, we set $M = m = 0.5$ [kg], $l = 1$ [m] and approximate $g \approx 10$ [m/s^2]. These choices lead to

$$\frac{Y(s)}{F(s)} = 2 \frac{(s-\sqrt{10})(s+\sqrt{10})}{s^2(s-\sqrt{20})(s+\sqrt{20})} \tag{9.6.5}$$

The presence of the NMP zero and the unstable pole lead to sensitivity peaks with lower bounds that can be computed from the theory presented above. Consider the schematic plots of sensitivity and complementary sensitivity given in Figure 9.3. In particular, recall the definitions of ω_ℓ, ω_h, and ϵ. We consider various choices for ω_ℓ and ω_h with $\epsilon = 0.1$:

- *$\omega_l = \sqrt{10}$ and $\omega_h = 100$–then equation (9.4.16) predicts $S_{max} \geq 432$. In this case, ω_h is much larger than the unstable pole, thus the large value for the bound results from $\omega_l = \sqrt{10}$, too close to the NMP zero.*

- *When $\omega_l = 1$ and $\omega_h = 100$, we have from (9.4.16) that $S_{max} \geq 16.7$, which is significantly lower than the previous case (although still very large), because now ω_l is much smaller than the NMP zero.*

- *If $\omega_l = 1$ and $\omega_h = \sqrt{20}$, we obtain from (9.5.5) that $T_{max} \geq 3171$, which is due to the fact that ω_h is too close to the unstable pole.*

- *If $\omega_l = 1$ and $\omega_h = 3$, we obtain from (9.5.5) that $T_{max} \geq 7.2 \times 10^5$. This huge lower bound originates from two facts: first, ω_h is lower than the unstable pole; second, ω_l and ω_h are very close.*

We can see from the sensitivity arguments just presented that the control of an inverted pendulum when only the cart position is measured is extremely difficult (and probably nearly impossible in practice). An alternative justification for this claim based on root-locus arguments, is given on the website.

The website also derives the following controller:

$$C(s) = \frac{p_3 s^3 + p_2 s^2 + p_1 s + p_o}{s^s + \ell_2 s^2 + \ell_1 s + \ell_o} \qquad (9.6.6)$$

where

$$
\begin{aligned}
p_o &= & -74.3 \\
p_1 &= & -472.8 \\
p_2 &= & 7213.0 \\
p_3 &= & 1690.5 \\
\ell_o &= & -8278.9 \\
\ell_1 &= & -2682.3 \\
\ell_2 &= & 41.5
\end{aligned}
$$

This controller results in the sensitivity and complementary sensitivity plots given in Figure 9.5.

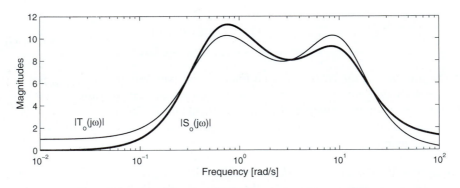

Figure 9.5. Sensitivities for the inverted pendulum

Of course, the plots in Figure 9.5 show very large peaks, indicating that the design is very sensitive to noise, disturbance, model error, and so forth. It can easily be *checked that, for this particular design (using $\epsilon = 0.1$), where $\omega_l \approx 0.03$ and $\omega_h \approx 60$, equation (9.4.16) predicts $S_{max} \geq 6.34$ and equation (9.5.5) predicts $T_{max} \geq 7.19$. Inspection of Figure 9.5 shows that these lower bounds are indeed exceeded by this particular design. The key point here is that these kinds of difficulties are inherent in the problem and cannot be circumvented by any control-system design. (The reader is invited to look at the associated step responses on the website and to try his or her own hand at designing a better SISO controller. The reader can perhaps appreciate the difficulty of the problem by trying to balance a broom on one's finger when one's eyes are closed!)*

□□□

Remark 9.3. *The reader might speculate that the difficulty encountered in the above example is easily resolved by building some virtual sensor for the missing angle measurements. This may well be a useful strategy to help in the design of the control law. However, at the end of the day, the resultant controller will be driven only by u and y–i.e., it reduces to a SISO feedback loop. Thus identical sensitivity issues apply!*

The reader might judge that these arguments contradict the spectacular success we had in subsection §8.8.1, where we suggested the use of a vertical sensor to estimate the exit thickness in a rolling mill. The essential difference is that, in the latter case, we used an additional measurement (force) to infer thickness–i.e., we changed the basic architecture of the control problem. This is not achieved in the case of the inverted pendulum when we try to estimate θ by using y, because no new measurements are involved.

□□□

9.7   Summary

- One class of design constraints covers those which hold at a particular frequency.

- Thus we can view the law $S(j\omega) = 1 - T(j\omega)$ on a frequency-by-frequency basis. It states that no single frequency can be removed from both the sensitivity, $S(j\omega)$, and the complementary sensitivity, $T(j\omega)$.

- There is, however, an additional class of design considerations, which results from so-called frequency-domain integral constraints; see Table 9.1 on page 261.

- This chapter explores the origin and nature of these integral constraints and derives their implications for control-system performance:

  - The constraints are a direct consequence of the requirement that all sensitivity functions must be stable.

○ Mathematically, this means that the sensitivities are required to be analytic in the right-half complex plane.

○ Results from analytic function theory then show that this requirement necessarily implies that weighted integrals of the frequency response necessarily evaluate to a constant.

○ Hence, if one designs a controller to have low sensitivity in a particular frequency range, then the sensitivity will necessarily increase at other frequencies–a consequence of the weighted integrals always being a constant; this phenomenon has also been called the *water-bed effect*. (Pushing down on the water bed in one area, raises it somewhere else.)

• These trade-offs show that systems become increasingly difficult to control as

○ unstable zeros become slower;

○ unstable poles become faster; or

○ time delays get bigger.

9.8   Further Reading

Bode's original work

Bode, H. (1945). *Network analysis and feedback amplifier design.* Van Nostrand, New York.

Frequency-domain constraints

Freudenberg, J.S. and Looze, D.P. (1985). Right-half plane poles and zeros and design tradeoffs in feedback systems. *IEEE Transactions on Automatic Control,* 30(6):555-565.

Freudenberg, J.S. and Looze, D.P. (1988). *Frequency Domain Properties of Scalar and Multivariable Feedback Systems.* Lecture Notes in Control and Information Sciences, Vol. 104, Springer-Verlag, New York.

Seron, M.M., Braslavsky, J.H., and Goodwin, G.C. (1997). *Fundamental limitations in filtering and control.* Springer-Verlag.

Zames, G. and Francis, B.A. (1983). Feedback, minimax sensitivity, and optimal robustness. *IEEE Transactions on Automatic Control,* 28(5):585-601.

Complex variable theory

Churchill, R.V. and Brown, J.W. (1990). *Complex Variables and Applications.* McGraw-Hill, New York, 5th edition.

Notation	Constraints	Interpretation
p_i RHP poles	$\displaystyle\int_0^\infty ln|S_o(j\omega)|d\omega = \pi \sum_{i=1}^{N} \Re\{p_i\}$	Areas of sensitivity below 1 (i.e. $ln|S_o| < 0$) must be matched by areas of sensitivity above 1.
c_i RHP zeros	$\displaystyle\int_0^\infty \frac{ln|T_o(j\omega)|}{\omega^2}d\omega = \pi \sum_{i=0}^{N} \frac{1}{c_i} - \frac{\pi}{2k_v}$	Areas of complementary sensitivity below 1 must be matched by areas of complementary sensitivity above 1.
$W(c_k,\omega)$ weighting function $B_p(c_k)$ Blaschke product	$\displaystyle 2\int_0^\infty ln|S_o(j\omega)|W(c_k,\omega)d\omega = -\pi ln|B_p(c_k)|$	The above sensitivity trade-off is focused on areas near RHP zeros.
$W(p_i,\omega)$ weighting function $B_z(p_i)$ Blaschke product	$\displaystyle 2\int_0^\infty ln|T_o(j\omega)|W(p_i,\omega)d\omega = -\pi ln|B_z(p_i)|$	The above complementary sensitivity trade-off is focused on areas near RHP poles.

Table 9.1. Integral constraints for SISO systems

Krzyz, J.G. (1971). *Problems in Complex Variable Theory.* Elsevier, New York.

Levinson, N. and Redheffer, R.M. (1970). *Complex Variables.* Holden-Day, San Francisco.

General design issues

Doyle, J.C., Francis, B.A., and Tannenbaum, A.R. (1992). *Feedback Control Theory.* Macmillan Publishing Company.

Skogestad, S. and Postlethwaite, I. (1996). *Multivariable Feedback Control: Analysis and Design.* Wiley, New York.

Horowitz, I. (1963). *Synthesis of Feedback Systems.* Academic Press.

Horowitz, I. and Shaked, U. (1975). Superiority of transfer function over state-variable methods in linear, time-invariant feedback systems design. *IEEE Transactions on Automatic Control,* 20:84-97.

9.9   Problems for the Reader

Problem 9.1. *Consider a feedback control loop where*

$$G_o(s)C(s) = \frac{9}{s(s+4)} \tag{9.9.1}$$

9.1.1 *Verify Lemma 9.1 on page 242 by directly computing the integral in (9.2.3).*

9.1.2 *Repeat for $G_oC(s) = \dfrac{17s + 100}{s(s+4)}$.*

Problem 9.2. *A plant model is given by*

$$G(s) = \frac{e^{-s}}{s+1} \tag{9.9.2}$$

Use a second-order Padé approximation for the delay. Then, using Lemma 9.5 and the result in Example 9.1, with $\epsilon = 0.1$ and $\omega_l = 3$ to derive a lower bound for the nominal sensitivity peak.

Problem 9.3. *The nominal model for a plant is given by*

$$G_o(s) = \frac{5(s-1)}{(s+1)(s-5)} \tag{9.9.3}$$

This plant has to be controlled in a feedback loop with one-degree-of-freedom. Assume exact inversion at $\omega = 0$ and step-like reference and disturbances.

9.3.1 *Assume that the loop must satisfy*

$$|S_o(j\omega)| < 0.2 \qquad for \ \ \omega \leq 2 \qquad |T_o(j\omega)| < 0.2 \qquad for \ \ \omega \geq 8 \tag{9.9.4}$$

*Using the results presented in this chapter (and the provided MATLAB routines **smax.m** and **tmax.m**), obtain the best estimate you can for a lower bound on the nominal sensitivity peak and the complementary sensitivity peak.*

9.3.2 *Why is the control of this nominal plant specially difficult? Discuss.*

Problem 9.4. *Assume that in a one-d.o.f. control loop for a plant with nominal model given by $G_o(s)$, unstable and nonminimum phase, a biproper controller is designed. Derive frequency-domain constraints for the nominal control sensitivity, S_{uo}.*

Chapter 10

ARCHITECTURAL ISSUES IN SISO CONTROL

10.1 Preview

The analysis in previous chapters has focused on basic feedback-loop properties and feedback-controller synthesis. In this chapter, we will extend the scope of the analysis to focus on further architectural issues that are aimed at achieving exact compensation of certain types of deterministic disturbances and exact tracking of particular reference signals. We will show that, when compared to relying solely on feedback, the use of feedforward and or cascade structures offers advantages in many cases. We also generalize the idea of integral action to more general classes of disturbance compensation.

10.2 Models for Deterministic Disturbances and References

Throughout this chapter, we will need to refer to certain classes of disturbances and reference signals. The particular signals of interest here are those that can be described as the output of a linear dynamic system having zero input and certain specific initial conditions. The simplest example of such a signal is a constant, which can be described by the model:

$$\dot{x}_d = 0 \; ; \quad x_d(0) \quad \text{given} \tag{10.2.1}$$

The generalization of this idea includes any disturbance that can be described by a differential equation of the form:

$$\frac{d^q d_g(t)}{dt^q} + \sum_{i=0}^{q-1} \gamma_i \frac{d^i d_g(t)}{dt^i} = 0 \tag{10.2.2}$$

The above model leads to the following expression for the Laplace transform of the disturbance:

$$D_g(s) = \frac{N_d(s)x_d(0)}{\Gamma_d(s)} \qquad (10.2.3)$$

where $\Gamma_d(s)$ is the *disturbance-generating polynomial* defined by

$$\Gamma_d(s) \triangleq s^q + \sum_{i=0}^{q-1} \gamma_i s^i \qquad (10.2.4)$$

This polynomial corresponds to the denominator of the Laplace transform of $d_g(t)$, and it can be factored according to the modes present in the disturbance. We illustrate by a simple example.

Example 10.1. *A disturbance takes the following form*

$$d_g(t) = K_1 + K_2 \sin(3t + K_3) \qquad (10.2.5)$$

where K_1, K_2, and K_3 are constants. Then the generating polynomial is given by

$$\Gamma_d(s) = s(s^2 + 9) \qquad (10.2.6)$$

Note that K_1, K_2, and K_3 are related to the initial state, $x_d(0)$, in the state space model.

We also characterize reference signals in a similar way to that used for disturbances, i.e., we assume that the reference originates as the output of a dynamic system having zero input and certain specific initial conditions. This means that the reference satisfies a homogeneous differential equation of the form

$$\frac{d^q r(t)}{dt^q} + \sum_{i=0}^{q-1} \gamma_i \frac{d^i r(t)}{dt^i} = 0 \qquad (10.2.7)$$

The above model leads to:

$$R(s) = \frac{N_r(s)x_r(0)}{\Gamma_r(s)} \qquad (10.2.8)$$

where $\Gamma_r(s)$ is the *reference-generating polynomial* defined by

$$\Gamma_r(s) \triangleq s^q + \sum_{i=0}^{q-1} \gamma_i s^i \qquad (10.2.9)$$

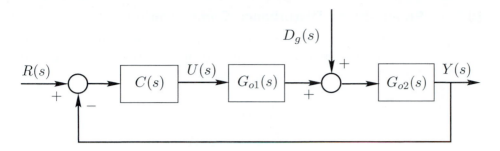

Figure 10.1. Control loop with a generalized disturbance

This polynomial corresponds to the denominator of the Laplace transform of $r(t)$, and it can be factored according to the modes present in the reference. For future use, we factor $\Gamma_r(s)$ as

$$\Gamma_r(s) = \Gamma_r^o(s)\Gamma_r^\infty(s) \tag{10.2.10}$$

where Γ_r^o is a polynomial with roots strictly inside the LHP and $\Gamma_r^\infty(s)$ is a poly-nomial with roots in the closed RHP, located at $s = \epsilon_i$, for $i = 1, 2, \dots, n_e$. For simplicity, we assume in the sequel that these roots are distinct.

10.3   Internal Model Principle for Disturbances

10.3.1   Disturbance Entry Points

In previous chapters, we distinguished between input and output disturbances so as to highlight the issues arising from the point at which disturbances enter the loop. Here, we will adopt a unified description encompassing both input and out-put disturbances. Specifically, for a nominal model $G_o(s)$ with input $U(s)$ and output $Y(s)$, we will assume that the disturbance $D_g(s)$ acts on the plant at some intermediate point, i.e., we model the output as follows:

$$Y(s) = G_{o2}(s)(G_{o1}(s)U(s) + D_g(s)) \qquad \text{where} \qquad G_o(s) = G_{o1}(s)G_{o2}(s) \tag{10.3.1}$$

This set-up is depicted in Figure 10.1

The input-disturbance case is included by setting $G_{o1}(s) = 1$ and $G_o(s) = G_{o2}(s)$. The output-disturbance case is covered by considering $G_{o2}(s) = 1$ and $G_o(s) = G_{o1}(s)$.

10.3.2 Steady-State Disturbance Compensation

We note that, for the generalized disturbance description given above, and assuming closed-loop stability, the nominal model output and controller output are given respectively by

$$Y(s) = S_o(s)G_{o2}(s)D_g(s) \tag{10.3.2}$$

$$U(s) = -S_{uo}G_{o2}(s)D_g(s) = \frac{T_o(s)}{G_{o1}(s)}D_g(s) \tag{10.3.3}$$

From equation (10.3.2) we observe that the effect of the disturbance on the model output vanishes asymptotically when at least one of the following conditions is satisfied:

- $d_g(t) \to 0$ as $t \to \infty$

- The polynomial $\Gamma_d(s)$ is a factor in the numerator of $S_o(s)G_{o2}(s)$

The first case is not interesting from the viewpoint of steady-state performance, although it will be considered in the context of transient performance.

When the roots of $\Gamma_d(s)$ lie in the closed RHP, then only the second condition guarantees zero steady-state error due to the disturbance. This requirement can be met if the factors of $\Gamma_d(s)$ are in the denominator of the controller $C(s)$ and/or in the denominator of $G_{o1}(s)$. Note that, for an input disturbance, i.e. $G_{o1}(s) = 1$, the only valid alternative is that $\Gamma_d(s)$ be a factor in the denominator of $C(s)$.

We observe that, when $\Gamma_d(s)$ is a factor in the denominator of $C(s)$, then the unstable components of $d_g(t)$ will be asymptotically compensated in both the nominal feedback loop and in the true loop, provided that the nominal loop is robustly stable. The reason for this strong property is that $S_o(s)$ will vanish at the unstable roots of $\Gamma_d(s)$, and this, combined with the robust stability property, ensures that $S(s)$ also vanishes at those roots.

We may thus conclude the following:

Steady-state disturbance compensation requires that the generating polynomial be included as part of the controller denominator. This is known as the *Internal Model Principle* (IMP). The roots of the generating polynomial, in particular the unstable ones, impose the same performance trade-offs on the closed loop as if those poles were part of the plant.

When the IMP is used, $T_o(s)$, and $T(s)$ are equal to one at the roots of $\Gamma_d(s)$. Thus, from equation (10.3.3), we see that the controller output $u(t)$ will, in general, contain the disturbances modes.

The IMP can be explicitly enforced during the synthesis stage. We next examine one way in which this can be achieved.

10.3.3   Pole Assignment

Recall the theory presented in section §7.2. For a nominal plant model and controller as given in (7.2.2), i.e., $G_o(s) = \frac{B_o(s)}{A_o(s)}$ and $C(s) = \frac{P(s)}{L(s)}$, we have that the IMP can be enforced if we factor $L(s)$ as $\Gamma_d(s)\bar{L}(s)$. This leads to a pole-assignment equation of the form

$$A_o(s)\Gamma_d(s)\bar{L}(s) + B_o(s)P(s) = A_{cl}(s) \tag{10.3.4}$$

where $L(s) = \Gamma_d(s)\bar{L}(s)$.

If the degree of $A_o(s)$ is n, then $A_{cl}(s)$ can be arbitrarily specified if and only if its degree is, at least, $2n - 1 + q$. Note that in subsection §7.2.2 this idea was introduced for the special case $\Gamma_d(s) = s$.

We illustrate the procedure by the following example.

Example 10.2. *Consider a nominal model $G_o(s) = \frac{3}{s+3}$ and an input disturbance $d_g(t) = K_1 + K_2\sin(2t + K_3)$. It is required to build a controller $C(s)$ such that the IMP is satisfied for this class of disturbances.*

We first note that $q = 3$, $\Gamma_d(s) = s(s^2 + 4)$, and $n = 1$. This means that $A_{cl}(s)$ should at least be of degree $n_c = 4$. Say we choose $A_{cl}(s) = (s^2 + 4s + 9)(s + 5)^2$. We then have that the controller should have the form

$$C(s) = \frac{\beta_3 s^3 + \beta_2 s^2 + \beta_1 s + \beta_0}{s(s^2 + 4)} \tag{10.3.5}$$

The corresponding pole-assignment equation becomes

$$s(s^2 + 4)(s + 3) + 3(\beta_3 s^3 + \beta_2 s^2 + \beta_1 s + \beta_0) = (s^2 + 4s + 9)(s + 5)^2 \tag{10.3.6}$$

*leading to $\beta_3 = \frac{14}{3}$, $\beta_2 = \frac{74}{3}$, $\beta_1 = \frac{190}{3}$, and $\beta_0 = 75$. (Use **paq.m**.)*
Equation (10.3.2), when applied to this example, leads to

$$Y(s) = \frac{3s(s^2 + 4)}{A_{cl}(s)}D_g(s) = \frac{3N_g(s)}{A_{cl}(s)} \tag{10.3.7}$$

where $N_g(s)$ is the numerator polynomial of $D_g(s)$. We have also used the fact that $G_{o2}(s) = G_o(s)$. Notice, however, the trade-off price that must be paid to achieve perfect cancellation of this particular disturbance: the disturbance-generating polynomial will necessarily appear as zeros of the sensitivity function and affect step-output disturbance response and complementary sensitivity accordingly.

An interesting observation is the following: Assume that, for this input disturbance, we had the following transfer functions for the nominal model and the controller.

$$G_o(s) = \frac{3}{s} \quad and \quad C(s) = \frac{\beta_3 s^3 + \beta_2 s^2 + \beta_1 s + \beta_0}{(s+3)(s^2+4)} \tag{10.3.8}$$

Note that the transfer function of the combined plant and controller, $G_o(s)C(s)$, has the same poles as above, but now the factor s appears in $G_o(s)$ rather than in $C(s)$.

The closed-loop polynomial, $A_{cl}(s)$, and the nominal sensitivity, $S_o(s)$, remain unchanged; however, now $G_{o2}(s) = \frac{3}{s}$, and thus

$$Y(s) = \frac{3(s+3)(s^2+4)}{A_{cl}(s)} D_g(s) = \frac{3(s+3)N_g(s)}{s A_{cl}(s)} \tag{10.3.9}$$

From here, it is evident that the constant mode of the disturbance remains in the output, leading to nonzero steady-state error. This is due to the fact that the IMP is satisfied for the disturbance mode represented by the factor $s^2 + 4$ but not for the one represented by the factor s.

□□□

10.3.4   Industrial Application: Roll-Eccentricity Compensation in Rolling Mills

A common technique used for gauge control in rolling mills (see §8.8.1) is to infer thickness from roll-force measurements. This is commonly called a BISRA gauge. However, these measurements are affected by roll eccentricity. We can explain why eccentricity affects the thickness by a physical argument, as follows: Normally, an increased force means the exit thickness has increased (thus pushing the rolls apart); however, if the rolls are eccentric, then, when the largest radius passes through the roll gap, the force increases, but the exit thickness actually decreases. Hence, a change in force is misinterpreted when eccentricity components are present.

A very common strategy for dealing with this problem is to model the eccentricity components as multiple sinusoids (ten sine waves per roll are typically used; with four rolls, this amounts to forty sinusoids). These sinusoids can be modeled as in (10.2.4), where

$$\Gamma_d(s) = \prod_{i=1}^{m}(s^2 + \omega_i^2) \tag{10.3.10}$$

The Internal Model Principle can then be used to remove the disturbance from the exit gauge. One such scheme is a patented technique known as AUSREC, which was developed, in collaboration with others, by one of the authors of this book.

The reader is referred to the Rolling-Mill example given on the book's website. There, it is shown how this form of controller is a very effective tool for dealing with the roll-eccentricity issue.

10.4   Internal Model Principle for Reference Tracking

For reference tracking, we consider the two-degree-of-freedom architecture shown in Figure 5.2 on page 124, with zero disturbances. Then tracking performance can be quantified through the following equations:

$$Y(s) = H(s)T_o(s)R(s) \tag{10.4.1}$$

$$E(s) = R(s) - Y(s) = (1 - H(s)T_o(s))R(s) \tag{10.4.2}$$

$$U(s) = H(s)S_{uo}(s)R(s) \tag{10.4.3}$$

The block with transfer function $H(s)$ (which must be stable, since it acts in open loop) is known as the *reference-feedforward transfer function*. This is usually referred to as the second-degree-of-freedom in control loops. We will now address the issues of steady-state and transient performance in this architecture.

If we are to use the Internal Model Principle for reference tracking, it suffices to set $H(s) = 1$ and then to include the reference-generating polynomial in the denominator of $C(s)G_o(s)$. This leads to $S_o(\epsilon_i) = 0$, where $\epsilon_i = 1, \ldots n_r$ are the poles of the reference-generating polynomial. This in turn implies that the achieved sensitivity, $S(\epsilon_i) = 0$ for $\epsilon_i = 1, \ldots n_r$. Hence, the achieved complementary sensitivity satisfies $T(\epsilon_i) = 1$, and, from (10.4.2), we have that, for $H(s) = 1$,

$$\lim_{t \to \infty} e(t) = 0 \tag{10.4.4}$$

We can summarize this as follows:

> To achieve robust tracking, the reference-generating polynomial must be in the denominator of the product $C(s)G_o(s)$, i.e., the Internal Model Principle also has to be satisfied for the reference. When the reference-generating polynomial and the disturbance-generating polynomial share some roots, then these common roots need be included only once in the denominator of $C(s)$ to simultaneously satisfy the IMP for both the reference and disturbance.

10.5   Feedforward

The use of the IMP, as outlined above, provides complete disturbance compensation and reference tracking in steady state; however, this leaves unanswered the issue of transient performance: how the system responds during the initial phase of the response following a change in the disturbance or reference signal.

We have seen that steady-state compensation holds independent of the precise plant dynamics, provided that the closed loop is stable; however, the transient response is a function of the system dynamics and is thus subject to the inherent performance trade-offs that we have studied elsewhere. In particular, the transient response is influenced by the zeros of the transfer function between the disturbance or reference injection point and the output, the poles of the transfer function between the output and the disturbance or reference injection point, and the location of the closed-loop poles.

The transient performance can be influenced in various ways. The most obvious way is to change the location of the closed-loop poles by changing the controller; however, in some cases, it may be possible to measure the disturbances or reference signals directly. Judicious use of these measurements in the control structure gives us additional mechanisms for affecting the transient performance.

We will now discuss reference feedforward and disturbance feedforward.

10.5.1   Reference Feedforward

We can use a two-degree-of-freedom architecture for reference tracking. We recall the model describing this architecture for reference signals given in equations (10.4.1) to (10.4.3). The essential idea of reference feedforward is to use $H(s)$ to invert $T_o(s)$ at certain key frequencies, so that $H(s)T_o(s) = 1$ at the poles of the reference model (i.e., at $\epsilon_i; i = 1, \ldots, n_e$). Note that, with this strategy, one can avoid using high-gain feedback to bring $T_o(\epsilon_i)$ to 1. This would have advantages with respect to robust stability. On the other hand, if we use $H(s)$ to achieve $H(\epsilon_i)T_o(\epsilon_i) = 1$, then the performance on the real plant will be sensitive to differences between the nominal sensitivity, $T_o(s)$, and the true or achieved sensitivity. This, in turn, implies sensitivity to differences between the true plant and the nominal model, $G_o(s)$.

Example 10.3. *Consider a plant having a nominal model given by*

$$G_o(s) = \frac{2}{s^2 + 3s + 2} \tag{10.5.1}$$

The control goal is to cause the plant output to follow, as closely as possible, a reference given by

$$r(t) = K_1 \sin(t) + K_2 + r_a(t) \tag{10.5.2}$$

where K_1 and K_2 are unknown constants and $r_a(t)$ is a signal with energy in the frequency band $[0,5][rad/s]$. It is also assumed that the presence of measurement noise prevents the closed-loop bandwidth from exceeding $3[rad/s]$.

It is required to design a control system to achieve zero steady-state control error and good dynamic tracking.

We first note that the reference has energy concentrated at $\omega = 1[rad/s]$ and $\omega = 0[rad/s]$. It also has energy distributed in the band $[0,5][rad/s]$. (Note that this lies outside the allowed closed-loop bandwidth, so a two-degree-of-freedom architecture will be needed.)

We also need to pay attention to relative-degree issues, because $C(s)$ and $H(s)$ must be proper.

The design requirements can be summarized as follows:

(i) The bandwidth of T_o must be at most $3[rad/s]$.

(ii) The bandwidth of HT_o should be at least $5[rad/s]$.

(iii) The feedback-controller transfer function $C(s)$ must have poles at $s = 0$ and at $s = \pm j[rad/s]$.

We first synthesize the feedback controller by using polynomial pole placement. We note that the closed-loop polynomial must then be chosen of degree equal to at least 6 to force poles at $0, \pm j1$ into $C(s)$. For simplicity, we choose to cancel the plant-model poles.

Thus, we choose

$$C(s) = \frac{P(s)}{L(s)} = \frac{(s^2 + 3s + 2)(\beta_2 s^2 + \beta_1 s + \beta_0)}{s(s^2 + 1)(s + \alpha)} \tag{10.5.3}$$

The closed-loop polynomial is chosen as $A_{cl}(s) = (s^2+3s+2)(s^2+3s+4)(s+1)^2$. This choice is made by trial-and-error, to limit the closed-loop bandwidth to $3[rad/s]$; this constraint forces slow poles into $A_{cl}(s)$. Note also that the factor $s^2 + 3s + 2$ has been introduced in $A_{cl}(s)$ to force the cancellation of open-loop poles.

After the simplifying of the factor s^2+3s+2, the pole-placement equation reduces to

$$s(s^2 + 1)(s + \alpha) + 2(\beta_2 s^2 + \beta_1 s + \beta_0) = (s^2 + 3s + 4)(s + 1)^2 \tag{10.5.4}$$

This leads to

$$\alpha = 5; \quad \beta_2 = 5; \quad \beta_1 = 3; \quad \beta_0 = 2 \tag{10.5.5}$$

from which we obtain

$$T_o(s) = \frac{10s^2 + 6s + 4}{(s^2 + 3s + 4)(s + 1)^2} \tag{10.5.6}$$

We next design the reference-feedforward block $H(s)$ to achieve an overall reference-tracking bandwidth of, at least, $5[rad/s]$. The relative degree of $T_o(s)$ is 2, so this is also the minimum relative degree for $H(s)T_o(s)$–otherwise $H(s)$ would be improper. Say we choose

$$H(s)T_o(s) = \frac{1}{(0.01s + 1)^2} \qquad (10.5.7)$$

which then exceeds the required $5[rad/s]$ bandwidth. This leads to

$$H(s) = \frac{(s^2 + 3s + 4)(s + 1)^2}{(10s^2 + 6s + 4)(0.01s + 1)^2} \qquad (10.5.8)$$

□□□

10.5.2   Disturbance Feedforward

We next show how feedforward ideas can be applied to disturbance rejection.

A structure for feedforward from a measurable disturbance is shown in Figure 10.2. Note that this is precisely the configuration used early in the book (§2.3, in the mould-level example) to compensate for changes in casting rate.

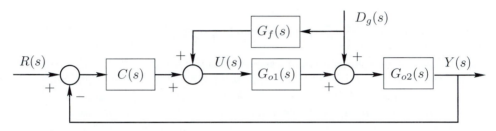

Figure 10.2. Disturbance-feedforward scheme

From Figure 10.2, we observe that the model output and the controller output response, both due to the disturbance, are given by:

$$Y_d(s) = S_o(s)G_{o2}(s)(1 + G_{o1}G_f(s))D_g(s) \qquad (10.5.9)$$
$$U_d(s) = -S_{uo}(s)(G_{o2}(s) + G_f(s))D_g(s) \qquad (10.5.10)$$

The proposed architecture has the following features:

(i) The feedforward-block transfer function $G_f(s)$ must be stable and proper, because it acts in open loop.

(ii) Equation (10.5.9) indicates that, ideally, the feedforward block should invert part of the nominal model–that is,

$$G_f(s) = -[G_{o1}(s)]^{-1} \qquad (10.5.11)$$

(iii) Usually $G_{o1}(s)$ will have a low-pass characteristic, so we should expect $G_f(s)$ to have a high-pass characteristic.

The use of disturbance feedforward is illustrated in the following example.

Example 10.4. *Consider a plant having a nominal model given by*

$$G_o(s) = \frac{e^{-s}}{2s^2 + 3s + 1} \qquad G_{o1}(s) = \frac{1}{s+1} \qquad G_{o2}(s) = \frac{e^{-s}}{2s+1} \qquad (10.5.12)$$

We will not discuss the feedback controller. (It actually uses a Smith-predictor architecture–see section §7.4.)

We assume that the disturbance $d_g(t)$ consists of infrequently occurring step changes. We choose $-G_f(s)$ as an approximation to the inverse of $G_{o1}(s)$–that is,

$$G_f(s) = -K \frac{s+1}{\beta s + 1} \qquad (10.5.13)$$

where β allows a trade-off to be made between the effectiveness of the feedforward and the size of the control effort. Note that K takes the nominal value 1.

A SIMULINK simulation was carried out for a loop with unit step reference at $t = 1$ and unit step disturbance at $t = 5$. Typical results are shown in Figure 10.3 for $\beta = 0.2$ and $K = 0$ (no feedforward) and $K = 1$ (full feedforward).

It can be seen from the figure that feedforward clearly improves the transient performance.

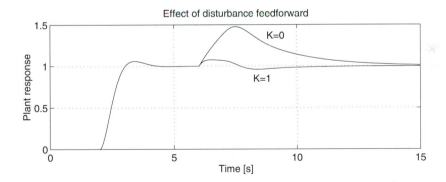

Figure 10.3. Control loop with ($K = 1$) and without ($K = 0$) disturbance feedforward

An interesting observation is that, in many cases, disturbance feedforward is beneficial even if only a static $G_f(s)$ is used–i.e., exact inversion of $G_{o1}(s)$ is only achieved at $\omega = 0$. The reader is encouraged to try this and other choices for $G_f(s)$

*and to evaluate the effects of those choices on, for instance, $u(t)$. The SIMULINK scheme for this example corresponds to file **distff.mdl**.*

Ideal disturbance feedforward requires inversion, so the main difficulties arise when $G_{o1}(s)$ has delays or NMP zeros, because this cannot be inverted. On the other hand, disturbance feedforward is especially helpful in certain plants, like the one in this example, where $G_{o2}(s)$ includes a significant time delay while $G_{o1}(s)$ does not. This occurs frequently in the process industries.

Disturbance feedforward can also be used in nonlinear plants, as is illustrated in the following example.

Example 10.5. *Consider a plant where the relationship between the input $u(t)$, disturbance $d_g(t)$, and output $y(t)$ is given by the model*

$$y(t) = G_{o2}\langle d_g(t) + w(t)\rangle \qquad with \quad w(t) \triangleq G_{o1}\langle u(t)\rangle \qquad (10.5.14)$$

where $G_{o2}\langle \circ \rangle$ is a linear time-invariant operator and $G_{o1}\langle \circ \rangle$ is a nonlinear operator such that

$$\frac{dw(t)}{dt} + (2 + 0.2u(t))w(t) = 2u(t) \qquad (10.5.15)$$

Assume that we want to use the disturbance-feedforward architecture shown in Figure 10.2. Determine a suitable G_f.

Solution

The ideal choice for G_f is a nonlinear operator such that

$$G_{o1}\langle G_f\langle d_g(t)\rangle\rangle = -d_g(t) \qquad i.e., \quad G_f\langle d_g(t)\rangle = G_{o1}^{-1}\langle -d_g(t)\rangle \qquad (10.5.16)$$

Thus

$$G_f\langle d_g(t)\rangle = G_{o1}^{-1}\langle -d_g(t)\rangle = \left[\frac{d[d_g(t)]}{dt} + 2d_g(t)\right]\frac{1}{-2 + 0.2d_g(t)} \qquad (10.5.17)$$

However, this operator cannot be implemented, because pure derivatives are required. To remove that constraint, a fast transient is introduced,

$$\tau\frac{dG_f\langle d_g(t)\rangle}{dt} + G_f\langle d_g(t)\rangle = \left[\frac{d[d_g(t)]}{dt} + 2d_g(t)\right]\frac{1}{-2 + 0.2d_g(t)} \qquad (10.5.18)$$

where $\tau \ll 1$.

The operator defined in equation (10.5.18) is used to compute the disturbance-feedforward error, $e_f(t)$, defined as

$$e_f(t) \stackrel{\triangle}{=} d_g(t) + w(t) = d_g(t) + G_{o1} \otimes G_f \langle d_g(t) \rangle \qquad (10.5.19)$$

This example illustrates how disturbance feedforward can be used in connection with nonlinear models; however, caution must be exercised, because stability of the feedforward mechanism depends on the magnitude of the signals involved. In this example, if $d_g(t)$ gets close to 10, saturation might occur, because the term $-2 + 0.2d_g(t)$ grows unboundedly.

Another issue of relevance in this architecture is the fact that disturbance measurement can include noise that would have a deleterious effect on the loop performance. This is illustrated in the following example.

Example 10.6. *Consider a plant with a generalized disturbance (as in Figure 10.1 on page 267), where*

$$G_{o1}(s) = \frac{1}{s+1}; \quad G_{o2}(s) = e^{-2s}; \quad D_g(s) = \frac{24\alpha}{s(s+3)(s+8)} \qquad (10.5.20)$$

where α is an unknown constant.

10.6.1 Design disturbance feedforward, assuming that the bandwidth of the complementary sensitivity cannot exceed 3[rad/s].

10.6.2 Assume now that the closed-loop bandwidth is unconstrained. If one uses only one-d.o.f., would it be possible to obtain similar performance to that obtained through the above feedforward design?

10.6.3 Assume that measurement noise in the band [2, 20][rad/s] is injected when measuring the disturbance. Discuss the effect of this noise on the plant output.

Solution

10.6.1 The disturbance-feedforward block $G_f(s)$ should be an approximation to the negative of the inverse of $G_{o1}(s)$. The disturbance has significant energy in the frequency band [0, 8][rad/s], so we need to have a good inverse in that band. Say we choose

$$G_f(s) = -\frac{10(s+1)}{s+10} \qquad (10.5.21)$$

We will not describe the design of the feedback controller $C(s)$, because it is not relevant to the current discussion. (Actually, the simulations use a Smith controller structure–see section §7.4.)

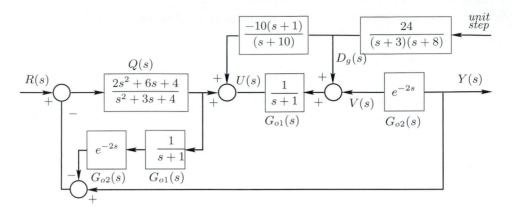

Figure 10.4. First- and third-d.o.f. design

The performance of the loop is shown in Figure 10.5. In this figure, the plant output is shown for a unit step reference at $t = 1[s]$ and a disturbance (with $\alpha = 1$) that starts at $t = 10[s]$.

10.6.2 *No. Because of the plant delay, the first two seconds of any disturbance change will appear without attenuation at the plant output. The reader is invited to confirm this by deleting the feedforward block in the SIMULINK diagram (in file **dff3.mdl**), corresponding to Figure 10.4.*

10.6.3 *The feedforward block with transfer function $G_f(s)$ has a high-pass characteristic, with a gain ranging from $10[dB]$ at $\omega = 3[rad/s]$ up to almost $19[dB]$ at $\omega = 20[rad/s]$. This will significantly amplify the disturbance-measurement noise.*

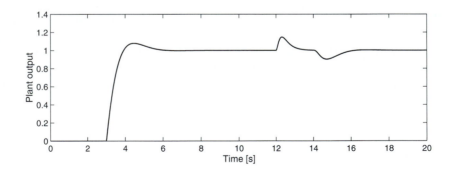

Figure 10.5. First- and third-d.o.f. performance

Remark 10.1. *Disturbance feedforward is sometimes referred to as the third-degree-of-freedom in a control loop.*

10.6   Industrial Applications of Feedforward Control

Feedforward control is generally agreed to be the single most useful concept in practical control-system design, beyond the use of elementary feedback ideas. Clearly, if one can measure up-stream disturbances, then, by feeding these forward, one can take anticipatory control action that preempts the disturbance affecting the process. The power of this idea has been illustrated in the classroom example given in Figure 10.3. A word of warning is that feedforward depends on the use of open-loop inverses and hence is susceptible to the impact of model errors. Thus, one usually needs to supplement feedforward control by some form of feedback control, so as to correct any miscalculation involved in the anticipatory control action inherent in feedforward.

Another use of feedforward control is to enlist other control inputs to help one achieve desirable control performance. Indeed, this is the first step towards multi-input, multi-output control (a topic that will be addressed in full in Parts VI, VII, and VIII). As an illustration of the power of feedforward control, we present the following industrial case study.

10.6.1   Hold-Up Effect in Reversing Mill, Revisited

Consider again the Rolling Mill problem discussed in section §8.7. There, we saw that the presence of imaginary-axis zeros was a fundamental limitation impeding the achievement of a rapid response between unloaded roll-gap position and exit thickness. We called this the *hold-up* effect. The physical cause of the problem is thickness-tension interactions. To verify this claim, the rolling-mill model was changed so that the tension remained constant. This is physically impossible to achieve, but it can be done in simulation to test the hypothesis. The hold-up effect disappears when there are no tension interactions. Also, as was argued in Chapter 8, the difficulty cannot be overcome without a radical change in the architecture of the control system, because the defect holds true for *any* single-input single-output controller, no matter how designed! However, if we go beyond the single-input single-output architecture, then one might be able to make progress. A clue to how this might be achieved is provided by the fact that there are actually three inputs available (roll gap, coiler current, and uncoiler current). Moreover, the imaginary-axis zeros appear only between the unloaded roll gap and the exit thickness. Hence, it seems feasible that one might be able to utilize the other inputs in some kind of feedforward arrangement to avoid the *hold-up* effect.

Motivated by this idea, we will describe a feedforward and feedback control scheme that is aimed at minimizing tension fluctuations.

To formulate the idea, we note that there is conservation of mass across the roll gap. This implies that

$$h_e(t)v_e(t) = h_i(t)v_i(t) \tag{10.6.1}$$

where $h_e(t)$, $h_i(t)$, $v_e(t)$ and $v_i(t)$ are the exit thickness, input thickness, exit velocity, and input velocity respectively. We note that the exit velocity is held (roughly) constant by another control system acting on the rolls. Hence, (10.6.1) can be expressed in incremental form as

$$\left(h_e^o + \Delta h_e(t)\right)v_e^o = \left(h_i^o + \Delta h_i(t)\right)\left(v_i^o + \Delta v_i(t)\right) \tag{10.6.2}$$

where h_e^o, v_e^o, h_i^o, and v_i^o are the nominal values. Also,

$$h_e^o v_e^o = h_i^o v_i^o \tag{10.6.3}$$

It is also known that, in the absence of tension fluctuations, the exit thickness is related to the roll-gap position by a static relationship of the form

$$\Delta h_e(t) = c_1\Delta\sigma(t) + c_2\Delta h_i(t) \tag{10.6.4}$$

where c_1, c_2 are constants related to the stiffness of the mill and $\Delta\sigma(t)$ is the change in the gap between the rolls without load.

Substituting (10.6.4), (10.6.3) into (10.6.2) and ignoring second-order terms gives

$$\Delta v_i(t) = \frac{1}{h_i^o}\left[v_e^o\Delta h_e(t) - v_i^o\Delta h_i(t)\right] \tag{10.6.5}$$

$$= \frac{1}{h_i^o}\left[c_1 v_0^o\Delta\sigma(t) + c_2 v_e^o\Delta h_i(t) - v_i^o\Delta h_i(t)\right]$$

Analysis of the physics of the problem indicates that tension fluctuations occur because the uncoiler rotational speed does not keep up with changes in the input velocity. This is avoided if we can make the percentage change in the roll angular velocity the same as the percentage change in input velocity–i.e., we require

$$\frac{\Delta v_i(t)}{v_i^o} = \frac{\Delta\omega_u(t)}{\omega_u^o} \tag{10.6.6}$$

where ω_u denotes the uncoiler angular velocity. Hence, from (10.6.5), we require

$$\Delta\omega_u = \frac{w_u^o}{v_i^o h_i^o}\left[c_1 v_0^o\Delta\sigma(t) + c_2 v_e^o\Delta h_i 8t) - v_i^o\Delta h_i(t)\right] \tag{10.6.7}$$

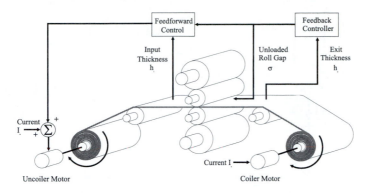

Figure 10.6. Feedforward controller for reversing mill

Now, a simple model for the uncoiler dynamics is

$$J_u \frac{d\omega_u(t)}{dt} = K_m i_u(t) \tag{10.6.8}$$

where $i_u(t)$ is the uncoiler current, J_u is the uncoiler inertia, and K_m is the motor-torque constant.

Substituting (10.6.8) into (10.6.7), we finally see that tension fluctuations would be avoided by choosing

$$i_u(t) = \frac{J_u \omega_u^o}{v_i^o h_i^o K_m} \left[c_1 v_0^o \frac{d\sigma(t)}{dt} + c_2 v_0^o \frac{dh_i(t)}{dt} - v_i^o \frac{dh_i(t)}{dt} \right] \tag{10.6.9}$$

Equation (10.6.9) is a feedforward signal linking (the derivatives of) the unloaded roll-gap position, $\sigma(t)$, and the input thickness, $h_i(t)$, to the uncoiler current.

This design has been tested on a practical mill and found to achieve a significant reduction in the hold-up effect. The core idea is now part of several commercially available thickness-control systems. The idea is shown schematically in Figure 10.6.

10.7   Cascade Control

Next, we turn to an alternative architecture for dealing with disturbances. The core idea is to feed back intermediate variables that lie between the disturbance injection point and the output. This gives rise to so-called *cascade control*.

A motivating example is shown in Figure 10.7. In this example, we assume that we want to control a variable $y(t)$ by manipulating the flow rate $q(t)$. The simplest architecture for achieving this is the one shown in part a) of Figure 10.7. Note that the controller output commands the valve opening; however, in this case, changes

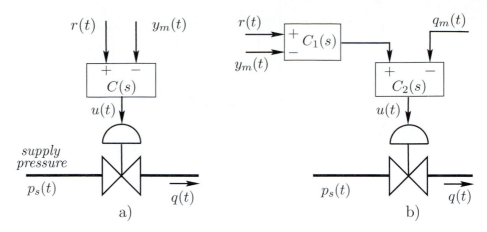

Figure 10.7. Example of application of cascade control

in the supply pressure $p_s(t)$ will yield different flow rates for the same value of $u(t)$ and thus affect the control goal.

An alternate cascade architecture is shown in part b) of Figure 10.7. A second loop has been introduced to control the flow rate $q(t)$. This loop requires the measurement of $q(t)$, denoted by $q_m(t)$ in the figure. Note that the first controller output provides the reference for the second loop.

The generalization of this idea has the structure shown in Figure 10.8.

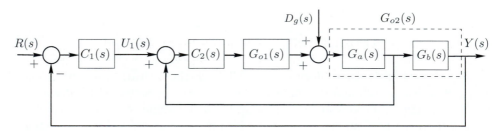

Figure 10.8. Cascade-control structure

In this architecture, there are two control loops, a primary loop with a primary controller, $C_1(s)$, and a secondary loop with a secondary controller, $C_2(s)$. The secondary controller is designed to attenuate the effect of the disturbance before it significantly affects the plant output $y(t)$.

The main benefits of cascade control are obtained under the following circumstances:

(i) when $G_a(s)$ contains significant nonlinearities that limit the loop performance;

or

(ii) when $G_b(s)$ limits the bandwidth in a basic control architecture.

The second scenario above occurs when $G_b(s)$ has NMP zeros and/or pure time delays, as shown in chapter 8. Then, the advantages of the secondary loop become apparent. These advantages can be quantified by noting, from Figure 10.8, that $C_1(s)$ must be designed to control an equivalent plant with input $U_1(s)$, output $Y(s)$, and disturbance $D_g(s)$:

$$Y(s) = G_{o2}(s)S_{o2}(s)D_g(s) + C_2(s)G_o(s)S_{o2}(s)U_1(s); \qquad G_o(s) = G_{o1}(s)G_{o2}(s) \tag{10.7.1}$$

where $S_{o2}(s)$ is the sensitivity function for the secondary loop. Note also that, if we denote by $T_{o2}(s)$ the complementary sensitivity in the secondary loop, then equation (10.7.1) can be rewritten as

$$Y(s) = G_{o2}(s)S_{o2}(s)D_g(s) + G_b(s)T_{o2}(s)U_1(s) \tag{10.7.2}$$

We can then compare this with the model for the original plant, as it appears in the loop shown in Figure 10.1 on page 267, where

$$Y(s) = G_{o2}(s)D_g(s) + G_o(s)U(s) \tag{10.7.3}$$

From this comparison, we see that, in the cascade architecture, the disturbance which $C_1(s)$ has to deal with is *precompensated* by the secondary-loop sensitivity. Thus, a better transient in the disturbance rejection might be expected.

Once the secondary controller $C_2(s)$ has been designed, the primary controller $C_1(s)$ can be designed so as to deal with the equivalent plant with transfer function

$$G_{oeq} \triangleq G_b(s)T_{o2}(s) \tag{10.7.4}$$

The benefits of this architecture are significant in many applications. An example is next presented to illustrate these benefits quantitatively.

Example 10.7. *Consider a plant having the same nominal model as in example 10.4 and assume that the measurement for the secondary loop is the input to $G_{o2}(s)$; then*

$$G_{o1}(s) = \frac{1}{s+1}; \qquad G_{o2}(s) = \frac{e^{-s}}{2s+1}; \qquad G_a(s) = 1; \qquad G_b(s) = G_{o2}(s) = \frac{e^{-s}}{2s+1} \tag{10.7.5}$$

We first choose the secondary controller to be a PI controller for which

$$C_2(s) = \frac{8(s+1)}{s} \qquad (10.7.6)$$

This controller has been chosen to obtain satisfactory disturbance compensation in the secondary loop. The complementary sensitivity in the secondary loop is then given by

$$T_{o2}(s) = \frac{8}{s+8} \qquad (10.7.7)$$

and the primary controller sees an equivalent plant with transfer function

$$G_{oeq}(s) = \frac{8e^{-s}}{2s^2 + 17s + 8} \qquad (10.7.8)$$

The primary controller can now be designed. Here again, we use a Smith-predictor controller–see section §7.4.

The disturbance-compensation performance is evaluated for a unit step disturbance. The results are shown in Figure 10.9. A unit step reference was applied at $t = 1$. Then, at $t = 4$, a unit step disturbance occurs. We can now compare these results with those shown in Figure 10.3 on page 275. Cascade control is clearly better than the basic one-degree-of-freedom control configuration and is comparable with that obtained from feedforward.

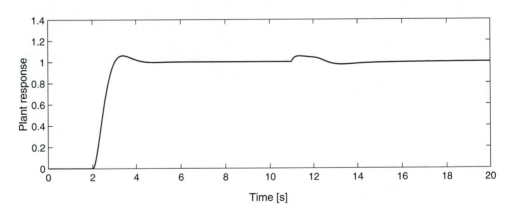

Figure 10.9. Disturbance rejection with a cascade-control loop

The main features of cascade control are the following:

(i) Cascade control is a feedback strategy.

(ii) A second measurement of a process variable is required; however, the disturbance itself does not need to be measured. Indeed, the secondary loop can be interpreted as having an observer to estimate the disturbance.

(iii) Measurement noise in the secondary loop must be considered in the design, because it can limit the achievable bandwidth in this loop.

(iv) Although cascade control (like feedforward) requires inversion, it can be made less sensitive to modeling errors by using the advantages of feedback.

$$\square\square\square$$

Remark 10.2. *Many practical control problems depend upon regulating a flow variable by using a valve. One solution is to assume knowledge of the valve characteristics and simply use the control law to open the valve to some position (which, via the valve characteristics, leads to the desired flow). However, this solution will be significantly affected by valve imperfections (stiction, smooth nonlinearities, etc.) and pressure changes (in the fluid whose flow is being regulated). This suggests a hierarchical control architecture in which the principal control law calls for a desired flow and then a secondary controller achieves this flow by a secondary loop in which the real flow is measured and fedback. (Compare the two architectures given in Figure 10.7). The cascade-controller architecture is usually preferable to the one-degree-of-freedom architecture. Of course, the negative side of the solution is that one needs to measure the actual flow resulting from the control action. In some applications, this cannot be easily achieved. Recall, for example, the Mould-Level Controller discussed in section §8.3.2, where it is not really feasible to measure the flow of steel.*

$$\square\square\square$$

Remark 10.3. *Another way of thinking about cascade control is that one uses fast acting feedback control loops to turn certain process outputs (or process variables) into "pseudo inputs" (or "pseudo manipulated variables"). For example, in the scenario discussed in Remark 10.2, the material flow becomes the pseudo input variable. The reason for doing this is that, it is frequently the case that the control of the overall process may be more readily formulated and solved in terms of the pseudo inputs than when the original inputs are utilized.*

$$\square\square\square$$

10.8   Summary

- This chapter focuses the discussion of the previous chapter on a number of special topics with high application value:

  o internal disturbance models–compensation for classes of references and disturbances

  o feedforward

  o cascade control

  o two-degree-of-freedom architectures

- Signal models

  o Certain classes of reference or disturbance signals can be modeled explicitly by their Laplace transforms:

Signal Type	Transform
Step	$\dfrac{1}{s}$
Ramp	$\dfrac{a_1 s + 1}{s^2}$
Parabola	$\dfrac{a_2 s^2 + a_1 s + 1}{s^3}$
Sinusoid	$\dfrac{a_1 s + 1}{s^2 + \omega^2}$

  o Such references (disturbances) can be asymptotically tracked (rejected) if and only if the closed loop contains the respective transform in the sensitivity S_0.

  o This is equivalent to having imagined the transforms as being (unstable) poles of the open-loop and stabilizing them with the controller.

  o In summary, the Internal Model Principle augments poles to the open-loop gain function $G_o(s)C(s)$. However, this implies that the same design trade-offs apply as if these poles had been in the plant to begin with.

  o Thus, internal model control is not cost-free but must be considered as part of the design trade-off considerations.

- Reference feedforward

  o A simple but very effective technique for improving responses to set-point changes is prefiltering the set-point (Figure 10.10).

  o This is the so-called two-degree-of-freedom (two-d.o.f.) architecture, because the prefilter H provides an additional design freedom. If, for example, there is significant measurement noise, then the loop must not be designed with too high a bandwidth. In this situation, reference tracking can be sped up with the prefilter.

○ Also, if the reference contains high-frequency components (such as step changes, for example), which anyhow are beyond the bandwidth of the loop, then one might as well filter them, so as not to excite uncertainties and actuators with them unnecessarily.

○ It is important to note, however, that design inadequacies in the loop (such as poor stability or performance) cannot be compensated by the prefilter. This is due to the fact that the prefilter does not affect the loop dynamics excited by disturbances.

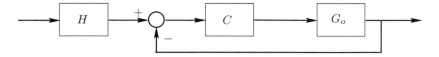

Figure 10.10. Two-degree-of-freedom architecture for improved tracking

• Disturbance feedforward

The trade-offs regarding sensitivities to reference, measurement noise, input, and output disturbances, discussed in the previous chapters, refer to the case when these disturbances are technically or economically not measurable.

Measurable disturbances can be compensated for explicitly (Figure 10.11), thus relaxing one of the trade-off constraints and giving the design more flexibility.

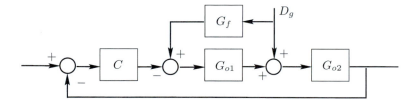

Figure 10.11. Disturbance-feedforward structure

• Cascade Control

○ Cascade control is a well-proven technique applicable when two or more systems feed sequentially into each other (Figure 10.12 on the next page).

○ All previously discussed design trade-offs and insights apply.

○ If the inner loop (C_2 in Figure 10.12) were not utilized, then the outer controller (C_1 in Figure 10.12) would–implicitly or explicitly–estimate y_1

as an internal state of the overall system $(G_{o1}G_{o2})$. This estimate, however, would inherit the model uncertainty associated with G_{o2}. Therefore, utilizing the available measurement of y_1 reduces the overall uncertainty, and one can achieve the associated benefits.

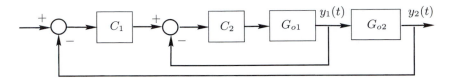

Figure 10.12. Cascade-control structure

10.9   Further Reading

Internal Model Principle

Francis, B.A. and Wonham, W. (1976). The Internal Model Principle of control theory. *Automatica*, 12:457-465.

Feedforward and cascade

Shinskey, F. (1998). *Process control systems: application, design and adjustment.* McGraw-Hill Book Company, New York, 3rd edition.

Stephanopoulos, G. (1984).   *Chemical process control: an introduction to theory and practice.* Prentice-Hall, Englewood Cliffs, N.J.

10.10   Problems for the Reader

Problem 10.1. *Compute the disturbance-generating polynomial, $\Gamma_d(s)$, for each of the following signals.*

(a) $3 + t$ (b) $2\cos(0.1t + \pi/7)$

(c) $3\left(\sin 1.5t\right)^3$ (d) $e^{-0.1t}\cos(0.2t)$

Problem 10.2. *In a feedback control loop of a minimum-phase stable plant, the reference and disturbances are combinations of sinewaves of frequencies 0, 1, and 2 [rad/s]. A set of possible complementary sensitivity transfer function includes the following:*

(a) $\dfrac{9}{s^2 + 4s + 9}$ (b) $\dfrac{25s^4 + 175s^3 + 425s^2 + 425s + 150}{s^5 + 25s^4 + 180s^3 + 425s^2 + 429s + 150}$

(c) $\dfrac{4s^2 + 20s + 16}{s^3 + 4s^2 + 21s + 16}$ (d) $\dfrac{30s^3 + 420s^2 + 1680s + 1920}{s^4 + 31s^3 + 424s^2 + 1684s + 1920}$

Analyze for each candidate whether the IMP is satisfied for all three frequencies.

Problem 10.3. *Consider a plant having a nominal model given by*

$$G_o(s) = \frac{-s + 8}{(s + 2)(s + 4)} \qquad (10.10.1)$$

Synthesize a controller to satisfy the IMP for $\omega = 0$ and $\omega = 2$ [rad/s].

Problem 10.4. *Consider a plant with input $u(t)$, disturbance $d_g(t)$, output $y(t)$, and nominal model given by*

$$Y(s) = G_b(s)V(s); \qquad where \qquad V(s) = G_a(s)\left(D_g(s) + G_{o1}(s)U(s)\right)$$

and where $v(t)$ is a measurable plant variable.
  For the plant given by

$$G_a(s) = e^{-0.2s}; \quad G_b(s) = \frac{e^{-0.5s}}{s+1}; \quad G_{o1}(s) = \frac{1}{s+1} \tag{10.10.2}$$

the nominal complementary sensitivity must satisfy

$$T_o(s) = \frac{4e^{-0.7s}}{s^2 + 3s + 4} \tag{10.10.3}$$

It is also known that the disturbance is a signal with abrupt changes.

10.4.1 Design a control loop including the first-d.o.f (i.e., a feedback controller) and the third-degree-of-freedom (i.e., disturbance feedforward). Hint: use Smith controllers.

10.4.2 Repeat the design, but using cascade control (measuring $v(t)$) instead of the third-d.o.f. You should aim to obtain the prescribed T_o and to achieve a degree of disturbance compensation as close as possible to that obtained in your solution to 10.4.1.

Problem 10.5. Consider Figure 10.2 on page 274, where the plant has a model given by

$$G_{o1}(s) = \frac{2}{s-2} \quad and \quad G_{o2}(s) = \frac{1}{s+1} \tag{10.10.4}$$

The reference is a constant signal and the disturbance satisfies $d_g(t) = \mathbf{K_d} + d_v(t)$, where $\mathbf{K_d}$ is constant and $d_v(t)$ is a signal with energy in the frequency band $[0,4][rad/s]$.

Design the first- (feedback controller $C(s)$) and third- (disturbance feedforward $G_f(s)$) d.o.f.. Pay special attention to the unstable nature of $G_{o1}(s)$. (This issue is not as simple as it might appear at first sight.)

Use the SIMULINK file **distffun.mdl** to evaluate your design.

Problem 10.6. Consider the cascade structure shown in Figure 10.8 on page 282, where

$$G_{o1}(s) = 1; \quad G_a(s) = \frac{1}{s}; \quad G_b(s) = \frac{e^{-s}}{s+1} \tag{10.10.5}$$

Assume that the reference is constant and that the disturbance $d_g(t)$ is as in Problem 10.5.

Design the primary and secondary controllers, under two different structural constraints for the secondary controller:

(i) $C_2(s)$ *must have a pole at the origin.*

(ii) $C_2(s)$ *has to be stable, i.e., all its poles must be in the open LHP.*

Compare the two designs. Discuss.

Problem 10.7. *Consider a feedback control loop with reference feedforward for a plant with nominal model given by*

$$G_o(s) = \frac{s+2}{(s+1)(s+3)} \tag{10.10.6}$$

The reference signal has significant energy only in the frequency band $[0, 10]$. However, due to noise constraints, the closed-loop bandwidth is restricted to $3[rad/s]$.
Design the feedback controller, $C(s)$, and the reference-feedforward transfer function, $H(s)$, such that good tracking is achieved.

Problem 10.8. *Closed-loop control has to be synthesized for a plant having nominal model $G_o(s) = \frac{-s+4}{(s+1)(s+4)}$, to achieve the following goals:*

(i) zero steady-state errors to a constant reference input;

(ii) zero steady-state errors for a sine-wave disturbance of frequency $0.25[rad/s]$;

(iii) a biproper controller transfer function, $C(s)$.

Use a pole-placement method to obtain a suitable controller $C(s)$.

Problem 10.9. *Consider a plant with nominal model*

$$G_o(s) = \frac{2}{(-s+1)(s+4)} \tag{10.10.7}$$

Design a one-d.o.f. control such that the feedback control loop tracks step references in the presence of measurement noise with energy in the frequency band $[5, 50][rad/s]$.

Chapter 11

DEALING WITH CONSTRAINTS

11.1 Preview

An ubiquitous problem in control is that all real actuators have limited authority. This implies that they are constrained in amplitude and/or rate of change. If one ignores this possibility, then serious degradation in performance can result in the event that the input reaches a constraint limit. This is clearly a very important problem. There are two ways of dealing with it:

(i) Reduce the performance demands so that a linear controller never violates the limits.

(ii) Modify the design to account for the limits.

Option (i) above is the more common strategy; however, it implies that either the actuator was oversized in the first place or one is unnecessarily compromising the performance. Anyway, we will see below that option (ii) is quite easy to achieve.

This chapter gives a first treatment of option (ii) based on modifying a given linear design. This will usually work satisfactorily for modest violations of the constraint (up to say 100%). If more serious violations of the constraints occur, then we would argue that the actuator has been undersized for the given application.

We will also show how the same ideas can be used to avoid simple kinds of state constraints.

Here, we assume that the control laws are biproper and of minimum phase. This will generally be true in SISO systems. Biproperness can be achieved by adding extra zeros if necessary. Techniques that do not depend on those assumptions will be described in Chapter 18.

Also, in a later chapter (Chapter 23), we will describe another technique for dealing with control and state constraints based on constrained optimal control theory. These latter techniques are generally called "Model Predictive Control" methods and can be thought of as more general versions of the ideas presented here.

11.2   Wind-Up

One very common consequence when an input hits a saturation limit is that the integrator in the controller (assuming it has one) will continue to integrate whilst the input is constrained. This means that the state of the integrator can reach an unacceptable high value leading to poor transient response. We illustrate by a simple example.

Example 11.1. *Consider the following nominal plant model:*

$$G_o(s) = \frac{2}{(s+1)(s+2)} \tag{11.2.1}$$

Say that the target complementary sensitivity is

$$T_o(s) = \frac{100}{s^2 + 13s + 100} \tag{11.2.2}$$

It is readily seen that this is achieved with the following controller:

$$C(s) = \frac{50(s+1)(s+2)}{s(s+13)} \tag{11.2.3}$$

A unit step reference is applied at $t = 1$, and a negative unit step-output disturbance occurs at $t = 10$. The plant input saturates when it is outside the range $[-3, 3]$. The plant output $y(t)$ is shown in Figure 11.1.

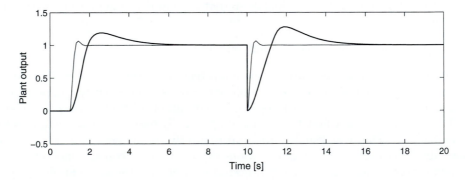

Figure 11.1. Loop performance with (thick line) and without (thin line) saturation at the plant input

We observe from Figure 11.1 that the plant output exhibits undesirable transient behavior that is inconsistent with the linear nominal bandwidth of approximately $10[rad/s]$. This deficiency originates from the saturation, because a unit step in the reference produces an instantaneous demanded change of 50 in the controller output and hence saturation occurs; a linear design procedure for $C(s)$ does not take this consequence into account.

☐☐☐

A simple ad-hoc mechanism for dealing with wind-up in PI controllers was described in subsection §8.8.3. We will show how this can be generalized below.

11.3   Anti-Wind-up Scheme

There are many alternative ways of achieving protection against wind-up. All of these methods rely on making sure that the states of the controller have two key properties:

(i) The state of the controller should be driven by the actual (i.e., constrained) plant input.

(ii) The states of the controller should have a stable realization when driven by the actual plant input.

This is particularly easy to achieve when the controller is biproper and of minimum phase. Say that the controller has transfer function $C(s)$; then we split this into the direct feedthrough term c_∞ and a strictly proper transfer function $\overline{C}(s)$:

$$C(s) = c_\infty + \overline{C}(s) \tag{11.3.1}$$

Then, consider the feedback loop shown in Figure 11.2.

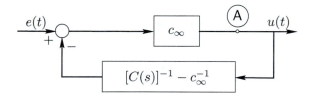

Figure 11.2. Feedback form of biproper controller

The transfer function from $e(t)$ to $u(t)$ in Figure 11.2 is readily seen to be

$$
\begin{aligned}
\frac{U(s)}{E(s)} &= \frac{c_\infty}{1 + ([C(s)]^{-1} - c_\infty^{-1})c_\infty} \tag{11.3.2}\\
&= \frac{c_\infty}{[C(s)]^{-1}c_\infty}\\
&= C(s)
\end{aligned}
$$

Also, because $[C(s)]^{-1}$ is stable (we are assuming for the moment that $C(s)$ is of minimum phase), the bottom block (driven by $u(t)$) in Figure 11.2 is stable. It also contains all of the dynamics of the controller. Hence, conditions (i) and (ii) listed above are satisfied.

We now redraw Figure 11.2 as in Figure 11.3.

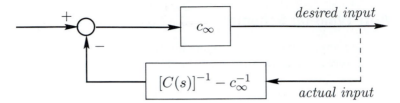

Figure 11.3. Desired and actual plant input

In the case of a limited input, all we now need to do is ensure that the correct relationship is achieved between the desired and actual input in Figure 11.3. If we denote by $\hat{u}(t)$ the desired input and $u(t)$ the actual input, then saturation and slew-rate limits can be described as follows.

Saturation

$$u(t) = Sat\langle\hat{u}(t)\rangle \triangleq \begin{cases} u_{max} & \text{if } \hat{u}(t) > u_{max}, \\\\ \hat{u}(t) & \text{if } u_{min} \le \hat{u}(t) \le u_{max}, \\\\ u_{min} & \text{if } \hat{u}(t) < u_{min}. \end{cases} \quad (11.3.3)$$

where $\hat{u}(t)$ is the unconstrained controller output and $u(t)$ is the effective plant input.

Slew-rate limit

$$\dot{u}(t) = Sat\langle\dot{\hat{u}}(t)\rangle \triangleq \begin{cases} \sigma_{max} & \text{if } \dot{\hat{u}}(t) > \sigma_{max}, \\\\ \dot{\hat{u}}(t) & \text{if } \sigma_{min} \le \dot{\hat{u}}(t) \le \sigma_{max}, \\\\ \sigma_{min} & \text{if } \dot{\hat{u}}(t) < \sigma_{min}. \end{cases} \quad (11.3.4)$$

Slew-rate limits can be modeled as shown in Figure 11.4, where an Euler approximation has been used to model the derivative.

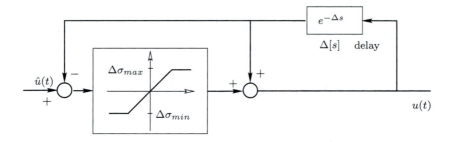

Figure 11.4. Slew-rate-limit model

Combined saturation and slew-rate limits can be modeled as in Figure 11.5.

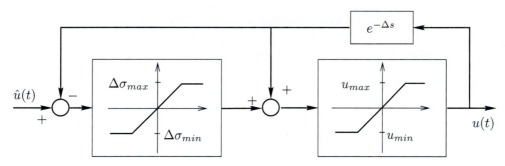

Figure 11.5. Combined saturation and slew-rate-limit model

Saturation and slew-rate limits can also apply to variables internal to the plant. In this case, they are known as state-saturation and state slew-rate limits, respectively. These issues will be explored later in this chapter.

Comparing Figure 11.3 with Figure 11.2, we see that all we need to do, to ensure that the lower block in Figure 11.3 is driven by the actual input, is to place the appropriate limiting circuit at point A in Figure 11.2.

This leads to the final implementation, as in Figure 11.6, where *Lim* denotes the appropriate limiting circuit (saturation, slew-rate limit, or both).

To illustrate the efficacy of the scheme, we repeat Example 11.1 on page 294.

Example 11.2. *Consider the same plant as in example 11.1, with identical reference and disturbance conditions. However, this time we implement the control loop as in Figure 11.6 on the following page.*

Upon running a simulation, the results are as shown in Figure 11.7, where the plant output has been plotted.

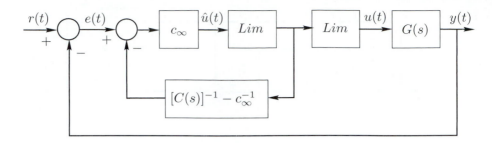

Figure 11.6. Simplified anti-wind-up control loop (C form)

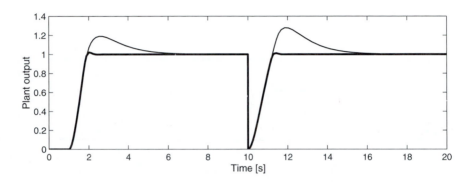

Figure 11.7. Loop performance with anti-wind-up controller (thick line) compared to performance achieved with no anti-wind-up feature (thin
line). The latter corresponds to the thick line in Figure 11.1.

*Figure 11.7 illustrates the dramatic improvement achieved by use of the proposed
anti-wind-up scheme.*

*The SIMULINK file **qawup.mdl** contains the schematic used in this simulation.
Note that the decomposition of $C(s)$ was carried out by using the utility MATLAB
program **awup.m** (included on the accompanying CD-ROM).*

□□□

Another example having slew-rate limits is described next.

Example 11.3. *Consider a plant having a linear model given by*

$$Y(s) = e^{-s} \left(\frac{1}{(s+1)^2} U(s) + D_g(s) \right)$$ (11.3.5)

where $U(s)$ and $D_g(s)$ are the Laplace transforms of the plant input and a generalized disturbance.

Assume that a PI controller with $K_P = 0.5$ and $T_r = 1.5[s]$, has been tuned for the linear operating range of this model, i.e., ignoring any nonlinear actuator dynamics.

If the input $u(t)$ cannot change at a rate faster than $0.2[s^{-1}]$, verify that implementation of the controller as in Figure 11.6 provides better performance than ignoring this slew-rate limitation.

Solution

We build a control loop with the controller structure shown in Figure 11.6 on the facing page, with Lim replaced by the slew-rate limiter in Figure 11.4.

$$c_\infty = K_p = 0.5; \qquad [C(s)]^{-1} - c_\infty^{-1} = -\frac{1}{K_p(T_r s + 1)} = -\frac{2}{(1.5s + 1)} \qquad (11.3.6)$$

A simulation was run with $r(t) = \mu(t - 1)$ and $d_g(t) = \mu(t - 20)$. The results of the simulation are shown in Figure 11.8.

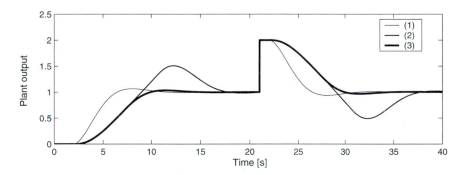

Figure 11.8. Performance of a PI control loop: when no slew-rate limitation exists (1); with slew-rate limitation but no compensation (2); and with anti-wind-up for slew-rate limitation (3).

In Figure 11.8, the curve labeled 1 describes the performance of the nominal design (no slew-rate limitation). Curve 2 describes the control performance for the plant with slew-rate limitation, but a linear controller. Finally, curve 3 describes the loop performance for the slew-rate-limited plant and the anti-wind-up PI controller. The comparison of curves 2 and 3 verifies the advantage of using the nonlinear implementation for the controller with anti-wind-up protection.

*The reader can evaluate the robustness of the scheme and other design ideas by using the SIMULINK file **slew1.mdl**.*

11.3.1   Interpretation in terms of Conditioning

We next give an alternative motivation for the structure developed in the last section. This alternative point of view is called *conditioning*. Here, we ask the following question: What *conditioned* set-point $\bar{r}$ would have avoided producing an input $\hat{u}$ beyond the limits of saturation in the first place?

In this subsection, we will show that the anti-wind-up strategy developed above is equivalent to conditioning, so that $u(t) = \hat{u}(t)$ at all times. Consider a controller having input $e(t)$ and output $u(t)$. We assume that $C(s)$ is biproper and can hence be expanded in terms of its strictly proper and feed-through terms as

$$C(s) = \overline{C}(s) + c_\infty \tag{11.3.7}$$

where $\overline{C}(s)$ is strictly proper and $c_\infty \neq 0$ is the high-frequency gain.

Let us assume that we have avoided saturation up to this point by changing $e(t)$ to $\bar{e}(t)$. Then, at the current time, we want to choose $\bar{e}$ so that

$$C\langle\bar{e}\rangle = u_{sat} = Sat\langle\overline{C}\langle\bar{e}\rangle + c_\infty e\rangle = \overline{C}\langle\bar{e}\rangle + c_\infty\bar{e} \tag{11.3.8}$$

Clearly, this requires that we choose $\bar{e}$ to be

$$\bar{e} = c_\infty^{-1}\left[Sat\langle\overline{C}\langle\bar{e}\rangle + c_\infty e\rangle - \overline{C}\bar{e}\right] \tag{11.3.9}$$

This can be represented as in Figure 11.9.

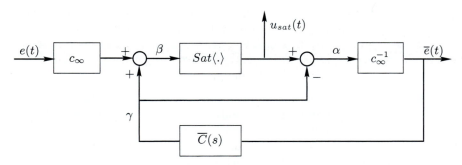

Figure 11.9. Conditioning equivalent for the anti-wind-up controller

To show that this is equivalent to the previous design, we note that, in Figure 11.9,

$$\gamma(t) = c_\infty^{-1}\overline{C}\langle u_{sat}(t) - \gamma(t)\rangle \Leftrightarrow c_\infty\gamma(t) = \overline{C}\langle u_{sat}(t)\rangle - \overline{C}\langle\gamma(t)\rangle \tag{11.3.10}$$

From that, upon using (11.3.7), we finally obtain

$$\gamma(t) = -c_\infty(C^{-1} - c_\infty^{-1})\langle u_{sat}(t)\rangle \qquad (11.3.11)$$

Also

$$\beta(t) = c_\infty e(t) + \gamma(t) = c_\infty \left(e(t) - (C^{-1} - c_\infty^{-1})\langle u_{sat}(t)\rangle\right) \qquad (11.3.12)$$

and

$$u_{sat}(t) = Sat\langle\beta(t)\rangle \qquad (11.3.13)$$

and so, by using (11.3.12), we finally obtain

$$u_{sat}(t) = Sat\left\langle c_\infty \left(e(t) - (C^{-1} - c_\infty^{-1})\langle u_{sat}(t)\rangle\right)\right\rangle \qquad (11.3.14)$$

Hence, the scheme in Figure 11.9 implements the same controller as that in Figure 11.6.

11.4   State Saturation

As a further illustration of the application of anti-wind-up procedures, we next show how they can be applied to maintain state limits.

Say we wish to place a constraint on an internal variable $z(t)$. (We obviously assume that this constraint does not impede the plant output from reaching the value prescribed by the reference, in steady state.) There might be many practical reasons why this may be highly desirable. For instance, we might be interested in whether a given process variable (or its rate of change) is bounded.

We consider a plant with nominal model given by

$$Y(s) = G_o(s)U(s); \qquad\qquad Z(s) = G_{oz}(s)U(s) \qquad (11.4.1)$$

We assume we are either able to measure $z(t)$ or estimate it from the available data $u(t)$ and $y(t)$, using some form of virtual sensor–i.e., we use an observer to construct an estimate $\hat{z}(t)$ for $z(t)$. This virtual sensor will depend on the input $u(t)$ and $y(t)$, and hence can be expressed by using Laplace transforms as

$$\hat{Z}(s) = T_{1z}(s)U(s) + T_{2z}(s)Y(s) \qquad (11.4.2)$$

We will now show how this estimate can be used to develop a strategy to achieve state constraints based on switching between two controllers. One of these controllers (the prime controller) is the standard controller aimed at achieving the main control goal: that the plant output $y(t)$ track a given reference, say $r_y(t)$. The task for the secondary controller is to keep the variable $z(t)$ within prescribed bounds. This is achieved by use of a secondary closed loop aimed at the regulation of the estimated state, $\hat{z}(t)$, using a fixed set-point.

Our strategy will be to switch between the primary and secondary controller. However, it can be seen that there is a strong potential for wind-up because one or other of the two controllers, at any given time, will be running in open-loop. We will thus implement this in anti-wind-up form.

For simplicity of presentation, we assume that a bound is set upon $|z(t)|$, i.e., that $z(t)$ is symmetrically bounded.

The general structure of the proposed control scheme is shown in Figure 11.10 on the next page.

This strategy has the following features:

(i) The switching between the two controllers operates according to a policy based on the value of $|z(t)|$. The switching is performed by the block $\mathcal{W}\langle.\rangle$

(ii) Both controllers have been implemented as in Figure 11.6. Thus, the prime (linear) controller has a transfer function, $C_y(s)$, given by

$$C_y(s) = \frac{c_{y\infty}}{1 + c_{y\infty}H_y(s)}; \qquad H_y(s) = [C_y(s)]^{-1} - c_{y\infty}^{-1} \qquad (11.4.3)$$

Analogously, the secondary (linear) controller has a transfer function, $C_z(s)$, given by

$$C_z(s) = \frac{c_{z\infty}}{1 + c_{z\infty}H_z(s)}; \qquad H_z(s) = [C_z(s)]^{-1} - c_{z\infty}^{-1} \qquad (11.4.4)$$

The controllers are designed so as to achieve satisfactory performance in the control of $y(t)$ and $z(t)$, respectively. We also assume that $C_y(s)$ and $C_z(s)$ are minimum-phase biproper transfer functions. This implies that $H_y(s)$ and $H_z(s)$ are stable transfer functions.

(iii) Plant-input saturation can be accounted for, through the block $Sat\langle\circ\rangle$. Anti-wind-up protection is given to both controllers by ensuring that each dynamic block (i.e., $H_y(s)$ and $H_z(s)$) is driven by the true plant input, no matter which controller is active in controlling the real plant.

One can envision different strategies for specifying the switching law. Two possibilities are considered below.

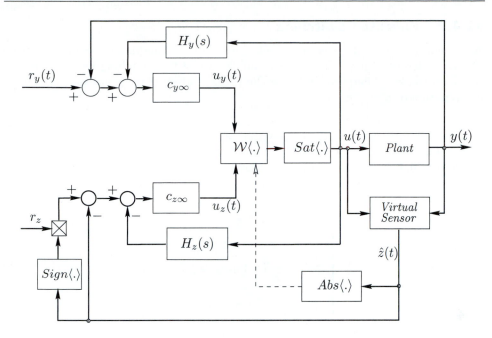

Figure 11.10. Switching strategy for state saturation

11.4.1 Substitutive Switching with Hysteresis

A simple approach is to transfer the generation of the real plant input, $u(t)$, from one controller to the other in such a way that, at any time, $u(t)$ is determined by either $u_y(t)$ or $u_z(t)$.

If we aim to keep $|z(t)|$ bounded by a known constant $z_{sat} > 0$, then this approach can be implemented by using a switch with hysteresis, where the switching levels, z_l and z_h, are chosen as $0 < z_l < z_h < z_{sat}$.

When the state controller is in charge, the loop is driven by a set-point z_{sp}, given by

$$z_{sp}(t) = \text{sign}\langle z(t)\rangle r_z \qquad \text{where} \qquad 0 < r_z < z_l < z_{sat} \qquad (11.4.5)$$

In this scheme, the secondary controller $C_z(s)$ takes over from the primary controller $C_y(s)$ when $|z(t)|$ grows beyond z_h. The control reverts to the original controller $C_y(s)$ when $|z(t)|$ falls below z_l. We observe that this latter transition is made possible by the choice (11.4.5).

11.4.2   Weighted Switching

A switching strategy that is an embellishment of the one described in subsection §11.4.1 is described next. It also relies on the use of the switching levels z_l and z_h, but with the key difference that now the (unsaturated) plant input $u(t)$ is a linear combination of $u_y(t)$ and $u_z(t)$:

$$u(t) = Sat\langle \lambda u_z(t) + (1 - \lambda)u_y(t)\rangle \qquad (11.4.6)$$

where $\lambda \in [0, 1]$ is a weighting factor. One way of determining λ would be

$$\lambda = \begin{cases} 0 & \text{for } |z(t)| \leq z_l \\[2ex] \dfrac{|z(t)| - z_l}{z_h - z_l} & \text{for } z_h > |z(t)| \geq z_l \\[2ex] 1 & \text{for } |z(t)| > z_h \end{cases} \qquad (11.4.7)$$

This strategy is illustrated in the following example.

Example 11.4. *Consider a plant with a model as in (11.4.1), where*

$$G_o(s) = \frac{16}{(s + 2)(s + 4)(s + 1)}; \qquad G_{oz}(s) = \frac{16}{(s + 2)(s + 4)} \qquad (11.4.8)$$

The reference is a square-wave of unity amplitude and frequency 0.3[rad/s]. It is desired that the state $z(t)$ not go outside the range $[-1.5; 1.5]$. Furthermore, the plant input saturates outside the range $[-2; 2]$.

For this plant, the primary and secondary controller are designed to be

$$C_y(s) = \frac{90(s + 1)(s + 2)(s + 4)}{16s(s^2 + 15s + 59)}; \qquad C_z(s) = \frac{16(3s + 10)(s + 4)}{s(s + 14)} \qquad (11.4.9)$$

The basic guidelines used to develop the above designs is to have the secondary control loop faster than the primary control loop, so that the state $z(t)$ can be quickly brought within the allowable bounds.

The virtual sensor for $z(t)$ is actually based on a linear observer designed to have poles to the left of -10. (See Chapter 18.) After some simple trials, the following parameters were chosen for the switching strategy.

$$z_l = 1.2; \qquad z_h = 1.5; \qquad z_{sp} = 0.9; \qquad (11.4.10)$$

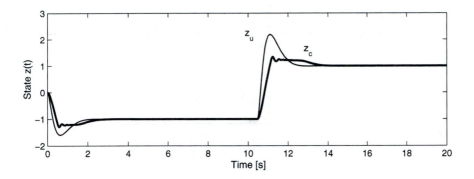

Figure 11.11. Process variable $z(t)$ with state-control saturation (z_c) and without state-control saturation (z_u)

The performance of the loop is shown in Figure 11.11, where a comparison is made between the evolution of the state $z(t)$ with and without the switching strategy:
The effect of the switching strategy on the plant output is shown in Figure 11.12.
Figure 11.12 shows a (predictably) slightly slower control of the plant output, due to the limitations imposed on the state $z(t)$.

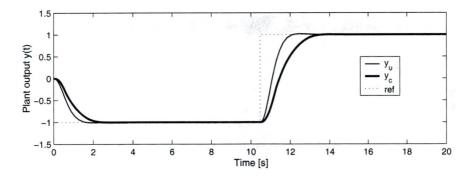

Figure 11.12. Plant output with (y_c) and without (y_u) state-control saturation

The evolution of the weighting factor $\lambda(t)$ is shown in Figure 11.13.
Figure 11.13 shows that the strategy uses a weight that does not reach its maximum value–i.e., the upper level, z_h, is never reached.

□□□

The reader is invited to try different designs, using the SIMULINK schematic in file **newss.mdl**. The observer in this schematic includes the possibility of step-like

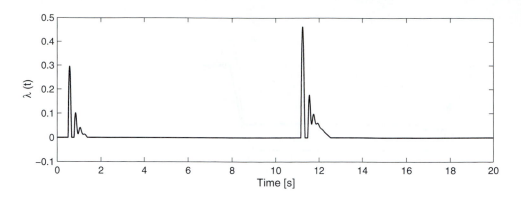

Figure 11.13. Weighting factor behavior

input disturbances.

11.5   Introduction to Model Predictive Control

Another class of strategies for dealing with constraints are the so-called *Model Predictive Control* (mpc) algorithms. The essential idea in these algorithms is to formulate controller design as an on-line receding-horizon optimization problem that is solved (usually by quadratic programming methods) subject to the given hard constraints. The idea has been widely used in the petrochemical industries, where constraints are prevalent. Until recently, mpc and anti-wind-up strategies were considered alternative and quite separate schemes for solving the same problem. However, it has recently been recognized (perhaps not unsurprisingly) that the schemes are very closely related and are sometimes identical. We will give a more complete treatment of mpc in Chapter 23.

11.6   Summary

- Constraints are ubiquitous in real control systems.

- There are two possible strategies for dealing with them:
  - Limit the performance so that the constraints are never violated.
  - Carry out a design with the constraints in mind.

- Here, we have given a brief introduction to the latter idea.

- A very useful insight is provided by the arrangement shown in Figure 11.2 on page 295, which can be set in a general inversion set-up:

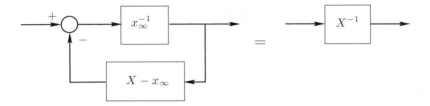

Figure 11.14. Implicit inversion X^{-1}

- X is a biproper, stable, minimum-phase transfer function.
- x_∞ is the high-frequency gain of $X : x_\infty = \lim_{s \to \infty} X(s)$.
- Given that X is biproper, then $x_\infty \neq 0$ and hence x_∞^{-1} is finite.
- The overall system is equivalent to X^{-1}.
- The fascinating and useful point about the implementation in Figure 11.14 is that the circuit inverts X *effectively*, but not *explicitly*.

- Here, we have used this idea to describe anti-wind-up mechanisms for achieving input saturation and slew-rate limits.

- We have also shown how state constraints might be achieved.

- In later chapters, we will use the same idea for many other purposes.

11.7   Further Reading

Saturating actuators

Bernstein, D. and Michel, A. (1995). A chronological bibliography on saturation actuation. *Int. J. of Robust and Nonlinear Control*, 5:375-380.

Hanus, R., Kinnaert, M., and Henrotle, J. (1987). Conditioning technique, a general anti-wind-up and bumpless transfer method. *Automatica*, 23(6):729-739.

Kothare, M., Campo, P., Morari, M., and Nett, C. (1994). A unified framework for the study of anti-wind-up designs. *Automatica*, 30(12):1869-1883.

Seron, M.M., Graebe, S.F., and Goodwin, G.C. (1994). All stabilizing controllers, feedback linearization and anti-wind-up: a unified review. In *Proceedings of the 1994 American Control Conference, Baltimore, Maryland*, 2:1685-1689.

Stability of anti-wind-up and switching schemes

De Doná, J.A., Feuer, A., and Goodwin, G.C. (1998). A high performance controller incorporating over-saturation of the input signal. *Technical Report, Department of*

E&CE, University of Newcastle, Australia.

Kapoor, N., Teel, A.R., and Daoutidis, P. (1996). On anti-integrator-windup and global asymptotic stability. In *Proceedings of the 13th World Congress of IFAC, Volume D*, pages 67-72.

Teel, A.R. (1998). Anti-wind-up for exponentially unstable linear systems. *Technical Report, CCEC,* University of California at Santa Barbara.

Model Predictive Control (MPC)

Garcia, C.E., Prett, D.M. and Morari, M. (1989). Model Predictive Control: Theory and practice–a survey. *Automatica,* 25(3):335-348.

Keerthi, S.S. and Gilbert, E.G. (1988). Optimal, infinite horizon feedback laws for a general class of constrained discrete-time systems: Stability and moving-horizon approximations. *Journal of Optimization Theory and Applications,* 57(2):265-293.

Mayne, D.Q. and Michalska, H. (1990). Receding horizon control of non-linear systems. *IEEE Transactions on Automatic Control,* 35(5):814-824.

Rawlings, J.B. and Muske, K.R. (1993) Model Predictive Control with linear models. *AIChE Journal,* 39(2):262-287.

Connections between MPC and anti-wind-up

De Doná, J.A., Goodwin, G.C., and Seron, M.M. (1998). Connections between Model Predictive Control and anti-wind-up strategies for dealing with saturating actuators. In *Proceedings of the 1999 European Control Conference.*

11.8 Problems for the Reader

Problem 11.1. *Consider the following controller transfer functions, $C(s)$.*

$$(a) \quad \frac{9(s+1)}{s} \qquad (b) \quad \frac{12(s^2+5s+4)}{s(s+3)}$$

$$(c) \quad \frac{8s+7}{s(s-4)} \qquad (d) \quad \frac{6(s-1)(s+7)}{s(s+8)}$$

Obtain, for each case, the form given in Figure 11.2 on page 295. Discuss, especially, cases (c) and (d).

Problem 11.2. *In a feedback loop, with plant-input saturation, the controller has a strictly proper transfer function given by*

$$C(s) = \frac{12(s+4)}{s(s+2)} \tag{11.8.1}$$

Because the use of the anti-wind-up strategy developed in this chapter requires a biproper *controller, someone suggest making this controller biproper by adding a large zero, leading to a modified transfer function $C_m(s)$ given by*

$$C_m(s) = \frac{12(s+4)(\tau s+1)}{s(s+2)} \tag{11.8.2}$$

where $\tau \ll 1$.
Analyze the proposal.

Problem 11.3. *Consider a proper controller (with integral action) having a transfer function, $C(s)$, given by*

$$C(s) = \frac{P(s)}{s\overline{L}(s)} = \frac{K_I}{s} + \frac{\overline{P}(s)}{\overline{L}(s)} \tag{11.8.3}$$

Someone suggests, as an anti-wind-up mechanism, freezing the integral action of the controller while the controller output is saturated.

11.3.1 *Implement this idea to deal with the same control problem proposed in example 11.1. Use SIMULINK file* **piawup.mdl**.

11.3.2 *Is this mechanism, in general, equivalent to that proposed in section §11.3?*

Problem 11.4. *Consider a plant with nominal model given by*

$$G_o(s) = \frac{1}{(s+1)^2} \tag{11.8.4}$$

The loop has a unit step reference and output disturbances showing abrupt changes with magnitude 0.5. The plant input, $u(t)$, is constrained to the interval $[-2, 2]$.

11.4.1 *Design a linear, biproper controller such that the closed loop is dominated by poles located at $-0.7\omega_n \pm j0.7\omega_n$.*

11.4.2 *Choose ω_n so that the plant input never saturates.*

11.4.3 *Choose ω_n so that the plant input saturates for a given combination of reference and disturbances. Evaluate the performance of the loop under this condition.*

11.4.4 *Implement the anti-wind-up mechanism, and compare the loop performance with the case having no special anti-wind-up mechanism.*

Problem 11.5. *Devise a circuit similar to that shown in Figure 11.5 that would allow you to constrain the acceleration (rate of change of the rate of change) of an input signal.*

Problem 11.6. *Extend your answer to Problem 11.5 so that you can simultaneously constrain the acceleration, the velocity, and the amplitude of the input.*

Problem 11.7. *Test your circuit by repeating Example 11.3, where $u(t)$ is required to satisfy acceleration, velocity, and amplitude constraints.*

Problem 11.8. *An alternative circuit that is sometimes used to protect against anti-wind-up is shown in Figure 11.15, where g is a static gain.*
    Explain the operation of this circuit.

Problem 11.9. *Test the circuit given in Problem 11.8, and compare its performance with the results given in, for example, Example 11.2.*

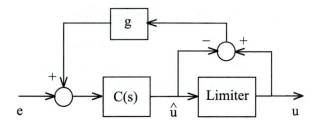

Figure 11.15. Alternative anti-wind-up scheme

Problem 11.10. *Say that $C(s)$ is a simple integral controller of the form*

$$C(s) = \frac{k}{s} \tag{11.8.5}$$

11.10.1 *Turn $C(s)$ into a biproper controller by adding a term $(\varepsilon s + 1)$ to the numerator, for ε small.*

11.10.2 *Write the modified biproper controller as*

$$[C'(s)]^{-1} = [C'_\infty]^{-1} + \{[C'(s)]^{-1} - [C'_\infty]^{-1}\} \tag{11.8.6}$$

11.10.3 *Hence, show that, when the limiter is operating, the circuit of Figure 11.6 leads to*

$$\hat{u} = u + \left\{ k\varepsilon e - \left[\frac{\varepsilon s}{\varepsilon s + 1}\right] u \right\} \tag{11.8.7}$$

Problem 11.11. *Consider again the simple integral controller given in Problem 11.10.*

11.11.1 *By placing the controller in the feedback circuit given in Figure 11.15, show that*

$$\hat{u} = u + \left\{ \frac{k\varepsilon}{(\varepsilon s + 1)} e - \left[\frac{\varepsilon s}{\varepsilon s + 1}\right] u \right\} \tag{11.8.8}$$

where $\varepsilon = \frac{1}{gk}$.

11.11.2 *Compare this result with the result found in Problem 11.10.*

Problem 11.12. *Yet another alternative scheme for anti-wind-up protection of simple controllers is to freeze the integrator output when the control is saturated.*

Compare this answer with the results found in Problems 11.10 and 11.11 for small ε.

Problem 11.13. *(This problem was suggested by a colleague from industry. It is a slightly harder problem.)*

11.13.1 *Consider the situation illustrated in Figure 11.16.*

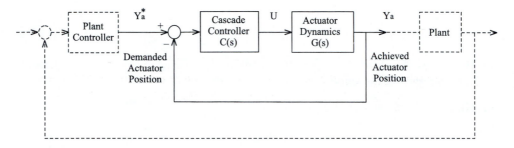

Figure 11.16. Cascaded actuator control

The actuator is required to have constrained amplitude and slew rate. However, the actuator is itself controlled via a cascaded loop whose output u drives the actuator system. Say that the transfer function from input u to actuator position y_a is given by

$$\frac{Y_a(s)}{U(s)} = \frac{b}{s^2 + a_1 s + a_o} \tag{11.8.9}$$

We assume that the states y_a and $\dot{y}_a$ are measured. Also, we note that constraints on y_a and $\dot{y}_a$ correspond to state constraints.

Say that the (cascade) loop bandwidth is w_b. Let $\hat{u}$ denote a demanded input, and choose $\Delta \simeq \frac{1}{5W_b}$.

Our objective is to construct an anti-wind-up circuit for the problem.

One way to proceed is as follows:

You can think of $b\hat{u} - a_1\dot{y}_a - a_0 y_a$ as a demand for actuator acceleration, $\ddot{y}_a$. Thus, if you form

$$\hat{\ddot{y}}_a = \dot{y}_a + \Delta[b\hat{u} - a_1\dot{y}_a - a_0 y_a] \tag{11.8.10}$$

then you can think of this as a prediction of demanded actuator velocity.
You can then constrain this velocity to produce $(\hat{\dot{y}}_a)_{cons}$.
Similarly,

$$\hat{y}_a = y_a + \Delta(\hat{\dot{y}}_a)_{cons} \tag{11.8.11}$$

can be thought of as a prediction of the demanded actuator position.
We can then constrain $\hat{y}_a$, to produce $(\hat{y}_a)_{cons}$.

The above steps can now be reversed to evaluate a control input that is consistent with the constraints on y_a and $\dot{y}_a$. Thus, reversing (11.8.11) gives

$$(\hat{\dot{y}}_a{}')_{cons} = \frac{1}{\Delta}[(\hat{y}_a)_{cons} - y_a] \tag{11.8.12}$$

Also, reversing (11.8.10) gives the allowed input $\hat{u}'$ as

$$\hat{u}' = \frac{[\frac{1}{\Delta}[(\hat{\dot{y}}_a{}')_{cons} - \dot{y}_a] + a_1\dot{y}_a + a_0 y_a]}{b} \tag{11.8.13}$$

Equations (11.8.10) to (11.8.13) give a special form of input limiter.
Build a SIMULINK form of this limiter.

11.13.2 *How would you combine the limiter described above with an anti-wind-up form of the controller $C(s)$ so that $\hat{u}$ is also limited?*

11.13.3 *Test your design by using the actuator transfer function*

$$G(s) = \frac{2}{s^2 + 2s + 2} \tag{11.8.14}$$

together with the controller

$$C(s) = \frac{4(s^2 + 2s + 2)}{s(s+4)} \tag{11.8.15}$$

Choose $\Delta = 0.1$, and examine the performance for a unit step input subject to the constraints

$$|\dot{y}_a| \leq 0.5 \tag{11.8.16}$$

$$|y_a| \leq 1 \tag{11.8.17}$$

$$|u| \leq 1.5 \tag{11.8.18}$$

(Aside: Actually, the above design can be viewed as a procedure for predicting two steps ahead, so as to "see" the effect of the current demanded control on the future states. If these states are constrained, then the input is modified so that the constraints are not violated. This leads to a special form of input limiter. The control law is then implemented in a feedback arrangement, as in Figure 11.6, so that the states of the controller are fed by the actual plant input that is the output of this limiter.

This is one possible way to include state constraints into anti-wind-up circuits. The key idea being used here is seen to be prediction of the future state response. Indeed, we will see in Chapter 23 that the idea described above, predicting ahead to evaluate the impact of the input on future state constraints, is the key idea in Model Predictive Control.)

Part IV

DIGITAL COMPUTER CONTROL

PREVIEW

So far in the book, we have treated only continuous-time (or analog) control systems. However, the reality is that almost all modern control systems will be implemented in a digital computer of one form or another. Thus, in this fourth part of the book, we will describe the essential differences between analog and digital control. Some authors devote entire books to this topic. This tends to highlight the differences between analog and digital control. However, our perspective is quite different. We want to show the reader that these topics actually have a remarkably close connection. We show, for example, that all properly formulated digital controllers converge to an underlying continuous controller as the sampling period goes to zero. We also point to the issues where digital and analog control are different–especially at slow sampling rates.

Chapter 12

MODELS FOR
SAMPLED-DATA SYSTEMS

12.1 Preview

Modern control systems are almost always implemented in a digital computer. It is thus important to have an appreciation of the impact of implementing a particular control law in digital form. In this chapter, we provide the fundamental modeling tools necessary to describe the sampled response of continuous-time plants.

The main topics included are

- Discrete-time signals

- Z-transforms and Delta-transforms

- Sampling and reconstruction

- Aliasing and anti-aliasing filters

- Sampled-data control systems

It is common, in many courses on control, to treat sampling at a much later point or even to move it to a separate course specifically devoted to digital control. However, our view is that there is so much in common between continuous and digital control that one can introduce sampling ideas early and then, essentially, present both ideas simultaneously.

12.2 Sampling

Computers work with sequences of numbers rather than continuous functions of time. To connect an analog plant to a computer, one therefore needs to sample the output of the system (thus converting a continuous-time function into a sequence of numbers). We will use the notation $\{f[k]\}$ to denote a sequence $f[0], f[1], f[2], \ldots$. When $\{f[k]\}$ arises from sampling a continuous-time signal, $f(t)$, at a fixed interval Δ, we will write

$$f[k] = f(k\Delta) \qquad\qquad k = 0, 1, 2, \ldots \qquad\qquad (12.2.1)$$

From the point of view of control, we are interested in, among other things, the process of coding a continuous-time signal into such a sequence of numbers (the sampling process) and of building a continuous-time signal from a sequence of numbers (the reconstruction process).

There will always be loss of information due to sampling. However, the extent of this loss depends on the sampling method and the associated parameters. For example, assume that a sequence of samples of a signal $f(t)$ is taken every Δ seconds; then the sampling frequency needs to be large in comparison with the maximum rate of change of $f(t)$. Otherwise, high-frequency components will be mistakenly interpreted as low frequencies in the sampled sequence. We illustrate by a simple example.

Example 12.1. *Consider the signal*

$$f(t) = 3\cos 2\pi t + \cos\left(20\pi t + \frac{\pi}{3}\right) \qquad\qquad (12.2.2)$$

We observe that, if the sampling period Δ is chosen equal to $0.1[s]$, then

$$f(k\Delta) = 3\cos(0.2k\pi) + \cos\left(2k\pi + \frac{\pi}{3}\right) \qquad\qquad (12.2.3)$$

$$= 3\cos(0.2k\pi) + 0.5 \qquad\qquad (12.2.4)$$

from which it is evident that the high-frequency component has been shifted to a constant, i.e., the high-frequency component appears as a signal of low frequency (here zero). This phenomenon is known as aliasing.

The result is depicted in Figure 12.1, where the sampled signal is described by a sequence of small circles.

We can see in Figure 12.1 that the sampled high-frequency component appears as a d.c. upwards shift of the low-frequency component, as predicted by (12.2.4).

□□□

To mitigate the effect of aliasing, the sampling rate must be high relative to the rate of change of the signals of interest. A typical rule of thumb is to require that the sampling rate be 5 to 10 times the bandwidth of the system. If this rule is violated, the observed samples can be a very poor reflection of the underlying continuous-time signal. This issue will be studied in detail later in the book.

Even when the above rule of thumb is followed, one needs to protect the sampling process from other contaminating signals with high-frequency content, such as noise. To keep these signals from being aliased to low frequencies, it is common practice to place an analog filter prior to the sampling process. This filter is called an anti-aliasing filter , because its purpose is to avoid aliasing of high-frequency noise signals to lower frequencies via the sampling process.

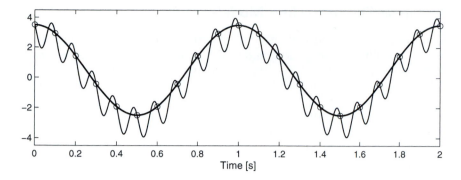

Figure 12.1. Aliasing effect when using low sampling rate.

12.3   Signal Reconstruction

The output of a digital controller is another sequence of numbers $\{u[k]\}$, which are the sample values of the intended control signal. These sample values need to be converted back to continuous-time functions before they can be applied to the plant. Usually, this is done by interpolating them into a *staircase* function $u(t)$, as illustrated in Figure 12.2.

Computer-controlled systems are typically analyzed at *the samples only*. (The danger in this approach, as well as a more general description, will be studied in Chapter 14.) In the simple *at-sample* only case, the plant description is transformed into one that relates the sampled input sequence $\{u[k]\}$ to the sampled output sequence $\{y[k]\}$. Thus, we need convenient ways of describing dynamic models that relate one sequence (the input) to another sequence (the output).

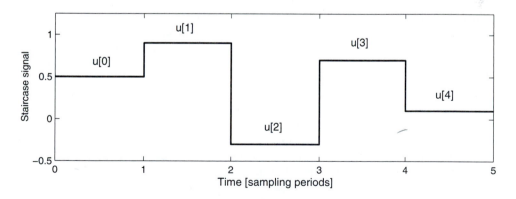

Figure 12.2. Staircase reconstruction

12.4   Linear Discrete-Time Models

A useful discrete-time model of the type referred to above is the linear version of the high-order difference-equation model of section §3.8. In the discrete case, this model takes the form

$$y[k+n] + \overline{a}_{n-1}y[k+n-1] + \cdots + \overline{a}_0 y[k] \qquad (12.4.1)$$
$$= \overline{b}_{n-1}u[k+n-1] + \cdots + \overline{b}_0 u[k]$$

Example 12.2. *We consider a so-called PI controller which generates the control sequence, $u[k]$, as the sum of a term proportional to the control error and a term proportional to the accumulation (integral) of the error. This can be modeled as*

$$\overline{e}[k+1] = \overline{e}[k] + c_1 e[k] \qquad (12.4.2)$$
$$u[k] = c_2 e[k] + \overline{e}[k] \qquad (12.4.3)$$

Shifting one step forward in (12.4.3) gives

$$u[k+1] = c_2 e[k+1] + \overline{e}[k+1] \qquad (12.4.4)$$

Subtracting (12.4.3) from (12.4.4) and using (12.4.2) gives the following high-order difference-equation model:

$$u[k+1] - u[k] = c_2\,(e[k+1] - e[k]) + \overline{e}[k+1] - \overline{e}[k] \qquad (12.4.5)$$
$$= c_2\,(e[k+1] - e[k]) + c_1 e[k]$$

In the notation of (12.4.1), we have $\overline{a}_0 = -1$, $\overline{b}_1 = c_2$, $\overline{b}_0 = c_1 - c_2$. Note that the extra term $c_2 e[k+1]$ appears on the right-hand side because, in this case, the controller is biproper. (The left- and right-hand sides of (12.4.5) have the same order.) Also note that, here, $\{u[k]\}$ is the output and $\{e[k]\}$ the input.

□□□

12.5   The Shift Operator

In writing down discrete-time models, we will find it convenient to use operator notation to denote the forward shift operation. Thus we define the
Forward Shift Operator as follows:

$$\boxed{q(f[k]) \triangleq f[k+1]} \qquad (12.5.1)$$

In terms of this operator, the model (12.4.1) becomes

$$q^n y[k] + \bar{a}_{n-1} q^{n-1} y[k] + \cdots + \bar{a}_0 y[k] = \bar{b}_m q^m u[k] + \cdots + \bar{b}_0 u[k] \qquad (12.5.2)$$

For a discrete-time system, it is also possible to have discrete state space models. In the shift domain, these models take the form

$$qx[k] = \mathbf{A}_q x[k] + \mathbf{B}_q u[k] \qquad (12.5.3)$$

$$y[k] = \mathbf{C}_q x[k] + \mathbf{D}_q u[k] \qquad (12.5.4)$$

where $\{x[k]\}$, $\{u[k]\}$, and $\{y[k]\}$ are the state, input, and output sequences, respectively.

The solution to equations (12.5.3) and (12.5.4) is easily evaluated by iterating (12.5.3).

$$x[k] = \mathbf{A}_q^k x(0) + \sum_{l=0}^{k-1} \mathbf{A}_q^l \mathbf{B}_q u[k - l - 1] \qquad (12.5.5)$$

$$y[k] = \mathbf{C}_q \mathbf{A}_q^k x(0) + \mathbf{C}_q \sum_{l=0}^{k-1} \mathbf{A}_q^l \mathbf{B}_q u[k - l - 1] \qquad (12.5.6)$$

Example 12.3. *Consider the discrete PI controller of Example 12.2 on the preceding page. Actually, (12.4.2) and (12.4.3) were already in discrete state space form where*

$$\mathbf{A}_q = 1; \qquad \mathbf{B}_q = c_1; \qquad \mathbf{C}_q = 1; \qquad \mathbf{D}_q = c_2 \qquad (12.5.7)$$

□□

12.6   Z-Transform

In the same way that Laplace Transforms turn differential equations into algebraic equations, we can use Z-transforms to turn difference equations into algebraic equations.

Consider a sequence $\{y[k]; \ k = 0, 1, 2, \dots\}$. Then the Z-transform pair associated with $\{y[k]\}$ is given by

$$\mathcal{Z}\left[y[k]\right] = Y(z) = \sum_{k=0}^{\infty} z^{-k} y[k] \qquad (12.6.1)$$

$$\mathcal{Z}^{-1}\left[Y(z)\right] = y[k] = \frac{1}{2\pi j} \oint z^{k-1} Y(z) dz \qquad (12.6.2)$$

where the contour of integration is a circle with center at the origin and with radius ρ.

$Y(z)$ is referred to as the Z-transform of $y(t)$. The transform pair is well-defined if there exists $\rho \in \mathbb{R}^+$ such that

$$|y[k]| < \rho^k \qquad\qquad \forall k \geq 0 \qquad\qquad (12.6.3)$$

Z-transforms of various common signals are shown in Table 12.1 on the facing page. Some useful properties of the Z-transform are summarized in Table 12.2 on page 326.

Example 12.4. *Consider the discrete PI controller of Example 12.2, and say that* $u[0] = 0$ *and* $e[k]$ *is a unit step applied at* $k = 0$. *Then, taking the Z-transform in (12.4.5) gives*

$$zU(z) - zu[0] - U(z) = c_2 \left(zE(z) - E(z) - ze[0]\right) + c_1 E(z)$$

Hence,[1]

$$U(z) = \frac{c_2 z + (c_1 - c_2)}{(z - 1)} E(z); \quad E(z) = \frac{z}{z - 1} \qquad (12.6.4)$$

$$\frac{U(z)}{z} = \frac{c_1}{(z - 1)^2} + \frac{c_2}{z - 1}; \ U(z) = \frac{c_1 z}{(z - 1)^2} + \frac{c_2 z}{z - 1} \qquad (12.6.5)$$

Hence, $u[k] = c_1 k + c_2 \,; \ k \geq 0$.

Z-transforms can be used to convert linear difference equations into algebraic equations.

12.7 Discrete Transfer Functions

Taking Z-transforms on each side of the high-order difference-equation model (12.5.2) leads to

$$A_q(z)Y_q(z) = B_q(z)U_q(z) + f_q(z, x_o) \qquad (12.7.1)$$

where $Y_q(z)$ and $U_q(z)$ are the Z-transforms of the sequences $\{y[k]\}$ and $\{u[k]\}$, respectively.

---

[1]Note that $u[0] = c_2 e[0]$.

$f[k]$	$\mathcal{Z}\left[f[k]\right]$	Region of convergence
1	$\dfrac{z}{z-1}$	$\lvert z \rvert > 1$
$\delta_K[k]$	1	$\lvert z \rvert > 0$
k	$\dfrac{z}{(z-1)^2}$	$\lvert z \rvert > 1$
k^2	$\dfrac{z(z-1)}{(z-1)^3}$	$\lvert z \rvert > 1$
a^k	$\dfrac{z}{z-a}$	$\lvert z \rvert > \lvert a \rvert$
ka^k	$\dfrac{az}{(z-a)^2}$	$\lvert z \rvert > \lvert a \rvert$
$\cos k\theta$	$\dfrac{z(z-\cos\theta)}{z^2-2z\cos\theta+1}$	$\lvert z \rvert > 1$
$\sin k\theta$	$\dfrac{z\sin\theta}{z^2-2z\cos\theta+1}$	$\lvert z \rvert > 1$
$a^k \cos k\theta$	$\dfrac{z(z-a\cos\theta)}{z^2-2az\cos\theta+a^2}$	$\lvert z \rvert > a$
$a^k \sin k\theta$	$\dfrac{az\sin\theta)}{z^2-2az\cos\theta+a^2}$	$\lvert z \rvert > a$
$k\cos k\theta$	$\dfrac{z(z^2\cos\theta-2z+\cos\theta)}{z^2-2z\cos\theta+1}$	$\lvert z \rvert > 1$
$\mu[k]-\mu[k-k_o],\quad k_o \in \mathbb{N}$	$\dfrac{1+z+z^2+\ldots+z^{k_o-1}}{z^{k_o-1}}$	$\lvert z \rvert > 0$

Table 12.1. Z-transform table

$f[k]$	$\mathcal{Z}\left[f[k]\right]$	Names
$\displaystyle\sum_{i=1}^{l} a_i f_i[k]$	$\displaystyle\sum_{i=1}^{l} a_i F_i(z)$	Partial fractions
$f[k+1]$	$zF(z) - zf(0)$	Forward shift
$\displaystyle\sum_{l=0}^{k} f[l]$	$\dfrac{z}{z-1}F(z)$	Summation
$f[k-1]$	$z^{-1}F(z) + f(-1)$	Backward shift
$y[k-l]\mu[k-l]$	$z^{-l}Y(z)$	Unit step
$kf[k]$	$-z\dfrac{dF(z)}{dz}$	
$\dfrac{1}{k}f[k]$	$\displaystyle\int_{z}^{\infty}\dfrac{F(\zeta)}{\zeta}d\zeta$	
$\displaystyle\lim_{k\to\infty} y[k]$	$\displaystyle\lim_{z\to 1}(z-1)Y(z)$	Final-value theorem
$\displaystyle\lim_{k\to 0} y[k]$	$\displaystyle\lim_{z\to\infty} Y(z)$	Initial value theorem
$\displaystyle\sum_{l=0}^{k} f_1[l]f_2[k-l]$	$F_1(z)F_2(z)$	Convolution
$f_1[k]f_2[k]$	$\dfrac{1}{2\pi j}\oint F_1(\zeta)F_2\left(\dfrac{z}{\zeta}\right)\dfrac{d\zeta}{\zeta}$	Complex convolution
$(\lambda)^k f_1[k]$	$F_1\left(\dfrac{z}{\lambda}\right)$	Frequency scaling

Table 12.2. Z-transform properties. Note that $F_i(z) = \mathcal{Z}\left[f_i[k]\right]$, that $\mu[k]$ denotes, as usual, a unit step, that $y[\infty]$ must be well defined, and that the convolution property holds (provided that $f_1[k] = f_2[k] = 0$ for all $k < 0$).

Note that

$$A_q(z) = z^n + a_{n-1}z^{n-1} + \cdots + a_o \qquad (12.7.2)$$
$$B_q(z) = b_m z^m + b_{m-1}z^{m-1} + \cdots + b_o \qquad (12.7.3)$$

and $f_q(z, x_o)$ is an initial-condition term, analogous to $f(s, x_o)$ in (4.5.1).

Equation (12.7.1) can be rewritten as

$$Y_q(z) = G_q(z)U_q(z) + \frac{f_q(z, x_o)}{A_q(z)} \qquad (12.7.4)$$

where

$$G_q(z) \triangleq \frac{B_q(z)}{A_q(z)} \qquad (12.7.5)$$

is called the *discrete (shift form) transfer function*. As in the continuous-time case, the transfer function uniquely determines the input-output behavior at the discrete sampling times, for zero initial conditions.

We can also use the Z-transform to obtain the transfer function corresponding to a discrete state space model in shift form. Taking Z-transforms in (12.5.3) and (12.5.4) and ignoring initial conditions yields

$$\mathbf{G_q}(z) = \mathbf{C_q}(z\mathbf{I} - \mathbf{A_q})^{-1}\mathbf{B_q} + \mathbf{D_q} \qquad (12.7.6)$$

The role of discrete transfer functions in describing the dynamic behavior of systems parallels that of transfer functions for continuous-time systems. In particular, the location of the poles (roots of $A_q(z)$) determines the natural modes of the system. Although continuous and discrete transfer functions share many common concepts, there are some special features of the discrete-time case, as is illustrated in the following two examples.

Example 12.5 (Poles at the origin–finite settling time). *Compute the unit step response of a discrete-time system with transfer function*

$$G_q(z) = \frac{0.5z^2 - 1.2z + 0.7}{z^3} \qquad (12.7.7)$$

Solution

The Z-transform of the system response, $y[k]$, to an input $u[k]$ is given by

$$Y_q(z) = \frac{0.5z^2 - 1.2z + 0.9}{z^3}U_q(z) = 0.5z^{-1}U_q(z) - 1.2z^{-2}U_q(z) + 0.9z^{-3}U_q(z)$$
$$(12.7.8)$$

Hence,

$$y[k] = 0.5u[k-1] - 1.2u[k-2] + 0.9u[k-3] \tag{12.7.9}$$

Then, when $u[k] = \mu[k]$, *the system response is given by*

$$y[k] = \begin{cases} 0 & k = 0 \\ 0.5 & k = 1 \\ -0.7 & k = 2 \\ 0.2 & \forall k \geq 3 \end{cases} \tag{12.7.10}$$

The key aspect of this system is that its step response has a finite settling time. This characteristic is due to the poles at the origin.

□□□

Example 12.6 (Stable negative real poles–ringing). *Find the unit step response of a system having the transfer function given by*

$$G_q(z) = \frac{0.5}{z + 0.8} \tag{12.7.11}$$

Solution

The Z-transform of the step response, $y[k]$, *is given by*

$$Y_q(z) = \frac{0.5}{z + 0.5} U_q(z) = \frac{0.5z}{(z + 0.5)(z - 1)} \tag{12.7.12}$$

Expanding in partial fractions (by means of MATLAB command **residue**), *we obtain*

$$Y_q(z) = \frac{z}{3(z-1)} - \frac{z}{3(z+0.5)} \iff y[k] = \frac{1}{3}\left(1 - (-0.5)^k\right)\mu[k] \tag{12.7.13}$$

Note that the response contains the term $(-0.5)^k$, which corresponds to an oscillating behavior (known as ringing). In discrete time, this can occur (as in the example) for a single negative real pole, whereas, in continuous time, a pair of complex-conjugate poles is necessary to produce this effect.

This behavior can be appreciated in Figure 12.3, where the step response (12.7.13) is shown.

□□□

12.8 Discrete Delta-Domain Models

The forward shift operator, q, defined in section §12.5, is the most commonly used discrete-time operator. However, in some applications, the forward shift operator can lead to difficulties. The reason for these difficulties is explained below.

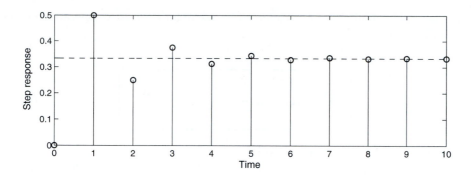

Figure 12.3. Unit step response of a system exhibiting ringing response

Although there is a clear resemblance between the differential-operator model of equation (4.2.4) and the shift-operator model of equation (12.5.2), there is a subtle (but far reaching) difference between these representations. To illustrate it, consider the first-order continuous-time equation

$$\rho y(t) + y(t) = \frac{dy(t)}{dt} + y(t) = u(t) \tag{12.8.1}$$

and the discretized shift-operator equation

$$a_2 q y(t_k) + a_1 y(t_k) = b_1 u(t_k) \tag{12.8.2}$$

Expanding the differential explicitly as a limiting operation in (12.8.1), we obtain

$$\lim_{\Delta \to 0} \left(\frac{y(t + \Delta) - y(t)}{\Delta} \right) + y(t) = u(t) \tag{12.8.3}$$

If we now compare (12.8.1) to the expanded form of (12.8.2),

$$a_2 y(t + \Delta) + a_1 y(t) = b_1 u(t); \qquad \text{where} \quad \Delta = t_{k+1} - t_k \tag{12.8.4}$$

we see that the fundamental difference between continuous and discrete time is tangibly captured by the limiting operation in (12.8.3). Beyond that, however, we notice that (12.8.3) is fundamentally based on a *relative* and *time-scaled* displacement $\frac{y(t+\Delta)-y(t)}{\Delta}$ from the absolute position $y(t)$, whereas (12.8.4) models the same dynamics by the *two absolute* positions $y(t)$ and $y(t+\Delta)$. In later chapters, we will see that this subtle difference prevents the discrete-time results based on shift operators from having any sensible relationship to the corresponding continuous-time results as $\Delta \to 0$.

This fundamental difficulty is avoided by use of an alternative operator, namely, the *Delta operator*:

$$\delta\left(f(k\Delta)\right) \triangleq \frac{f((k+1)\Delta) - f(k\Delta)}{\Delta} \qquad (12.8.5)$$

where Δ denotes the sampling period.

For sampled signals, an important feature of this operation is the observation that

$$\lim_{\Delta \to 0} [\delta\{f(k\Delta)\}] = \rho(f(t)) \qquad (12.8.6)$$

that is, as the sampling time goes to zero, the δ operator tends to the differential operator. Note, however, that *no approximations* will be involved in employing the delta operator for finite sampling periods, because we will derive *exact* model descriptions relevant to this operator at the given sampling rate.

Example 12.7. *Consider the PI controller of Example 12.4 on page 324, and let Δ denote the sampling period. Dividing both sides of (12.4.5) by Δ leads to the following exact delta-domain form of the controller:*

$$\delta u(k\Delta) = c_2 \delta e(k\Delta) + \frac{c_1}{\Delta} e(k\Delta) \qquad (12.8.7)$$

□□□

In delta form, the general discrete model (12.4.1) can be expressed as

$$\delta^n y[k] + a'_{n-1}\delta^{n-1}y[k] + \cdots + a'_0 y[k] = b'_m \delta^m u[k] + \cdots + b'_0 u[k] \qquad (12.8.8)$$

Note that there is a simple one-to-one relationship between the coefficients $(\bar{a}_i, \bar{b}_i)$ in (12.4.1) and (a'_i, b'_i) in (12.8.8).

Discrete-time state space models can also be described in the delta domain. The appropriate state space model format is

$$\delta x[k] = \mathbf{A}_\delta x[k] + \mathbf{B}_\delta u[k] \qquad (12.8.9)$$
$$y[k] = \mathbf{C}_\delta x[k] + \mathbf{D}_\delta u[k] \qquad (12.8.10)$$

We can readily convert a discrete state model given in shift form to one in delta form (and vice versa). For example, the model (12.5.3) and (12.5.4) becomes

$$\frac{qx[k] - x[k]}{\Delta} = \frac{(\mathbf{A}_q - \mathbf{I})}{\Delta} x[k] + \frac{\mathbf{B}_q}{\Delta} u[k] \qquad (12.8.11)$$

$$y[k] = \mathbf{C}_q x[k] + \mathbf{D}_q u[k]$$

This model is as in (12.8.9) and (12.8.10), where

$$\mathbf{A}_\delta = \frac{\mathbf{A}_q - \mathbf{I}}{\Delta}; \quad \mathbf{B}_\delta = \frac{\mathbf{B}_q}{\Delta}; \quad \mathbf{C}_\delta = \mathbf{C}_q; \quad \mathbf{D}_\delta = \mathbf{D}_q \qquad (12.8.12)$$

Example 12.8. *For the PI controller of Example 12.7 on the preceding page, we have*

$$\mathbf{A}_\delta = 0, \quad \mathbf{B}_\delta = \frac{c_1}{\Delta}, \quad \mathbf{C}_\delta = 1, \quad \mathbf{D}_\delta = c_2 \qquad (12.8.13)$$

□□□

The solution of (12.8.9), (12.8.10) is easily seen to be

$$x[k] = (I + \mathbf{A}_\delta \Delta)^k x[0] + \sum_{l=0}^{k-1} \Delta (I + \mathbf{A}_\delta \Delta)^l \mathbf{B}_\delta u[k - l - 1] \qquad (12.8.14)$$

$$y[k] = \mathbf{C}_\delta (I + \mathbf{A}_\delta \Delta)^k x[0] + \mathbf{C}_\delta \sum_{l=0}^{k-1} \Delta (I + \mathbf{A}_\delta \Delta)^l \mathbf{B}_\delta u[k - l - 1] + \mathbf{D}_\delta u[k]$$

$$(12.8.15)$$

12.9   Discrete Delta-Transform

As can be seen by comparing the results in Table 12.1 against those in Table 4.1, expressions in Laplace and Z-transform do not exhibit an obvious structural equivalence. Intuitively, we would expect such an equivalence to exist when the discrete sequence is obtained by sampling a continuous-time signal. In particular, one would expect that the Laplace Transform should approach the Z-transform as the sampling frequency increases. We will show that this indeed happens if we use the alternative delta operator. To show how this occurs, consider the sequence $\{y[k] = y(k\Delta)\}$ arising from sampling a continuous-time signal $y(t)$ every Δ seconds.

Recall that the Laplace Transform is given by

$$\mathcal{L}[y(t)] = \int_{0^-}^{\infty} e^{-st} y(t) dt \qquad (12.9.1)$$

It is natural to seek a discrete version of this Laplace Transform by using the Riemann sum

$$Y_\delta(s) = \sum_{k=0}^{\infty} e^{-st_k} y(t_k)\Delta \qquad\qquad t_k = k\Delta \qquad\qquad (12.9.2)$$

For reasons to be motivated presently, it is also desirable to invoke the isomorphic change of argument

$$e^{s\Delta} \triangleq 1 + \gamma\Delta \qquad\qquad (12.9.3)$$

$$\gamma = \frac{e^{s\Delta} - 1}{\Delta} \qquad\qquad (12.9.4)$$

In terms of the argument γ, we then define the *Discrete Delta-Transform* pair

$$\mathcal{D}\left[y(k\Delta)\right] \triangleq Y_\delta(\gamma) = \sum_{k=0}^{\infty}(1+\gamma\Delta)^{-k}y(k\Delta)\Delta \qquad\qquad (12.9.5)$$

$$\mathcal{D}^{-1}\left[Y_\delta(\gamma)\right] = y(k\Delta) = \frac{1}{2\pi j}\oint (1+\gamma\Delta)^{k-1}Y_\delta(\gamma)d\gamma \qquad\qquad (12.9.6)$$

The Discrete Delta-Transform can be related to the Z-transform by noting that

$$Y_\delta(\gamma) = \Delta Y_q(z)\Big|_{z=\Delta\gamma+1} \qquad\qquad (12.9.7)$$

where $Y_q(z) = \mathcal{Z}\left[y(k\Delta)\right]$. Conversely,

$$Y_q(z) = \frac{1}{\Delta}Y_\delta(\gamma)\Big|_{\gamma=\frac{z-1}{\Delta}} \qquad\qquad (12.9.8)$$

Equations (12.9.7) and (12.9.8) allow us to derive a table of Delta-Transforms from the corresponding Z-transforms–see Table 12.3. Properties of the Delta-Transform are given in Table 12.4.

A key property of Delta-Transforms is that they converge to the associated Laplace Transform as $\Delta \to 0$–that is,

$$\lim_{\Delta \to 0} Y_\delta(\gamma) = Y(s)\Big|_{s=\gamma} \qquad\qquad (12.9.9)$$

Next, we illustrate this by a simple example.

$f[k] \qquad (k \geq 0)$	$\mathcal{D}\left[f[k]\right]$	Region of Convergence
1	$\dfrac{1 + \Delta\gamma}{\gamma}$	$\left\|\gamma + \dfrac{1}{\Delta}\right\| > \dfrac{1}{\Delta}$
$\dfrac{1}{\Delta}\delta_K[k]$	1	$\|\gamma\| < \infty$
$\mu[k] - \mu[k-1]$	$\dfrac{1}{\Delta}$	$\|\gamma\| < \infty$
k	$\dfrac{1 + \Delta\gamma}{\Delta\gamma^2}$	$\left\|\gamma + \dfrac{1}{\Delta}\right\| > \dfrac{1}{\Delta}$
k^2	$\dfrac{(1 + \Delta\gamma)(2 + \Delta\gamma)}{\Delta^2\gamma^3}$	$\left\|\gamma + \dfrac{1}{\Delta}\right\| > \dfrac{1}{\Delta}$
$e^{\alpha\Delta k} \qquad \alpha \in \mathbb{C}$	$\dfrac{1 + \Delta\gamma}{\gamma - \frac{e^{\alpha\Delta}-1}{\Delta}}$	$\left\|\gamma + \dfrac{1}{\Delta}\right\| > \dfrac{e^{\alpha\Delta}}{\Delta}$
$ke^{\alpha\Delta k} \qquad \alpha \in \mathbb{C}$	$\dfrac{(1 + \Delta\gamma)e^{\alpha\Delta}}{\Delta\left(\gamma - \frac{e^{\alpha\Delta}-1}{\Delta}\right)^2}$	$\left\|\gamma + \dfrac{1}{\Delta}\right\| > \dfrac{e^{\alpha\Delta}}{\Delta}$
$\sin(\omega_o\Delta k)$	$\dfrac{(1 + \Delta\gamma)\omega_o\mathrm{sinc}(\omega_o\Delta)}{\gamma^2 + \Delta\phi(\omega_o,\Delta)\gamma + \phi(\omega_o,\Delta)}$ where $\mathrm{sinc}(\omega_o\Delta) = \dfrac{\sin(\omega_o\Delta)}{\omega_o\Delta}$ and $\phi(\omega_o,\Delta) = \dfrac{2(1 - \cos(\omega_o\Delta))}{\Delta^2}$	$\left\|\gamma + \dfrac{1}{\Delta}\right\| > \dfrac{1}{\Delta}$
$\cos(\omega_o\Delta k)$	$\dfrac{(1 + \Delta\gamma)(\gamma + 0.5\Delta\phi(\omega_o,\Delta))}{\gamma^2 + \Delta\phi(\omega_o,\Delta)\gamma + \phi(\omega_o,\Delta)}$	$\left\|\gamma + \dfrac{1}{\Delta}\right\| > \dfrac{1}{\Delta}$

Table 12.3. Delta-Transform table

$f[k]$	$\mathcal{D}\,[f[k]]$	Names
$\displaystyle\sum_{i=1}^{l} a_i f_i[k]$	$\displaystyle\sum_{i=1}^{l} a_i F_i(\gamma)$	Partial fractions
$f_1[k+1]$	$(\Delta\gamma + 1)(F_1(\gamma) - f_1[0])$	Forward shift
$\dfrac{f_1[k+1] - f_1[k]}{\Delta}$	$\gamma F_1(\gamma) - (1 + \gamma\Delta)f_1[0]$	Scaled difference
$\displaystyle\sum_{l=0}^{k-1} f[l]\Delta$	$\dfrac{1}{\gamma}F(\gamma)$	Reimann sum
$f[k-1]$	$(1 + \gamma\Delta)^{-1}F(\gamma) + f[-1]$	Backward shift
$f[k-l]\mu[k-l]$	$(1 + \gamma\Delta)^{-l}F(\gamma)$	
$kf[k]$	$-\dfrac{1+\gamma\Delta}{\Delta}\dfrac{dF(\gamma)}{d\gamma}$	
$\dfrac{1}{k}f[k]$	$\displaystyle\int_{\gamma}^{\infty}\dfrac{F(\zeta)}{1+\zeta\Delta}d\zeta$	
$\displaystyle\lim_{k\to\infty} f[k]$	$\displaystyle\lim_{\gamma\to 0}\gamma F(\gamma)$	Final-value theorem
$\displaystyle\lim_{k\to 0} f[k]$	$\displaystyle\lim_{\gamma\to\infty}\dfrac{\gamma F(\gamma)}{1+\gamma\Delta}$	Initial value theorem
$\displaystyle\sum_{l=0}^{k-1} f_1[l]f_2[k-l]\Delta$	$F_1(\gamma)F_2(\gamma)$	Convolution
$f_1[k]f_2[k]$	$\dfrac{1}{2\pi j}\displaystyle\oint F_1(\zeta)F_2\left(\dfrac{\gamma-\zeta}{1+\zeta\Delta}\right)\dfrac{d\zeta}{1+\zeta\Delta}$	Complex convolution
$(1+a\Delta)^k f_1[k]$	$F_1\left(\dfrac{\gamma-a}{1+a\Delta}\right)$	

Table 12.4. Delta-Transform properties. Note that $F_i(\gamma) = \mathcal{D}\,[f_i[k]]$, that $\mu[k]$ denotes, as usual, a unit step, that $f[\infty]$ must be well defined, and that the convolution property holds (provided that $f_1[k] = f_2[k] = 0$ for all $k < 0$).

Example 12.9. *Say that $\{y[k]\}$ arises from sampling, at period Δ, a continuous-time exponential $e^{\beta t}$. Then*

$$y[k] = e^{\beta k \Delta} \tag{12.9.10}$$

and, from Table 12.3,

$$Y_\delta(\gamma) = \frac{1 + \gamma \Delta}{\gamma - \left[\frac{e^{\beta \Delta} - 1}{\Delta}\right]} \tag{12.9.11}$$

In particular, note that as $\Delta \to 0, Y_\delta(\gamma) \to \dfrac{1}{\gamma - \beta}$ –the Laplace Transform of $e^{\beta t}$.

□□□

For our purposes, the key property of Delta-Transforms is the following:

The Delta-Transform can be used to convert difference equations into algebraic equations. The Delta-Transform also provides a smooth transition from discrete to continuous time as the sampling rate increases.

Historically, discrete systems analysis began with the use of the Delta-Transform. Emphasis later moved to the Z-transform. More recently, there has been a movement back to the Delta-Transform. This has been especially motivated by the availability of faster computers, which allow smaller sampling periods; in that case, the Delta form has major numerical and conceptual advantages, as we show below. Indeed, we conjecture that the reason that digital control has been seen as a subject separate from continuous control is, in part, the misunderstandings that were associated with the wide spread use of Z-Transforms and shift operators. These conceptual problems are resolved by the Delta Operation, which makes it clear that digital and continuous are actually quite close.

12.10   Discrete Transfer Functions (Delta Form)

Applying the discrete Delta-Transform and its difference rule to the high-order difference equation in delta form (12.8.8), we obtain

$$A_\delta(\gamma)Y_\delta(\gamma) = B_\delta(\gamma)U_\delta(\gamma) + f_\delta(\gamma, x_o) \tag{12.10.1}$$

where $Y_\delta(\gamma)$ and $U_\delta(\gamma)$ are the Delta-transforms of the sequences $\{y[k]\}$ and $\{u[k]\}$, respectively.

Here, analogously to (12.7.2) and (12.7.3),

$$A_\delta(\gamma) = \gamma^n + a'_{n-1}\gamma^{n-1} + \cdots + a'_o \qquad (12.10.2)$$

$$B_\delta(\gamma) = b'_m\gamma^m + b'_{m-1}\gamma^{m-1} + \cdots + b'_o \qquad (12.10.3)$$

and $f_\delta(\gamma, x_o)$ is an initial-condition term, analogous to $f(s, x_o)$ in (4.5.1) and to $f_q(z, x_o)$ in (12.7.1).

Equation (12.10.1) can be rewritten as

$$Y_\delta(\gamma) = G_\delta(\gamma)U_\delta(\gamma) + \frac{f_\delta(\gamma, x_o)}{A_\delta(\gamma)} \qquad (12.10.4)$$

where

$$G_\delta(\gamma) \triangleq \frac{B_\delta(\gamma)}{A_\delta(\gamma)} \qquad (12.10.5)$$

is called the *discrete (delta-form) transfer function*.

We can also use Delta-transforms to derive the transfer function corresponding to the delta-domain state space model given in (12.8.9) and (12.8.10). Taking the Delta-transforms and ignoring initial conditions, yields

$$\mathbf{G}_\delta(\gamma) = \mathbf{C}_\delta(\gamma\mathbf{I} - \mathbf{A}_\delta)^{-1}\mathbf{B}_\delta + \mathbf{D}_\delta \qquad (12.10.6)$$

12.11   Transfer Functions and Impulse Responses

For each of the three transforms introduced so far (namely, the Laplace Transform, the Z-transform and the Delta-Transform), there exists a characteristic signal, which when transformed, yields a value equal to one. These signals are as follows:

$\delta_D(t)$ (the Dirac-delta function) for the Laplace Transform;
$\delta_K[k]$ (the unit impulse, or Kronecker delta) for the Z-transform;
$\frac{1}{\Delta}\delta_K[k]$ (the scaled unit impulse) for the Delta-Transform.

We then see that the transfer function, in each of the three domains, is the corresponding transformation of the response of the system to the particular characteristic signal, with zero initial conditions.

12.12   Discrete System Stability

12.12.1   Relationship to Poles

We have seen that the response of a discrete system (in the shift operator) to an input $U(z)$ has the form

$$Y(z) = G_q(z)U(z) + \frac{f_q(z, x_o)}{(z - \alpha_1)(z - \alpha_2) \cdots (z - \alpha_n)} \qquad (12.12.1)$$

where $\alpha_1 \cdots \alpha_n$ are the poles of the system.

We then know, via a partial-fraction expansion, that $Y(z)$ can be written as

$$Y(z) = \sum_{j=1}^{n} \frac{\beta_j z}{z - \alpha_j} + \text{terms depending on } U(z) \qquad (12.12.2)$$

where, for simplicity, we have assumed nonrepeated poles.

The corresponding time response is

$$y[k] = \beta_j \left[\alpha_j\right]^k + \text{ terms depending on the input} \qquad (12.12.3)$$

Stability requires that $[\alpha_j]^k \to 0$, which is the case if $|\alpha_j| < 1$. Hence, stability requires the poles to have magnitude less than 1, i.e., to lie inside a unit circle centered at the origin.

12.12.2   Delta-Domain Stability

Actually, the delta domain is simply a shifted and scaled version of the Z-Domain, as is clear from, for example, (12.9.7) and (12.9.8). It follows that the delta-domain boundary is a circle of radius $\frac{1}{\Delta}$ centered on $-\frac{1}{\Delta}$ in the γ domain. Note again the close connection between the continuous s-domain and discrete δ-domain: The δ-stability region approaches the s-stability region (OLHP) as $\Delta \to 0$.

12.13   Discrete Models for Sampled Continuous Systems

As was mentioned in section §12.2, the controller in most modern control systems is implemented in a digital computer, whereas the process itself evolves in continuous time.

Thus, our objective in this section is to obtain discrete-time models that link the *sampled output* of a continuous-time system to the *sampled input*. We first recall how a digital controller is connected to the continuous-time plant.

A typical way of making this interconnection is shown in Figure 12.4.

The analog-to-digital converter (A/D in the figure) implements the process of sampling (usually at some fixed interval Δ). Anti-aliasing filters could also be introduced before the samples are taken. The digital-to-analog converter (D/A in the figure) interpolates the discrete control action into a function suitable for application to the plant input. In practice, this is usually achieved by holding the discrete-time signal over the sample period. This is known as a zero-order hold. It

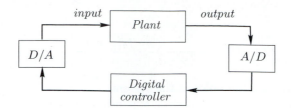

Figure 12.4. Digital control of a continuous-time plant

leads to the piecewise-constant or staircase input shown earlier in Figure 12.2 on page 321.

When a zero-order hold is used to construct $u(t)$, then

$$u(t) = u[k] \qquad \text{for} \quad k\Delta \leq t < (k+1)\Delta \qquad (12.13.1)$$

Discrete-time models typically relate the sampled signal $y[k]$ to the sampled input $u[k]$. Also, a digital control usually evaluates $u[k]$ on the basis of $y[j]$ and $r[j]$, where $\{r(k\Delta)\}$ is the reference sequence and $j \leq k$.

12.13.1 Using Continuous Transfer-Function Models

We observe that the generation of the staircase signal $u(t)$ from the sequence $\{u[k]\}$ can be modeled as in Figure 12.5.

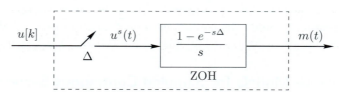

Figure 12.5. Zero-order hold

In Figure 12.5, the *impulse sampler* generates a Dirac sequence $u^s(t)$ given by

$$u^s(t) = \sum_{k=-\infty}^{\infty} u[k]\delta(t - k\Delta) \qquad (12.13.2)$$

This sequence, when processed through the zero-order-hold block, ZOH, yields the staircase signal $u(t)$. It is important to emphasize that the system in Figure 12.5 makes sense only when considered as a whole; the impulse sampler does not have any physical meaning.

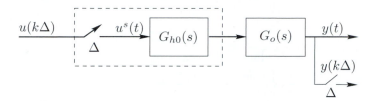

Figure 12.6. Discrete-time equivalent model with zero-order hold

We can now see that the digital controller drives an equivalent discrete-time system, as shown in Figure 12.6.

In Figure 12.6, we consider the discrete-time system with input $u[k]$ and output $y[k] = y(k\Delta)$. We know (per section §12.11) that the transfer function of a discrete-time system, in Z-transform form, is the Z-transform of the output (the sequence $\{y[k]\}$) when the input, $u[k]$, is a Kronecker delta, with zero initial conditions. We also have, from Figure 12.6, that if $u[k] = \delta_K[k]$, then $u^s(t) = \delta(t)$. If we denote by $H_{oq}(z)$ the transfer function from $U_q(z)$ to $Y_q(z)$, we then have the following result.

The discrete transfer function in Figure 12.6 satisfies

$$H_{oq}(z) = \mathcal{Z}\,[\text{the sampled impulse response of } G_{h0}(s)G_o(s)] \qquad (12.13.3)$$
$$= (1 - z^{-1})\mathcal{Z}\,[\text{the sampled step response of } G_o(s)] \qquad (12.13.4)$$

(The second line of the above expression can also be derived by noting that the Kronecker delta is equivalent to a step up at $k = 0$, followed by a delayed step down at $k = 1$).

The transfer function $H_{oq}(z)$ is sometimes expressed as $[G_{h0}G_o]_q$. This should be interpreted as in (12.13.3).

The discrete-transfer function is usually known as the *pulse-transfer function* of the continuous-time system. This terminology arises from the fact that a Dirac delta applied at the input of the zero-order sample-and-hold device translates into a pulse (unit height and width equal to $\Delta[s]$) at the input of the original continuous-time system.

Example 12.10. *Consider the d.c. servomotor problem of Example 3.4 on page 47.*
The continuous-time transfer function is

$$G_o(s) = \frac{b_0}{s(s + a_0)} \qquad (12.13.5)$$

Using (12.13.3), we see that

$$H_{oq}(z) = \frac{(z-1)}{z} \mathcal{Z} \left\{ \frac{b_0}{a_0}(k\Delta) - \frac{b_0}{a_0^2} + \frac{b_0}{a_0^2}e^{-\delta k} \right\} \tag{12.13.6}$$

$$= \frac{(z-1)}{a_0^2} \left\{ \frac{a_0 b_0 z \Delta}{(z-1)^2} - \frac{b_0 z}{z-1} + \frac{b_0}{z - e^{-a_0\Delta}} \right\}$$

$$= \frac{\left(b_0 a_0 \Delta + b_0 e^{-a_0\Delta} - b_0\right)z - b_0 a_0 \Delta e^{-a_0\Delta} - b_0 e^{-a_0\Delta} + b_0}{a_0^2(z-1)(z - e^{-a_0\Delta})}$$

□ □ □

12.14   Using Continuous State Space Models

Next, we show how to derive a discrete state space model when a zero-hold input is applied to a continuous-time plant (described in state space form). Consider a continuous-time plant in which the input to the plant, $u(t)$, is generated by a zero-order hold from an input sequence $\{u[k]\}$. This implies that

$$u(t) = u[k] \qquad \text{for} \quad k\Delta \leq t < (k+1)\Delta \tag{12.14.1}$$

If the plant is described by the continuous-time state space model

$$\frac{dx(t)}{dt} = \mathbf{A}x(t) + \mathbf{B}u(t) \tag{12.14.2}$$

$$y(t) = \mathbf{C}x(t) \tag{12.14.3}$$

then, by the solution formula (3.7.2), the sampled state response is given by

$$x((k+1)\Delta) = e^{\mathbf{A}\Delta}x(k\Delta) + \int_0^\Delta e^{\mathbf{A}(\Delta-\tau)}\mathbf{B}u(\tau)d\tau \tag{12.14.4}$$

and, using (12.14.1), we can write

$$x((k+1)\Delta) = \mathbf{A}_q x(k\Delta) + \mathbf{B}_q u(k\Delta) \tag{12.14.5}$$

where

$$\mathbf{A}_q = e^{\mathbf{A}\Delta} \tag{12.14.6}$$

$$\mathbf{B}_q = \int_0^\Delta e^{\mathbf{A}(\Delta-\tau)}\mathbf{B}d\tau \tag{12.14.7}$$

Also, the output is

$$y(k\Delta) = \mathbf{C}_q x(k\Delta) \qquad \text{where} \quad \mathbf{C}_q = \mathbf{C} \tag{12.14.8}$$

12.14.1   Shift Form

As usual, the discrete-time model (12.14.5)-(12.14.8) can be expressed compactly by using the forward shift operator, q, as follows:

$$qx[k] = \mathbf{A}_q x[k] + \mathbf{B}_q u[k] \tag{12.14.9}$$
$$y[k] = \mathbf{C}_q x[k] \tag{12.14.10}$$

where

$$\mathbf{A}_q \triangleq e^{\mathbf{A}\Delta} = \sum_{k=0}^{\infty} \frac{(\mathbf{A}\Delta)^k}{k!} \tag{12.14.11}$$

$$\mathbf{B}_q \triangleq \int_0^\Delta e^{\mathbf{A}(\Delta-\tau)} \mathbf{B} d\tau = \mathbf{A}^{-1}\left[e^{\mathbf{A}\Delta} - I\right] \quad \text{if} \quad \mathbf{A} \quad \text{is nonsingular} \tag{12.14.12}$$

$$\mathbf{C}_q \triangleq \mathbf{C} \tag{12.14.13}$$

$$\mathbf{D}_q \triangleq \mathbf{D} \tag{12.14.14}$$

12.14.2   Delta Form

Alternatively, if we define $t_k \triangleq k\Delta$, then (12.14.4)-(12.14.8) can be expressed in the delta-operator form as

$$\delta x(t_k) = \mathbf{A}_\delta x(t_k) + \mathbf{B}_\delta u(t_k) \tag{12.14.15}$$
$$y(t_k) = \mathbf{C}_\delta x(t_k) + \mathbf{D}_\delta u(t_k) \tag{12.14.16}$$

where $\mathbf{C}_\delta = \mathbf{C}_q = \mathbf{C}$, $\mathbf{D}_\delta = \mathbf{D}_q = \mathbf{D}$ and

$$\mathbf{A}_\delta \triangleq \frac{e^{\mathbf{A}\Delta} - I}{\Delta} \tag{12.14.17}$$

$$\mathbf{B}_\delta \triangleq \Omega \mathbf{B} \tag{12.14.18}$$

$$\Omega = \frac{1}{\Delta}\int_0^\Delta e^{\mathbf{A}\tau} d\tau = I + \frac{\mathbf{A}\Delta}{2!} + \frac{\mathbf{A}^2\Delta^2}{3!} + \dots \tag{12.14.19}$$

12.14.3   Some Comparisons of Shift and Delta Forms

There are several advantages over (12.14.9) associated with the formulation (12.14.15). For example, we can readily see that, for the delta form, we have

$$\lim_{\Delta \to 0} \mathbf{A}_\delta = \mathbf{A} \qquad\qquad \lim_{\Delta \to 0} \mathbf{B}_\delta = \mathbf{B} \tag{12.14.20}$$

where $(\mathbf{A}, \mathbf{B})$ are the underlying continuous values.

This coincides with the intuition that a discrete-time model should converge to the underlying continuous-time model as the sampling period goes to zero. Similarly, the left-hand side of (12.14.15) could be thought of as a finite difference approximation to the continuous derivative. These intuitions are lost in the formulation of (12.14.11), because

$$\lim_{\Delta \to 0} \mathbf{A}_q = I \qquad\qquad \lim_{\Delta \to 0} \mathbf{B}_q = 0 \qquad (12.14.21)$$

Other advantages of the delta format include the following:

- It can be shown that, under rapid sampling conditions, (12.14.15) is numerically better conditioned than (12.14.9).

- More fundamentally, (12.14.9) models the absolute displacement of the state vector in one interval, whereas the underlying continuous model is based on differentials, that is, infinitesimal increments of the state. This characteristic is better captured by (12.14.15), which also describes the increment of the state in one time step.

12.15 Frequency Response of Sampled-Data Systems

The idea of frequency response also applies to sampled-data systems. Consider a continuous-time system having a zero-order hold at the input. The input to the sample-and-hold device and the system output are then signals sampled every Δ seconds. Consider, now, a sine-wave input given by

$$u(k\Delta) = \sin(\omega k\Delta) = \sin\left(2\pi k \frac{\omega}{\omega_s}\right) = \frac{1}{2j}\left(e^{j2\pi k\frac{\omega}{\omega_s}} - e^{-j2\pi k\frac{\omega}{\omega_s}}\right) \qquad (12.15.1)$$

where $\omega_s = \frac{2\pi}{\Delta}$.

Following the same procedure as in the continuous-time case (section §4.9), we see that the system output response to the input (12.15.1) is

$$y(k\Delta) = \alpha(\omega)\sin(\omega k\Delta + \phi(\omega)) \qquad (12.15.2)$$

where

$$H_q(e^{j\omega\Delta}) = \alpha(\omega)e^{j\phi(\omega)} \qquad (12.15.3)$$

or

$$H_\delta \left(\frac{e^{j\omega\Delta} - 1}{\Delta} \right) = \alpha(\omega)e^{j\phi(\omega)} \tag{12.15.4}$$

and where $H_q(z)$ or $H_\delta(\gamma)$ is the discrete-system transfer function in z or δ form, respectively.

The frequency response of a sampled-data system can also be depicted by using Bode diagrams. However, $H_q(e^{j\omega\Delta})$ is a nonrational function of ω, so some of the simple rules for drawing Bode diagrams for continuous-time systems do not apply here. Today, this difficulty has lost some of its traditional significance, because modern computer packages provide a direct way to represent the frequency response of discrete-time systems graphically.

A characteristic feature of the frequency response of a sampled-data system is its periodicity (in ω). This property arises from the fact that $e^{j\omega\Delta}$ is periodic in ω with period $2\pi/\Delta$. The periodicity naturally holds for the magnitude as well as for the phase of the frequency response. To illustrate this idea, Figure 12.7 shows the frequency response of a sampled-data system having transfer function

$$H_q[z] = \frac{0.3}{z - 0.7} \tag{12.15.5}$$

Another feature of particular interest is that the sampled-data frequency response converges to its continuous counterpart as $\Delta \to 0$, and hence much insight can be obtained simply by looking at the continuous version. This is exemplified below.

Example 12.11. *Consider the two systems shown in Figure 12.8 on the next page. Compare the frequency responses of the systems in the range $[0, \omega_s]$.*

Solution

For System 1, the frequency response is given by

$$H(j\omega) \triangleq \frac{Y(j\omega)}{U(j\omega)} = \frac{a}{j\omega + a} \tag{12.15.6}$$

For System 2, the frequency response is given by

$$H_q(e^{j\omega\Delta}) \triangleq \frac{Y_q(e^{j\omega\Delta})}{U_q(e^{j\omega\Delta})} = \mathcal{Z}\left\{ G_{h0}(s)\frac{a}{s+a} \right\}\bigg|_{z=e^{j\omega\Delta}} = \frac{1 - e^{-a\Delta}}{e^{j\omega\Delta} - e^{-a\Delta}} \tag{12.15.7}$$

Note that, if $\omega \ll \omega_s$ and $a \ll \omega_s$–i.e., $\omega\Delta \ll 1$ and $a\Delta \ll 1$–then we can use a first-order Taylor's series approximation for the exponentials $e^{-a\Delta}$ and $e^{j\omega\Delta}$, leading to

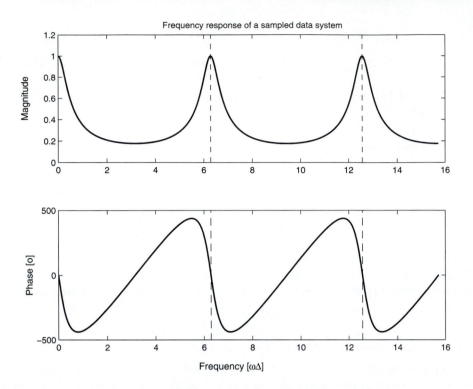

Figure 12.7. Periodicity in the frequency response of sampled-data systems

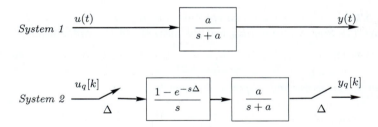

Figure 12.8. Continuous and sampled-data systems

$$H_q(j\omega\Delta) \approx \frac{1 - 1 + a\Delta}{1 + j\omega\Delta - 1 + a\Delta} = \frac{a}{j\omega + a} = H(j\omega) \qquad (12.15.8)$$

Hence, both frequency responses will asymptotically (in ω_s) approach each other. To illustrate this asymptotic behavior, an error function $H_e(j\omega)$ is defined as

$$H_e(j\omega) = H(j\omega) - H_q[e^{j\omega}] \qquad (12.15.9)$$

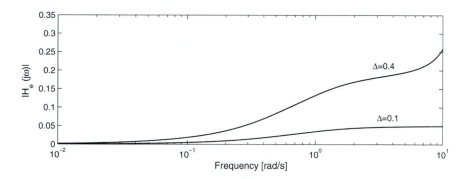

Figure 12.9. Asymptotic behavior of a sampled-data transfer function

Figure 12.9 displays the magnitude of the frequency response of $H_e(j\omega)$, for two different values of the sampling period Δ and for $a = 1$. We see that, for Δ small, the continuous and discrete frequency responses are very close indeed.

A complete analysis of this issue, in the framework of digital control, will be carried out in Chapter 14.

□□□

12.16 Summary

- Very few plants encountered by the control engineer are digital; most are continuous. That is, the control signal applied to the process, as well as the measurements received from the process, are usually continuous time.

- Modern control systems, however, are almost exclusively implemented on digital computers.

- Compared to the historical analog-controller implementation, the digital computer provides

  o much greater ease of implementing complex algorithms,

  o convenient (graphical) man-machine interfaces,

  o logging, trending, and diagnostics of the internal controller, and

  o flexibility to implement filtering and other forms of signal-processing operations.

- Digital computers operate with sequences in time, rather than continuous functions in time.

- Therefore,

    ○ input signals to the digital controller–notably, process measurements– must be sampled;

    ○ outputs from the digital controller–notably, control signals–must be interpolated from a digital sequence of values to a continuous function in time.

- Sampling (see Figure 12.10) is carried out by A/D (analog-to-digital) converters.

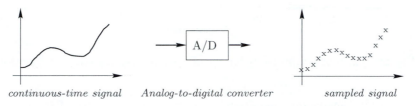

continuous-time signal   Analog-to-digital converter   sampled signal

Figure 12.10. The result of sampling

- The converse, reconstructing a continuous-time signal from digital samples, is carried out by D/A (digital-to-analog) converters. There are different ways of interpolating between the discrete samples, but the so-called zero-order hold (see Figure 12.11) is by far the most common.

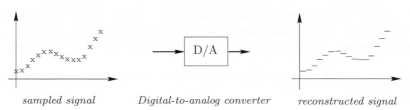

sampled signal   Digital-to-analog converter   reconstructed signal

Figure 12.11. The result of reconstruction

- When sampling a continuous-time signal,

    ○ an appropriate sampling rate must be chosen;

    ○ an anti-aliasing filter (low-pass) should be included to avoid frequency folding.

- Analysis of digital systems relies on discrete-time versions of the continuous operators.

- The chapter has introduced two discrete operators:

    - the shift operator, q, defined by $qx[k] \triangleq x[k+1]$
    - the δ-operator, δ, defined by $\delta x[k] \triangleq \dfrac{x[k+1] - x[k]}{\Delta}$

- Thus, $\delta = \dfrac{q-1}{\Delta}$, or $q = \delta\Delta + 1$.

- Thanks to this conversion possibility, the choice is largely based on preference and experience. Comparisons are outlined below.

- The shift operator, q,

    - is the traditional operator;
    - is the operator many engineers feel more familiar with;
    - is used in the majority of the literature.

- The δ-operator, δ, has the following advantages:

    - it emphasizes the link between continuous and discrete systems (resembles a differential);
    - δ-expressions converge to familiar continuous expressions as $\Delta \to 0$, which is intuitive;
    - it is vastly superior numerically at fast sampling rates, when properly implemented.

- Analysis of digital systems relies on discrete-time versions of the continuous operators.

    - The discrete version of the differential operator is a difference operator.
    - The discrete version of the Laplace Transform is either the Z-transform (associated with the shift operator) or the γ-transform (associated with the δ-operator).

- With the help of these operators,

    - continuous-time differential-equation models can be converted to discrete-time difference-equation models;
    - continuous-time transfer or state space models can be converted to discrete-time transfer or state space models in either the shift or δ operators.

12.17   Further Reading

Sampling

Feuer, A. and Goodwin, G.C. (1996). *Sampling in Digital Signal Processing and Control*. Birkäusser Boston, Cambridge, Mass.

Kwakernaak, H. and Sivan, R. (1991). *Modern Signals and Systems*. Prentice-Hall, Englewood Cliffs, N.J.

Oppenheim, A.V., Wilsky, A.S., and Hamid, N.S. (1997). *Signals and Systems*. Prentice-Hall, Upper Saddle River, N.J., 2nd edition.

Shift operator and Z-transform

Åström, K. and Wittenmark, B. (1990). *Computer Controlled Systems. Theory and Design*. Prentice-Hall, Englewood Cliffs, N.J., 2nd edition.

Cadzow, J. (1985). *Signals, Systems and Transforms*. Prentice-Hall, Englewood Cliffs, N.J.

Delta-Transform

Feuer, A. and Goodwin, G.C. (1996). *Sampling in Digital Signal Processing and Control*. Birkäusser Boston, Cambridge, Mass.

Goodwin, G.C, Middleton, R.H., and Poor, H.V. (1992). High-speed digital signal processing and control. Invited paper in: *IEEE Proceedings*, 80(2):240-259.

Middleton, R.H. and Goodwin, G.C. (1990). *Digital Control and Estimation. A Unified Approach*. Prentice-Hall, Englewood Cliffs, N.J.

12.18   Problems for the Reader

Problem 12.1. *Compute the Z-transform and the discrete Delta-Transform for the discrete sequences that result from sampling the following signals at* 1 *[Hz].*

(a) $\mu(t) - \mu(t - 3)$ (b) $e^{-0.1t}\cos(0.5t + \pi/4)$

(c) $t^2 e^{-0.25t}$ (d) $te^{-0.1t}\cos(t)$

Problem 12.2. *Consider the following recursive equations describing the relationship between the input* $u[k]$ *and the output* $y[k]$ *in different discrete-time (sampled-data) systems.*

(a) $y[k] - 0.8y[k - 1] = 0.4u[k - 2]$ (12.18.1)

(b) $y[k] - 0.5y[k - 1] + 0.06y[k - 2] = 0.6u[k - 1] + 0.3u[k - 2]$ (12.18.2)

(c) $y[k] - 0.5y[k - 1] + 0.64y[k - 2] = u[k - 1]$ (12.18.3)

12.2.1 *For each case, determine the transfer function.*

12.2.2 *From the above result, compute the response of each system to a unit Kronecker delta.*

Problem 12.3. *Determine the step response of the discrete-time systems that have the following transfer functions in Z-form:*

$$\frac{z - 0.5}{z^2(z + 0.5)} \qquad \frac{z}{z^2 - 0.5z + 1} \qquad (12.18.4)$$

$$\frac{1}{(z - 0.6)^2} \qquad \frac{z + 1}{z^2 - 0.4z + 0.64} \qquad (12.18.5)$$

Problem 12.4. *Consider a continuous-time system having the transfer function*

$$G(s) = \frac{4}{s^2 + 2.4s + 4} \qquad (12.18.6)$$

Find a sampling rate such that the pulse transfer function $[GG_{h0}]_q(z)$ *has only one pole.*

Problem 12.5. *Assume that, in Figure 12.6 on page 339, $G_o(s)$ is given by*

$$G_o(s) = \frac{2}{(s+1)(s+2)} \qquad (12.18.7)$$

12.5.1 *Compute the Delta-Transform of the transfer function from $u[k]$ to $y[k]$, $H_{o\delta}(\gamma)$, as a function of the sampling interval Δ.*

12.5.2 *Verify that, if we make $\Delta \to 0$, then*

$$\lim_{\Delta \to 0} H_{o\delta}(\gamma) \Big|_{\gamma=s} = G_o(s) \qquad (12.18.8)$$

Problem 12.6. *The output $y(t)$ of a continuous-time system having a unit step input is sampled every $1[s]$. The expression for the sequence $\{y[k]\}$ is given by*

$$y[k] = 0.5 - 0.5(0.6)^k \qquad \forall k \geq 0 \qquad (12.18.9)$$

12.6.1 *Determine $Y_q(z)$.*

12.6.2 *Determine the transfer function from $U_q(z)$ to $Y_q(z)$.*

12.6.3 *From the above result, derive the difference equation linking $\{y[k]\}$ to $\{u[k]\}$.*

Problem 12.7. *Consider the setup shown in Figure 12.6 on page 339, where the sampling interval is $\Delta = 0.5[s]$. Assume that the transfer function from $U_q(z)$ to $Y_q(z)$ is given by*

$$H_{oq}(z) = \frac{z - 0.5}{(z - 0.8)(z - 0.2)} \qquad (12.18.10)$$

12.7.1 *Find $G_o(s)$. (Try MATLAB command **d2c**.)*

12.7.2 *Explain why the above solution is not unique and provide some alternative expressions for $G_o(s)$ that also yield (12.18.10).*

Problem 12.8. *Consider the setup shown in Figure 12.6 on page 339, where*

$$G(s) = \frac{e^{-s}}{s + 0.8} \qquad (12.18.11)$$

Find the transfer function from $U_q(z)$ to $Y_q(z)$, first for $\Delta = 1[s]$ and then for $\Delta = 0.75[s]$.

Problem 12.9. *The transfer function of a sampled-data system (in delta form) is given by*

$$G_\delta(\gamma) = \frac{\gamma + 0.5}{(\gamma - 0.1)(\gamma + 0.8)} \qquad (12.18.12)$$

12.9.1 *If $\Delta = 3.5[s]$, is the system stable?*

12.9.2 *Find the corresponding Z-transform transfer function for $\Delta = 3.5[s]$.*

12.9.3 *Repeat parts (12.9.1) and (12.9.2) for $\Delta = 1.5[s]$.*

Problem 12.10. *Consider a continuous-time and a discrete-time low-pass filter having transfer functions $H(s)$ and $H_\delta(\gamma)$, respectively. Assume that, for the digital filter, the sampling frequency, ω_s, is chosen equal to $25[rad/s]$, and that*

$$H(s) = \frac{1}{s^2 + \sqrt{2}s + 1} \qquad H_\delta(\gamma) = \frac{1}{\gamma^2 + \sqrt{2}\gamma + 1} \qquad (12.18.13)$$

12.10.1 *Compare the frequency responses for the filters in the frequency range $[0; \quad 3\omega_s]$. Comment your findings.*

12.10.2 *Can this digital filter act as an anti-aliasing filter?*

Problem 12.11. *Consider two transfer functions, $G_1(s)$ and $G_2(s)$, given by*

$$G_1(s) = \frac{1}{s+1} \qquad and \qquad G_2(s) = \frac{2}{s+2} \qquad (12.18.14)$$

Compare the frequency responses of the following two pulse transfer functions:

$$[G_{ho}G_1G_2]_q(z) \qquad and \qquad [G_{ho}G_1]_q(z)[G_{ho}G_2]_q(z) \qquad (12.18.15)$$

for two different sampling periods: $\Delta = 0.05[s]$ and $\Delta = 0.5[s]$. Discuss the general problem raised by this example.

Chapter 13

DIGITAL CONTROL

13.1 Preview

Models for discrete-time systems have been described in Chapter 12. There, we saw that digital and continuous systems were actually quite close. Hence, it is often true that digital responses approach the corresponding continuous response as the sampling period goes to zero. For this reason, in the remainder of the book we will present continuous and discrete ideas in parallel. The purpose of the current chapter is to provide a smooth transition to this latter work by highlighting the special issues associated with digital control. In particular, the chapter covers

- why one cannot simply treat digital control as if it were exactly the same as continuous control, and

- how to carry out designs for digital-control systems so that the *at-sample* response is treated exactly.

13.2 Discrete-Time Sensitivity Functions

The reader is asked to recall the results presented in Chapter 12 on models for sampled-data systems.

We assume (as will almost always be the case) that the plant operates in continuous time, whereas the controller is implemented in digital form. Having the controller implemented in digital form introduces several constraints into the problem:

(a) the controller *sees* the output response only at the sample points,

(b) an anti-aliasing filter will usually be needed (see section §12.2) prior to the output-sampling process, to avoid folding of high-frequency signals (such as noise) onto lower frequencies, where they will be misinterpreted, and

(c) the continuous plant input bears a simple relationship to the (sampled) digital controller output–e.g., via a zero-order-hold device.

A key idea from Chapter 12 is that, if one is interested only in the *at-sample* response, these samples can be described by discrete-time models in either the shift or delta operator. For example, consider the sampled-data control loop shown in Figure 13.1.

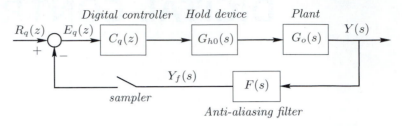

Figure 13.1. Sampled-data control loop

Note that, in the figure, we have used the Laplace variable s to describe continuous-time transfer functions and the z-transform variable z to describe the digital controller.

If we focus on only the sampled response, then it is straightforward to derive an equivalent discrete model for the *at-sample* response of the hold-plant-anti-aliasing filter combination. This can be done either via transfer function or state space methods. For convenience, we here use the transfer-function form, which we recall from (12.13.3).

$$[FG_oG_{h0}]_q(z) \triangleq \mathcal{Z}\{\text{sampled impulse response of } F(s)G_o(s)G_{h0}(s)\} \qquad (13.2.1)$$

$$[G_oG_{h0}]_q(z) \triangleq \mathcal{Z}\{\text{sampled impulse response of } G_o(s)G_{h0}(s)\} \qquad (13.2.2)$$

Given the equivalent discrete-plant transfer function, we can immediately write down other relevant closed-loop transfer functions:

The discrete sensitivity function is

$$S_{oq}(z) = \frac{E_q(z)}{R_q(z)} = \frac{1}{\left(1 + C_q(z)[FG_oG_{h0}]_q(z)\right)} \qquad (13.2.3)$$

The discrete complementary sensitivity function is

$$T_{oq}(z) = \frac{Y_{fq}(z)}{R_q(z)} = \frac{C_q(z)[FG_oG_{h0}]_q(z)}{\left(1 + C_q(z)[FG_oG_{h0}]_q(z)\right)} \qquad (13.2.4)$$

So far, this all seems directly analogous to the continuous case (and, indeed, it is). Of course, the transfer functions above capture only the sampled responses. More will be said about this in Chapter 14, when we investigate the intersample response of a digital-control loop. The term *intersample response* refers to the actual–but unobservable (at the computer)–continuous-time response of the underlying process.

13.3   Zeros of Sampled-Data Systems

As shown in Chapter 8, the open-loop zeros of a system have a profound impact on achievable closed-loop performance. The importance of an understanding of the zeros in discrete-time models is therefore not surprising. It turns out that there exist some subtle issues here, as we now investigate.

If we use shift-operator models, then it is difficult to see the connection between continuous- and discrete-time models. However, if we use the equivalent delta-domain description, then it is clear that discrete transfer functions converge to the underlying continuous-time descriptions. In particular, the relationship between continuous and discrete (delta-domain) poles is revealed by equation (12.14.17). Specifically, we see that

$$p_i^\delta = \frac{e^{p_i \Delta} - 1}{\Delta}; \qquad i = 1, \dots n \qquad (13.3.1)$$

where p_i^δ and p_i denote the discrete (delta-domain) poles and the continuous-time poles, respectively. In particular, we see that $p_i^\delta \to p_i$ as $\Delta \to 0$.

However, the relationship between continuous and discrete zeros is more complex. Perhaps surprisingly, all discrete-time systems turn out to have relative degree 1, irrespective of the relative degree of the original continuous system.[1]

Hence, if the continuous system has n poles and $m(< n)$ zeros, then the corresponding discrete system will have n poles and $(n - 1)$ zeros.

In view of the convergence of the discrete zeros to the continuous zeros, we (somewhat artificially) divide the discrete zeros into two sets.

1. **System zeros**: $z_1^\delta, \cdots, z_m^\delta$ having the property that

$$\lim_{\Delta \to 0} z_i^\delta = z_i \qquad i = 1, \dots, m \qquad (13.3.2)$$

where z_i^δ are the discrete-time zeros (expressed in the delta domain for convenience) and z_i are the zeros of the underlying continuous-time system.

---

[1]Exceptions arise when sampling continuous-time systems either with pure time delays or with NMP zeros (and a particular choice of sampling period).

2. **Sampling zeros**: $z_{m+1}^\delta, \cdots z_{n-1}^\delta$ having the property that

$$\lim_{\Delta \to 0} \left| z_i^\delta \right| = \infty \qquad i = m+1, \ldots, n-1 \qquad (13.3.3)$$

Of course, if $m = n - 1$ in the continuous-time system, then there are no sampling zeros. Also, note that, as the sampling zeros tend to infinity, they contribute to the continuous relative degree.

Example 13.1. *Consider the continuous-time servo system of Example 3.4 on page 47, having continuous-transfer function*

$$G_o(s) = \frac{1}{s(s+1)} \qquad (13.3.4)$$

where $n = 2$, $m = 0$. Then we anticipate that discretizing would result in one sampling zero, which we verify as follows.

With a sampling period of 0.1 seconds, the exact shift-domain digital model is

$$G_{oq}(z) = K \frac{z - z_o^q}{(z-1)(z-\alpha_o)} \qquad (13.3.5)$$

where $K = 0.0048$, $z_o^q = -0.967$, and $\alpha_o = 0.905$.
The corresponding exact delta-domain digital model is

$$G_\delta(\gamma) = \frac{K'(\gamma - z_o^\delta)}{\gamma(\gamma - \alpha_o')} \qquad (13.3.6)$$

where $K' = 0.0048$, $z_o^\delta = -19.67$, and $\alpha_o = -0.9516$.
The location of the shift- and delta-domain sampling zeros, as a function of the sampling interval Δ, are given, respectively, by

$$z_o^q = \frac{(1+\Delta)e^{-\Delta} - 1}{\Delta + e^{-\Delta} - 1} \qquad and \qquad z_o^\delta = \frac{e^{-\Delta} - 1}{\Delta + e^{-\Delta} - 1} \qquad (13.3.7)$$

from which it is straightforward to verify that, for very small Δ, $z_o^q \to (-1)^+$ and $z_o^\delta \to \infty$. Also, for very large Δ, $z_o^q \to 0^-$ and $z_o^\delta \to 0^-$. These evolutions are shown in Figure 13.2.

□□□

In the control of discrete-time systems, special care needs to be taken with the sampling zeros. For example, these zeros can be nonminimum phase even if the

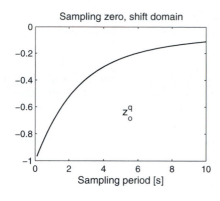
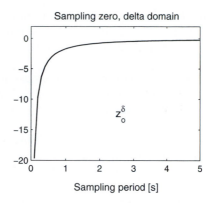

Figure 13.2. Location of sampling zero with different sampling periods (Example 13.1)

original continuous system is minimum phase. Consider, for instance, the minimum-phase continuous-time system having transfer function given by

$$G_o(s) = \frac{s+4}{(s+1)^3} \tag{13.3.8}$$

For this system, the shift-domain zeros of $[G_oG_{h0}]_q(z)$ for two different sampling periods are

$$\Delta = 2[s] \quad \Rightarrow \text{ zeros at } -0.6082 \text{ and } -0.0281$$
$$\Delta = 0.5[s] \quad \Rightarrow \text{ zeros at } -1.0966 \text{ and } 0.1286$$

Note that, for $\Delta = 0.5[s]$, the pulse-transfer function has a zero outside the stability region.

13.4   Is a Dedicated Digital Theory Really Necessary?

We have seen in section §13.3 that, under reasonable conditions, the discrete-time model of a system will converge, as the sampling rate increases, to the underlying continuous-time model.

This inevitably leads to the question: *Do we really need to worry about a separate theory of digital control?* It is intuitively reasonable that the constraints labeled (a), (b), (c) in section §13.2 are almost certainly negligible if one samples quickly enough. Also, we see from section §13.3 that discrete (delta-domain) poles and zeros converge to the continuous locations as $\Delta \to 0$. Although these observations are absolutely correct, the sampling rate needed is probably somewhat higher than might be used in practice. Practical sampling rates in current use are typically 5 to 10 times the closed-loop bandwidth, whereas to make the impact of sampling truly negligible would require sampling rates an order of magnitude higher.

One therefore needs to think about digital control in its own right. Three possible design options are the following:

1) Design the controller in continuous time, discretize the result for implementation, and ensure that the sampling constraints do not significantly affect the final performance.

2) Work in discrete time by doing an exact analysis of the *at-sample* response, and ensure that the intersample response is not too surprising.

3) Carry out an exact design by optimizing the continuous response with respect to the (constrained) digital controller.

The first two of these options will be analyzed in the following sections.

13.5 Approximate Continuous Designs

Here, we explore the idea of simply carrying out a normal continuous-time design and then mapping the resultant controller into the discrete domain.

To illustrate the effect of simple continuous-to-discrete mappings in control-system design, we mention three methods drawn from the digital signal-processing literature.

1. Simply take a continuous-time controller expressed in terms of the Laplace variable s, and then replace every occurrence of s by the corresponding delta-domain operator γ. This leads to the following digital-control law:

$$\overline{C}_1(\gamma) = C(s)\big|_{s=\gamma} \tag{13.5.1}$$

where $C(s)$ is the transfer function of the continuous-time controller and where $\overline{C}_1(\gamma)$ is the resultant transfer function of the discrete-time controller in delta form.

2. Convert the controller to a zero-order-hold discrete equivalent. This is called a *step-invariant transformation*. This leads to

$$\overline{C}_2(\gamma) = \mathcal{D}\,[\text{sampled impulse response of } \{C(s)G_{h0}(s)\}] \tag{13.5.2}$$

where $C(s)$, $G_{h0}(s)$ and $\overline{C}_2(\gamma)$ are the transfer functions of the continuous-time controller, zero-order hold, and resultant discrete-time controller, respectively.

3. We could use a more sophisticated mapping from s to γ. For example, we could carry out the following transformation, commonly called a *bilinear transformation with prewarping*. We first let

$$s = \frac{\alpha\gamma}{\frac{\Delta}{2}\gamma + 1} \iff \gamma = \frac{s}{\alpha - \frac{\Delta}{2}s} \tag{13.5.3}$$

The discrete controller is then defined by

$$\overline{C}_3(\gamma) = C(s)|_{s = \frac{\alpha\gamma}{\frac{\Delta}{2}\gamma+1}} \tag{13.5.4}$$

We next choose α so as to make the frequency responses of the two controllers match at some desired frequency, say ω^*. For example, one might choose ω^* as the frequency at which the continuous-time sensitivity function has its maximum value.

Now, we recall from Chapter 12 that the discrete frequency response at ω^* is evaluated by using $\gamma = \left(e^{j\omega^*\Delta} - 1\right)/\Delta$, and the continuous frequency response is obtained by replacing s by $j\omega^*$. Hence, to equate the discrete and continuous frequency responses at ω^*, we require that α satisfy

$$j\omega^* = \frac{\alpha\left[\frac{e^{j\omega^*\Delta}-1}{\Delta}\right]}{\frac{\Delta}{2}\left[\frac{e^{j\omega^*\Delta}-1}{\Delta}\right] + 1} \tag{13.5.5}$$

The solution is

$$\alpha = \frac{\omega^*\Delta}{2}\left[\frac{\sin\omega^*\Delta}{1 - \cos\omega^*\Delta}\right] = \frac{\omega^*\Delta}{2}\cot\left[\frac{\omega^*\Delta}{2}\right] \tag{13.5.6}$$

We illustrate by the following examples.

Example 13.2. *A plant has a nominal model given by*

$$G_o(s) = \frac{1}{(s-1)^2} \tag{13.5.7}$$

13.2.1 Synthesize a continuous-time PID controller such that the dominant closed-loop poles are the roots of the polynomial $s^2 + 3s + 4$.

13.2.2 Using the above result, obtain a discrete-time PID controller. Assume that the sampling frequency can be as large as required and that a zero-order hold is present at the plant input.

13.2.3 Using SIMULINK, compare the response to a unit step reference for both the continuous- and the discrete-time loop.

Solution

13.2.1 The closed-loop characteristic polynomial $A_{cl}(s)$ is chosen as

$$A_{cl}(s) = (s^2 + 3s + 4)(s^2 + 10s + 25) \qquad (13.5.8)$$

where the factor $s^2 + 10s + 25$ has been added to ensure that the degree of $A_{cl}(s)$ is 4, which is the minimum degree required for an arbitrarily chosen $A_{cl}(s)$.

Upon solving the pole-assignment equation, we obtain $P(s) = 88s^2 + 100s + 100$ and $\bar{L}(s) = s + 15$. This leads to the following PID controller:

$$C(s) = \frac{88s^2 + 100s + 100}{s(s + 15)} \qquad (13.5.9)$$

13.2.2 We can use high sampling rates, so a straightforward procedure to obtain a discrete-time PID controller is to replace s by γ in (13.5.9):

$$C_\delta(\gamma) = \frac{88\gamma^2 + 100\gamma + 100}{\gamma(\gamma + 15)} \qquad (13.5.10)$$

or, in Z-transform form,

$$C_q(z) = \frac{88z^2 - 166z + 79}{(z - 1)(z + 0.5)} \qquad (13.5.11)$$

where we have assumed a sampling interval $\Delta = 0.1$.

13.2.3 The continuous- and the discrete-time loops are simulated with SIMULINK for a unit step reference at $t = 1$ and a unit step input disturbance at $t = 10$. The difference of the plant outputs is shown in Figure 13.3 on the facing page.

*The reader is invited to verify that if a smaller sampling period is used, say $\Delta = 0.01$, then the difference would be barely noticeable. Both loops are included in the SIMULINK scheme in file **lcodi.mdl**.*

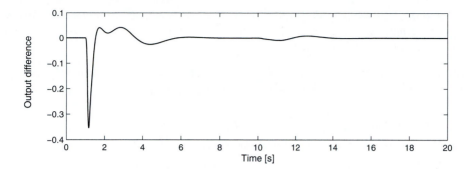

Figure 13.3. Difference in plant outputs due to discretization of the controller
(sampling period =0.1 [s])

However, the fact that none of the ad-hoc transformations listed above is entirely
satisfactory with more modest sampling rates is illustrated in the following example.

Example 13.3. *The system nominal transfer function is given by*

$$G_o(s) = \frac{10}{s(s+1)} \qquad (13.5.12)$$

and the continuous-time controller is

$$C(s) = \frac{0.416s + 1}{0.139s + 1} \qquad (13.5.13)$$

Replace this controller by a digital controller with $\Delta = 0.157[s]$ *preceded by a
sampler and followed by a ZOH, using each of the three approximations outlined
above. Test the unit step response for each such approximation.*

Solution

1. *Replacing s by* γ *in* $C(s)$, *we get*

$$\overline{C}_1(\gamma) = \frac{0.416\gamma + 1}{0.139\gamma + 1} \qquad (13.5.14)$$

2. *The ZOH equivalent of* $C(s)$ *is*

$$\overline{C}_2(\gamma) = \frac{0.694\gamma + 1}{0.232\gamma + 1} \qquad (13.5.15)$$

3. *For the bilinear mapping with prewarping, we first look at the continuous-time sensitivity function*

$$S_o(j\omega) = \frac{1}{1 + C(j\omega)G(j\omega)} \qquad (13.5.16)$$

We find that $|S_o(j\omega)|$ has a maximum at $\omega^* = 5.48\,[rad/s]$.

Now, using the formula (13.5.6), we obtain $\alpha = 0.9375$.

With this value of α, we find the approximation

$$\overline{C}_3(\gamma) = C(s)\big|_{s = \frac{\alpha\gamma}{\frac{\Delta}{2}\gamma+1}} = \frac{0.4685\gamma + 1}{0.2088\gamma + 1} \qquad (13.5.17)$$

The closed-loop unit step responses obtained with the continuous-time controller $C(s)$ and the three discrete-time controllers $\overline{C}_1(\gamma), \overline{C}_2(\gamma)$, and $\overline{C}_3(\gamma)$ are presented in Figure 13.4 on the next page.

□□□

We see from the figure that none of the approximations exactly reproduces the closed-loop response obtained with the continuous-time controller. Actually, for this example, we see that simple substitution appears to give the best result and that there is not much to be gained by fancy methods here. However, it would be dangerous to draw general conclusions from this one example.

The above example shows the difficulty of obtaining discrete-time control laws by ad-hoc means. We therefore proceed to examine discrete-time and sampled-data control-system design.

13.6   At-Sample Digital Design

The next option we explore is that of doing an exact digital control-system design *for the sampled response.*

We recall that the sampled response is exactly described by appropriate discrete-time models (expressed in either the shift or delta operation).

Many possible design strategies could be used. We briefly outline some of the time- and frequency-domain options.

13.6.1   Time-Domain Design

Any algebraic technique (such as pole assignment) has an immediate digital counterpart. Essentially all that is needed is to work with z (or γ) instead of the Laplace variable s and to keep in mind the different region for closed-loop stability. Recall the following regions for closed-loop stability:

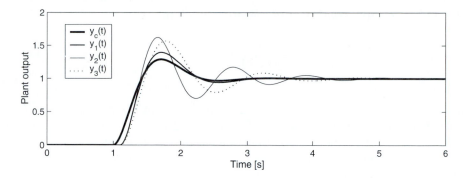

Figure 13.4. Performance of different control designs: continuous time $(y_c(t))$, simple substitution $(y_1(t))$, step invariance $(y_2(t))$, and bilinear transformation $(y_3(t))$

Continuous: $\Re\{s\} < 0$

Discrete z-Domain: $|z| < 1$

Discrete δ-Domain: $\left|\delta + \frac{1}{\Delta}\right| < \frac{1}{\Delta}$

As already discussed in Chapter 7, in specifying the closed-loop poles, we fix the characteristics of the closed-loop natural modes. Thus, the general approaches for pole assignment described in Chapter 7 can be applied, with obvious modifications, in digital-control synthesis. However, discrete-time systems have the appealing feature of being able to exhibit finite settling time (impossible to achieve exactly in continuous-time systems). This is achieved when all the sensitivity transfer-function poles are located at $z = 0$ or, equivalently, at $\gamma = -\frac{1}{\Delta}$ (per Example 12.5 on page 327). Several synthesis approaches have been suggested to obtain digital-control loops with finite settling time (to step references), measured *at sample times*. We briefly examine two of those approaches for the following framework.

Consider a continuous-time plant having a sampled transfer function given by

$$G_{oq}(z) = [G_o G_{h0}]_q(z) = \frac{B_{oq}(z)}{A_{oq}(z)} \qquad (13.6.1)$$

and a digital controller with transfer function given by

$$C_q(z) = \frac{P_q(z)}{L_q(z)} \qquad (13.6.2)$$

$$(13.6.3)$$

where $A_{oq}(z)$, $B_{oq}(z)$, $L_q(z)$, and $P_q(z)$ are polynomials in z, with $A_{oq}(z)$ and $L_q(z)$ monic. The degrees of $A_{oq}(z)$ and $B_{oq}(z)$ are n and m ($m < n$), respectively. We further assume that the open loop has *accumulation effect* (integral action)–that is, that $A_{oq}(z)L_q(z)$ has at least one root at $z = 1$, to ensure zero steady-state errors to step references and output disturbances.

13.6.2 Minimal Prototype

The basic idea in this control-design strategy is to achieve zero error at the sample points in the minimum number of sampling periods, for step references and step-output disturbances (with zero initial conditions). This implies that the complementary sensitivity must be of the form

$$T_o(z) = \frac{p(z)}{z^l} \qquad (13.6.4)$$

where $l \in \mathbb{N}$, $p(z)$ is a polynomial of degree $< l$, and $p(1) = 1$. This last condition ensures that $T_{oq}(1) = 1$, which is necessary and sufficient to achieve the zero steady-state requirement.

The controller synthesis can then be carried out by using pole-assignment techniques. (See section §7.2.) We will consider two cases.

Case 1 The plant sampled transfer function, $G_{oq}(z)$, is assumed to have all its poles and zeros strictly inside the stability region. Then the controller can cancel the numerator and the denominator of $G_{oq}(z)$, and the pole-assignment equation becomes

$$L_q(z)A_{oq}(z) + P_q(z)B_{oq}(z) = A_{clq}(s) \qquad (13.6.5)$$

where

$$L_q(z) = (z - 1)B_{oq}(z)\overline{L}_q(z) \qquad (13.6.6)$$
$$P_q(z) = K_o A_{oq}(z) \qquad (13.6.7)$$
$$A_{clq}(s) = z^{n-m}B_{oq}(z)A_{oq}(z) \qquad (13.6.8)$$

Note that the polynomial degrees have been chosen according to the theory presented in section §7.2.

Using (13.6.6)–(13.6.8) in (13.6.5) and simplifying, we obtain

$$(z - 1)\overline{L}_q(z) + K_o = z^{n-m} \qquad (13.6.9)$$

Equation (13.6.9) can now be solved for K_o by evaluating the expression at $z = 1$. This leads to $K_o = 1$ and to a controller and a complementary sensitivity given by

$$C_q(z) = [G_{oq}(z)]^{-1} \frac{1}{z^{n-m} - 1}; \qquad \text{and} \qquad T_o(z) = \frac{1}{z^{n-m}} \qquad (13.6.10)$$

We illustrate this case with an example.

Example 13.4. *Consider a continuous-time plant with transfer function*

$$G_o(s) = \frac{50}{(s+2)(s+5)} \qquad (13.6.11)$$

Synthesize a minimum prototype controller with sampling period $\Delta = 0.1[s]$.

Solution

The sampled transfer function is given by

$$G_{oq}(z) = \frac{0.0398(z + 0.7919)}{(z - 0.8187)(z - 0.6065)} \qquad (13.6.12)$$

Notice that $G_{oq}(z)$ is stable and of minimum phase, with $m = 2$ and $n = 3$. Upon applying (13.6.10), we have that

$$C_q(z) = \frac{25.124(z - 0.8187)(z - 0.6065)}{(z - 1)(z + 0.7919)} \qquad \text{and} \qquad T_{oq}(z) = \frac{1}{z} \quad (13.6.13)$$

The performance of the resultant control loop is evaluated for a unit step reference at $t = 0.1[s]$. The plant output is shown in Figure 13.5.

We see that the sampled response settles into exactly one sample period. This is as expected, because $T_{oq}(z) = z^{-1}$. However, Figure 13.5 illustrates one of the weaknesses of minimal-prototype control: perfect tracking is guaranteed only at the sampling instants. Indeed, we see a very substantial intersample ripple! We will analyze the cause of this problem in the next chapter. Another drawback of this approach is the large control effort required: The controller is biproper, as it reacts instantaneously to the step reference with an initial value equal to 48.73 times the magnitude of the step.

□□□

Case 2 The plant is assumed to be of minimum phase and stable, except for a pole at $z = 1$, i.e., $A_{oq}(z) = (z - 1)\overline{A}_{oq}(z)$. In this case, the minimal-prototype idea does not require that the controller have a pole at $z = 1$. Thus, equations (13.6.6) to (13.6.8) become

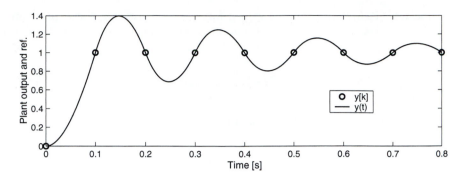

Figure 13.5. Plant output for a unit step reference and a minimal-prototype digital control; stable plant

$$L_q(z) = B_{oq}(z)\overline{L}_q(z) \tag{13.6.14}$$

$$P_q(z) = K_o\overline{A}_{oq}(z) \tag{13.6.15}$$

$$A_{clq}(z) = z^{n-m}B_{oq}(z)\overline{A}_{oq}(z) \tag{13.6.16}$$

Equation (13.6.9) holds, as in Case 1. Thus, $K_o = 1$, and

$$C_q(z) = [G_{oq}(z)]^{-1}\frac{1}{z^{n-m}-1} = \frac{\overline{A}_{oq}(z)}{B_{oq}(z)}\frac{z-1}{z^{n-m}-1} \tag{13.6.17}$$

$$= \frac{\overline{A}_{oq}(z)}{B_{oq}(z)(z^{n-m-1}+z^{n-m-2}+z^{n-m-3}+\ldots+z+1)} \tag{13.6.18}$$

$$T_{oq}(z) = \frac{1}{z^{n-m}} \tag{13.6.19}$$

This case is illustrated in the following example.

Example 13.5. *Consider the servo system of Example 3.4 on page 47. Recall that its transfer function is given by*

$$G_o(s) = \frac{1}{s(s+1)} \tag{13.6.20}$$

Synthesize a minimal-prototype controller having sampling period $\Delta = 0.1[s]$.

Solution

From Example 13.1 on page 356, we have that the system pulse transfer function (with $\Delta = 0.1[s]$) is given by (13.3.5), i.e.,

$$G_{oq}(z) = 0.0048\frac{z + 0.967}{(z-1)(z-0.905)} \tag{13.6.21}$$

Thus,

$$B_{oq}(z) = 0.0048(z + 0.967) \qquad and \quad \overline{A}_{oq}(z) = z - 0.905 \tag{13.6.22}$$

and, by using (13.6.17),

$$C_q(z) = 208.33\frac{z - 0.905}{z + 0.967} \tag{13.6.23}$$

$$T_{oq}(z) = \frac{1}{z} \tag{13.6.24}$$

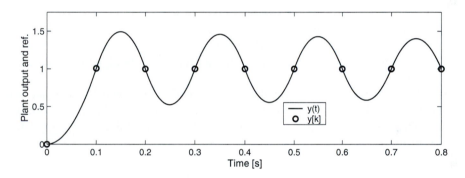

Figure 13.6. Plant output for a unit step reference and a minimal-prototype digital control; plant with integration

The performance of the resultant control loop is evaluated for a unit step reference at $t = 0.1[s]$. The plant output is shown in Figure 13.6.

Again we see that the sampled response settles into one sample period. However, Figure 13.6 also verifies the main characteristics of minimal-prototype control: nonzero errors in the intersample response and large control magnitudes.   □□□

The remarkable common feature in both cases is the intersample response: there exists a *lightly damped oscillation of frequency equal to half the sampling frequency*. We will examine the reasons for this in Chapter 14.

Further insights into the minimal-prototype approach can be obtained by analyzing the behavior of the controller output $u[k]$. From (13.6.4), we have that

$$Y_q(z) = T_{oq}(z) = \frac{p(z)}{z^l}R_q(z) \iff U_q(z) = [G_{oq}(z)]^{-1}Y_q(z) = \frac{A_{oq}(z)p(z)}{z^lB_{oq}(z)}R_q(z)$$

$$(13.6.25)$$

For the above examples, $p(z) = 1$, and l is equal to the relative degree of $G_{oq}(z)$ (this will generally be the case). Thus, the natural modes in $u[k]$ will depend on the location of the zeros of $G_{oq}(z)$, i.e., the roots of $B_{oq}(z)$ (including the sampling zeros!). This implies that $u[k]$ *does not settle to its steady-state value in finite time*. Because the sampling zeros generally have negative real values, this transient will include oscillatory modes; this phenomenon (illustrated in Example 12.6 on page 328) is known as *ringing*, and it is undesirable due to its wearing effect on the actuators.

13.6.3   Minimum-Time Dead-Beat Control

The basic idea in dead-beat control design is similar to that in the minimal-prototype case: to achieve zero error at the sample points in a finite number of sampling periods for step references and step-output disturbances (and with zero initial conditions). However, in this case we add the requirement that, for this sort of reference and disturbance, the controller output $u[k]$ also reach its steady-state value in the same number of intervals.

Consider a digital-control loop with step reference and in which $y[k]$ must reach its steady-state value (with zero control error at the sample points) in n sampling instants, where n is the degree of the denominator polynomial. Then $Y(z)$ must have the form

$$Y_q(z) = \frac{w(z)}{z^n}R_q(z) \iff T_{oq}(z) = \frac{w(z)}{z^n} \qquad (13.6.26)$$

$$w(z) = w_n z^n + w_{n-1}z^{n-1} + \ldots + w_1 z + w_0 \qquad (13.6.27)$$

To achieve perfect steady-state tracking at the sample points (for step references and step-output disturbances) we need $T_{oq}(1) = 1$, i.e.,

$$w(1) = \sum_{i=0}^{n} w_i = 1 \qquad (13.6.28)$$

We also require that the controller output $u[k]$ reach its steady-state value in n sampling periods. This condition allows us to compute[2] the polynomial $w(z)$, because

---

[2]Note that, to have $S_{uoq}(z)$ proper, the degree of $w(z)$ must be, at most, equal to the degree of $B_{oq}(z)$.

$$U_q(z) = [G_{oq}(z)]^{-1}Y_q(z) = S_{uoq}(z)R_q(z) = \frac{A_{oq}(z)w(z)}{B_{oq}(z)z^n}R_q(z) \qquad (13.6.29)$$

Now, to have $u[k]$ settle to its steady-state value in n steps, it follows that $B_{oq}(z)$ cannot appear in the denominator of the term in the last equality in (13.6.29). Hence, we must choose $w(z)$ to satisfy

$$w(z) = \alpha B_{oq}(z) \qquad \text{where} \quad \alpha = \frac{1}{B_{oq}(1)} \qquad (13.6.30)$$

Finally, we see that

$$U_q(z) = \frac{\alpha A_{oq}(z)}{z^n}R_q(z) \qquad (13.6.31)$$

which is achieved by the following control law:

$$C_q(z) = \frac{\alpha A_{oq}(z)}{z^n - \alpha B_{oq}(z)} \qquad (13.6.32)$$

The dead-beat control is illustrated in the following example.

Example 13.6. *Consider the servo system of Examples 3.4 on page 47 and 13.5 on page 366, which has the transfer function*

$$G_o(s) = \frac{1}{s(s+1)} \qquad (13.6.33)$$

Synthesize a minimum-time dead-beat control with sampling interval $\Delta = 0.1[s]$.

Solution

From Example 13.5 on page 366, we have that the system pulse transfer function (with $\Delta = 0.1[s]$) is given by (13.6.21):

$$G_{oq}(z) = 0.0048\frac{z + 0.967}{(z-1)(z-0.905)} \implies \alpha = 105.49 \qquad (13.6.34)$$

Hence, applying (13.6.32), we have that

$$C_q(z) = \frac{\alpha A_{oq}(z)}{z^n - \alpha B_{oq}(z)} = \frac{105.49z - 95.47}{z + 0.4910} \qquad (13.6.35)$$

The performance of this controller is shown in Figure 13.7 for a unit step reference applied at time $t = 0$.

Observe that the continuous-time plant output, $y(t)$, reaches its steady-state value after two sampling periods, as expected. Furthermore, the intersample response is now quite acceptable. To evaluate the control effort, we apply (13.6.31) to obtain

$$U_q(z) = \frac{\alpha A_{oq}(z)}{z^n} R_q(z) = 105.49 \frac{(z-1)(z-0.905)}{z^2} R_q(z) \qquad (13.6.36)$$

Solving, we obtain the control sequence: $u[0] = 105.49$, $u[1] = -95.47$, and $u[k] = 0 \quad \forall k \geq 2$. Note that the zero value is due to the fact that the plant exhibits integration. Also observe the magnitudes of the first two sample values of the control signal; they run the danger of saturating the actuator in practical applications. In this respect, dead-beat control is not different from minimal prototype or continuous-time control; fast control (with respect to the plant bandwidth) will always demand large control magnitudes (as happens in this example)–this is a trade-off that does not depend on the control architecture or control philosophy. (See Chapter 8.)

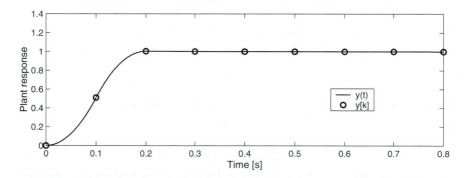

Figure 13.7. Minimum-time dead-beat control for a second-order plant

Remark 13.1. *The controller presented above has been derived for stable plants or plants with at most one pole at the origin, because cancellation of $A_{oq}(z)$ is allowed. However, the dead-beat philosophy can also be applied to unstable plants, provided that dead-beat is attained in more than n sampling periods. To do this, we simply use pole assignment and place all of the closed-loop poles at the origin.*

□□□

13.6.4   Digital-Control Design by Pole Assignment

Minimal-prototype and dead-beat approaches are particular applications of pole assignment. If, for example, one needs to reduce the control effort demanded by the dead-beat design, we can resort to pole-assignment synthesis by shifting the poles at the origin to less demanding positions by moving them towards $(1; 0)$. With this strategy, we are trading control speed (loop bandwidth) against control effort (magnitude of $u[k]$). This idea is applied to the servo problem in the next example.

Example 13.7 (General pole-assignment synthesis). *Consider the same servo system used in the previous examples. It has continuous-time transfer function given by (13.6.33) and pulse-transfer function given by (13.6.34) (for $\Delta = 0.1[s]$).*

A pole-assignment synthesis is carried out with a closed-loop polynomial $A_{clq}(z) = (z - 0.905)(z - 0.8)(z - 0.6)$. If we compare this choice with the dead-beat synthesis case, we observe that we are still canceling the pole at $z = 0.905$; this time, however, the other closed-loop poles have been shifted from the origin to the positions $z = 0.8$ and $z = 0.6$. The pole-assignment equation is then

$$
\begin{aligned}
A_{clq}(z) &= (z - 1)(z - 0.905)L_q(z) + 0.0048(z + 0.967)P_q(z) \\
&= (z - 0.905)(z - 0.6)(z - 0.8)
\end{aligned}
\tag{13.6.37}
$$

where

$$
L_q(z) = z - \alpha_0 \qquad P_q(z) = \beta_1(z - 0.905) \Longrightarrow C_q(z) = \frac{\beta_1(z - 0.905)}{z - \alpha_0} \tag{13.6.38}
$$

Once equations (13.6.37) and (13.6.38) are solved, we obtain

$$
C_q(z) = \frac{8.47(z - 0.905)}{z - 0.44} \Longrightarrow S_{uoq}(z) = \frac{8.47(z - 0.905)(z - 1)}{(z - 0.6)(z - 0.8)} \tag{13.6.39}
$$

The reader can verify that, with this controller, the loop response to a unit step reference settles within 2% of its final value in approximately 20[s]. Notice that this is ten times the settling time obtained with the minimum-time dead-beat control strategy. However, the trade-off becomes evident when we use the control sensitivity $S_{uoq}(z)$ in (13.6.39) to compute the control effort for the same reference. For this case, the control sequence is

$\{u[k]\} = \{8.47, 4.19, 1.80, 0.51, -0.15, -0.45, -0.56, -0.57, -0.52, \ldots, 0\}.$

This compares very favorably with the control effort demanded in the dead-beat design. The trade-off between response time and control energy in this and the previous example is completely in line with the trade-offs derived in Chapter 8.

A further example of digital design by pole assignment is presented next.

Example 13.8. *Consider a continuous-time plant having a nominal model given by*

$$
G_o(s) = \frac{1}{(s + 1)(s + 2)} \tag{13.6.40}
$$

Design a digital controller $C_q(z)$ that achieves a loop bandwidth of approximately $3[rad/s]$. The loop must also yield zero steady-state error for constant references.

Solution

*We choose to carry out the design by using the delta-transform and then to transform
the resulting controller, $C_\delta(\gamma)$, into the Z form, $C_q(z)$. The sampling interval, Δ,
is chosen to be equal to $0.1[s]$. (Note that, with this choice, the sampling frequency
is significantly higher than the required bandwidth.)*

*We first use the MATLAB program **c2del.m** (on the accompanying CD-ROM)
to obtain the discrete-transfer function in delta form, representing the combination
of the continuous-time plant and the zero-order-hold mechanism. This yields*

$$\mathcal{D}\{G_{ho}(s)G_o(s)\} = \frac{0.0453\gamma + 0.863}{\gamma^2 + 2.764\gamma + 1.725} \tag{13.6.41}$$

We next choose the closed-loop polynomial $A_{cl\delta}(\gamma)$ to be equal to

$$A_{cl\delta}(\gamma) = (\gamma + 2.5)^2(\gamma + 3)(\gamma + 4) \tag{13.6.42}$$

*Note that a fourth-order polynomial was chosen, because we have to force integral
action in the controller. The resulting pole-assignment equation has the form*

$$(\gamma^2 + 2.764\gamma + 1.725)\gamma\overline{L}_\delta(\gamma) + (0.0453\gamma + 0.863)P_\delta(\gamma) = (\gamma + 2.5)^2(\gamma + 3)(\gamma + 4) \tag{13.6.43}$$

*The MATLAB program **paq.m** is then used to solve this equation, leading to
$C_\delta(\gamma)$, which is finally transformed into $C_q(z)$. The delta and shift controllers are
given by*

$$C_\delta(\gamma) = \frac{29.1\gamma^2 + 100.0\gamma + 87.0}{\gamma^2 + 7.9\gamma} = \frac{P_\delta(\gamma)}{\gamma\overline{L}_\delta(\gamma)} \qquad and \tag{13.6.44}$$

$$C_q(z) = \frac{29.1z^2 - 48.3z + 20.0}{(z - 1)(z - 0.21)} \tag{13.6.45}$$

*The design is evaluated by simulating a square-wave reference with the SIMULINK
file **dcpa.mdl**. The results are shown in Figure 13.8.*

*The reader is encouraged to use the SIMULINK file **dcpa.mdl** to check that the
bandwidth of the loop (use the command* dlinmod*) exceeds the required value. It is
also of interest to evaluate how the locations of the closed-loop poles change when
the sampling period is changed (without modifying $C_q(z)$).*

13.7 Internal Model Principle for Digital Control

Most of the ideas presented in previous chapters carry over to digital systems. One
simply needs to take account of such issues as the different stability domains and
model types.

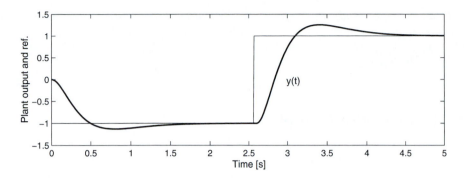

Figure 13.8. Performance of a digital-control loop

As an illustration, we consider here IMP applied to digital control. Say that the disturbance is modeled by a discrete model having disturbance-generating polynomial $\Gamma_{dq}(z)$ (or $\Gamma'_{d\delta}(\gamma)$ for delta operators). $\Gamma_{dq}(z)$ will typically have poles on the unit circle; $\Gamma'_{d\delta}(\gamma)$ will typically have poles on the shifted circle $\gamma : |\gamma + \frac{1}{\Delta}| = \frac{1}{\Delta}$. The IMP is then satisfied by including $\Gamma_{dq}(z)$ or $\Gamma'_{d\delta}(\gamma)$ in the denominator of the controller transfer function. For example, discrete integral action is achieved by including $(\Gamma_{dq}(z) = z - 1)$ or $(\Gamma'_{d\delta}(\gamma) = \gamma)$ in the denominator of the controller.

We illustrate by a simple example.

Example 13.9. *A continuous-time plant is being digitally controlled. The plant output should track a sinusoidal reference of frequency $0.2[rad/s]$, in the presence of step disturbances.*

Determine the polynomial (to be included in the transfer-function denominator of the digital controller) necessary to achieve zero steady-state errors. Assume that the sampling period is $1[s]$.

Solution

We first note that the sampled reference is $r[k] = K_1 \sin(0.2k + K_2)$, where K_1 and K_2 are unknown constants. Thus, the reference-generating polynomials, in shift and delta form, correspond to the denominators of the Z-transform and the delta-transform of $r[k]$, respectively. We then obtain

$$\Gamma_{rq}(z) = z^2 - 1.96z + 1; \qquad \Gamma_{r\delta}(\gamma) = \gamma^2 + 0.04\gamma + 0.04 \qquad (13.7.1)$$

Similarly, the disturbance-generating polynomials are given by

$$\Gamma_{dq}(z) = z - 1; \qquad \Gamma_{d\delta}(\gamma) = \gamma \qquad (13.7.2)$$

To obtain zero steady-state errors, the IMP must be satisfied. This requires that the denominator of the controller must include either the factor $\Gamma_{rq}(z)\Gamma_{dq}(z) = (z^2 - 1.96z + 1)(z - 1)$ (for the shift form) or the factor $\Gamma_{r\delta}(\gamma)\Gamma_{d\delta}(\gamma) = (\gamma^2 + 0.04\gamma + 0.04)\gamma$ (for the delta form).

□□□

13.7.1 Repetitive Control

An interesting special case of the Internal Model Principle in digital control occurs with periodic signals. It is readily seen that any periodic signal of period N_p samples can be modeled by a discrete-time model (in shift-operator form) by using a generating polynomial given by

$$\Gamma_{dq}(q) = \left(q^{N_p} - 1\right) \tag{13.7.3}$$

Hence, any N_p period reference signal can be exactly tracked (at least at the sample points) by including $\Gamma_{dq}(q)$ in the denominator of the controller. This idea is the basis of a technique known as *repetitive control* aimed at causing a system to *learn* how to carry out a repetitive (periodic) task.

This idea has found application in many areas–e.g., robotics. The authors of this book have used the idea in real-world applications in the steel industry.

The core idea as set out above is simple and intuitively reasonable. In applications, it is usually necessary to modify the scheme to meet practical requirements. One such issue is that of robustness.

Incorporation of (13.7.3) in the denominator of the controller ensures that the complementary sensitivity function is exactly one at frequencies

$$\omega_i = \frac{i2\pi}{N_p\Delta}; \qquad\qquad i = 0, 1, \ldots, N_p - 1 \tag{13.7.4}$$

where Δ is the sampling interval. The above ideas are illustrated with the following example.

Example 13.10. *Consider a continuous-time plant with nominal transfer function $G_o(s)$ given by*

$$G_o(s) = \frac{2}{(s + 1)(s + 2)} \tag{13.7.5}$$

Assume that this plant has to be digitally controlled with sampling interval $\Delta = 0.2[s]$ in such a way that the plant output tracks a periodic reference, $r[k]$, given by

$$r[k] = \sum_{i=0}^{\infty} r_T[k - 10i] \iff R_q(z) = R_{Tq}(z)\frac{z^{10}}{z^{10} - 1} \tag{13.7.6}$$

where $\{r_T[k]\} = \{0.0; 0.1; 0.25; 0.6; 0.3; 0.2; -0.1; -0.3; -0.4; 0.0\}$ and [3] $R_{Tq}(z) = \mathcal{Z}[r_T[k]]$.

Synthesize the digital control that achieves zero steady-state at-sample errors.

Solution

From (13.7.6), we observe that the reference-generating polynomial, $\Gamma_{qr}(z)$, is given by $z^{10} - 1$. Thus, the IMP leads to the following controller structure:

$$C_q(z) = \frac{P_q(z)}{L_q(z)} = \frac{P_q(z)}{\overline{L}_q(z)\Gamma_{qr}(z)} \qquad (13.7.7)$$

We then apply pole assignment with the closed-loop polynomial chosen as $A_{clq}(z) = z^{12}(z - 0.2)$. The solution of the Diophantine equation yields

$$\begin{aligned} P_q(z) = &13.0z^{11} + 11.8z^{10} - 24.0z^9 + 19.7z^8 - 16.1z^7 + 13.2z^6 - \\ &- 10.8z^5 + 8.8z^4 - 7.2z^3 + 36.4z^2 - 48.8z + 17.6 \end{aligned} \qquad (13.7.8)$$

$$L_q(z) = (z^{10} - 1)(z + 0.86) \qquad (13.7.9)$$

Figure 13.9, shows the performance of the resulting digital control loop.

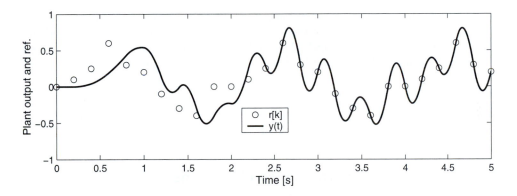

Figure 13.9. Repetitive control

In Figure 13.9, we verify that, after a transient period, the plant output, $y(t)$ follows exactly the periodic reference, at the sampling instants. Notice the dangers of a purely at-sample analysis and design. The intersample behavior of this example can be predicted with the techniques for hybrid systems that will be presented in Chapter 14.

□□□

---

[3]Note that $R_{Tq}(z)$ is a polynomial in z^{-1}.

Perfect tracking in steady state, for reference with high-frequency harmonics, could compromise the robustness of the nominal design. This can be appreciated by introducing a 0.02 [s] unmodelled time delay in the control loop designed in Example 13.10. For this case, the loop behavior is depicted in Figure 13.10.

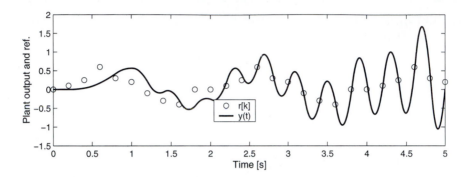

Figure 13.10. Repetitive control loop in the presence of an unmodeled time delay

This behavior can be readily understood. In particular, perfect tracking requires T_o to be 1 at all frequencies of interest. However, we know from Theorem 5.3 on page 146 that robust stability usually requires that $|T_o|$ be reduced at high frequencies. This can be achieved in several ways. For example, one could redefine the internal model in (13.7.3) to include only the frequency components up to some maximum frequency determined by robustness issues–we could use[4]

$$\Gamma_{dq}(q) = (q-1) \prod_{i=1}^{N_{max}} \left(q^2 - 2\cos\left(\frac{i2\pi}{N_p}\right) q + 1 \right); \qquad N_{max} \leq \frac{N_p - 1}{2} \quad (13.7.10)$$

13.8   Fundamental Performance Limitations

The results presented in Chapters 8 and 9 on fundamental design trade-offs also extend in a straightforward way to the discrete-time case.

Analogous results to those presented in subsection §8.6.5, for example, can be derived for discrete-time systems.

Thus, the time-domain properties can be derived from a lemma equivalent to Lemma 4.1 on page 81, which is stated as follows.

Lemma 13.1. *Let $H(z)$ be a rational function of z and analytic for $|z| > \rho$. Also, let the corresponding discrete-time function be*

---

[4]Recall that the roots of the polynomial in (13.7.3) are $e^{jl\theta}$, $l = 0, 1, \ldots, N_p - 1$, where $\theta = \frac{2\pi}{N_p}$.

$$H(z) = \mathcal{Z}\{h[k]\} \tag{13.8.1}$$

Then, for any z_o such that $|z_o| > \rho$, we have that

$$\sum_{k=0}^{\infty} h[k](z_o)^{-k} = \lim_{z \to z_o} H(z) \tag{13.8.2}$$

Proof

From the definition of the Z-transform we have that, for all z in the region of convergence of the transform, i.e., for $|z| > \rho$,

$$H(z) = \sum_{k=0}^{\infty} h[k]z^{-k} \tag{13.8.3}$$

The result follows, because z_o is in the region of convergence of the transform.

□□□

The application of this lemma leads to the same conclusions as those in the continuous-time case. This is due to the fact that, in both cases, the key elements are as follows:

(i) The sensitivity function must vanish at the open-loop unstable poles, and the complementary sensitivity function is equal 1 at those values.

(ii) The complementary sensitivity function must vanish at the NMP zeros, and the sensitivity function must be equal to 1 at those values.

(iii) We use a cumulative measure of the signals of interest. This leads to the same line of argument regarding compensation of positive and negative accumulated values.

To illustrate this parallel, the discrete-time counterpart of Lemma 8.3 on page 213 is formulated below.

Lemma 13.2. *Consider a feedback control loop having stable closed-loop poles located inside a circle centered at the origin and having radius ρ, for some $\rho < 1$. Also assume that the controller $C(z) = \frac{P(z)}{L(z)}$ has at least one pole at (1,0). Then, for a plant zero z_o and a plant pole η_o satisfying $|z_o| > \rho$ and $|\eta_o| > \rho$, respectively, the following hold:*

(i) *For a positive unit reference step or a negative unit step-output disturbance,*
we have

$$\sum_{k=0}^{\infty} e[k](z_o)^{-k} = \frac{1}{1 - z_o^{-1}} \tag{13.8.4}$$

$$\sum_{k=0}^{\infty} e[k](\eta_o)^{-k} = 0 \tag{13.8.5}$$

(ii) *For a positive unit step reference and for z_o outside the unit disk, we have*

$$\sum_{k=0}^{\infty} y[k](z_o)^{-k} = 0 \tag{13.8.6}$$

(iii) *For a negative unit step input disturbance, we have*

$$\sum_{k=0}^{\infty} e[k](z_o)^{-k} = \frac{L(\eta_o)}{\left(1 - \eta_o^{-1}\right) P(\eta_o)} \tag{13.8.7}$$

Proof

The proof depends on the key fact that the poles and zeros that we consider are in
the region of convergence of the transform–i.e., their magnitudes are larger than
that of all closed-loop poles.

(i) *In this case, the control error satisfies*

$$E(z) = S_{oq}(z) \left(R_q(z) - D_o(z) \right) \tag{13.8.8}$$

with either $R_q(z) = (1 - z^{-1})^{-1}$ or $D_o(z) = (1 - z^{-1})^{-1}$. We also note that
$S_{oq}(z_o) = 1$ and $S_{oq}(\eta_o) = 0$. The result then follows from Lemma 13.1 on
page 376, with $h[k]$ replaced by $e[k]$.

(ii) *In this case, the plant output satisfies*

$$Y_q(z) = T_{oq}(z) R_q(z) \tag{13.8.9}$$

with $R_q(z) = (1 - z^{-1})^{-1}$. We also note that $T_{oq}(z_o) = 0$. The result then
follows from Lemma 13.1 on page 376, with $h[k]$ replaced by $y[k]$.

(iii) In this case, the control error satisfies

$$E(z) = -S_{ioq}(z)D_i(z) \tag{13.8.10}$$

with $-D_i(z) = (1 - z^{-1})^{-1}$. *We also note that* $S_{i0}(z_o) = 0$ *and* $S_{i0}(\eta_o) = \frac{L(\eta_o)}{P(\eta_o)}$. *The result then follows from Lemma 13.1 on page 376, with* $h[k]$ *replaced by* $e[k]$.

□□□

It is left as a task for the reader to establish the discrete-time counterparts to other results presented in subsection §8.6.5. These discrete-time constraints, however, require careful interpretation. For example, the condition (13.8.6) does not necessarily imply that the system exhibits undershoot, because the sum on the left-hand side can be equal to zero if $z_o \in \mathbb{R}^-$, even if $y[k]$ never changes sign. This is the case when z_o is a sampling zero.

It is also possible to extend the frequency-domain results of Chapter 9 to the discrete case. The reader is encouraged to view Appendix C on the web at this point.

To give the flavor of the discrete results, we present the following lemma, which is the discrete version of Lemma 9.2 on page 244.

Lemma 13.3. *Consider a one-d.o.f. discrete stable control loop with the open-loop rational transfer function* $H_{oq}(z)$.

Assume that $H_{oq}(z)$ *has q open-loop poles outside the unit disk, located at* $\zeta_1, \zeta_2, \ldots, \zeta_q$. *Then the nominal sensitivity function satisfies*

$$\frac{1}{2\pi} \int_0^{2\pi} \ln|S_o(e^{j\omega})|d\omega = \frac{1}{\pi} \int_0^{\pi} \ln|S_o(e^{j\omega})|d\omega = \sum_{i=1}^{q} \ln|\zeta_i| \tag{13.8.11}$$

Proof

The proof follows from a straightforward application of Jensen's formula for the unit disc (Theorem C.11). Note that $h(z)$ must be replaced by $S_o(z)$, which by definition is a biproper stable quotient of two monic polynomials, $K_f = 1$ and $m = \bar{m} \longleftrightarrow m' = 0$.

□□□

Equation (13.8.11) is very similar to (9.2.9); both indicate the need to balance the low-sensitivity region (negative logarithm) with the high-sensitivity region (positive logarithm). They also show that the existence of large unstable open-loop poles shift this balance significantly towards sensitivities larger than 1. The main difference is that, in the discrete-time case, the sensitivity compensation must be achieved in a *finite* frequency band, $[0, 2\pi]$.

Other parallel results for discrete-time systems are more or less direct applications of Lemma C.2 and Theorem C.11.

The above result is illustrated in the following example.

Example 13.11. *A plant with nominal model* $G_o(s) = \frac{13}{s^2 - 4s + 13}$ *is to be digitally controlled. The sampling interval is* Δ *and a zero-order hold is used. Assume that the nominal sensitivity must satisfy*

$$|S_o(e^{j\omega\Delta})| \leq \epsilon < 1 \qquad \forall \omega \leq \frac{\omega_s}{4} = \frac{\pi}{2\Delta} \qquad (13.8.12)$$

Use Lemma 13.3 on the preceding page to determine a lower bound for the sensitivity peak S_{max}.

Solution

We note that the nominal model has two unstable poles, located at $p_{1,2} = 2 \pm j3$. *When the discrete model for the plant is obtained, these unstable poles map into* $\zeta_{1,2} = e^{(2\pm j3)\Delta}$. *We then apply Lemma 13.3, using normalized frequency. This leads to*

$$\int_0^\pi \ln|S_o(e^{j\omega})|d\omega = 4\pi\Delta \qquad (13.8.13)$$

If we split the integration interval as $[0, \pi] = [0, \frac{\pi}{2}] \cup (\frac{\pi}{2}, \pi]$, *then*

$$\ln S_{max} > 4\Delta - \ln(\epsilon) \qquad (13.8.14)$$

This bound becomes smaller if the sampling frequency increases. The reader is invited to investigate the use of Theorem C.10 to obtain a tighter bound for S_{max}.

13.9   Summary

- There are a number of ways of designing digital-control systems:

  ○ Design in continuous time, then discretize the controller prior to implementation;

  ○ Model the process by a digital model, and carry out the design in discrete time.

- Continuous-time design can be discretized for implementation:

  ○ Continuous-time signals and models are utilized for the design.

- Prior to implementation, the controller is replaced by an equivalent discrete-time version.

- *Equivalent* means to simply map s to δ (where δ is the delta operator).

- Caution must be exercised: The analysis was carried out in continuous time, and the expected results are therefore based on the assumption that the sampling rate is high enough to mask sampling effects.

- If the sampling interval is chosen carefully, in particular with respect to the open- and closed-loop dynamics, then the results should be acceptable.

- Discrete design can be based on a discretized process model:

  - First, the model of the continuous process is discretized.

  - Then, on the basis of the discrete process, a discrete controller is designed and implemented.

  - Caution must be exercised with so-called intersample behavior: The analysis is based entirely on the behavior as observed at discrete points in time, but the process has a continuous behavior between sampling instances.

  - Problems can be avoided by refraining from designing solutions that appear feasible in a discrete-time analysis but are known to be unachievable in a continuous-time analysis (such as removing nonminimum-phase zeros from the closed loop!).

- The following rules of thumb will help avoid intersample problems if a purely digital design is carried out.

  - Sample 10 times the desired closed-loop bandwidth.

  - Use simple anti-aliasing filters to avoid excessive phase shift.

  - Never try to cancel or otherwise compensate for discrete sampling zeros.

  - Always check the intersample response.

13.10   Further Reading

General

Åström, K. and Wittenmark, B. (1990). *Computer controlled systems. Theory and design.* Prentice-Hall, Englewood Cliffs, N.J., 2nd edition.

Feuer, A. and Goodwin, G.C. (1996). *Sampling in Digital Signal Processing and Control.* Birkäusser Boston, Cambridge, Mass.

Franklin, G.F., Powell, J.D., and Workman, M. (1990). *Digital control of dynamic systems*. Addison-Wesley, Reading, Mass., 2nd edition.

Jury, E. (1958). *Sample Data Control Systems*. Wiley, New York.

Kuo, B.C. (1992). *Digital control systems*. Oxford Series in Electrical and Computer Engineering, 2nd edition.

Ogata, K. (1987). *Discrete-time Control Systems*. Prentice-Hall, Englewood Cliffs, N.J.

Zeros of sampled-data systems

Åström, K., Hagander, P., and Sternby, J. (1984). Zeros of sample-data systems. *Automatica*, 20(1):31-38.

Weller, S.R., Moran, W., Ninness, B.M., and Pollington, A. (1997). Sampling zeros and the Euler-Fröbenius polynomials. In *Proceedings of the 36th IEEE CDC*, pages 1471-1476.

Difficulties of digital control

Keller, J. and Anderson, B. (1992). A new approach to the discretization of continuous-time controllers. *IEEE Transactions on Automatic Control*, 37(2):214-233.

Repetitive control

Longman, R.W. and Lo, C-P. (1997). Generalized holds, ripple attenuation, and tracking additional outputs in learning control. *Journal of Guidance, Control, and Dynamics*, 20(6):1207-1214.

Middleton, R.H., Goodwin, G.C., and Longman, R.W. (1989). A method for improving the dynamic accuracy of a robot performing a repetitive task. *International Journal of Robotics Research*, 8(5):67-74.

Ryu, Y.S. and Longman, R.W. (1994). Use of anti-reset windup in integral control based learning and repetitive control. In *Proceedings of IEEE International Conference on Systems, Man and Cybernetics*, San Antonio, TX, 3:2617-2622.

13.11   Problems for the Reader

Problem 13.1. *The frequency response of the zero-order sample-and-hold device is computed from*

$$G_{ho}(j\omega) = \frac{1 - e^{-j\omega\Delta}}{j\omega} \tag{13.11.1}$$

13.1.1 *Plot the magnitude of this frequency response for different values of Δ.*

13.1.2 *What is the relationship between the results obtained and the staircase nature of the plant input $u(t)$?*

Problem 13.2. *A continuous-time plant has a transfer function given by*

$$G_o(s) = \frac{1}{(s+1)^2(s+2)} \tag{13.11.2}$$

13.2.1 *Compute the location of the sampling zeros for $\Delta = 0.2[s]$.*

13.2.2 *How do those zeros evolve when we vary Δ in the range $[0.02; 2]$?*

Problem 13.3. *A continuous-time plant has a transfer function given by*

$$G_o(s) = \frac{-s+1}{(s+2)(s+1)} \tag{13.11.3}$$

13.3.1 *Is there any sampling frequency at which no zero appears in the pulse-transfer function?*

13.3.2 *Synthesize a minimal-prototype controller for $\Delta = 0.5[s]$. Evaluate the control-loop performance for a unit step-output disturbance.*

Problem 13.4. *A continuous-time plant has a nominal model having a transfer function given by*

$$G_o(s) = \frac{1}{(s+2)(s+1)} \tag{13.11.4}$$

13.4.1 *Synthesize a minimum-time dead-beat control with $\Delta = 0.2[s]$. Evaluate the control-loop performance.*

13.4.2 *Assume that the* true *transfer function has an additional pole at $s = -8$ (without affecting the d.c. gain). Evaluate the robustness of the original synthesis. Do you still have dead-beat performance?*

Problem 13.5. *Consider the minimum dead-beat control presented in subsection §13.6.3.*

13.5.1 *Prove that, in general, there is no dead-beat behavior for input step disturbances.*

13.5.2 *Show that a dead-beat control (for input step disturbances) that settles in $2n$ sampling intervals can be synthesized by using pole-assignment techniques, where n is the number of the plant-model poles.*

13.5.3 *Apply your results to synthesize a dead-beat controller for a plant having the nominal model*

$$G_o(s) = \frac{4}{s^2 + 3s + 4} \tag{13.11.5}$$

Problem 13.6. *Consider the following plant nominal models:*

(a) $\dfrac{9}{s^2 + 4s + 9}$ (b) $\dfrac{2}{-s + 2}$

(c) $\dfrac{-s + 8}{(s + 2)(s + 3)}$ (d) $\dfrac{-s + 4}{(-s + 1)(s + 4)}$

13.6.1 *For each case, synthesize a controller providing dead-beat control (for step-output disturbances) in the minimum number of sampling periods. Use $\Delta = 0.1\,[s]$.*

13.6.2 *Discuss the difficulties you found in cases (b), (c), and (d), and suggest a general procedure to synthesize dead-beat control for systems with unstable poles and nonminimum-phase zeros.*

Problem 13.7. *Consider a plant having the nominal model*

$$G_o(s) = \frac{e^{-s}}{s + 0.5} \tag{13.11.6}$$

13.7.1 *Choose* $\Delta = 0.2$ *[s], and synthesize a digital controller such that the control error* $e[k]$*, for a step input disturbance, decays faster than* $(0.5)^k$*.*

13.7.2 *Using a Smith predictor, design a continuous-time controller that achieves similar performance. Compare and discuss.*

Problem 13.8. *Extend the anti-wind-up mechanism described in Chapter 11 to digital control loops having biproper digital controllers.*

Problem 13.9. *Build the anti-wind-up form of Figure 11.6 on page 298 for the following controllers.*

$$(a)\quad \frac{2z-1}{z-0.5}\qquad (b)\quad \frac{z(z-0.2)}{z^2-0.8z+0.5}$$

$$(c)\quad \frac{(z-0.4)^3}{z^3}\qquad (d)\quad \frac{z+1.2}{z-0.8}$$

Analyze the difficulties arising in case (d).

Problem 13.10. *Consider a plant having the nominal model*

$$G_o(s) = \frac{e^{-s}}{(2s+1)(4s+1)} \qquad (13.11.7)$$

13.10.1 *Using the Cohen-Coon tuning method, as a first iteration point, find a continuous-time PID controller,* $C(s)$*.*

13.10.2 *Build a delta-form digital controller by taking* $C_\delta(\gamma) = C(\gamma)$*, and compare the performance of the continuous and the discrete control loop, for step reference and step input disturbances, with sampling interval* $\Delta = 0.1$ *[s].*

13.10.3 *Repeat for* $\Delta = 1$ *[s]. Compare and discuss.*

Problem 13.11. *Consider a digital feedback control loop for which the open-loop transfer function is given by*

$$C_q(z)\,[G_{h0}G_o]_q\,(z) = \frac{K(z-\alpha)}{(z-1)(z-0.7)} \qquad (13.11.8)$$

13.11.1 *Fix $\alpha = 0.4$, and study the closed-loop stability by using Nyquist theory.*

13.11.2 *Fix $K = 2$, and study the closed-loop stability by using root locus.*

Problem 13.12. *Consider a digital-control loop for which*

$$[G_{h0}G_o]_q(z) = \frac{0.1}{z - 0.9} \tag{13.11.9}$$

Assume that a digital controller, $C_q(z)$, is designed to achieve a complementary sensitivity given by

$$T_{oq}(z) = \frac{1 - p_o}{z - p_o} \qquad with \quad 0 < p_o < 1 \tag{13.11.10}$$

13.12.1 *Determine the control sensitivity $S_{uoq}(z)$.*

13.12.2 *Use SIMULINK to verify that a fast control loop–one with a small p_o–leads to large magnitudes in the controller output, $u[k]$.*

13.12.3 *Set $p_o = 0.1$, and draw, on the same graph, the magnitudes of the frequency response of the plant and the complementary sensitivity.*

Problem 13.13. *In a digital-control loop of a continuous-time plant, we have that*

$$\{G_{h0}G_o\}_q(z) = \frac{z - \alpha}{(z - 0.2)(z - 0.6)} \tag{13.11.11}$$

13.13.1 *Prove that, if $\alpha > 1$, then the plant output will always exhibit undershoot for a step reference.*

13.13.2 *Will the above be true for $\alpha < -1$?*

Problem 13.14. *Consider the digital control of a loop, with $\Delta = 0.1$ [s], for a plant having the nominal model*

$$G_o(s) = \frac{3}{(-s + 1)(s + 3)} \tag{13.11.12}$$

Synthesize a digital controller that provides perfect steady-state at-sample tracking for a reference $r(t) = 2 + \cos(2\pi t)$.

Chapter 14

HYBRID CONTROL

14.1 Preview

Chapter 13 gives a traditional treatment of digital control based on analyzing the *at-sample* response. Generally, we found that this was a simple and problem-free approach to digital-control design. However, at several points we warned the reader that the resultant continuous response could contain nasty surprises if certain digital controllers were implemented on continuous systems. The purpose of this chapter is to analyze this situation and to explain the following:

- why the continuous response can appear very different from that predicted by the at-sample response

- how to avoid these difficulties in digital control

The general name for this kind of analysis, in which we mix digital control and continuous responses, is *hybrid control*.

14.2 Hybrid Analysis

In this chapter, we will examine what causes the unexpected differences between continuous and sampled responses to occur. We will achieve this by analyzing the underlying continuous-time behavior. We will use the term *hybrid analysis* to describe formulations of this type, which allow one to employ digital control of a continuous-time process in a unified framework.

14.3 Models for Hybrid Control Systems

A hybrid control loop containing both continuous- and discrete-time elements is shown in Figure 14.1 on the following page.

To carry out a hybrid analysis of this loop, we will need to mix continuous- and discrete-time signals and systems.

Using the notation of section §12.13, we denote the discrete equivalent transfer function of the combination { zero-order hold + Continuous Plant + Filter } as $[FG_oG_{h_0}]_q$. From (12.13.3), we have

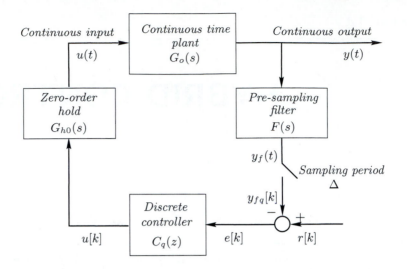

Figure 14.1. Sampled-data control loop (block form)

$$[FG_oG_{h0}]_q = \mathcal{Z}\left\{\text{sampled impulse response of } F(s)G_o(s)G_{h_o}(s)\right\} \qquad (14.3.1)$$

In this section, we will need to mix Z-transform and Laplace transform. We will use a subscript q to distinguish the former.

We also associate a fictitious staircase function, $\hat{y}_f(t)$, with the sequence $\{y_f[k]\}$, where

$$\hat{y}_f(t) = \sum_{k=0}^{\infty} y_f[k]\left(\mu(t - k\Delta) - \mu(t - (k+1)\Delta)\right) \qquad (14.3.2)$$

where $\mu(t - \tau)$ is a unit step function starting at τ. We illustrate the connection between $y_f(t)$, $y_f[k]$, and $\hat{y}_f(t)$ for a special case in Figure 14.2 on the next page.

We also note that, because of the zero-order hold, $u(t)$ is already a staircase function:

$$u(t) = \hat{u}(t) = \sum_{k=0}^{\infty} u[k]\left(\mu(t - k\Delta) - \mu(t - (k+1)\Delta)\right) \qquad (14.3.3)$$

The reason for introducing $\hat{y}_f(t)$ is that it has a Laplace Transform (similarly for $\hat{u}(t)$). For example, the Laplace Transform of $\hat{u}(t)$ can be related to the Z-transform of $\{u[k]\}$ as follows:

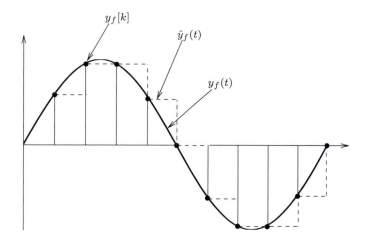

Figure 14.2. Connections between $y_f(t)$, $y_f[k]$, and $\hat{y}_f(t)$ for $y_f(t) = \sin(2\pi t)$, $\Delta = 0.1$

$$\hat{U}(s) = \mathcal{L}\left\{\hat{u}(t)\right\} = \int_0^\infty e^{-st}\hat{u}(t)dt \tag{14.3.4}$$

$$= \int_0^\infty e^{-st} \sum_{k=0}^\infty u[k]\left(\mu(t - k\Delta) - \mu(t - (k+1)\Delta)\right)dt$$

Interchanging the order of summation and integration, we have

$$U(s) = \hat{U}(s) = \sum_{k=0}^\infty u[k]\left(\frac{e^{-k\Delta s} - e^{-(k+1)\Delta s}}{s}\right) \tag{14.3.5}$$

$$= \sum_{k=0}^\infty u[k]e^{-k\Delta s}\left[\frac{1 - e^{-\Delta s}}{s}\right]$$

$$= U_q\left(e^{\Delta s}\right)G_{ho}(s)$$

where $U_q(z)$ is the Z-transform of $\{u[k]\}$. It is clear that

$$Y(s) = G_o(s)\hat{U}(s) \tag{14.3.6}$$

We also know that $Y_{fq}(z)$ is related to $U_q(z)$ and to the sampled reference input $R_q(z)$ via standard discrete transfer functions, i.e.,

$$U_q(z) = C_q(z)\left[R_q(z) - Y_{fq}(z)\right] \tag{14.3.7}$$

Multiplying both sides by $G_{ho}(s)$ and setting $z = e^{s\Delta}$ gives

$$\left[G_{ho}(s)U_q(e^{s\Delta})\right] = -C_q(e^{s\Delta})G_{ho}(s)Y_{fq}(e^{s\Delta}) \qquad (14.3.8)$$
$$+ C(e^{s\Delta})G_{ho}(s)R_q(e^{s\Delta})$$

and, by using (14.3.5) for $\hat{U}(s)$ and $\hat{Y}_f(s)$, we obtain

$$\hat{U}(s) = -C_q(e^{s\Delta})\hat{Y}_f(s) + C_q(e^{s\Delta})G_{ho}(s)R_q(e^{s\Delta}) \qquad (14.3.9)$$

Similarly, we can see that

$$\hat{Y}_f(s) = [FG_oG_{h0}]_q (e^{s\Delta})\hat{U}(s) \qquad (14.3.10)$$

Hence, for analysis purposes, we can redraw the loop in Figure 14.1 on page 388 as in Figure 14.3 on the next page, where all discrete functions (with subscript q) are evaluated at $e^{s\Delta}$ and all other functions are evaluated at s.

Figure 14.3 on the facing page describes the hybrid system containing both discrete and continuous signals. This diagram can be used to make various hybrid calculations.

For example,

The Laplace transform of the continuous output of the hybrid loop is given by

$$Y(s) = \left[\frac{C_q(e^{s\Delta})G_o(s)G_{h0}(s)}{1 + C_q(e^{s\Delta})[FG_oG_{h0}]_q(e^{s\Delta})}\right] R_q(e^{s\Delta}) \qquad (14.3.11)$$

Remark 14.1. *Although the continuous-transfer function $G_o(s)$ in Figure 14.3 on the next page appears to be in open loop, feedback is actually provided around it by the discrete loop. Thus feedback will ensure that unstable modes in $G_o(s)$ are stabilized.*

Remark 14.2. *Note that, even when the reference input is a pure sinusoid, the continuous-time output will not, in general, be sinusoidal. This is because $R_q(e^{j\omega_o})$ is a periodic function and hence it follows from (14.3.11) that $Y(j\omega)$ will have components at $\{\omega = \omega_o + \frac{2k\pi}{\Delta}; k = \ldots, -1, 0, 1, \ldots\}$.*

□□□

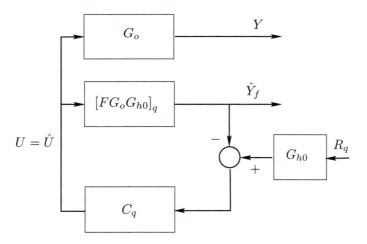

Figure 14.3. Transfer-function form of a sampled-data control loop

14.4   Analysis of Intersample Behavior

The starting point for analyzing intersample behavior is the set of results given in subsection §14.3 for the continuous-time output of a hybrid loop. Here, we work with the filtered output, $y_f(t)$.

From those results we have that

$$Y_f(s) = \frac{C_q(e^{s\Delta})F(s)G_o(s)G_{h0}(s)}{1 + C_q(e^{s\Delta})\left[FG_oG_{h0}\right]_q(e^{s\Delta})} R_q(e^{s\Delta}) \qquad (14.4.1)$$

Also, we recall that the sampled output response is given by

$$Y_{fq}(e^{s\Delta}) = T_{oq}(e^{s\Delta})R_q(e^{s\Delta}) \qquad (14.4.2)$$

where $T_{oq}(z)$ is the shift-domain complementary sensitivity:

$$T_{oq}(z) = \frac{Y_{fq}(z)}{R_q(z)} = \frac{C_q(z)\left[FG_oG_{h0}\right]_q(z)}{\left(1 + C_q(z)\left[FG_oG_{h0}\right]_q(z)\right)} \qquad (14.4.3)$$

Also, the staircase approximation to the sampled output is given by

$$\hat{Y}_f(s) = G_{h0}(s)Y_{fq}(e^{s\Delta}) \tag{14.4.4}$$

From equations (14.4.1) and (14.4.4), the ratio of the continuous-time output response to the staircase form of the sampled output response is given by

$$\frac{Y_f(s)}{\hat{Y}_f(s)} = \frac{F(s)G_o(s)}{[FG_oG_{h0}]_q (e^{s\Delta})} \tag{14.4.5}$$

For the moment, let us ignore the effect of the anti-aliasing filter. (This is reasonable, because one usually designs this filter to be somewhat transparent to the dynamics anyway).

Then one sees from (14.4.5) that the ratio of the continuous-time output response to the staircase form of the sampled output response depends on the ratio

$$\Theta(s) = \frac{G_o(s)}{[G_oG_{h0}]_q (e^{s\Delta})} \tag{14.4.6}$$

As shown in section §13.3, the discrete transfer function $[G_oG_{h0}]_q$ will typically have sampling zeros. The effect of those zeros will be particularly significant at or near half the sampling frequency. Hence, the ratio $\Theta(s)$ given in (14.4.6) can be expected to become large in the vicinity of half the sampling frequency. We illustrate this feature by considering the servo system in Example 13.5 on page 366.

Example 14.1. *Compare the continuous-time and discrete-time responses for Examples 13.5 on page 366 and 13.6 on page 369.*

Solution

The magnitude of the ratio $\Theta(j\omega)$ for Example 13.5 on page 366 is shown in Figure 14.4 on the facing page. We see from this figure that the ratio is 1 at low frequencies, but at $\omega = \frac{\pi}{\Delta}$ [rad/s] there is a ratio of approximately $23 : 1$ between the frequency contents of the continuous-time response and that of the staircase form of the sampled output.

We next use the above analysis to compare the continuous frequency response of the minimal-prototype design and the dead-beat design for this plant.

(a) **Minimal-prototype design** *We recall that this design canceled the sampling zero and led to $T_{oq}(z) = z^{-1}$, which is an all-pass transfer function. Hence, a sampled sine-wave input in the reference leads to a sampled sine-wave output of the same magnitude. However, Figure 14.4 on the next page predicts that*

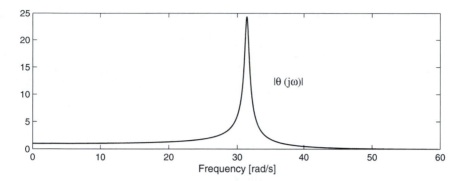

Figure 14.4. Frequency response of $\Theta(j\omega)$, $\Delta = 0.1$

the corresponding continuous output will have 23 times more amplitude for a sinusoidal frequency $\omega = \frac{\pi}{\Delta}$ [rad/s]. The reason for this peak is easily seen. In particular, the minimal prototype cancels the sampling zero in the discrete system. However, this sampling zero is near $\omega = \frac{\pi}{\Delta}$ [rad/s]. Hence, it follows from (14.4.6) that the continuous-time output must have significant energy at $\omega = \frac{\pi}{\Delta}$ [rad/s].

(b) Minimum-time dead-beat design *By way of contrast, the minimum-time dead-beat design of Example 13.6 on page 369 did not cancel the sampling zeros and led to the following discrete-time complementary sensitivity functions:*

$$T_{oq}(z) = \frac{B_{oq}(z)}{B_{oq}(1)z^2} = \frac{0.5083z + 0.4917}{z^2} \qquad (14.4.7)$$

The magnitude of the frequency response of this complementary sensitivity is shown in Figure 14.5 on the next page. We see that, in this case, the discrete-time gain drops dramatically at $\omega = \frac{\pi}{\Delta}$ [rad/s] and, hence, although Figure 14.4 still applies with respect to $\Theta(j\omega)$, there is now little discrete-time response at $\frac{\pi}{\Delta}$ [rad/s] to yield significant intersample ripple.

We observe that this design makes no attempt to compensate for the sampling zero, and hence there are no unpleasant differences between the sampled response and the full continuous-time response.

□□□

14.5   Repetitive Control Revisited

We recall the repetitive controller described in subsection §13.7.1. There, we found that a digital-control system could be designed to track (at the sample points) any

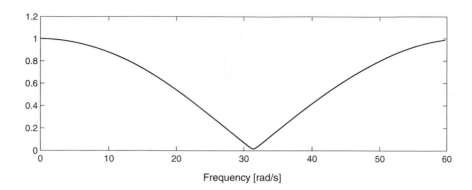

Figure 14.5. Frequency response of the complementary sensitivity for a minimum-time dead-beat design

arbitrary periodic reference. However, we found that this involved having infinite loop gain at high frequencies (relative to the sampling period), and this was undesirable due to robustness effects. We therefore suggested modifying the idea so that only frequency components up to some maximum frequency were exactly tracked.

Another reason for not using the idealized form of repetitive control comes from intersample response issues. We have seen in subsection §13.7.1 that significant intersample response will occur if one tries to enforce a significant discrete response at frequencies approaching π/Δ. However, this is exactly what the idealized form of the repetitive controller aims to do. Thus, if a repetitive controller is to be applied to a continuous-time system, then it is again a good idea to limit the bandwidth over which exact tracking occurs.

14.6   Poisson Summation Formula

Finally, we present a result that frequently is useful in the context of hybrid control. In particular, we wish to evaluate explicitly the Z-transform of a sequence $\{f(k\Delta)\}$ obtained by sampling, at a fixed interval Δ, a continuous-time signal $f(t)$. We use $F(s)$ and $F_q(z)$ to denote the Laplace transform of $f(t)$ and the Z-transform of $\{f(k\Delta)\}$, respectively:

$$F(s) = \int_o^\infty f(t)e^{-st}dt \tag{14.6.1}$$

$$F_q(z) = \sum_{k=0}^\infty f(k\Delta)z^{-k} \tag{14.6.2}$$

Then, subject to various regularity conditions, we have

$$F_q(e^{s\Delta}) = \frac{1}{\Delta} \sum_{k=-\infty}^{\infty} F\left(s + jk\frac{2\pi}{\Delta}\right) + \frac{f(0)}{2} \qquad (14.6.3)$$

This result is known as the *Poisson summation formula* . This formula is useful in the analysis of hybrid control systems. For example, it shows that the frequency response of a sampled signal is the superposition of infinitely many copies of the corresponding continuous-time frequency response.

To establish the result, we write

$$\frac{1}{\Delta} \sum_{k=-\infty}^{\infty} F\left(s + jk\frac{2\pi}{\Delta}\right) = \frac{1}{\Delta} \lim_{n \longrightarrow \infty} \sum_{k=-n}^{n} \int_0^{\infty} f(t)e^{-\left(s+jk\frac{2\pi}{\Delta}\right)t}dt \qquad (14.6.4)$$

$$= \frac{1}{\Delta} \lim_{n \longrightarrow \infty} \int_0^{\infty} f(t)D_n\left(\frac{\pi t}{\Delta}\right)e^{-st}dt \qquad (14.6.5)$$

where $D_n\left(\frac{\pi t}{\Delta}\right)$ is the special function given by

$$D_n\left(\frac{\pi t}{\Delta}\right) = \sum_{k=-n}^{n} e^{-jk\frac{\pi t}{\Delta}} = \frac{\sin\left(\frac{(2n+1)\pi t}{\Delta}\right)}{\sin\left(\frac{\pi t}{\Delta}\right)} \qquad (14.6.6)$$

The function $D_n\left(\frac{\pi t}{\Delta}\right)$ is very frequently employed in proofs relating to convergence of Fourier transforms. It is known as the *Dirichlet kernel*. To give an idea what this function looks like, we plot it in Figure 14.6 for the case $n = 8$.

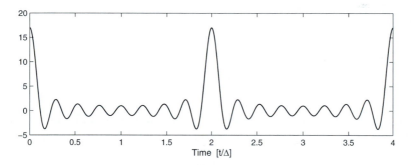

Figure 14.6. Dirichlet kernel ($n = 8$)

An interesting fact that can readily be established is that, for all n,

$$\int_0^{\frac{\Delta}{2}} D_n\left(\frac{\pi t}{\Delta}\right) = \frac{\Delta}{2} \qquad (14.6.7)$$

As $n \to \infty$, it can be seen that integrating (a not too bizarre) function multiplied by $D_n\left(\frac{\pi t}{\Delta}\right)$ will yield the values of the function at $t = \Delta, 2\Delta, \ldots$ scaled by Δ, and the value of the function at $t = 0$ scaled by $\frac{\Delta}{2}$.

Thus, we might envision that (14.6.5) simply becomes

$$\frac{1}{\Delta} \lim_{n \to \infty} \int_0^\infty f(t) D_n\left(\frac{\pi t}{\Delta}\right) e^{-st} = \frac{f(0)}{2} + \sum_{k=1}^{\infty} f(k\Delta) e^{-sk\Delta} \qquad (14.6.8)$$

$$= F_q\left(e^{-s\Delta}\right) - \frac{f(0)}{2} \qquad (14.6.9)$$

This essentially establishes (14.6.3).

Remark 14.3. *The above development is somewhat heuristic, but a formal proof of the result (under very general conditions) is given in the references given at the end of the chapter. Actually if $f(t)$ is such that $e^{-\sigma t} f(t)$ is of uniform bounded variation for some $\sigma \in \mathbb{R}$, then (14.6.3) holds for every s such that $\Re\{s\} > \sigma$.*

14.7   Summary

- Hybrid analysis allows one to mix continuous- and discrete-time systems properly.

- Hybrid analysis should always be utilized when design specifications are particularly stringent and one is trying to push the limits of the fundamentally achievable.

- The ratio of the magnitude of the continuous-time frequency content at frequency ω to the frequency content of the staircase form of the sampled output is

$$\Theta(s) = \frac{G_o(s)}{[G_o G_{h0}]_q\left(e^{s\Delta}\right)}$$

- The preceding formula allows one to explain apparent differences between the sampled and the continuous response of a digital-control system.

- Sampling zeros typically cause $[G_o G_{h0}]_q\left(e^{j\omega\Delta}\right)$ to fall in the vicinity of $\omega = \frac{\pi}{\Delta}-$ that is, $|\Theta(j\omega)|$ increases at these frequencies.

- It is therefore usually necessary to ensure that the discrete complementary sensitivity has been reduced significantly below 1 by the time the folding frequency, $\frac{\pi}{\Delta}$, is reached.

- This is often interpreted by saying that the closed-loop bandwidth should be 20%, or less, of the folding frequency.

- In particular, it is never a good idea to carry out a discrete design that either implicitly or explicitly cancels sampling zeros; this will inevitably lead to significant intersample ripple.

14.8   Further Reading

Intersample issues

Araki, M., Ito, Y., and Hagiwara, T. (1996). Frequency response of sample-data systems. *Automatica*, 32(4):483-497.

Chen, T. and Francis, B.A. (1995). *Optimal Sample-Data Control Systems*. Springer-Verlag.

Feuer, A. and Goodwin, G.C. (1996). *Sampling in Digital Signal Processing and Control*. Birkäusser Boston, Cambridge, Mass.

Goodwin, G.C. and Salgado, M.E. (1994). Frequency-domain sensitivity function for continuous-time systems under sample data control. *Automatica*, 30(8):1263-1270.

Fundamental limitations in hybrid control systems

Braslavsky, J.H., Middleton, R.H., and Freudenberg, J.S. (1995). Frequency response of generalized sampled-data hold function. *Proceedings of the 34th CDC*, New Orleans, LA, 4:3596-3601.

Freudenberg, J.S., Middleton, R.H., and Braslavsky, J.H. (1995). Inherent design limitations for linear sampled-data feedback systems. *International Journal of Control*, 61(6):1387-1421.

Goodwin, G.C. and Salgado, M.E. (1994). Frequency domain sensitivity function for continuous-time systems under sample data control. *Automatica*, 30(8):1263-1270.

Poisson sampling formula

Braslavsky, J.H., Meinsma, G., Middleton, R.H., and Freudenberg, J.S. (1997). On a key sampling formula relating the Laplace and Z-transforms. *Systems and Control Letters*, 29(4):181-190.

14.9 Problems for the Reader

Problem 14.1. *Consider a plant having a nominal model with transfer function given by*

$$G_o(s) = \frac{-s+2}{(s+2)(s+4)} \tag{14.9.1}$$

Find an expression for the location of the zero for the sampled-data function $[GoG_{h0}]_q(z)$, as a function of the sampling interval Δ.

Problem 14.2. *Assume that $[GoG_{h0}]_q(z)$ is given by*

$$[GoG_{h0}]_q(z) = \frac{(z+1.2)}{(z-0.5)^2(z-0.8)} \tag{14.9.2}$$

If $G_o(s)$ is the transfer function of a third-order system, find $G_o(s)$ for $\Delta = 0.1$, $\Delta = 0.5$, and $\Delta = 1\ [s]$.

Problem 14.3. *Evaluate the function $\Theta(j\omega)$ as in (14.4.6) for the functions and sampling intervals indicated below.*

$$G_o(s) = \frac{1}{(s-1)(s+2)} \qquad \Delta = 0.10$$

$$G_o(s) = \frac{e^{-0.4s}}{(s+0.5)(s+0.25)} \qquad \Delta = 0.20$$

$$G_o(s) = \frac{1}{s^2+0.1s+1} \qquad \Delta = 0.25$$

Problem 14.4. *A digital feedback control loop has to be designed for a plant having the nominal model*

$$G_o(s) = \frac{2}{(10s+1)(5s+1)} \tag{14.9.3}$$

A sample interval of 0.5 [s] is used. However, instead of using the standard zero-order hold, a first order hold is used. The impulse response of this hold appears in Figure 14.7.

*Synthesize a controller achieving zero error **at the sampling instants** in minimum time, for constant references. Discuss your results.*

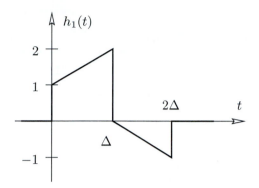

Figure 14.7. First-order hold impulse response

Problem 14.5. *Assume that a digital controller is needed for a plant having a nominal model given by*

$$G_o(s) = \frac{-0.25s + 1}{(0.25s + 1)(s + 1)} \tag{14.9.4}$$

14.5.1 *Synthesize a digital controller such that the loop tracks, as closely as possible, a triangular-wave reference of period 10Δ, where the sampling interval is $\Delta = 0.05$.*

14.5.2 *Simulate and evaluate your design. Discuss.*

Problem 14.6. *An idea that has been described in the control-theory literature is to use a more sophisticated hold function than a simple zero-order hold. We recall from subsection §12.13.1 that a zero-order hold can be modeled as follows:*

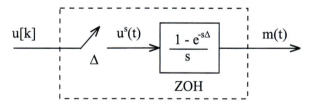

Figure 14.8. Zero-order hold

This can be generalized by simply replacing the Laplace Transform of the impulse response of the zero-order hold by a more general transfer function, as shown in Figure 14.9.

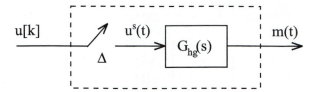

Figure 14.9. Generalized hold

14.6.1 *Use the above ideas to show that the discrete model corresponding to a continuous plant $G_o(s)$ with generalized hold $G_{hg}(s)$ is given by*

$$H_{oq}(z) = \mathcal{Z}\left[\text{the sampled impulse response of } G_{hg}(s)G_o(s)\right] \qquad (14.9.5)$$

14.6.2 *One way of forming a generalized hold is via piecewise-constant functions. Thus, consider the pulse response shown in Figure 14.10.*

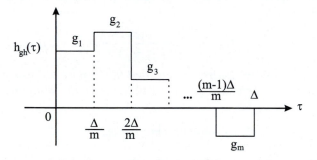

Figure 14.10.

Show that, for this generalized hold, we have

$$G_{hg}(s) = \sum_{k=1}^{m} \frac{e^{-(k-1)\frac{\Delta}{m}s} g_k \left[1 - e^{\frac{-s\Delta}{m}}\right]}{s} \qquad (14.9.6)$$

Problem 14.7. *Consider the generalized hold described in (14.9.6). Say that this is used with a continuous-time model (in state space form):*

$$\dot{x}(t) = Ax(t) + Bm(t) \qquad (14.9.7)$$
$$y(t) = Cx(t) \qquad (14.9.8)$$

Show that the corresponding discrete-time model takes the form

$$x[k + 1] = A_q x[k] + B_{gq} u[k] \qquad (14.9.9)$$
$$y[k] = C_q x[k] \qquad (14.9.10)$$

where

$$A_q = e^{A\Delta} \qquad (14.9.11)$$

$$B_{gq} = \sum_{i=1}^{m} g_i \Gamma_i \qquad (14.9.12)$$

where

$$\Gamma_i = \int_{\frac{(i-1)\Delta}{m}}^{\frac{i\Delta}{m}} e^{A(\Delta - \tau)} B \, d\tau \qquad (14.9.13)$$

Problem 14.8. *Generalized holds can actually be used to shift the discrete zeros arbitrarily. As an illustration, consider the continuous system.*

$$\mathbf{A} = \begin{bmatrix} -1 & 0 \\ 0 & -2 \end{bmatrix}, \qquad \mathbf{B} = \begin{bmatrix} 1 \\ 1 \end{bmatrix}, \qquad \mathbf{C} = \begin{bmatrix} 2 & -3 \end{bmatrix} \qquad (14.9.14)$$

14.8.1 *Obtain the corresponding discrete-transfer function with ZOH and $\Delta = 0.1$.*

14.8.2 *Show that this discrete model is nonminimum phase with zero at 1.10573 and poles at 0.90484 and 0.81873.*

14.8.3 *Use a generalized hold as in (14.9.6), with $m = 2$, to obtain a discrete-time model having zero at 0.90484. This zero cancels the stable pole and leads to the discrete-transfer function*

$$G_{gq}(z) = \frac{0.1}{z - 0.81873} \qquad (14.9.15)$$

Problem 14.9. *Problems 14.6 to 14.8 suggest that one can remove the effects of discrete nonminimum-phase zeros by use of a generalized hold. Hence, one should* *be able to increase the bandwidth of a closed-loop system well beyond that which would result from the use of a ZOH.*

Explain why this is not a good idea (in general) by examining the difference between the sampled and continuous output using (14.4.6). (Actually, it is a bit subtle. The continuous system appears to have less output response than the discrete system. Thus, the discrete system response must be produced by folding (of higher-frequency components to lower frequencies) due to sampling.)

Problem 14.10. *Illustrate the idea captured in Problem 14.9 by designing a minimal-prototype controller for $G_{qq}(z)$ given in (14.9.15).*

Simulate the discrete closed-loop response and compare it with the corresponding continuous system response using the same hold function and controller.

Discuss the result.

Part V

ADVANCED SISO CONTROL

PREVIEW

In this fifth part of the book, we move onto some slightly more advanced topics in SISO control. In particular, we show how to parameterize *all* linear time-invariant controllers that stabilize a given linear system. We then use this parameterization to introduce optimization-based methods for control-system design. We then turn to alternative methods for control-system design based on state space models. These methods provide us with a smooth transition to multivariable control systems, which will be the subject of the remaining parts of the book. Finally, we introduce some basic ideas related to the control of nonlinear systems.

Chapter 15

SISO CONTROLLER PARAMETERIZATIONS

15.1 Preview

Up to this point in the book, we have seen many different methods for designing controllers of different types. On reading all of this, one might be tempted to ask whether there is some easy way that one could specify *all* possible controllers that, at least, stabilized a given system. This sounds, at first hearing, a formidable task. However, we will show in this chapter that it is actually quite easy to give a relatively straightforward description of all stabilizing controllers for both open-loop stable and unstable linear plants. This leads to an affine parameterization of all possible nominal sensitivity functions. This affine structure, in turn, gives valuable insights into the control problem and opens the door to various optimization-based strategies for design. The main ideas presented in this chapter include the following:

- motivation for the affine parameterization and the idea of open-loop inversion

- affine parameterization and Internal Model Control

- affine parameterization and performance specifications

- PID synthesis using the affine parameterization

- control of time-delayed plants and affine parameterization; connections with the Smith controller

- interpolation to remove undesirable open-loop poles

15.2 Open-Loop Inversion Revisited

In previous chapters, we have discussed the basic nature of control design, the modeling problem, key issues in feedback control loops, performance limitations, and various synthesis methods. Underlying these topics there is a set of basic concepts, including the key notion that control implicitly and explicitly depends on plant-model inversion. Thus, inversion is a convenient framework within which to develop an alternative discussion of the control-design problem.

In open-loop control, the input, $U(s)$, can be generated from the reference signal, $R(s)$, by a transfer function, $Q(s)$. This leads to an input-output transfer function of the following form:

$$T_o(s) = G_o(s)Q(s) \qquad (15.2.1)$$

This simple formula is the basis of the affine parameterization to be discussed in this chapter. It highlights the fundamental importance of inversion: $T_o(j\omega)$ will be 1 only at those frequencies where $Q(j\omega)$ inverts the model. Note that this is consistent with the prototype solution to the control problem described in section §2.5.

A key point is that (15.2.1) is affine in $Q(s)$. On the other hand, with a conventional feedback controller, $C(s)$, the closed-loop transfer function has the form

$$T_o(s) = \frac{G_o(s)C(s)}{1 + G_o(s)C(s)} \qquad (15.2.2)$$

The above expression is nonlinear in $C(s)$. This makes tuning $C(s)$ to achieve desired closed-loop properties difficult. We are thus motivated to see whether one could retain the simplicity of (15.2.1) in a more general feedback setting. Actually, comparing (15.2.1) and (15.2.2), we see that the former affine relationship holds if we simply could parameterize $C(s)$ in the following fashion:

$$Q(s) = \frac{C(s)}{1 + C(s)G_o(s)} \qquad (15.2.3)$$

This is the essence of the idea presented in this chapter. In particular, we will use the idea of inversion first to design $Q(s)$ in (15.2.1) and then to use (15.2.3) to determine the corresponding value of $C(s)$.

15.3   Affine Parameterization: The Stable Case

15.3.1   The Parameterization

Our starting point will be the relationship (15.2.3) between $Q(s)$ and $C(s)$. We can invert this relationship to express $C(s)$ in terms of $Q(s)$ and $G_o(s)$:

$$C(s) = \frac{Q(s)}{1 - Q(s)G_o(s)} \qquad (15.3.1)$$

This is known as the Youla parameterization of all stabilizing controllers for stable plants. The result is formalized in the following lemma.

Lemma 15.1 (Affine parameterization for stable systems). *Consider a plant having a stable nominal model $G_o(s)$ controlled in a one-d.o.f. feedback architecture with a proper controller. Then the nominal loop (Figure 5.1 on page 122) is internally stable if and only if $Q(s)$ is any stable proper transfer function, when the controller transfer function $C(s)$ is parameterized as in (15.3.1).*

Proof

We first note that the four sensitivity functions, defined in (5.3.1) to (5.3.4), can be written, by using (15.3.1), as

$$T_o(s) = Q(s)G_o(s) \qquad (15.3.2)$$
$$S_o(s) = 1 - Q(s)G_o(s) \qquad (15.3.3)$$
$$S_{io}(s) = (1 - Q(s)G_o(s))G_o(s) \qquad (15.3.4)$$
$$S_{uo}(s) = Q(s) \qquad (15.3.5)$$

Necessity and sufficiency can then be established as follows.

Necessity

From equation (15.3.5), we immediately conclude that stability of $Q(s)$ is necessary for internal stability.

Sufficiency

Equations (15.3.2) to (15.3.5), together with the assumption regarding stability of $G_o(s)$, imply that stability of $Q(s)$ is sufficient to ensure the stability of the four sensitivity functions and, hence, to ensure the internal stability of the loop.

□□□

The key point about the parameterization (15.3.1) is that it describes *all* possible stabilizing linear time-invariant controllers for the given linear time-invariant plant $G_o(s)$. All that we need to do (in the light of Lemma 15.1) is ensure that $Q(s)$ is chosen to be a stable transfer function. The above parameterization can be made explicit if the feedback loop is redrawn as in Figure 15.1.

This description for the controller can also be used to give expressions for the achieved (or true) sensitivities. Using equations (5.9.15) to (5.9.19) from Chapter 5, we obtain

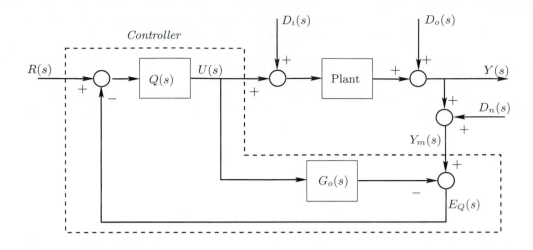

Figure 15.1. Youla's parameterization of all stabilizing controllers for stable plants

$$S(s) = (1 - Q(s)G_o(s))S_\Delta(s) = S_o(s)S_\Delta(s) \qquad (15.3.6)$$

$$T(s) = Q(s)G_o(s)(1 + G_\Delta(s))S_\Delta(s) \qquad (15.3.7)$$

$$S_i(s) = (1 - Q(s)G_o(s))(1 + G_\Delta(s))S_\Delta(s)G_o(s) \qquad (15.3.8)$$

$$S_u(s) = Q(s)S_\Delta(s) \qquad (15.3.9)$$

$$S_\Delta(s) = \frac{1}{1 + Q(s)G_\epsilon(s)} \qquad (15.3.10)$$

where $G_\epsilon(s)$ and $G_\Delta(s)$ are the additive and multiplicative modeling errors, respectively, as defined in section §4.12.

15.3.2   Design Considerations

We see, from equations (15.3.2) to (15.3.5), that, by use of $Q(s)$, we can shape *one* of the four nominal sensitivities. The remaining three are then, of course, specified by this choice. A common objective is to initially focus on the shaping of $S_o(s)$. A typical requirement is that $|S_o(j\omega)|$ be small at low frequencies and then rise to 1 at high frequencies. The latter requirement is employed so as to reduce $|T_o(j\omega)|$ at high frequencies, which is usually required to ensure that high-frequency measurement noise is rejected by the control loop and to provide robustness against modeling errors. Under these circumstances, it would seem that a reasonable choice for $Q(s)$

might be

$$Q(s) = F_Q(s)[G_o(s)]^{-1} \tag{15.3.11}$$

where $[G_o(s)]^{-1}$ is the exact inverse of $G_o(s)$. Not unexpectedly, we see that inversion plays a central role in this prototype solution.

The transfer function $F_Q(s)$ plays a key role in controller design, as we shall see.

Although the design proposed in (15.3.11) is a useful starting point, it will usually have to be refined further to accommodate more detailed design considerations. In particular, we will investigate the following issues:

- nonminimum-phase zeros

- model relative degree

- disturbance rejection

- control effort

- robustness

15.3.3   Nonminimum-Phase Zeros

Recall that, provided $G_o(s)$ is stable, $Q(s)$ need only be stable to ensure closed-loop stability. However, this implies that, if $G_o(s)$ contains NMP zeros, then they cannot be included in $[G_o(s)]^{-1}$ in equation (15.3.11). One might therefore think of replacing (15.3.11) by

$$Q(s) = F_Q(s)G_o^i(s) \tag{15.3.12}$$

where $G_o^i(s)$ is a stable approximation to $[G_o(s)]^{-1}$. For example, if one factors $G_o(s)$ as:

$$G_o(s) = \frac{B_{os}(s)B_{ou}(s)}{A_o(s)} \tag{15.3.13}$$

where $B_{os}(s)$ and $B_{ou}(s)$ are the stable and unstable factors in the numerator, respectively, with $B_{ou}(0) = 1$, then a suitable choice for $G_o^i(s)$ would be

$$G_o^i(s) = \frac{A_o(s)}{B_{os}(s)} \tag{15.3.14}$$

15.3.4   Model Relative Degree

To have a proper controller, it is necessary that $Q(s)$ be proper. Thus, in view of (15.3.12) and (15.3.14), it is necessary that the shaping filter, $F_Q(s)$, have relative degree at least equal to the negative of that of $[G_o^i(s)]^{-1}$. Conceptually, this can be achieved by including factors of the form $(\tau s + 1)^{n_d}$ $(\tau \in \mathbb{R}^+)$ in the denominator. In this form, n_d is chosen so as to make $Q(s)$ at least biproper, and τ should be chosen to meet the necessary design trade-offs.

15.3.5   Disturbance Rejection

(a) Steady-state errors

Steady-state errors due to input and output disturbances can be reduced to zero if $Q(j\omega)$ is the exact inverse of $G_o(j\omega)$ in all frequency bands where the input or output disturbance has significant energy. This will yield zero sensitivity and zero input sensitivity in those bands. The affine parameterization can be generalized to include the case when the input and output-disturbance energies are concentrated at certain known frequencies. Specifically, we have, from $C(s) = Q(s)\,(1 - Q(s)G_o(s))^{-1}$, that the controller has integral action if and only if $Q(0) = [G_o(0)]^{-1}$. Note, however, that this way of formulating the disturbance-rejection problem does not capture the remaining solution space in terms of the degrees of freedom available to the designer. Such a parameterization is provided by the following result.

Lemma 15.2. *Consider a stable model $G_o(s)$ with (input or output or both) disturbance at zero frequency. Then a one-d.o.f. control loop giving zero steady-state tracking error is stable if and only if the controller $C(s)$ can be expressed as in (15.3.1), where $Q(s)$ satisfies*

$$Q(s) = s\overline{Q}(s) + [G_o(0)]^{-1}Q_a(s) \qquad (15.3.15)$$

where $\overline{Q}(s)$ is any stable transfer function and $Q_a(s)$ is any stable transfer function that satisfies $Q_a(0) = 1$.

Proof

- Sufficiency

  We see that, if $\overline{Q}(s)$ and $Q_a(s)$ are stable, then so is $Q(s)$, and this implies that the loop is stable. Also, we see that (15.3.15) implies that $C(s)$ contains an integrator.

- Necessity

  Consider a controller that stabilizes the model in closed loop and also yields zero steady-state error at d.c. This is equivalent to saying that the nominal complementary sensitivity is 1 at d.c. From Equation (15.3.2), we see that *this is also equivalent to having $Q(0) - [G_o(0)]^{-1} = 0$, i.e., $Q(s) - [G_o(0)]^{-1}$ is an arbitrary stable function that has a zero at $s = 0$. A characterization of all such transfer functions is*

$$Q(s) - [G_o(0)]^{-1} = s \left[\overline{Q}(s) + \frac{Q_b(s)}{s} \right] \qquad where \qquad Q_b(0) = 0 \quad (15.3.16)$$

Equation (15.3.16) shows that any $Q(s)$ giving zero steady-state error for constant reference or disturbances or both, can be written as

$$Q(s) = s\overline{Q}(s) + [G_o(0)]^{-1}[1 + G_o(s)Q_b(s)] \qquad where \qquad Q_b(0) = 0$$
$$(15.3.17)$$

The result follows with

$$Q_a(s) = 1 + G_o(s)Q_b(s) \qquad (15.3.18)$$

□□□

Remark 15.1. *We observe that the simplest choice in (15.3.15) is $Q_a(s) = 1$.*

Lemma 15.3. *Consider a stable model $G_o(s)$, and assume that the input disturbance has frequency components at $\omega_1, \omega_2, \dots, \omega_l$. Then a one-d.o.f. control loop yields zero steady-state tracking errors if and only if the controller $C(s)$ can be expressed as in (15.3.1), where $Q(s)$ satisfies*

$$Q(s) = \frac{N_Q(s)}{D_Q(s)} = \frac{N_1(s) \prod_{i=1}^{l}(s^2 + \omega_i^2) + N_2(s)}{D_Q(s)} \qquad (15.3.19)$$

where $N_Q(s), N_1(s), N_2(s)$, and $D_Q(s)$ are real polynomials in s with $D_Q(s)$ stable and

$$N_2(j\omega_i) = D_Q(j\omega_i)[G_o(j\omega_i)]^{-1} \qquad where \qquad i = 1, 2, \dots, l \qquad (15.3.20)$$

Proof

- Sufficiency

  We see that, because $D_Q(s)$ is a stable polynomial, $Q(s)$ is stable, and this implies that the loop is stable. Also, we see from (15.3.19) and (15.3.20) that $Q(j\omega_i) = [G_o(j\omega_i)]^{-1}$, and thus $C(\pm j\omega_i) = \infty$ for $i = 1, 2, \dots l$–i.e., perfect model inversion is achieved at this set of frequencies.

- Necessity

    Consider a controller that stabilizes the model in closed loop and also yields zero steady-state error for disturbances at frequencies $\omega = \omega_1, \omega_2, \ldots, \omega_l$.

    Denote the numerator of $Q(s)$ by $N_Q(s)$ and its denominator by $D_Q(s)$. If we divide $N_Q(s)$ by the polynomial $P_d(s) \triangleq \prod_{i=1}^{l}(s^2 + \omega_i^2)$, we obtain a polynomial result $\overline{N}_1(s)$ and a remainder $\overline{N}_2(s)$:

    $$\frac{N_Q(s)}{P_d(s)} = \overline{N}_1(s) + \frac{\overline{N}_2(s)}{P_d(s)} \tag{15.3.21}$$

    From this, (15.3.19) follows upon identifying $\overline{N}_1(s)$ as $N_1(s)$ and $\overline{N}_2(s)$ as $N_2(s)$.

    Furthermore, the zero steady-state error condition is equivalent to saying that the nominal input sensitivity is 0 at $s = \pm j\omega_i$ for $i = 1, 2, \ldots l$. From equation (15.3.4), we see that this is also equivalent to having $Q(\pm j\omega_i) = [G_o(\pm j\omega_i)]^{-1}$, which requires that (15.3.20) be satisfied.

                                                                                        □□□

Remark 15.2. *We note that in (15.3.19) a possible choice for $N_2(s)$ is*

$$N_2(s) = \overline{D}_Q(s) \sum_{i=1}^{2l} (\tau j\omega_i + 1)^{2l-1} [G_o(j\omega_i)]^{-1} \prod_{k \in \Omega_{2l,i}} \frac{s - j\omega_k}{j\omega_i - j\omega_k} \tag{15.3.22}$$

where $\tau > 0$, $\omega_{l+i} = -\omega_i$ for $i = 1, 2, \ldots, l$, $\Omega_{2l,i} \triangleq \{1, 2, \ldots, 2l\} - \{i\}$ and

$$D_Q(s) = \overline{D}_Q(s)(\tau s + 1)^{2l-1} \tag{15.3.23}$$

We can then parameterize $Q(s)$ as

$$Q(s) = \overline{Q}(s) \prod_{i=1}^{l}(s^2 + w_i^2) + \sum_{i=1}^{2l} [G_o(j\omega_i)]^{-1} \prod_{k \in \Omega_{2l,i}} \frac{s - j\omega_k}{j\omega_i - j\omega_k} \tag{15.3.24}$$

The above lemmas allow us to parameterize the control problem satisfying steady-state constraints while preserving the affine structure of the sensitivity functions in the design variable $\overline{Q}(s)$. We will make use of these parameterizations in Chapter 16 in the context of control-system design via optimization.

(b) Disturbance-rejection trade-offs

Focusing on disturbances and noise only, we see from Chapter 5 that the nominal output response is given by

$$
\begin{aligned}
Y_o(s) &= -\,T_o(s)D_n(s) + S_o(s)D_o(s) + S_{io}(s)D_i(s) \\
&= -\,Q(s)G_o(s)D_n(s) + (1 - Q(s)G_o(s))D_o(s) \\
&\quad + (1 - Q(s)G_o(s))G_o(s)D_i(s)
\end{aligned}
\tag{15.3.25}
$$

With the choices (15.3.12) and (15.3.14), we see that

$$
\begin{aligned}
Y_o(s) &= -\,(F_Q(s)B_{ou}(s))D_n(s) + (1 - F_Q(s)B_{ou}(s))D_o(s) \\
&\quad + (1 - F_Q(s)B_{ou}(s))G_o(s)D_i(s)
\end{aligned}
\tag{15.3.26}
$$

As we know, the first two transfer functions on the right-hand side sum to 1: $S_o(s) + T_o(s) = 1$.

A more subtle trade-off occurs between D_o and D_i. We can use $F_Q(s)$ to cancel any desired open-loop poles in $S_i(s)$, but then these poles necessarily occur as zeros in S_o, as shown in subsection §8.6.3. In view of the impact of both poles and zeros on performance, as discussed in Chapter 8, we see that one has a definite trade-off between dealing with input disturbances and dealing with output disturbances. For example, if one has a slow pole in $G_o(s)$, then this appears either as a slow pole in $S_{io}(s)$, which leads to a long settling time, or as a slow zero in $S_o(s)$, with a consequential sensitivity peak. This will be discussed further in sections §15.5 and §15.6 below.

The case in which disturbance and measurement noise are present and the plant is of nonminimum phase is illustrated in the next example.

Example 15.1 (Nonminimum-phase plant). *Consider a one-d.o.f. control loop where the plant has a nominal model given by*

$$
G_o(s) = \frac{-s + 2}{(s + 2)(s + 1)}
\tag{15.3.27}
$$

Assume that there are output disturbances, which are infrequent abrupt changes. The measurement noise is a signal with significant energy only in the frequency region above 5[rad/s].

Design a controller $C(s)$, by using the affine-parameterization approach, such that the controller output $u(t)$ does not contain significant noise components.

Solution

Considering only the nature of the output disturbance, we require that

- *the controller include integration (i.e., $Q(0) = [G_o(0)]^{-1} = 1$) to ensure zero steady-state error, and that*

- *the closed-loop bandwidth be as large as possible, to provide fast disturbance compensation.*

However, the noise requirement sets an upper bound for the loop bandwidth, say $\omega_c = 5[rad/s]$. We note that the noise component in the controller output is given by (15.3.5). We can then use filter synthesis theory, because $F_Q(s)$ must be a low-pass filter with cut-off frequency $\omega_c = 5[rad/s]$. After iterating with different filter types (Butterworth, Tchebyshev, and elliptic) of different orders, a fourth-order Butterworth filter was chosen:

$$F_Q(s) = \frac{625}{s^4 + 13.0656s^3 + 85.3553s^2 + 326.6407s + 625} \tag{15.3.28}$$

*The reader is encouraged to test other choices for $F_Q(s)$ by using the SIMULINK diagram in file **qaff1.mdl**.*

□□□

15.3.6   Control Effort

We see from (15.3.3) and (15.3.1) that, if we achieve $S_o = 0$ at a given frequency (e.g., $QG_o = 1$), then we have infinite gain in the controller C at the same frequency. For example, say the plant is of minimum phase; then we could choose $G_o^i(s) = G_o^{-1}$. However, by using (15.3.12), we would then have

$$C(s) = \frac{F_Q(s)G_o^i(s)}{1 - F_Q(s)} \tag{15.3.29}$$

By way of illustration, say that we choose

$$F_Q(s) = \frac{1}{(\tau s + 1)^r} \tag{15.3.30}$$

Then the high-frequency gain of the controller, K_{hfc}, and the high-frequency gain of the model, K_{hfg}, are related by

$$K_{hfc} = \frac{1}{\tau^r K_{hfg}} \tag{15.3.31}$$

Thus, as we make $F_Q(s)$ faster (i.e., τ becomes smaller), we see that K_{hfc} increases. This, in turn, implies that the control energy will increase. This consequence can be appreciated from the fact that, under the assumption that $G_o(s)$ is of minimum phase and is stable, we have that

$$S_{uo}(s) = Q(s) = \frac{[G_o(s)]^{-1}}{(\tau s + 1)^r} \tag{15.3.32}$$

15.3.7 Robustness

The issue of modeling errors was introduced in section §4.12. A key observation was that modeling errors are usually significant at high frequencies. This observation was central to the analysis of robustness carried out in section §5.9. A fundamental result in that analysis was that, in order to ensure robustness, the closed-loop bandwidth should be such that the frequency response $|T_o(j\omega)|$ rolls off before the effects of modeling errors become significant.

Now say that we choose $Q(s)$ as in (15.3.12); then

$$T_o(s) = F_Q(s)B_{ou}(s) \qquad (15.3.33)$$

Thus, in the framework of the affine parameterization under discussion here, the robustness requirement can be satisfied if $F_Q(s)$ reduces the gain of $T_o(j\omega)$ at high frequencies. This is usually achieved by including appropriate poles in $F_Q(s)$. Of course, reducing $|T_o(j\omega)|$ to a value $\ll 1$ beyond some frequency necessarily means that $S_o(s)$ tends to 1 beyond the same frequency.

15.3.8 Choice of Q. Summary for the Case of Stable Open-Loop Poles

We have seen that a prototype choice for $Q(s)$ is simply the inverse of the open-loop plant transfer function $G_o(s)$. However, this "ideal" solution needs to be modified in practice to account for the following:

- **Nonminimum-phase zeros**. Internal stability precludes the cancellation of these zeros, so they must appear in $T_o(s)$. This implies that the gain of $Q(s)$ must be reduced at these frequencies to avoid poor transient response, as discussed in Chapter 8.

- **Relative degree**. Excess poles in the model must necessarily appear as a lower bound for the relative degree of $T_o(s)$, because $Q(s)$ must be proper to ensure that the controller $C(s)$ is proper.

- **Disturbance trade-offs**. Whenever we roll T_o off to satisfy measurement-noise rejection, we necessarily increase sensitivity to output disturbances at that frequency. Also, slow open-loop poles must appear either as poles of $S_{io}(s)$ or as zeros of $S_o(s)$, and in either case there is a performance penalty.

- **Control energy**. Plants are typically low pass. Hence, any attempt to make $Q(s)$ close to the model inverse necessarily gives a high-pass transfer function from $D_o(s)$ to $U(s)$. This will lead to large input signals and can lead to controller saturation.

- **Robustness**. Modeling errors usually become significant at high frequencies; hence, to retain robustness, it is necessary to attenuate T_o (and, hence, Q) at these frequencies.

15.4 PID Synthesis by using the Affine Parameterization

In this section, we will illustrate the application of the affine parameterization by using it to develop synthesis strategies for PI and PID controllers when all-open poles are stable (and hence acceptable as closed-loop poles).

That PID design is possible by selecting $Q(s)$ is no surprise; (15.3.1) covers all stabilizing controllers for an open-loop stable plant model, so it must also include the PID architecture as applied to stable plants.

15.4.1 Plant Models for PID Control

In the sequel, we will consider the following models:

$$G_o(s) = \frac{K_o}{\nu_o s + 1} \qquad \text{first order} \qquad (15.4.1)$$

$$G_o(s) = \frac{K_o e^{-s\tau_o}}{\nu_o s + 1} \qquad \text{time-delayed first order} \qquad (15.4.2)$$

$$G_o(s) = \frac{K_o}{s^2 + 2\zeta_o\omega_o s + \omega_o^2} \qquad \text{relative degree 2 resonant} \qquad (15.4.3)$$

$$G_o(s) = \frac{K_o(b_o s + 1)}{s^2 + 2\zeta_o\omega_o s + \omega_o^2}; \quad b_o > 0 \quad \text{relative degree 1 resonant} \qquad (15.4.4)$$

$$G_o(s) = \frac{K_o(-b_o s + 1)}{s^2 + 2\zeta_o\omega_o s + \omega_o^2}; \quad b_o > 0 \quad \text{nonminimum-phase resonant} \qquad (15.4.5)$$

In the above models, all coefficients are assumed to be positive.

Together, these models cover many systems met in industrial applications. In particular, (15.4.2) usually occurs in applications with transport delays and (15.4.3)–(15.4.5) typically occur in electromechanical systems having resonant structures.

15.4.2 First-Order Models

In this subsection, we consider the model (15.4.1):

$$G_o(s) = \frac{K_o}{\nu_o s + 1}$$

We employ the affine-synthesis methodology. There are no unstable zeros, so the model is exactly invertible. We then choose

$$G_o^i(s) = [G_o(s)]^{-1} = \frac{\nu_o s + 1}{K_o} \qquad (15.4.6)$$

In order for $Q(s)$ to be biproper, $F_Q(s)$ must have relative degree 1–for example,

$$F_Q(s) = \frac{1}{\alpha s + 1} \qquad (15.4.7)$$

Hence,

$$Q(s) = F_Q(s)G_o^i(s) = \frac{\nu_o s + 1}{K_o(\alpha s + 1)} \qquad (15.4.8)$$

and the controller (15.3.1) becomes

$$C(s) = \frac{Q(s)}{1 - Q(s)G_o(s)} = \frac{\nu_o s + 1}{K_o \alpha s} = \frac{\nu_o}{K_o \alpha} + \frac{1}{K_o \alpha s} \qquad (15.4.9)$$

which is a PI controller with

$$K_P = \frac{\nu_o}{K_o \alpha} \qquad K_I = \frac{1}{K_o \alpha} \qquad (15.4.10)$$

With these controller parameters, the nominal complementary sensitivity becomes

$$T_o(s) = Q(s)G_o(s) = F_Q(s) = \frac{1}{\alpha s + 1} \qquad (15.4.11)$$

where α becomes a tuning parameter: choosing α smaller makes the loop faster, a larger value for α slows the loop down. With this controller, output disturbances are rejected by the nominal sensitivity function

$$S_o(s) = 1 - T_o(s) = 1 - F_Q(s) = \frac{\alpha s}{\alpha s + 1} \qquad (15.4.12)$$

Again, a small value for α rejects the output disturbance faster than a large value for α. This effect is shown in Figure 15.2 on the following page, where the response to an output step disturbance is shown for $\alpha = 0.1, 0.5, 1.0$, and 2.0.

However, as discussed in Chapter 8, α cannot be chosen arbitrarily small, because of actuator limitations and robustness considerations. Still, performing the experiment that leads to Figure 6.6 on page 167 (and provided that $t_0 \approx t_1$), the relationships in (15.4.10) provide a simple method for PI-controller design with a single easily tuned parameter α. Furthermore, if the time constant ν_o or gain K_o happens to be time varying in a known way, the controller can be easily adapted, because K_P and K_I in (15.4.10) are explicitly expressed in terms of these values.

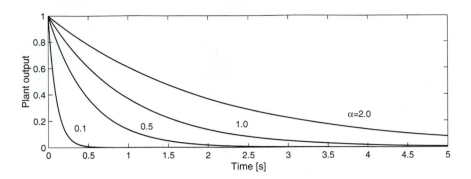

Figure 15.2. Effect of α on output-disturbance rejection

15.4.3   Second-Order Models with Medium Damping

Next, consider the design of PID controllers for second-order models of the form (15.4.3) to (15.4.5). Throughout this subsection, we assume that the plant is medium-well-damped. The exact meaning of medium versus light damping, regarding design, depends on the size of the prevalent modeling errors; typically, however, the design in this subsection is applicable for damping factors larger than, say, 0.6.

We consider initially the relative degree 2 model (15.4.3).

Because there are no unstable zeros, we can choose

$$G_o^i(s) = [G_o(s)]^{-1} = \frac{s^2 + 2\zeta_o\omega_o s + \omega_o^2}{K_o} \qquad (15.4.13)$$

In order to ensure biproperness of $Q(s) = F_Q(s)G_o^i(s)$, $F_Q(s)$ must have relative degree 2–typically,

$$F_Q(s) = \frac{1}{\alpha_2 s^2 + \alpha_1 s + 1} \qquad (15.4.14)$$

Then the equivalent unity feedback controller becomes

$$C(s) = \frac{Q(s)}{1 - Q(s)G_o(s)} = \frac{F_Q(s)G_o^i(s)}{1 - F_Q(s)G_o^i(s)} = \frac{s^2 + 2\zeta_o\omega_o s + \omega_o^2}{K_o(\alpha_2 s^2 + \alpha_1 s)} \qquad (15.4.15)$$

and the equivalent PID controller has the proportional gain:

$$K_P = \frac{2\zeta_o\omega_o\alpha_1 - \alpha_2\omega_o^2}{K_o\alpha_1^2} \qquad (15.4.16)$$

the integral gain

$$K_I = \frac{\omega_o^2}{K_o\alpha_1} \tag{15.4.17}$$

the derivative gain

$$K_D = \frac{\alpha_1^2 - 2\zeta_o\omega_o\alpha_1\alpha_2 + \alpha_2^2\omega_o^2}{K_o\alpha_1^3} \tag{15.4.18}$$

and the derivative time constant

$$\tau_D = \frac{\alpha_2}{\alpha_1} \tag{15.4.19}$$

It is again useful to parameterize the nominal closed-loop complementary sensitivity function directly in terms of the desired closed-loop natural frequency ω_{cl} and damping ψ_{cl}, by choosing

$$\alpha_2 = \frac{1}{\omega_{cl}^2}; \qquad\qquad \alpha_1 = \frac{2\psi_{cl}}{\omega_{cl}} \tag{15.4.20}$$

that is,

$$T_o(s) = F_Q(s) = \frac{\omega_{cl}^2}{s^2 + 2\psi_{cl}\omega_{cl}s + \omega_{cl}^2} \tag{15.4.21}$$

Then, inserting (15.4.20) into (15.4.16) to (15.4.19) yields the following PID gains:

$$K_P = \frac{4\zeta_o\psi_{cl}\omega_o\omega_{cl} - \omega_o^2}{4K_o\psi_{cl}^2} \tag{15.4.22}$$

$$K_I = \frac{\omega_o^2\omega_{cl}}{2K_o\psi_{cl}} \tag{15.4.23}$$

$$K_D = \frac{4\psi_{cl}^2\omega_{cl}^2 - 4\zeta_o\omega_o\zeta_{cl}\omega_{cl} + \omega_o^2}{8K_o\psi_{cl}^3\omega_{cl}} \tag{15.4.24}$$

$$\tau_D = \frac{1}{2\psi_{cl}\omega_{cl}} \tag{15.4.25}$$

Compared to the classical PID techniques of Chapter 6, this model-based approach has several advantages. In particular, the PID gains (15.4.22)–(15.4.25) are explicit functions of the desired closed-loop model (15.4.21) and so allows systematic trade-off decisions. They are also explicit in the model parameters, which allow gain-scheduling to changing parameters without additional effort.

Similar expressions can be derived for the relative degree 1 model (15.4.4), by using the inverse,

$$G_o^i = [G_o(s)]^{-1} = \frac{s^2 + 2\zeta_o\omega_o s + \omega_o^2}{K_o(b_o s + 1)} \tag{15.4.26}$$

and the relative degree 1 filter,

$$F_Q(s) = \frac{1}{\alpha s + 1} \tag{15.4.27}$$

For the nonminimum-phase model (15.4.5), an appropriate stable inverse approximation is given by

$$G_o^i(s) = \frac{s^2 + 2\zeta_o\omega_o s + \omega_o^2}{K_o} \tag{15.4.28}$$

because the unstable zero cannot be inverted. To ensure biproperness of $Q(s)$, $F_Q(s)$ must have relative degree 2, and so (15.4.14) is again an appropriate choice.

15.4.4   Lightly Damped Models

In this subsection, we consider plant models of the type (15.4.3), with a very small value for the damping factor ζ_o. We must warn the reader that the control of lightly damped systems is an inherently difficult problem. The aim of this subsection is to discuss the main issues associated with the problem and to highlight the difficulties involved in solving it.

The problem with very lightly damped systems is the sensitivity at the resonance peaks, which have increasingly higher gains for lower damping ratios. This characteristic results in severe performance degradation even for small errors in the model parameters, when the closed loop is beyond the location of the open-loop resonance. To better appreciate the nature of the problem, we apply the general model-based controller (15.3.1) to the true plant model $G(s)$. As usual, the achieved complementary sensitivity is

$$T(s) = \frac{G(s)C(s)}{1 + G(s)C(s)} = \frac{G(s)Q(s)}{1 + Q(s)(G(s) - G_o(s))} \tag{15.4.29}$$

To avoid cluttering the analysis by including other plant limitations, such as nonminimum-phase zeros, we assume, for clarity, that the model is exactly invertible; thus, we can choose

$$G_o^i(s) = [G_o(s)]^{-1} \tag{15.4.30}$$

and so

$$Q(s) = F_Q(s)[G_o(s)]^{-1} \tag{15.4.31}$$

This yields

$$T(s) = \frac{F_Q(s)G(s)}{F_Q(s)(G(s) - G_o(s)) + G_o(s)} = \frac{F_Q(s)G(s)}{F_Q(s)G(s) + (1 - F_Q(s))G_o(s)} \tag{15.4.32}$$

We recall now that the usual control objective of set-point tracking requires $T(j\omega)$ to be unity at the relevant frequencies. Equation (15.4.32) highlights that this can be achieved in two alternative ways. One obvious way is first to use a very good model at those frequencies ($G_\epsilon = G - G_o$ small), then to shape the filter $F_Q(s)$ to be close to unity at those frequencies.

Note also that $T(j\omega)$ is also close to unity at those frequencies where $|G_o(j\omega)(1 - F_Q(j\omega))|$ is small. If the nominal model damping is close to that of the true system, then the bandwidth of $F_Q(s)$ must be very large to achieve this. A large bandwidth $F_Q(s)$ would have severe consequences regarding loop robustness. An alternative strategy would be to use a well-damped model, even if the true system is not well-damped. Caution has to be exercised in this approach, because we have to ensure robust stability.

These arguments suggest that, at important tracking frequencies, one should choose F_Q and G_o such that $|G_o(j\omega)(1 - F_Q(j\omega))|$ is small.

To see the implications for lightly damped systems, we return to the model (15.4.3) and assume that the true system is of the form

$$G(s) = \frac{K}{s^2 + 2\zeta\omega_n s + \omega_n^2} \tag{15.4.33}$$

We assume also that $F_Q(s)$ is the same as in the previous subsection:

$$F_Q(s) = \frac{\omega_{cl}^2}{s^2 + 2\psi_{cl}\omega_{cl}s + \omega_{cl}^2} \tag{15.4.34}$$

We consider a nominal model where there is no error in the damping factor ($\zeta_o = \zeta$). We also assume that the d.c. gain in both the true system and the nominal model is 1–i.e., $K = \omega_n^2$ and $K_o = \omega_o^2$. If we measure the frequency in units of ω_o, we obtain

$$G_o(s) = \frac{1}{s^2 + 2\zeta s + 1} \tag{15.4.35}$$

$$G(s) = \frac{x^2}{s^2 + 2\zeta xs + x^2} \tag{15.4.36}$$

$$F_Q(s) = \frac{v^2}{s^2 + 2\psi_{cl}vs + v^2} \tag{15.4.37}$$

$$G_\epsilon(s) = (x-1)s\frac{(x+1)s + 2\zeta x}{(s^2 + 2\zeta s + 1)(s^2 + 2\zeta xs + x^2)} \tag{15.4.38}$$

$$G_\Delta(s) = (x-1)s\frac{(x+1)s + 2\zeta x}{s^2 + 2\zeta xs + x^2} \tag{15.4.39}$$

where s is now the normalized Laplace variable, and

$$x \triangleq \frac{\omega_n}{\omega_o}; \qquad\qquad v \triangleq \frac{\omega_{cl}}{\omega_o} \tag{15.4.40}$$

.

To illustrate the nature and magnitude of modeling error due to a mismatch between ω_n and ω_o–i.e., when $x \neq 1$–we calculate and plot $|G_\epsilon(j\omega)|$ as a function of the normalized frequency, for $x = 0.9$ and $x = 1.1$. The results are shown in Figure 15.3.

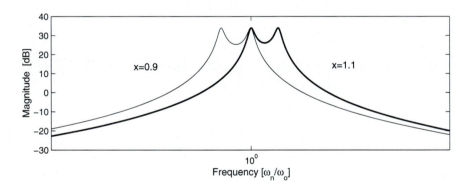

Figure 15.3. Additive modeling error for a lightly damped system ($\zeta = 0.01$)

Figure 15.3 shows that the magnitude of the error has a peak of approximately $33[dB]$ in both cases. However, the similarity between the cases for $x = 0.9$ (overestimating the natural frequency) and $x = 1.1$ (underestimating the natural frequency) is deceptive. To have a more accurate picture of the effects of modeling errors,

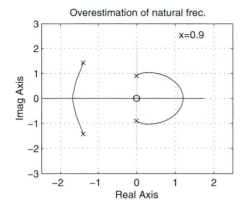
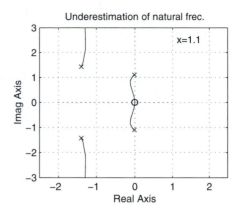

Figure 15.4. Root-locus analysis for $S_\Delta(s)$ over- and underestimating the natural frequency

we recall the expression for the achieved or true sensitivity (15.3.6). If, in that expression, we substitute (15.3.10), (15.3.12), and (15.4.30), we obtain

$$S(s) = (1 - Q(s)G_o(s))S_\Delta(s) = \frac{S_o(s)}{1 + F_Q(s)G_\Delta(s)} \tag{15.4.41}$$

Thus, because the nominal model is stable, robust stability is achieved if and only if the Nyquist plot of $F_Q(j\omega)G_\Delta(j\omega)$ does not encircle the point $(-1,0)$ in the complex plane. This is equivalent to requiring the stability of an equivalent closed loop in which the open-loop transfer function is given by

$$F_Q(s)G_\Delta(s) = \frac{v^2}{s^2 + 2\psi_{cl}vs + v^2}(x - 1)s\frac{(x + 1)s + 2\zeta x}{s^2 + 2\zeta xs + x^2} \tag{15.4.42}$$

Figure 15.4 shows two root-locus diagrams. They correspond to the cases $x = 0.9$ and $x = 1.1$; in both cases, $\zeta = 0.01$, $\psi_{cl} = 0.7$, and $v = 2$. We see that, for the first case (overestimation), the true loop will be unstable even for small errors. By contrast, underestimation always yields a stable loop. The result for the latter case must be interpreted with caution, because only mismatch in the natural frequency has been considered. However, from this analysis, it seems sensible, when facing uncertainty in the value of the true natural frequency, to choose the nominal natural frequency equal to the lower bound known.

The above analysis highlights one of the main difficulties in the control of lightly damped systems: the extreme sensitivity to modeling errors. Another major problem arises in the presence of input disturbances when the model poles are canceled.

The uncontrollable poles appear in the input sensitivity and the associated modes are lightly damped.

15.4.5 Models with Time Delays using Padé Approximation

We next turn to first-order models with time delay, as in (15.4.2). The main difficulty in this type of system arises from the factor $e^{-s\tau_o}$, which is noninvertible and irrational. Initially, we will replace the time-delay term by its so-called first-order Padé approximation, given by

$$e^{-s\tau_o} \approx \frac{2 - s\tau_o}{2 + s\tau_o} \qquad (15.4.43)$$

If we substitute (15.4.43) in (15.4.2), we obtain the rational-model approximation

$$G_o(s) \approx G_{op}(s) \triangleq \frac{(-\tau_o K_o)s + 2K_o}{(\tau_o \nu_o)s^2 + (\tau_o + 2\nu_o)s + 2} \qquad (15.4.44)$$

This model is stable but of nonminimum phase, so we use an approximate inverse, given by

$$G_{op}^i = \frac{(\tau_o \nu_o)s^2 + (\tau_o + 2\nu_o)s + 2}{2K_o} \qquad (15.4.45)$$

A biproper $Q(s)$ is obtained if $F_Q(s)$ is chosen to have relative degree 2, say

$$F_Q(s) = \frac{1}{\alpha_2 s^2 + \alpha_1 s + 1} \qquad (15.4.46)$$

The unity feedback controller is then given by

$$C_p(s) = \frac{Q(s)}{1 - Q(s)G_{op}(s)} = \frac{F_Q(s)G_{op}^i(s)}{1 - F_Q(s)G_{op}^i(s)G_{op}(s)}$$

$$= \frac{(\tau_o \nu_o)s^2 + (\tau_o + 2\nu_o)s + 2}{(2K_o \alpha_2)s^2 + (2K_o \alpha_1 + \tau_o K_o)s} \qquad (15.4.47)$$

which, according to Lemma 7.2 on page 187, is of PID type (7.3.2), with

$$K_P = \frac{2\tau_o\alpha_1 + 4\nu_o\alpha_1 + \tau_o^2 + 2\tau_o\nu_o - 4\alpha_2}{4K_o\alpha_1^2 + 4K_o\alpha_1\tau_o + \tau_o^2 K_o} \qquad (15.4.48)$$

$$K_I = \frac{2}{2K_o\alpha_1 + \tau_o K_o} \qquad (15.4.49)$$

$$K_D = \frac{(2K_o\alpha_1 + \tau_o K_o)^2(\tau_o\nu_o) - (2K_o\alpha_2)(2K_o\alpha_1 + \tau_o K_o)(\tau_o + 2\nu_o) + 8K_o^2\alpha_2^2}{(2K_o\alpha_1 + \tau_o K_o)^3}$$
$$(15.4.50)$$

$$\tau_D = \frac{2\alpha_2}{2\alpha_1 + \tau_o} \qquad (15.4.51)$$

Observe, once again, that the PID gains are given directly in terms of the model parameters (15.4.2) and the design filter (15.4.46). The rather involved and nonlinear nature of these expressions hints at the origins of why trial-and-error tuning of a PID controller for time-delayed systems has traditionally been known to be cumbersome. The expressions given here, however, lead to a straightforward design.

Another advantage of these explicit expressions is that time delays are frequently time varying. This is the case, for example, where the time delay is due to a transportation speed that varies with production requirements. If the speed is measurable, the time delay can be computed on-line, and the PID gains (15.4.48) to (15.4.51) can be adapted correspondingly.

An alternative approach is the Smith controller, presented in the next section. The advantage of the Smith controller over the PID controller (15.4.48) to (15.4.51) is that the Padé approximation is avoided; this tends to be an advantage if the time delay τ_o is comparable to (or larger than) the dominant closed-loop time constant. However, the resulting controller is not strictly of PID type, because it introduces a time-delay element in the parallel model.

15.5   Affine Parameterization for Systems Having Time Delays

A classical method for dealing with pure time delays was to use a *dead-time* compensator. This idea was introduced by Otto Smith in the 1950's. Here, we give this a modern interpretation via the affine parameterization.

Smith's controller is based upon two key ideas: affine synthesis, and the recognition that delay characteristics cannot be inverted. The structure of the traditional Smith controller can be obtained from the scheme in Figure 15.6, which is a particular case of the general scheme shown in Figure 15.1 on page 410.

In Figure 15.5, the nominal model is given by the product of a time-delay factor and a rational transfer function:

$$G_o(s) = e^{-s\tau}\overline{G}_o(s) \qquad (15.5.1)$$

Then, from (15.3.2), the nominal complementary sensitivity is

$$T_o(s) = e^{-s\tau}\overline{G}_o(s)Q(s) \tag{15.5.2}$$

Equation (15.5.2) suggests that $Q(s)$ must be designed by considering only the rational part of the model, $\overline{G}_o(s)$, because the delay cannot be inverted. To carry out the design, the procedures and criteria discussed in the previous sections can be used. In particular, we need an approximate (stable, causal, and proper) inverse for $G_o(s) = e^{-s\tau}\overline{G}_o(s)$. The delay has no causal inverse, so we seek an approximate inverse for $\overline{G}_o(s)$. This can be achieved directly, as in (15.3.14). Alternatively, one can use the idea of feedback to generate a stable inverse. Thus, we might conceive of evaluating $Q(s)$ by

$$Q(s) = \frac{C(s)}{1 + C(s)\overline{G}_o(s)} \tag{15.5.3}$$

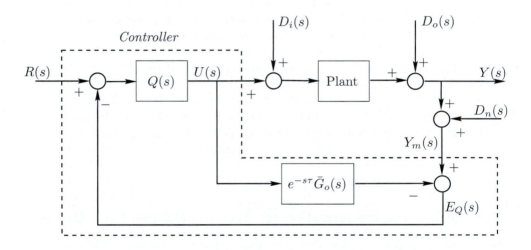

Figure 15.5. Smith's controller (Q form)

When $|C(j\omega)|$ is large, then $Q(j\omega) \approx [\overline{G}_o(j\omega)]^{-1}$.

If $Q(s)$ is implemented via (15.5.3), then the structure in Figure 15.6 on the next page is obtained. This is the traditional form for the Smith controller, as was presented in section §7.4. However, the form given in Figure 15.5 is equally valid.

Remark 15.3. *It is sometimes said that one should not use Smith controllers because of their sensitivity to the modeling errors. However, Lemma 15.1 on page 409 shows that the structure given in Figure 15.5 covers all stabilizing controllers. Hence,*

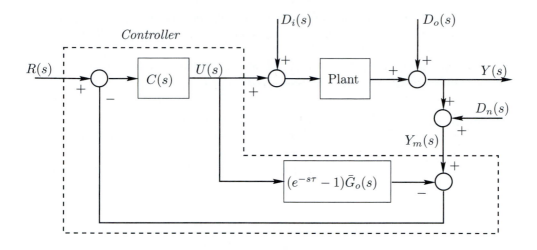

Figure 15.6. Smith's controller (traditional form)

the only question remaining is how to choose $Q(s)$. In this context, there are in-escapable bandwidth-limitation issues associated with robustness. The reader is in-vited to reread subsection §8.6.2.

   The ideas presented above are applied in the following example.

Example 15.2. *Consider a plant with a nominal model given by*

$$G_o(s) = \frac{e^{-s}}{2s+1} \tag{15.5.4}$$

   Synthesize a controller that provides good reference tracking in the frequency band $[0,1][rad/s]$.

Solution

We use the structure shown in Figure 15.5, and thus pay attention only to the rational part of $G_o(s)$. A simple approach is to make

$$Q(s) = \frac{2s+1}{s^2 + 1.3s + 1} \tag{15.5.5}$$

which leads to

$$T_o(s) = \frac{e^{-s}}{s^2 + 1.3s + 1} \tag{15.5.6}$$

*The performance of this design can be evaluated, via simulation, by using SIMULINK file **qaff2.mdl**. Using this schematic, the reader can, among other things, assess the effect of modeling errors.*

A more elaborate solution can be obtained if cancellation of the plant pole is unacceptable, as we explore in the next section.

□□□

Example 15.3 (Rolling-mill thickness control). *We recall, from Example 8.3 on page 208, that there is an inescapable measurement delay associated with the use of a downstream X-ray gauge to measure thickness in a rolling mill. Hence, if one wishes to use this measurement for feedback control, then a Smith-predictor architecture of the type shown in Figure 15.6 could well be appropriate. Two practical comments:*

(i) The delay τ is actually a function of strip velocity, and thus it may be desirable to schedule the delay in the controller on the basis of strip speed.

(ii) Robustness issues precludes the use of excessively high bandwidth for the controller $C(s)$.

□□□

15.6   Undesirable Closed-Loop Poles

15.6.1   Interpolation Constraints

Up to this point, it has been implicitly assumed that all open-loop plant poles were stable and hence could be *tolerated* in the closed-loop input sensitivity function $S_{io}(s)$. In practice, we need to draw a distinction between *stable* poles and *desirable* poles. For example, a lightly damped resonant pair might well be stable but is probably undesirable. Say the open-loop plant contains some undesirable (including unstable) poles. The only way to remove poles from the complementary sensitivity is to choose $Q(s)$ to contain these poles as zeros. This results in cancellation of these poles from the product $Q(s)G_o(s)$ and hence from $S_o(s)$ and $T_o(s)$. (See equations (15.3.1) and (15.3.2).) However, we see from (15.3.4) that the canceled poles might still appear as poles of the nominal input sensitivity $S_{io}(s)$, depending on the zeros of $1 - Q(s)G_o(s)$–(i.e., the zeros of $S_o(s)$). To eliminate these poles from $S_{io}(s)$, we need to ensure that the offending poles are also zeros of $[1 - Q(s)G_o(s)]$. The following lemma summarizes the above observations.

Lemma 15.4 (Interpolation constraints to avoid undesirable poles). *Consider a nominal feedback control loop with one-d.o.f., and assume that $G_o(s)$ contains undesirable (including unstable) open-loop poles. We then have the following:*

(a) Each of the sensitivity functions $T_o(s)$, $S_o(s)$, $S_{io}(s)$, and $S_{uo}(s)$ will have no undesirable poles if and only if the controller $C(s)$ is expressed as in (15.3.1), where $Q(s)$ satisfies the following constraints:

(i) $Q(s)$ *is proper stable and has only desirable poles.*

(ii) *Any undesirable poles of $G_o(s)$ are zeros of $Q(s)$ having at least the same multiplicity as $G_o(s)$.*

(iii) *Any undesirable poles of $G_o(s)$ are zeros of $1 - Q(s)G_o(s)$ having at least the same multiplicity as $G_o(s)$.*

(b) *When conditions (ii) and (iii) are satisfied, then all resultant unstable pole-zero cancellations in $C(s)$ induced by (15.3.1) should be performed analytically, prior to implementation.*

Proof

Part (a) follows immediately from equations (15.3.2) to (15.3.5).

Part (b) avoids undesirable (unstable) pole-zero cancellation inside the controller.

□□□

Remark 15.4. *In connection with part (b) of the above result, the reader should note the difference between an analytical and an implementation pole-zero cancellation. This can be better understood by considering an example which involves the following three transfer functions:*

$$H(s) = \frac{4}{(s+1)(s+4)}; \qquad H_1(s) = \frac{4}{(s+1)(-s+2)}; \qquad H_2(s) = \frac{(-s+2)}{(s+4)}$$
$$(15.6.1)$$

We observe that $H(s) = H_1(s)H_2(s)$. If the cancellation is performed prior to implementation of the transfer function $H(s)$ (analytic cancellation), no problem arises. However, if we cascade the implementation of $H_1(s)$ with the implementation of $H_2(s)$, internal instability arises, because then the cancellation will appear in the implementation.

This distinction explains why the structure in Figure 15.1 on page 410 cannot be used to implement a controller for unstable systems. Instead, the pole-zero cancellations inherent in the interpolation constraints given in Lemma 15.4 on the preceding page must be performed prior to implementation.

□□□

We illustrate Lemma 15.4 on the facing page by a simple example.

Example 15.4. *Consider the nominal model $G_o(s)$ given by*

$$G_o(s) = \frac{6}{(s+1)(s+6)} \qquad (15.6.2)$$

Assume that the measurement noise limits the closed-loop bandwidth to $\omega = 10[rad/s]$. Under these conditions, a possible choice for $Q(s)$ is

$$Q(s) = F_Q(s)\frac{(s+1)(s+6)}{6} \qquad where \qquad F_Q(s) = 1000\frac{\beta_1 s + 1}{(s^2 + 14s + 100)(s+10)} \tag{15.6.3}$$

where β_1 is a free coefficient to be chosen later.

The relative degree of $F(s)$ has been chosen equal to 2, to make $Q(s)$ biproper. $T_o(s) = F_Q(s)$, so the choice of $F_Q(0) = 1$ ensures exact inversion at $\omega = 0$–i.e., $C(s)$ will have a pole at the origin.

We see that the controller is actually given by

$$C(s) = \frac{Q(s)}{1 - Q(s)G_o(s)} = \frac{F_Q(s)[G_o(s)]^{-1}}{1 - F_Q(s)} = \frac{(\beta_1 s + 1)(s+1)(s+6)}{6s(s^2 + 24s + 240 - 1000\beta_1)} \tag{15.6.4}$$

Next, let us hypothesize that simply canceling the slow plant pole at $s = -1$ in $S_o(s)$ is inadequate, because this would lead to a (relatively) slow pole in $S_{io}(s)$, which, in turn, would produce a slow response to input disturbances. To avoid this outcome, it is desirable to further constrain $Q(s)$ so that $s = -1$ is a zero of $S_o(s)$. To achieve this, we seek a value of β_1 such that $S_o(-1) = 0$ ($\Rightarrow T_o(-1) = F_Q(-1) = 1$). Using (15.6.3), we require

$$F_Q(-1) = \frac{1000}{783}(1 - \beta_1) = 1 \Longrightarrow \beta_1 = \frac{217}{1000} \tag{15.6.5}$$

We observe that, with this value for β_1, the denominator of $C(s)$ in (15.6.4) can be factored as $6s(s+23)(s+1)$. We then see that $(s+1)$ cancels in the numerator and denominator of the controller, leading to

$$C(s) = \frac{(217s + 1000)(s+6)}{6000s(s+23)} \tag{15.6.6}$$

From this last equation, we see that only the pole at $s = -6$ will be uncontrollable from the reference and controllable from the input disturbance.

15.6.2   PID Design Revisited

We return to the design of subsection §15.4.2, where a PI controller was synthesized for a first-order plant. We found that the design of subsection §15.4.2 (based on canceling the open-loop poles in $C(s)$) gave excellent output-disturbance rejection. In chemical processes, however, disturbances are frequently better modeled as occurring at the *input* to the system. We then recall that the input-disturbance response $Y_d(s)$ is given by

$$Y_d(s) = S_{io}(s)D_i(s) \tag{15.6.7}$$
$$S_{io}(s) = S_o(s)G_o(s) \tag{15.6.8}$$

Hence, when any plant pole is canceled in the controller, it remains controllable from the input disturbance and is still observable at the output. Thus, the transient component in the input-disturbance response will have a mode associated with that pole. Canceling *slow* poles thus has a deleterious effect on the response. For example, consider again the design given in (15.4.6), (15.4.7). The response to a unit step *input* disturbance is illustrated in Figure 15.7. Time has been expressed in units of ν_o, and three cases have been considered. Each curve has been obtained for a different value of α (measured relative to ν_o). These values have been chosen to be $1, 2$ and 5. In each case, the plant model d.c. gain K_o has been set to 1. We see that the response has a long *tail* and that changing α only scales the response without changing its shape.

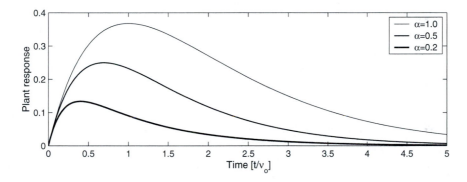

Figure 15.7. Input-disturbance rejection with plant-pole cancellation, for different values of α

The origin of this problem is the cancellation of a pole in $G_o(s)$ with a zero in $C(s)$. As shown in section §15.6, the only way to remove the pole from $S_{io}(s)$ is to choose $F_Q(s)$ in such a way that the offending pole is a zero of $S_o(s) = 1 - Q(s)G_o(s)$–that is, we require

$$S_o(-a) = 0 \Longrightarrow T_o(-a) = F_Q(-a) = 1 \qquad \text{where} \qquad a \stackrel{\triangle}{=} \frac{1}{\nu_o} \tag{15.6.9}$$

The following result shows how this can be achieved:

Lemma 15.5. *Consider the plant model (15.4.1) and the control scheme shown in Figure 15.1 on page 410, where $Q(s) = [G_o(s)]^{-1} F_Q(s)$. Then a PI controller which does not cancel the plant pole is obtained as*

$$C(s) = K_P + \frac{K_I}{s} \qquad \text{with} \qquad K_P = \frac{2\psi_{cl}\omega_{cl}\nu_o - 1}{K_o}; \qquad K_I = \frac{\nu_o}{K_o}\omega_{cl}^2$$

$$(15.6.10)$$

where ψ_{cl} and ω_{cl} are chosen to obtain a closed-loop characteristic polynomial given by

$$A_{cl}(s) = \left(\frac{s}{\omega_{cl}}\right)^2 + 2\psi_{cl}\left(\frac{s}{\omega_{cl}}\right) + 1 \qquad (15.6.11)$$

Proof

We first parameterize $F_Q(s)$ as

$$F_Q(s) = \frac{\beta_1 s + 1}{\alpha_2 s^2 + \alpha_1 s + 1} \qquad (15.6.12)$$

where, to ensure that (15.6.11) is satisfied, we choose

$$\alpha_2 = \frac{1}{\omega_{cl}^2}; \qquad \alpha_1 = 2\frac{\psi_{cl}}{\omega_{cl}} \qquad (15.6.13)$$

Then, we note that

$$S_o(s) = 1 - F_Q(s) = \frac{(\alpha_2 s + \alpha_1 - \beta_1)s}{\alpha_2 s^2 + \alpha_1 s + 1} \qquad (15.6.14)$$

This leads to

$$C(s) = \frac{T_o(s)}{S_o(s)}[G_o(s)]^{-1} = \frac{(\beta_1 s + 1)(\nu_o s + 1)}{K_o s(\alpha_2 s + \alpha_1 - \beta_1)} \qquad (15.6.15)$$

Thus, to obtain a PI controller that removes the open-loop pole at $-\nu_o^{-1}$ from the input-disturbance response, it is sufficient that

$$F_Q(-a) = \frac{-a\beta_1 + 1}{a^2\alpha_2 - a\alpha_1 + 1} = 1 \qquad (15.6.16)$$

This leads to the choice

$$a = \frac{\alpha_1 - \beta_1}{\alpha_2} \qquad (15.6.17)$$

or

$$\beta_1 = \alpha_1 - \frac{\alpha_2}{\nu_o} = \frac{2\psi_{cl}\omega_{cl}\nu_o - 1}{\nu_o\omega_{cl}^2} \qquad (15.6.18)$$

The result follows upon noting that (15.6.15) leads to

$$C(s) = \frac{F_Q(s)}{G_o(s)(1 - F_Q(s))} = \frac{\beta_1\nu_o}{\alpha_2 K_o} + \frac{\nu_o}{\alpha_2 K_o}\frac{1}{s} \qquad (15.6.19)$$

□□□

Lemma 15.5 on the facing page allows one to reduce the design problem to the choice of one parameter. To do that, we first do a time-scaling by ν_o; this implies a frequency scaling by a factor of $[\nu_o]^{-1}$. If we choose a damping factor $\psi_{cl} = 0.7$, then, using (15.6.10), we have that

$$G_o(s)C(s) = \frac{s(1.4v - 1) + v^2}{s(s + 1)} \qquad (15.6.20)$$

where v is the normalized value for ω_{cl} and s has been also scaled by ω_{cl}.

From (15.6.20), it is evident that the choice of one parameter (ω_{cl}) determines the loop performance. We recall that ω_{cl} is a measure of the closed-loop bandwidth, in units of the plant-model bandwidth; it thus becomes a meaningful tuning parameter.

To illustrate the application of the above result, we repeat the simulations in Figure 15.7 on page 433, but this time using the controller given by (15.6.10). Also, (15.6.11) is

$$\left(\frac{s}{\omega_{cl}}\right)^2 + 2\psi_{cl}\left(\frac{s}{\omega_{cl}}\right) + 1 = (\beta_1 s + 1) + s\chi_o(\nu_o s + 1) \qquad (15.6.21)$$

The response to a unit step input disturbance is shown in Figure 15.8 for three different (normalized) values of ω_{cl}. These values are $1, 2$, and 5.

In contrast to the results in Figure 15.7, when cancellation was used, we now see that the time response to an input disturbance can be changed both in shape and in scale. The difference can be appreciated by examining Figures 15.8 on the following page and 15.7 on page 433 for the case when a value of 5 is chosen for the ratio between the closed loop and the plant bandwidths. Note, however, that $(\nu_o s + 1)$ now appears in the numerator of $S_o(s)$. $S_{io}(s) = S_o(s)G_o(s)$, so it is inescapable that every open-loop pole appears either as a pole in S_{io} or as a zero in $S_o(s)$.

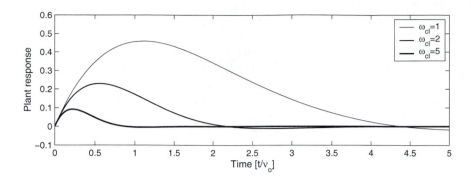

Figure 15.8. Input-disturbance rejection without plant-pole cancellation

15.6.3 Integrating Models

In this subsection we consider a first-order model with integrator:

$$G_o(s) = \frac{K_o}{s(\nu_o s + 1)} \tag{15.6.22}$$

A free integrator, of the type appearing in this model, frequently models mass accumulation, the rotational angle of a motor shaft, or the position of a moving object. Of course an integration is an undesirable (indeed, unstable) closed-loop pole, and thus we need to apply the methodology explained in section 15.6.

Recall from Lemma 15.4 on page 430 that the parameterization (15.3.1) is still valid, but $Q(s)$ must satisfy additional interpolation constraints. In this case, there are only two constraints, which originate from the presence of the pole at the origin. In particular, internal stability requires that $Q(s)$ and $1 - Q(s)G_o(s)$ both must have a zero at the origin.

Because (15.6.22) has no unstable zero, we can choose

$$G_o^i(s) = \frac{s(\nu_o s + 1)}{K_o} \tag{15.6.23}$$

which requires, for properness of $Q(s)$, that $F_Q(s)$ be of relative degree at least 2. Also, with this choice, $Q(s) = F_Q(s)G_o^i(s)$ automatically satisfies one of the constraints ($Q(0) = 0$). The second constraint is satisfied if $F_Q(s)$ is chosen such that $F_Q(0) = 1$. A valid choice for $F_Q(s)$ is, for instance, the one given in (15.4.46). This leads to

$$C(s) = \frac{F_Q(s)[G_o(s)]^{-1}}{1 - F_Q(s)} = \frac{s(\nu_o s + 1)}{K_o s(\alpha_2 s + \alpha_1)} \tag{15.6.24}$$

Again we notice that the controller (15.6.24) is obtained after an unstable pole-zero cancellation. This cancellation must be carried out analytically, i.e., prior to implementation; see Remark 15.4 on page 431.

From a design point of view, (15.6.24) still has the shortcoming of not rejecting constant input disturbances. This happens because $S_{io}(s)$ does not have a zero at the origin, even though $S_o(s)$ does. This is remedied if the choice of $F_Q(s)$ forces the sensitivity $S_o(s)$ to have two poles at the origin; then $S_{io}(s)$ will have one zero at the origin. The simplest choice for $F_Q(s)$ leading to this desired result is

$$F_Q(s) = \frac{F_N(s)}{F_D(s)} = \frac{\alpha_1 s + 1}{\alpha_3 s^3 + \alpha_2 s^2 + \alpha_1 s + 1} \tag{15.6.25}$$

Observe that the coefficients in the numerator coincide with the coefficients of the same power of s in the denominator. The resulting sensitivity function is then

$$S_o(s) = 1 - F_Q(s) = \frac{s^2(\alpha_3 s + \alpha_2)}{\alpha_3 s^3 + \alpha_2 s^2 + \alpha_1 s + 1} \tag{15.6.26}$$

As a result, the input sensitivity becomes

$$S_{io}(s) = S_o(s)G_o(s) = \frac{K_o s(\alpha_3 s + \alpha_2)}{(\alpha_3 s^3 + \alpha_2 s^2 + \alpha_1 s + 1)(\nu_o s + 1)} \tag{15.6.27}$$

Hence, constant input disturbances are indeed rejected with zero steady-state error. The associated unity feedback controller is given by

$$C(s) = \frac{(\nu_o \alpha_1)s^2 + (\nu_o + \alpha_1)s + 1}{K_o s(\alpha_3 s + \alpha_2)} \tag{15.6.28}$$

which is once again a PID controller with

$$K_P = \frac{(\nu_o + \alpha_1)\alpha_2 - \alpha_3}{K_o \alpha_2^2} \tag{15.6.29}$$

$$K_I = \frac{1}{K_o \alpha_2} \tag{15.6.30}$$

$$K_D = \frac{\alpha_2^2 \alpha_1 \nu_o - \alpha_3 \alpha_2 \nu_o - \alpha_3 \alpha_2 \alpha_1 + \alpha_3^2}{K_o \alpha_2^2} \tag{15.6.31}$$

$$\tau_D = \frac{\alpha_3}{\alpha_2} \tag{15.6.32}$$

$$\tag{15.6.33}$$

As in the previous cases, the PID parameters (15.6.29) to (15.6.32) are given explicitly in terms of the parameters of the integrating model (15.6.22) and the design filter (15.6.25). This filter, in turn, is designed by first selecting the closed-loop characteristic polynomial $\alpha_3 s^3 + \alpha_2 s^2 + \alpha_1 s + 1$, which then induces the zero of $\alpha_1 s + 1$. We recall that a general third-order characteristic polynomial can be obtained by combining a complex-conjugate pair and a simple real pole–i.e., we can write

$$F_D(s) = \left(\frac{1}{\omega_{cl}^2} s^2 + \frac{2\psi_{cl}}{\omega_{cl}} s + 1 \right) (\alpha_c s + 1) \qquad (15.6.34)$$

leading to

$$F_N(s) = \left(\frac{2\psi_{cl} + \alpha_c \omega_{cl}}{\omega_{cl}} s + 1 \right) \qquad (15.6.35)$$

Typically the single real pole is placed at the same distance from the origin as the complex pair, that is,

$$\alpha_c = \frac{1}{\omega_{cl}} \qquad (15.6.36)$$

and the damping factor is chosen as $\psi_{cl} = 0.7$; thus, the third-order filter, and hence the PID controller, are parameterized in terms of a single convenient tuning parameter, ω_{cl}.

Nominally, if the true system equals the model, the complementary sensitivity is given by $F_Q(s)$. We see that there is a zero located at

$$z_o = -\frac{\omega_{cl}}{2\psi_{cl} + 1} \qquad (15.6.37)$$

Hence, the closed-loop characteristic polynomial induces a zero that is slower (smaller) the smaller ω_{cl} is, and slower zeros produce larger overshoots, as was discussed in Chapter 8. Observe that the overshoot-producing zero of $F_Q(s)$ was induced by the desire to obtain zero steady-state error in response to a step disturbance at the input of an integrating system.

15.7 Affine Parameterization: The Unstable Open-Loop Case

In the examples given above, we went to some trouble to ensure that the poles of all closed-loop sensitivity functions (especially the input-disturbance sensitivity, S_{io}) lay in desirable regions of the complex plane. In this section, we will simplify this procedure by considering a general design problem in which the open-loop plant can have one (or many) poles in undesirable regions. We will not give a formal definition of an undesirable region, but, heuristically, one would expect this to include all unstable poles, stable poles that were "close" to the origin (i.e., those leading to slow transients), and lightly damped poles (i.e., those leading to ringing transients).

In the last section, we found that extra interpolation constraints on $Q(s)$ were needed to eliminate undesirable poles from the input sensitivity $S_{io}(s)$. In the design examples presented to date, we have chosen $Q(s)$ so as to account explicitly for the interpolation constraints. However, this is a tedious task, and one is led to ask the following question: *Can we reparameterize $C(s)$ in such a way that the interpolation constraints given in Lemma 15.4 on page 430 are automatically satisfied?* The answer is yes, and the solution is described in the following lemma.

Lemma 15.6 (Affine parameterization undesirable open-loop poles). *Consider a one-d.o.f. control loop for the plant with nominal model $G_o(s) = \frac{B_o(s)}{A_o(s)}$. We assume that $B_o(s)$ and $A_o(s)$ are coprime polynomials and that $G_o(s)$ might contain undesirable poles (including unstable poles).*

Then the nominal closed loop will be internally stable and all sensitivity functions will contain only desirable poles if and only if $C(s)$ is parameterized by

$$
C(s) = \frac{\dfrac{P(s)}{E(s)} + Q_u(s)\dfrac{A_o(s)}{E(s)}}{\dfrac{L(s)}{E(s)} - Q_u(s)\dfrac{B_o(s)}{E(s)}}
\tag{15.7.1}
$$

where

(a) $Q_u(s)$ is a proper stable transfer function having desirable poles, and

(b) $P(s)$ and $L(s)$ are polynomials satisfying the following pole-assignment equation:

$$
A_o(s)L(s) + B_o(s)P(s) = E(s)F(s)
\tag{15.7.2}
$$

where $E(s)$ and $F(s)$ are polynomials of suitable degrees which have zeros lying in the desirable region of the complex plane but otherwise are arbitrary.

Proof

For simplicity, we assume that all poles in $A_o(s)$ lie in an undesirable region, (See Remark 15.7 on page 441, for the more general case in which there exists a mixture of poles.) Without loss of generality, we write $Q(s)$ as the ratio of two polynomials

$$Q(s) = \frac{\tilde{P}(s)}{\tilde{E}(s)} \tag{15.7.3}$$

giving

$$C(s) = \frac{Q(s)}{1 - Q(s)G_o(s)} = \frac{\tilde{P}(s)A_o(s)}{\tilde{E}(s) - P(s)B_o(s)} \tag{15.7.4}$$

From equation (15.7.3), we see that necessary and sufficient conditions for the interpolation constraints of Lemma 15.4 on page 430 to be satisfied are as follows:

(i) the zeros of $\tilde{E}(s)$ lie in the desirable region;

(ii) $A_o(s)$ is a factor of $\tilde{P}(s)$, i.e., there exists a $\overline{P}(s)$ such that $\tilde{P}(s) = A_o(s)\overline{P}(s)$; and

(iii) $A_o(s)$ is a factor of $[1 - Q(s)G(s)]$.

However,

$$1 - Q(s)G_o(s) = 1 - \frac{\tilde{P}(s)B_o(s)}{\tilde{E}(s)A_o(s)} \tag{15.7.5}$$

$$= 1 - \frac{\overline{P}(s)B_o(s)}{\tilde{E}(s)}$$

$$= \frac{\tilde{E}(s) - \overline{P}(s)B_o(s)}{\tilde{E}(s)}$$

Hence, to satisfy (iii), there must exist a polynomial $\overline{L}(s)$ such that

$$\tilde{E}(s) - \overline{P}(s)B_o(s) = \overline{L}(s)A_o(s) \tag{15.7.6}$$

or

$$\overline{L}(s)A_o(s) + \overline{P}(s)B_o(s) = E(s)F(s) \tag{15.7.7}$$

where we have set $\tilde{E}(s) = E(s)F(s)$.

We recognize (15.7.7) as the standard pole-assignment equation of Chapter 7. Thus, by choosing the orders of $E(s), F(s), \overline{L}(s), \overline{P}(s)$, we know that a unique solution $L(s), P(s)$ can be found.

Given one solution $(L(s), P(s))$ to (15.7.7), it is a standard result of algebra that any other solution $(\overline{L}(s), \overline{P}(s))$ can be expressed as follows:

$$\frac{\overline{L}(s)}{E(s)} = \frac{L(s)}{E(s)} - Q_u(s)\frac{B_o(s)}{E(s)} \tag{15.7.8}$$

$$\frac{\overline{P}(s)}{E(s)} = \frac{P(s)}{E(s)} + Q_u(s)\frac{A_o(s)}{E(s)} \tag{15.7.9}$$

where $Q_u(s)$ is a stable proper transfer function having no undesirable poles. (Check that the solutions satisfy (15.7.7).)

Substituting (15.7.7) into (15.7.4) shows that

$$C(s) = \frac{\overline{P}(s)}{\overline{L}(s)} \tag{15.7.10}$$

Finally, using (15.7.8) and (15.7.9) in (15.7.10), we see that any controller satisfying the desired conditions can be parameterized as in (15.7.1), (15.7.2).

□□□

Remark 15.5. *Equation (15.7.2) provides a link between the affine parameterized result and the pole-assignment methods studied in Chapter 7.*

□□□

Remark 15.6. *Actually, Lemma 15.6 on page 439 simply gives an automatic way of parameterizing $Q(s)$ so that the interpolation constraints given in Lemma 15.4 on page 430 are satisfied. Indeed, substituting (15.7.7), (15.7.8), (15.7.9) into (15.7.3), we find that the original $Q(s)$ is now constrained to the form*

$$Q(s) = \frac{A_o(s)}{F(s)}\left[\frac{P(s)}{E(s)} + Q_u(s)\frac{A_o(s)}{E(s)}\right] \tag{15.7.11}$$

where $Q_u(s)$ has desirable poles. It is then verified that this form for $Q(s)$ automatically ensures that the interpolation constraints (i) to (iii) of Lemma 15.4 on page 430 hold.

□□□

Remark 15.7. *If $A_o(s)$ has a mixture of desirable and undesirable poles, then we can write*

$$A_o(s) = A_d(s)A_u(s) \tag{15.7.12}$$

where $A_u(s)$ contains the undesirable poles. In this case, we can write $E(s) = A_d(s)\overline{E}(s)$, and (15.7.2) becomes

$$A_d(s)A_u(s)L(s) + B_o(s)P(s) = A_d(s)\overline{E}(s)F(s) \qquad (15.7.13)$$

Clearly, this equation requires the existence of a $\tilde{P}(s)$ such that $P(s) = \tilde{P}(s)A_d(s)$, and hence (15.7.2) reduces to

$$A_u(s)L(s) + B_o(s)\tilde{P}(s) = \overline{E}(s)F(s) \qquad (15.7.14)$$

When $A_o(s)$ contains only desirable poles, then $A_o(s) = E(s), A_u(s) = 1$, and $\tilde{E}(s) = 1$, and we can take $L(s) = F(s)$ and $\tilde{P}(s) = 0$.

The result in Lemma 15.6 on page 439 then reduces to the one given in Lemma 15.1 on page 409. Also, we clearly have

$$Q(s) = Q_u(s) \qquad (15.7.15)$$

$\square\square\square$

We illustrate the application of Lemma 15.6 on page 439 by the following example.

Example 15.5. *Consider a nominal plant model*

$$G_o(s) = \frac{s-4}{(s-1)(s+4)} \qquad (15.7.16)$$

Say that we require all closed-loop poles to lie to the left of -0.5 in the complex plane. We also require that the controller include integral action.

(a) Find a particular controller satisfying these conditions.

(b) Parameterize all controllers satisfying this condition.

Solution

(a) We note that the open-loop pole at -4 lies in the desirable region. Also, the controller is required to have integral action, so $\overline{L}(s)$ in (15.7.10) must be of the form $s\tilde{L}(s)$.

Then equation (15.7.14) becomes

$$s(s-1)\tilde{L}(s) + (s-4)\tilde{P}(s) = \overline{E}(s)F(s) \qquad (15.7.17)$$

For a unique solution, we choose the degrees of $\overline{E}(s)F(s)$, $\tilde{L}(s)$, and $\tilde{P}(s)$ as 3, 1, and 1, respectively. To be specific, we choose

$$\overline{E}(s)F(s) = \left(s^2 + 4s + 9\right)(s + 10) \tag{15.7.18}$$

We find

$$\tilde{L}(s) = s + \frac{263}{6}; \qquad \tilde{P}(s) = -\frac{1}{6}(173s + 135) \tag{15.7.19}$$

Hence, a particular solution is

$$C(s) = -\frac{(173s + 135)(s + 4)}{s(6s + 263)} \tag{15.7.20}$$

(b) All possible solutions can be expressed as in (15.7.1), and after some simplifications, we have

$$C(s) = \frac{\left[-\dfrac{(173s + 135)}{6(s^2 + 4s + 9)}\right] + Q_u(s)\left[\dfrac{(s - 1)}{(s^2 + 4s + 9)}\right]}{\left[\dfrac{s(6s + 263)}{6(s^2 + 4s + 9)(s + 4)}\right] - Q_u(s)\left[\dfrac{(s - 4)}{(s^2 + 4s + 9)(s + 4)}\right]} \tag{15.7.21}$$

where $Q_u(s)$ is any proper transfer function having poles in the desirable region.

□□□

The parameterization in Equation (15.7.1) leads to the following parameterized version of the nominal sensitivities:

$$S_o(s) = \frac{A_o(s)L(s)}{E(s)F(s)} - Q_u(s)\frac{B_o(s)A_o(s)}{E(s)F(s)} \tag{15.7.22}$$

$$T_o(s) = \frac{B_o(s)P(s)}{E(s)F(s)} + Q_u(s)\frac{B_o(s)A_o(s)}{E(s)F(s)} \tag{15.7.23}$$

$$S_{io}(s) = \frac{B_o(s)L(s)}{E(s)F(s)} - Q_u(s)\frac{(B_o(s))^2}{E(s)F(s)} \tag{15.7.24}$$

$$S_{uo}(s) = \frac{A_o(s)P(s)}{E(s)F(s)} + Q_u(s)\frac{(A_o(s))^2}{E(s)F(s)} \tag{15.7.25}$$

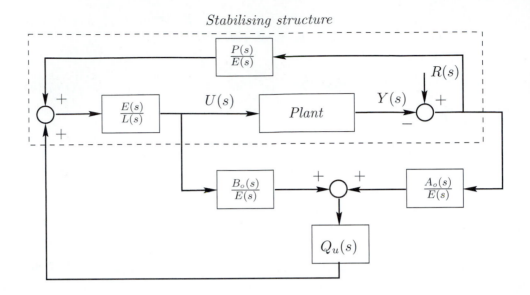

Figure 15.9. Q parameterization for unstable plants

The controller parameterization developed above can also be described in block-diagram form. The equation for $C(s)$ directly implies that the controller is as in Figure 15.9.

We observe that, by superposition, the input signal $U(s)$ in Figure 15.9 can be written as the sum of two signals, one coming from the upper loop via $\frac{P(s)}{L(s)}$ and the other via $Q_u(s)$.

It is also interesting to examine the class of all stabilizing controllers for a presta-bilized plant. Thus, consider the configuration shown in Figure 15.10 on the next page. Note that the prestabilized plant has the closed-loop transfer function $\frac{B_o(s)}{F(s)}$, where $A_o(s)L(s) + B_o(s)P(s) = E(s)F(s)$. Hence, Figure 15.10 on the facing page corresponds to Figure 15.1 on page 410, where $G_o(s)$ has been replaced by the prestabilized plant.

A simple calculation shows that the equivalent unity feedback controller in Figure 15.10 is

$$\overline{C}(s) = \frac{Q_x(s)E(s)F(s) + P(s)(F(s) - Q_x(s)B_o(s))}{L(s)(F(s) - Q_x(s)B_o(s))} \tag{15.7.26}$$

By using the expression $A_o(s)L(s) + B_o(s)P(s) = E(s)F(s)$, the above expres-sion can be simplified to

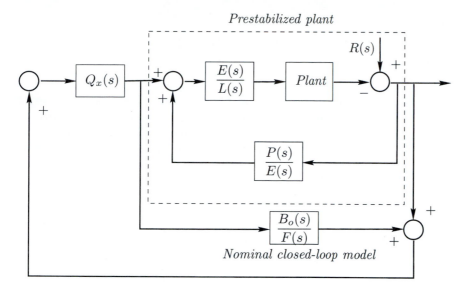

Figure 15.10. Q interpretation for a prestabilized plant

$$\overline{C}(s) = \frac{Q_x(s)A_o(s)L(s) + P(s)F(s)}{L(s)F(s) - L(s)Q_x(s)B_o(s)} \tag{15.7.27}$$

We then have the following result:

Lemma 15.7. *Consider the control structures shown in Figures 15.9 and 15.10.*

(i) If $Q_x(s)$ is stable, then Figure 15.10 can always be redrawn as in Figure 15.9 on the preceding page, where $Q_u(s)$ takes the particular value

$$Q_u(s) = \frac{Q_x(s)L(s)}{F(s)} \tag{15.7.28}$$

(ii) If $\frac{Q_u(s)}{L(s)}$ is stable, then Figure 15.9 on the facing page can be redrawn as in Figure 15.10, where $Q_x(s)$ takes the particular value

$$Q_x(s) = \frac{F(s)Q_u(s)}{L(s)} \tag{15.7.29}$$

Proof

Equating $\overline{C}(s)$ to $C(s)$, we find that the relationship between $Q_u(s)$ and $Q_x(s)$ is given by (15.7.28). Then the result follows.

□□□

Remark 15.8. *Part (i) of the above result is unsurprising; the loop in Figure 15.10 is clearly stable for $Q_x(s)$ stable, so, by Lemma 15.6 on page 439, the controller can be expressed as in Figure 15.9 on page 444 for some stable $Q_u(s)$. The converse, given in part (ii), is more interesting, because it shows that there exist structures of the type shown in Figure 15.9 on page 444 that cannot be expressed as in Figure 15.10. However, inspection of equation (15.7.22) indicates that low sensitivity is achieved if $Q_u(s)$ is chosen close to $\frac{L(s)}{B_o(s)}$ over the frequency band of interest. Thus, it may be reasonable to choose $Q_u(s)$ so that $\frac{Q_u(s)}{L(s)}$ is stable. In this case, part (ii) of Lemma 15.7 shows that it is possible to prestabilize the plant and then use the simple representation of Lemma 15.1.*

15.8 Discrete-Time Systems

Throughout this chapter, we have illustrated the affine parameterization by using the case of continuous-time plants. However, all the methods are algebraically based, so they can be extended immediately to discrete-time systems. As a simple illustration, consider the following example:

Example 15.6. *Consider a discrete-time control loop, where the plant model is*

$$G_{oq}(z) = \frac{B_{oq}(z)}{A_{oq}(z)} \tag{15.8.1}$$

and $A_{oq}(z)$ has at least one root outside the open unit disk.

Determine the equivalent form of Lemma 15.2 on page 412 for discrete-time open-loop unstable systems.

Solution

We first note that the affine parameterization in the unstable discrete-time case leads to

$$T_{oq}(z) = \frac{B_{oq}(z)P_q(z)}{E_q(z)F_q(z)} + Q_q(z)\frac{B_{oq}(z)A_{oq}(z)}{E_q(z)F_q(z)} \tag{15.8.2}$$

where

$$A_{oq}(z)L_q(z) + B_{oq}(z)P_q(z) = E_q(z)F_q(z) \tag{15.8.3}$$

and where we assume that $B_{oq}(z)$ does not vanish on the unit circle. Then, with a constant input disturbance, zero steady-state tracking error can be achieved only if

$$S_{ioq}(z)|_{z=1} = 0 \Longrightarrow L_q(1) - Q_q(1)B_{oq}(1) = 0 \qquad (15.8.4)$$

Thus, the discrete counterpart of (15.3.15) is

$$Q_q(z) = (z-1)\overline{Q}_q(z) + \frac{L_q(1)}{B_{oq}(1)}Q_{aq}(z) \qquad (15.8.5)$$

where $\overline{Q}_q(z)$ is any stable transfer function and $Q_{aq}(z)$ is any stable transfer function that, additionally, satisfies $Q_{aq}(1) = 1$.

Note that the delta form of equation (15.8.5) is given by

$$Q_\delta(\gamma) = \gamma\overline{Q}_\delta(\gamma) + \frac{L_\delta(0)}{B_{o\delta}(0)}Q_{a\delta}(\gamma) \qquad (15.8.6)$$

where $\overline{Q}_\delta(\gamma)$ is any stable transfer function and $Q_{a\delta}(\gamma)$ is any stable transfer function that, additionally, satisfies $Q_{a\delta}(0) = 1$.

□□□

A final point in connection with digital Q synthesis is that *sampling zeros must always be considered as undesirable closed-loop pole locations* (whether they are stable or unstable). This follows from the points made in Chapter 14 relating to intersample response.

15.9   Summary

- The previous part of the book established that closed-loop properties are interlocked in a network of trade-offs. Hence, tuning for one property automatically affects other properties. This necessitates an understanding of the interrelations and conscious trade-off decisions.

- The fundamental laws of trade-off presented in previous chapters allow one to identify unachievable specifications and to establish where further effort is warranted or wasted.

- However, when pushing a design maximally towards a subtle trade-off, the earlier formulation of the fundamental laws falls short because it is difficult to push the performance of a design by tuning in terms of controller numerator and denominator: The impact on the trade-off-determining sensitivity poles and zeros is very nonlinear, complex, and subtle.

- This shortcoming raises the need for an alternative controller representation that

○ allows one to design more explicitly in terms of the quantities of interest (the sensitivities),

○ makes stability explicit, and

○ makes the impact of the controller on the trade-offs explicit.

• This need is met by the affine parameterization, also known as *Youla parameterization.*

• Summary of results for stable systems:

○ $C = Q(1 - QG_o)^{-1}$, where the design is carried out by designing the transfer function Q.

○ *Nominal sensitivities*:

$$T_o = QG_o \tag{15.9.1}$$
$$S_o = 1 - QG_o \tag{15.9.2}$$
$$S_{io} = (1 - QG_o)\, G_o \tag{15.9.3}$$
$$S_{uo} = Q \tag{15.9.4}$$

○ *Achieved sensitivities* [1]:

$$S_\Delta = \frac{1}{1 + QG_oG_\Delta} = \frac{1}{1 + QG_\epsilon} \tag{15.9.5}$$
$$T = QGS_\Delta \tag{15.9.6}$$
$$S = S_oS_\Delta \tag{15.9.7}$$
$$S_i = GS_oS_\Delta \tag{15.9.8}$$
$$S_u = QS_\Delta \tag{15.9.9}$$

• Observe the following advantages of the affine parameterization:

○ Nominal stability is explicit.

○ The known quantity G_o and the quantity sought by the control engineer (Q) occur in the highly insightful relation $T_o = QG_o$ (multiplicative in the frequency domain); whether a designer chooses to work in this quantity from the beginning or prefers to start with a synthesis technique and then convert, the simple multiplicative relation QG_o provides deep insights into the trade-offs of a particular problem and provides a very direct means of pushing the design by shaping Q.

---

[1] See the definitions of modeling errors in section §3.9.

○ The sensitivities are affine in Q, which is a great advantage for synthesis techniques relying on numerical minimization of a criterion. (See Chapter 16 for a detailed discussion of *optimization* methods that exploit this parameterization.)

- The following points are important to avoid some common misconceptions:

  ○ The associated trade-offs are not a consequence of the affine parameterization: they are perfectly general, and they hold for any linear time-invariant controller, including LQR, PID, pole placement based, and H_∞.

  ○ We have used the affine parameterization to make the general trade-offs more visible and to provide a direct means for the control engineer to make trade-off decisions; this should not be confused with synthesis techniques that make particular choices in the affine parameterization to synthesize a controller.

  ○ The fact that Q must approximate the inverse of the model at frequencies where the sensitivity is meant to be small is perfectly general and highlights the fundamental importance of inversion in control. This does *not* necessarily mean that the *controller*, C, must contain this approximate inverse as a factor and should not be confused with the pros and cons of that particular design choice.

- PI and PID design based on affine parameterization:

  ○ PI and PID controllers are traditionally tuned in terms of their parameters.

  ○ However, systematic design, trade-off decisions, and deciding whether a PI(D) is sufficient, is significantly easier in the model-based affine structure.

  ○ Inserting a first-order model into the affine structure automatically generates a PI controller.

  ○ Inserting a second-order model into the Q-structure automatically generates a PID controller.

  ○ All trade-offs and insights of the previous chapters also apply to PID-based control loops.

  ○ Whether a PI(D) is sufficient for a particular process is directly related to whether a first- (second-) order model can approximate the process well up to the frequencies where performance is limited by other factors such as delays, actuator saturations, sensor noise, or fundamentally unknown dynamics.

  ○ The first- and second-order models are easily obtained from step-response models (Chapter 3).

Smith predictor

Åström, K. (1977). Frequency domain properties of Otto Smith regulator. *International Journal of Control*, 26:307-314.

Smith, O. (1958). *Feedback Control Systems*. McGraw-Hill, New York.

Affine parameterization

Desoer, C., Liu, R., Murray, J., and Saeks, R. (1980). Feedback systems design: The fractional representation approach to analysis and synthesis. *IEEE Transactions on Automatic Control*, 25(3):399-412.

Youla, D., Jabr, H., and Bongiorno, J. (1976). Modern Wiener-Hopf design of optimal controllers. Part II: The multivariable case. *IEEE Transactions on Automatic Control*, 21(3):319-338.

15.11   Problems for the Reader

Problem 15.1. *For a plant having nominal model*

$$G_o(s) = 2\frac{-s+15}{(s+5)(s+10)} \qquad (15.11.1)$$

describe the class of controllers $Q(s)$ that provide zero steady-state error for constant references.

Problem 15.2. *For the same plant as in problem 15.1, find a $Q(s)$ such that the complementary sensitivity has dominant poles located at $-2 \pm j1.5$.*

Problem 15.3. *Given a plant with nominal model $G_o(s) = (s+1)^{-2}$, characterize the class of controllers that provide zero steady-state errors for sinusoidal references of frequency 0.5 [rad/s].*

Problem 15.4. *Consider an unstable plant having nominal model $G_o(s) = (s - 1)^{-1}$. Describe, with the simplest $P(s)$ and $L(s)$, the class of all stabilizing controllers. Hint: note that this plant can be stabilized by using a simple proportional controller.*

Problem 15.5. *Consider a plant having a model given by*

$$G_o(s) = \frac{2(-0.1s+1)}{(s+1)(s+3)} \qquad (15.11.2)$$

Assume that the output of this plant has to be regulated to a constant value in the presence of an input disturbance. Further assume that the input disturbance has a d.c. component and a variable component, with energy distributed in the frequency band $[0,4]$ [rad/s].

15.5.1 *Synthesize $Q(s)$, by using cancellation of the stable dynamics of $G_o(s)$. The dominant component in the denominator of $T_o(s)$ must be $s^2 + 6s + 16$.*

15.5.2 *Test your design by using the SIMULINK schematic in file **nmpq.mdl**.*

15.5.3 *Redesign your controller to improve disturbance compensation, without significantly affecting the performance tracking of a step reference.*

Problem 15.6. *Consider the IMC structure shown in Figure 15.1 on page 410 and your final design for problem 15.5 on the page before.*
Assume now that the true model for the plant is given by

$$G(s) = G_o(s)\frac{1}{\tau s + 1} = \frac{2(-0.1s + 1)}{(s+1)(s+3)(\tau s + 1)} \qquad (15.11.3)$$

Analyze the performance of your design for $\tau = 0.05$ and $\tau = 0.2$. Redesign your controller, if necessary.

Problem 15.7. *Consider a discrete-time controller having a transfer function given by*

$$C_q(z) = \frac{z - 0.5}{z - 1} \qquad (15.11.4)$$

Find a general characterization for the transfer function of all discrete-time plants that are stabilized by this controller. (Hint: use, in reverse form, the parameterization of all stabilizing controllers for unstable plants–i.e., interchange the roles of plant and controller.)

Problem 15.8. *Consider a continuous-time plant having a model given by*

$$G_o(s) = \frac{e^{-0.4s}}{s + 1} \qquad (15.11.5)$$

Assume that one desires to control this plant digitally, through a zero-order sample and hold, with sampling interval $\Delta = 0.1[s]$; then we require a discrete-time model, which is given (per 12.6 on page 339) by

$$G_{h0}G_o(z) = \frac{1 - e^{-0.1}}{z^4(z - e^{-0.1})} \qquad (15.11.6)$$

Use the IMC architecture to design $Q_q(z)$ in such a way that the closed-loop modes are at least as fast as $(0.6)^k$. Note that this requirement has also to be satisfied when compensating input disturbances.

Problem 15.9. *Consider a plant having a nominal model given by*

$$G_o(s) = \frac{-\alpha s + 1}{(s+1)(s+2)} \qquad where \quad \alpha \in \mathbb{R}^+ \qquad (15.11.7)$$

15.9.1 *Propose a method to synthesize PID controllers, along the lines followed in section §15.4. Evaluate your methodology, via simulation, for $\alpha \in [0.1, 20]$. Integrate into your analysis the ideas presented in Chapter 8.*

15.9.2 *Compare your results with those obtained from the Cohen-Coon tuning strategy, and evaluate the results for the same range of values for α.*

Problem 15.10. *Assume that an unstable plant has a transfer function $G_o(s) = B_o(s)/A_o(s)$ such that the polynomial $A_o(s) + B_o(s)$ has all its roots with negative real parts.*

Parameterize all stabilizing controllers for this plant, as a function of $A_o(s)$ and $B_o(s)$.

Problem 15.11. *Develop an anti-wind-up mechanism for the IMC architecture, one that parallels the structure developed in Chapter 11.*

Chapter 16

CONTROL DESIGN BASED
ON OPTIMIZATION

16.1 Preview

Thus far, we have seen that design constraints arise from a number of different sources:

- structural plant properties, such as NMP zeros or unstable poles;

- disturbances–their frequency content, point of injection, and measurability;

- architectural properties and the resulting algebraic laws of trade-off; and

- integral constraints and the resulting integral laws of trade-off.

The subtlety as well as complexity of the emergent trade-off web, into which the designer needs to ease a solution, motivates interest in what is known as *criterion-based* control design or *optimal* control theory: the aim here is to capture the control objective in a mathematical criterion and solve it for the controller that (depending on the formulation) maximizes or minimizes it.

Three questions arise:

(1) Is optimization of the criterion mathematically feasible?

(2) How good is the resulting controller?

(3) Can the constraint of the trade-off web be circumvented by optimization?

Question (1) has an affirmative answer for a number of criteria. In particular, quadratic formulations tend to favor tractability. Also, the affine parameterization of Chapter 15 is a key enabler, because it renders the sensitivity functions affine in the sought variable, Q.

The answer to question (2) has two facets: (a) How good is the controller as measured by the criterion? Answer: it is optimal by construction; but (b) How good is the resulting control-loop performance as measured by the original performance specifications? Answer: as good, or as poor, as the criterion in use is at

457

capturing the design intention and active trade-offs. A poorly formulated criterion will simply yield a controller that *optimally* implements the poor design. However, when selected well, a design criterion can synthesize a controller that would have been difficult to conceive of by the techniques covered thus far; this is particularly true for multivariable systems, covered in the next part of the book.

Question (3) is simply answered by *no*: all linear time-invariant controllers, whether they were synthesized by trial-and-error, pole assignment, or optimization, are subject to the same fundamental trade-off laws.

16.2 Optimal Q (Affine) Synthesis

16.2.1 General Ideas

Chapter 15 showed how Q synthesis could be used to control stable and unstable plants. The main difficulties in the methods arise from the need to ensure internal stability while, at the same time, achieving acceptable loop performance. The stability requirement forces certain structures on the sensitivity functions. These make it hard to satisfy simultaneously both structural requirements and prescribed performance specifications. An approach alternative to those presented above can be formulated by first specifying a performance target for a loop property–e.g., the complementary sensitivity. Then, a function $Q(s)$ can be found, within the class of all $Q(s)$ that ensure internal stability, in such a way that the *distance* from the achieved complementary sensitivity to the target is minimized. This approach is facilitated by the fact, already observed, that the nominal sensitivity functions are affine functions of $Q(s)$. The specific measure used for the distance defines the particular optimization procedure to be used. There exists a wide variety of possible choices. We will illustrate by using a particular approach based on quadratic optimization. This has been chosen since it is simple and can be presented without the need for extensive mathematical background beyond that already covered.

16.2.2 Quadratic Optimal Q Synthesis

Assume that a target function $H_o(s)$ is chosen for the complementary sensitivity $T_o(s)$. We have seen in Chapter 15 that, if we are given *some* stabilizing controller $C(s) = P(s)/L(s)$, then *all* stabilizing controllers can be expressed as in (15.7.1) for stable $Q_u(s)$. Also, the nominal complementary sensitivity function is then given be (15.7.23):

$$T_o(s) = H_1(s) + Q_u(s)V(s) \qquad (16.2.1)$$

where $H_1(s)$ and $V(s)$ are stable transfer functions of the form

$$H_1(s) = \frac{B_o(s)P(s)}{E(s)F(s)}; \qquad V(s) = \frac{B_o(s)A_o(s)}{E(s)F(s)} \qquad (16.2.2)$$

Let $\mathcal{S}$ denote the set of all real rational stable transfer functions; then the quadratic optimal synthesis problem can be stated as follows:

Problem   (Quadratic optimal synthesis problem). *Find $Q_u^o(s) \in \mathcal{S}$ such that*

$$Q_u^o(s) = \arg \min_{Q_u(s) \in \mathcal{S}} \left\| H_o - T_o \right\|_2^2 = \arg \min_{Q_u(s) \in \mathcal{S}} \left\| H_o - H_1 - Q_u V \right\|_2^2 \quad (16.2.3)$$

---

where the quadratic norm, also called the $\mathcal{H}_2-norm$, of a function[1] $X(s)$ is defined as

$$\left\| X \right\|_2 = \left[\frac{1}{2\pi} \int_{-\infty}^{\infty} X(j\omega)X(-j\omega)d\omega \right]^{\frac{1}{2}} \quad (16.2.4)$$

Remark 16.1. *The optimization approach can also be used in the synthesis of discrete-time controllers. In this case, the quadratic norm of a function $X_q(z)$ is defined as*

$$\left\| X_q \right\|_2 = \left[\frac{1}{2\pi} \int_0^{2\pi} X_q(e^{j\omega})X_q(e^{-j\omega})d\omega \right]^{\frac{1}{2}} \quad (16.2.5)$$

□□□

To solve this problem, we first need a preliminary result that is an extension of Pythagoras' theorem.

Lemma 16.1. *Let $\mathcal{S}_o \subset \mathcal{S}$ be the set of all real strictly proper stable rational functions, and let $\mathcal{S}_o^{\perp}$ be the set of all real strictly proper rational functions that are analytical for $\Re\{s\} \leq 0$. Furthermore assume that $X_s(s) \in \mathcal{S}_o$ and $X_u(s) \in \mathcal{S}_o^{\perp}$. Then*

$$\left\| X_s + X_u \right\|_2^2 = \left\| X_s \right\|_2^2 + \left\| X_u \right\|_2^2 \quad (16.2.6)$$

Proof

$$\left\| X_s + X_u \right\|_2^2 = \left\| X_s \right\|_2^2 + \left\| X_u \right\|_2^2 + 2\Re \left[\frac{1}{2\pi} \int_{-\infty}^{\infty} X_s(j\omega)X_u(-j\omega)d\omega \right] \quad (16.2.7)$$

but

---

[1]We here assume that $X(s)$ is a real function of s.

$$\frac{1}{2\pi} \int_{-\infty}^{\infty} X_s(j\omega)X_u(-j\omega)d\omega = \frac{1}{2\pi j} \oint_C X_s(s)X_u(-s)ds \qquad (16.2.8)$$

where C is the closed contour shown in Figure C.4 in Appendix C. We note that $X_s(s)X_u(-s)$ is analytic on the closed region encircled by C. We then see, upon applying Theorem C.7, that the integral in (16.2.8) is zero, and the result follows.

□□□

In what follows, we will need to split a general function $X(s)$ into a stable part $X_s(s)$ and an unstable part $X_u(s)$. We can do this via a partial-fraction expansion. The stable poles and their residues constitute the stable part.

We note that the cost function given in (16.2.3) has the general form

$$Q_u^o(s) = \arg \min_{Q_u(s) \in \mathcal{S}} \left\| W(s) - Q_u(s)V(s) \right\|_2^2 \qquad (16.2.9)$$

where $W(s) = H_o(s) - H_1(s)$, $H_o(s)$ is the target complementary sensitivity, and $H_1(s)$ is as in (16.2.2).

The solution to the above problem is then described in the next lemma.

Lemma 16.2. *Provided that $V(s)$ has no zeros on the imaginary axis, then*

$$\arg \min_{Q_u(s) \in \mathcal{S}} \left\| W(s) - Q_u(s)V(s) \right\|_2^2 = (V_m(s))^{-1}[V_a(s)^{-1}W(s)]_s \qquad (16.2.10)$$

where

$$V(s) = V_m(s)V_a(s) \qquad (16.2.11)$$

such that $V_m(s)$ is a factor with poles and zeros in the open LHP and $V_a(s)$ is an all-pass factor with unity gain, and where $[X]_s$ denotes the stable part of X.

Proof

Substituting (16.2.11) into (16.2.10) gives

$$\left\| W - Q_u V_m V_a \right\|_2^2 = \left\| V_a(V_a^{-1}W - Q_u V_m) \right\|_2^2 = \left\| V_a^{-1}W - Q_u V_m \right\|_2^2 \qquad (16.2.12)$$

where we used the fact that $|V_a|$ is constant (because V_a is all pass).

We then decompose $V_a^{-1}(s)W(s)$ into a stable term, $[V_a^{-1}(s)W(s)]_s$, and an unstable term, $[V_a^{-1}(s)W(s)]_u$ –i.e., we write

$$V_a^{-1}(s)W(s) = [V_a^{-1}(s)W(s)]_s + [V_a^{-1}(s)W(s)]_u \qquad (16.2.13)$$

With the decomposition (16.2.13), we obtain

$$\left\| V_a^{-1} W - Q_u V_m \right\|_2^2 = \left\| [V_a^{-1} W]_u + [V_a^{-1} W]_s - Q_u V_m \right\|_2^2 \qquad (16.2.14)$$

If we now apply Lemma 16.1 on page 459, we obtain

$$\left\| W - Q_u V_m V_a \right\|_2^2 = \left\| [V_a^{-1} W]_u \right\|_2^2 + \left\| [V_a^{-1} W]_s - Q_u V_m \right\|_2^2 \qquad (16.2.15)$$

From that, the optimal $Q_u(s)$ is seen to be

$$Q_u^o(s) = (V_m(s))^{-1} [(V_a(s))^{-1} W(s)]_s \qquad (16.2.16)$$

Remark 16.2. *The factorization (16.2.11) is also known factorization in the literature as an inner-outer.*

Remark 16.3. *The solution will be proper only either if V has relative degree zero or if both V has relative degree one and W has relative degree of at least one. However, improper solutions can readily be turned into approximate proper solutions by adding an appropriate number of fast poles to $Q_u^o(s)$.*

□□□

Returning to the problem posed in equation (16.2.3), we see that Lemma 16.2 provides an immediate solution, by setting

$$W(s) = H_o(s) - H_1(s) \qquad (16.2.17)$$
$$V(s) = V_m(s) V_a(s) \qquad (16.2.18)$$

We note that, for this case, the poles of $[V_a^{-1}(s) W(s)]$ are the poles of $W(s)$ (because this is a stable transfer function here).

Remark 16.4. *The above procedure can be modified to include a weighting function $\Omega(j\omega)$. In this framework, the cost function is now given by*

$$\left\| (H_o - T_o) \Omega \right\|_2^2 \qquad (16.2.19)$$

No additional difficulty arises, because it is enough to simply redefine $V(s)$ and $W(s)$ to convert the problem into the form of equation (16.2.9).

Remark 16.5. *It is also possible to restrict the solution space to satisfy additional design specifications. For example, forcing an integration is achieved by parameterizing $Q(s)$ as in (15.3.15) and by introducing a weighting function $\Omega(s) = s^{-1}$. ($H_o(0) = 1$ is also required.) This does not alter the affine nature of $T_o(s)$ on the unknown function. Hence, the synthesis procedure developed above can be applied, provided that we first redefine the function, $V(s)$ and $W(s)$.*

Remark 16.6. *A MATLAB program to find the solution for the quadratic optimal Q synthesis problem is included on the accompanying CD-ROM in file* **oph2.m**.

The synthesis procedure is illustrated in the following two examples.

Example 16.1 (Unstable plant). *Consider a plant with nominal model*

$$G_o(s) = \frac{2}{(s-1)(s+2)} \tag{16.2.20}$$

Assume that the target function for $T_o(s)$ is given by

$$H_o(s) = \frac{9}{s^2 + 4s + 9} \tag{16.2.21}$$

We first choose the observer polynomial $E(s) = (s+4)(s+10)$ and the controller polynomial $F(s) = s^2 + 4s + 9$. Note that, for simplicity, $F(s)$ was chosen to be equal to the denominator of $H_o(s)$.

We then solve the pole-assignment equation $A_o(s)L(s) + B_o(s)P(s) = E(s)F(s)$ to obtain the prestabilizing control law expressed in terms of $P(s)$ and $L(s)$. (Use the routine **paq.m***, provided on the accompanying CD-ROM.) The resultant polynomials are*

$$P(s) = 115s + 270; \qquad L(s) = s^2 + 17s + 90 \tag{16.2.22}$$

Now consider any *controller from the class of stabilizing control laws as parameterized in (15.7.1). The quadratic cost function is then as in (16.2.9), where*

$$W(s) = H_o(s) - \frac{B_o(s)P(s)}{E(s)F(s)} = \frac{9s^2 - 104s - 180}{E(s)F(s)} \tag{16.2.23}$$

$$V(s) = \frac{B_o(s)A_o(s)}{E(s)F(s)} = V_a(s)V_m(s) = \left[\frac{s-1}{s+1}\right]\left[\frac{2(s+2)(s+1)}{E(s)F(s)}\right] \tag{16.2.24}$$

Consequently

$$[V_a^{-1}(s)W(s)]_s = \left(\frac{1}{7}\right)\frac{5s^3 + 158s^2 + 18s - 540}{E(s)F(s)} \tag{16.2.25}$$

The optimal $Q_u(s)$ is then obtained from (16.2.16):

$$Q_u^o(s) = (V_m(s))^{-1}[(V_a(s))^{-1}W(s)]_s = \left(\frac{1}{14}\right)\frac{5s^3 + 158s^2 + 18s - 540}{(s+1)(s+2)} \tag{16.2.26}$$

We observe that $Q_u^o(s)$ is improper. However, we can approximate it by a sub-optimal (but proper) transfer function, $\tilde{Q}(s)$, by adding one fast pole to $Q_u^o(s)$:

$$\tilde{Q}(s) = Q_u^o(s)\frac{1}{\tau s + 1} \qquad where \qquad \tau \ll 1 \qquad (16.2.27)$$

The reader is encouraged to investigate the resultant controller and the associated qualities of the closed loop–see Problem 16.11.

□□□

Example 16.2 (Nonminimum-phase plant). *Consider a plant with nominal model*

$$G_o(s) = \frac{-3s + 18}{(s+6)(s+3)} \qquad (16.2.28)$$

It is required to synthesize, by using $\mathcal{H}_2$ optimization, a one-d.o.f. control loop having the target function

$$H_o(s) = \frac{16}{s^2 + 5s + 16} \qquad (16.2.29)$$

and to provide exact model inversion at $\omega = 0$.

Solution

We adopt the strategy suggested in Remark 16.5 on page 461. The cost function is thus defined as

$$J(Q) = \left\| (H_o(s) - (s\overline{Q}(s) + [G_o(0)]^{-1})G_o(s))\Omega(s) \right\|_2^2 \qquad where \qquad \Omega(s) = \frac{1}{s} \qquad (16.2.30)$$

Then the cost function takes the form

$$J(Q) = \left\| W - \overline{Q}V \right\|_2^2 \qquad (16.2.31)$$

where

$$V(s) = G_o(s) = \frac{-s+6}{s+6}\frac{3}{s+3}; \qquad W(s) = \frac{3s^2 + 13s + 102}{(s^2 + 5s + 16)(s^2 + 9s + 16)} \qquad (16.2.32)$$

We first note that

$$V_a(s) = \frac{-s+6}{s+6}; \qquad V_m(s) = \frac{3}{s+3} \qquad (16.2.33)$$

The optimal $\overline{Q}(s)$ can then be obtained by using (16.2.16):

$$\overline{Q}^o(s) = \frac{0.1301s^2 + 0.8211s + 4.6260}{s^2 + 5s + 16} \qquad (16.2.34)$$

from this $Q^o(s)$ can be obtained as $Q^o(s) = s\overline{Q}^o(s) + 1$. One fast pole has to be added to make this function proper.

The reader is again encouraged to investigate the resultant controller and the associated qualities of the closed loop–see Problem 16.12.

□□□

16.3   Robust Control Design with Confidence Bounds

We next show briefly how optimization methods can be used to change a nominal controller so that the resultant performance is more robust against model errors. Methods for doing this range from ad-hoc procedures to sophisticated optimization methods that preserve stability in the presence of certain types of model error. Our objective is to give the flavor of these methods.

16.3.1   Statistical Confidence Bounds

We have argued in section §3.9 that no model can give an exact description of a real process. This realization is the key driving force for the modern idea of robust control design. Robust control-design methods include $\mathcal{H}_\infty$ methods, $\mathcal{L}_1$ methods, and $\mathcal{H}_2$ methods. Here, we will use the latter approach, because it falls nicely into the framework described above.

Our starting point will be to assume the existence of statistical confidence bounds on the modeling error. Obtaining these confidence bounds is a research area in its own right, but approximations can be obtained from uncertainty estimates provided by standard system identification methods. The interested reader is referred to the literature cited at the end of the chapter.

We will show how these statistical confidence bounds can be used for robust control design. The essential idea of the procedure is to modify the nominal controller so as to minimize the expected variation (variance) of the actual system performance from an a-priori given desired performance.

We assume that we are given a nominal frequency response, $G_o(j\omega)$, together with a statistical description of the associated errors of the form

$$G(j\omega) = G_o(j\omega) + G_\epsilon(j\omega) \qquad (16.3.1)$$

where $G(j\omega)$ is the true (but unknown) frequency response and $G_\epsilon(j\omega)$, as usual, represents the additive modeling error. We assume that G_ϵ possesses the following probabilistic properties:

$$\mathcal{E}\{G_\epsilon(j\omega)\} = 0 \qquad (16.3.2)$$

$$\mathcal{E}\{G_\epsilon(j\omega)G_\epsilon(-j\omega)\} = \alpha(j\omega)\alpha(-j\omega) = \tilde{\alpha}^2(\omega) \qquad (16.3.3)$$

In (16.3.3), $\alpha(s)$ is the stable, minimum-phase spectral factor. Also, in (16.3.3), $\tilde{\alpha}$ is the given measure of the modeling error. In order to use this measure for control design, we require that it be a rational function.

16.3.2   Robust Control Design

Based on the nominal model $G_o(j\omega)$, we assume that a design is carried out that leads to acceptable nominal performance. This design will typically account for the usual control-design issues such as nonminimum-phase behavior, the available input range, and unstable poles. Let us say that this has been achieved with a nominal controller C_o and that the corresponding nominal sensitivity function is S_o. Of course, the true plant is assumed to satisfy (16.3.1)–(16.3.3); hence, the value S_o will not be achieved in practice, because of the variability of the achieved sensitivity, S, from S_o. Let us assume, to begin, that the open-loop system is *stable*. (The more general case will be treated in section §16.3.6.) We can thus use the simple form of the parameterization of all stabilizing controllers to express C_o and S_o in terms of a stable parameter Q_o:

$$C_o(s) = \frac{Q_o(s)}{1 - G_o(s)Q_o(s)} \qquad (16.3.4)$$

$$S_o(s) = 1 - G_o(s)Q_o(s) \qquad (16.3.5)$$

The achieved sensitivity, S_1, when the nominal controller C_o is applied to the true plant is given by

$$S_1(s) = \frac{S_o(s)}{1 + Q_o(s)G_\epsilon(s)} \qquad (16.3.6)$$

Our proposal for robust design now is to adjust the controller so that the *distance* between the resulting achieved sensitivity, S_2, and S_o is minimized. If we change Q_o to Q and hence C_o to C, then the achieved sensitivity changes to

$$S_2(s) = \frac{1 - G_o(s)Q(s)}{1 + G_\epsilon(s)Q(s)} \qquad (16.3.7)$$

where

$$C(s) = \frac{Q(s)}{1 - G_o(s)Q(s)} \tag{16.3.8}$$

and

$$S_2(s) - S_o(s) = \frac{1 - G_o(s)Q(s)}{1 + G_\epsilon(s)Q(s)} - (1 - G_o(s)Q_o(s)) \tag{16.3.9}$$

Observe that S_2 denotes, the sensitivity achieved when the plant is G and the controller is parameterized by Q, and S_o denotes the sensitivity achieved when the plant is G_o and the controller is parameterized by Q_o. Unfortunately, $(S_2 - S_o)$ is a nonlinear function of Q and G_ϵ; however, in the next section, we show that this problem can be remedied by using a weighted sensitivity error.

16.3.3   Frequency Weighted Errors

In place of minimizing some measure of the sensitivity error given in (16.3.9), we instead consider a weighted version, with $W_2 = 1 + G_\epsilon Q$. Thus, consider

$$
\begin{aligned}
W_2(s)(S_2(s) - S_o(s)) = & \ (1 - G_o(s)Q(s)) - (1 - G_o(s)Q_o(s))(1 + G_\epsilon(s)Q(s)) \\
= & \ -G_o(s)\tilde{Q}(s) - S_o(s)Q_o(s)G_\epsilon(s) - S_o(s)\tilde{Q}(s)G_\epsilon(s).
\end{aligned}
\tag{16.3.10}
$$

where $\tilde{Q}(s) = Q(s) - Q_o(s)$ is the desired adjustment in $Q_o(s)$ to account for $G_\epsilon(s)$.

Before proceeding, let us interpret (16.3.10).

Lemma 16.3. *The weighted sensitivity error has bias*

$$\mathcal{E}\left\{W_2(j\omega)\left(S_2(j\omega) - S_o(j\omega)\right)\right\} = -G_o(j\omega)\tilde{Q}(j\omega) \tag{16.3.11}$$

and variance

$$
\begin{aligned}
\mathcal{E}\left\{|W_2(j\omega)\left(S_2(j\omega) - S_o(j\omega)\right)|^2\right\} = & \ |G_o(j\omega)|^2|\tilde{Q}(j\omega)|^2 + \\
& \ |S_o(j\omega)Q_o(j\omega) + S_o(j\omega)\tilde{Q}(j\omega)|^2\tilde{\alpha}^2(\omega)
\end{aligned}
\tag{16.3.12}
$$

Proof

To establish (16.3.11) and (16.3.12), we observe that

$$W_2(s)(S_2(s) - S_o(s)) = -G_o(s)\tilde{Q}(s) - S_o(s)Q_o(s)G_\epsilon(s) - S_o(s)\tilde{Q}(s)G_\epsilon(s). \tag{16.3.13}$$

The result follows upon using the properties of G_ϵ.

□□□

The procedure that we now propose for choosing $\tilde{Q}$ is to find the value that minimizes

$$
J = \|W_2(S - S_o)\|_2^2 = \int_{-\infty}^{\infty} \mathcal{E}\left\{|W_2(j\omega)\left(S_2(j\omega) - S_o(j\omega)\right)|^2\right\} d\omega
$$

$$
= \int_{-\infty}^{\infty} |G_o(j\omega)|^2 |\tilde{Q}(j\omega)|^2 + |S_o(j\omega)Q_o(j\omega) + S_o(j\omega)\tilde{Q}(j\omega)|^2 \tilde{\alpha}^2(\omega) d\omega \quad (16.3.14)
$$

Remark 16.7. *This loss function has intuitive appeal. The first term on the right-hand side represents the* bias *error. It can be seen that this term is zero if $\tilde{Q} = 0$ (i.e., we leave the controller unaltered). The second term in (16.3.14) represents the* variance *error. This term is zero if $\tilde{Q} = -Q_o$–i.e., if we choose open-loop control. These observations suggest that there are two extreme cases. For $\tilde{\alpha} = 0$ (no model uncertainty), we leave the controller unaltered; as $\tilde{\alpha} \to \infty$ (large model uncertainty), we choose open-loop control, which clearly is robust for the case of an open-loop stable plant.*

Remark 16.8. *One might ask the significance of the weighting function W_2. For $|G_\epsilon(j\omega)Q(j\omega)| << 1$, we have $|W_2(j\omega)| \approx 1$. However, $|G_\epsilon(j\omega)Q(j\omega)| < 1$ is a sufficient condition for robust stability, and thus the weighting should not significantly affect the result. (See also Remark 16.10 on the next page.)*

□□□

Lemma 16.4. *Suppose that*

(i) G_o is strictly proper with no zeros on the imaginary axis and

(ii) $\mathcal{E}\{G_\epsilon(j\omega)G_\epsilon(-j\omega)\}$ has a spectral factorization as in (16.3.3).

Then $\alpha(s)\alpha(-s)S_o(s)S_o(-s) + G_o(s)G_o(-s)$ has a spectral factor, which we label H, and the optimal $\tilde{Q}$ is given by

$$
\tilde{Q}^{opt}(s) = \arg \min_{\tilde{Q}(s)\in\mathcal{S}} \|W_2(S_2 - S_o)\|_2
$$

$$
= -\frac{1}{H(s)} \times \text{ stable part of } \frac{\alpha(s)\alpha(-s)S_o(s)S_o(-s)Q_o(s)}{H(-s)} \quad (16.3.15)
$$

Proof

First, we prove that

$$
F(s) \stackrel{\triangle}{=} \alpha(s)\alpha(-s)S_o(s)S_o(-s) + G_o(s)G_o(-s) \quad (16.3.16)
$$

has a spectral factor. By hypothesis, $\alpha(s)\alpha(-s)$ has a spectral factor, B, so

$$F(s) = B(s)B(-s)S_o(s)S_o(-s) + G_o(s)G_o(-s) \qquad (16.3.17)$$

$B(\infty) \neq 0$ and, by hypothesis, $G_o(\infty) = 0$ and $S_o(\infty) = 1$, so

$$F(\infty) = B(\infty)^2 > 0 \qquad (16.3.18)$$

and, because G_o has no zeros on the imaginary axis, clearly F does not either. Finally, it is straightforward to prove that $F(-s) = F(s)$, so it follows that F has a spectral factor, which we label H.

By completing the squares, equation (16.3.14) can be written as

$$J = \int_{-\infty}^{\infty} \left| H(j\omega)\tilde{Q}(j\omega) + \frac{\tilde{\alpha}^2(\omega)|S_o(j\omega)|^2 Q_o(j\omega)}{H(-j\omega)} \right|^2 d\omega +$$

$$\int_{-\infty}^{\infty} \tilde{\alpha}^2(\omega)|S_o(j\omega)|^2 |Q_o(j\omega)|^2 \left(1 - \frac{\tilde{\alpha}^2(\omega)|S_o|^2(j\omega)}{|H|^2(j\omega)} \right) d\omega. \qquad (16.3.19)$$

Now,

$$1 - \frac{\alpha(s)\alpha(-s)S_o(s)S_o(-s)}{H(s)H(-s)} = \frac{G_o(s)G_o(-s)}{H(s)H(-s)} \qquad (16.3.20)$$

has relative degree at least two, so we see that the second term on the right-hand side of (16.3.19) is finite. Furthermore, this term is independent of $\tilde{Q}$, we can minimize the cost by minimizing the first term on the right-hand side:

$$\left\| \underbrace{H(s)}_{V(s)} \tilde{Q}(s) + \underbrace{\frac{\alpha(s)\alpha(-s)S_o(s)S_o(-s)Q_o(s)}{H(-s)}}_{W(s)} \right\|_2^2 \qquad (16.3.21)$$

H is a spectral factor, so $V = H$ has no zeros on the imaginary axis and is outer. This also means that W has no poles on the imaginary axis. By Lemma 16.2, the solution of the optimization problem is as stated.

$\square\square\square$

Remark 16.9. *The value of $\tilde{Q}$ found in Lemma 16.4 gives an optimal, in the sense of (16.3.14), trade-off between the bias error (the first term in (16.3.14)) and the variance term (the second term in (16.3.14)).*

Remark 16.10. *A final check on robust stability (which is not automatically guaranteed by the algorithm) requires us to check that $|G_\epsilon(j\omega)||Q(j\omega)| < 1$ for all ω and all "likely" values of $G_\epsilon(j\omega)$. To make this more explicit, let us assume that $G_\epsilon(j\omega)$ has a gaussian distribution with mean zero and covariance $\overline{P}(\omega)$. (Note that* *this would arise naturally when the model is estimated in the presence of gaussian measurement noise; for details, see the references on identification given at the end of the chapter.)*

Let us denote the real and imaginary components of $G_\epsilon(j\omega)$ by $\tilde{g}_R(\omega)$ and $\tilde{g}_I(\omega)$, respectively. Also, let $\tilde{g}(\omega) = (\tilde{g}_R(\omega), \tilde{g}_I(\omega))^T$. Finally let the covariance of $\tilde{g}(\omega)$ be $P(\omega)$. Then it is a well-known result from statistics that $\tilde{g}(\omega)^T P(\omega)^{-1} \tilde{g}(\omega)$ has a Chi-squared distribution with two degrees of freedom. By the use of this distribution, we can find a scalar β such that 99% (say) of the values of $\tilde{g}(\omega)$ satisfy

$$\tilde{g}(\omega)^T P(\omega)^{-1} \tilde{g}(\omega) \le \beta \qquad (16.3.22)$$

If we now check that $|Q(j\omega)| < [\beta \lambda_{\max} P(\omega)]^{-1}$, for all ω, then this ensures that $|G_\epsilon(j\omega)||Q(j\omega)| < 1$ with probability 0.99. To verify this claim, we note that, from Equation (16.3.22), 99% of values of $\tilde{g}(\omega)$ satisfy the following stronger condition:

$$(\tilde{g}^T \tilde{g}) \le \beta \lambda_{\max} P(\omega) \qquad (16.3.23)$$

Hence, if

$$|Q(j\omega)| < [\beta \lambda_{\max} P(\omega)]^{-1} \qquad (16.3.24)$$

then, by combining (16.3.23) and (16.3.24), we see that the probability that $|\tilde{g}^T \tilde{g}||Q(j\omega)| < 1$ is 0.99, as required. In principle, it is necessary to check (16.3.24) for all ω. However, a reasonable practical compromise is to check (16.3.24) for each discrete ω used in the original quadratic optimization.

□□□

Remark 16.11. *Of course, the procedure outlined in Remark 16.10 allows one to check only whether a given design, $Q(j\omega)$, is consistent with robust stability at a given probabilistic confidence level. If the test fails, then one could artificially increase the uncertainty in the design phase so as to shift the balance to a more conservative outcome. Alternatively, it is possible to include Equation (16.3.24) as a set of side constraints in the quadratic optimization procedure itself, but we will not pursue this option further here.*

□□□

Remark 16.12. *As we have seen in Chapter 15, it usually is necessary to make a distinction between stable closed-loop poles and desirable closed-loop poles. The reader is asked to note the remark on input-disturbance response in Chapter 15, page 425. It is possible to modify the algorithm presented here to ensure that all poles lie in desirable regions. For example, a preliminary transformation $s \to s' = s + \beta$ with $\beta > 0$ can be used to transform the region to the left of $-\beta$ to the LHP.*

16.3.4   Incorporating Integral Action

The methodology given in section §16.3.3 can be extended to include integral action. Assuming that Q_o provides this property, the final controller will do so as well, if $\tilde{Q}$ has the form

$$\tilde{Q}(s) = s\tilde{Q}'(s) \tag{16.3.25}$$

with $\tilde{Q}'$ strictly proper. (See Lemma 15.2 on page 412.)

There are a number of ways to enforce this constraint. A particularly simple way is to follow Remark 16.5 and change the cost function to

$$J' = \int_{-\infty}^{\infty} \frac{\mathcal{E}\left\{|W_2(j\omega)|^2|S_2(j\omega) - S_o(j\omega)|^2\right\}}{|jw|^2}\, d\omega \tag{16.3.26}$$

Lemma 16.5. *Suppose that*

(i) G_o is strictly proper with no zeros on the imaginary axis and

(ii) $\mathcal{E}\{G_\epsilon(j\omega)G_\epsilon(-j\omega)\}$ has a spectral factorization as in (16.3.3).

Then $\alpha(s)\alpha(-s)S_o(s)S_o(-s) + G_o(s)G_o(-s)$ has a spectral factor, which we label H, and

$$\arg\min_{\tilde{Q}(s)\in\mathcal{S}} J' = -\frac{s}{H(s)} \times \text{ stable part of } \frac{\alpha(s)\alpha(-s)S_o(s)S_o(-s)Q_o(s)}{sH(-s)} \tag{16.3.27}$$

Proof

First of all, the existence of H follows from Lemma 16.4. Second, by using (16.3.25), we have

$$J' = \int_{-\infty}^{\infty} \left| H(j\omega)\tilde{Q}'(j\omega) + \frac{\tilde{\alpha}^2(\omega)|S_o(j\omega)|^2Q_o(j\omega)}{j\omega H(-j\omega)} \right|^2 d\omega +$$
$$\int_{-\infty}^{\infty} \frac{a|S_o(j\omega)|^2|Q_o(j\omega)|^2}{\omega^2}\left(1 - \frac{a|S_o|^2}{|H|^2}\right) d\omega \tag{16.3.28}$$

From the assumption that C_o has integral action, it follows that S_o has a zero at zero, and because

$$\frac{\alpha(s)\alpha(-s)S_o(s)S_o(-s)Q_o(s)Q_o(-s)}{-s^2}\left(1 - \frac{\alpha(s)\alpha(-s)S_o(s)S_o(-s)}{H(s)H(-s)}\right)$$

has relative degree at least two, it follows that the second term in the above integral is finite. This term is independent of $\tilde{Q}'$, so we can minimize the cost by minimizing *the first term on the right-hand side, namely,*

$$\left\|\underbrace{H(s)}_{V(s)}\tilde{Q}'(s) + \underbrace{\frac{\alpha(s)\alpha(-s)S_o(s)S_o(-s)Q_o(s)}{sH(-s)}}_{W}\right\|_2^2.$$

H is a spectral factor, so $V = H$ has no zeros on the imaginary axis and is outer. This also means that W has no poles on the imaginary axis. By Lemma 16.2, the optimal $\tilde{Q}' \in \mathcal{RH}\infty$ is

$$-\frac{1}{H(s)} \times \textit{ stable part of } \left\{\frac{\alpha(s)\alpha(-s)S_o(s)S_o(-s)Q_o(s)}{sH(-s)}\right\} \qquad (16.3.29)$$

W is strictly proper, so it follows that $\tilde{Q}'$ is as well and hence that $\tilde{Q}(s) = s\tilde{Q}'(s)$ is proper, as required.

16.3.5   A Simple Example

Consider a first-order system having constant variance for the model error in the frequency domain:

$$G_o(s) = \frac{1}{\tau_o s + 1} \qquad (16.3.30)$$

$$Q_o(s) = \frac{\tau_o s + 1}{\tau_{cl} s + 1} \qquad (16.3.31)$$

$$S_o(s) = \frac{\tau_{cl} s}{\tau_{cl} s + 1} \qquad (16.3.32)$$

$$\tilde{\alpha}^2(\omega) = \epsilon > 0 \qquad \forall \omega \qquad (16.3.33)$$

Note that choosing $\tilde{\alpha}$ to be a constant is an approximation to the case where the identification is performed by means of a finite impulse response model with a white-noise type of input and a white-noise disturbance.

(a) Without integral-action constraint

In this case, with a_1 and a_2 appropriate functions of τ_o, τ_{cl}, and ϵ, we can write

$$\begin{aligned}
H(s)H(-s) &= \frac{1}{1 - \tau_o^2 s^2} + \frac{\epsilon(-\tau_{cl}^2 s^2)}{1 - \tau_{cl}^2 s^2}\\
&= \frac{1 - \tau_{cl}^2(1 + \epsilon)s^2 + \epsilon\tau_{cl}^2\tau_o^2 s^4}{(1 - \tau_o^2 s^2)(1 - \tau_{cl}^2 s^2)}\\
&= \frac{(1 + \sqrt{a_1}s)(1 + \sqrt{a_2}s)(1 - \sqrt{a_1}s)(1 - \sqrt{a_2}s)}{(1 + \tau_o s)(1 + \tau_{cl}s)(1 - \tau_o s)(1 - \tau_{cl}s)} \qquad (16.3.34)
\end{aligned}$$

Then there exist A_1, A_2, A_3, and A_4, also appropriate functions of τ_o, τ_{cl}, and ϵ, so that

$$\frac{\alpha(s)\alpha(-s)S_o(s)S_o(-s)}{H(-s)}Q_o(s) = \frac{(1-\tau_o s)(1-\tau_{cl}s)}{(1-\sqrt{a_1}s)(1-\sqrt{a_2}s)}\frac{\epsilon(-\tau_{cl}^2 s^2)(1+\tau_o s)}{(1-\tau_{cl}s)(1+\tau_{cl}s)^2}$$

$$= A_o + \frac{A_1}{1-\sqrt{a_1}s} + \frac{A_2}{1-\sqrt{a_2}s} + \frac{A_3}{(1+\tau_{cl}s)^2} + \frac{A_4}{1+\tau_{cl}s}$$

$$(16.3.35)$$

The optimal $\tilde{Q}$ is then

$$\tilde{Q}(s) = -\frac{(1+\tau_o s)(1+\tau_{cl}s)}{(1+\sqrt{a_1}s)(1+\sqrt{a_2}s)}\left[A_o + \frac{A_3}{(1+\tau_{cl}s)^2} + \frac{A_4}{1+\tau_{cl}s}\right] \qquad (16.3.36)$$

To examine this example numerically, we take $\tau_o = 1$, $\tau_{cl} = 0.5$, and $\epsilon = 0.4$. Then, from (16.3.34)-(16.3.36), we obtain the optimal $\tilde{Q}$ as

$$\tilde{Q}(s) = -\frac{0.316s^3 + 1.072s^2 + 1.285s + 0.529}{0.158s^3 + 0.812s^2 + 1.491s + 1.00} \qquad (16.3.37)$$

It is interesting to investigate how this optimal $\tilde{Q}$ contributes to the reduction of the loss function (16.3.14). If $\tilde{Q}(s) = 0$, then

$$J = \int_{-\infty}^{\infty} |S_o(j\omega)Q_o(j\omega)|^2 \epsilon d\omega = \infty,$$

and if the optimal $\tilde{Q}$ given by (16.3.37) is used, then the total error is $J = 4.9$, which has a *bias* error of

$$\int_{-\infty}^{\infty} |G_o(j\omega)\tilde{Q}|^2 d\omega = 4.3$$

and a *variance* error of

$$\int_{-\infty}^{\infty} |S_o(j\omega)Q_o(j\omega) + S_o(j\omega)\tilde{Q}(j\omega)|^2 \epsilon d\omega = 0.6.$$

(b) With integral-action constraint

Following subsection §16.3.4, we write

$$\frac{\alpha(s)\alpha(-s)S_o(s)S_o(-s)}{H(-s)}Q_o(s) = \frac{(1-\tau_o s)(1-\tau_{cl}s)}{(1-\sqrt{a_1}s)(1-\sqrt{a_2}s)}\frac{\epsilon(-\tau_{cl}^2 s)(1+\tau_o s)}{(1-\tau_{cl}s)(1+\tau_{cl}s)^2}$$

$$= \frac{B_1}{1-\sqrt{a_1}s} + \frac{B_2}{1-\sqrt{a_2}s} + \frac{B_3}{(1+\tau_{cl}s)^2} + \frac{B_4}{1+\tau_{cl}s}$$

Then, from (16.3.25) and Lemma 16.5, the optimal $\tilde{Q}$ is given by

$$\tilde{Q}(s) = -\frac{s(1 + \tau_o s)(1 + \tau_{cl} s)}{(1 + \sqrt{a_1} s)(1 + \sqrt{a_2} s)} \left[\frac{B_3}{(1 + \tau_{cl} s)^2} + \frac{B_4}{1 + \tau_{cl} s} \right] \qquad (16.3.38)$$

For the same set of process parameters as above, we obtain the optimal $\tilde{Q}$ as

$$\tilde{Q}(s) = -\frac{s(0.184 s^2 + 0.411 s + 0.227)}{0.158 s^3 + 0.812 s^2 + 1.491 s + 1.00} \qquad (16.3.39)$$

and the Q for controller implementation is simply

$$Q(s) = Q_o(s) + \tilde{Q}(s) = \frac{(0.265 s + 1)(s + 1)}{0.316 s^2 + 0.991 s + 1} \qquad (16.3.40)$$

Note that the open-loop pole at -1 has been canceled. If this were not desirable (on account of input-disturbance considerations), then the suggestion made in Remark 16.12 could be used. For this example, if $\hat{Q} = 0$, then the loss function shown in (16.3.26) is $J' = 1.4$. Using the optimal $\tilde{Q}$ of (16.3.39), the total error is reduced to $J' = 0.94$, where the *bias* error term is 0.22 and the *variance* error term is 0.72.

(c) Closed-loop system-simulation results

For the same process parameters as above, we now examine how the robust controller given by (16.3.40) copes with plant uncertainty by simulating closed-loop responses with different processes, and we compare the results for the cases when Q_o in (16.3.31) is used. We choose the following three different plants.
Case 1

$$G_1(s) = \frac{1}{s + 1} = G_o(s) \qquad (16.3.41)$$

Case 2

$$G_2(s) = \frac{1.3 e^{-0.3}}{0.5 s + 1} \qquad (16.3.42)$$

Case 3

$$G_3(s) = \frac{0.5}{0.5 s + 1} \qquad (16.3.43)$$

The frequency responses of the three plants are shown in Figure 16.1 on the next page; they are within the statistical confidence bounds centered at $G_o(j\omega)$ and have standard deviation of $\sqrt{0.4}$. Note that the difference between these 3 plants is much larger than might be expected in any practical situation. We have deliberately chosen the example to highlight the issues involved.

Figures 16.2, 16.3, and 16.4 show the closed-loop responses of the three plants for a unit set-point change, controlled by using C and C_o. The results are discussed next.

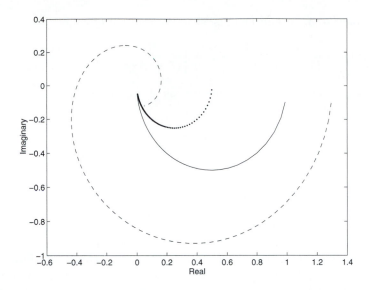

Figure 16.1. Plant frequency response: Case 1 (solid); case 2 (dashed); case 3 (dotted)

- **Case 1**. $G_1(s) = G_o(s)$, so the closed-loop response based on Q_o for this case is the desired response, as specified. However, there is a difference in responses between the desired closed-loop system and the robust control system, which has accounted for uncertainty in the model. The existence of $\tilde{Q}$ causes degradation in the *nominal* closed-loop performance, but this degradation is reasonably small, as can be seen from the closeness of the closed-loop responses; in particular, the control signal is seen to be less aggressive. This is the price one pays for including a robustness margin aimed at decreasing sensitivity to modeling errors.

- **Case 2**. There is a large model error between $G_2(s)$ and $G_o(s)$, shown in Figure 16.1. This corresponds to the so called *dangerous* situation in robust control. It is seen from Figure 16.3 on the next page that, without the compensation of optimal $\tilde{Q}$, the closed-loop system is at the edge of instability. However, $\tilde{Q}$ stabilizes the closed-loop system and achieves acceptable closed-loop performance in the presence of this large model uncertainty.

- **Case 3**. Although there is a large model error between $G_3(s)$ and $G_o(s)$ in the low-frequency region, this model error is less likely to cause instability of the closed-loop system (though more likely to result in a slower closed-loop response). Figure 16.4 on page 476, illustrates that the closed-loop response speed when using the optimal $\tilde{Q}$ is indeed slower than the response speed from

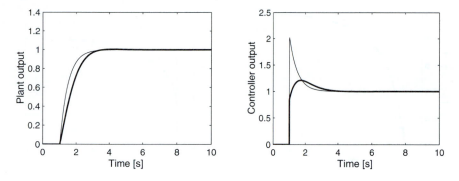

Figure 16.2. Closed-loop responses for case 1: when using Q_o (thin line), and when using optimal Q (thick line)

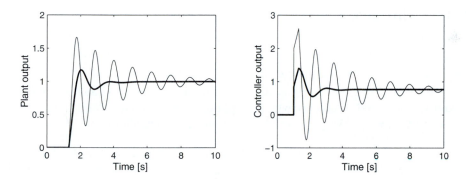

Figure 16.3. Closed-loop responses for case 2: when using Q_o (thin line), and when using optimal Q (thick line)

Q_o, but the difference is small. It is interesting to note that, in this case, the control signal from the compensated robust closed-loop system is a smoother response than the one from Q_o.

Remark 16.13. *We do not want to give the impression that one must always carry out a formal, criterion-based robust design. We have carefully chosen this example to highlight the issues. However, in practice, careful nominal design performed with an eye on robustness issues is likely to yield an adequate solution, especially because precise bounds for model errors will be difficult to come by.*   □□□

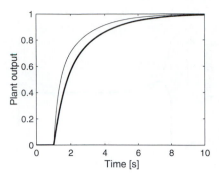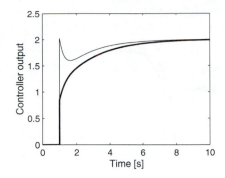

Figure 16.4. Closed-loop responses for case 3: when using Q_o (thin line), and when using optimal Q (thick line)

16.3.6   Unstable Plant

We next briefly show how the robust design method can be extended to the case of an unstable open-loop plant. As before, we denote the nominal model by $G_o(s) = \frac{B_o(s)}{A_o(s)}$, the nominal controller by $C_o(s) = \frac{P(s)}{L(s)}$, the nominal sensitivity by S_o, the sensitivity when the plant remains at $G_o(s)$ (but the controller changes to $C(s)$) by S_1, and the achieved sensitivity when the plant changes to $G(s)$ (and the controller changes to C) by S_2. We parameterize the modified controller by:

$$C(s) = \frac{\frac{P(s)}{E(s)} + \frac{A_o(s)}{E(s)}Q(s)}{\frac{L(s)}{E(s)} - \frac{B_o(s)}{E(s)}Q(s)} \tag{16.3.44}$$

where $Q(s)$ is a stable proper transfer function.

It then follows, as in Chapter 15, that

$$S_o(s) = \frac{A_o(s)L(s)}{A_o(s)L(s) + B_o(s)P(s)} \qquad T_o(s) = \frac{B_o(s)P(s)}{A_o(s)L(s) + B_o(s)P(s)} \tag{16.3.45}$$

$$S_1(s) = S_o(s)\left(1 - \frac{B_o(s)Q(s)}{L(s)}\right) \qquad T_1(s) = T_o(s) + \frac{S_o(s)B_o(s)Q(s)}{L(s)} \tag{16.3.46}$$

$$S_2(s) = \frac{S_1(s)}{1 + T_1(s)G_\Delta(s)} = \frac{S_1(s)}{1 + T_1(s)\frac{A_o(s)}{B_o(s)}G_\epsilon}$$

$$= \frac{S_o(s) - \frac{A_o(s)B_o(s)Q(s)}{A_o(s)L(s)+B_o(s)P(s)}}{1 + \left(\frac{A_o(s)P(s)}{A_o(s)L(s)+B_o(s)P(s)} + \frac{A_o(s)^2 Q(s)}{A_o(s)L(s)+B_o(s)P(s)}\right)G_\epsilon(s)} \tag{16.3.47}$$

where $G_\Delta(s)$ and $G_\epsilon(s)$ denote, as usual, the MME and AME, respectively.

As before, we used a weighted measure of $S_2(s) - S_o(s)$, where the weight is now chosen as

$$W_2(s) = (1 + T_1(s)G_\Delta(s)) \tag{16.3.48}$$

In this case,

$$W_2(s)[S_2(s) - S_o(s)] = -\frac{A_o(s)B_o(s)Q(s)}{A_o(s)L(s) + B_o(s)P(s)}$$
$$- \frac{L(s)\left[P(s) + A_o(s)Q(s)\right]}{[A_o(s)L(s) + B_o(s)P(s)]^2}A_o(s)^2 G_\epsilon(s) \tag{16.3.49}$$

A point that will be appreciated by those who have studied identification is that identification for unstable plants typically uses fractional models. Thus, the nominal model would typically be expressed as $\frac{N_o}{D_o}$, where

$$N_o(s) = \frac{B_o(s)}{E(s)} \tag{16.3.50}$$

$$D_o(s) = \frac{A_o(s)}{E(s)} \tag{16.3.51}$$

for some stable polynomial $E(s)$. Also, the true plant would normally be represented as $\frac{N(s)}{D(s)}$, where

$$N(s) = N_o(s) + \frac{B_\epsilon(s)}{E(s)} \tag{16.3.52}$$

$$D(s) = D_o(s) + \frac{A_\epsilon(s)}{E(s)} \tag{16.3.53}$$

Under these conditions, the additive modeling error $G_\epsilon(s)$ would take the form

$$G_\epsilon(s) = \frac{N(s)}{D(s)} - \frac{N_o(s)}{D_o(s)} = \frac{B_o(s) + B_\epsilon(s)}{A_o(s) + A_\epsilon(s)} - \frac{B_o(s)}{A_o(s)} \tag{16.3.54}$$

$$\simeq \frac{A_o(s)B_\epsilon(s)}{A_o(s)^2} - \frac{B_o(s)A_\epsilon(s)}{A_o(s)^2} \tag{16.3.55}$$

Substituting (16.3.54) into (16.3.49) yields

$$W_2(s)[S_2(s) - S_o(s)] = -\frac{A_o(s)B_o(s)Q(s)}{A_o(s)L(s) + B_o(s)P(s)}$$
$$- \frac{L(s)\left[P(s) + A_o(s)Q(s)\right]}{[A_o(s)L(s) + B_o(s)P(s)]^2}\left(A_o(s)B_\epsilon(s) - B_o(s)A_\epsilon(s)\right) \tag{16.3.56}$$

We can then proceed essentially as in the open-loop stable case.

Remark 16.14. *The comments made in Remark 16.12 also apply here.*

16.4   Cheap Control Fundamental Limitations

Consider the standard single-input single-output feedback control loop shown, for example, in Figure 5.1 on page 122. We will be interested in minimizing the quadratic cost associated with the output response expressed by:

$$J = \frac{1}{2} \int_0^{\infty} y(t)^2 dt \qquad (16.4.1)$$

Note that, no account is taken here of the "size" of the control effort. Hence, this class of problem, is usually called "cheap control." It is obviously impractical to allow arbitrarily large control signals. However, by not restricting the control effort, we obtain a benchmark against which other, more realistic , scenarios can be judged. Thus, these results give a fundamental limit to the achievable performance. We will consider two types of disturbances, namely

(i) impulsive measurement noise $(d_m(t) = \delta(t))$, and

(ii) a step-output disturbance $(d_o(t) = \mu(t))$.

We then have the following result that expresses the connection between the minimum achievable value for the cost function (16.4.1) and the open-loop properties of the system.

Theorem 16.1. *Consider the SISO feedback control loop as in Figure 5.1 on page 122 and the cheap control cost function (16.4.1). Then,*

(i) For impulsive measurement noise, the minimum value for the cost (16.4.1) is

$$J^* = \sum_{i=1}^{N} p_i \qquad (16.4.2)$$

where $p_i, \ldots, p_N$, denote the open-loop plant poles in the right-half plane, and

(ii) For a step-output disturbance, the minimum value for the cost (16.4.1) is

$$J^* = \sum_{i=1}^{M} \frac{1}{c_i} \qquad (16.4.3)$$

where $c_i, \ldots, c_M$ denote the open-loop plant zeros in the right-half plane.

Proof

(i) *The transfer function from measurement noise to output is $\{-T(s)\}$. Hence, for impulsive noise, the Laplace Transform of the output is*

$$Y(s) = -T(s) \tag{16.4.4}$$

Hence, using Parseval's Theorem (Theorem 4.1 on page 96), the cost function (16.4.1) becomes

$$J = \frac{1}{4\pi} \int_{-\infty}^{\infty} |T_o(j\omega)|^2 d\omega \tag{16.4.5}$$

We next use the Q-parameterization of all stabilizing controllers to write $T_o(s)$ as in (15.7.23), i.e.,

$$T_o(s) = \frac{B_o(s)P(s) + Q_u(s)B_o(s)A_o(s)}{A^*(s)} \tag{16.4.6}$$

where $A^*(s) = A_o(s)L_o(s) + B_o(s)P(s)$

Hence, the cost function takes the form:

$$J = \frac{1}{4\pi} \int_{-\infty}^{\infty} |W(j\omega) - Q_u(j\omega)V(j\omega)|^2 d\omega \tag{16.4.7}$$

where

$$W(s) = \frac{B_o(s)P(s)}{A^*(s)} \; ; \quad V(s) = \frac{B_o(s)A_o(s)}{A^*(s)} \tag{16.4.8}$$

By using Equations (16.2.15) and (16.2.16) from Lemma 16.2 on page 460, we see that the optimal controller yields

$$J^* = \frac{1}{4\pi} \int_{-\infty}^{\infty} |(V_a(j\omega)^{-1}W(j\omega))_u|^2 d\omega \tag{16.4.9}$$

Noting that $B(s)P(s) = A^*(s)$ at a zero of $A(s)$ and by using the Cauchy Integral Formula (Theorem C.8, Appendix C) we have

$$J^* = \sum_{i=1}^{N} p_i \qquad (16.4.10)$$

where p_i are the open-loop right-half plane poles.

(ii) Similarly to part (i).

□□□

The above result implies the minimum achievable output energy resulting from impulsive measurement noise is equal to the sum of the right-half plane poles and that the minimum achievable output energy resulting from a unit-step output disturbance is equal to the sum of the inverses of right-half plane zeros. The result is heuristically reasonable. For example, to compensate for an output disturbance, we need to invert the plant. The difficulty of doing this is a function of right-half plane plant zeros. Similarly, dealing with impulsive measurement noise, requires us to make minimal control perturbations so as to keep the plant stable. This is a function of right-half plane plant poles.

16.5   Frequency-Domain Limitations Revisited

We saw earlier in Chapter 9 (Lemma 9.2 on page 244 and Lemma 9.4 on page 247) that the sensitivity and complementary sensitivity functions satisfied the following integral equations in the frequency domain

(i)

$$\frac{1}{\pi} \int_0^\infty \ln |S_o(j\omega)| + \frac{k_h}{2} = \sum_{i=1}^{N} p_i \qquad (16.5.1)$$

where k_h denotes $\lim_{s \to \infty} sH_{ol}(s)$ and $H_{ol}(s)$ is the open-loop transfer function.

(ii)

$$\frac{1}{\pi} \int_0^\infty \frac{1}{\omega^2} \ln |T_o(j\omega)| + \frac{1}{2k_v} = \sum_{i=1}^{M} \frac{1}{c_i} \qquad (16.5.2)$$

where $k_v = \lim_{s \to 0} sH_{ol}(s)$.

There is clearly a remarkable consistency between the right-hand sides of (16.5.1), (16.4.2), (16.5.2), and (16.4.3). This is not a coincidence as shown in the following result:

Theorem 16.2. *Consider the standard SISO control loop on Figure 5.1 on page 122, in which the open-loop transfer function, $H_{ol}(s)$ is strictly proper and $H_{ol}(0)^{-1} = 0$ (i.e., there is integral action), then*

(i) *For impulse measurement noise, the following inequality holds:*

$$\frac{1}{2}\int_0^\infty y(t)^2 \geq \frac{k_h}{2} + \frac{1}{\pi}\int_0^\infty \ln|S_o(j\omega)|d\omega = \sum_{i=1}^N p_i \qquad (16.5.3)$$

where $p_i, \ldots, p_N$ denote the plant right-half plane poles.

(ii) *For a unit-step output disturbance, then*

$$\frac{1}{2}\int_0^\infty y(t)^2 \geq \frac{1}{2k_v} + \frac{1}{\pi}\int_0^\infty \ln|T_o(j\omega)|\frac{d\omega}{\omega^2} = \sum_{i=1}^M \frac{1}{c_i} \qquad (16.5.4)$$

where $c_i, \ldots, c_M$ denote the plant right-half plane zeros.

Proof

Actually, the result follows immediately by comparing (16.5.1), (16.4.2), (16.5.2), and (16.4.3). However, a direct proof is also possible as we show below:

(i) *We know that the transfer function from measurement noise to plant output is given by $-T_o(s)$ where $T_o(s)$ is the complementary sensitivity function. For impulsive noise, the Laplace Transform of the output is given by*

$$Y(s) = -T_o(s) \qquad (16.5.5)$$

Hence, by Parseval's Theorem (Theorem 4.1 on page 96):

$$\frac{1}{2}\int_0^\infty y(t)^2 dt = \frac{1}{2\pi}\int_0^\infty |T_o(j\omega)|^2 d\omega \qquad (16.5.6)$$

Adding and subtracting $\frac{1}{\pi}\int_0^\infty \Re\{T_o(j\omega)\}d\omega$ gives

$$\frac{1}{2}\int_0^\infty y(t)^2 dt = \frac{1}{2\pi}\int_{-\infty}^\infty T_o(j\omega)d\omega \qquad (16.5.7)$$
$$+ \frac{1}{2\pi}\int_0^\infty \left[-2\Re\{T_o(j\omega)\} + |T_o(j\omega)|^2\right]d\omega$$

where we have used the fact that $\text{Im}\{T_o(j\omega)\}$ *is an odd function.*

Next, we use the fact that $\ln(1 + x) \leq x$. *Hence,*

$$\frac{1}{2}\int_0^\infty y(t)^2 dt \geq \frac{1}{2\pi}\int_{-\infty}^\infty T_o(j\omega)d\omega \tag{16.5.8}$$

$$+ \frac{1}{2\pi}\int_0^\infty \ln\left(1 - 2\Re\{T_o(j\omega)\} + |T_o(j\omega)|^2\right) d\omega$$

$$= \frac{1}{2\pi}\int_{-\infty}^\infty T_o(j\omega)d\omega + \frac{1}{2\pi}\int_0^\infty \ln|1 - T_o(j\omega)|^2 d\omega$$

Now, using techniques similar to those employed in the proof of Lemma 9.3 on page 246, and noting that, for large values of s, $T(s) \simeq \frac{k_h}{s}$, we have that

$$\frac{1}{2\pi}\int_{-\infty}^\infty T_o(j\omega)d\omega = \frac{k_h}{2} \tag{16.5.9}$$

Also, we know that $S_o + T_o = 1$. The result follows by using (9.2.3).

(ii) *This can be established along the same lines as part (i), by using (9.3.11), and on noting that the Laplace Transform of the output in response to a unit-step output disturbance is*

$$Y(s) = \frac{S_o(s)}{s} \tag{16.5.10}$$

where $S_o(s)$ is the sensitivity function.

□□□

16.6   Summary

- Optimization can often be used to assist with certain aspects of control-system design.

- The answer provided by an optimization strategy is only as good as the question that has been asked–that is, how well the optimization criterion captures the relevant design specifications and trade-offs.

- Optimization needs to be employed carefully: keep in mind the complex web of trade-offs involved in all control-system design.

- Quadratic optimization is a particularly simple strategy and leads to a closed-form solution.

- Quadratic optimization can be used for optimal Q synthesis.

- We have also shown that quadratic optimization can be used effectively to formulate and solve robust control problems when the model uncertainty is specified in the form of a frequency-domain probabilistic error.

- Within this framework, the robust controller *biases* the nominal solution so as to create conservatism, in view of the expected model uncertainty, while attempting to minimize affecting the achieved performance.

- This can be viewed as a formal way of achieving the bandwidth reduction that was discussed earlier as a mechanism for providing a robustness *gap* in control-system design.

16.7   Further Reading

Optimization

Doyle, J.C., Glover, K., Khargonekar, P.P., and Francis, B.A. (1989). State space solutions to standard H_2 and H_∞ control problems. *IEEE Transactions on Automatic Control*, 34(8):831-847.

Youla, D., Jabr, H., and Bongiorno, J. (1976). Modern Wiener-Hopf design of optimal controllers. Part I. *IEEE Transactions on Automatic Control*, 21(2):3-13.

Zhou, K., Doyle, J.C., and Glover, K. (1996). *Robust and Optimal Control*. Prentice-Hall, Upper Saddle River, N.J.

Robust control

Dorato, P.E. (1987). *Robust Control*. IEEE Press, New York.

Dorato, P.E., Tempo, R., and Muscato, G.M. (1993). Bibliography on robust control. *Automatica*, 29(1):201-213.

Doyle, J.C., Francis, B.A., and Tannenbaum, A.R. (1992). *Feedback control Theory*. Macmillan Publishing Company.

Francis, B.A. (1987). *A Course on H_∞ Control Theory*. Lecture Notes in Control and Information Sciences, Vol. 8, Springer-Verlag, New York.

Francis, B.A. and Zames, G. (1984). On H_∞ optimal sensitivity theory for SISO feedback systems. *IEEE Transactions on Automatic Control*, 29(1):9-16.

Graebe, S.F. (1994). Control of an Unknown Plant: A Benchmark of New Format. *Automatica*, 30(4):567-575.

Morari, M. and Zafiriou, E. (1989). *Robust process control*. Prentice-Hall, Englewood Cliffs, N.J.

Zhou, K. and Doyle, J.C. (1998). *Essentials of Robust Control.* Prentice-Hall, Upper Saddle River, N.J.

Zhou, K., Doyle, J.C., and Glover, K. (1996). *Robust and Optimal Control.* Prentice-Hall, Upper Saddle River, N.J.

Identification

Bohlin, T. and Graebe, S.F. (1995). Issues in nonlinear stochastic grey box identification. *International Journal of Adaptive Control and Signal Processing*, 9(6):465-490.

Goodwin, G.C. and Payne, R.L. (1977). *Dynamic System.* Academic Press, New York.

Ljung, L. (1999). *System Identification. Theory for the User.* Prentice-Hall, Englewood Cliffs, N.J. 2nd edition.

Statistical confidence bounds accommodating for undermodeling

Goodwin, G.C., Gevers, M., and Ninness, B.M. (1992). Quantifying the error in estimated transfer functions with application to model order selection. *IEEE Transactions on Automatic Control*, 37(7):913-928.

Goodwin, G.C., Braslavsky, J.H., and Seron, M.M. (1999). Non-stationary stochastic embedding for transfer function estimation. In *Proceedings of the 14th IFAC World Congress*, Beijing, China.

Robust design and stochastic errors

Boyd, S.P., Balakrishnan, V., Barratt, C.H., Khraishi, N., Li, X., Meyer, D. and Norman, S.A. (1988). A new CAD method and associated architectures for linear controllers. *IEEE Transactions on Automatic Control*, 33(3):268-283.

Goodwin, G.C. and Miller, D.E. (1998). *Robust performance optimization based on stochastical model errors: The stable open-loop case.* Technical report, Dept. of Electrical and Computer Engineering, University of Newcastle, Australia.

Goodwin, G.C., Wang, L., and Miller, D.E. (1999). Bias-variance tradeoff issues in robust controller design using statistical confidence bounds. In *Proceedings of the 14th IFAC World Congress*, Beijing, China.

Cheap control fundamental limitations

Qui, L. and Davison, E.J. (1993). Performance limitations for nonminimum phase systems in the servomechanism problem. *Automatica*, 29(2):337-349.

Saberi, A. and Sannuti, P. (1987). Cheap and singular controls for linear quadratic regulators. *IEEE Transactions on Automatic Control*, 32(3):208-219.

Middleton, R.H. and Braslavsky J.H. (2000). *On the relationship between logarithmic sensitivity integrals and limiting optimal control problems.* Technical report, Dept. of Electrical and Computer Engineering, University of Newcastle, Australia.

16.8   Problems for the Reader

Problem   16.1. *Consider the quadratic optimization problem*

$$\min_{Q(s)\in\mathcal{S}} \left\| W(s) - Q(s)V(s) \right\|_2^2 \qquad (16.8.1)$$

where $V(s)$ and $W(s)$ are stable and proper transfer functions. Solve the optimiza-tion problem (i.e., find the minimizing, stable, and (possibly) improper $Q(s)$) for the following pairs:

	Case 1	Case 2	Case 3
$V(s)$	$\dfrac{1}{(s+1)}$	$\dfrac{(-s+1)^2}{s^2+4s+2}$	$\dfrac{1}{s+2}$
$W(s)$	$\dfrac{s+3}{s+5}$	$\dfrac{2}{(s+1)^2}$	$\dfrac{(-s+1)}{(s+6)(s+2)}$

Problem   16.2. *Consider a plant having a nominal model with transfer function given by*

$$G_o(s) = \frac{-s+2}{(s+2)(s+4)} \qquad (16.8.2)$$

It is required to design a one-d.o.f. control loop to track a reference with nonzero mean and significant energy in the frequency band $[0,3][rad/s]$.

*Synthesize $Q(s)$ as the optimal quadratic norm inverse of $G_o(s)$. (Hint: use an appropriate weighting function and set the problem up as a standard quadratic-minimization problem; then use MATLAB program **oph2.m**.)*

Problem   16.3. *Consider a plant having a nominal model $G_o(s)$. The comple-mentary sensitivity target is $H_o(s)$. Assume that*

$$G_o(s) = 4\frac{-s+1}{(s+1)(s+8)} \quad and \quad H_o(s) = \frac{b^2}{s^2+1.3bs+b^2} \qquad (16.8.3)$$

16.3.1 *Compute the optimal (quadratic) $Q(s)$ for $b = 0.5$.*

16.3.2 *Repeat for $b = 6$.*

16.3.3 *Compare the two solutions, and discuss them in connection with the funda-mental design limitations arising from NMP zeros.*

Problem 16.4. *Consider a plant having a nominal model $G_o(s)$. The comple-mentary sensitivity target is $H_o(s)$. Assume that*

$$G_o(s) = 4\frac{s+4}{(s+1)(s+8)} \quad and \quad H_o(s) = \frac{4}{s^2 + 2.6s + 4} \quad (16.8.4)$$

Compute the optimal (quadratic) $Q(s)$, with the additonal constraint that its poles should have real parts less or equal than -2. Hint: make the substitution $v = s+2$ and then solve the equivalent optimization problem in the complex variable v.

Problem 16.5. *Find the best (in a quadratic sense) stable approximation for the unstable transfer function*

$$F(s) = \frac{4}{(-s+2)(s+3)} \quad (16.8.5)$$

in the frequency band $[0; 4]$ $[rad/s]$. To do that, set the optimization problem as

$$\min_{X(s) \in S} \left\| W(s)\left(1 - [F(s)]^{-1}X(s)\right) \right\|_2^2 \quad (16.8.6)$$

where $X(s)$ is the optimal stable approximation and $W(s)$ is a suitable weighting function.

Problem 16.6. *A disturbance-feedforward scheme (as in Figure 10.2 on page 274) is to be used in the two-d.o.f. control of a plant with nominal model*

$$G_{o1}(s) = \frac{-s+3}{(s+1)(s+4)} \quad (16.8.7)$$

It is also known that the disturbance $d_g(t)$ is a signal with energy distributed in the frequency band $[0, 4][rad/s]$.

Design the feedforward block $G_f(s)$ by using quadratic optimization–i.e., set the problem up so as to find a transfer function $G_f^o(s)$ such that

$$G_f^o(s) = \arg\min_{X \in S} \|(1 - G_{o1}X)\Omega\|_2 \quad (16.8.8)$$

where S is the set of all real rational and stable transfer functions and $\Omega(s)$ is a suitable weighting function that takes into account the disturbance bandwidth.

Problem 16.7. *Consider a plant with nominal model*

$$G_o(s) = \frac{2}{(-s+1)(s+4)} \tag{16.8.9}$$

Design a one-d.o.f. controller such that the feedback control loop tracks step references in the presence of measurement noise with energy in the frequency band $[5, 50]$ $[rad/s]$.

Problem 16.8. *A feedback control loop is designed for a plant with nominal model*

$$G_o(s) = \frac{2}{(s+1)(s+2)} \tag{16.8.10}$$

so as to compensate for an input disturbance with nonzero mean and a variable component with energy in the frequency band $[0, 5]$ $[rad/s]$.

16.8.1 *Synthesize a nominal controller $Q_o(s)$ by using quadratic optimization.*

16.8.2 *Use the robust control strategy presented in this chapter to design a robust controller, $Q(s)$, if we additionally know that the frequency response of the additive modeling error satisfies*

$$\mathcal{E}\{G_\epsilon(j\omega)G_\epsilon(-j\omega)\} = \tilde{\alpha}^2(\omega) = \frac{2\omega^2}{(\omega^2+64)(\omega^2+1)} \tag{16.8.11}$$

Problem 16.9. *Build a quadratic optimization Q synthesis theory, similar to that developed in subsection §16.2.2, for the sampled-data control case. To that end, assume that you want to synthesize a digital controller $Q_q(z)$ for a continuous-time plant, such that the complementary sensitivity $T_{oq}(z)$ is at a minimum distance, in quadratic norm, to a prescribed design target $H_{oq}(z)$.*

Problem 16.10. *Apply the methodology developed in Problem 16.9 to a case where the plant has a nominal model given by*

$$G_o(s) = \frac{-s+2}{(s+1)(s+2)} \tag{16.8.12}$$

The digital controller output is applied to the plant through a zero-order sample-and-hold device, and the plant output is sampled every 0.2 $[s]$.

Synthesize the controller $Q_q(z)$ that minimizes $\left\| H_{oq} - T_{oq} \right\|_2$, where the complementary sensitivity target, $H_{oq}(z)$, is given by

$$H_{oq}(z) = \frac{0.1(z + 0.7)}{z^2 - 1.2z + 0.45} \qquad (16.8.13)$$

Problem 16.11. *Recall Example 16.1. Compute the resulting complementary sensitivity $T_o(s)$, and compare it to the target value $H_o(s)$. Compare the frequency responses, and simulate the step responses. Discuss.*

Problem 16.12. *Repeat Problem 16.11 for Example 16.2.*

Chapter 17

LINEAR STATE SPACE
MODELS

17.1 Preview

We have seen that there are many alternative model formats that can be used for linear dynamic systems. In simple SISO problems, any representation is probably as good as any other. However, as we move to more complex problems (especially multivariable problems), it is desirable to use special model formats. One of the most flexible and useful structures is the state space model. As we saw in Chapter 3, this model takes the form of a coupled set of *first-order* differential (or difference) equations. This model format is particularly useful with regard to numerical computations.

State space models were briefly introduced in Chapter 3. Here, we will examine linear state space models in a little more depth for the SISO case. Note, however, that many of the ideas will carry over directly to the multivariable case presented later. In particular, we will study

- similarity transformations and equivalent state representations,
- state space model properties:
    - controllability, reachability, and stabilizability,
    - observability, reconstructability, and detectability,
- special (canonical) model formats.

The key tools used in studying linear state space methods are linear algebra and vector space methods. The reader is encouraged to briefly review these concepts as a prelude to reading this chapter.

17.2 Linear Continuous-Time State Space Models

As we have seen in Chapter 3, a continuous-time linear time-invariant state space model takes the form

$$\dot{x}(t) = \mathbf{A}x(t) + \mathbf{B}u(t) \qquad\qquad x(t_o) = x_o \qquad (17.2.1)$$
$$y(t) = \mathbf{C}x(t) + \mathbf{D}u(t) \qquad\qquad\qquad\qquad\quad (17.2.2)$$

where $x \in \mathbb{R}^n$ is the state vector, $u \in \mathbb{R}^m$ is the control signal, $y \in \mathbb{R}^p$ is the output, $x_o \in \mathbb{R}^n$ is the state vector at time $t = t_o$ and $\mathbf{A}, \mathbf{B}, \mathbf{C}$, and $\mathbf{D}$ are matrices of appropriate dimensions.

Equation (17.2.1) is called the state equation, (17.2.2) is called the output equation.

17.3   Similarity Transformations

It is readily seen that the definition of the state of a system is nonunique. Consider, for example, a linear transformation of $x(t)$ to $\overline{x}(t)$ defined as

$$\overline{x}(t) = \mathbf{T}^{-1}x(t) \qquad\qquad x(t) = \mathbf{T}\,\overline{x}(t) \qquad (17.3.1)$$

where $\mathbf{T}$ is any nonsingular matrix, called a similarity transformation. If (17.3.1) is substituted into (17.2.1) and (17.2.2), the following alternative state description is obtained:

$$\dot{\overline{x}}(t) = \overline{\mathbf{A}}\,\overline{x}(t) + \overline{\mathbf{B}}\,u(t) \qquad\qquad \overline{x}(t_o) = \mathbf{T}^{-1}x_o \qquad (17.3.2)$$
$$y(t) = \overline{\mathbf{C}}\,\overline{x}(t) + \overline{\mathbf{D}}\,u(t) \qquad\qquad\qquad\qquad (17.3.3)$$

where

$$\overline{\mathbf{A}} \triangleq \mathbf{T}^{-1}\mathbf{A}\mathbf{T} \quad \overline{\mathbf{B}} \triangleq \mathbf{T}^{-1}\mathbf{B} \quad \overline{\mathbf{C}} \triangleq \mathbf{C}\mathbf{T} \quad \overline{\mathbf{D}} \triangleq \mathbf{D} \qquad (17.3.4)$$

The model $\{(17.3.2), (17.3.3)\}$ is an equally valid description of the system. Indeed, there are an infinite number of (equivalent) ways of expressing the state space model of a given system. All equivalent models are related via a similarity transform as in (17.3.4). We shall see below that certain choices for the transformation $\mathbf{T}$ make various aspects of the system response easier to visualize and compute.

As an illustration, say that the matrix $\mathbf{A}$ can be diagonalized by a similarity transformation $\mathbf{T}$; then

$$\overline{\mathbf{A}} = \mathbf{\Lambda} \triangleq \mathbf{T}^{-1}\mathbf{A}\mathbf{T} \qquad (17.3.5)$$

where, if $\lambda_1, \lambda_2, \ldots, \lambda_n$ are the eigenvalues of $\mathbf{A}$, then

$$\mathbf{\Lambda} = \operatorname{diag}(\lambda_1, \lambda_2, \ldots, \lambda_n) \qquad (17.3.6)$$

Because $\mathbf{\Lambda}$ is diagonal, we have

$$\overline{x}_i(t) = e^{\lambda_i(t-t_o)}\overline{x}_o + \int_{t_o}^{t} e^{\lambda_i(t-\tau)}\overline{b}_i u(\tau) d\tau \tag{17.3.7}$$

where the subscript i denotes the ith component of the state vector. Equation (17.3.7) gives a direct connection between the system output and the eigenvalues of the matrix $\mathbf{A}$. Note that, in the case of complex-conjugate eigenvalues, $\overline{x}_i(t)$ will also be complex, although the output $y(t)$ is necessarily real:

$$y(t) = \sum_{i=1}^{n} \overline{c}_i e^{\lambda_i(t-t_o)}\overline{x}_o + \sum_{i=1}^{n} \overline{c}_i \int_{t_o}^{t} e^{\lambda_i(t-\tau)}\overline{b}_i u(\tau) d\tau + \mathbf{D}u(t) \tag{17.3.8}$$

Example 17.1. *Consider a system with a state space description given by*

$$\mathbf{A} = \begin{bmatrix} -4 & -1 & 1 \\ 0 & -3 & 1 \\ 1 & 1 & -3 \end{bmatrix}; \quad \mathbf{B} = \begin{bmatrix} -1 \\ 1 \\ 0 \end{bmatrix}; \quad \mathbf{C} = \begin{bmatrix} -1 & 1 & 0 \end{bmatrix} \quad \mathbf{D} = 0 \tag{17.3.9}$$

*Using the MATLAB command **eig**, we find that the eigenvalues of $\mathbf{A}$ are -5, -3, and -2. The matrix $\mathbf{A}$ can be diagonalized by using a similarity transformation matrix $\mathbf{T}$ whose columns are a set of linearly independent eigenvectors of $\mathbf{A}$. The matrix $\mathbf{T}$ can also be obtained by using the MATLAB command **eig**, which yields*

$$\mathbf{T} = \begin{bmatrix} 0.8018 & 0.7071 & 0.0000 \\ 0.2673 & -0.7071 & 0.7071 \\ -0.5345 & -0.0000 & 0.7071 \end{bmatrix} \tag{17.3.10}$$

Using (17.3.4) and (17.3.10), we obtain the similar state space description given by

$$\overline{\mathbf{A}} = \mathbf{\Lambda} = \begin{bmatrix} -5 & 0 & 0 \\ 0 & -3 & 0 \\ 0 & 0 & -2 \end{bmatrix}; \quad \overline{\mathbf{B}} = \begin{bmatrix} 0.0 \\ -1.414 \\ 0.0 \end{bmatrix}; \quad (17.3.11)$$

$$\overline{\mathbf{C}} = \begin{bmatrix} -0.5345 & -1.4142 & 0.7071 \end{bmatrix} \quad \overline{\mathbf{D}} = 0 \quad (17.3.12)$$

17.4   Transfer Functions Revisited

As might be expected, the transfer function of a system does not depend on the basis used for the state space. Consider, for example, the alternative description given in (17.3.2) to (17.3.4).

The solution to equation (17.3.3) from using Laplace Transforms is given by

$$Y(s) = [\overline{\mathbf{C}}(s\mathbf{I} - \overline{\mathbf{A}})^{-1}\overline{\mathbf{B}} + \overline{\mathbf{D}}]U(s) + \overline{\mathbf{C}}(s\mathbf{I} - \overline{\mathbf{A}})^{-1}\overline{x}(0) \quad (17.4.1)$$

$$= [\mathbf{CT}(s\mathbf{I} - \mathbf{T}^{-1}\mathbf{AT})^{-1}\mathbf{T}^{-1}\mathbf{B} + \mathbf{D}]U(s) + \mathbf{CT}(s\mathbf{I} - \mathbf{T}^{-1}\mathbf{AT})^{-1}\mathbf{T}^{-1}x(0) \quad (17.4.2)$$

$$= [\mathbf{C}(s\mathbf{I} - \mathbf{A})^{-1}\mathbf{B} + \mathbf{D}]U(s) + \mathbf{C}(s\mathbf{I} - \mathbf{A})^{-1}x(0) \quad (17.4.3)$$

which is identical to (4.5.10), as anticipated.

The previous analysis shows that different choices of state variables lead to different internal descriptions of the model, but to the same input-output model, because

$$\overline{\mathbf{C}}(s\mathbf{I} - \overline{\mathbf{A}})^{-1}\overline{\mathbf{B}} + \overline{\mathbf{D}} = \mathbf{C}(s\mathbf{I} - \mathbf{A})^{-1}\mathbf{B} + \mathbf{D} \quad (17.4.4)$$

for any nonsingular $\mathbf{T}$.

Hence, given a transfer function, there exist infinitely many input-output equivalent state space models.

This is consistent with physical understanding, as is illustrated in the following example.

Example 17.2. *Consider again the d.c. motor of Example 3.4 on page 47. Physically, the angle of the shaft is an objective observation, which is clearly independent of the units used to measure the other state: angular speed. If $x_2(t)$ is a measurement of the angular speed in one unit, and $\overline{x}_2(t)$ one in another unit, with conversion factor $\overline{x}_2(t) = \alpha x_2(t)$, then*

$$\bar{x}(t) = \begin{bmatrix} \bar{x}_1(t) \\ \bar{x}_2(t) \end{bmatrix} = \mathbf{T}^{-1} \begin{bmatrix} x_1(t) \\ x_2(t) \end{bmatrix} = \mathbf{T}^{-1}x(t) \qquad where \quad \mathbf{T}^{-1} = \begin{bmatrix} 1 & 0 \\ 0 & \alpha \end{bmatrix}$$

$$(17.4.5)$$

The input-output relation is, both physically and mathematically, independent of this transformation.

□□□

In this example, we transformed from one set of physical states to another set of physical states (namely, the same quantities measured in different units). A common application of such similarity transformations is to convert to a state vector such that the **A** matrix has particular properties; the associated states can then lose their physical meaning (if they had one), but the original state can always be recovered by the inverse transformation.

For example, we can carry out a transformation to highlight the set of *natural frequencies* of the system that correspond to the eigenvalues of the matrix **A**. We note that

$$\det(s\mathbf{I} - \mathbf{A}) = \det(s\mathbf{I} - \mathbf{T}^{-1}\mathbf{A}\mathbf{T}) \qquad (17.4.6)$$

which verifies that the eigenvalues are invariant with respect to a similarity transformation.

Example 17.3. *Consider the 4-tuple (**A**,**B**,**C**,**D**) used in Example 17.1 on page 493. The transfer function $G(s)$, obtained from (4.5.13), is given by*

$$G(s) = \frac{2s^2 + 14s + 20}{s^3 + 10s^2 + 31s + 30} \qquad (17.4.7)$$

Note that the computation of $G(s)$ can be done by using MATLAB command ***ss2tf***.

□□□

Remark 17.1. *We observe that $G(s)$ can be expressed as*

$$G(s) = \frac{\mathbf{C}\,Adj(s\mathbf{I} - \mathbf{A})^{-1}\mathbf{B}}{\det(s\mathbf{I} - \mathbf{A})} \qquad (17.4.8)$$

*Hence, the poles of $G(s)$ are eigenvalues of **A**–i.e., they are roots of the characteristic polynomial of the matrix **A***

$$\det(s\mathbf{I} - \mathbf{A}) = \prod_{i=1}^{n}(s - \lambda_i) \qquad (17.4.9)$$

where $\lambda_1, \lambda_2, \dots, \lambda_n$ are the eigenvalues of $\mathbf{A}$.

17.5 From Transfer Function to State Space Representation

We have seen above how to go from a state space description to the corresponding
transfer function. The converse operation leads to the following question: *given a
transfer function $G(s)$, how can a state representation for this system be obtained?*
In this section, we present one answer to this question.

Consider a transfer function $G(s)$, as in (4.5.3) to (4.5.5). We can then write

$$Y(s) = \sum_{i=1}^{n} b_{i-1}V_i(s) \qquad \text{where} \qquad V_i(s) = \frac{s^{i-1}}{A(s)}U(s) \qquad (17.5.1)$$

We note from the above definitions that

$$v_i(t) = \mathcal{L}^{-1}[V(s)] = \frac{dv_{i-1}(t)}{dt} \qquad \text{for} \quad i = 1, 2, \dots, n \qquad (17.5.2)$$

We can then choose, as state variables, $x_i(t) = v_i(t)$, which, with the use of
(17.5.1), lead to

$$\mathbf{A} = \begin{bmatrix} 0 & 1 & 0 & \cdots & 0 & 0 \\ 0 & 0 & 1 & \cdots & 0 & 0 \\ \vdots & \vdots & \vdots & & \vdots & \vdots \\ -a_0 & -a_1 & -a_2 & \cdots & -a_{n-2} & -a_{n-1} \end{bmatrix}; \quad \mathbf{B} = \begin{bmatrix} 0 \\ 0 \\ \vdots \\ 0 \\ 1 \end{bmatrix} \qquad (17.5.3)$$

$$\mathbf{C} = \begin{bmatrix} b_0 & b_1 & b_2 & \cdots & b_{n-1} \end{bmatrix} \quad \mathbf{D} = 0 \qquad (17.5.4)$$

This is illustrated by the following example.

Example 17.4. *Consider building a state space description by using (17.5.1) to
(17.5.4) for the transfer function in (17.4.7). This leads to*

$$\mathbf{A} = \begin{bmatrix} 0 & 1 & 0 \\ 0 & 0 & 1 \\ -30 & -31 & -10 \end{bmatrix} ; \quad \mathbf{B} = \begin{bmatrix} 0 \\ 0 \\ 1 \end{bmatrix} \quad (17.5.5)$$

$$\mathbf{C} = \begin{bmatrix} 20 & 14 & 2 \end{bmatrix} \quad \mathbf{D} = 0 \quad (17.5.6)$$

As an exercise for the reader, we suggest the problem of finding similarity transformations to convert this description into the other two given previously for the same system, in equations (17.3.9) and (17.3.11).

Of course, a similar procedure applies to discrete-time models. For example, consider the transfer function $G_q(z)$ given by

$$Y_q(z) = G_q(z)U_q(z) \quad (17.5.7)$$

$$G_q(z) = \frac{B_q(z)}{A_q(z)} = \frac{b_{n-1}z^{n-1} + b_{n-2}z^{n-2} + \ldots + b_1 z + b_0}{z^n + a_{n-1}z^{n-1} + \ldots + a_1 z + a_0} \quad (17.5.8)$$

This can be expressed as

$$Y_q(z) = \sum_{i=1}^{n} b_{i-1} V_i(z) \qquad \text{where} \qquad V_i(z) = \frac{z^{i-1}}{A_q(z)} U_q(z) \quad (17.5.9)$$

From the above definitions, we have that

$$v_i[k] = \mathcal{Z}^{-1}\left[V(z)\right] = q v_{i-1}[k] \qquad \text{for} \qquad i = 1, 2, \ldots, n \quad (17.5.10)$$

where q is the forward shift operator.

We can then choose, as state variables, $x_i[k] = v_i[k]$, which lead to

$$\mathbf{A}_q = \begin{bmatrix} 0 & 1 & 0 & \cdots & 0 & 0 \\ 0 & 0 & 1 & \cdots & 0 & 0 \\ \vdots & \vdots & \vdots & & \vdots & \vdots \\ -a_0 & -a_1 & -a_2 & \cdots & -a_{n-2} & -a_{n-1} \end{bmatrix} ; \quad \mathbf{B}_q = \begin{bmatrix} 0 \\ 0 \\ \vdots \\ 0 \\ 1 \end{bmatrix} \quad (17.5.11)$$

$$\mathbf{C}_q = \begin{bmatrix} b_0 & b_1 & b_2 & \cdots & b_{n-1} \end{bmatrix} ; \qquad \mathbf{D}_q = 0 \quad (17.5.12)$$

17.6 Controllability and Stabilizability

An important question that lies at the heart of control using state space models is whether we can steer the state via the control input to certain locations in the state space. Technically, this property is called controllability or reachability. A closely related issue is that of stabilizability. We begin with controllability.

> The issue of controllability concerns whether a given initial state x_o can be steered to the origin in finite time using the input $u(t)$.

Formally, we have the following:

Definition 17.1. *A state x_o is said to be controllable if there exists a finite interval $[0,T]$ and an input $\{u(t), t \in [0,T]\}$ such that $x(T) = 0$. If all states are controllable, then the system is said to be completely controllable.*

□□□

Remark 17.2. *A related concept is that of reachability. This concept is sometimes used in discrete-time systems. It is formally defined as follows:*

Definition 17.2. *A state $\bar{x} \neq 0$ is said to be reachable (from the origin) if, given $x(0) = 0$, there exist a finite time interval $[0,T]$ and an input $\{u(t), t \in [0,T]\}$ such that $x(T) = \bar{x}$. If all states are reachable, the system is said to be completely reachable.*

□□□

For continuous, time-invariant, linear systems, there is no distinction between complete controllability and complete reachability. However, the following example illustrates that there is a subtle difference in discrete time.
Consider the following shift-operator state space model:

$$x[k+1] = 0 \tag{17.6.1}$$

This system is obviously completely controllable: the state immediately goes to the origin. However, no nonzero state is reachable.

□□□

In view of the subtle distinction between controllability and reachability in discrete time, we will use the term "controllability" in the sequel to cover the stronger of the two concepts. The discrete-time proofs for the results presented below are a little easier. We will thus prove the results on the following discrete-time (delta-domain) model:

$$\delta x[k] = \mathbf{A}_\delta x[k] + \mathbf{B}_\delta u[k] \tag{17.6.2}$$
$$y[k] = \mathbf{C}_\delta x[k] + \mathbf{D}_\delta u[k] \tag{17.6.3}$$

where $x \in \mathbb{R}^n, u \in \mathbb{R}^m$ and $y \in \mathbb{R}^l$.

For simplicity of notation, we will drop the subscript δ from the matrices $\mathbf{A}_\delta, \mathbf{B}_\delta, \mathbf{C}_\delta,$ and $\mathbf{D}_\delta$ in the remainder of this section.

Our next step will be to derive a simple algebraic test for controllability that can easily be applied to a given state space model. In deriving this result, we will use a result from linear algebra known as the Cayley-Hamilton Theorem. For the convenience of the reader, we now briefly review this result.

Theorem 17.1 (Cayley-Hamilton theorem). *Every matrix satisfies its own characteristic equation–i.e., if*

$$\det(s\mathbf{I} - \mathbf{A}) = s^n + a_{n-1}s^{n-1} + \ldots + a_0 \qquad (17.6.4)$$

then

$$\mathbf{A}^n + a_{n-1}\mathbf{A}^{n-1} + \ldots + a_0\mathbf{I} = 0 \qquad (17.6.5)$$

Proof

Let

$$Adj(s\mathbf{I} - \mathbf{A}) = \mathbf{B_n}s^n + \mathbf{B_{n-1}}s^{n-1} + \ldots + \mathbf{B_0} \qquad (17.6.6)$$

Now, $[\det(s\mathbf{I} - \mathbf{A})]\mathbf{I} = (Adj[s\mathbf{I} - \mathbf{A}])(s\mathbf{I} - \mathbf{A})$, *so we have*

$$(s^n + a_{n-1}s^{n-1} + \ldots + a_0)\mathbf{I} = (\mathbf{B_n}s^n + \mathbf{B_{n-1}}s^{n-1} + \ldots + \mathbf{B_0})(s\mathbf{I} - \mathbf{A}) \quad (17.6.7)$$

Equating coefficients gives

$$\mathbf{B_n} = \mathbf{0} \qquad (17.6.8)$$

$$\mathbf{B_{n-1}} = \mathbf{I} \qquad (17.6.9)$$

$$-\mathbf{B_{n-1}}\mathbf{A} + \mathbf{B_{n-2}} = a_{n-1}\mathbf{I} \qquad (17.6.10)$$

$$-\mathbf{B_{n-2}}\mathbf{A} + \mathbf{B_{n-3}} = a_{n-2}\mathbf{I} \qquad (17.6.11)$$

$$\vdots$$

$$-\mathbf{B_1}\mathbf{A} + \mathbf{B_0} = a_1\mathbf{I} \qquad (17.6.12)$$

$$-\mathbf{B_0}\mathbf{A} = a_0\mathbf{I} \qquad (17.6.13)$$

Successively eliminating the $\mathbf{B}'s$ *gives (17.6.5).*

□□□

Armed with the foregoing result, we can now easily establish the following result on controllability.

Theorem 17.2. *Consider the state space model (17.6.2)-(17.6.3).*

i) *The set of all controllable states is the range space of the controllability matrix* $\boldsymbol{\Gamma}_c[\mathbf{A}, \mathbf{B}]$, *where*

$$\boldsymbol{\Gamma}_c[\mathbf{A}, \mathbf{B}] \triangleq \begin{bmatrix} \mathbf{B} & \mathbf{AB} & \mathbf{A^2B} & \cdots & \mathbf{A^{n-1}B} \end{bmatrix} \tag{17.6.14}$$

ii) *The model is completely controllable if and only if* $\boldsymbol{\Gamma}_c[\mathbf{A}, \mathbf{B}]$ *has full row rank.*

Proof

The solution of (17.6.2) is

$$x[N] = (\mathbf{I} + \mathbf{A}\Delta)^N x[0] + \Delta \sum_{i=1}^{N} (\mathbf{I} + \mathbf{A}\Delta)^{i-1} \mathbf{B} u[N - i] \tag{17.6.15}$$

and so

$$x[N] = (\mathbf{I} + \mathbf{A}\Delta)^N x[0] + \Delta \begin{bmatrix} \mathbf{B} & \mathbf{AB} & \mathbf{A^2B} & \cdots & \mathbf{A^{N-1}B} \end{bmatrix} \boldsymbol{\Xi}\, \varsigma \tag{17.6.16}$$

where

$$\boldsymbol{\Xi} \triangleq \begin{bmatrix} \mathbf{I} & \mathbf{I} & \mathbf{I} & \cdots & \mathbf{I} \\ 0 & \Delta\mathbf{I} & 2\Delta\mathbf{I} & & \\ 0 & 0 & \Delta^2\mathbf{I} & & \\ \vdots & \vdots & & \ddots & \vdots \\ 0 & 0 & 0 & \cdots & \Delta^{N-1}\mathbf{I} \end{bmatrix} \qquad \varsigma \triangleq \begin{bmatrix} u[N-1] \\ u[N-2] \\ \vdots \\ u[0] \end{bmatrix} \tag{17.6.17}$$

Because the matrix $\boldsymbol{\Xi}$ *is nonsingular, the result follows from the Cayley-Hamilton theorem (Theorem 17.1 on the preceding page).*

□□□

Example 17.5. *Consider the state space model*

$$\mathbf{A} = \begin{bmatrix} -3 & 1 \\ -2 & 0 \end{bmatrix} ; \quad \mathbf{B} = \begin{bmatrix} 1 \\ -1 \end{bmatrix} \tag{17.6.18}$$

Check for complete controllability.

Solution

The controllability matrix is given by

$$\Gamma_c[\mathbf{A}, \mathbf{B}] = [\mathbf{B}, \mathbf{A}\mathbf{B}] = \begin{bmatrix} 1 & -4 \\ -1 & -2 \end{bmatrix} \tag{17.6.19}$$

Clearly, rank $\Gamma_c[\mathbf{A}, \mathbf{B}] = 2$; thus, the system is completely controllable

□□□

Example 17.6. *Repeat for*

$$\mathbf{A} = \begin{bmatrix} -1 & 1 \\ 2 & 0 \end{bmatrix} ; \quad \mathbf{B} = \begin{bmatrix} 1 \\ -1 \end{bmatrix} \tag{17.6.20}$$

Solution

$$\Gamma_c[\mathbf{A}, \mathbf{B}] = [\mathbf{B}, \mathbf{A}\mathbf{B}] = \begin{bmatrix} 1 & -2 \\ -1 & 2 \end{bmatrix} \tag{17.6.21}$$

Rank $\Gamma_c[\mathbf{A}, \mathbf{B}] = 1 < 2$; thus, the system is not completely controllable.

□□□

Remark 17.3. *Although we have derived the above result by using the delta model, it holds equally well for shift and/or continuous-time models.*

Remark 17.4. *We see that controllability is a black and white issue: a model either is completely controllable or it is not. Clearly, to know that something is uncontrollable is a valuable piece of information. However, to know that something is controllable really tells us nothing about the degree of controllability, i.e., about the difficulty that might be involved in achieving a certain objective. The latter issue lies at the heart of the fundamental design trade-offs in control that were the subject of Chapters 8 and 9.*

Remark 17.5. *We notice that controllability is a system property that does not depend on the particular choice of state variables. This can be easily seen as follows:*

- *Consider a transformation defined by the nonsingular matrix $\mathbf{T}$, for example, $\overline{x}(t) = \mathbf{T}^{-1}x(t)$, yielding*

$$\overline{\mathbf{A}} = \mathbf{T}^{-1}\mathbf{A}\mathbf{T} \tag{17.6.22}$$

$$\overline{\mathbf{A}}^{\,i} = \mathbf{T}^{-1}\mathbf{A}^{i}\mathbf{T} \tag{17.6.23}$$

$$\overline{\mathbf{B}} = \mathbf{T}^{-1}\mathbf{B} \tag{17.6.24}$$

$$\overline{\mathbf{A}}^{\,i}\overline{\mathbf{B}} = \mathbf{T}^{-1}\mathbf{A}^{i}\mathbf{B} \tag{17.6.25}$$

- *From equations (17.6.22) to (17.6.25), it can be seen that*

$$\mathbf{\Gamma}_c[\,\overline{\mathbf{A}}, \overline{\mathbf{B}}\,] = \mathbf{T}^{-1}\mathbf{\Gamma}_c[\mathbf{A}, \mathbf{B}] \tag{17.6.26}$$

which implies that $\mathbf{\Gamma}_c[\,\overline{\mathbf{A}}, \overline{\mathbf{B}}\,]$ *and* $\mathbf{\Gamma}_c[\mathbf{A}, \mathbf{B}]$ *have the same rank.*

If a system is not completely controllable, it can be decomposed into a controllable and a completely uncontrollable subsystem, as explained below.

Lemma 17.1. *Consider a system having* $rank\{\mathbf{\Gamma}_c[\mathbf{A}, \mathbf{B}]\} = k < n$; *then there exists a similarity transformation* T *such that* $\overline{x} = \mathbf{T}^{-1}x$,

$$\overline{\mathbf{A}} = \mathbf{T}^{-1}\mathbf{A}\mathbf{T}; \qquad\qquad \overline{\mathbf{B}} = \mathbf{T}^{-1}\mathbf{B} \tag{17.6.27}$$

and $\overline{\mathbf{A}}, \overline{\mathbf{B}}$ *have the form*

$$\overline{\mathbf{A}} = \begin{bmatrix} \overline{\mathbf{A}}_c & \overline{\mathbf{A}}_{12} \\ 0 & \overline{\mathbf{A}}_{nc} \end{bmatrix}; \qquad\qquad \overline{\mathbf{B}} = \begin{bmatrix} \overline{\mathbf{B}}_c \\ 0 \end{bmatrix} \tag{17.6.28}$$

where $\overline{\mathbf{A}}_c$ *has dimension* k *and* $(\overline{\mathbf{A}}_c, \overline{\mathbf{B}}_c)$ *is completely controllable.*

Proof

Let $\mathbf{T}_1$ *be any basis for the range space of* $\mathbf{\Gamma}_c$. *Choose* $\mathbf{T}_2$ *arbitrarily, except that* $\mathbf{T} = [\mathbf{T}_1\ \mathbf{T}_2]$ *must be nonsingular. Define*

$$\mathbf{T}^{-1} = \begin{bmatrix} \mathbf{S}_1 \\ \mathbf{S}_2 \end{bmatrix} \tag{17.6.29}$$

Then $\mathbf{T}^{-1}\mathbf{T} = \mathbf{I}$ *implies*

$$\mathbf{S}_1 \mathbf{T}_1 = \mathbf{I} \qquad \mathbf{S}_1 \mathbf{T}_2 = 0 \qquad \mathbf{S}_2 \mathbf{T}_1 = 0 \qquad \mathbf{S}_2 \mathbf{T}_2 = \mathbf{I} \qquad (17.6.30)$$

After the applying of the transformation (17.6.27), the transformed system has the form:

$$\overline{\mathbf{B}} = \begin{bmatrix} \mathbf{S}_1 \\ \mathbf{S}_2 \end{bmatrix} \mathbf{B} = \begin{bmatrix} \mathbf{S}_1 \mathbf{B} \\ \mathbf{0} \end{bmatrix} \qquad (17.6.31)$$

$$\overline{\mathbf{A}} = \begin{bmatrix} \mathbf{S}_1 \\ \mathbf{S}_2 \end{bmatrix} \mathbf{A} [\mathbf{T}_1 \ \mathbf{T}_2] = \begin{bmatrix} \mathbf{S}_1 \mathbf{A} \mathbf{T}_1 & \mathbf{S}_1 \mathbf{A} \mathbf{T}_2 \\ 0 & \mathbf{S}_2 \mathbf{A} \mathbf{T}_2 \end{bmatrix} \qquad (17.6.32)$$

The zero entries in (17.6.31) follow, because the values of B belong to the range space of $\mathbf{\Gamma}_c$ and $\mathbf{S}_2 \mathbf{T}_1 = 0$. Similarly, the zero elements in (17.6.32) are a consequence of the fact that the columns of $\mathbf{A}\mathbf{T}_1$ belong to the range space of $\mathbf{\Gamma}_c$ by the Cayley-Hamilton theorem and that $S_2 \mathbf{T}_1 = 0$. This establishes (17.6.28). Furthermore, we see that

$$\begin{bmatrix} \overline{\mathbf{B}}_c & \overline{\mathbf{A}}_c \overline{\mathbf{B}}_c & \cdots & \overline{\mathbf{A}}_c{}^{n-1} \overline{\mathbf{B}}_c \\ 0 & \cdots & \cdots & 0 \end{bmatrix} = \begin{bmatrix} \mathbf{\Gamma}_c [\ \overline{\mathbf{A}}_c, \overline{\mathbf{B}}_c \ ] & \overline{\mathbf{A}}_c{}^k \overline{\mathbf{B}}_c & \cdots & \overline{\mathbf{A}}_c{}^{n-1} \overline{\mathbf{B}}_c \\ 0 & & \cdots & \cdots & 0 \end{bmatrix}$$

$$\qquad (17.6.33)$$

$$= \mathbf{T}^{-1} \mathbf{\Gamma}_c [\mathbf{A}, \mathbf{B}] \qquad (17.6.34)$$

Hence,

$$rank\{\mathbf{\Gamma}_c [\ \overline{\mathbf{A}}_c, \overline{\mathbf{B}}_c \ ]\} = rank\{\mathbf{\Gamma}_c [\mathbf{A}, \mathbf{B}]\} = k \qquad (17.6.35)$$

□□□

The above result has important consequences regarding control. To appreciate this, express the (transformed) state and output equations in partitioned form as

$$\delta \begin{bmatrix} \overline{x}_c[k] \\ \overline{x}_{nc}[k] \end{bmatrix} = \begin{bmatrix} \overline{\mathbf{A}}_c & \overline{\mathbf{A}}_{12} \\ \mathbf{0} & \overline{\mathbf{A}}_{nc} \end{bmatrix} \begin{bmatrix} \overline{x}_c[k] \\ \overline{x}_{nc}[k] \end{bmatrix} + \begin{bmatrix} \overline{\mathbf{B}}_c \\ \mathbf{0} \end{bmatrix} u[k] \qquad (17.6.36)$$

$$y[k] = \begin{bmatrix} \overline{\mathbf{C}}_c & \overline{\mathbf{C}}_{nc} \end{bmatrix} \begin{bmatrix} \overline{x}_c[k] \\ \overline{x}_{nc}[k] \end{bmatrix} + \mathbf{D}u[k] \qquad (17.6.37)$$

A pictorial representation of these equations is shown in Figure 17.1.

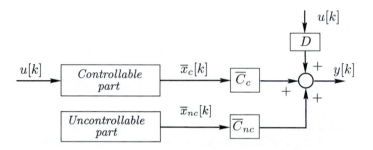

Figure 17.1. Controllable-uncontrollable decomposition

Equations (17.6.36) and (17.6.37) and Figure 17.1 show that caution must be exercised when controlling a system (or designing a controller with a model that is not completely controllable), because the output has a component $\overline{\mathbf{C}}_{nc}\overline{x}_{nc}[k]$ that does not depend on the manipulated input $u[k]$. For example, if $\overline{x}_{nc}[k]$ grows unboundedly (one or more eigenvalues of $\overline{\mathbf{A}}_{nc}$ are outside the stability region), then the plant cannot be stabilized by feedback by using the input $u[k]$. Nevertheless there are situations in which the deliberate use of models with uncontrollable subsets is beneficial; see Remark 17.6 on page 507.

Definition 17.3. *The controllable subspace of a state space model is composed of all states generated through every possible linear combination of the states in $\overline{x}_c$. The stability of this subspace is determined by the location of the eigenvalues of $\overline{\mathbf{A}}_c$.*

Definition 17.4. *The uncontrollable subspace of a state space model is composed of all states generated through every possible linear combination of the states in $\overline{x}_{nc}$. The stability of this subspace is determined by the location of the eigenvalues of $\overline{\mathbf{A}}_{nc}$.*

Definition 17.5. *A state space model is said to be "stabilizable" if its uncontrollable subspace is stable.*

A fact that we will find useful in what follows is that, if the system given by equations (17.6.2) and (17.6.3) is completely controllable, there exist similarity transformations that convert it into special forms, known as *canonical forms*. This is established in the following two lemmas.

Lemma 17.2. *Consider a completely reachable state space model for a SISO system. Then there exists a similarity transformation that converts the state space model into the following controllability-canonical form:*

$$
\mathbf{A}' = \begin{bmatrix} 0 & 0 & \cdots & 0 & -\alpha_0 \\ 1 & 0 & \cdots & 0 & -\alpha_1 \\ 0 & 1 & \cdots & 0 & -\alpha_2 \\ \vdots & \vdots & \ddots & \vdots & \vdots \\ 0 & 0 & \cdots & 1 & -\alpha_{n-1} \end{bmatrix} \qquad \mathbf{B}' = \begin{bmatrix} 1 \\ 0 \\ 0 \\ \vdots \\ 0 \end{bmatrix} \qquad (17.6.38)
$$

where
$$\lambda^n + \alpha_{n-1}\lambda_{n-1} + \cdots + \alpha_1\lambda + \alpha_0 = \det(\lambda\mathbf{I} - \mathbf{A}) \text{ is the characteristic polynomial}$$
of $\mathbf{A}$.

Proof

We first note that $\mathbf{\Gamma}_c(\mathbf{A}', \mathbf{B}') = I$. Using this result and equation (17.6.26), we have that the similarity transformation should be given by $\mathbf{T} = \mathbf{\Gamma}_c(\mathbf{A}, \mathbf{B})$. We now verify that this transformation yields (17.6.38). By comparing the first column in both sides of equation (17.6.26), it is clear that this choice of T transforms B into B', as given by (17.6.38). We next note that

$$
\mathbf{\Gamma}_c[\mathbf{A}, \mathbf{B}]\mathbf{A}' = \begin{bmatrix} \mathbf{A}\mathbf{B} & \mathbf{A}^2\mathbf{B} & \cdots & \mathbf{A}^{n-1}\mathbf{B} & -(\alpha_0 I + \alpha_1\mathbf{A} + \cdots + \alpha_{n-1}\mathbf{A}^{n-1})\mathbf{B} \end{bmatrix}
$$
$$(17.6.39)$$

By applying the Cayley-Hamilton Theorem, we have that $-(\alpha_0 I + \alpha_1\mathbf{A} + \cdots + \alpha_{n-1}\mathbf{A}^{n-1}) = \mathbf{A}^n$. Then,

$$
\mathbf{\Gamma}_c[\mathbf{A}, \mathbf{B}]\mathbf{A}' = \mathbf{A}\mathbf{\Gamma}_c[\mathbf{A}, \mathbf{B}] \Rightarrow \mathbf{\Gamma}_c[\mathbf{A}, \mathbf{B}]\mathbf{A}'\mathbf{\Gamma}_c[\mathbf{A}, \mathbf{B}]^{-1} = \mathbf{A} \qquad (17.6.40)
$$

which completes the proof.

□□□

Lemma 17.3. *Consider a completely controllable state space model for a SISO system. Then there exists a similarity transformation that converts the state space model into the following controller-canonical form:*

$$
\mathbf{A}'' = \begin{bmatrix} -\alpha_{n-1} & -\alpha_{n-2} & \cdots & -\alpha_1 & -\alpha_0 \\ 1 & 0 & \cdots & 0 & 0 \\ 0 & 1 & \cdots & 0 & 0 \\ \vdots & \vdots & \ddots & \vdots & \vdots \\ 0 & 0 & \cdots & 1 & 0 \end{bmatrix} \qquad \mathbf{B}'' = \begin{bmatrix} 1 \\ 0 \\ 0 \\ \vdots \\ 0 \end{bmatrix} \qquad (17.6.41)
$$

where

$$\lambda^n + \alpha_{n-1}\lambda_{n-1} + \cdots + \alpha_1\lambda + \alpha_0 = \det(\lambda\mathbf{I} - \mathbf{A}) \text{ is the characteristic polynomial}$$

of $\mathbf{A}$.

Proof

Using (17.6.26) and the controllability property of the SISO system, we note that there exists a unique similarity transformation to take the model (17.6.41) into the model (17.6.38). The appropriate similarity transformation is

$$x = \mathbf{\Gamma}_c[\mathbf{A}, \mathbf{B}]\mathbf{M}x'' \qquad (17.6.42)$$

where

$$
\mathbf{M} \triangleq \begin{bmatrix} 1 & \alpha_{n-1} & \cdots & \alpha_2 & \alpha_1 \\ & \ddots & & & \alpha_2 \\ & & \ddots & & \vdots \\ & & & \ddots & \alpha_{n-1} \\ & & & & 1 \end{bmatrix} \qquad (17.6.43)
$$

The similarity transformation gives

$$\mathbf{A}'' = \mathbf{M}^{-1}\mathbf{\Gamma}_c^{-1}[\mathbf{A}, \mathbf{B}]\mathbf{A}\mathbf{\Gamma}_c[\mathbf{A}, \mathbf{B}]\mathbf{M} \qquad\qquad \mathbf{B}'' = \mathbf{M}^{-1}\mathbf{\Gamma}_c^{-1}[\mathbf{A}, \mathbf{B}]\mathbf{B} \quad (17.6.44)$$

or

$$\mathbf{A}'' = \mathbf{M}^{-1}\mathbf{A}'\mathbf{M} \qquad\qquad \mathbf{B}'' = \mathbf{M}^{-1}\mathbf{B}' \qquad (17.6.45)$$

where $\mathbf{A}' = \mathbf{\Gamma}_c^{-1}[\mathbf{A},\mathbf{B}]\mathbf{A}\mathbf{\Gamma}_c[\mathbf{A},\mathbf{B}]$, $\mathbf{B}' = \mathbf{\Gamma}_c^{-1}[\mathbf{A},\mathbf{B}]\mathbf{B}$ *is in controllability form, in view of Lemma 17.2 on page 505. From the definition of* $\mathbf{M}$, *it can be seen that* $\mathbf{M}^{-1}$ *is also an upper-triangular matrix with unit diagonal elements. Thus*

$$\mathbf{B}'' = \mathbf{M}^{-1}\mathbf{B}' = \begin{bmatrix} 1 \\ 0 \\ \vdots \\ 0 \end{bmatrix} \qquad (17.6.46)$$

The rest of the proof follows upon separately computing $\mathbf{M}\mathbf{A}''$ *and* $\mathbf{A}'\mathbf{M}$ *and verifying that the two matrices are equal. Because* $\mathbf{M}$ *is a nonsingular matrix, the result follows.*

□□□

Remark 17.6. *Finally, we remark that, as we have seen in Chapter 10, it is very common indeed to employ uncontrollable models in control-system design. This is because they are a convenient way of describing various commonly occurring disturbances. For example, a constant disturbance can be modeled by the following state space model:*

$$\dot{x}_d = 0 \qquad (17.6.47)$$

which is readily seen to be nonstabilizable.

More generally, disturbances are frequently described by models of the form

$$\xi\langle x_d \rangle = \mathbf{A}_d x_d \qquad (17.6.48)$$

where ξ *is an appropriate operator (differential, shift, delta) and* $\mathbf{A}_d$ *is a matrix that typically, has roots on the stability boundary.*

These models can be combined with other models to include the disturbances in the total system description. For example, a system with an input disturbance might be modeled with $\mathbf{A}'$ *or* $\mathbf{B}'$, *given by*

$$\mathbf{A}' = \begin{bmatrix} \mathbf{A} & \mathbf{B} \\ \mathbf{0} & \mathbf{A}_d \end{bmatrix} ; \qquad \mathbf{B}' = \begin{bmatrix} \mathbf{B} \\ \mathbf{0} \end{bmatrix} \qquad (17.6.49)$$

17.7   Observability and Detectability

Consider again the state space model {(17.6.2), (17.6.3)}.

In general, the dimension of the observed output, y, can be less than the dimension of the state, x. However, one might conjecture that, if one observed the output over some nonvanishing time interval, then this might tell us something about the state. The associated properties are called observability (or reconstructability). A related issue is that of detectability. We begin with observability.

Observability is concerned with the issue of what can be said about the state when one is given measurements of the plant output.

A formal definition is as follows:

Definition 17.6. *The state $x_o \neq 0$ is said to be "unobservable" if, given $x(0) = x_o$, and $u[k] = 0$ for $k \geq 0$, then $y[k] = 0$ for $k \geq 0$. The system is said to be completely observable if there exists no nonzero initial state that it is unobservable.*

Remark 17.7. *A concept related to observability is that of reconstructability. This concept is sometimes used in discrete-time systems. Reconstructability is concerned with what can be said about $x(T)$, on the basis of the past values of the output, i.e., $y[k]$ for $0 \leq k \leq T$. For linear time-invariant continuous-time systems, the distinction between observability and reconstructability is unnecessary. However, the following example illustrates that, in discrete time, the two concepts are different. Consider*

$$x[k+1] = 0 \qquad\qquad x[0] = x_o \qquad\qquad (17.7.1)$$
$$y[k] = 0 \qquad\qquad\qquad\qquad (17.7.2)$$

This system is clearly reconstructible for all $T \geq 1$, because we know for certain that $x[T] = 0$ for $T \geq 1$. However, it is completely unobservable, because $y[k] = 0$ $\forall k$, irrespective of the value of x_o.

□□□

In view of the subtle difference between observability and reconstructability, we will use the term "observability" in the sequel to cover the stronger of the two concepts.

A test for observability of a system is established in the following theorem.

Theorem 17.3. *Consider the the state model {(17.6.2) and (17.6.3)}.*

i) The set of all unobservable states is equal to the null space of the observability matrix $\mathbf{\Gamma}_o[\mathbf{A}, \mathbf{C}]$, where

$$\mathbf{\Gamma}_o[\mathbf{A},\mathbf{C}] \triangleq \begin{bmatrix} \mathbf{C} \\ \mathbf{CA} \\ \vdots \\ \mathbf{CA}^{n-1} \end{bmatrix} \tag{17.7.3}$$

ii) *The system is completely observable if and only if* $\mathbf{\Gamma}_o[\mathbf{A},\mathbf{C}]$ *has full column rank* n.

Proof

The result holds for both continuous and discrete time; here, we will focus on the discrete-time (delta-operator domain) proof. For zero input, the solution of equation (17.6.2) is

$$x[k+1] = (I + \Delta\mathbf{A})x[k] \tag{17.7.4}$$

Hence, for $N > n - 1$,

$$Y_N \triangleq \begin{bmatrix} y[0] \\ y[1] \\ y[2] \\ \vdots \\ y[N] \end{bmatrix} = \begin{bmatrix} \mathbf{C} \\ \mathbf{C}(\mathbf{I}+\Delta\mathbf{A}) \\ \mathbf{C}(\mathbf{I}+\Delta\mathbf{A})^2 \\ \vdots \\ \mathbf{C}(\mathbf{I}+\Delta\mathbf{A})^N \end{bmatrix} x[0] \tag{17.7.5}$$

$$= \begin{bmatrix} \mathbf{I} & 0 & 0 & \cdots & 0 \\ \mathbf{I} & \Delta\mathbf{I} & 0 & \cdots & 0 \\ \mathbf{I} & 2\Delta\mathbf{I} & \Delta^2\mathbf{I} & \cdots & 0 \\ \vdots & \vdots & \vdots & \ddots & \vdots \\ & & & & \Delta^N\mathbf{I} \end{bmatrix} \begin{bmatrix} \mathbf{C} \\ \mathbf{CA} \\ \mathbf{CA}^2 \\ \vdots \\ \mathbf{CA}^N \end{bmatrix} x[0] \tag{17.7.6}$$

The first matrix in (17.7.6) is nonsingular, and hence Y_N is zero if and only if $x[0]$ lies in the null space of $[\mathbf{C}^T \ \mathbf{A}^T\mathbf{C}^T \ \dots (\mathbf{A}^N)^T\mathbf{C}^T]^T$. From the Cayley-Hamilton Theorem, this is the same as the null space of $\mathbf{\Gamma}_o[\mathbf{A},\mathbf{C}]$.

□□□

As for controllability, the above result also applies to continuous-time and discrete (shift) operator models.

Example 17.7. *Consider the following state space model:*

$$\mathbf{A} = \begin{bmatrix} -3 & -2 \\ 1 & 0 \end{bmatrix}; \quad \mathbf{B} = \begin{bmatrix} 1 \\ 0 \end{bmatrix}; \quad \mathbf{C} = \begin{bmatrix} 1 & -1 \end{bmatrix} \tag{17.7.7}$$

Check on observability.

Solution

$$\mathbf{\Gamma}_o[\mathbf{A},\mathbf{C}] = \begin{bmatrix} \mathbf{C} \\ \mathbf{CA} \end{bmatrix} = \begin{bmatrix} 1 & -1 \\ -4 & -2 \end{bmatrix} \tag{17.7.8}$$

Hence, rank $\mathbf{\Gamma}_o[\mathbf{A},\mathbf{C}] = 2$, and the system is completely observable.

□□□

Example 17.8. *Repeat for*

$$\mathbf{A} = \begin{bmatrix} -1 & -2 \\ 1 & 0 \end{bmatrix}; \quad \mathbf{B} = \begin{bmatrix} 1 \\ 0 \end{bmatrix}; \quad \mathbf{C} = \begin{bmatrix} 1 & -1 \end{bmatrix} \tag{17.7.9}$$

Solution

$$\mathbf{\Gamma}_o[\mathbf{A},\mathbf{C}] = \begin{bmatrix} 1 & -1 \\ -2 & -2 \end{bmatrix} \tag{17.7.10}$$

Hence, rank $\mathbf{\Gamma}_o[\mathbf{A},\mathbf{C}] = 1 < 2$, and the system is not completely observable.

□□□

We see a remarkable similarity between the results in Theorem 17.2 on page 499 and in Theorem 17.3 on page 508. We can formalize this as follows:

Theorem 17.4 (Duality). *Consider a state space model described by the 4-tuple* $(\mathbf{A}, \mathbf{B}, \mathbf{C}, \mathbf{D})$. *Then the system is completely controllable if and only if the dual system* $(\mathbf{A}^T, \mathbf{C}^T, \mathbf{B}^T, \mathbf{D}^T)$ *is completely observable.*

Proof

Immediate from Theorems 17.2 on page 499 and 17.3 on page 508.

□□□

The above theorem can often be used to go from a result on controllability to one on observability, and vice versa. For example, the dual of Lemma 17.1 on page 502 is the following:

Lemma 17.4. *If* $rank\{\mathbf{\Gamma}_o[\mathbf{A}, \mathbf{C}]\} = k < n$, *there exists a similarity transformation* $\mathbf{T}$ *such that with* $\overline{x} = \mathbf{T}^{-1}x$, $\overline{\mathbf{A}} = \mathbf{T}^{-1}\mathbf{A}\mathbf{T}$, $\overline{\mathbf{C}} = \mathbf{C}\mathbf{T}$, *then* $\overline{\mathbf{C}}$ *and* $\overline{\mathbf{A}}$ *take the form*

$$
\overline{\mathbf{A}} = \begin{bmatrix} \overline{\mathbf{A}}_o & 0 \\ \overline{\mathbf{A}}_{21} & \overline{\mathbf{A}}_{no} \end{bmatrix} \qquad \overline{\mathbf{C}} = \begin{bmatrix} \overline{\mathbf{C}}_o & 0 \end{bmatrix} \tag{17.7.11}
$$

where $\overline{\mathbf{A}}_o$ *has dimension* k *and the pair* $(\overline{\mathbf{C}}_o, \overline{\mathbf{A}}_o)$ *is completely observable.*

Proof

We consider the pair $(\mathbf{A}', \mathbf{B}') \triangleq (\mathbf{A}^T, \mathbf{C}^T)$, *and apply Lemma 17.1 on page 502. The result follows by duality.*

□□□

The above result has a relevance similar to that of the controllability property and the associated decomposition. To appreciate this, we apply the dual of Lemma 17.1 on page 502 to express the (transformed) state and output equations in partitioned form as

$$
\delta \begin{bmatrix} \overline{x}_o[k] \\ \overline{x}_{no}[k] \end{bmatrix} = \begin{bmatrix} \overline{\mathbf{A}}_o & 0 \\ \overline{\mathbf{A}}_{21} & \overline{\mathbf{A}}_{n0} \end{bmatrix} \begin{bmatrix} \overline{x}_o[k] \\ \overline{x}_{no}[k] \end{bmatrix} + \begin{bmatrix} \overline{\mathbf{B}}_o \\ \overline{\mathbf{B}}_{no} \end{bmatrix} u[k] \tag{17.7.12}
$$

$$
y[k] = \begin{bmatrix} \overline{\mathbf{C}}_o & 0 \end{bmatrix} \begin{bmatrix} \overline{x}_o[k] \\ \overline{x}_{no}[k] \end{bmatrix} + \mathbf{D}u[k] \tag{17.7.13}
$$

A pictorial description of these equations is shown in Figure 17.2 on the following page.

Equations (17.7.12) and (17.7.13), together with Figure 17.2, show that problems can arise when solely using the output to control the plant, or designing a controller from a model, which is not completely observable. This is due to the fact that the output does not provide *any* information regarding $\overline{x}_{no}[k]$. For example, if $\overline{x}_{nc}[k]$ grows unboundedly, then the plant cannot be stabilized by output feedback.

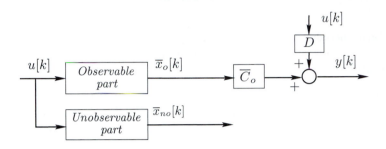

Figure 17.2. Observable-unobservable decomposition

Definition 17.7. *The observable subspace of a plant is composed of all states generated through every possible linear combination of the states in $\overline{x}_o$. The stability of this subspace is determined by the location of the eigenvalues of $\overline{\mathbf{A}}_o$.*

Definition 17.8. *The unobservable subspace of a plant is composed of all states generated through every possible linear combination of the states in $\overline{x}_{no}$. The stability of this subspace is determined by the location of the eigenvalues of $\overline{\mathbf{A}}_{no}$.*

Definition 17.9. *A plant is said to be "detectable" if its unobservable subspace is stable.*

In section §17.6 we remarked that noncontrollable (indeed nonstabilizable) models are frequently used in control-system design. This is not true for nondetectable models. Essentially all models used in the sequel can be taken to be detectable, without loss of generality.

There are also duals of the canonical forms given in Lemmas 17.2 on page 505 and 17.3 on page 506. For example, the dual of Lemma 17.3 on page 506 is

Lemma 17.5. *Consider a completely observable SISO system given by (17.6.2) and (17.6.3). Then there exists a similarity transformation that converts the model to the observer-canonical form*

$$\delta x'[k] = \begin{bmatrix} -\alpha_{n-1} & 1 & & \\ \vdots & & \ddots & \\ \vdots & & & 1 \\ -\alpha_0 & 0 & & 0 \end{bmatrix} x'[k] + \begin{bmatrix} b_{n-1} \\ \vdots \\ \vdots \\ b_0 \end{bmatrix} u[k] \qquad (17.7.14)$$

$$y[k] = \begin{bmatrix} 1 & 0 & \cdots & 0 \end{bmatrix} x'[k] + \mathbf{D} u[k] \qquad (17.7.15)$$

Proof

We note that $(\mathbf{C}, \mathbf{A})$ observable $\Leftrightarrow (\mathbf{A}^T, \mathbf{C}^T)$ controllable. Hence, we consider the dual system

$$\delta x[k] = \mathbf{A}^T x[k] + \mathbf{C}^T u[k] \qquad (17.7.16)$$

$$y[k] = \mathbf{B}^T x[k] + \mathbf{D} u[k] \qquad (17.7.17)$$

and convert this to controller form as in Lemma 17.3 on page 506. Finally, a second application of duality gives the form $\{(17.7.14),(17.7.15)\}$.

□□□

Remark 17.8. *We note that $\{(17.7.14),(17.7.15)\}$ is another way of writing an nth-order difference equation. The continuous time version of $\{(17.7.14),(17.7.15)\}$ is an alternative way of writing an nth-order differential equation.*

17.8   Canonical Decomposition

Further insight into the structure of linear dynamical systems is obtained by considering those systems that are only partially observable or controllable. These systems can be separated into completely observable and completely controllable systems.

The two results of Lemmas 17.1 on page 502 and 17.4 on page 511 can be combined as follows.

Theorem 17.5 (Canonical Decomposition Theorem). *Consider a system described in state space form. Then, there always exists a similarity transformation $\mathbf{T}$ such that the transformed model for $\bar{x} = \mathbf{T}^{-1} x$ takes the form*

$$\overline{\mathbf{A}} = \begin{bmatrix} \overline{\mathbf{A}}_{co} & 0 & \overline{\mathbf{A}}_{13} & 0 \\ \overline{\mathbf{A}}_{21} & \overline{\mathbf{A}}_{22} & \overline{\mathbf{A}}_{23} & \overline{\mathbf{A}}_{24} \\ 0 & 0 & \overline{\mathbf{A}}_{33} & 0 \\ 0 & 0 & \overline{\mathbf{A}}_{34} & \overline{\mathbf{A}}_{44} \end{bmatrix} ; \quad \overline{\mathbf{B}} = \begin{bmatrix} \overline{\mathbf{B}}_1 \\ \overline{\mathbf{B}}_2 \\ 0 \\ 0 \end{bmatrix} ; \quad \overline{\mathbf{C}} = \begin{bmatrix} \overline{\mathbf{C}}_1 & 0 & \overline{\mathbf{C}}_2 & 0 \end{bmatrix}$$

$$(17.8.1)$$

where

i) *The subsystem* $[\overline{\mathbf{A}}_{co}, \overline{\mathbf{B}}_1, \overline{\mathbf{C}}_1]$ *is both completely controllable and completely observable and has the same transfer function as the original system. (See Lemma 17.6 on the next page.)*

ii) *The subsystem*

$$\begin{bmatrix} \overline{\mathbf{A}}_{co} & 0 \\ \overline{\mathbf{A}}_{21} & \overline{\mathbf{A}}_{22} \end{bmatrix}, \begin{bmatrix} \overline{\mathbf{B}}_1 \\ \overline{\mathbf{B}}_2 \end{bmatrix}, \begin{bmatrix} \overline{\mathbf{C}}_1 & 0 \end{bmatrix} \qquad (17.8.2)$$

is completely controllable.

iii) *The subsystem*

$$\begin{bmatrix} \overline{\mathbf{A}}_{co} & \overline{\mathbf{A}}_{13} \\ 0 & \overline{\mathbf{A}}_{33} \end{bmatrix}, \begin{bmatrix} \overline{\mathbf{B}}_1 \\ 0 \end{bmatrix}, \begin{bmatrix} \overline{\mathbf{C}}_1 & \overline{\mathbf{C}}_2 \end{bmatrix} \qquad (17.8.3)$$

is completely observable.

Proof

Consequence of Lemmas 17.1 on page 502 and 17.4 on page 511.

□□□

The canonical decomposition described in Theorem 17.5 on the preceding page leads to

Lemma 17.6. *Consider the transfer-function matrix* $\mathbf{H}(s)$ *given by*

$$Y(s) = \mathbf{H}(s)U(s) \qquad (17.8.4)$$

Then

$$\mathbf{H} = \mathbf{C}(s\mathbf{I} - \mathbf{A})^{-1}\mathbf{B} + \mathbf{D} = \overline{\mathbf{C}}_1(s\mathbf{I} - \overline{\mathbf{A}}_{co})^{-1}\overline{\mathbf{B}}_1 + \mathbf{D} \qquad (17.8.5)$$

where $\overline{\mathbf{C}}_1$, $\overline{\mathbf{A}}_{co}$, *and* $\overline{\mathbf{B}}_1$ *are as in equations (17.8.1).*

Proof

Straightforward from the definitions of the matrices, followed by the application of the formula for inversion of triangular partitioned matrices.

□□□

Remark 17.9. *Lemma 17.6 shows that the uncontrollable and the unobservable parts of a linear system do not appear in the transfer function. Conversely, given a transfer function, it is possible to generate a state space description that is both completely controllable and observable. We then say that this state description is a minimal realization of the transfer function. As mentioned earlier, nonminimal models are frequently used in control-system design to include disturbances.*

Remark 17.10. *If* $\mathbf{M}$ *is any square matrix, and we denote by* $\Lambda\{\mathbf{M}\}$ *the set of eigenvalues of* $\mathbf{M}$, *then*

$$\Lambda\{\overline{\mathbf{A}}\} = \Lambda\{\overline{\mathbf{A}}_{co}\} \cup \Lambda\{\overline{\mathbf{A}}_{22}\} \cup \Lambda\{\overline{\mathbf{A}}_{33}\} \cup \Lambda\{\overline{\mathbf{A}}_{44}\} \qquad (17.8.6)$$

where

$\Lambda\{\overline{\mathbf{A}}\}$ *Eigenvalues of the system*

$\Lambda\{\overline{\mathbf{A}}_{co}\}$ *Eigenvalues of the controllable and observable subsystem*

$\Lambda\{\overline{\mathbf{A}}_{22}\}$ *Eigenvalues of the controllable but unobservable subsystem*

$\Lambda\{\overline{\mathbf{A}}_{33}\}$ *Eigenvalues of the uncontrollable but observable subsystem*

$\Lambda\{\overline{\mathbf{A}}_{44}\}$ *Eigenvalues of the uncontrollable and unobservable subsystem*

    Although the definition of the system properties refers to the states in certain subsystems, it is customary to use the same terminology for the eigenvalues and natural modes associated with those subsystems. For example, if a controllable but unobservable continuous-time subsystem has two distinct eigenvalues, λ_a *and* λ_b, *then it is usually said that the modes $e^{\lambda_a t}$ and $e^{\lambda_b t}$ and the eigenvalues λ_a and λ_b are controllable but unobservable.*

Remark 17.11. *We observe that controllability for a given system depends on the structure of the input ports: where, in the system, the manipulable inputs are applied. Thus, the states of a given subsystem might be uncontrollable for one given input but completely controllable for another. This distinction is of fundamental importance in control-system design, because not all plant inputs can be manipulated (consider, for example, disturbances) to steer the plant to reach certain states.*

    Similarly, the observability property depends on which outputs are being considered. Certain states may be unobservable from a given output, but they may be completely observable from some other output. This also has a significant impact on output-feedback control systems, because some states might not appear in the plant output being measured and fed back. However, they could appear in crucial internal variables and thus be important to the control problem.

17.9   Pole-Zero Cancellation and System Properties

The system properties described in the previous sections are also intimately related to issues of pole-zero cancellations. The latter issue was discussed in Chapter 15 (in particular, Remark 15.4 on page 431), when we made a distinction between analytic (or exact) cancellations and implementation cancellations. To facilitate the subsequent development, we introduce the following test, which is useful for studying issues of controllability and observability.

Lemma 17.7 (PBH Test). *Consider a state space model* $(\mathbf{A}, \mathbf{B}, \mathbf{C})$.

(i) *The system is not completely observable if and only if there exist a nonzero vector* $x \in \mathbb{C}^n$ *and a scalar* $\lambda \in \mathbb{C}$ *such that*

$$\mathbf{A}x = \lambda x \tag{17.9.1}$$
$$\mathbf{C}x = 0$$

(ii) *The system is not completely controllable if and only if there exist a nonzero vector* $x \in \mathbb{C}^n$ *and a scalar* $\lambda \in \mathbb{C}$ *such that*

$$x^T \mathbf{A} = \lambda x^T \tag{17.9.2}$$
$$x^T \mathbf{B} = 0$$

Proof

(i) **If:** *The given conditions imply*

$$
\begin{bmatrix} \mathbf{C} \\ \mathbf{CA} \\ \vdots \\ \mathbf{CA}^{n-1} \end{bmatrix} x = 0 \tag{17.9.3}
$$

This implies that the system is not completely observable.

Only if: *Let us assume the system is not completely observable. It can then be decomposed as in* $\{(17.7.12), (17.7.13)\}$. *Now, choose* λ *as an eigenvalue of* $\overline{\mathbf{A}}_{no}$, *and define*

$$
x = \begin{bmatrix} 0 \\ \overline{x}_{no} \end{bmatrix} \tag{17.9.4}
$$

where $\overline{x}_{n0} \neq 0$ *is an eigenvector of* $\overline{\mathbf{A}}_{n0}$ *corresponding to* λ. *Clearly, we have*

$$
\begin{bmatrix} \overline{\mathbf{A}}_0 & \mathbf{0} \\ \overline{\mathbf{A}}_{21} & \overline{\mathbf{A}}_{no} \end{bmatrix} \begin{bmatrix} 0 \\ \overline{x}_{no} \end{bmatrix} = \lambda \begin{bmatrix} 0 \\ \overline{x}_{no} \end{bmatrix} \tag{17.9.5}
$$

and

$$
\begin{bmatrix} \overline{\mathbf{C}}_o & 0 \end{bmatrix} \begin{bmatrix} 0 \\ \overline{x}_{no} \end{bmatrix} = 0 \tag{17.9.6}
$$

(ii) Similar to part (i), using properties of controllability.

□□□

We will next use the preceding result to study the system properties of cascaded systems.

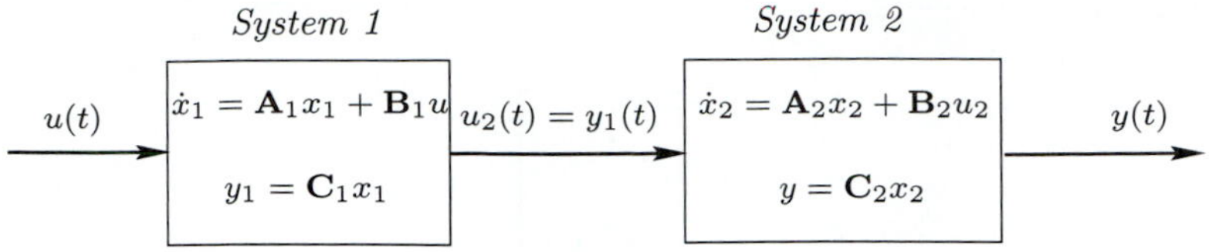

Figure 17.3. Pole-zero cancellation

Consider the cascaded system in Figure 17.3.

We assume that $u(t), u_2(t), y_1(t), y(t) \in \mathbb{R}$, that both subsystems are minimal, and that

System 1   has a zero at α and pole at β,

System 2   has a pole at α and zero at β.

Then the combined model has the property that

(a) the system pole at β is not observable from Y, and

(b) the system pole at α is not controllable from u.

To establish this, we note that the combined model has the state space description

$$\begin{bmatrix} \mathbf{A}_1 & 0 \\ \mathbf{B}_2\mathbf{C}_1 & \mathbf{A}_2 \end{bmatrix} \; ; \; \begin{bmatrix} \mathbf{B}_1 \\ 0 \end{bmatrix} \; ; \; \begin{bmatrix} 0 & \mathbf{C}_2 \end{bmatrix} \qquad (17.9.7)$$

To establish (a), let $\lambda = \beta$ and choose $x \neq 0$ defined as follows:

$$x = \begin{bmatrix} x_1^T & x_2^T \end{bmatrix}^T \qquad (17.9.8)$$

$$0 = (\mathbf{A}_1 - \beta\mathbf{I}) x_1 \qquad (17.9.9)$$

$$x_2 = (\beta\mathbf{I} - \mathbf{A}_2)^{-1} \mathbf{B}_2\mathbf{C}_1 x_1 \qquad (17.9.10)$$

Then, clearly,

$$\begin{bmatrix} \mathbf{0} & \mathbf{C}_2 \end{bmatrix} \begin{bmatrix} x_1 \\ x_2 \end{bmatrix} = \mathbf{C}_2 x_2 \qquad (17.9.11)$$

$$= \mathbf{C}_2 \left(\beta \mathbf{I} - \mathbf{A}_2 \right)^{-1} \mathbf{B}_2 \mathbf{C}_1 x_1$$
$$= 0 \qquad \text{because } \beta \text{ is a zero of System 2.}$$

Also from $\{(17.9.8), (17.9.10)\}$,

$$\begin{bmatrix} \mathbf{A}_1 & 0 \\ \mathbf{B}_2\mathbf{C}_1 & \mathbf{A}_2 \end{bmatrix} \begin{bmatrix} x_1 \\ x_2 \end{bmatrix} = \beta \begin{bmatrix} x_1 \\ x_2 \end{bmatrix} \qquad (17.9.12)$$

Thus, by using the PBH test, the composite system is not completely observable. Part (b) is established similarly, by choosing x so that

$$x^T = \begin{bmatrix} x_1^T & x_2^T \end{bmatrix} \qquad (17.9.13)$$
$$0 = x_2^T \left(\alpha \mathbf{I} - \mathbf{A}_2 \right)$$
$$x_1^T = x_2^T \mathbf{B}_2 \mathbf{C}_1 \left(\alpha \mathbf{I} - \mathbf{A}_1 \right)^{-1}$$

Then

$$\begin{bmatrix} x_1^T & x_2^T \end{bmatrix} \begin{bmatrix} \mathbf{B}_1 \\ 0 \end{bmatrix} = x_2^T \mathbf{B}_2 \mathbf{C}_1 \left(\alpha \mathbf{I} - \mathbf{A}_1 \right) \mathbf{B}_1 = 0 \qquad (17.9.14)$$

$$\begin{bmatrix} x_1^T & x_2^T \end{bmatrix} \begin{bmatrix} \mathbf{A}_1 & 0 \\ \mathbf{B}_2\mathbf{C}_1 & \mathbf{A}_2 \end{bmatrix} = \alpha \begin{bmatrix} x_1^T & x_2^T \end{bmatrix} \qquad (17.9.15)$$

Thus, by using the PBH test, the composite system is not completely controllable.

□□□

17.10   Summary

- State variables are system internal variables, upon which a full model for the system behavior can be built. The state variables can be ordered in a state vector.

- Given a linear system, the choice of state variables is not unique–however,

  ○ the minimal dimension of the state vector is a system invariant,

  ○ there exists a nonsingular matrix that defines a similarity transformation between any two state vectors, and

  ○ any designed system output can be expressed as a linear combination of the state variables and the inputs.

- For linear, time-invariant systems, the state space model is expressed in the following equations:

continuous-time systems

$$\dot{x}(t) = \mathbf{A}x(t) + \mathbf{B}u(t) \tag{17.10.1}$$
$$y(t) = \mathbf{C}x(t) + \mathbf{D}u(t) \tag{17.10.2}$$

discrete-time systems, shift form

$$x[k+1] = \mathbf{A}_q x[k] + \mathbf{B}_q u[k] \tag{17.10.3}$$
$$y[k] = \mathbf{C}_q x[k] + \mathbf{D}_q u[k] \tag{17.10.4}$$

discrete-time systems, delta form

$$\delta x[k] = \mathbf{A}_\delta x[k] + \mathbf{B}_\delta u[k] \tag{17.10.5}$$
$$y[k] = \mathbf{C}_\delta x[k] + \mathbf{D}_\delta u[k] \tag{17.10.6}$$

- Stability and natural response characteristics of the system can be studied from the eigenvalues of the matrix $\mathbf{A}$ $(\mathbf{A}_q, \mathbf{A}_\delta)$.

- state space models facilitate the study of certain system properties that are paramount in the solution to the control-design problem. These properties relate to the following questions:

  ○ By proper choice of the input u, can we steer the system state to a desired state (point value)? (controllability)

  ○ If some states are uncontrollable, will these states generate a time-decaying component? (stabilizability)

  ○ If one knows the input, $u(t)$, for $t \geq t_0$, can we infer the state at time $t = t_0$ by measuring the system output, $y(t)$, for $t \geq t_0$? (observability)

  ○ If some of the states are unobservable, do these states generate a time-decaying signal? (detectability)

- Controllability tells us about the feasibility of attempting to control a plant.

- Observability tells us about whether it is possible to know what is happening inside a given system by observing its outputs.

- The above system properties are system invariants. However, changes in the number of inputs, in their injection points, in the number of measurements, and in the choice of variables to be measured can yield different properties.

- A transfer function can always be derived from a state space model.

- A state space model can be built from a transfer-function model. However, only the completely controllable and observable part of the system is described in that state space model. Thus *the transfer-function model might be only a partial description of the system.*

- The properties of individual systems do not necessarily translate unmodified to composed systems. In particular, given two systems completely observable and controllable, their cascaded connection

  - is not completely observable if a pole of the first system coincides with a zero of the second system (pole-zero cancellation),

  - is not detectable if the pole-zero cancellation affects an unstable pole,

  - is not completely controllable if a zero of the first system coincides with a pole of the second system (zero-pole cancellation), and

  - is not stabilizable if the zero-pole cancellation affects a NMP zero.

- This chapter provides a foundation for the design criterion that states that one should never attempt to cancel unstable poles and zeros.

17.11   Further Reading

Linear state space models

Chen, C-T. (1984).   *Introduction to Linear System Theory.*   Holt, Rinehart and Winston.

Kailath, T. (1980). *Linear Systems.* Prentice-Hall, Englewood Cliffs, N.J.

Ogata, K. (1967). *State Space Analysis of Control Systems.* Prentice-Hall, Englewood Cliffs, N.J.

Rosenbrock, H. (1970). *State Space and Multivariable Theory.* Wiley, New York.

Schultz, D. and Melsa, J. (1967).   *State Function and Linear Control Systems.* McGraw-Hill, New York.

Wiberg, D. (1971).   *Theory and Problems of State Space and Linear Systems.* McGraw-Hill, New York.

Controllability, observability and canonical forms

Gilbert, E.G. (1963).   Controllability and observability in multivariable control systems. *SIAM Journal of Control and Optimization*, 1:128-151.

Kailath, T. (1980). *Linear Systems*. Prentice-Hall, Englewood Cliffs, N.J.

Kalman, R.E., Ho, Y.C., and Narendra, K.S. (1962).   Controllability of linear dynamical systems. *Contrib. Differ. Equations*, 1(2):189-213.

Luenberger, D. (1967).   Canonical forms for linear multivariable systems. *IEEE Transactions on Automatic Control*, 12(6):290-293.

Poles and zeros

Brockett, R. (1965). Poles, zeros, and feedback: State space interpretation. *IEEE Transactions on Automatic Control*, 10:129-135, April.

Kailath, T. (1980). *Linear Systems*. Prentice-Hall, Englewood Cliffs, N.J.

Observers

Kwakernaak, H. and Sivan, R. (1972). *Linear Optimal Control Systems*. Wiley-Interscience, New York.

Luenberger, D. (1964).   Observing the state of a linear system.   *IEEE Trans. Military Electr.*, 8:74-80.

Luenberger, D. (1971).   An introduction to observers.   *IEEE Transactions on Automatic Control*, 16(6):596-602.

17.12   Problems for the Reader

Problem 17.1. *Consider the following differential equations describing system input-output relationships:*

$$\frac{d^2y(t)}{dt^2} + 4\frac{dy(t)}{dt} + 5y(t) = 2u(t) \tag{17.12.1}$$

$$\frac{d^3y(t)}{dt^3} - 2y(t) = u(t) \tag{17.12.2}$$

$$\frac{d^2y(t)}{dt^2} - 4\frac{dy(t)}{dt} + 3y(t) = \frac{du(t)}{dt} + 2u(t) \tag{17.12.3}$$

$$\frac{d^2y(t)}{dt^2} + 4\frac{dy(t)}{dt} + 5y(t) = -\frac{d^2u(t)}{dt^2} + \frac{du(t)}{dt} + 2u(t) \tag{17.12.4}$$

For each case, build a steady-state representation.

Problem 17.2. *Build a state space representation for a system having a unit step response given by*

$$y(t) = 2 - 3e^{-t} + e^{-t} - te^{-t} \tag{17.12.5}$$

Problem 17.3. *Build state space representations for the systems with the following transfer functions:*

(a) $\dfrac{9}{s^2 + 4s + 9}$ (b) $\dfrac{2}{(-s+2)^2}$

(c) $\dfrac{-s+8}{(s+2)(s+3)}$ (d) $\dfrac{-s+4}{s+4}$

(e $\dfrac{z-0.5}{(z+0.5)(z-0.4)}$ (f) $\dfrac{0.4}{z^2(z-0.8)}$

Problem 17.4. *Consider a one-d.o.f. control loop, with controller $C(s)$ and nominal plant model $G_o(s)$. The state space models for plant and controller are $(\mathbf{A_o}, \mathbf{B_o}, \mathbf{C_o}, 0)$ and $(\mathbf{A_c}, \mathbf{B_c}, \mathbf{C_c}, \mathbf{D_c})$, respectively.*

17.4.1 *Prove that, for a biproper controller, it is necessary and sufficient that $\mathbf{D_c}$ be a nonzero scalar.*

17.4.2 *Build a state space representation for the whole loop, one having the reference as the input and the plant output as the output of the composite system.*

Problem 17.5. *A continuous-time system has a state space model given by*

$$\mathbf{A} = \begin{bmatrix} -3 & 1 \\ 0 & -2 \end{bmatrix}; \qquad \mathbf{B} = \begin{bmatrix} 1 \\ \epsilon \end{bmatrix} \qquad (17.12.6)$$

$$\mathbf{C} = \begin{bmatrix} c_1 & c_2 \end{bmatrix}; \qquad \mathbf{D} = 0 \qquad (17.12.7)$$

17.5.1 *Compute the system transfer function.*

17.5.2 *Verify that, as $\epsilon \to 0$, the transfer function has a near pole-zero cancellation. Interpret this in terms of controllability.*

17.5.3 *Build a dual example with respect to observability.*

Problem 17.6. *A continuous-time system has a state space model given by*

$$\mathbf{A} = \begin{bmatrix} -3 & 1 \\ -1 & 2 \end{bmatrix}; \qquad \mathbf{B} = \begin{bmatrix} 1 \\ -1 \end{bmatrix} \qquad (17.12.8)$$

$$\mathbf{C} = \begin{bmatrix} 1 & 0 \end{bmatrix}; \qquad \mathbf{D} = 0 \qquad (17.12.9)$$

17.6.1 *Determine whether the system is stable.*

17.6.2 *Compute the transfer function.*

17.6.3 *Investigate the system properties (i.e., controllability, stabilizability, observability, and detectability).*

Problem 17.7. *A discrete-time system has an input-output model given by*

$$\delta^2 y[k] + 5\delta y[k] + 6y[k] = \delta u[k] + u[k] \qquad (17.12.10)$$

Build a state space model.

Problem 17.8. *Consider a generic state space description* $(\mathbf{A}, \mathbf{B}, \mathbf{C}, \mathbf{D})$, *where* $\mathbf{B} = \mathbf{0}$, $\mathbf{D} = \mathbf{0}$.

17.8.1 *Determine a pair of matrices* $\mathbf{A}$ *and* $\mathbf{C}$ *and an initial state* $x(0)$ *such that* $y(t) = 5$, $\forall t \geq 0$.

17.8.2 *Repeat for* $y(t) = -3 + 2\sin(0.5t + \pi/3)$.

17.8.3 *Try to generalize the problem.*

Problem 17.9. *Consider the cascaded system in Figure 17.3 on page 518. Assume that the transfer functions are given by*

$$G_1(s) = \frac{2s+3}{(s-1)(s+5)}; \qquad G_2(s) = \frac{2(s-1)}{s^2+4s+9} \qquad (17.12.11)$$

Prove that, at the output of System 1, there exists an unbounded mode e^t *that cannot be detected at the output of the cascaded system (i.e., at the output of System 2).*

Problem 17.10. *Consider the same cascaded system in Figure 17.3 on page 518. Assume now that the systems have been connected in reverse order:*

$$G_1(s) = \frac{2(s-1)}{s^2+4s+9}; \qquad G_2(s) = \frac{2s+3}{(s-1)(s+5)} \qquad (17.12.12)$$

Prove that, for nonzero initial conditions, the output of System 2 includes an unbounded mode e^t *that is not affected by the input,* $u(t)$, *to the cascaded system.*

Problem 17.11. *Consider a system composed of a continuous-time plant and a zero-order hold at its input. The plant transfer function is given by*

$$G_o(s) = \frac{e^{-0.5s}}{s+0.5} \qquad (17.12.13)$$

Assume that the input and the output of this composed system are sampled every $\Delta[s]$.

17.11.1 *Build a discrete-time state space model for the sampled system, for* $\Delta = 0.5[s]$.

17.11.2 *Repeat for $\Delta = 0.75[s]$.*

17.11.3 *Assume that one would like to build a continuous-time state space model for $G_o(s)$. What would be the main difficulty?*

Problem 17.12. *Consider again Problem 12.4 on page 349. Interpret the apparent loss of a pole at certain sampling rates by constructing a discrete-time state space model of the original continuous-time system and then checking for observability and controllability.*

Chapter 18

SYNTHESIS VIA STATE SPACE METHODS

18.1 Preview

Here, we will give a state space interpretation to many of the results described earlier in Chapters 7 and 15. In a sense, this will duplicate the earlier work. Our reason for doing so, however, is to gain additional insight into linear-feedback systems. Also, it will turn out that the alternative state space formulation carries over more naturally to the multivariable case.

Results to be presented here include the following:

- pole assignment by state-variable feedback

- design of observers to reconstruct missing states from available output measurements

- combining state feedback with an observer

- transfer-function interpretation

- dealing with disturbances in state-variable feedback

- reinterpretation of the affine parameterization of all stabilizing controllers

18.2 Pole Assignment by State Feedback

We will begin by examining the problem of closed-loop pole assignment. For the moment, we make a simplifying assumption that all of the system states are measured. We will remove this assumption in section §18.4. We will also assume that the system is completely controllable. The following result then shows that the closed-loop poles of the system can be arbitrarily assigned by feeding back the state through a suitably chosen constant-gain vector.

527

Lemma 18.1. *Consider the state space nominal model*

$$\dot{x}(t) = \mathbf{A_o}x(t) + \mathbf{B_o}u(t) \tag{18.2.1}$$
$$y(t) = \mathbf{C_o}x(t) \tag{18.2.2}$$

Let $\bar{r}(t)$ denote an external signal.

Then, provided that the pair $(\mathbf{A_o}, \mathbf{B_o})$ is completely controllable, there exists state-variable feedback of the form

$$u(t) = \bar{r} - \mathbf{K}x(t) \tag{18.2.3}$$
$$\mathbf{K} \triangleq [k_0, k_1, \dots, k_{n-1}] \tag{18.2.4}$$

such that the closed-loop characteristic polynomial is $A_{cl}(s)$, where $A_{cl}(s)$ is an arbitrary polynomial of degree n.

Proof

The model is completely controllable, so it can be transformed into controller form (see Chapter 17), by using a similarity transformation:

$$x = \mathbf{T}x_c \Longrightarrow \mathbf{A}_c \triangleq \mathbf{T}^{-1}\mathbf{A_o}\mathbf{T}; \qquad \mathbf{B}_c \triangleq \mathbf{T}^{-1}\mathbf{B_o}; \qquad \mathbf{C}_c \triangleq \mathbf{C_o}\mathbf{T} \tag{18.2.5}$$

where

$$\dot{x}_c(t) = \mathbf{A}_c x_c(t) + \mathbf{B}_c u(t) \tag{18.2.6}$$
$$y(t) = \mathbf{C}_c x_c(t) \tag{18.2.7}$$

and

$$\mathbf{A}_c = \begin{bmatrix} -a_{n-1} & \cdots & \cdots & -a_0 \\ 1 & & & 0 \\ & \ddots & & \vdots \\ & & 1 & 0 \end{bmatrix} \qquad \mathbf{B}_c = \begin{bmatrix} 1 \\ 0 \\ \vdots \\ 0 \end{bmatrix} \tag{18.2.8}$$

Now, if one uses state feedback of the form

$$u(t) = -\mathbf{K_c}x_c(t) + \bar{r}(t) \tag{18.2.9}$$

where

$$\mathbf{K_c} \triangleq \begin{bmatrix} k^c_{n-1} & k^c_{n-2} & \cdots & k^c_0 \end{bmatrix} \tag{18.2.10}$$

then the closed-loop state space representation becomes

$$\dot{x}_c(t) = \begin{bmatrix} -a_{n-1} & \cdots & \cdots & -a_0 \\ 1 & & & \\ & \ddots & & \\ & & 1 & 0 \end{bmatrix} x_c(t) + \begin{bmatrix} 1 \\ 0 \\ \vdots \\ 0 \end{bmatrix} (-\mathbf{K_c}x_c(t) + \bar{r}(t)) \tag{18.2.11}$$

$$\dot{x}_c(t) = \begin{bmatrix} -a_{n-1} - k^c_{n-1} & \cdots & \cdots & -a_0 - k^c_0 \\ 1 & & & \\ & \ddots & & \\ & & 1 & 0 \end{bmatrix} x_c(t) + \begin{bmatrix} 1 \\ 0 \\ \vdots \\ 0 \end{bmatrix} \bar{r}(t) \tag{18.2.12}$$

Thus, the closed-loop poles satisfy

$$s^n + (a_{n-1} + k^c_{n-1})s^{n-1} + \ldots (a_0 + k^c_0) = 0 \tag{18.2.13}$$

It is therefore clear that, by choice of $\mathbf{K_c}$, the closed-loop polynomial, and there-fore the nominal closed-loop poles, can be assigned arbitrarily. We can also express the above result in terms of the original state space description by reversing the transformation (18.2.5). This leads to

$$u(t) = -\mathbf{K_c}\mathbf{T}^{-1}x(t) + \bar{r}(t) = -\mathbf{K}x(t) + \bar{r}(t) \qquad where \quad \mathbf{K} = \mathbf{K_c}\mathbf{T}^{-1} \tag{18.2.14}$$

□□□

Note that state feedback does not introduce additional dynamics in the loop, because the scheme is based only on proportional feedback of certain system vari-ables. We can easily determine the overall transfer function from $\bar{r}(t)$ to $y(t)$. It is given by

$$\frac{Y(s)}{R(s)} = \mathbf{C_o}(s\mathbf{I} - \mathbf{A_o} + \mathbf{B_oK})^{-1}\mathbf{B_o} = \frac{\mathbf{C_o}Adj\{s\mathbf{I} - \mathbf{A_o} + \mathbf{B_oK}\}\mathbf{B_o}}{F(s)} \qquad (18.2.15)$$

where

$$F(s) \triangleq \det\{s\mathbf{I} - \mathbf{A_o} + \mathbf{B_oK}\} \qquad (18.2.16)$$

and *Adj* stands for *adjoint matrix*.

We can further simplify the expression given in equation (18.2.15). To do this, we will need to use the following two results from Linear Algebra.

Lemma 18.2 (Matrix inversion lemma). *Consider three matrices,* $\mathbf{A} \in \mathbb{C}^{n \times n}$, $\mathbf{B} \in \mathbb{C}^{n \times m}$ *and* $\mathbf{C} \in \mathbb{C}^{m \times n}$. *Then, if* $\mathbf{A} + \mathbf{BC}$ *is nonsingular, we have that*

$$(\mathbf{A} + \mathbf{BC})^{-1} = \mathbf{A}^{-1} - \mathbf{A}^{-1}\mathbf{B}\left(\mathbf{I} + \mathbf{CA}^{-1}\mathbf{B}\right)^{-1}\mathbf{CA}^{-1} \qquad (18.2.17)$$

Proof

Multiply both sides by $(\mathbf{A} + \mathbf{BC})$.

□□□

Remark 18.1. *In the case for which* $\mathbf{B} = g \in \mathbb{R}^n$ *and* $\mathbf{C}^T = h \in \mathbb{R}^n$, *equation (18.2.17) becomes*

$$\left(\mathbf{A} + gh^T\right)^{-1} = \left(\mathbf{I} - \mathbf{A}^{-1}\frac{gh^T}{1 + h^T\mathbf{A}^{-1}g}\right)\mathbf{A}^{-1} \qquad (18.2.18)$$

Lemma 18.3. *Given a matrix* $W \in \mathbb{R}^{n \times n}$ *and a pair of arbitrary vectors* $\phi_1 \in \mathbb{R}^n$ *and* $\phi_2 \in \mathbb{R}^n$, *then, provided that* W *and* $W + \phi_1\phi_2^T$ *are nonsingular,*

$$Adj(W + \phi_1\phi_2^T)\phi_1 = Adj(W)\phi_1 \qquad (18.2.19)$$
$$\phi_2^T Adj(W + \phi_1\phi_2^T) = \phi_2^T Adj(W) \qquad (18.2.20)$$

Proof

Using the definition of matrix inverse, and applying the matrix inversion lemma (18.2) to the inverse of $W + \phi_1\phi_2^T$, *we have that*

$$(W + \phi_1\phi_2^T)^{-1}\phi_1 = \frac{Adj(W + \phi_1\phi_2^T)}{\det(W + \phi_1\phi_2^T)}\phi_1 \qquad (18.2.21)$$

$$= W^{-1}\phi_1 - \frac{W^{-1}\phi_1\phi_2^T W^{-1}\phi_1}{1 + \phi_2^T W^{-1}\phi_1} \qquad (18.2.22)$$

$$= \frac{W^{-1}\phi_1}{1 + \phi_2^T W^{-1}\phi_1} \qquad (18.2.23)$$

It is also known that

$$W^{-1} = \frac{Adj(W)}{\det(W)} \tag{18.2.24}$$

$$1 + \phi_2^T W^{-1} \phi_1 = \det(1 + \phi_2^T W^{-1} \phi_1) = \det(I + W^{-1} \phi_1 \phi_2^T) \tag{18.2.25}$$

$$= \frac{\det(W + \phi_1 \phi_2^T)}{\det(W)} \tag{18.2.26}$$

We first use (18.2.24) and (18.2.26) on the right-hand side of (18.2.23). Then, by comparing the resulting expression with the right-hand side of (18.2.21), the result (18.2.19) is obtained. The proof of (18.2.20) follows along the same lines.

□□□

Application of Lemma 18.3 on the facing page to equation (18.2.15) leads to

$$\mathbf{C_o} Adj\{s\mathbf{I} - \mathbf{A} + \mathbf{B_o K}\}\mathbf{B_o} = \mathbf{C_o} Adj\{s\mathbf{I} - \mathbf{A_o}\}\mathbf{B_o} \tag{18.2.27}$$

We then see that the right-hand side of the above expression is the numerator $B_o(s)$ of the nominal model, $G_o(s)$. Hence, state feedback assigns the closed-loop poles to a prescribed position, while the zeros in the overall transfer function remain the same as those of the plant model.

State feedback encompasses the essence of many fundamental ideas in control design and lies at the core of many design strategies. However, this approach requires that all states be measured. In most cases, this is an unrealistic requirement. For that reason, the idea of observers is introduced next, as a mechanism for estimating the states from the available measurements.

18.3 Observers

In the previous section, we found that state-variable feedback could assign the closed-loop poles arbitrarily. Frequently, however, not all states can be measured. We thus introduce the idea of an observer. This is basically a mechanism to estimate the unmeasured states from the available output measurements. This is a special form of *virtual* sensor.

We consider again the state space model

$$\dot{x}(t) = \mathbf{A_o}x(t) + \mathbf{B_o}u(t) \tag{18.3.1}$$

$$y(t) = \mathbf{C_o}x(t) \tag{18.3.2}$$

A general linear observer then takes the form

$$\dot{\hat{x}}(t) = \mathbf{A_o}\hat{x}(t) + \mathbf{B_o}u(t) + \mathbf{J}(y(t) - \mathbf{C_o}\hat{x}(t)) \tag{18.3.3}$$

where the matrix $\mathbf{J}$ is called the observer gain and $\hat{x}(t)$ is the *state estimate*. Note that, for the particular choice $\mathbf{J} = 0$, the observer (18.3.3) degenerates into an open-loop model of (18.3.1).

The term

$$\nu(t) \triangleq y(t) - \mathbf{C_o}\hat{x}(t) \tag{18.3.4}$$

is known as the *innovation process*. For nonzero $\mathbf{J}$ in (18.3.3), $\nu(t)$ represents the feedback error between the observation and the predicted model output.

The following result shows how the observer gain $\mathbf{J}$ can be chosen such that the error, $\tilde{x}(t)$, defined as

$$\tilde{x}(t) \triangleq x(t) - \hat{x}(t) \tag{18.3.5}$$

can be made to decay at any desired rate.

Lemma 18.4. *Consider the state space model (18.3.1)-(18.3.2) and an associated observer of the form (18.3.3).*

Then the estimation error $\tilde{x}(t)$, defined by (18.3.5), satisfies

$$\dot{\tilde{x}}(t) = (\mathbf{A_o} - \mathbf{J}\mathbf{C_o})\tilde{x}(t) \tag{18.3.6}$$

Moreover, provided the model is completely observable, then the eigenvalues of $(\mathbf{A_o} - \mathbf{J}\mathbf{C_o})$ can be arbitrarily assigned by choice of $\mathbf{J}$.

Proof

Subtracting (18.3.1) and (18.3.3) and using (18.3.2) and (18.3.5) gives (18.3.6).

Also, if the model is completely observable, then there always exists a similarity transformation

$$x = \overline{\mathbf{T}}x_o \implies \overline{\mathbf{A}} \triangleq \overline{\mathbf{T}}^{-1}\mathbf{A_o}\overline{\mathbf{T}}; \qquad \overline{\mathbf{B}} \triangleq \overline{\mathbf{T}}^{-1}\mathbf{B_o}; \qquad \overline{\mathbf{C}} \triangleq \mathbf{C_o}\overline{\mathbf{T}} \tag{18.3.7}$$

that takes the system state description to observer canonical form. (See Chapter 17). It is

$$\dot{x}_o(t) = \overline{\mathbf{A}}x_o(t) + \overline{\mathbf{B}}u(t) \tag{18.3.8}$$

$$y(t) = \overline{\mathbf{C}}x_o(t) \tag{18.3.9}$$

where

$$
\overline{\mathbf{A}} = \begin{bmatrix} -a_{n-1} & 1 & & \\ -a_{n-2} & & \ddots & \\ \vdots & & & 1 \\ -a_0 & 0 & \cdots & 0 \end{bmatrix} ; \qquad \overline{\mathbf{B}} = \begin{bmatrix} b_{n-1} \\ b_{n-2} \\ \vdots \\ b_0 \end{bmatrix} ; \qquad \overline{\mathbf{C}} = \begin{bmatrix} 1 & 0 & \cdots & 0 \end{bmatrix}
$$

$$(18.3.10)$$

In this form, the observer becomes

$$
\dot{\hat{x}}_o(t) = \overline{\mathbf{A}}\hat{x}_o(t) + \overline{\mathbf{B}}u(t) + \mathbf{J}_o(y(t) - \overline{\mathbf{C}}\hat{x}_o(t)) \tag{18.3.11}
$$

Now, let

$$
\mathbf{J_o} \triangleq \begin{bmatrix} j_{n-1}^o & j_{n-2}^o & \cdots & j_0^o \end{bmatrix}^T \tag{18.3.12}
$$

The error then becomes

$$
\tilde{x}_o(t) = x_o(t) - \hat{x}_o(t) \tag{18.3.13}
$$

and that leads to

$$
\dot{\tilde{x}}_o(t) = (\overline{\mathbf{A}} - \mathbf{J_o C_o})\tilde{x}_o(t) = \begin{bmatrix} -a_{n-1} - j_{n-1}^o & 1 & & \\ -a_{n-2} - j_{n-2}^o & & \ddots & \\ \vdots & & & 1 \\ -a_0 - j_0^o & 0 & \cdots & 0 \end{bmatrix} \tilde{x}_o(t) \tag{18.3.14}
$$

The error dynamics are therefore governed by the roots of the following characteristic equation:

$$
\det(s\mathbf{I} - \overline{\mathbf{A}} - \mathbf{J_o}\overline{\mathbf{C}}) = s^n + (a_{n-1} + j_{n-1}^o)s^{n-1} + \ldots (a_0 + j_0^o) = 0 \tag{18.3.15}
$$

Hence, by choice of $\mathbf{J_o}$, the error dynamics can be arbitrarily assigned.
  Finally, by transforming the state back to the original representation, we can see that, in Equation (18.3.3), we require that

$$\mathbf{J} = \overline{\mathbf{T}}\mathbf{J_o} \tag{18.3.16}$$

□□□

Example 18.1 (Tank-level estimation). *As a simple application of a linear observer to estimate states, we consider the problem of two coupled tanks in which only the height of the liquid in the second tank is actually measured but where we are also interested in estimating the height of the liquid in the first tank. We will design a virtual sensor for this task.*

A system of this type is also discussed on the book's web page, where a photograph of a real system is given.

A schematic diagram is shown in Figure 18.1.

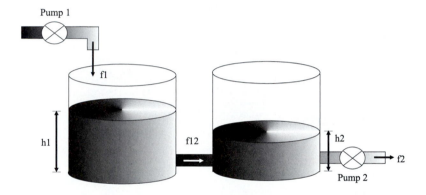

Figure 18.1. Schematic diagram of two coupled tanks

Water flows into the first tank through pump 1 a rate $f_i(t)$ that obviously affects the height of the water in tank 1 (denoted by $h_1(t)$). Water flows out of tank 1 into tank 2 at a rate $f_{12}(t)$, affecting both $h_1(t)$ and $h_2(t)$. Water then flows out of tank 2 at a rate f_e controlled by pump 2.

Given this information, the challenge is to build a virtual sensor (or observer) to estimate the height of liquid in tank 1 from measurements of the height of liquid in tank 2 and the flows $f_1(t)$ and $f_2(t)$.

Before we continue with the observer design, we first make a model of the system. The height of liquid in tank 1 can be described by the equation

$$\frac{dh_1(t)}{dt} = \frac{1}{A}(f_i(t) - f_{12}(t)) \tag{18.3.17}$$

Similarly, $h_2(t)$ is described by

$$\frac{dh_2(t)}{dt} = \frac{1}{A}(f_{12}(t) - f_e) \qquad (18.3.18)$$

The flow between the two tanks can be approximated by the free-fall velocity for the difference in height between the two tanks:

$$f_{12}(t) = \sqrt{2g(h_1(t) - h_2(t))} \qquad (18.3.19)$$

Now, if we measure the liquid heights in the tanks in % (where 0% is empty and 100% is full), we can convert the flow rates into equivalent values in % per second (where $f_1(t)$ is the equivalent flow into tank 1 and $f_2(t)$ is the equivalent flow out of tank 2). The model for the system is then

$$\frac{d}{dt}\begin{bmatrix} h_1(t) \\ h_2(t) \end{bmatrix} = \begin{bmatrix} -K\sqrt{h_1(t) - h_2(t)} \\ K\sqrt{h_1(t) - h_2(t)} \end{bmatrix} + \begin{bmatrix} 1 & 0 \\ 0 & -1 \end{bmatrix}\begin{bmatrix} f_1(t) \\ f_2(t) \end{bmatrix} \qquad (18.3.20)$$

where

$$K = \frac{\sqrt{2g}}{A} = 0.26 \qquad (18.3.21)$$

We can linearize this model for a nominal steady-state height difference (or operating point). Let

$$h_1(t) - h_2(t) = \Delta h(t) = H + h_d(t) \qquad (18.3.22)$$

Now,

$$\begin{aligned} K\sqrt{h_1(t) - h_2(t)} &= K\sqrt{\Delta h} \qquad (18.3.23) \\ &= K\sqrt{H + h_d(t)} \\ &= K\sqrt{H}\sqrt{1 + \frac{h_d(t)}{H}} \\ &\approx K\sqrt{H}\left(1 + \frac{h_d(t)}{2H}\right) \end{aligned}$$

because

$$\sqrt{1 + \epsilon} \approx 1 + \frac{\epsilon}{2} \qquad (18.3.24)$$

and so

$$K\sqrt{H}\left(1+\frac{h_d(t)}{2H}\right) = K\sqrt{H} + \frac{K}{2\sqrt{H}}h_d(t) \tag{18.3.25}$$

$$= K\sqrt{H} + \frac{K}{2\sqrt{H}}(h_1(t) - h_2(t) - H)$$

$$= K\sqrt{H} + \frac{K}{2\sqrt{H}}(h_1(t) - h_2(t)) - \frac{K}{2\sqrt{H}}H$$

$$= \frac{K\sqrt{H}}{2} + \frac{K}{2\sqrt{H}}(h_1(t) - h_2(t))$$

This yields the following linear model:

$$\frac{d}{dt}\begin{bmatrix}h_1(t)\\h_2(t)\end{bmatrix} = \begin{bmatrix}-k & k\\k & -k\end{bmatrix}\begin{bmatrix}h_1(t)\\h_2(t)\end{bmatrix} + \begin{bmatrix}1 & 0\\0 & -1\end{bmatrix}\begin{bmatrix}f_1(t) - \frac{K\sqrt{H}}{2}\\f_2(t) + \frac{K\sqrt{H}}{2}\end{bmatrix} \tag{18.3.26}$$

where

$$k = \frac{K}{2\sqrt{H}} \tag{18.3.27}$$

We are assuming that $h_2(t)$ can be measured and $h_1(t)$ cannot, so we set $C = \begin{bmatrix}0 & 1\end{bmatrix}$ and $D = \begin{bmatrix}0 & 0\end{bmatrix}$. The resulting system is both controllable and observable (as you can easily verify). Now we wish to design an observer

$$J = \begin{bmatrix}J_1\\J_2\end{bmatrix} \tag{18.3.28}$$

to estimate the value of $h_2(t)$. The characteristic polynomial of the observer is readily seen to be

$$s^2 + (2k + J_2)s + J_2k + J_1k \tag{18.3.29}$$

so we can choose the observer poles; that choice gives us values for J_1 and J_2. If we assume that the operating point is $H = 10\%$, then $k = 0.0411$. If we wanted poles at $s = -0.9291$ and $s = -0.0531$, then we would calculate that $J_1 = 0.3$ and $J_2 = 0.9$. If we wanted two poles at $s = -2$, then $J_2 = 3.9178$ and $J_1 = 93.41$.

The equation for the final observer is then

$$
\frac{d}{dt}\begin{bmatrix} \hat{h}_1(t) \\ \hat{h}_2(t) \end{bmatrix} = \begin{bmatrix} -k & k \\ k & -k \end{bmatrix}\begin{bmatrix} \hat{h}_1(t) \\ \hat{h}_2(t) \end{bmatrix} + \begin{bmatrix} 1 & 0 \\ 0 & -1 \end{bmatrix}\begin{bmatrix} f_1(t) - \frac{K\sqrt{H}}{2} \\ f_2(t) + \frac{K\sqrt{H}}{2} \end{bmatrix} + J(h_2(t) - \hat{h}_2(t))
$$

$$(18.3.30)$$

$\square\square\square$

Notice that the above observer allows us to estimate the missing height $h_1(t)$ by using measurements of $h_2(t)$ and knowledge of the flows $f_1(t)$ and $f_2(t)$. It thus provides a *virtual* sensor for $h_2(t)$.

The reader can interactively experiment with this observer on the book's web page.

18.4   Combining State Feedback with an Observer

A reasonable conjecture arising from the last two sections is that it would be a good idea, in the presence of unmeasurable states, to proceed by estimating these states via an observer and then to complete the feedback control strategy by feeding back these estimates in lieu of the true states. Such a strategy is indeed very appealing, because it separates the task of observer design from that of controller design. A-priori, however, it is not clear how the observer poles and the state feedback interact. Nor is it clear how they affect the closed loop. The following theorem shows that the resultant closed-loop poles are the combination of the observer and the state-feedback poles.

Theorem 18.1 (Separation theorem). *Consider the state space model (18.3.1)-(18.3.2), and assume that it is completely controllable and completely observable. Consider also an associated observer of the form (18.3.3) and state-variable feedback of the form (18.2.3), where the state estimates are used in lieu of the true states:*

$$u(t) = \bar{r}(t) - \mathbf{K}\hat{x}(t) \tag{18.4.1}$$

$$\mathbf{K} \triangleq \begin{bmatrix} k_0 & k_1 & \dots & k_{n-1} \end{bmatrix} \tag{18.4.2}$$

Then

(i) the closed-loop poles are the combination of the poles from the observer and the poles that would have resulted from using the same feedback on the true states–Specifically, the closed-loop polynomial $A_{cl}(s)$ is given by

$$A_{cl}(s) = \det(s\mathbf{I} - \mathbf{A_o} + \mathbf{B_o}\mathbf{K})\det(s\mathbf{I} - \mathbf{A_o} + \mathbf{J}\mathbf{C_o}) \qquad (18.4.3)$$

(ii) *The state-estimation error $\tilde{x}(t)$ cannot be controlled from the external signal $\bar{r}(t)$.*

Proof

(i)

We note from Lemma 18.4 on page 532 that $x(t)$, $\hat{x}(t)$, and $\tilde{x}(t)$ satisfy

$$\dot{\tilde{x}}(t) = (\mathbf{A_o} - \mathbf{J}\mathbf{C_o})\tilde{x}(t) \qquad (18.4.4)$$

$$\dot{x}(t) = \mathbf{A_o}x(t) + \mathbf{B_o}[\bar{r}(t) - \mathbf{K}\hat{x}(t)] \qquad (18.4.5)$$

$$= \mathbf{A_o}x(t) + \mathbf{B_o}[\bar{r}(t) - \mathbf{K}(x(t) - \tilde{x}(t))] \qquad (18.4.6)$$

Equations (18.4.4) and (18.4.6) can be written jointly as

$$\frac{d}{dt}\begin{bmatrix} x(t) \\ \tilde{x}(t) \end{bmatrix} = \begin{bmatrix} \mathbf{A_o} - \mathbf{B_o}\mathbf{K} & \mathbf{B_o}\mathbf{K} \\ 0 & \mathbf{A_o} - \mathbf{J}\mathbf{C_o} \end{bmatrix}\begin{bmatrix} x(t) \\ \tilde{x}(t) \end{bmatrix} + \begin{bmatrix} \mathbf{B_o} \\ 0 \end{bmatrix}\bar{r}(t) \qquad (18.4.7)$$

$$y(t) = \begin{bmatrix} \mathbf{C_o} & 0 \end{bmatrix}\begin{bmatrix} x(t) \\ \tilde{x}(t) \end{bmatrix} \qquad (18.4.8)$$

The result follows upon noting that the eigenvalues of a block triangular matrix are equal to the combination of the eigenvalues of the diagonal blocks.

(ii)

The structure of Equation (18.4.7) immediately implies that the state-estimation error $\tilde{x}(t)$ cannot be controlled from the external signal $\bar{r}(t)$, as can be readily checked by using the controllability matrix. This observation implies that the state-estimation error cannot be improved by manipulating $\bar{r}(t)$.

□□□

The above theorem makes a very compelling case for the use of state-estimation feedback. However, the reader is cautioned that the location of closed-loop poles is only one among many factors that come into control-system design. Indeed, we shall see later that state-estimate feedback is not a panacea, because it is subject to the same issues of sensitivity to disturbances, model errors, etc. as all feedback solutions. Indeed, all of the schemes turn out to be essentially identical. These connections are the subject of the next section.

18.5   Transfer-Function Interpretations

In the material presented above, we have developed a seemingly different approach to SISO linear control-systems synthesis. This could leave the reader wondering what the connection is between this and the transfer-function ideas presented earlier in Chapter 7. We next show that these two methods are actually different ways of expressing the same result.

18.5.1   Transfer-Function Form of Observer

We first give a transfer-function interpretation to the observer. We recall from (18.3.3) that the state space observer takes the form

$$\dot{\hat{x}}(t) = \mathbf{A_o}\hat{x}(t) + \mathbf{B_o}u(t) + \mathbf{J}(y(t) - \mathbf{C_o}\hat{x}(t)) \tag{18.5.1}$$

where $\mathbf{J}$ is the observer gain and $\hat{x}(t)$ is the state estimate.

A transfer-function interpretation for this observer is given in the following lemma.

Lemma 18.5. *The Laplace transform of the state estimate in (18.5.1) has the following properties:*

(a) *The estimate can be expressed in transfer-function form as:*

$$\hat{X}(s) = (s\mathbf{I} - \mathbf{A_o} + \mathbf{JC_o})^{-1}(\mathbf{B_o}U(s) + \mathbf{J}Y(s)) = T_1(s)U(s) + T_2(s)Y(s) \tag{18.5.2}$$

where $T_1(s)$ and $T_2(s)$ are the following two stable transfer functions:

$$T_1(s) \overset{\triangle}{=} (s\mathbf{I} - \mathbf{A_o} + \mathbf{JC_o})^{-1}\mathbf{B_o} \tag{18.5.3}$$

$$T_2(s) \overset{\triangle}{=} (s\mathbf{I} - \mathbf{A_o} + \mathbf{JC_o})^{-1}\mathbf{J} \tag{18.5.4}$$

Note that $T_1(s)$ and $T_2(s)$ have in common the denominator

$$E(s) \overset{\triangle}{=} \det(s\mathbf{I} - \mathbf{A_o} + \mathbf{JC_o}) \tag{18.5.5}$$

(b) *The estimate is related to the input and initial conditions by*

$$\hat{X}(s) = (s\mathbf{I} - \mathbf{A_o})^{-1}\mathbf{B_o}U(s) - \frac{f_0(s)}{E(s)} \tag{18.5.6}$$

where $f_0(s)$ is a polynomial vector in s with coefficients depending linearly on the initial conditions of the error $\tilde{x}(t)$.

(c) *The estimate is unbiased in the sense that*

$$T_1(s) + T_2(s)G_o(s) = (s\mathbf{I} - \mathbf{A_o})^{-1}\mathbf{B_o} \qquad (18.5.7)$$

where $G_o(s)$ is the nominal plant model.

Proof

(a)

Equation (18.5.2) is obtained directly by applying the Laplace transform to equation (18.5.1), and using the definition for $T_1(s)$ and $T_2(s)$. Zero initial conditions are assumed.

(b)

Equation (18.5.6) is derived by noting that

$$\hat{X}(s) = X(s) - \tilde{X}(s) \qquad (18.5.8)$$

by using the fact that $X(s) = (s\mathbf{I} - \mathbf{A_o})^{-1}\mathbf{B_o}U(s)$, and by also using Equation (18.4.7) to find the expression for $\tilde{X}(s)$.

(c)

Using the fact that $G_o(s) = \mathbf{C_o}(s\mathbf{I} - \mathbf{A_o})^{-1}\mathbf{B_o}$ and using Equations (18.5.3) and (18.5.4), we have that

$$T_1(s) + T_2(s)G_o(s) = (s\mathbf{I} - \mathbf{A_o} + \mathbf{JC_o})^{-1}(\mathbf{B_o} + \mathbf{JC_o}(s\mathbf{I} - \mathbf{A_o})^{-1}\mathbf{B_o}) \qquad (18.5.9)$$
$$= (s\mathbf{I} - \mathbf{A_o} + \mathbf{JC_o})^{-1}(s\mathbf{I} - \mathbf{A_o} + \mathbf{JC_o})(s\mathbf{I} - \mathbf{A_o})^{-1}\mathbf{B_o}$$
$$(18.5.10)$$

from which the result follows.

□□□

Part (b) of the above lemma is known as the property of *asymptotic unbiasedness* of the estimate. It shows that the transfer function from U to $\hat{X}$ is the same as that from U to X. Thus, the only difference (after transients have decayed) between the true state and the estimated state is due to effects we have not yet included (e.g., noise, disturbances,and model errors).

18.5.2   Transfer-Function Form of State-Estimate Feedback

We next give a transfer-function interpretation to the interconnection of an observer with state-variable feedback, as was done in section §18.4. The key result is described in the following lemma.

Lemma 18.6.   *(a) The state-estimate feedback law (18.4.1) can be expressed in transfer-function form as:*

$$\frac{L(s)}{E(s)}U(s) = -\frac{P(s)}{E(s)}Y(s) + \overline{R}(s) \qquad (18.5.11)$$

where $E(s)$ is the polynomial defined in equation (18.5.5) and

$$\frac{L(s)}{E(s)} = 1 + \mathbf{K}T_1(s) = \frac{\det(s\mathbf{I} - \mathbf{A_o} + \mathbf{JC_o} + \mathbf{B_oK})}{E(s)} \qquad (18.5.12)$$

$$\frac{P(s)}{E(s)} = \mathbf{K}T_2(s) = \frac{\mathbf{K}Adj(s\mathbf{I} - \mathbf{A_o})\mathbf{J}}{E(s)} \qquad (18.5.13)$$

$$\frac{P(s)}{L(s)} = \mathbf{K}[s\mathbf{I} - \mathbf{A_o} + \mathbf{JC_o} + \mathbf{B_oK}]^{-1}\mathbf{J} \qquad (18.5.14)$$

where $\mathbf{K}$ is the feedback gain and $\mathbf{J}$ is the observer gain.

(b) The closed-loop characteristic polynomial is

$$A_{cl}(s) = \det(s\mathbf{I} - \mathbf{A_o} + \mathbf{B_oK})\det(s\mathbf{I} - \mathbf{A_o} + \mathbf{JC_o}) \qquad (18.5.15)$$

$$= F(s)E(s) = A_o(s)L(s) + B_o(s)P(s) \qquad (18.5.16)$$

(c) The transfer function from $\overline{R}(s)$ to $Y(s)$ is given by

$$\frac{Y(s)}{\overline{R}(s)} = \frac{B_o(s)E(s)}{A_o(s)L(s) + B_o(s)P(s)} \qquad (18.5.17)$$

$$= \frac{B_o(s)}{\det(s\mathbf{I} - \mathbf{A_o} + \mathbf{B_oK})}$$

$$= \frac{B_o(s)}{F(s)}$$

where $B_o(s)$ and $A_o(s)$ are the numerator and denominator of the nominal loop, respectively. $P(s)$ and $L(s)$ are the polynomials defined in (18.5.12) and (18.5.13), respectively. $F(s)$ is the polynomial defined in (18.2.16).

Proof

(a) Substituting (18.5.2) into (18.4.1) yields

$$U(s) = -\mathbf{K}[T_1(s)U(s) + T_2(s)Y(s)] + \overline{R}(s)$$
$$\iff (1 + \mathbf{K}T_1(s))U(s) = -\mathbf{K}T_2(s)Y(s) + \overline{R}(s) \quad (18.5.18)$$

We next note that

$$1 + \mathbf{K}(s\mathbf{I} - \mathbf{A_o} + \mathbf{JC_o})^{-1}\mathbf{B_o} = \det(\mathbf{I} + (s\mathbf{I} - \mathbf{A_o} + \mathbf{JC_o})^{-1}\mathbf{B_o}\mathbf{K}) \quad (18.5.19)$$
$$= \det\big([s\mathbf{I} - \mathbf{A_o} + \mathbf{JC_o}]^{-1}(s\mathbf{I} - \mathbf{A_o} + \mathbf{JC_o} + \mathbf{B_o}\mathbf{K})\big) \quad (18.5.20)$$
$$\mathbf{K}Adj(s\mathbf{I} - \mathbf{A_o} + \mathbf{JC_o})\mathbf{J} = \mathbf{K}Adj(s\mathbf{I} - \mathbf{A_o})\mathbf{J} \quad (18.5.21)$$

where (18.5.21) was obtained by application of Lemma 18.3 on page 530.

Finally, result 18.5.14 follows after applying Lemma 18.3 twice to show that

$$\mathbf{K}Adj(s\mathbf{I} - \mathbf{A_o})\mathbf{J} = \mathbf{K}Adj(s\mathbf{I} - \mathbf{A_o} + \mathbf{JC_o} + \mathbf{B_o}\mathbf{K})\mathbf{J} \quad (18.5.22)$$

(b) By using (18.4.3), the closed-loop polynomial is given as

$$A_{cl}(s) = \det(s\mathbf{I} - \mathbf{A_o} + \mathbf{B_o}\mathbf{K})\det(s\mathbf{I} - \mathbf{A_o} + \mathbf{JC_o}) \quad (18.5.23)$$
$$= F(s)E(s) = A_o(s)L(s) + B_o(s)P(s) \quad (18.5.24)$$

(c) From Equation (18.5.11) and the nominal model $G_o(s) = B_o(s)/A_o(s)$, we have that

$$\frac{\overline{Y}(s)}{\overline{R}(s)} = \frac{B_o(s)E(s)}{A_o(s)L(s) + B_o(s)P(s)} \quad (18.5.25)$$

Then the result follows by using (18.5.16) to simplify the expression (18.5.25).

$\square\square\square$

The foregoing lemma shows that polynomial pole assignment and state-estimate feedback lead to the same result. Thus, the only difference is in the terms of implementation. Actually, in the form that we have presented it, the state space formulation makes certain assumptions about the form of the polynomial solution. In particular, we see from equations (18.5.2) to (18.5.4) that the observer is based on strictly proper transfer functions. According to Remark 7.2 on page 184, this requires that the closed-loop polynomial be chosen of degree at least $2n$.

Note that the cancellation of the factor $\det(s\mathbf{I} - \mathbf{A_o} + \mathbf{JC_o})$ in (18.5.17) is consistent with the fact that the observer states are uncontrollable from the external signal $\bar{r}(t)$.

The combination of observer and state-estimate feedback has some simple interpretations in terms of a standard feedback loop. A first possible interpretation derives directly from Equation (18.5.11), by an expression of the controller output as

$$U(s) = \frac{E(s)}{L(s)}\left(\overline{R}(s) - \frac{P(s)}{E(s)}Y(s)\right) \qquad (18.5.26)$$

This is graphically depicted in part (a) of Figure 18.2 on the following page. We see that this is a two-degree-of-freedom control loop. A standard one-degree-of-freedom loop can be obtained if we generate $\bar{r}(t)$ from the loop reference $r(t)$ as follows:

$$\overline{R}(s) = \frac{P(s)}{E(s)}R(s) \qquad (18.5.27)$$

We then have that (18.5.26) can be written as

$$U(s) = \frac{P(s)}{L(s)}(R(s) - Y(s)) \qquad (18.5.28)$$

This corresponds to the one-degree-of-freedom loop shown in part (b) of Figure 18.2 on the next page.

Note that, in part (a) of Figure 18.2 on the following page, the transfer function from $\overline{R}(s)$ to $Y(s)$ is given by (18.5.17), whereas, in part (b) of the same figure, the transfer function from $R(s)$ to $Y(s)$ is given by

$$\frac{Y(s)}{R(s)} = \frac{B_o(s)P(s)}{E(s)F(s)} \qquad (18.5.29)$$

where $F(s)$ is given in (18.2.16). Here E(s) appears in the denominator due to the particular choice of $\overline{R}(s)$, which reintroduces the polynomial above.

Remark 18.2. *Note that (18.5.14) says that the feedback controller can be implemented as a system defined, in state space form, by the $4 - tuple$ ($\mathbf{A_o} - \mathbf{JC_o} - \mathbf{B_o K}, \mathbf{J}, \mathbf{K}, 0$). (MATLAB provides a special command, **reg**, to obtain the transfer-function form.)*

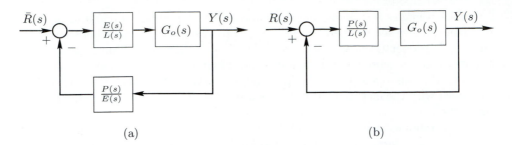

Figure 18.2. Separation theorem in standard loop forms

18.5.3   Transfer Function for Innovation Process

We finally give an interpretation to the innovation process introduced in (18.3.4). Recall that

$$\nu(t) = y(t) - \mathbf{C_o}\hat{x}(t) \qquad (18.5.30)$$

This equation can also be expressed in terms of Laplace transfer functions by using (18.5.2) to (18.5.4) as

$$
\begin{aligned}
E_\nu(s) = \mathcal{L}\left[\nu(t)\right] &= Y(s) - \mathbf{C_o}[T_1(s)U(s) + T_2(s)Y(s)] \\
&= (1 - \mathbf{C_o}T_2(s))Y(s) - \mathbf{C_o}T_1U(s)
\end{aligned}
\qquad (18.5.31)
$$

We can use the above result to express the innovation process $\nu(t)$ in terms of the original plant transfer function. In particular, we have the next lemma.

Lemma 18.7. *Consider the state space model (18.3.1)-(18.3.2) and the associated nominal transfer function $G_o(s) = B_o(s)/A_o(s)$. (We recall that state space description matrices are set in bold face, to distinguish them from polynomials defined with the same symbol.) Then the innovations process, $\nu(t)$, can be expressed as*

$$E_\nu(s) = \frac{A_o(s)}{E(s)}Y(s) - \frac{B_o(s)}{E(s)}U(s) \qquad (18.5.32)$$

where $E(s)$ is the polynomial defined in (18.5.5), called the observer characteristic polynomial.

Proof

We see that the term multiplying $Y(s)$ in (18.5.31) is given by

$$1 - \mathbf{C_o}T_2(s) = 1 - \mathbf{C_o}(s\mathbf{I} - \mathbf{A_o} + \mathbf{JC_o})^{-1}\mathbf{J} \tag{18.5.33}$$

$$= \det(I - (s\mathbf{I} - \mathbf{A_o} + \mathbf{JC_o})^{-1}\mathbf{JC_o}) \tag{18.5.34}$$

$$= \det((s\mathbf{I} - \mathbf{A_o} + \mathbf{JC_o})^{-1})\det(s\mathbf{I} - \mathbf{A_o} + \mathbf{JC_o} - \mathbf{JC_o}) \tag{18.5.35}$$

$$= \frac{\det(s\mathbf{I} - \mathbf{A_o})}{\det(s\mathbf{I} - \mathbf{A_o} + \mathbf{JC_o})} \tag{18.5.36}$$

$$= \frac{A_o(s)}{E(s)} \tag{18.5.37}$$

Similarly, the term multiplying $U(s)$ in (18.5.31) is the negative of

$$\mathbf{C_o}T_1(s) = \mathbf{C_o}(s\mathbf{I} - \mathbf{A_o} + \mathbf{JC_o})^{-1}\mathbf{B_o} \tag{18.5.38}$$

$$= \mathbf{C_o}\frac{Adj(s\mathbf{I} - \mathbf{A_o} + \mathbf{JC_o})}{\det(s\mathbf{I} - \mathbf{A_o} + \mathbf{JC_o})}\mathbf{B_o} \tag{18.5.39}$$

$$= \mathbf{C_o}\frac{Adj(s\mathbf{I} - \mathbf{A_o})}{\det(s\mathbf{I} - \mathbf{A_o} + \mathbf{JC_o})}\mathbf{B_o} \tag{18.5.40}$$

$$= \frac{B_o(s)}{E(s)} \tag{18.5.41}$$

where we used Lemma 18.3 on page 530 to derive (18.5.40) from (18.5.39).

□□□

For the moment, the above result is simply an interesting curiosity. However, we will show that this result has very interesting implications. For example, we use it in the next section to reinterpret the parameterization of all stabilizing controllers given earlier in Chapter 15.

18.6   Reinterpretation of the Affine Parameterization of all Stabilizing Controllers

We recall the parameterization of all stabilizing controllers given in section §15.7– see especially Figure 15.9 on page 444. In the sequel, we take $R(s) = 0$, because the reference signal can always be reintroduced if desired. We note that the input $U(s)$ in Figure 15.9 on page 444 satisfies

$$\frac{L(s)}{E(s)}U(s) = -\frac{P(s)}{E(s)}Y(s) + Q_u(s)\left[\frac{B_o(s)}{E(s)}U(s) - \frac{A_o(s)}{E(s)}Y(s)\right] \tag{18.6.1}$$

We can connect this result to state-estimate feedback and innovations feedback from an observer by using the results of section §18.5. In particular, we have the next lemma.

Lemma 18.8. *The class of all stabilizing linear controllers can be expressed in state space form as*

$$U(s) = -\mathbf{K}\hat{X}(s) - Q_u(s)E_\nu(s) \tag{18.6.2}$$

where $\mathbf{K}$ *is a state-feedback gain,* $\hat{X}(s)$ *is a state estimate provided by any stable linear observer, and* $E_\nu(s)$ *denotes the corresponding innovation process.*

Proof

The result follows immediately upon comparing (18.6.1) *with* (18.5.32) *and* (18.5.26).
$$\square\square\square$$

This alternative form of the class of all stabilizing controllers is shown in Figure 18.3.

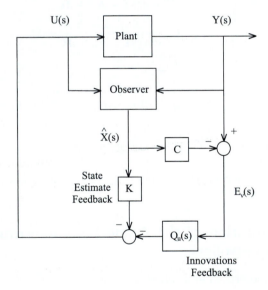

Figure 18.3. State-estimate feedback interpretation of all stabilizing controllers

18.7   State Space Interpretation of Internal Model Principle

A generalization of the above ideas on state-estimate feedback is the Internal Model Principle (IMP) described in Chapter 10. We next explore the state space form of IMP from two alternative perspectives.

(a) Disturbance-estimate feedback

One way that the IMP can be formulated in state space is to assume that we have a general deterministic *input disturbance* $d(t)$ with a generating polynomial $\Gamma_d(s)$.

We then proceed by building an observer so as to generate a model state estimate $\hat{x}_o(t)$ and a disturbance estimate $\hat{d}(t)$. These estimates can then be combined in a control law of the form

$$u(t) = -\mathbf{K_o}\hat{x}_o(t) - \hat{d}(t) + r\bar{(t)} \qquad (18.7.1)$$

We observe that this law ensures asymptotic cancellation of the input disturbance from the input signal, provided that the estimate of the disturbance, $\hat{d}(t)$, is stable and unbiased. We will show below that the control law (18.7.1) automatically ensures that the polynomial $\Gamma_d(s)$ appears in the denominator, $L(s)$, of the corresponding transfer-function form of the controller.

Before examining the general situation, we first present a simple example to illustrate the key ideas.

Example 18.2. *Consider a linear plant with input $u(t)$, input disturbance $d(t)$, and measurable output $y(t)$, where*

$$Y(s) = \frac{1}{s+1}\left(U(s) + D(s)\right) \qquad (18.7.2)$$

and where $d(t)$ is a "constant" but "unknown" signal. Use an observer to estimate $d(t)$; then design a feedback control law as in (18.7.1).

Solution

The basic idea is to model $d(t)$ as the output of a particular system, one that reflects our knowledge regarding the nature of the disturbance.

Once the overall model is built, the states in the disturbance-generating model can be observed from $y(t)$, and thus the disturbance can be estimated.

A combined state representation for a constant disturbance is given by

$$\begin{bmatrix} \dot{x}_1(t) \\ \dot{x}_2(t) \end{bmatrix} = \begin{bmatrix} -1 & 1 \\ 0 & 0 \end{bmatrix} \begin{bmatrix} x_1(t) \\ x_2(t) \end{bmatrix} + \begin{bmatrix} 1 \\ 0 \end{bmatrix} u(t); \qquad \begin{bmatrix} y(t) \\ d(t) \end{bmatrix} = \begin{bmatrix} 1 & 0 \\ 0 & 1 \end{bmatrix} \begin{bmatrix} x_1(t) \\ x_2(t) \end{bmatrix} \qquad (18.7.3)$$

where the state $x_1(t)$ is the plant state and $x_2(t)$ is the state in the disturbance-generating model.

Next, the state estimate, $\hat{x}(t)$, and the disturbance estimate, $\hat{d}(t)$, are given by

$$\hat{x}(t) = \mathcal{L}^{-1}\left[T_1(s)U(s) + T_2(s)Y(s)\right] \tag{18.7.4}$$

$$\hat{d}(t) = \hat{x}_2(t) = \begin{bmatrix} 0 & 1 \end{bmatrix}\hat{x}(t) \tag{18.7.5}$$

where $T_1(s)$ and $T_2(s)$ are the transfer functions defined in (18.5.3) and (18.5.4), respectively.

If the observer gain is $\mathbf{J} = \begin{bmatrix} j_1 & j_2 \end{bmatrix}^T$, then the observer polynomial defined in (18.5.5) is

$$E(s) = s^2 + (1 + j_1)s + j_2 \tag{18.7.6}$$

As usual, j_1 and j_2 are chosen to make $E(s)$ stable.

Once $T_1(s)$ and $T_2(s)$ have been computed, we have that

$$\hat{d}(t) = \hat{x}_2(t) = \mathcal{L}^{-1}\left[\frac{j_2(s+1)}{E(s)}Y(s) + \frac{-j_2}{E(s)}U(s)\right] \tag{18.7.7}$$

The fact that $\hat{d}(t)$ tends to $d(t)$ as $t \to \infty$ is easily proven by substituting (18.7.2) in (18.7.7) and considering that the disturbance satisfies

$$D(s) = \frac{\beta}{s} \qquad \text{for some unknown } \beta \text{ to be estimated} \tag{18.7.8}$$

This yields

$$\hat{d}(t) = \mathcal{L}^{-1}\left[\frac{j_2}{s^2 + (1 + j_1)s + j_2}\frac{\beta}{s}\right] \tag{18.7.9}$$

From this, upon applying the Final-Value Theorem, we have that

$$\lim_{t \to \infty} \hat{d}(t) = \beta \tag{18.7.10}$$

The transient in $\hat{d}(t)$ is a linear combination of the observer modes. Also,

$$\hat{x}(t) = \mathcal{L}^{-1}\left[\frac{(sj_1 + j_2)}{E(s)}Y(s) + \frac{s}{E(s)}U(s)\right] \tag{18.7.11}$$

We see that the transfer function from $U(s)$ to $\hat{X}(s)$ is $\frac{1}{s+1}$, as required. Also, choosing K to place the controllable poles at -2 gives $K = 1$, and the final control law becomes

$$u(t) = -\hat{x}(t) - \hat{d}(t) \tag{18.7.12}$$

In terms of transfer functions, we see that this control is of the form

$$U(s) = -\left[\frac{(sj_1 + j_2)}{E(s)}Y(s) + \frac{s}{E(s)}U(s)\right] - \left[\frac{j_2(s+1)}{E(s)}Y(s) - \frac{j_2}{E(s)}U(s)\right]$$
$$\tag{18.7.13}$$

This simplifies to

$$U(s) = -\left[\frac{(j_1 + j_2)s + 2j_2}{s(s + j_1 + 2)}\right]Y(s) \tag{18.7.14}$$

Thus, as expected, the controller incorporates integral action and the closed-loop characteristic polynomial is $(s + 2)(s^2 + (j_1 + 1)s + j_2)$.

□□□

Returning to the general case, we begin by considering a composite state description, which includes the plant-model state, as in (18.2.1)-(18.2.2), and the disturbance model state:

$$\dot{x}_d(t) = A_d x_d(t) \tag{18.7.15}$$
$$d(t) = C_d x_d(t) \tag{18.7.16}$$

We note that the corresponding 4-tuples that define the partial models are $(\mathbf{A}_o, \mathbf{B}_o, \mathbf{C}_o, 0)$ and $(\mathbf{A}_d, 0, \mathbf{C}_d, 0)$ for the plant and disturbance, respectively. For the combined state $x(t) = \begin{bmatrix} x_o^T(t) & x_d^T(t) \end{bmatrix}^T$, we have

$$\dot{x}(t) = \mathbf{A}x(t) + \mathbf{B}u(t) \qquad \text{where} \qquad \mathbf{A} = \begin{bmatrix} \mathbf{A}_o & \mathbf{B}_o\mathbf{C}_d \\ 0 & \mathbf{A}_d \end{bmatrix} \qquad \mathbf{B} = \begin{bmatrix} \mathbf{B}_o \\ 0 \end{bmatrix}$$
$$\tag{18.7.17}$$

The plant-model output is given by

$$y(t) = \mathbf{C}x(t) \qquad \text{where} \qquad \mathbf{C} = \begin{bmatrix} \mathbf{C}_o & 0 \end{bmatrix} \tag{18.7.18}$$

Note that this composite model will, in general, be observable but *not* controllable (on account of the disturbance modes). Thus, we will only attempt to stabilize *the plant modes*, by choosing $\mathbf{K}_o$ so that $(\mathbf{A}_o - \mathbf{B}_o\mathbf{K}_o)$ is a stability matrix.

The observer and state-feedback gains can then be partitioned as

$$\mathbf{J} = \begin{bmatrix} \mathbf{J_o} \\ \mathbf{J_d} \end{bmatrix}; \qquad \mathbf{K} = \begin{bmatrix} \mathbf{K_o} & \mathbf{K_d} \end{bmatrix} \tag{18.7.19}$$

When the control law (18.7.1) is used, then, clearly, $\mathbf{K_d} = \mathbf{C}_d$. We thus obtain

$$s\mathbf{I} - \mathbf{A} = \begin{bmatrix} s\mathbf{I} - \mathbf{A}_o & -\mathbf{B}_o\mathbf{C}_d \\ 0 & s\mathbf{I} - \mathbf{A}_d \end{bmatrix}; \qquad \mathbf{JC} = \begin{bmatrix} \mathbf{J}_o\mathbf{C}_o & 0 \\ \mathbf{J_d}\mathbf{C}_o & 0 \end{bmatrix}; \qquad \mathbf{BK} = \begin{bmatrix} \mathbf{B}_o\mathbf{K_o} & \mathbf{B}_o\mathbf{C}_d \\ 0 & 0 \end{bmatrix} \tag{18.7.20}$$

The final control law is thus seen to correspond to the following transfer function:

$$C(s) = \frac{P(s)}{L(s)} = \begin{bmatrix} \mathbf{K_o} & \mathbf{K_d} \end{bmatrix} \begin{bmatrix} s\mathbf{I} - \mathbf{A}_o + \mathbf{B}_o\mathbf{K}_o + \mathbf{J}_o\mathbf{C}_o & 0 \\ \mathbf{J_d}\mathbf{C}_o & s\mathbf{I} - \mathbf{A}_d \end{bmatrix}^{-1} \begin{bmatrix} \mathbf{J}_o \\ \mathbf{J}_d \end{bmatrix} \tag{18.7.21}$$

From this, we see that the denominator of the control law in polynomial form is

$$L(s) = \det(s\mathbf{I} - \mathbf{A}_o + \mathbf{J}_o\mathbf{C}_o + \mathbf{B}_o\mathbf{K_o}) \det(s\mathbf{I} - \mathbf{A}_d) \tag{18.7.22}$$

Using (18.7.15) in (18.7.22), we finally see that $\Gamma_d(s)$ is indeed a factor of $L(s)$, as in the polynomial form of IMP.

(b) Forcing the Internal Model Principle via additional dynamics

Another method of satisfying the Internal Model Principle in state space is to filter the system output by passing it through the disturbance model. To illustrate this, say that the system is given by

$$\dot{x}(t) = \mathbf{A}_o x(t) + \mathbf{B}_o u(t) + \mathbf{B}_o d_i(t) \tag{18.7.23}$$
$$y(t) = \mathbf{C}_o x(t) \tag{18.7.24}$$
$$\dot{x}_d(t) = \mathbf{A}_d x_d(t) \tag{18.7.25}$$
$$d_i(t) = \mathbf{C}_d x_d(t) \tag{18.7.26}$$

We then modify the system by passing the system output through the following filter:

$$\dot{x}'(t) = \mathbf{A}_d^T x'(t) + \mathbf{C}_d^T y(t) \tag{18.7.27}$$

where observability of $(\mathbf{C}_d, \mathbf{A}_d)$ implies controllability of $(\mathbf{A}_d^T, \mathbf{C}_d^T)$. We then estimate $x(t)$ using a standard observer, ignoring the disturbance, leading to

$$\dot{\hat{x}}(t) = \mathbf{A}_o \hat{x}(t) + \mathbf{B}_o u(t) + \mathbf{J}_o(y(t) - \mathbf{C}_o \hat{x}(t)) \qquad (18.7.28)$$

The final control law is then obtained by feeding back both $\hat{x}(t)$ and $x'(t)$, to yield

$$u(t) = -\mathbf{K}_o \hat{x}(t) - \mathbf{K}_d x'(t) \qquad (18.7.29)$$

where $\begin{bmatrix} \mathbf{K}_o & \mathbf{K}_d \end{bmatrix}$ is chosen to stabilize the composite system (18.7.23), (18.7.24), (18.7.27).

Indeed, the results in section §17.9 establish that the cascaded system is completely controllable, provided, that the original system does not have a zero coinciding with any eigenvalue of $\mathbf{A}_d$.

The resulting control law is finally seen to have the following transfer function:

$$C(s) = \frac{P(s)}{L(s)} = \begin{bmatrix} \mathbf{K}_o & \mathbf{K}_d \end{bmatrix} \begin{bmatrix} s\mathbf{I} - \mathbf{A}_o + \mathbf{B}_o \mathbf{K}_o + \mathbf{J}_o \mathbf{C}_o & \mathbf{B}_o \mathbf{K}_o \\ \mathbf{0} & s\mathbf{I} - \mathbf{A}_d^T \end{bmatrix}^{-1} \begin{bmatrix} \mathbf{J}_o \\ \mathbf{C}_d^T \end{bmatrix}$$
$$(18.7.30)$$

The denominator polynomial is thus seen to be

$$L(s) = \det(s\mathbf{I} - \mathbf{A}_o + \mathbf{J}_o \mathbf{C}_o + \mathbf{B}_o \mathbf{K}_o) \det(s\mathbf{I} - \mathbf{A}_d) \qquad (18.7.31)$$

and we see again that $\Gamma_d(s)$, as defined in (10.2.4), is a factor of $L(s)$ as required. Indeed, one sees that there is an interesting connection between the results in (18.7.21) and in (18.7.30). They are simply alternative ways of achieving the same result.

18.8 Trade-Offs in State Feedback and Observers

In section §18.2, it was shown that, under the assumption of controllability and by a suitable choice of the feedback gain $\mathbf{K}$, the closed-loop poles could be assigned to any desired set of locations. However, if the closed-loop modes are chosen to be much faster than those of the plant, then the gain $\mathbf{K}$ will be large, leading to large magnitudes for the plant input $u(t)$. We have thus restated the known trade-off between speed and the required input energy.

A similar (dual) problem arises in state estimation. To illustrate this point, consider the state space model given by (18.3.1)-(18.3.2), and assume that measurement noise, $v(t)$, is present, so that the model becomes

$$\dot{x}(t) = \mathbf{A}_o x(t) + \mathbf{B}_o u(t) \qquad (18.8.1)$$
$$y(t) = \mathbf{C}_o x(t) + v(t) \qquad (18.8.2)$$

Then the state estimate and the estimation error are, respectively, given by

$$\dot{\hat{x}}(t) = \mathbf{A_o}\hat{x}(t) + \mathbf{B_o}u(t) + \mathbf{JC_o}(x(t) - \hat{x}(t)) + \mathbf{J}v(t) \qquad (18.8.3)$$

$$\dot{\tilde{x}}(t) = (\mathbf{A_o} - \mathbf{JC_o})\tilde{x}(t) - \mathbf{J}v(t) \qquad (18.8.4)$$

We then see, on applying the Laplace transform to (18.8.4), that

$$\tilde{X}(s) = [s\mathbf{I} - \mathbf{A_o} + \mathbf{JC_o}]^{-1}\tilde{x}(0) - [s\mathbf{I} - \mathbf{A_o} + \mathbf{JC_o}]^{-1}\mathbf{J}V(s) \qquad (18.8.5)$$

Equation (18.8.5) reveals a trade-off in the observer design. We see that, if $\mathbf{J}$ is chosen so as to have the eigenvalues of $\mathbf{A_o} - \mathbf{JC_o}$ well into the left-half plane, we will quickly eliminate the effect of the initial error $\tilde{x}(0)$. However, this will almost certainly require a large value for $\mathbf{J}$. We then see that the second term on the right-hand side of Equation (18.8.5) will usually enhance the effect of the measurement noise, because this is usually a high-frequency signal.

The problem is compounded if, in addition to measurement noise, we also have unmodeled disturbances, i.e., signals that appear in additive form in the state equation (18.8.1).

The above analysis suggests that, in many real cases, the problem of observer design requires a compromise to be made between speed of response and noise immunity.

We thus see that both state feedback and observer design require compromises to be made between conflicting requirements. One way to formulate the resolution of these compromises is to pose them as optimization problems, with a suitable cost function which balances the competing requirements. A particularly nice cost function to use is one with quadratic terms, because this leads to simplifications in the solution of the problem. Thus, quadratic optimal control theory and quadratic optimal-filter theory have become widely used tools in feedback design.

The theoretical foundation of quadratic optimal control and filter design will be presented later in Chapter 22 of Part VII.

18.9   Dealing with Input Constraints in the Context of State-Estimate Feedback

Finally, we give a state space interpretation to the anti-wind-up schemes presented in Chapter 11.

We remind the reader of the two conditions placed on an anti-wind-up implementation of a controller, given in section §11.3:

 (i) the states of the controller should be driven by the actual plant input;

 (ii) the states should have a stable realization when driven by the actual plant input.

The above requirements are easily met in the context of state-variable feedback. This leads to the anti-wind-up scheme shown in Figure 18.4.

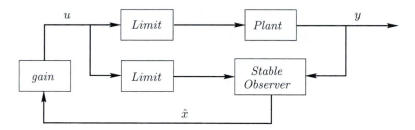

Figure 18.4. Anti-wind-up scheme

In Figure 18.4, the state $\hat{x}$ should also include estimates of disturbances. Actually, to achieve a one-degree-of-freedom architecture for reference injection, then all one need do is subtract the reference prior to feeding the plant output into the observer.

We thus see that anti-wind-up has a particularly simple interpretation in state space.

18.10   Summary

- We have shown that controller synthesis via pole placement can also be presented in state space form:

   Given a model in state space form, and given desired locations of the closed-loop poles, it is possible to compute a set of constant gains, one gain for each state, such that feeding back the states through the gains results in a closed loop with poles in the prespecified locations.

- Viewed as a theoretical result, this insight complements the equivalence of transfer function and state space models which an equivalence of achieving pole placement by synthesizing a controller either as transfer function via the Diophantine equation or as constant-gain state-variable feedback.

- Viewed from a practical point of view, implementing this controller would require sensing the value of each state. Due to physical, chemical, and economic constraints, however, one hardly ever has actual measurements of all system states available.

- This raises the question of alternatives to actual measurements and introduces the notion of so-called *observers*, sometimes also called *soft sensors*, *virtual sensors*, *filters*, or *calculated data*.

- The purpose of an observer is to infer the value of an unmeasured state from other states that are correlated with it and that *are* being measured.

- Observers have a number of commonalities with control systems:

  - they are dynamical systems;
  - they can be treated in either the frequency or the time domain;
  - they can be analyzed, synthesized, and designed;
  - they have performance properties, such as stability, transients, and sensitivities;
  - these properties are influenced by the pole/zero patterns of their sensitivities.

- State estimates produced by an observer are used for several purposes:

  - constraint monitoring;
  - data logging and trending;
  - condition and performance monitoring;
  - fault detection;
  - feedback control.

- To implement a synthesized state-feedback controller as discussed above, one can use state-variable estimates from an observer in lieu of unavailable measurements; the emergent closed-loop behavior is due to the interaction between the dynamical properties of system, controller, *and observer*.

- The interaction is quantified by the third-fundamental result presented in this chapter: the nominal poles of the overall closed loop are the union of the observer poles and the closed-loop poles induced by the feedback gains if all states could be measured. This result is also known as the *separation theorem*.

- Recall that controller synthesis is concerned with how to compute a controller that will give the emergent closed loop a particular property, the *constructed property*.

- The main focus of the chapter is on synthesizing controllers that place the closed-loop poles in chosen locations; this is a particular constructed property that allows certain design insights to be achieved.

- There are, however, other useful constructed properties as well.

- Examples of constructed properties for which there exist synthesis solutions:

  - to arrive at a specified system state in minimal time with an energy constraint

  - to minimize the weighted square of the control error and energy consumption

  - to achieve minimum-variance control

- One approach to synthesis is to cast the constructed property into a so-called *cost-functional*, *objective function* or *criterion*, which is then minimized numerically.

  - This approach is sometimes called optimal control, because one optimizes a criterion.

  - One must remember, however, that the result cannot be better than the criterion.

  - *Optimization* shifts the primary engineering task from explicit controller design to criterion design, which then generates the controller automatically.

  - Both approaches have benefits, including personal preference and experience.

18.11   Further Reading

General

Chen, C-T. (1984).   *Introduction to Linear System Theory.*   Holt, Rinehart and Winston.

Doyle, J.C., Francis, B.A., and Tannenbaum, A.R. (1992).   *Feedback Control Theory.*   Macmillan Publishing Company.

Kwakernaak, H. and Sivan, R. (1972).   *Linear Optimal Control Systems.*   Wiley-Interscience, New York.

Fortmann, T. and Hitz, K. (1977).   *An Introduction to Linear Control Systems.*   Marcel Dekker, New York.

Filter and Observer Design

Seron, M.M., Braslavsky, J.H., and Goodwin, G.C. (1997).   *Fundamental limitations in filtering and control.*   Springer-Verlag.

18.12   Problems for the Reader

Problem 18.1. *A state space model for a linear system is given by* $(\mathbf{A_o}, \mathbf{B_o}, \mathbf{C_o}, 0)$, *where*

$$\mathbf{A_o} = \begin{bmatrix} -3 & 2 \\ 0 & -1 \end{bmatrix}; \qquad \mathbf{B_o} = \begin{bmatrix} -1 \\ 1 \end{bmatrix}; \qquad \mathbf{C_o}^T = \begin{bmatrix} -1 \\ \alpha \end{bmatrix} \qquad (18.12.1)$$

18.1.1 *Determine, if it exists, a state-feedback gain* $\mathbf{K}$ *such that the closed-loop poles are located at* -5 *and* -6.

18.1.2 *Determine, if it exists, a state-feedback gain* $\mathbf{K}$ *such that the closed-loop natural modes have the form* $\beta_1 e^{-2t} \cos(0.5t + \beta_2)$.

18.1.3 *Find the range of values of* α *for which the system is completely observable.*

18.1.4 *Choose a particular value of* α *that makes the system completely observable, and build an observer, whose error decays faster than* e^{-5t}.

Problem 18.2. *Assume a measurable signal* $f(t)$ *has the form* $f(t) = \alpha_0 + \beta_1 \sin(2t) + \beta_2 \cos(2t)$, *where* α_0, β_1, *and* β_2 *are unknown constants.*

18.2.1 *Build a state space model to generate the signal* $f(t)$. *Note that, in this model, the initial state is a linear function of* α_0, β_1, *and* β_2.

18.2.2 *Using that model, build an observer to estimate* α_0, β_1, *and* β_2. *The estimation error should be negligible after* 2 $[s]$.

Problem 18.3. *Consider a discrete-time linear system having its transfer function given by*

$$H_q(z) = \frac{z + 0.8}{(z - 0.8)(z - 0.6)} \qquad (18.12.2)$$

18.3.1 *Build a state space model for this system.*

18.3.2 *Design an observer for the system state that achieves zero estimation error in exactly 3 time units.*

Problem 18.4. *Assume you need to estimate a constant signal which is measured in noise. Further, assume that the noise energy is concentrated around* 0.5 $[rad/s]$.

Design an observer to estimate the unknown constant. Try an observer with its pole located at $s = -5$; *then, repeat the design, but this time with the pole located at* $s = -0.1$. *Compare and discuss your results. (Model the noise as a pure sine wave of frequency* 0.5 $[rad/s]$.)

Problem   18.5. *Consider a plant with input $u(t)$, input disturbance $d(t)$, and output $y(t)$. The nominal model of this plant is*

$$Y(s) = G_o(s)(U(s) + D(s)) \qquad \text{where} \quad G_o(s) = \frac{2}{(s+2)} \qquad (18.12.3)$$

18.5.1 *If the disturbance is constant but unknown, build an observer that generates estimates for the disturbance and for the model state. The observer polynomial should be chosen as $E(s) = s^2 + 8s + 40$.*

18.5.2 *Use the above result to implement state-estimate feedback control in the form*

$$u(t) = -k_1 \hat{x}_1(t) - \hat{d}(t) + \bar{r}(t) \qquad (18.12.4)$$

where $\hat{x}_1(t)$ is the plant state estimate and $\hat{d}(t)$ is the disturbance estimate. Choose a convenient value for k_1.

18.5.3 *Compute $P(s)$ and $L(s)$ as defined in Lemma 18.6 on page 541, and set your final control loop in the form shown in part (b) of Figure 18.2 on page 544. Analyze the resulting closed-loop polynomial.*

Problem   18.6. *A continuous-time plant having transfer function $G_o(s)$ has to be controlled digitally. To that end, a zero-order sample and hold with a sampling interval Δ is implemented. Assume that*

$$G_o(s) = \frac{e^{-0.5s}}{s+1} \qquad \text{and} \quad \Delta = 0.25 \qquad (18.12.5)$$

18.6.1 *Build a state space model for the sampled-data transfer function.*

18.6.2 *Build an observer for the state. The estimation error should decay at a rate on the order of $(0.5)^k$.*

Problem   18.7. *Consider a plant having a linear model $G_o(s)$ controlled in a one-d.o.f. architecture by controller $C(s)$, where*

$$G_o(s) = \frac{2}{(s+1)^2} \qquad \text{and} \quad C(s) = \frac{2(s+1)^2}{s(s+3)(0.01s+1)} \qquad (18.12.6)$$

Find, if possible, an observer vector gain $\mathbf{J}$ and a controller vector gain $\mathbf{K}$ such that the above controller can be interpreted as the outcome of a design based on state-estimate feedback.

Problem 18.8. *A linear plant has a nominal model given by $G_o(s) = 6(s+3)^{-2}$.*

18.8.1 *Build a control loop that uses state-estimate feedback. The loop should achieve zero steady-state error for constant input disturbances, and the closed-loop poles each should have real part less than or equal to -4.*

18.8.2 *Design a controller by using polynomial pole placement, such that the loop satisfies the same requirements as above. Compare and discuss.*

Problem 18.9. *Consider a system having input $u(t)$, output $y(t)$, and a transfer function*

$$G_o(s) = \frac{-s+8}{(s+8)(-s+2)} \tag{18.12.7}$$

18.9.1 *Build a state space model.*

18.9.2 *Design an observer to estimate the signal $y(t) - 2\dot{y}(t)$. The estimation error must decay at least as fast as e^{-2t}.*

Chapter 19

INTRODUCTION TO NONLINEAR CONTROL

19.1 Preview

With the exception of our treatment of actuator limits, all previous material in the book has been aimed at linear systems. This is justified by the fact that most real-world systems exhibit (near) linear behavior within a limited operating range and by the significantly enhanced insights available in the linear case. However, one occasionally meets a problem on which the nonlinearities are so important that they cannot be ignored. This chapter is intended to give a brief introduction to nonlinear control. Our objective is not to be comprehensive but to simply give some basic extensions of linear strategies that might allow a designer to make a start on a nonlinear problem. As far as possible, we will build on the linear methods so as to benefit maximally from linear insights. We also give a *taste* of more rigorous nonlinear theory in subsection §19.10.3, so as to give the reader an appreciation for this fascinating and evolving subject. However, this section should probably be skipped on a first reading; it depends on more advanced mathematical ideas.

19.2 Linear Control of a Nonlinear Plant

Before turning to nonlinear control-system design, per se, an initial question that the reader might reasonably ask is what happens if a linear controller is applied to a nonlinear plant. We know from experience that this must be a reasonable strategy in many cases because one knows that all real plants exhibit some (presumably) mild form of nonlinearity and yet almost all real-world controllers are based on linear designs.

We saw in section §3.10 that one can obtain an approximate linear model for a nonlinear system by linearization techniques. The residual modeling error is, of course, a nonlinear operator. We will thus examine here the impact of nonlinear additive modeling errors on the performance of a linear controller.

To carry out this analysis, we have to combine linear and nonlinear descriptions in the analysis. The transfer-function concept is thus no longer adequate. We will therefore base the analysis on operators (transformations).

We define a nonlinear dynamic operator f as a mapping from one function space to another. Thus, f maps functions of time into other functions of time. Say, for example, that the time functions have finite energy; then we say that they belong to the space L_2, and we then say that f maps functions from L_2 into L_2. As in section §2.5, we will use the symbol y (without brackets) to denote an element of a function space; i.e., $y \in L_2$ is of the form $\{y(t) : \mathbb{R} \to \mathbb{R}\}$. We also use the notation $y = f\langle u \rangle$ to represent the mapping (transformation) from u to y via f.

To keep track of the different components of the loop, the controller is represented by a linear operator C and the nominal plant by the linear operator G_o; the modeling error will be characterized by an additive *nonlinear* operator G_ϵ.

We consider a reference input and input disturbance, because both can be treated in a consistent fashion.

The nominal and the true one-degree-of-freedom loops are shown in Figure 19.1.

Starting from the loops shown in Figure 19.1, we will obtain expressions for the following error signals:

- Plant output error $y_\epsilon = y - y_o$

- Control output error $u_\epsilon = u - u_o$

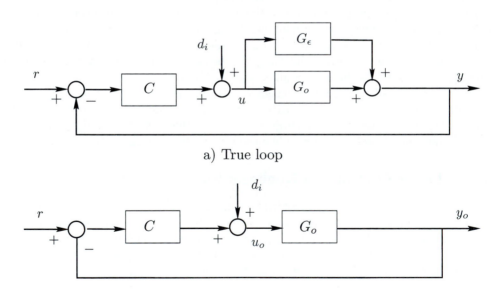

a) True loop

b) Nominal loop

Figure 19.1. True and nominal control loops

19.2.1   The Effect of Modeling Errors

To simplify the notation, we begin by grouping the external signals at the input by defining

$$d_i{}' = C\langle r \rangle + d_i \tag{19.2.1}$$

Note that this is possible because superposition holds for the linear part of the loop in Figure 19.1 a).

From Figure 19.1 a), we see that

$$y = (G_o + G_\epsilon)\langle u \rangle \tag{19.2.2}$$
$$u = -C\langle y \rangle + d_i{}' \tag{19.2.3}$$

Substituting (19.2.3) into (19.2.2) yields

$$y = (G_o + G_\epsilon)\langle -C\langle y \rangle + d_i{}' \rangle \tag{19.2.4}$$
$$= -G_o\langle C\langle y \rangle \rangle + G_o\langle d_i{}' \rangle + G_\epsilon\langle -C\langle y \rangle + d_i{}' \rangle \tag{19.2.5}$$

Similarly, for the nominal loop in Figure 19.1 b), we have

$$y_o = -G_o\langle C\langle y_o \rangle \rangle + G_o\langle d_i{}' \rangle \tag{19.2.6}$$
$$u_o = -C\langle y_o \rangle + d_i{}' \tag{19.2.7}$$

Subtracting (19.2.6) from (19.2.5) yields

$$y_\epsilon = -G_o\langle C\langle y_\epsilon \rangle \rangle + G_\epsilon\langle -C\langle y \rangle + d_i{}' \rangle \tag{19.2.8}$$

As usual, we define linear nominal sensitivities via

$$S_o\langle \circ \rangle = \frac{1}{1 + G_o\langle C\langle \circ \rangle \rangle} \tag{19.2.9}$$

$$S_{uo}\langle \circ \rangle = \frac{C\langle \circ \rangle}{1 + G_o\langle C\langle \circ \rangle \rangle} \tag{19.2.10}$$

Hence, rearranging the linear terms in (19.2.8) and using (19.2.9), we have

$$y_\epsilon = S_o\langle G_\epsilon\langle -C\langle y \rangle + d_i{}' \rangle \rangle \tag{19.2.11}$$

Then, replacing y by $y_o + y_\epsilon$ we have

$$y_\epsilon = S_o\langle G_\epsilon\langle -C\langle y_o + y_\epsilon \rangle + d_i{}' \rangle \rangle \tag{19.2.12}$$

and, using the linear result in (19.2.6), we have

$$y_\epsilon = S_o \langle G_\epsilon \langle -C\langle y_\epsilon \rangle + S_o \langle d_i{'} \rangle \rangle \rangle \tag{19.2.13}$$

Similarly, subtracting (19.2.7) from (19.2.3) yields

$$u_\epsilon = -C\langle y_\epsilon \rangle \tag{19.2.14}$$

Equations (19.2.13) and (19.2.14) can be shown schematically, as in Figure 19.2.

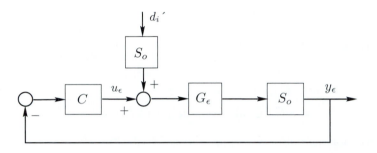

Figure 19.2. Error feedback loop

Reintroducing r and d_i via equation (19.2.1) and using $\{(19.2.9),\ (19.2.10)\}$ leads to the final representation, as in Figure 19.3.

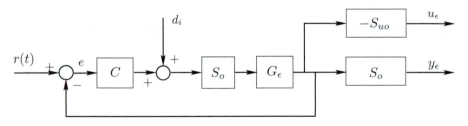

Figure 19.3. Equivalent signal loop describing the effects of modeling errors

Figure 19.3 is a compact way of depicting the effect of unmodeled plant nonlinearities on the robustness and performance of the feedback loop. Stability, robustness, and performance robustness can be studied by using this representation. For example, in section §19.10.3, we will show that stability is retained in the presence of sufficiently *small* nonlinear undermodeling. Of course, nonlinear systems are generally very difficult to analyze. However, some simplifying assumptions allow us to interpret Figure 19.3. For example, it is usual for G_ϵ to have the property that its output will be small when its input is small. In that case, Equations (19.2.13) and (19.2.14) suggest that errors in the feedback loop due to nonlinear modeling errors depend on whether

- $|S_o(j\omega)|$ is small in the frequency band where the input disturbance is significant, and

- $|S_{uo}(j\omega)|$ is small in the frequency band where the reference is significant.

These requirements might conflict. For example, if $|S_o(j\omega)|$ is required to be small in a frequency band that significantly exceeds the plant bandwidth, then $|S_{uo}(j\omega)|$ will almost certainly be large outside the plant bandwidth. This situation probably means that linear control is not a suitable strategy in this case.

Remark 19.1. *Of course, Figure 19.3 is also valid in the case of linear modeling errors. Thus, for example, the figure could be used to rederive Theorem 5.3 on page 146. Also, the figure could be used to gain insight into robust performance in both the linear and nonlinear cases.*

Example 19.1. *Consider the nonlinear plant having its state space model given by*

$$\frac{dx_1(t)}{dt} = x_2(t) + \big(x_2(t)\big)^3 \tag{19.2.15}$$

$$\frac{dx_2(t)}{dt} = -2x_1(t) - 3x_2(t) + u(t) + 0.1\big(x_1(t)\big)^2 u(t) \tag{19.2.16}$$

$$y(t) = x_1(t) \tag{19.2.17}$$

Assume that a linear controller is designed for this plant, and also assume that the design has been based upon a small-signal linearized model. This linearized model has been obtained for the operating point determined by a constant input $u(t) = u_Q = 2$, and the (linear) design achieves nominal sensitivities given by

$$T_o(s) = \frac{9}{s^2 + 4s + 9} \qquad S_o(s) = 1 - T_o(s) = \frac{s(s+4)}{s^2 + 4s + 9} \tag{19.2.18}$$

Analyze the performance of the (linear) design when used with the true (nonlinear) plant.

Solution

*We first need to obtain the small-signal linearized model. This is done by using the SIMULINK schematic **softpl1.mdl**. With a step input equal to 2, we have that the operating point is given by*

$$x_{1Q} = y_Q \overset{\triangle}{=} \lim_{t \to \infty} x_1(t) = 1.13 \qquad and \qquad x_{2Q} \overset{\triangle}{=} \lim_{t \to \infty} x_2(t) = 0 \qquad (19.2.19)$$

*The values for u_Q, x_{1Q}, and x_{2Q} are then used, in conjunction with the MAT-LAB commands **linmod** and ss2tf, to obtain the linearized model $G_o(s)$, which is given by*

$$G_o(s) = \frac{1.13}{s^2 + 3.0s + 1.55} \qquad (19.2.20)$$

To achieve the desired $T_o(s)$, we have that the controller transfer function must be given by

$$C(s) = [G_o(s)]^{-1}\frac{T_o(s)}{S_o(s)} = 7.96\frac{s^2 + 3.0s + 1.55}{s(s+4)} \qquad (19.2.21)$$

We then use SIMULINK to implement the block diagram shown in Figure 19.3, where $S_{uo}(s)$ is given by

$$S_{uo}(s) = [G_o(s)]^{-1}T_o(s) = 7.96\frac{s^2 + 3.0s + 1.55}{s^2 + 4s + 9} \qquad (19.2.22)$$

A simulation is run with a reference, $r(t)$, which has a mean value equal to 2 and a superimposed square-wave of amplitude 0.3. A pulse is added as an input disturbance. Figure 19.4 on the next page shows the plant output error, $y_\epsilon(t)$.

*The reader is encouraged to evaluate the performance of the linear design under other operating conditions and under other choice for $T_o(s)$. This is facilitated by the SIMULINK schematic in file **amenl.mdl**.*

□□□

We see from the above analysis that linear designs can perform well on nonlinear plants, provided that the nonlinearity is *sufficiently small*. However, as performance demands grow, one is inevitably forced to carry out an inherently nonlinear design. A first step in this direction is described in the next section, where we still use a linear controller, but a different linear controller is chosen at different operating points (or in different regions of the state space).

19.3   Switched Linear Controllers

One useful strategy for dealing with nonlinear systems is to split the state space up into small regions inside which a localized linear model gives a reasonable approximation to the response. One can then design a set of fixed linear controllers–one for each region. Two issues remain to be solved:

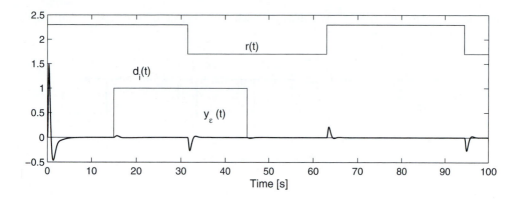

Figure 19.4. Effects of linear modeling of a nonlinear plant on the performance of linear-based control design

(i) how to know which region one is in, and

(ii) how to transfer between controllers.

The first problem above is often resolved if there exists some measured variable that is a key indicator of the dynamics of the system. These variables can be used to *schedule* the controllers. For example, in high-performance aircraft control, Mach number and altitude are frequently used as scheduling variables.

The second problem requires that each controller run in a stable fashion regardless of whether it is *in charge* of the plant. This can be achieved by the anti-wind-up strategies described in Chapter 11 or section §18.9.

An alternative architecture for an anti-wind-up scheme is shown in Figure 19.5. Here, a controller $(\bar{C}_i(s))$ is used to cause the ith controller output to track the true plant input. To describe how this might be used, let us say that we have k_c controllers and that switching between controllers is facilitated by a device that we call a *supervisor*. Thus, depending on the state of the system at any given time instant, the *supervisor* may select one of the k_c controllers to be the *active controller*. The other $k_c - 1$ controllers will be in *standby mode*.

We wish to arrange *bumpless transfer* between the controllers as directed by the supervisor. "Bumpless" transfer refers to the requirement that no *bump* occur in the control signal when switching from one controller to the others. For linear controllers having a similar structure, this can be achieved by allowing all controllers to share a common set of states and to simply switch the output gains attached to these states. However, in general, the controllers can have different structures, including different state dimensions, and inclusion of nonlinear elements. To cope with this general situation, we place each of the standby controllers under separate feedback control, so that their outputs track the output of the active controller. This is illustrated in Figure 19.5.

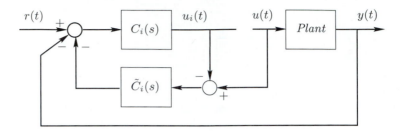

Figure 19.5. Bumpless controller transfer

In Figure 19.5, $u(t)$ denotes the active controller output (this controller is not shown in the diagram), and $C_i(s)$ denotes a standby controller having reference signal $r(t)$ for the particular plant output $y(t)$. In the figure, $\tilde{C}_i(s)$ denotes the controller for the ith controller, $C_i(s)$. Thus, in a sense, $\tilde{C}_i(s)$ is a *controller-controller*. This controller-controller is relatively easy to design: the *plant* in this case is well known, because it is actually the ith controller. For this controller-controller loop, the signals $r(t)$ and $y(t)$ act as disturbances.

In the special case, when the ith controller is biproper and of minimum phase, it turns out that perfect tracking of the active controller's past outputs is possible. Indeed, this is precisely what the arrangement in Figure 11.6 on page 298 ensures.

Examples of the application of the general structure in Figure 19.5:

- saturating actuators (where the supervisor responds to input saturation)

- state-constraining controllers (where the supervisor responds to imminent changes of state constraint violation)–see section §11.4

- gain scheduling (where the supervisor responds to certain measured variables– e.g., aircraft altitude or Mach number–to *schedule* the controller)

- adaptive control (where the supervisor responds to estimated model parameters)

A final point is that the active control need not be restricted to the output of one of the k_c controllers but can be any control signal, including manual actions or a combination of all controller outputs. Thus, for example, we might have

$$u(t) = \sum_{j=1}^{k_c} \lambda_j(t)u_j(t) \tag{19.3.1}$$

where $0 \leq \lambda_i(t) \leq 1$ and $\sum_{j=1}^{k_c} \lambda_j = 1$, $\forall t$.

This can be used when the supervisor makes *soft* switching decisions by generating a confidence weight attached to each possible controller. These weights might, for example, be the probability that each controller is the correct one, or some other set-membership function.

The reader will observe that we have already used this controller-mixing idea in connection with state constraints, in section §11.4.

The third problem raised at the beginning of this section is partially addressed by nonlinear robustness analysis of the type presented in section §19.2. Another related issue is that one really needs to analyze the effect of the time-varying linear controllers. A first step in this direction will be given in section §19.10.

19.4   Control of Systems with Smooth Nonlinearities

When the nonlinear features are significant, the above switching could lead to a conservative design. If this is the case, inherently nonlinear strategies should be considered in the design. To highlight some interesting features of nonlinear problems, we will begin by considering a simple scenario in which the model is both stable and stably invertible. We will later examine more general problems.

In the analysis to follow, we will need to describe interconnections of linear and nonlinear systems; thus, as we did in Chapter 5, we will resort to the use of operators. One of these operators, in the continuous-time case, will be the Heaviside operator ρ.

19.5   Static Input Nonlinearities

We will first examine the nonlinear control problem as an extension of the ideas in Chapter 15.

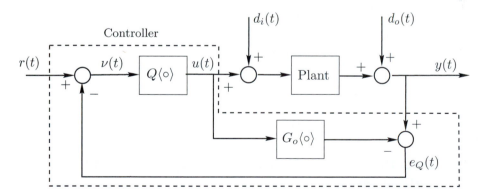

Figure 19.6. IMC architecture for control of (smooth) nonlinear systems

Consider the architecture shown in Figure 19.6. We observe that output disturbances have the same injection point that the reference does. This makes analysis
more straightforward than for input disturbances. In particular, using the assumption of open-loop stability for the plant and $d_i(t) \equiv 0$, we see that the error $e_Q(t)$
will converge (in the nominal case) to the value of the output disturbance.

We also recall, from Chapter 15, that, in the linear case of stable and minimum-
phase plants, $Q(s)$ was chosen as

$$Q(s) = F_Q(s)[G_o(s)]^{-1} \tag{19.5.1}$$

where $F_Q(s)$ was a suitable shaping filter, which ensures the properness of $Q(s)$ and
indirectly determines the associated trade-offs. The simplest choice was to make
$[F_Q(s)]^{-1}$ a stable polynomial of degree equal to the relative degree of $G_o(s)$. In
the linear case, this leads to the result

$$y(t) = F_Q \langle \nu(t) \rangle \tag{19.5.2}$$

The same basic principle can be applied to nonlinear systems. The simplest
situation occurs when the plant has a static input nonlinearity, and there is only an
output disturbance; then the nominal model output is given by

$$y(t) = G_o \langle u(t) \rangle + d_o(t) = \overline{G}_o \langle \phi(u) \rangle + d_o(t) \tag{19.5.3}$$

where $\overline{G}_o$ is a linear operator and $\phi(\ )$ is a static invertible function of u. Introducing $\overline{F}_Q$ as a linear stable operator of appropriate relative degree, the nonlinear
version of (19.5.1) becomes

$$Q\langle \circ \rangle = \phi^{-1} \left(\overline{G}_o^{-1} \langle \overline{F}_Q \langle \circ \rangle \rangle \right) \tag{19.5.4}$$

Hence,

$$u(t) = \phi^{-1} \left(\overline{G}_o^{-1} \langle \overline{F}_Q \langle \nu(t) \rangle \rangle \right) \tag{19.5.5}$$

Notice that ϕ^{-1} appears adjacent to ϕ in the plant.

Thus, Equation (19.5.5) represents an approximate inverse for the nonlinear
plant and leads directly to

$$y(t) = \overline{F}_Q \langle \nu(t) \rangle + d_o(t) \tag{19.5.6}$$

19.6 Smooth Dynamic Nonlinearities for Stable and Stably Invertible Models

For this class of systems, we essentially can use the IMC architecture shown in
Figure 19.6. The parallel model is easily constructed. We will thus focus attention
on the approximate inverse in $Q\langle \circ \rangle$.

We consider a class of nonlinear systems in which the nonlinear features are not necessarily restricted to the plant input. In particular, we consider a plant having a nominal model of the form

$$\rho x(t) \triangleq \frac{dx(t)}{dt} = f(x) + g(x)u(t) \tag{19.6.1}$$

$$y(t) = h(x) \tag{19.6.2}$$

where $x \in \mathbb{R}^n$, $y \in \mathbb{R}$, and $u \in \mathbb{R}$. We assume that $f(x)$, $g(x)$, and $h(x)$ are smooth mappings, in the sense that their first k derivatives, with respect to x, are well-defined, where k is at least equal to the *relative degree of the nonlinear system*. This latter concept is defined next.

Definition 19.1. *Consider a nonlinear system described by (19.6.1) and (19.6.2). Then the relative degree of the system is the minimum value $\eta \in \mathbb{N}$ such that*

$$\rho^{\eta} y(t) = \beta_{\eta}(x) + \alpha_{\eta}(x)u(t) \tag{19.6.3}$$

where ρ is the Heaviside operator and $\alpha_{\eta}(x)$ is not identically zero.

Example 19.2. *Consider a system having the model*

$$\rho x_1(t) = -x_1(t) - 0.1\big(x_2(t)\big)^2 \tag{19.6.4}$$

$$\rho x_2(t) = -2x_1(t) - 3x_2(t) - 0.125\big(x_2(t)\big)^3 + \Big[1 + 0.1\big(x_1(t)\big)^2\Big]u(t) \tag{19.6.5}$$

$$y(t) = x_1(t) \tag{19.6.6}$$

Then, if we compute the first- and second-time derivatives of $y(t)$, we obtain

$$\rho y(t) = \rho x_1(t) = -x_1(t) - 0.1\big(x_2(t)\big)^2 \tag{19.6.7}$$

$$\begin{aligned}\rho^2 y(t) = &\,1.4x_1(t) + 0.6x_2(t) + 0.1x_2(t)^2 + 0.25x_2(t)^3 \\ &- \big[0.2 + 0.02x_1(t)^2\big]u(t)\end{aligned} \tag{19.6.8}$$

from which we see that the relative degree of the system is equal to 2.

□□□

For the system described in equations (19.6.1) and (19.6.2), having relative degree η, we have that, if the ith derivative of $y(t)$, $i = 0, 1, 2, \ldots, \eta$ takes the form

$$\rho^i y(t) = \beta_i(x) + \alpha_i(x)u(t), \tag{19.6.9}$$

then $\alpha_i(x) \equiv 0$ for $i = 0, 1, 2, \ldots, \eta - 1$. Consider now the operator polynomial $p(\rho) = \sum_{i=0}^{\eta} p_i \rho^i$; then we see that

$$p(\rho)y(t) = b(x) + a(x)u(t) \qquad (19.6.10)$$

where

$$b(x) = \sum_{i=0}^{\eta} p_i \beta_i(x) \qquad \text{and} \qquad a(x) = p_\eta \alpha_\eta(x) \qquad (19.6.11)$$

We next show how the design methods of Chapter 15 might be extended to the nonlinear case. We recall that, in the linear case, we needed to invert the plant model (after using $F(s)$ to bring it to biproper form). The extension of these ideas to the nonlinear case is actually quite straightforward. In particular, we see from (19.6.10) that, provided that $a(x)$ is never zero for any state x in the operation region, the approximate inverse for the plant is obtained simply by setting $p(\rho)y(t)$ equal to $\nu(t)$ in (19.6.10). This leads (after changing the argument of the equation) to the following result for $u(t)$:

$$u(t) = \big(a(x)\big)^{-1} \left(\nu(t) - b(x)\right) = Q\langle \nu \rangle \qquad (19.6.12)$$

where $\nu(t)$ is defined in Figure 19.6 on page 567. Then, combining (19.6.10) and (19.6.12), we have that

$$p(\rho)y(t) = \nu(t) \iff y(t) = [p(\rho)]^{-1}\langle \nu \rangle \qquad (19.6.13)$$

This means that if we define

$$\overline{F}_Q\langle \circ \rangle \overset{\triangle}{=} [p(\rho)]^{-1}\langle \circ \rangle \qquad (19.6.14)$$

then

$$y(t) = \overline{F}_Q\langle \nu\langle t \rangle \rangle \qquad (19.6.15)$$

Finally, to obtain a perfect inverse at d.c., we set $p_0 = 1$. The control strategy (19.6.12) is commonly known as *input-output feedback linearization*, because, as is seen from (19.6.15), it leads to a linear closed-loop system from the input $\nu(t)$ to the output $y(t)$.

The reader will see that the above procedure is formally analogous to the linear-system case described in Chapter 15. As pointed out previously, however, there are stability, internal-stability, and relative-degree requirements that must be satisfied. These correspond to the nonlinear generalization of the linear stability and properness requirements on Q.

A remaining issue is how to implement $Q\langle \circ \rangle$ as described in (19.6.12), because it depends on the plant state, which is normally unavailable. Following the same

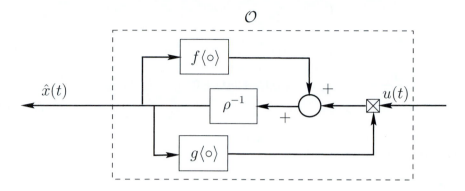

Figure 19.7. Nonlinear observer

philosophy as in the linear case, we can estimate the state by means of a nonlinear state observer. For the moment, we are assuming that the plant is open-loop stable, so we can use an open-loop observer. (More general cases are examined in subsection §19.8.1.) An open-loop observer is driven only by the plant input $u(t)$; its structure is shown in Figure 19.7, where ρ^{-1} denotes the integral operator.

With this observer, we can use Equation (19.6.12) to build the Q block, as shown in Figure 19.8. The observer is represented by the block labeled $\mathcal{O}$.

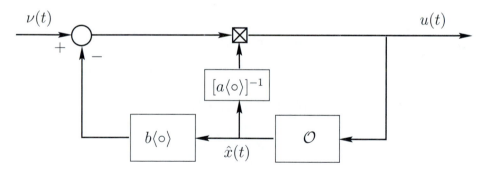

Figure 19.8. Implementation of the Q block

We now illustrate this concept by two simple examples.

Example 19.3. *Consider the nonlinear plant having a state space model given by*

$$\frac{dx_1(t)}{dt} = x_2(t) + \big(x_2(t)\big)^3 \tag{19.6.16}$$

$$\frac{dx_2(t)}{dt} = -2x_1(t) - 3x_2(t) + u(t) + 0.1\big(x_1(t)\big)^2 u(t) \tag{19.6.17}$$

$$y(t) = x_1(t) \tag{19.6.18}$$

Build a nonlinear controller based on the results of section §19.6. The desired bandwidth is approximately 3[rad/s].

Solution

To use the IMC structure shown in Figure 19.6, we first need to compute an approximate inverse for the plant, along the lines of section §19.6. We thus have, from (19.6.16) to (19.6.18), that

$$\rho y(t) = x_2(t) + \big(x_2(t)\big)^3 \tag{19.6.19}$$

$$\rho^2 y(t) = \big(1 + 3(x_2(t))^2\big)\frac{dx_2(t)}{dt} \tag{19.6.20}$$

$$= -(1 + 3x_2^2(t))(2x_1(t) + 3x_2(t)) + (1 + 3(x_2(t))^2)(1 + 0.1(x_1(t))^2)u(t) \tag{19.6.21}$$

We then have to choose the operator $F\langle\circ\rangle$ to achieve the desired bandwidth. To do that, we choose the operator polynomial $p(\rho)$ as

$$p(\rho) = (\rho^2 + 4\rho + 9)/9 \tag{19.6.22}$$

and we can then obtain $a(x)$ and $b(x)$ in Equation (19.6.10) as

$$a(x) = \frac{(1 + 3(x_2(t))^2)(1 + 0.1(x_1(t))^2)}{9} \tag{19.6.23}$$

$$b(x) = \frac{7x_1(t) + x_2(t) - 5\big(x_2(t)\big)^3 - 6x_1(t)(x_2(t))^2}{9} \tag{19.6.24}$$

Finally, the control loop in Figure 19.6 on page 567 can be implemented with $Q\langle\circ\rangle$, as in Figure 19.8 on the page before. The plant is stable, so the required (nonlinear) state observer can be built by using the (nonlinear) state space plant model.

To assess the performance of this design, we compare it with the linear design based upon a linearized plant model. (See Example 19.1 on page 563.) In this case, we use the same operating point as in Example 19.1–the one determined by $u_Q = 2$. This leads to the linear model

$$G_o(s) = \frac{1.13}{s^2 + 3.0s + 1.55} \tag{19.6.25}$$

The linear controller is thus designed to achieve the prescribed bandwidth. We now choose an appropriate complementary sensitivity, say

$$T_o(s) = \frac{9}{s^2 + 4s + 9} \tag{19.6.26}$$

which leads to

$$C(s) = [G_o(s)]^{-1} \frac{T_o(s)}{S_o(s)} = 7.96 \frac{s^2 + 3.0s + 1.55}{s(s+4)} \tag{19.6.27}$$

The linear and nonlinear control loops are simulated by using SIMULINK. The reference signal is chosen as the sum of a constant, of value 2, and a square-wave, of amplitude equal to 1.

The results of the simulation are shown in Figure 19.9. This figure shows the superior tracking performance of the nonlinear design.

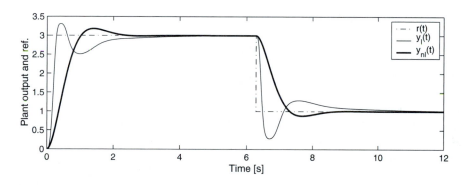

Figure 19.9. Tracking performance of linear (y_l) and nonlinear (y_{nl}) control designs for a nonlinear plant

Both control loops are in SIMULINK file **softloop1.mdl**. *The reader can use this file to investigate stability of both loops with different references, disturbance-compensation performance, robustness to model errors in the nonlinear case, effect of input saturation, and so forth.*

□□□

Example 19.4 (pH neutralization). *pH control is an extremely difficult problem in practical situations, because of the large dynamic range needed in the controller.*

To deal with this issue, it is often necessary to make structural changes to the physical setup–e.g., by providing additional mixing tanks and multiple reagent valves. The interested reader is referred to, for instance, the extensive discussion of this problem on the web page for the book. Here, we will consider a rather idealized form of the problem. We use this as a vehicle to illustrate nonlinear IMC methods rather than as a practical solution to the pH control problem.

It can be shown, from elementary mass-balance considerations, that an appropriate state space model for pH neutralization of the strong-acid–strong-base type is given by

$$\frac{dc_o(t)}{dt} = \frac{u(t)}{V}\left(c_u - c_o(t)\right) + \frac{q}{V}\left(c_i + d(t) - c_o(t)\right) \tag{19.6.28}$$

$$\frac{dp_m(t)}{dt} = \frac{1}{\alpha}\left(p_o(t) - p_m(t)\right) \tag{19.6.29}$$

$$p_o(t) = -\log\left[\sqrt{0.25\left(c_o(t)\right)^2 + 10^{-14}} + 0.5c_o(t)\right] \tag{19.6.30}$$

where

c_i, c_o, c_u :   *excess of hydrogen in inlet stream, effluent and control acid stream, respectively*
q :   *flow rate of inlet stream*
V :   *tank volume*
d :   *disturbance in the inlet stream*
u :   *flow rate of control acid stream*
p_o, p_m :   *true pH and measured pH respectively, of effluent*

The purpose of the control system is to regulate the effluent pH, $p_o(t)$, to a set-point value, by controlling the flow rate, $u(t)$, of the control acid stream. Note that the process itself is a first-order nonlinear system, with the sensor introducing a lag and second state.

We next observe that Equations (19.6.28) to (19.6.30) correspond to the state Equation (19.6.1), where $x_1(t) = c_o(t)$ and $x_2(t) = p_m(t)$. Also, equation (19.6.2) corresponds to $y(t) = x_2(t)$.

Then, the relative degree of this nonlinear system is 2, except when $c_o = c_u$. In practice, this point of singularity can be avoided by appropriate choice of c_u.

Using the theory presented in section §19.6, with

$$p(\rho) = \alpha\beta\rho^2 + (\alpha + \beta)\rho + 1 \tag{19.6.31}$$

we have that an approximate inverse is obtained, for $d = 0$, if

$$u(t) = \frac{V \ln(10)}{\beta(c_o(t) - c_u)}\sqrt{(c_o(t))^2 + 4\times 10^{-14}}(\nu(t) - p_o(t)) + \frac{q(c_i - c_o(t))}{c_o(t) - c_u} \tag{19.6.32}$$

The implementation of this inverse requires that c_o and p_o be replaced by their estimates, $\hat{c}_o$ and $\hat{p}_o$, which are obtained by using a nonlinear observer.

The above design has been implemented in a SIMULINK schematic (file phloop.mdl). To evaluate the above design strategy for this pH control problem, a simulation was run with the following data:

V : $83.67\ [lt]$
q : $1[lt/min]$
c_i : -10^{-3}
c_u : 10^{-4}
α : $1\ [s]$
β : $0.5\ [s]$

In this case, there was no control until $t = 2\ [s]$, when a step set-point was introduced to regulate the effluent pH to 7. (Note that the initial effluent pH is equal to 11.) The results are shown in Figure 19.10.

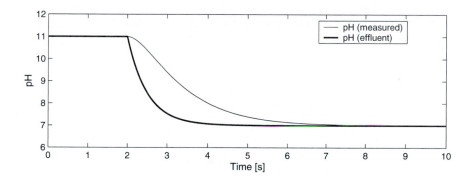

Figure 19.10. pH control by using a nonlinear control-design strategy–the effluent pH (thick line) and the measured pH (thin line) are shown

19.7   Disturbance Issues in Nonlinear Control

We have seen, in Chapter 8, that disturbances need special attention in control-system design. In particular, we found that there were subtle issues regarding the differences between input and output disturbances. In the nonlinear case, these same issues arise, but there is an extra complication arising from nonlinear behavior. The essence of the difficulty is captured in Figure 19.11.

In the linear case, the two strategies give the same result: $c_1 = c_2$. However, in the nonlinear case, $c_1 \neq c_2$, in general. For example, if $f(a) = a^2$, then $c_1 = a^2 + b^2$ whereas $c_2 = (a + b)^2$. The implications of this observation are that, whereas in

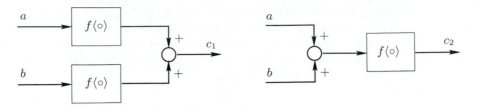

Figure 19.11. Nonlinear operators

the linear case one can freely move disturbances from input to output via linear transformations, this does not hold, in general, in the nonlinear case.

To illustrate, consider the set-up of Figure 19.12.

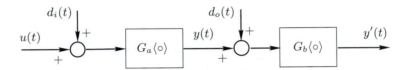

Figure 19.12. Disturbances in nonlinear systems

In Figure 19.12, $G_a\langle\circ\rangle$ and $G_b\langle\circ\rangle$ represent nonlinear dynamics, $u(t)$, $y(t)$, and $y'(t)$ represent signals, and $d_i(t)$ and $d_o(t)$ represent input and output disturbances, respectively. In this set-up, $G_b\langle\circ\rangle$ could, for example, represent the measurement system.

We will address the case of input and output disturbances separately.

(i) Input disturbances

Here, we can operate on $y'(t)$ and $u(t)$ to estimate $d_i(t)$, then cancel this at the input to the plant by action of feedback control. This leads to the strategy illustrated in Figure 19.13.

In Figure 19.13, H_{ba} and H_a are *approximate* inverses for $G_b\langle G_a\langle\circ\rangle\rangle$ and $G_a\langle\circ\rangle$, respectively. In particular, if the operator to be inverted is as in $\{(19.6.1), (19.6.2)\}$, then the approximate inverse is given by $\{(19.6.10), (19.6.12)\}$. Finally, F in Figure 19.13 is a filter introduced to avoid an algebraic loop.

(ii) Output disturbance

Here we can operate on $y'(t)$ and $u(t)$ to estimate $d_o(t)$, which is combined with the reference $r(t)$ and passed through an (approximate) inverse for $G_a\langle\circ\rangle$ so as to cancel the disturbance and cause $y(t)$ to approach $r(t)$. This leads to the strategy

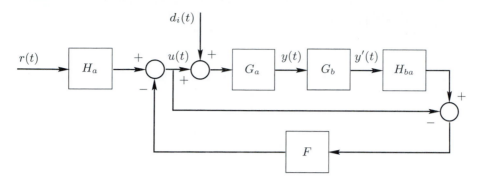

Figure 19.13. Control strategy for input disturbances

illustrated in Figure 19.14.

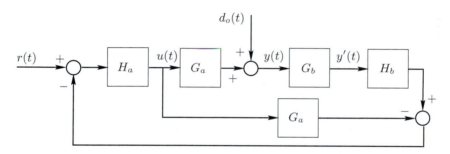

Figure 19.14. Control strategy for output disturbances

In Figure 19.14, H_b and H_a are *approximate* inverses for G_b and G_a, respectively, generated by the same methodology used for the input-disturbance case. The reader can verify that, in the linear case, one can commute the various inverses around the loops in Figures 19.13 and 19.14 to obtain identical results. However, the comments made in relation with Figure 19.11 indicate that this will not hold in the nonlinear case.

We illustrate by continuing Example 19.4 on page 573.

Example 19.5. *We consider the pH problem of Example 19.4, where G_a is given by (19.6.28) and (19.6.30) (with $y = p_o$), whilst G_b is given by (19.6.29) (with $y' = p_m$). We consider again the polynomial in (19.6.31), which we rewrite here as*

$$p(\rho) = f_2\rho^2 + f_1\rho + 1 \qquad (19.7.1)$$

By using (19.7.1), (19.6.32) becomes

$$u(t) = \frac{V \ln(10)\sqrt{(\hat{c}_o(t))^2 + 4 \times 10^{-14}}}{f_2\alpha(\hat{c}_o(t) - c_u)} \left[\alpha^2(\nu(t) - \hat{y}'(t)) - (\hat{y}'(t) - \hat{y}(t))(f_2 - \alpha f_1)\right] +$$

$$\frac{q(c_i(t) - \hat{c}_o(t))}{\hat{c}_o(t) - c_u} \quad (19.7.2)$$

where ν is the driving input to the inverse and $\hat{y}'$, $\hat{y}$, and $\hat{c}_o$ are state estimates generated by an open-loop observer, as in Figures 19.7 and 19.8 on page 571. For general values of f_1 and f_2 in (19.7.1), (19.7.2) generates H_{ba}. By choosing $f_1 = \alpha + \beta$, $f_2 = \alpha\beta$, where α is the sensor time constant in (19.6.29) and β is a free parameter, (19.7.2) generates H_a. Finally, H_b is easily chosen to be

$$H_b(s) = \frac{\alpha s + 1}{\tau s + 1}; \qquad \tau \ll \alpha \quad (19.7.3)$$

For the purpose of simulation, we choose $f_1 = 0.2$, $f_2 = 0.01$ for H_{ba}, $\beta = 0.5$ for H_a, and $\tau = 0.2$ for H_b. Finally, the filter F in Figure 19.13 on the page before was chosen as

$$F(s) = \frac{1}{0.5s + 1} \quad (19.7.4)$$

To illustrate the nonlinear characteristics of the problem, we consider several cases with set-point $r = 7.5$.

Case 1

Configuration (i)–(Figure 19.13) input-disturbance design with

(a) *input disturbance–a step of magnitude 0.005 at $t = 20$*

(b) *output disturbance–a step of magnitude -1.8886 at $t = 20$*

Note that both disturbances give the same steady-state perturbation to pH. The results are shown in Figure 19.15.
Notice that the response to the input disturbance is satisfactory, whereas the response to the output disturbance is poor. Of course, this configuration was predicated on the assumption that the disturbance was actually at the input.

Case 2

Configuration (ii)–(Figure 19.14) output-disturbance design with

(a) *input disturbance–a step of magnitude 0.005 at $t = 20$*

(b) *output disturbance–a step of magnitude -1.886 at $t = 20$*

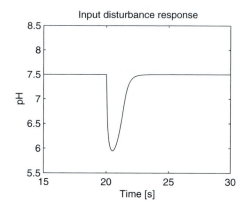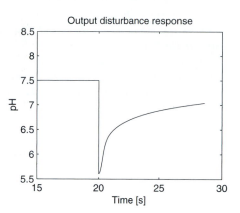

Figure 19.15. Response to input (left) and output (right) disturbances in a control loop designed for *input*-disturbance compensation ($d_i = 0.005$, $d_o = -1.886$) [Case 1]

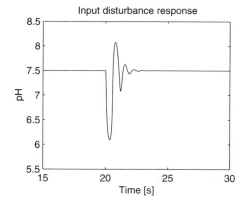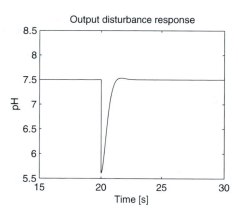

Figure 19.16. Response to input (left) and output (right) disturbances in a control loop designed for *output*-disturbance compensation ($d_i = 0.005$, $d_o = -1.886$) [Case 2]

The results are shown in Figure 19.16.

Notice that, in this case, the response to the output disturbances is satisfactory, whereas the response to the input disturbances is poor. Of course, this configuration was predicated on the assumption that the disturbance was actually at the output.

Two further cases (3 and 4) were considered. These mirror cases 1 and 2, save that the disturbances were increased in magnitude: to 0.05 (for the input disturbance), and to -2.8276 (for the output disturbance). The results are shown in Figures 19.17 and 19.18.

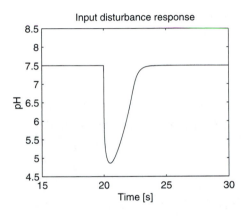 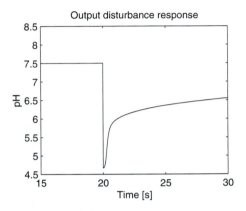

Figure 19.17. Response to input (left) and output (right) disturbances in a control loop designed for *input*-disturbance compensation ($d_i = 0.05$, $d_o = -2.8276$) [Case 3]

Notice that the same general comments apply as in Cases 1 and 2. Also, comparing Figures 19.15 and 19.16 with Figures 19.17 and 19.18, we see strong evidence of nonlinear behavior: the very nature of the responses changes with the "magnitude" of the disturbances.

This example highlights the point that, in nonlinear systems, certain linear insights no longer hold, because the principle of superposition is not valid. Thus, particular care needs to be exercised. We have seen that input and output disturbances need to be carefully considered and that the magnitude of signals plays a role. This is all part of the excitement associated with the control of nonlinear systems.

19.8   More General Plants with Smooth Nonlinearities

To highlight key issues, the discussion above has focused on plants that are both stable and stably invertible. We next examine briefly the more general situation. By analogy with the linear case (for example, in section §18.4), one might conceive

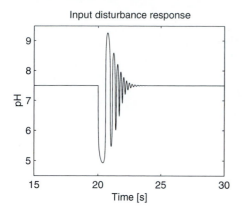
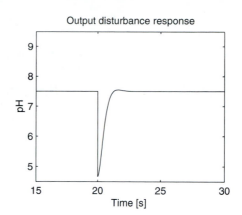

Figure 19.18. Response to input (left) and output (right) disturbances in a control loop designed for *output*-disturbance compensation ($d_i =$ 0.05, $d_o = -2.8276$) [Case 4]

of solving the general problem by a combination of a (nonlinear) observer and (nonlinear) state-estimate feedback. We found this to be a very profitable strategy in the linear case. For example, we found in section §18.4 that the closed-loop poles are simply the union of the dynamics of the observer and state feedback, considered separately. Unfortunately, this does not hold in the nonlinear case where, *inter alia*, there will generally be interaction between the observer and state feedback. This makes the nonlinear problem much more difficult; indeed, it is still the subject of ongoing research. We warn the reader that the methods described below will work only locally, because of the linear approximations involved in the deviations.

19.8.1   Nonlinear Observer

Consider a plant of the form given in (19.6.1) and (19.6.2). Say that we are given an estimate $\hat{x}(t)$ of the state at time t. We will use linearization methods to see how we might propagate this estimate.

The linearized forms of (19.6.1) and (19.6.2) about $\hat{x}(t)$ are, respectively,

$$\rho x(t) \approx f(\hat{x}) + \left.\frac{\partial f}{\partial x}\right|_{\hat{x}} [x(t) - \hat{x}(t)] + g(\hat{x})\, u(t) + \left.\frac{\partial g}{\partial x}\right|_{\hat{x}} [x(t) - \hat{x}(t)]\, u(t)$$

$$y(t) \approx h(\hat{x}) + \left.\frac{\partial h}{\partial x}\right|_{\hat{x}} [x(t) - \hat{x}(t)]$$

For notational convenience, let

$$\mathbf{A} = \left.\frac{\partial f}{\partial x}\right|_{\hat{x}} + \left.\frac{\partial g}{\partial x}\right|_{\hat{x}} u(t)$$

$$\mathbf{B} = g(\hat{x}) - \left.\frac{\partial g}{\partial x}\right|_{\hat{x}} \hat{x}(t)$$

$$\mathbf{C} = \left.\frac{\partial h}{\partial x}\right|_{\hat{x}}$$

$$\mathbf{D} = h(\hat{x}) - \left.\frac{\partial h}{\partial x}\right|_{\hat{x}} \hat{x}(t)$$

$$\mathbf{E} = f(\hat{x}) - \left.\frac{\partial f}{\partial x}\right|_{\hat{x}} \hat{x}(t)$$

where $\mathbf{A}, \mathbf{B}, \mathbf{C}, \mathbf{D}, \mathbf{E}$ are time-varying and depend on $\hat{x}, u$. The linearized model is then

$$\rho x = \mathbf{A}x + \mathbf{B}u + \mathbf{E}$$
$$y = \mathbf{C}x + \mathbf{D}$$

(19.8.1)

This suggests the following linearized observer:

$$\rho \hat{x} = \mathbf{A}\hat{x} + \mathbf{B}u + \mathbf{E} + \mathbf{J}(y - \mathbf{C}\hat{x} - \mathbf{D})$$

Before proceeding, we note that there are two common ways of designing the observer gain $\mathbf{J}$. One way is to take fixed nominal values for $\mathbf{A}, \mathbf{B}, \mathbf{C}, \mathbf{D}, \mathbf{E}$ (say $\mathbf{A_o}, \mathbf{B_o}, \mathbf{C_o}, \mathbf{D_o}, \mathbf{E_o}$) and use these to design a single fixed-value $\mathbf{J_o}$ by using linear observer theory. In more severe nonlinear problems, one can design a different $\mathbf{J}$ at every instant of time, one based on the current values of $\mathbf{A}, \mathbf{B}, \mathbf{C}, \mathbf{D}, \mathbf{E}$, which depend on the current estimate $\hat{x}(t)$. For example, if quadratic optimization (see Chapter 22) is used to design $\mathbf{J}$, then this leads to the so-called *Extended Kalman Filter* (EKF).

Substituting for $\mathbf{A}, \mathbf{B}, \mathbf{C}, \mathbf{D}, \mathbf{E}$ leads to the following compact representation for the observer:

$$\rho \hat{x} = f(\hat{x}) + g(\hat{x})u + \mathbf{J}[y - h(\hat{x})]$$

This result is intuitively appealing, because the nonlinear observer so obtained constitutes an open-loop nonlinear model with (linear) feedback gain multiplying the difference between the actual observations, y, and those given by the nonlinear model $h(\hat{x})$.

The reader is referred to the book's website, where this nonlinear observer is applied to the problem of estimating the level of water in a coupled-tank problem.

19.8.2   Nonlinear Feedback Design

There are myriad possibilities here. For example, if the system has a stable inverse, then one could use feedback linearization, as in (19.6.12), with x replaced by $\hat{x}$ generated, say, as in subsection §19.8.1. Another possible control algorithm, related to linearized state feedback, is described next.

We return to the linearized model (19.8.1) and add integral action by defining an additional state

$$\rho x_2 = r - y$$

where r is the reference input. The composite linearized model is then

$$\rho \begin{bmatrix} x \\ x_2 \end{bmatrix} = \begin{bmatrix} \mathbf{A} & 0 \\ -\mathbf{C} & 0 \end{bmatrix} \begin{bmatrix} x \\ x_2 \end{bmatrix} + \begin{bmatrix} \mathbf{B} \\ 0 \end{bmatrix} u + \begin{bmatrix} \mathbf{E} \\ -\mathbf{D} \end{bmatrix}$$

We then design state-estimate feedback as

$$u = - \begin{bmatrix} \mathbf{K}_1 & \mathbf{K}_2 \end{bmatrix} \begin{bmatrix} x \\ x_2 \end{bmatrix}$$

More will be said about this (linear) strategy in section §22.4 of Chapter 21.

As in the case of the observer, there are several ways we could design the feedback gain $\begin{bmatrix} \mathbf{K}_1 & \mathbf{K}_2 \end{bmatrix}$. Three possibilities are as follows:

1. Base the design on a fixed set of nominal values for $\mathbf{A}, \mathbf{B}, \mathbf{C}, \mathbf{D}, \mathbf{E}$.

2. Select a set of representative state values and carry out a design for each corresponding $\mathbf{A}, \mathbf{B}, \mathbf{C}, \mathbf{D}, \mathbf{E}$. Then, use some form of supervisor and anti-wind-up switch to select the appropriate controller. (See section §19.3 for details of how this might be done.)

3. Use a time-varying gain designed based on $\mathbf{A}, \mathbf{B}, \mathbf{C}, \mathbf{D}, \mathbf{E}$, as calculated for the current state estimate.

19.9   Nonsmooth Nonlinearities

The control strategies described above require the nonlinearities to be "smooth"– i.e., differentiable and invertible. In many practical cases, one meets nonsmooth nonlinearities–e.g., deadzones or stiction.–See, for example, the Mould-Level Controller described in Chapter 8.

These kinds of nonlinearity are very difficult to deal with because of the inherent problems of modeling them. (Note that we resorted to simple dither in subsection §8.8.4 to deal with stiction.) In some cases, it is possible to model the nonlinear element, and then corrective action can be taken. We illustrate by reference to deadzones.

A deadzone is a phenomenon normally found in practice in connection with mechanical actuators. Small deadzones can be ignored, because they will not produce significant performance deterioration. However, when the size is significant, some form of compensation is necessary.

For our purposes here, it suffices to define a deadzone $m = \mathcal{DZ}\langle u \rangle$ by

$$m(t) = \mathcal{DZ}\langle u(t) \rangle \triangleq \begin{cases} 0 & \text{if } \epsilon > u(t) > -\epsilon, \\ u(t) - \epsilon & \text{if } u(t) > \epsilon, \\ u(t) + \epsilon & \text{if } u(t) < -\epsilon. \end{cases} \tag{19.9.1}$$

where ϵ is a positive number that defines the nonlinearity.

Note that a deadzone is a nonsmooth nonlinearity–it has a discontinuous first derivative. However, there exists a precompensator for this nonlinearity, as shown in Figure 19.19.

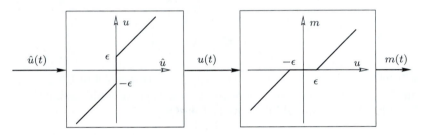

Figure 19.19. Deadzone precompensation

From Figure 19.19, it is straightforward to verify that $m(t) = \hat{u}(t)$. The advantage of this compensation is illustrated in the following example, where the size of the deadzone has been exaggerated to emphasize the difference.

Example 19.6. *Consider a plant having a model given by*

$$Y(s) = \frac{2}{s} \left(\frac{4}{(s+2)} M(s) + D_g(s) \right) \qquad with \quad m(t) = \mathcal{DZ}\langle u \rangle \tag{19.9.2}$$

We assume that the deadzone is characterized by $\epsilon = 0.5$. Also, $D_g(s)$ denotes the Laplace transform of a generalized disturbance, $d_g(t)$.

A one-d.o.f. control is designed, considering the linear part of the model. A pole-placement strategy, with integration in the controller, is used with a closed-loop polynomial $A_{cl}(s) = (s^2 + 3s + 4)(s + 5)^2$, leading to the controller

$$C(s) = \frac{4.625s^2 + 14.375s + 12.5}{s(s+11)} \qquad (19.9.3)$$

The loop is simulated with and without deadzone compensation. The reference is a square-wave of frequency 0.1[rad/s] and unit amplitude. The disturbance is taken to be $d_g(t) = \mu(t-10)$. The results are shown in Figure 19.20.

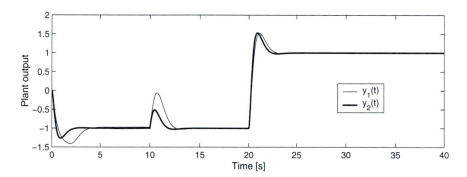

Figure 19.20. Effect of deadzone compensation in the loop performance

In Figure 19.20, the graph labeled $y_1(t)$ describes the plant output without dead-zone compensation. The graph labeled $y_2(t)$ describes the plant output when dead-zone compensation, of the form shown in Figure 19.19 on the preceding page, is used. The comparison of $y_1(t)$ and $y_2(t)$ shows the benefits of deadzone compensation.

Note that the performance difference is less significant for the step in the reference at $t = 10\pi[s]$. The reason for this is that, because of the nonzero disturbance, the steady-state input to the deadzone is no longer zero, unlike the case at $t = 0$. Thus, the effect of the deadzone is basically equivalent to a change in the d.c. gain of the process.

The reader is invited to test other design ideas by using the SIMULINK file **dead1.mdl.** *In particular, the reader is encouraged to modify the schematic, implementing the controller in the form required to add anti-wind-up control in the presence of input saturation (as in Figure 11.6 on page 298).*

□□□

19.10   Stability of Nonlinear Systems

Finally, we briefly address the issue of stability of nonlinear feedback systems. We summarize two formulations of nonlinear stability–namely, Lyapunov methods and function-space methods.

19.10.1   Lyapunov Stability

The basic idea of Lyapunov stability is to show that there exists a positive definite function (similar to energy) of the states that is decreasing along trajectories of the system. The positive definite function is usually called a *Lyapunov function*. The difficult part of applying the Lyapunov method is finding a suitable Lyapunov function; once this is done, stability follows immediately.

To give some further details, we first define the idea of global asymptotic stability for nonlinear systems. We will do this for discrete-time systems, but analogous results hold for continuous systems where differences are replaced by derivatives.

Definition 19.2. *Consider a discrete-time system of the form*

$$x[k+1] = f(x[k]); \qquad x[k_o] = x_o \qquad (19.10.1)$$

We say that the system is globally asymptotically stable, if, for all initial states $x[k_o]$ and for any $\epsilon > 0$, there exists a T such that $\| x[k_o + \tau] \| < \epsilon$, for all $\tau \geq T$.

□□□

Given the above definition, then testing for global stability is facilitated if we can find a function $V(x) \in \mathbb{R}$ (a Lyapunov function) having the following properties:

(i) $V(x)$ is a positive definite function of x: i.e., $V(x) > 0$ for all $x \neq 0, V(x)$ is continuous and is a strictly increasing function of $|x|$, and $V(x)$ is radially unbounded–i.e., $|V(x)| \to \infty$ for all $||x|| \to \infty$.

(ii) V is decreasing along trajectories of (19.10.1)–that is,

$$-(V(f(x)) - V(x)) \qquad \text{is positive definite} \qquad (19.10.2)$$

We then have, the following theorem, due to Lyapunov.

Theorem 19.1 (Lyapunov Stability). *The null solution of (19.10.1) is globally asymptotically stable if there exists a Lyapunov function for the system satisfying properties (i) and (ii) above.*

Proof

Because the system is time invariant, we can take $k_o = 0$. Also, for $||x(0)|| < \infty, V(x(0)) < \infty$. We argue by contradiction, and assume the theorem is false; that is, that, given some $\epsilon > 0$, there does not exist a T such that $||x(k)|| < \epsilon$ for all $k \geq T$.

Under these conditions, there exists an unbounded sequence of integers $S = \{k_i\}$ such that $||x(k_i)|| \geq \epsilon$ for all $k_i \epsilon S$. Using (19.10.2), we then know that

$$V(f(x(k_i))) - V(x(k_i)) < -\epsilon_2 \qquad (19.10.3)$$

for some $\epsilon_2 > 0$. Hence, if we add enough of these terms together, we can cause the sum to fall below $-V(x(0))$. Let N denote the index of the last time instant used in this sum. Then

$$\sum_{\substack{k_i \epsilon S \\ k_i \leq N}} V(f(x(k_i))) - V(x(k_i)) < -V(x(0)) \qquad (19.10.4)$$

Along trajectories of (19.10.1), *we have*

$$V(x(N)) = V(x(0)) + \sum_{k=0}^{N-1} V(x(k+1)) - V(x(k)) \qquad (19.10.5)$$

$$\leq V(x(0)) + \sum_{\substack{k_i \epsilon S \\ k_i \leq N}} V(x(k_i + 1)) - V(x(k_i))$$

$$= V(x(0)) + \sum_{\substack{k_i \epsilon S \\ k_i \leq N}} V(f(x(k_i))) - V(x(k_i))$$

$$\leq 0$$

where we have used (19.10.4)

However, this contradicts the positive definiteness of $V(x)$.

The result follows.

□□□

An application of Lyapunov methods will be given in Chapter 23, where we use this approach to prove stability of a general nonlinear-model predictive-control algorithm.

19.10.2   Circle Criterion

The Lyapunov approach to nonlinear stability is a powerful tool. The main difficulty, however, is in finding a suitable Lyapunov function. One class of problems for which an elegant solution to the issue of nonlinear stability exists is that of a feedback system comprising a linear dynamic block together with static (or memoryless) nonlinear feedback. This is often called the Lur'e problem–see Figure 19.21.

In the sequel, we will need to make use of the following result from Optimal Control Theory, that we state without proof.

Lemma 19.1 (Kalman-Yacubovitch Lemma). *Given a stable single-input single-output linear system, ($\mathbf{A}$,$\mathbf{B}$,$\mathbf{C}$,$\mathbf{D}$), with ($\mathbf{A}$,$\mathbf{B}$) controllable, and given a real vector, v, and scalars $\gamma \geq 0$ and $\varepsilon > 0$, and a positive definite matrix $\mathbf{Q}$, then there exists a positive definite matrix $\mathbf{P}$ and a vector q such that*

$$\mathbf{A}^T \mathbf{P} + \mathbf{P}\mathbf{A} = -qq^T - \varepsilon\mathbf{Q} \qquad (19.10.6)$$

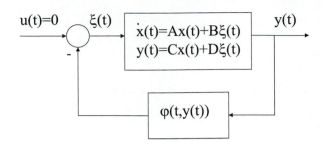

Figure 19.21. The single-input single-output Lur'e problem

and

$$\mathbf{PB} - v = \gamma^{\frac{1}{2}}q \qquad (19.10.7)$$

if and only if ε is small enough, and the scalar function

$$H(s) = \gamma + 2v^T(s\mathbf{I} - \mathbf{A})^{-1}\mathbf{B} \qquad (19.10.8)$$

satisfies

$$\Re\{H(j\omega)\} > 0, \qquad \text{for all } \omega \qquad (19.10.9)$$

where $\Re\{\cdot\}$ denotes the real part.

□□□

We then have the following stability result:

Theorem 19.2 (Circle Criterion). *Consider the Lur'e system illustrated in Figure 19.21. Provided,*

(i) the linear system $\dot{x} = \mathbf{A}x + \mathbf{B}\xi$; $y = \mathbf{C}x$ is stable, completely controllable, completely observable, and has a Nyquist plot that lies strictly to the right of $-\frac{1}{k}$, $k > 0$, and

(ii) the nonlinearity $\varphi(t, y)$ belongs to the sector $(0, k)$ in the sense that

$$0 \leq y\phi(t, y) \leq ky^2 \qquad \forall y \in \mathbb{R}, \forall t \geq 0 \qquad (19.10.10)$$

then the feedback loop of Figure 19.21 is globally asymptotically stable.

Proof

We consider the Lypunov function

$$V(x) = x^T \mathbf{P} x \qquad (19.10.11)$$

Calculating the derivative of V along the solutions of the system in Figure 19.21.

$$\begin{aligned} \dot{V}(x) &= \dot{x}^T \mathbf{P} x + x^T \mathbf{P} \dot{x} \\ &= x^T (\mathbf{A}^T \mathbf{P} + \mathbf{P} \mathbf{A}) x + 2 x^T \mathbf{P} \mathbf{B} \xi \end{aligned} \qquad (19.10.12)$$

Now it is easy to see that the following equality is true:

$$k \xi x^T \mathbf{C}^T + (1 + k \mathbf{D}) \xi^2 - \xi (ky + \xi) = 0 \qquad (19.10.13)$$

Subtracting (19.10.13) from (19.10.12) we obtain

$$\begin{aligned} \dot{V}(x) =& x^T (\mathbf{A}^T \mathbf{P} + \mathbf{P} \mathbf{A}) x + 2 \xi x^T (\mathbf{P} \mathbf{B} - \frac{1}{2} k \mathbf{C}^T) \\ &- (1 + k \mathbf{D}) \xi^2 + \xi (ky + \xi) \end{aligned} \qquad (19.10.14)$$

Now, since φ satisfies (19.10.10), we have that

$$0 \leq \frac{\phi(t, y)}{y} \leq k, \qquad \text{for all } y \neq 0 \qquad (19.10.15)$$

Equation (19.10.15) implies that $\varphi(t, y)$ and $ky - \varphi(t, y)$ always have the same sign, i.e.,

$$\varphi(t, y)[ky - \varphi(t, y)] \geq 0 \qquad \text{for all } y \in \mathbb{R} \qquad (19.10.16)$$

Recalling that–(see Figure 19.21)

$$\xi = -\varphi(t, y) \qquad (19.10.17)$$

and substituting into (19.10.16) we obtain

$$\xi(ky + \xi) \leq 0 \qquad (19.10.18)$$

Substituting (19.10.18) into (19.10.14) yields

$$\dot{V}(x) \leq x^T(\mathbf{A}^T\mathbf{P} + \mathbf{P}\mathbf{A})x + 2\xi x^T(\mathbf{P}\mathbf{B} - \frac{1}{2}k\mathbf{C}^T) - (1 + k\mathbf{D})\xi^2 \qquad (19.10.19)$$

Now condition (i) of the statement implies that

$$\Re\{[\mathbf{C}(s\mathbf{I} - \mathbf{A})^{-1}\mathbf{B} + \mathbf{D}] + \frac{1}{k}\} \geq 0; \quad s = j\omega \qquad (19.10.20)$$

or, because $k > 0$,

$$\Re\{k[\mathbf{C}(s\mathbf{I} - \mathbf{A})^{-1}\mathbf{B} + \mathbf{D}] + 1\} \geq 0; \quad s = j\omega \qquad (19.10.21)$$

We then define $\gamma = (1 + k\mathbf{D})$ and $v = \frac{1}{2}k\mathbf{C}^T$. We note that (19.10.21) then implies

$$\Re\{\gamma + 2v^T(s\mathbf{I} - \mathbf{A})^{-1}\mathbf{B}\} \geq 0; \quad s = j\omega \qquad (19.10.22)$$

that is, condition (i) is equivalent to (19.10.9) of the Kalman-Yacubovitch Lemma.

Also, we note that part (i) of the conditions evaluated at $\omega = \infty$ yields $\gamma = (1 + k\mathbf{D}) > 0$.

Hence, from the Kalman-Yacubovitch Lemma, we have that given a positive definite matrix $\mathbf{Q}$, there exists a positive definite matrix $\mathbf{P}$ and a vector q such that

$$\mathbf{A}^T\mathbf{P} + \mathbf{P}\mathbf{A} = -qq^T - \varepsilon\mathbf{Q} \qquad (19.10.23)$$

and

$$\mathbf{P}\mathbf{B} - \frac{1}{2}k\mathbf{C}^T = (1 + k\mathbf{D})^{\frac{1}{2}}q \qquad (19.10.24)$$

for any $\varepsilon > 0$.

Substituting (19.10.23), (19.10.24) into (19.10.19) yields

$$\begin{aligned}
\dot{V}(x) \leq &- \varepsilon x^T\mathbf{Q}x - x^Tqq^Tx \qquad (19.10.25)\\
&- 2\xi x^T(1 + k\mathbf{D})^{\frac{1}{2}}q - (1 + k\mathbf{D})\xi^2\\
= &- \varepsilon x^T\mathbf{Q}x - (x^Tq - (1 + k\mathbf{D})^{\frac{1}{2}}\xi)^2\\
\leq &- \varepsilon x^T\mathbf{Q}x
\end{aligned}$$

The result then follows from the continuous version of Theorem 19.2 on page 588.

□□□

The previous result can be extended to the case when the nonlinearity lies in a sector (k_1, k_2) in the sense that

$$k_1 y^2 \leq y\varphi(t, y) \leq k_2 y^2 \tag{19.10.26}$$

For example, we have the following corollary for the case $0 < k_1 < k_2$:

Corollary 19.1. *Consider the Lur'e system illustrated in Figure 19.21 with $G(s) = \mathbf{C}(s\mathbf{I} - \mathbf{A})^{-1}\mathbf{B} + \mathbf{D}$. Provided*

(i) The linear system ($\dot{x} = \mathbf{A}x + \mathbf{B}\xi, y = \mathbf{C}x$) has n_c unstable poles, is completely controllable, completely observable and has a Nyquist plot that does not enter a circle of centre $\frac{-(k_1+k_2)}{2k_1 k_2}$ and radius $\frac{(k_2-k_1)}{2k_1 k_2}$ but encircles it n_c times counterclockwise. Then, the loop is globally asymptotically stable.

Proof

We consider the loop transformation shown in Figure 19.22. In this case, we define

$$\hat{G}(s) = \frac{G(s)}{1 + k_1 G(s)} \tag{19.10.27}$$

and

$$\hat{\varphi}(y) = \varphi(y) - k_1 y \tag{19.10.28}$$

We can see that $\hat{\varphi}(y)$ satisfies Equation (19.10.10) with $k = (k_2 - k_1)$. We can thus apply Theorem 19.2 on page 588 provided $\hat{G}(s)$ is stable and has a Nyquist plot that lies to the right of $\frac{-1}{(k_2-k_1)}$. Arguing in a similar fashion to equations (19.10.20) and (19.10.21), we have

$$\Re\{(k_2 - k_1)\hat{G}(j\omega) + 1\} \geq 0 \tag{19.10.29}$$

Using (19.10.27) in (19.10.29) and writing $G(j\omega) = \alpha + j\beta$, we have

$$\frac{(1 + k_2\alpha)(1 + k_1\alpha) + k_1 k_2 \beta^2}{(1 + k_1\alpha)^2 + \beta^2 k_1^2} > 0 \tag{19.10.30}$$

The denominator of (19.10.30) is always greater than zero. Hence (19.10.30) implies

$$\left(\alpha + \frac{1}{k_1}\right)\left(\alpha + \frac{1}{k_2}\right) + \beta^2 > 0 \tag{19.10.31}$$

This implies that the Nyquist plot of $G(s)$ must lie outside the circle given in the corollary statement. Finally, the n_c counterclockwise encirclements ensures that $\hat{G}(s)$ is stable. The result follows from Theorem 19.2.

□□□

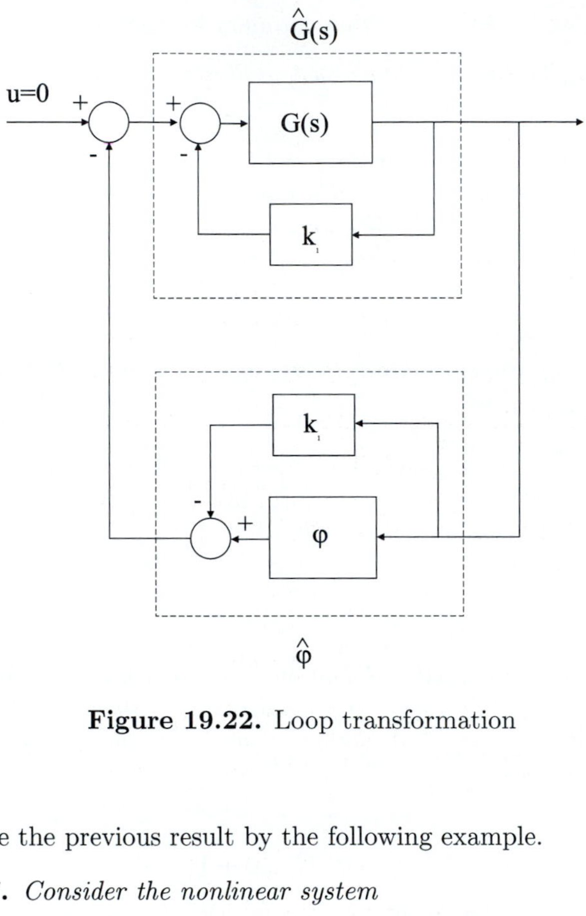

Figure 19.22. Loop transformation

We illustrate the previous result by the following example.

Example 19.7. *Consider the nonlinear system*

$$\dot{x}_1 = 10x_1 - 10x_2 \tag{19.10.32}$$

$$\dot{x}_2 = 16.925x_1 - 16x_2 + 0.1\tan^{-1}(x_2) - 0.1u \tag{19.10.33}$$

$$y = x_2 \tag{19.10.34}$$

Show that this system is globally asymptotically stable.

Solution

The system can be put into the Lur'e structure, with

$$\mathbf{A} = \begin{bmatrix} 10 & -10 \\ 16.925 & -16 \end{bmatrix}; \quad \mathbf{B} = \begin{bmatrix} 0 \\ -0.1 \end{bmatrix} \tag{19.10.35}$$

$$\mathbf{C} = \begin{bmatrix} 0 & 1 \end{bmatrix} \tag{19.10.36}$$

$$\varphi(y) = \tan^{-1}(y) \tag{19.10.37}$$

It is readily seen that

$$0 \le y\varphi(y) \le y^2 \tag{19.10.38}$$

Also, it is readily verified that the Nyquist plot of (19.10.35), (19.10.36) lies to the right of the point −1. The result follows from Theorem 19.2 on page 588.

□□□

19.10.3   Input-Output Stability via Function-Space Methods

Here we use the nonlinear-operator approach introduced in section §19.2. We will use χ to denote a Banach space[1] and we will use f to denote a nonlinear operator on χ, i.e., a mapping from its domain $\mathcal{D}(f) \subset \chi$ into χ. The domain and range of f are defined as follows:

$$\mathcal{D}(f) \triangleq \{\overline{x} \in \chi : f\langle\overline{x}\rangle \in \chi\} \tag{19.10.39}$$

$$\mathcal{R}(f) \triangleq \{f\langle\overline{x}\rangle : \overline{x} \in \mathcal{D}(f)\} \tag{19.10.40}$$

We also define $f^{-1}(Y)$ as the preimage set of Y through f:

$$f^{-1}(Y) = \{x \in \mathcal{D}(f) : f\langle x\rangle \in Y\} \tag{19.10.41}$$

With the above definitions, we have the following:

Definition 19.3. *We say that an operator is input-output stable if its domain is χ.*

Definition 19.4. *We say that an operator is input-output unstable if its domain is a strict subset of χ.*

Definition 19.5. *We say that an operator is input-output nonminimum phase if the closure of its range is a strict subset of χ.*

---

[1] A normed vector space is said to be a Banach space if all Cauchy sequences converge to limit points in the space.

Remark 19.2. *To provide motivation for the above concept of a nonminimum-phase nonlinear operator, we specialize to the case of linear time-invariant systems. In this case, if f has a nonminimum-phase zero, then the range of such an operator is the set of signals in χ that have a zero at the same frequency, i.e., a strict subset of χ.*

We will also need to define an operator gain. We will use the Lipschitz gain, which is defined as follows:

$$||f||_L \overset{\triangle}{=} \sup\left\{ \frac{|f\langle x\rangle - f\langle y\rangle|}{|x - y|} : x, y \in \mathcal{D}(f), x \neq y \right\} \tag{19.10.42}$$

where $|\circ|$ is the norm on χ.

Definition 19.6. *We say that f is Lipschitz stable if $\mathcal{D}(f) = \chi$ and $||f||_L < \infty$.*

We next define inversion of an operator.

Definition 19.7. *A Lipschitz-stable operator f is Lipschitz invertible if there is a Lipschitz-stable operator f^{-1} such that $f^{-1}\langle f\langle\circ\rangle\rangle = \mathbf{I}$ (the identity operator).*

□□□

Remark 19.3. *Note that it is clearly necessary for $\mathcal{R}(f)$ to be χ for f to be Lipschitz invertible; thus, a nonminimum-phase operator is not Lipschitz invertible.*

The following mathematical result is now stated without proof. (For a proof, see Martin (1976).)

Theorem 19.3. *Let χ be a Banach space and let f be Lipschitz stable. Suppose that $||f||_L < 1$; then $(\mathbf{I} + f)$ is Lipschitz invertible and*

$$||(\mathbf{I} + f)^{-1}||_L \leq (1 - ||f||_L)^{-1} \tag{19.10.43}$$

□□□

The above result can be used to analyze certain stability questions in nonlinear feedback systems. To illustrate, we will use it to extend the robust stability theorem of Chapter 5 to the nonlinear case. In particular, we have

Theorem 19.4. *Consider the set-up described in section §19.2. In particular, consider the equivalent signal loop given in Figure 19.2 on page 562, where G_ϵ is a nonlinear operator describing the effect of additive nonlinear modeling errors. A sufficient condition for robust (Lipschitz) stability is that g be Lipschitz stable and*

$$||g||_L < 1 \tag{19.10.44}$$

where g is the forward-path nonlinear operator in Figure 19.3:

$$g\langle\circ\rangle = G_\epsilon\left\langle S_o\langle C\langle\circ\rangle\rangle\right\rangle \tag{19.10.45}$$

Proof

We first note that in Figure 19.3, with r denoting all inputs lumped at the reference,

$$e = r - g\langle e \rangle \qquad (19.10.46)$$

Hence,

$$(\mathbf{I} + g)\langle e \rangle = r \qquad (19.10.47)$$

Now, using assumption (19.10.44) and Theorem 19.3, we know that $\mathbf{I} + g$ is Lipschitz invertible. (Hence, from Remark 19.3, $\mathcal{R}(\mathbf{I} + g) = \chi$ and $\mathcal{D}(\mathbf{I} + g) = \chi$.) The result follows from definition 19.7.

□□□

As a straightforward application of the above result, we can rederive Theorem 5.3 on page 146. In particular, in the case of linear operators in L_2 the Lipschitz gain reduces to the H_∞ norm.[2] Thus, in the linear case, condition (19.10.44) becomes

$$\sup_{\omega \in \mathbb{R}} |G_\epsilon(j\omega) S_o(j\omega) C(j\omega)| < 1 \qquad (19.10.48)$$

Note that this is identical to the condition given earlier in Equation (5.6.6).

19.11   Generalized Feedback Linearization for nonstability-Invertible Plants

We recall from section §19.6 that the feedback-linearization scheme in essence brings $p(\rho)y(t)$ to the set point signal $\nu(t)$–see Equation (19.6.13)–where $p(\rho)$ is a differential operator of degree equal to the relative degree of the nonlinear system. A drawback of the scheme, however, was that it cancelled the zero dynamics and hence required that the system have a stable inverse. Here we will show how the scheme can be generalized to cover classes of nonstable-invertible systems. In particular, we note that the basic feedback-linearization scheme achieves

$$p(\rho)y(t) = \nu(t) \qquad (19.11.1)$$

However, a difficulty in the nonstably-invertible case is that the corresponding input will not be bounded. By focusing temporarily on the input, it seems desirable to match (19.11.1) by some similar requirement on the input. Thus, we might ask that the input satisfy a linear dynamic model of the form

$$\ell(\rho)u(t) = u_s \qquad (19.11.2)$$

---

[2]The H_∞ norm of a transfer function $F(s)$ is defined as $||F||_\infty = \sup_{\omega \in \mathbb{R}} |F(j\omega)|$.

where $\ell(0) = 1$ and u_s is the steady-state input needed to achieve $y(t)$ equal to the steady-state value, ν_s, of $\nu(t)$.

Of course, (19.11.1) and (19.11.2) will, in general, not be simultaneously compatible. This suggests that we might determine the input by combining (19.11.1) and (19.11.2) in some way. For example, we could determine the input as that value of $u(t)$ that satisfies a linear combination of (19.11.1) and (19.11.2) of the form

$$(1 - \lambda)(p(\rho)y(t) - \nu(t)) + (\lambda)(\ell(\rho)u(t) - u_s) = 0 \qquad (19.11.3)$$

where $0 \leq \lambda \leq 1$.

Remark 19.4. *Equation (19.11.3) may give the false impression that it corresponds to a linear feedback policy. However, $p(\rho)y(t)$ actually encodes the future response of the system. (It has this exact interpretation in the discrete-time case.) Moreover, this "future" response is, in general, a nonlinear function of the state and control actions. Hence, in the nonlinear case, (19.11.3) actually corresponds to causal nonlinear feedback. Of course, in the special linear case, the "future" response can be evaluated as a linear function of states and control actions. Thus, in this particular case, (19.11.3) does correspond to a causal linear feedback policy. The latter case is further discussed below.*

□□□

Remark 19.5. *Clearly the control law implicitly defined by (19.11.3) is more general than the basic feedback-linearizing scheme. Indeed, it is easily seen that it leads to an internally stable control loop in the following special cases:*

(i) all stably invertible systems (whether or not they are stable)–take $\lambda = 0$,

(ii) all stable systems (whether on not they are stably invertible)–take $\lambda = 1$, and

(iii) arbitrary linear systems. To verify the latter claim, we note that if the open loop plant has linear transfer function $\frac{\mathbf{B}(s)}{\mathbf{A}(s)}$, then Equation (19.11.3) is equivalent to

$$(1 - \lambda)(\frac{p(\rho)\mathbf{B}(\rho)}{\mathbf{A}(\rho)}u(t) - \nu(t)) + (\lambda)(\ell(\rho)u(t) - u_s) = 0 \qquad (19.11.4)$$

This shows that the closed-loop characteristic equation is

$$(1 - \lambda)p(s)\mathbf{B}(s) + (\lambda)\ell(s)\mathbf{A}(s) = 0 \qquad (19.11.5)$$

Then by pole assignment principles it can be seen that there exists values of $\lambda, p,$ and ℓ which will stabilize any given linear system–see Chapter 7. (Take degree (p) equal to degree ($\mathbf{A}$) and degree (ℓ) equal to degree ($\mathbf{B}$).)

The above discussion suggests that the control law (19.11.3) has the potential to stabilize a broad class of nonlinear system which includes the class of stably invertible systems to which the basic feedback-linearizing scheme is applicable.

□□□

To develop the control law implicitly defined in (19.11.3), we introduce a dummy input $\bar{u}(t)$ defined by

$$\ell(\rho)u(t) = \bar{u}(t) \tag{19.11.6}$$

Say that $\ell(\rho)$ has degree h, and that the nonlinear system has relative degree m. Then the nonlinear system between $\bar{u}(t)$ and $y(t)$ will have relative degree $m + h$. Hence, if we use and $(m + h)$ degree operator, $p(\rho)$, then $p(\rho)y(t)$ will depend explicitly on $\bar{u}(t)$. Hence, following the development that led to (19.11.3), we can write:

$$p(\rho)y = b(x) + a(x)\bar{u} \tag{19.11.7}$$

Substituting into (19.11.3) gives the following nonlinear feedback control law:

$$\bar{u} = \frac{(1 - \lambda)(\nu - b) + \lambda u_s}{(1 - \lambda)a + \lambda} \tag{19.11.8}$$

Remark 19.6. *Of course, the success of the preceding control law depends on being able to make a judicious choice for $\ell(\rho)$ and $p(\rho)$. Equation (19.11.5) suggests that one approach would be to design $\ell(\rho), p(\rho)$ via pole assignment using a linearized model for the systems.*

□□□

This control law does not cancel the zero dynamics unless $\lambda = 0$. Clearly for $\lambda \to 0$, Equation (19.11.3) reduces to the feedback-linearizing control law (19.6.12) and for $\lambda \to 1$, $u(t)$ becomes the open-loop control policy $\ell(\rho)u(t) = u_s$. Because when $\lambda \to 1$, the control law becomes open loop, then it follows that one class of systems that this scheme will certainly handle is all open-loop stable nonlinear systems whether or not they are stably invertible.

To illustrate how the system might be used in practice, we will restrict attention in the sequel to open-loop stable systems so that, we can use an open-loop observer for the states. We follow the general philosophy introduced in Figure 19.14 on page 577 to estimate output disturbances (and reference input). This is illustrated in Figure 19.23.

An interesting property of the above scheme is that it incorporates integral action, i.e., the plant output $y(t)$–see Figure 19.23–goes to the reference $r(t)$ in steady state provided a stable steady state exists. This can be seen as follows:

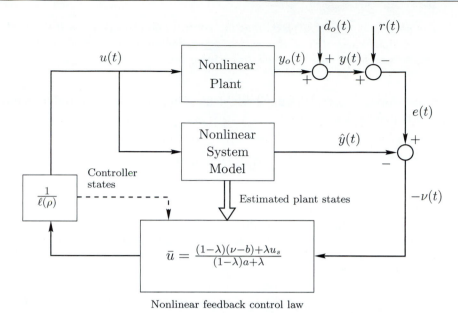

Figure 19.23. Generalized feedback linerization for open-loop stable plant

Provided a stable steady state exists, then $\nu(t)$ will converge to a constant (say ν_s). We have organized the control law so that, u_s, is chosen as the input that causes the model output to equal ν_s. If we denote the steady nonlinear gain of the model by $g(\cdot)$, then $\nu_s = g(u_s)$. Then, at steady state, the control law (19.11.3) becomes

$$(1 - \lambda)(g(u) - \nu_s) + (\lambda)(u - u_s) = 0 \qquad (19.11.9)$$

and clearly $u = u_s$ is the solution to this equation. This implies that, in steady state, $\hat{y}(t)$ in Figure 19.23 is ν_s. It then immediately follows that $e(t) = 0$ i.e., $y(t) = r(t)$, irrespective of the relationship between the model and the nonlinear plant. (Provided of course that a steady state is reached.)

We illustrate the above scheme by the following example.

Example 19.8. *Consider the nonlinear system*

$$\dot{x}_1 = 10x_1 - 10x_2 \qquad (19.11.10)$$

$$\dot{x}_2 = 16.925x_1 - 16x_2 - 0.1(u - \tan^{-1} x_2) \qquad (19.11.11)$$

$$y = x_2 + d_o \qquad (19.11.12)$$

where d_o represents a constant output disturbance.

(1) Evaluate the zero dynamics and hence show that the system is not stably invertible.

(2) Show that the system is open-loop stable.

(3) Test the basic feedback-linearizing control law with $p(\rho) = 0.2222\rho + 1$.

(4) Design a generalized feedback-linearization control law to cancel the unmeasured output disturbance, d_o, and cause the output, y, to track a constant reference signal r.

(5) Evaluate your design via simulation.

(6) Test the robustness of your design by multiplying the d.c. gain of the true plant by 2 without changing the model.

(7) Extend the algorithm to include an anti-wind-up scheme to deal with an input slew-rate limitation.

Solution

(1) The zero dynamics can be evaluated by setting $y = 0$ with $d_o = 0$. This leads to

$$\dot{x}_1 = 10x_1 \qquad\qquad (19.11.13)$$

Clearly the above zero dynamics are unstable indicating that the plant does not have a stable inverse.

(2) The system was studied in subsection §19.10.2 where it was shown to be open-loop stable.

(3) We follow the procedure of section §19.6. The system has relative degree 1. Also,

$$y = x_2 \qquad\qquad (19.11.14)$$
$$\dot{y} = 16.925x_1 - 16x_2 + 0.1\tan^{-1}x_2 - 0.1u \qquad\qquad (19.11.15)$$

Hence, in terms of the notation of section §19.6, we have

$$b = 0.2222(16.925x_1 - 16x_2 + 0.1\tan^{-1}x_2) + x_2 \qquad\qquad (19.11.16)$$
$$a = -0.02222 \qquad\qquad (19.11.17)$$

The system was then simulated as in Figure 19.23 with $\lambda = 0$ and $\ell(\rho) = 1$. A step reference change was applied. The results are shown in Figure 19.24. Note that the output response follows the desired trajectory, however, the input grows without bound. The latter outcome is a result of the nonstable invertibility of the system.

(4) *We follow the suggestion made in Remark 19.6 and linearize the system. This suggests the choice of $p(\rho)$ given in Part (3) of the problem, together with $\ell(\rho) = 1$. This leads to the control law of Equation (19.11.8) with $\bar{u} = u$ and 'a' and 'b' as in Equations (19.11.16) and (19.11.17).*

(5) *The system was simulated as in Figure 19.23 for different values of λ and for a unit reference step at $t = 1$ and a step disturbance of 0.5 at $t = 4$. The results are shown in Figure 19.25. Note that as λ decreases, so the response becomes faster and the undershoot increases. Of course, there is a lower limit on λ consistent with u remaining bounded.*

(6) *The response is shown in Figure 19.26 for the choice $\lambda = 0.2$. Note that there is no steady state error even though there is a mismatch of $2 : 1$ between the d.c. gain of the plant and model.*

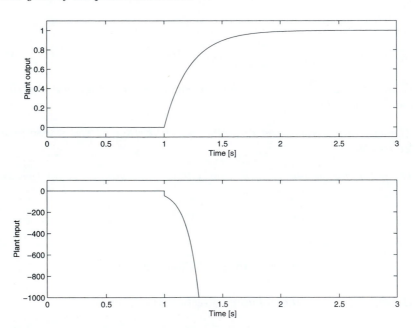

Figure 19.24. Simulation of basic feedback-linearization scheme

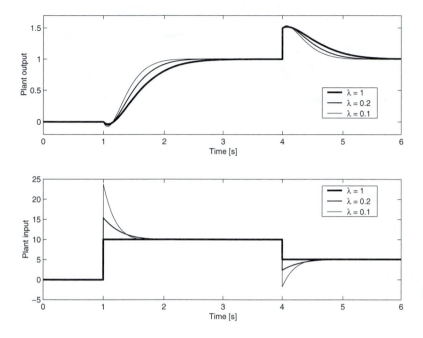

Figure 19.25. Simulation of the generalized feedback-linearization scheme

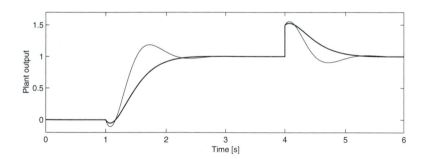

Figure 19.26. Evaluating the robustness of the generalized feedback-linearization scheme: (the thick line) is the nominal performance and (the thin line) is the response with model/plant gain mismatch

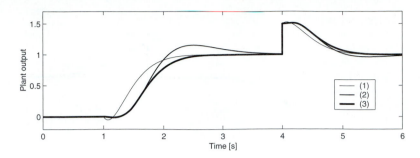

Figure 19.27. Simulation of anti-wind-up strategy: (1)–simulation of the generalized feedback-linearization strategy with $\lambda = 0.2$ when no slew-rate limitation applies, (2)–with slew-rate limitation but no compensation, and (3)–with anti-wind-up for slew-rate limitation

(7) We note that all states of the controller are contained in the parallel model used in the controller. Hence, in the light of the discussion of anti-wind-up strategies in Chapter 11, it seems that all we need do is limit the input fed to both the plant and parallel model. The efficacy of this suggestion is shown in Figure 19.27 where the slew rate constraint $|\dot{u}| \leq 20$ has been applied. We see that the nonlinear control with anti-wind-up leads to a very satisfactory solution without undesirable transients.

□□□

Remark 19.7. *So as not to leave a false impression, the only reason we required open-loop stability in the preceding example was so that an open-loop observer could be used for the system states. If the states were directly measured, or estimatable by some means, then the method would also be applicable to certain unstable open-loop systems.*

□□□

The preceding example illustrates that the generalized feedback-linearizing control law handles a wider class of nonlinear systems than the basic feedback-linearizing scheme. There remain systems however, that are not treatable within this generalized framework. The essential difficulty is that the design has a constrained structure. To deal with more difficult problems, and alternative scheme will be described in Chapter 23. There we will investigate an algorithm that incorporates measures of the future response. This, so-called, nonlinear Model Predictive Control scheme, will handle essentially all nonlinear systems at the cost of a significant increase in complexity. To avoid this complexity we suggest that it is usually desirable to, at least, begin by trying simpler schemes such as the basic feedback-linearization strategy or the generalized feedback-linearizing scheme, we have just described.

19.12   Summary

- So far, the book has emphasized linear systems and controllers.

- This chapter generalizes the scope to include various types of nonlinearities.

- A number of properties that are very helpful in linear control are not–or not directly–applicable to the nonlinear case.

  - *Frequency analysis*: The response to a sinusoidal signal is not necessarily a sinusoid; therefore, frequency analysis, Bode plots, etc., cannot be carried over directly from the linear case.

  - *Transfer functions*: The notion of transfer functions, poles, zeros, and their respective cancellation is not directly applicable.

  - *Stability* becomes more involved.

  - *Inversion*: It was highlighted in Chapter 15, on controller parameterizations, that, regardless of whether the controller contains the inverse of the model as a factor and regardless of whether one inverts the model explicitly, control is fundamentally linked to the ability to invert. Numerous nonlinear functions encountered, however, are not invertible (such as saturations, for example).

  - *Superposition* does not apply; that is: the effects of two signals (such as set-point and disturbance) acting on the system individually cannot simply be summed (superimposed) to determine the effect of the signals acting simultaneously on the system.

  - *Commutativity* does not apply.

- As a consequence, the mathematics for nonlinear control become more involved, solutions and results are not as complete, and intuition can fail more easily than in the linear case.

- Nevertheless, nonlinearities are frequently encountered and are a very important consideration.

- Smooth static nonlinearities at input and output

  - are frequently a consequence of nonlinear actuator and sensor characteristics

  - are the easiest form of nonlinearities to compensate

  - can be compensated by applying the inverse function to the relevant signal, thus obtaining a linear system in the precompensated signals (Use caution, however, with points singular such as division by zero, for particular signal values.)

- Nonsmooth nonlinearities cannot, in general, be exactly compensated or linearized.

- This chapter applies a nonlinear generalization of the affine parameterization of Chapter 15 to construct a controller that generates a feedback-linearizing controller if the model is smoothly nonlinear with stable inverse

- Nonlinear stability can be investigated by using a variety of techniques. Two common strategies are

  ○ Lyapunov methods;

  ○ function-space methods.

- Extensions of linear robustness analysis to the nonlinear case are possible

- There also exist nonlinear sensitivity limitations that mirror those for the linear case.

19.13   Further Reading

Nonlinear control

Isidori, A. (1995). *Nonlinear Control Systems*. Springer-Verlag, 3rd edition.

Khalil, H.K. (1996). *Nonlinear Systems*. Prentice-Hall, Upper Saddle River, NJ, 3rd edition.

Seron, M.M., Goodwin, G.C., and Graebe, S.F. (1995). Control system design issues for unstable linear systems having saturated input. *IEE Proceedings Part D*, 142(4):335-344.

Sepulchre, R., Janković, M. and Kokotović, P. (1997). *Constructive Nonlinear Control*. Springer.

Vidyasagar, M. (1993). *Nonlinear Systems Analysis*. Prentice-Hall, Englewood Cliffs, N.J., 2nd edition.

Smooth nonlinearities and disturbances

Graebe, S.F., Seron, M.M., and Goodwin, G.C. (1996). Nonlinear tracking and input-disturbance rejection with application to pH control. *Journal of Process Control.* 6(2-3):195-202.

Multiple controllers and bumpless transfer

Graebe, S.F. and Ahlen, A. (1996). Dynamic transfer among alternative controllers and its relation to anti-wind-up controller design. *IEEE Transactions on Control Systems Technology*, 4(1):92-99.

Graebe, S.F. and Ahlen, A. (1996). Bumpless Transfer. In *The Control Handbook*, W.S. Levine ed., p.381. CRC Press.

Rodríguez, J.A., Romagnoli, J.A., and Goodwin, G.C. (1998). Supervisory multiple model control.   In *Proceedings of the 5th IFAC Sytmposium on Dynamics and Control of Process Systems*, Corfu, Greece.

Rodríguez, J.A., Villanueva, H., Goodwin, G.C., Romagnoli, J.A., and Crisafulli, S. (1997).   Dynamic bumpless transfer mechanism.   In *Proceedings of the I.E. Aust. Control '97 Conference*, Sydney.

Rodríguez, J.A., Romagnoli, J.A., and Goodwin, G.C. (1998). Model-based switching policies for nonlinear systems.   In *Proceedings of the 1999 European Control Conference*.

Lyapunov-based design

Krstić, M., Kanellakopoulos, I., and Kokotović, P. (1995). *Nonlinear and adaptive control design*. John Wiley and Sons.

Stability

Desoer, C., and Vidyasagar, M.(1975). *Feedback systems: Input-output properties*. Academic Press.

Martin, Jr, R. (1976). *Nonlinear operations and differential equations in Banach spaces*. John Wiley and Sons Inc., New York.

Sastry, S. (1999). *Nonlinear systems*. Springer-Verlag, New York.

Nonlinear performance trade-offs

Seron, M.M., Braslavsky, J.H., and Goodwin, G.C. (1997). *Fundamental limitations in filtering and control*, Chapter 13, Springer-Verlag, Berlin.

Gain scheduling

Shamma, J. (2000). Linearization and gain scheduling. In *Control System Fundamentals*, W.S. Levine ed., CRC Press.

Robust analysis

Sandberg, J. (1965). An observation concerning the application of the contraction mapping fixed-point theorem and a result concerning the norm-boundedness of solutions of nonlinear functional equations. *Bell Systems Technical Journal*, 44:1809-1812.

Zames, G. (1966). On the input-output stability of time-varying nonlinear feedback systems. Part I. *IEEE Transactions on Automatic Control*, 11:228-238.

Zames, G. (1966). On the input-output stability of time-varying nonlinear feedback systems. Part II. *IEEE Transactions on Automatic Control*, 11:465-476.

19.14   Problems for the Reader

Problem 19.1. *Consider a nonlinear plant having a model given by*

$$\frac{d^2y(t)}{dt^2} + 7\frac{dy(t)}{dt} + 12y(t) = u(t) + 0.5\,(u(t))^3 \qquad (19.14.1)$$

Use the affine parameterization and the theory developed in this chapter for smooth input nonlinearities, and design a controller to achieve a closed-loop bandwidth approximately equal to 5 [rad/s].

Problem 19.2. *Consider a nonlinear function $y = f_D(u)$ given by*

$$y = \begin{cases} 0 & \text{if } |u| < D, \\ u & \text{if } |u| \geq D \end{cases} \qquad (19.14.2)$$

where $D \in \mathbb{R}$.

19.2.1 Determine a suitable nonlinear function $u = g(r)$ such that $g\langle\circ\rangle$ is a good approximate inverse of $f_D\langle\circ\rangle$.

19.2.2 Using the above result, design a one-d.o.f. controller for a plant having a model given by

$$-y(t) = G_{olin}\langle f_D\langle u\rangle\rangle \qquad \text{with} \qquad G_{olin}(s) = \frac{1}{s(s+1)} \qquad (19.14.3)$$

where $D = 0.3$.

19.2.3 Evaluate the tracking performance of your design for sinusoidal references of different amplitudes and frequencies.

19.2.4 Evaluate the disturbance-compensation performance of your design on step input disturbances of different magnitudes.

Problem 19.3. *Consider a nonlinear plant having a model given by*

$$\frac{d^2y(t)}{dt^2} + \left(1 + 0.2\sin\big(y(t)\big)\right)\frac{dy(t)}{dt} + 0.5y(t) = 3u(t) - sign\big(u(t)\big) \qquad (19.14.4)$$

19.3.1 Build a state space model for this system, and find its relative degree.

19.3.2 Find an approximate inverse for this plant. Assume that this approximate inverse must have reasonable accuracy in the frequency band $[0, 0.5][rad/s]$.

Problem 19.4. *Consider the same plant as in problem 19.3 and its approximate inverse.*

19.4.1 Design a nonlinear controller, based on the results presented in subsection §19.6, to track sine waves of frequency in the band $[0, 0.5][rad/s]$.

19.4.2 Compare the performance of this design with that of a loop using a linear controller. (See problem 3.9.)

Problem 19.5. *A discrete-time linear system has the state space model*

$$\begin{bmatrix} x_1[k+1] \\ x_2[k+1] \end{bmatrix} = \begin{bmatrix} 0.8 & 0.3 \\ 0.2 & 0.5 \end{bmatrix} \begin{bmatrix} x_1[k] \\ x_2[k] \end{bmatrix} \tag{19.14.5}$$

19.5.1 *Determine which of the functions below qualifies to be a Lyapunov function.*

(a) $x_1[k] + (x_2[k])^2$ (b) $(x_2[k])^2$

(c) $3(x_1[k])^2 + (x_2[k])^2$ (d) $(x_1[k] + x_2[k])^2 + (x_2[k])^2$

19.5.2 *Choose an appropriate Lyapunov function, and prove that the above system is globally asymptotically stable.*

Part VI

MIMO CONTROL ESSENTIALS

PREVIEW

In this part of the book, we turn to systems having multiple inputs and/or multiple outputs. Of course most (if not all) systems met in practice will be of this type. Fortunately, it is frequently the case that the inputs and outputs can be grouped into pairs and treated as if they were separate single-input single-output (SISO) problems. Indeed, this is why we have treated SISO control loops in such detail in earlier parts of the book. However, in other cases, nonnegligible interactions between the multiple inputs and outputs can occur. In this case, one really has no option but to tackle control design as a genuine multi-input multi-output (MIMO) problem. This is the topic of this part of the book.

Chapter 20

ANALYSIS OF MIMO CONTROL LOOPS

20.1 Preview

In previous chapters, we have focused on single-input single-output problems. Other signals in the control loop were considered to be disturbances. However, it frequently happens that what we have designated as disturbances in a given control loop are signals originating in other loops, and vice-versa. This phenomenon is known as *interaction* or *coupling*. In some cases, interaction can be ignored, either because the coupling signals are weak or because a clear time-scale or frequency-scale separation exists. However, in other cases it can be necessary to consider all signals simultaneously. This leads us to consider multi-input, multi-output (or MIMO) architectures.

Throughout this chapter, we will adhere to the convention used in the rest of the book: use boldface type to denote matrices.

20.2 Motivational Examples

All real-world systems comprise multiple interacting variables. The reader is asked to think of examples from personal experience: For example, one tries to increase the flow of water in a shower by turning on the hot tap, but then the temperature goes up; one wants to spend more time on holiday, but then one needs to spend more time at work to earn more money; the government tries to bring inflation down by reducing government expenditure, but then unemployment goes up; and so on. A more physical example is provided by the following:

Example 20.1 (Ammonia plant). *A typical industrial plant aimed at producing ammonia from natural gas is the Kellogg Process. In an integrated chemical plant of this type, there will be hundreds (possibly thousands) of variables that interact to some degree. Even if one focuses on one particular process unit–e.g., the ammonia-synthesis converters–one still ends up with 5 to 10 highly coupled variables. A typical ammonia-synthesis converter is shown in Figure 20.1. The process is exothermic;*

thus, the temperature rises across each catalyst bed. It is then cooled by mixing from the quench flows. Many measurements will typically be made–e.g., the temperature on either side of each bed.

The nature of the interactions can be visualized as follows. Say one incrementally opens quench valve 1; then all other flows will be affected, the temperature in zone 1 will drop, this will pass down the converter from bed to bed; as the reaction progressively slows, the heat exchanger will move to a different operating point and, finally, the temperature of the feed into the top of the converter will be affected. Thus, in the end, all variables will respond to the change in a single manipulated variable.

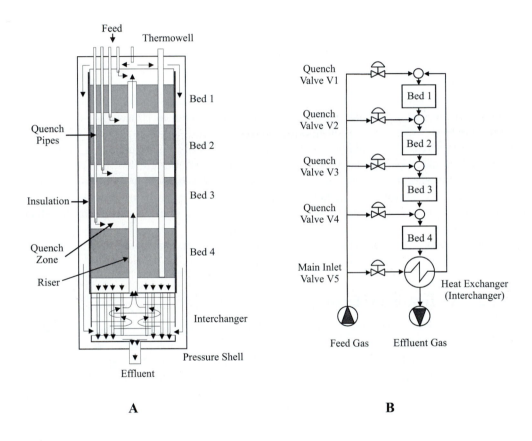

Figure 20.1. Ammonia-synthesis converter

To emphasize the wide range of physical systems exhibiting multivariable interactions, we consider a second example.

Example 20.2 (Small aircraft). *A small aircraft is ordered to maintain its present speed and course, but to climb to a significantly higher altitude; the new altitude is reached by manipulating the elevators; manipulating this control variable in isolation brings one output (altitude) to specification, but it inherently introduces an error into another, previously correct, output (the speed drops); thus, one control variable disperses its effects onto several outputs. To counteract dispersion and to change altitude without affecting speed, the elevator control action should be accompanied by just the right amount of additional throttle action.*

Consider now the same small aircraft ready for take-off; with all control surfaces neutral, a step change in the control variable throttle will accelerate the aircraft, wing lift will build, and a change will occur in the output altitude. If, however (as is routinely done to build safe lift-off speed), the control variable tail elevator is simultaneously activated downwards, then the step change in throttle will accelerate the aircraft down the runway, but without effect on altitude (until the elevator is released!). Thus, a step change in the control variable throttle is temporarily blocked from affecting the output altitude.

Obviously, these kinds of interaction are complex to understand and, as a result, they make control-system design *interesting* (to say the least). Of course, one could attempt using several SISO control loops, but this might not prove satisfactory. For example, in the ammonia-synthesis plant one could try controlling T_1, T_3, T_5, and T_7 by manipulating the four quench valves with individual PID controllers. However, this turns out to be a somewhat nontrivial task, on account of the associated interactions.

The reader may be left wondering *how interaction of the type described in the above examples affects the performance of controllers.*

20.3   Models for Multivariable Systems

Most of the ideas presented in early parts of the book apply (albeit with some slight enhancements) to multivariable systems. The main difficulty in the MIMO case is that we have to work with matrix, rather than scalar transfer functions. This means that care needs to be taken with such issues as the order in which transfer functions appear. (In general matrices do not commute.)

We review below the ideas of state space models, transfer functions, and matrix fraction descriptions (MFD's).

20.3.1   State Space Models, Revisited

Linear MIMO systems can be described by using the state space ideas presented in Chapter 17. The only change is the extension of the dimensions of inputs and outputs to vectors. In particular, if the system has input vector $u(t) \in \mathbb{R}^m$ and output vector $y(t) \in \mathbb{R}^m$, then its state space model can be written as

$$\dot{x}(t) = \mathbf{A}x(t) + \mathbf{B}u(t) \qquad\qquad x(t_o) = x_o \qquad\qquad (20.3.1)$$

$$y(t) = \mathbf{C}x(t) + \mathbf{D}u(t) \qquad\qquad\qquad\qquad\qquad (20.3.2)$$

where $x \in \mathbb{R}^n$ is the state vector, $x_o \in \mathbb{R}^n$ is the state vector at time $t = t_o$, $\mathbf{A} \in \mathbb{R}^{n \times n}, \mathbf{B} \in \mathbb{R}^{n \times m}, \mathbf{C} \in \mathbb{R}^{m \times n}$, and $\mathbf{D} \in \mathbb{R}^{m \times m}$.

We ask the reader to review the notions of system properties presented in sections §17.6 and §17.7, which also hold in the MIMO case.

Of course, if we start with a SISO system and make it multivariable, by adding extra inputs or outputs, then such system properties as controllability, stabilizability, observability, and detectability could also change. This can be appreciated by noting, for example, that, if we add inputs, then we change the matrix $\mathbf{B}$ (by adding columns), and we thus have the potential to reduce the dimension of the uncontrollable (unreachable) subspace. Similarly, by adding outputs, we change the matrix $\mathbf{C}$ (by adding rows), and the dimension of the unobservable subspace can be affected.

20.3.2   Transfer-Function Models, Revisited

As in section §17.4, it is straightforward to convert a state space model to a transfer-function model.

The Laplace transforming equations (20.3.1) and (20.3.2) lead to

$$sX(s) - x(0) = \mathbf{A}X(s) + \mathbf{B}U(s) \qquad\qquad (20.3.3)$$

$$Y(s) = \mathbf{C}X(s) + \mathbf{D}U(s) \qquad\qquad (20.3.4)$$

From here, upon taking $x(0) = 0$, we have that

$$Y(s) = (\mathbf{C}(s\mathbf{I} - \mathbf{A})^{-1}\mathbf{B} + \mathbf{D})U(s) \qquad\qquad (20.3.5)$$

The *matrix transfer function* $\mathbf{G}(s)$ is then defined by

$$\mathbf{G}(s) \stackrel{\triangle}{=} \mathbf{C}(s\mathbf{I} - \mathbf{A})^{-1}\mathbf{B} + \mathbf{D} \qquad\qquad (20.3.6)$$

Note that, if the dimension of the unobservable and/or uncontrollable subspace is nonzero, then the matrix $\mathbf{G}(s)$ will be an incomplete description of the system, insofar as it describes only the input-output properties with zero initial conditions.

For future reference, we will use $G_{ik}(s)$ to denote the transfer function from the kth component of $U(s)$ to the ith component of $Y(s)$. Then $\mathbf{G}(s)$ can be expressed as

$$\mathbf{G}(s) = \begin{bmatrix} G_{11}(s) & G_{12}(s) & \cdots & G_{1k}(s) & \cdots & G_{1m}(s) \\ G_{21}(s) & G_{22}(s) & \cdots & G_{2k}(s) & \cdots & G_{2m}(s) \\ \vdots & \vdots & \cdots & \vdots & \cdots & \vdots \\ G_{i1}(s) & G_{i2}(s) & \cdots & G_{ik}(s) & \cdots & G_{im}(s) \\ \vdots & \vdots & \cdots & \vdots & \cdots & \vdots \\ G_{m1}(s) & G_{m2}(s) & \cdots & G_{mk}(s) & \cdots & G_{mm}(s) \end{bmatrix} \tag{20.3.7}$$

Definition 20.1. *A transfer-function matrix $\mathbf{G}(s)$ is said to be a "proper matrix" if every one of its elements is a proper transfer function.*

A useful observation is that, when the kth component of $u(t)$ is an impulse, $G_{ik}(s)$ is the Laplace transform of the ith component of $y(t)$. This motivates the following definition.

Definition 20.2. *The "impulse response matrix" of the system, $\mathbf{g}(t)$, is the inverse Laplace transform of the transfer-function matrix $\mathbf{G}(s)$. For future reference, we express $\mathbf{g}(t)$ as*

$$\mathbf{g}(t) = \begin{bmatrix} g_{11}(t) & g_{12}(t) & \cdots & g_{1k}(t) & \cdots & g_{1m}(t) \\ g_{21}(t) & g_{22}(t) & \cdots & g_{2k}(t) & \cdots & g_{2m}(t) \\ \vdots & \vdots & \cdots & \vdots & \cdots & \vdots \\ g_{i1}(t) & g_{i2}(t) & \cdots & g_{ik}(t) & \cdots & g_{im}(t) \\ \vdots & \vdots & \cdots & \vdots & \cdots & \vdots \\ g_{m1}(t) & g_{m2}(t) & \cdots & g_{mk}(t) & \cdots & g_{mm}(t) \end{bmatrix} = \mathcal{L}^{-1}\left[\mathbf{G}(s)\right] \tag{20.3.8}$$

where

$$g_{ik}(t) = \mathcal{L}^{-1}\left[G_{ik}(s)\right] \tag{20.3.9}$$

Note that, for a state space model, with $\mathbf{D} = \mathbf{0}$, then $\mathbf{g}(t) = \mathbf{C}e^{\mathbf{A}t}\mathbf{B}$.

20.3.3   Matrix Fraction Descriptions

Clearly, all matrix transfer descriptions comprise elements having numerator and denominator polynomials. These matrices of rational functions of polynomials can be factorized in various ways.

Consider an $(n \times m)$ transfer-function matrix $\mathbf{G}(s)$. Let $d_i^r(s)$ denote the least common multiple of the denominator polynomials in the ith row. Also, let $e_i(s)$ denote a Hurwitz polynomial of the same degree as $d_i^r(s)$. Then it is clear that we can write

$$\mathbf{G}(s) = \left[\bar{G}_D(s)\right]^{-1}\left[\bar{G}_N(s)\right] \tag{20.3.10}$$

where

$$\bar{\mathbf{G}}_D(s) = \begin{bmatrix} \dfrac{d_1^r(s)}{e_1(s)} & & \\ & \ddots & \\ & & \dfrac{d_m^r(s)}{e_m(s)} \end{bmatrix} \tag{20.3.11}$$

$$\bar{\mathbf{G}}_N(s) = \begin{bmatrix} \dfrac{n_{11}(s)}{e_1(s)} & \cdots & \dfrac{n_{1m}(s)}{e_1(s)} \\ \vdots & & \\ \dfrac{n_{m1}(s)}{e_m(s)} & \cdots & \dfrac{n_{mm}(s)}{e_m(s)} \end{bmatrix} \tag{20.3.12}$$

where $n_{11}(s)\ldots n_{mm}(s)$ are polynomials. Clearly, $\bar{\mathbf{G}}_D(s)$ and $\bar{\mathbf{G}}_N(s)$ are stable proper transfer functions.

Equations (20.3.10), (20.3.11), and (20.3.12) describe a special form of Left Matrix Fraction Description (LMFD) for $\mathbf{G}(s)$.

To obtain a Right Matrix Fraction Description (RMFD), we need to be a little cautious; in general, matrices do not commute.

Hence, let $d_i^c(s)$ denote the least common multiple of the denominator polynomials in the ith column of $\mathbf{G}(s)$. Also, let $e_i'(s)$ denote a Hurwitz polynomial of the same degree as $d_i^c(s)$. Then we can write

$$\mathbf{G}(s) = \left[G_N(s)\right]\left[G_D(s)\right]^{-1} \tag{20.3.13}$$

where

$$\mathbf{G}_D(s) = \begin{bmatrix} \dfrac{d_1^c(s)}{e_1'(s)} & & \\ & \ddots & \\ & & \dfrac{d_m^c(s)}{e_m'(s)} \end{bmatrix} \tag{20.3.14}$$

$$\mathbf{G}_N(s) = \begin{bmatrix} \dfrac{n_{11}'(s)}{e_1'(s)} & \cdots & \dfrac{n_{1m}'(s)}{e_m'(s)} \\ \vdots & & \\ \dfrac{n_{m1}'(s)}{e_1'(s)} & \cdots & \dfrac{n_{mm}'(s)}{e_m'(s)} \end{bmatrix} \tag{20.3.15}$$

where $n'_{11}(s)\ldots n'_{mm}(s)$ are polynomials. Again, $\mathbf{G}_D(s)$ and $\mathbf{G}_N(s)$ are stable proper transfer functions.

Equations (20.3.13), (20.3.14), and (20.3.15) describe a special form of Right Matrix Fraction Description (RMFD) for $\mathbf{G}(s)$.

20.3.4   Connection Between State Space Models and MFD's

Actually, a RMFD and LMFD can be obtained from a state space description of a given system by designing stabilizing state-variable feedback and an observer, respectively. To illustrate this, consider the state space model

$$\dot{x}(t) = \mathbf{A}x(t) + \mathbf{B}u(t) \tag{20.3.16}$$
$$y(t) = \mathbf{C}x(t) \tag{20.3.17}$$

We assume that the state space model is stabilizable. If not, we must first eliminate nonstabilizable modes.

Let $u(t) = -\mathbf{K}x(t) + w(t)$ be stabilizing feedback. The system can then be written as follows, by adding and subtracting $\mathbf{BK}x(t)$:

$$\dot{x}(t) = (\mathbf{A} - \mathbf{BK})x(t) + \mathbf{B}w(t) \tag{20.3.18}$$
$$y(t) = \mathbf{C}x(t) \tag{20.3.19}$$
$$w(t) = u(t) + \mathbf{K}x(t) \tag{20.3.20}$$

We can express these equations, in the Laplace-transform domain with zero initial conditions, as

$$U(s) = (\mathbf{I} - \mathbf{K}[s\mathbf{I} - \mathbf{A} + \mathbf{BK}]^{-1}\mathbf{B})W(s) \tag{20.3.21}$$
$$Y(s) = \mathbf{C}[s\mathbf{I} - \mathbf{A} + \mathbf{BK}]^{-1}\mathbf{B}W(s) \tag{20.3.22}$$

These equations have the form

$$U(s) = \mathbf{G_D}(s)W(s); \qquad Y(s) = \mathbf{G_N}(s)W(s); \qquad Y(s) = \mathbf{G_N}(s)[\mathbf{G_D}(s)]^{-1}U(s) \tag{20.3.23}$$

where $\mathbf{G_N}(s)$ and $\mathbf{G_D}(s)$ are the following two *stable* transfer-function matrices:

$$\mathbf{G_N}(s) = \mathbf{C}[s\mathbf{I} - \mathbf{A} + \mathbf{BK}]^{-1}\mathbf{B} \tag{20.3.24}$$
$$\mathbf{G_D}(s) = \mathbf{I} - \mathbf{K}[s\mathbf{I} - \mathbf{A} + \mathbf{BK}]^{-1}\mathbf{B} \tag{20.3.25}$$

We see from (20.3.23) to (20.3.25) that $(\mathbf{G_N}(s), \mathbf{G_D}(s))$ is a RMFD.

Similarly, we can use an observer to develop a LMFD. We assume that the state space model is detectable. If not, we must first eliminate nondetectable modes. Then, consider the following observer description of (20.3.16) and (20.3.17):

$$\dot{\hat{x}}(t) = \mathbf{A}\hat{x}(t) + \mathbf{B}u(t) + \mathbf{J}(y(t) - \mathbf{C}\hat{x}(t)) \tag{20.3.26}$$

$$y(t) = \mathbf{C}\hat{x}(t) + \nu(t) \tag{20.3.27}$$

We can express these equations in the Laplace domain as

$$\Phi(s) \overset{\triangle}{=} \mathcal{L}[\nu(t)] = (\mathbf{I} - \mathbf{C}[s\mathbf{I} - \mathbf{A} + \mathbf{JC}]^{-1}\mathbf{J})Y(s) - \mathbf{C}[s\mathbf{I} - \mathbf{A} + \mathbf{JC}]^{-1}\mathbf{B}U(s) \tag{20.3.28}$$

We know that, for a stable observer, $\nu(t) \to 0$ exponentially fast, hence, in steady state, we can write

$$\overline{\mathbf{G}}_{\mathbf{D}}(s)Y(s) = \overline{\mathbf{G}}_{\mathbf{N}}(s)U(s) \tag{20.3.29}$$

where

$$\overline{\mathbf{G}}_{\mathbf{N}}(s) = \mathbf{C}(s\mathbf{I} - \mathbf{A} + \mathbf{JC})^{-1}\mathbf{B} \tag{20.3.30}$$

$$\overline{\mathbf{G}}_{\mathbf{D}}(s) = \mathbf{I} - \mathbf{C}(s\mathbf{I} - \mathbf{A} + \mathbf{JC})^{-1}\mathbf{J} \tag{20.3.31}$$

Hence $(\overline{\mathbf{G}}_{\mathbf{N}}(s), \overline{\mathbf{G}}_{\mathbf{D}}(s))$ is a LMFD for the system.

The RMFD and LMFD developed above have the following interesting property:

Lemma 20.1. *There always exist a RMFD and a LMFD for a system having the following coprime factorization property:*

$$\begin{bmatrix} \overline{\mathbf{C}}_{\mathbf{D}}(s) & \overline{\mathbf{C}}_{\mathbf{N}}(s) \\ -\overline{\mathbf{G}}_{\mathbf{N}}(s) & \overline{\mathbf{G}}_{\mathbf{D}}(s) \end{bmatrix} \begin{bmatrix} \mathbf{G}_{\mathbf{D}}(s) & -\mathbf{C}_{\mathbf{N}}(s) \\ \mathbf{G}_{\mathbf{N}}(s) & \mathbf{C}_{\mathbf{D}}(s) \end{bmatrix} =$$

$$\begin{bmatrix} \mathbf{G}_{\mathbf{D}}(s) & -\mathbf{C}_{\mathbf{N}}(s) \\ \mathbf{G}_{\mathbf{N}}(s) & \mathbf{C}_{\mathbf{D}}(s) \end{bmatrix} \begin{bmatrix} \overline{\mathbf{C}}_{\mathbf{D}}(s) & \overline{\mathbf{C}}_{\mathbf{N}}(s) \\ -\overline{\mathbf{G}}_{\mathbf{N}}(s) & \overline{\mathbf{G}}_{\mathbf{D}}(s) \end{bmatrix} = \mathbf{I} \tag{20.3.32}$$

where $\mathbf{C}_{\mathbf{N}}(s)$, $\mathbf{C}_{\mathbf{D}}(s)$, $\overline{\mathbf{C}}_{\mathbf{N}}(s)$, *and* $\overline{\mathbf{C}}_{\mathbf{D}}(s)$ *are stable transfer functions defined as follows:*

$$\mathbf{C}_{\mathbf{N}}(s) = \mathbf{K}[s\mathbf{I} - \mathbf{A} + \mathbf{BK}]^{-1}\mathbf{J} \tag{20.3.33}$$

$$\mathbf{C}_{\mathbf{D}}(s) = \mathbf{I} + \mathbf{C}[s\mathbf{I} - \mathbf{A} + \mathbf{BK}]^{-1}\mathbf{J} \tag{20.3.34}$$

$$\overline{\mathbf{C}}_{\mathbf{N}}(s) = \mathbf{K}[s\mathbf{I} - \mathbf{A} + \mathbf{JC}]^{-1}\mathbf{J} \tag{20.3.35}$$

$$\overline{\mathbf{C}}_{\mathbf{D}}(s) = \mathbf{I} + \mathbf{K}[s\mathbf{I} - \mathbf{A} + \mathbf{JC}]^{-1}\mathbf{B} \tag{20.3.36}$$

and $\mathbf{G}_{\mathbf{N}}(s)$, $\mathbf{G}_{\mathbf{D}}(s)$, $\overline{\mathbf{G}}_{\mathbf{N}}(s)$, *and* $\overline{\mathbf{G}}_{\mathbf{D}}(s)$ *are defined as in (20.3.24)-(20.3.25) and (20.3.30)-(20.3.31).*

Proof

We use the same construction as in (20.3.18) but apply it to the observer (20.3.26). This leads to

$$\dot{\hat{x}}(t) = (\mathbf{A} - \mathbf{B}\mathbf{K})\hat{x}(t) + \mathbf{B}v'(t) + \mathbf{J}\nu(t) \tag{20.3.37}$$

where

$$v'(t) = u(t) + \mathbf{K}\hat{x}(t) \qquad or \qquad u(t) = -\mathbf{K}\hat{x}(t) + v'(t) \tag{20.3.38}$$
$$\nu(t) = y(t) - \mathbf{C}\hat{x}(t) \qquad or \qquad y(t) = \mathbf{C}\hat{x}(t) + \nu(t) \tag{20.3.39}$$

Hence, we can write

$$\begin{bmatrix} U(s) \\ Y(s) \end{bmatrix} = \begin{bmatrix} \mathbf{G_D}(s) & -\mathbf{C_N}(s) \\ \mathbf{G_N}(s) & \mathbf{C_D}(s) \end{bmatrix} \begin{bmatrix} V'(s) \\ \Phi(s) \end{bmatrix} \tag{20.3.40}$$

where $V'(s) = \mathcal{L}[v'(t)]$, $\Phi(s) = \mathcal{L}[\nu(t)]$. Also, we have

$$\begin{bmatrix} V'(s) \\ \Phi(s) \end{bmatrix} = \begin{bmatrix} \overline{\mathbf{C}}_\mathbf{D}(s) & \overline{\mathbf{C}}_\mathbf{N}(s) \\ -\overline{\mathbf{G}}_\mathbf{N}(s) & \overline{\mathbf{G}}_\mathbf{D}(s) \end{bmatrix} \begin{bmatrix} U(s) \\ Y(s) \end{bmatrix} \tag{20.3.41}$$

The result follows.

□□□

20.3.5   Poles and Zeros of MIMO Systems

The reader will recall that, in the SISO case, the performance of control systems was markedly dependent on the location of open-loop zeros. Thus, it would seem to be important to extend the notion of zeros to the MIMO case. As a first guess at how this might be done, one might conjecture that the zeros could be defined in terms of the numerator polynomials in (20.3.12) or (20.3.15). Although this can be done, it turns out that this does not adequately capture the "blocking" nature of zeros that was so important in the SISO case. An alternative description of zeros that does capture this "blocking" property is to define zeros of a MIMO transfer function as those values of s that make the matrix $\mathbf{G}(s)$ lose rank. This means that there exists at least one nonzero constant vector v (zero right direction) such that

$$\mathbf{G}(c)v = 0 \tag{20.3.42}$$

and at least one nonzero constant vector w (zero left direction) such that

$$w^T \mathbf{G}(c) = 0 \qquad (20.3.43)$$

where $s = c$ is one of the zeros of $\mathbf{G}(s)$. The vectors v and w^T are part of null spaces generated by the columns or rows of $\mathbf{G}(c)$, respectively. Note that the number of linearly independent vectors that satisfy (20.3.42) depends on the rank loss of $\mathbf{G}(s)$ when evaluated at $s = c$. This number is known as the *geometric multiplicity* of the zero, and it is equal to the dimension of the null space generated by the columns of $\mathbf{G}(s)$.

An additional insight into the effect of zeros is obtained by considering a particular system input $U(s)$ given by

$$U(s) = v \frac{1}{s - c} \qquad (20.3.44)$$

where v is a vector that satisfies (20.3.42). Note that (20.3.44) implies that the input contains only one forcing mode, e^{ct}. We then see that the system output $Y(s)$ is given by

$$Y(s) = \mathbf{G}(c)v \frac{1}{s - c} + Y_{ad}(s) \qquad (20.3.45)$$

where $Y_{ad}(s)$ is the Laplace transform of the natural component of the response. If we now apply (20.3.42), we observe that the forcing mode e^{ct} is blocked by the system; it does not appear in the output. This property of zeros gives rise to the term *transmission zeros*.

System zeros as defined above are not always obvious by looking at the transfer function. This is illustrated in the following example.

Example 20.3. *Consider the matrix transfer function*

$$\mathbf{G}(s) = \begin{bmatrix} \dfrac{4}{(s+1)(s+2)} & \dfrac{-1}{(s+1)} \\[3mm] \dfrac{2}{(s+1)} & \dfrac{-1}{2(s+1)(s+2)} \end{bmatrix} \qquad (20.3.46)$$

It is difficult to tell by inspection where its zeros are. However, it turns out there is one zero at $s = -3$, as can be readily seen by noting that

$$\mathbf{G}(-3) = \begin{bmatrix} 2 & \dfrac{1}{2} \\[3mm] -1 & -\dfrac{1}{4} \end{bmatrix} \qquad (20.3.47)$$

which clearly has rank 1.

□□□

Example 20.4 (Quadruple-tank apparatus). *A very interesting piece of laboratory equipment based on four coupled tanks has recently been described by Karl Hendrix Johansson. (See references given at the end of the chapter.) A photograph and discussion of such a system built at the University of Newcastle, are given in the web page.*

A schematic diagram is given in Figure 20.2.

Figure 20.2. Schematic of a quadruple-tank apparatus

Physical modeling leads to the following (linearized) transfer function linking (u_1, u_2) *with* (y_1, y_2).

$$\mathbf{G}(s) = \begin{bmatrix} \dfrac{3.7\gamma_1}{62s + 1} & \dfrac{3.7(1 - \gamma_2)}{(23s + 1)(62s + 1)} \\ \dfrac{4.7(1 - \gamma_1)}{(30s + 1)(90s + 1)} & \dfrac{4.7\gamma_2}{90s + 1} \end{bmatrix} \qquad (20.3.48)$$

Where γ_1 *and* $(1 - \gamma_1)$ *represent the proportion of the flow from pump 1 that goes into tanks 1 and 4, respectively (similarly for* γ_2 *and* $(1 - \gamma_2)$*). Actually, the time constants change a little with the operating point, but this does not affect us here.*

The system has two multivariable zeros that satisfy $\det(\mathbf{G}(s)) = 0$:

$$(23s + 1)(30s + 1) - \eta = 0 \qquad where \quad \eta = \frac{(1 - \gamma_1)(1 - \gamma_2)}{\gamma_1 \gamma_2} \qquad (20.3.49)$$

A simple root-locus argument shows that the system is of nonminimum phase for $\eta > 1$, i.e., for $0 < \gamma_1 + \gamma_2 < 1$, and of minimum phase for $\eta < 1$, i.e., for $1 < \gamma_1 + \gamma_2 < 2$.

Also, the zero direction associated with a zero $z > 0$ satisfies (20.3.43). It then follows that, if γ_1 is small, the zero is associated mostly with the first output, whilst if γ_1 is close to 1, then the zero is associated mostly with the second output.

□□□

It is also possible to relate zeros to the properties of a coprime factorization of a given transfer-function matrix. In particular, if we write

$$\mathbf{G}(s) = \mathbf{G}_N(s)[\mathbf{G}_D(s)]^{-1} = [\bar{\mathbf{G}}_D(s)]^{-1}[\bar{\mathbf{G}}_N(s)] \qquad (20.3.50)$$

where $\mathbf{G}_N(s), \mathbf{G}_D(s), \bar{\mathbf{G}}_D(s), \bar{\mathbf{G}}_N(s)$ satisfy the coprime identity given in (20.3.32), then the zeros of $\mathbf{G}(s)$ correspond to those values of s where either $\mathbf{G}_N(s)$ or $\bar{\mathbf{G}}_N(s)$ (or both) lose rank.

Similarly, we can define the poles of $\mathbf{G}(s)$ as those values of s where $\mathbf{G_D}(s)$ or $\bar{\mathbf{G}}_{\mathbf{D}}(s)$, or both, loose rank.

20.3.6   Smith-McMillan Forms

Further insight into the notion of plant poles and zeros is provided by the Smith-McMillan form of the transfer function. This result depends on further mathematical ideas that are outside the core scope of this book. We therefore leave the interested reader to follow up the ideas in Appendix B.

20.4   The Basic MIMO Control Loop

In this section, we will develop the transfer functions involved in the basic MIMO control loop.

Unless otherwise stated, the systems we consider will be square (the input vector has the same number of components as the output vector). Also, all transfer-function matrices under study will be assumed to be nonsingular almost everywhere, which means that these matrices will be singular only at a finite set of "zeros."

We consider the control of a plant having a nominal model (with an output $Y(s)$) given by

$$Y(s) = D_o(s) + \mathbf{G_o}(s)(U(s) + D_i(s)) \qquad (20.4.1)$$

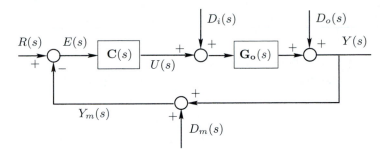

Figure 20.3. MIMO feedback loop

where $Y(s)$, $U(s)$, $D_i(s)$, and $D_o(s)$ are the vectors of dimension m corresponding to the output, input, input disturbance, and output-disturbance, respectively. $\mathbf{G_o}(s)$ is an $m \times m$ matrix (the nominal transfer function) with entries $G_{ik}(s)$.

We will consider the same basic feedback structure as in the SISO case, i.e., the structure shown in Figure 20.3.

We will emphasize the multivariable nature of these systems, by using boldface to denote matrix transfer functions.

The nominal MIMO control loop in Figure 20.3 can be described, as in the SISO case, by certain key transfer functions. In particular, we define

$\quad S_o(s) \quad : \quad$ the (matrix) transfer function connecting $D_o(s)$ to $Y(s)$
$\quad T_o(s) \quad : \quad$ the (matrix) transfer function connecting $R(s)$ to $Y(s)$
$\quad S_{uo}(s) \quad : \quad$ the (matrix) transfer function connecting $R(s)$ to $U(s)$
$\quad S_{io}(s) \quad : \quad$ the (matrix) transfer function connection $D_i(s)$ to $Y(s)$

As in the SISO case, we call S_o, T_o, S_{uo}, S_{io} the nominal sensitivity, nominal complementary sensitivity, nominal input sensitivity, and nominal input-disturbance sensitivity, respectively.

Expressions for these quantities can be obtained readily from Figure 20.3. However, care must be taken to respect the vector nature of all signals. The result of this analysis is

$$Y(s) = \mathbf{T_o}(s)R(s) - \mathbf{T_o}(s)D_m(s) + \mathbf{S_o}(s)D_o(s) + \mathbf{S_{io}}(s)D_i(s) \qquad (20.4.2)$$
$$U(s) = \mathbf{S_{uo}}(s)R(s) - \mathbf{S_{uo}}(s)D_m(s) - \mathbf{S_{uo}}(s)D_o(s) - \mathbf{S_{uo}}(s)\mathbf{G_o}(s)D_i(s)$$
$$\qquad (20.4.3)$$
$$E(s) = \mathbf{S_o}(s)R(s) - \mathbf{S_o}(s)D_m(s) - \mathbf{S_o}(s)D_o(s) - \mathbf{S_{io}}(s)D_i(s) \qquad (20.4.4)$$

where

$$\mathbf{S_o}(s) = [\mathbf{I} + \mathbf{G_o}(s)\mathbf{C}(s)]^{-1} \tag{20.4.5}$$

$$\mathbf{T_o}(s) = \mathbf{G_o}(s)\mathbf{C}(s)[\mathbf{I} + \mathbf{G_o}(s)\mathbf{C}(s)]^{-1} = [\mathbf{I} + \mathbf{G_o}(s)\mathbf{C}(s)]^{-1}\mathbf{G_o}(s)\mathbf{C}(s) \tag{20.4.6}$$

$$= \mathbf{I} - \mathbf{S_o}(s) \tag{20.4.7}$$

$$\mathbf{S_{uo}}(s) = \mathbf{C}(s)[\mathbf{I} + \mathbf{G_o}(s)\mathbf{C}(s)]^{-1} = \mathbf{C}(s)\mathbf{S_o}(s) = [\mathbf{G_o}(s)]^{-1}\mathbf{T_o}(s) \tag{20.4.8}$$

$$\mathbf{S_{io}}(s) = [\mathbf{I} + \mathbf{G_o}(s)\mathbf{C}(s)]^{-1}\mathbf{G_o}(s) = \mathbf{G_o}(s)[\mathbf{I} + \mathbf{C}(s)\mathbf{G_o}(s)]^{-1} = \mathbf{S_o}(s)\mathbf{G_o}(s) \tag{20.4.9}$$

From Equations (20.4.5)-(20.4.9), we can also derive the true, or achievable, matrix sensitivity functions by replacing the transfer function of the nominal model $\mathbf{G_o}(s)$ by the transfer function of the true system $\mathbf{G}(s)$. This leads to

$$\mathbf{S}(s) = [\mathbf{I} + \mathbf{G}(s)\mathbf{C}(s)]^{-1} \tag{20.4.10}$$

$$\mathbf{T}(s) = \mathbf{G}(s)\mathbf{C}(s)[\mathbf{I} + \mathbf{G}(s)\mathbf{C}(s)]^{-1} = [\mathbf{I} + \mathbf{G}(s)\mathbf{C}(s)]^{-1}\mathbf{G}(s)\mathbf{C}(s) \tag{20.4.11}$$

$$= \mathbf{I} - \mathbf{S}(s) \tag{20.4.12}$$

$$\mathbf{S_u}(s) = \mathbf{C}(s)[\mathbf{I} + \mathbf{G}(s)\mathbf{C}(s)]^{-1} = [\mathbf{G}(s)]^{-1}\mathbf{T}(s) \tag{20.4.13}$$

$$\mathbf{S_i}(s) = [\mathbf{I} + \mathbf{G}(s)\mathbf{C}(s)]^{-1}\mathbf{G}(s) = \mathbf{G}(s)[\mathbf{I} + \mathbf{C}(s)\mathbf{G}(s)]^{-1} = \mathbf{S}(s)\mathbf{G}(s) \tag{20.4.14}$$

Remark 20.1. *Note that, because matrix products, in general, do not commute, special care must be exercised when manipulating the above equations.*

Remark 20.2. *Note that $\mathbf{S_o}(s) + \mathbf{T_o}(s) = \mathbf{I}$ and $\mathbf{S}(s) + \mathbf{T}(s) = \mathbf{I}$. These are the multivariable versions of* (5.3.5).

20.5   Closed-Loop Stability

We next extend the notions of stability, described in section §5.4 for the SISO case, to the MIMO case.

We say that a MIMO system is stable if all its poles are strictly inside the stability region (the LHP for continuous-time systems, and the unit disk for discrete-time shift-operator systems). However, as in the SISO case, interconnection of systems, such as in the feedback loop in Figure 20.3, can yield hidden unstable modes, i.e., internal instability. This is due to potential unstable pole-zero cancellations.

We mirror Lemma 5.1 on page 128 with the following:

Lemma 20.2. *Consider the nominal control loop in Figure 20.3. Then the nominal loop is internally stable if and only if the four sensitivity functions defined in (20.4.5)-(20.4.9) are stable.*

Proof

If all input signals, i.e., $r(t)$, $d_i(t)$, $d_o(t)$, and $d_n(s)$, are bounded, then we see from (20.4.2)-(20.4.4) that stability of the four sensitivity functions is sufficient to have bounded outputs $y(t)$, $u(t)$, and $e(t)$. We also see that this condition is necessary, because $\mathbf{S_{uo}}(s)\mathbf{G_o}(s) = [\mathbf{G_o}(s)]^{-1}\mathbf{T_o}(s)\mathbf{G_o}(s)$ is stable if and only if $\mathbf{T_o}(s)$ is stable.

□□□

20.5.1   Stability in MFD Form

Stability can also be expressed by using matrix fraction descriptions (MFDs).

Consider RMFD and LMFD descriptions for the plant and the controller:

$$\mathbf{G_o}(s) = \mathbf{G_{oN}}(s)[\mathbf{G_{oD}}(s)]^{-1} = \left[\overline{\mathbf{G}}_{\mathbf{oD}}(s)\right]^{-1}\overline{\mathbf{G}}_{\mathbf{oN}}(s) \qquad (20.5.1)$$

$$\mathbf{C}(s) = \mathbf{C_N}(s)[\mathbf{C_D}(s)]^{-1} = \left[\overline{\mathbf{C}}_{\mathbf{D}}(s)\right]^{-1}\overline{\mathbf{C}}_{\mathbf{N}}(s) \qquad (20.5.2)$$

Then, the transfer functions appearing in (20.4.2) to (20.4.4) can be rewritten as

$$\mathbf{S_o}(s) = \mathbf{C_D}(s)\left[\overline{\mathbf{G}}_{\mathbf{oD}}(s)\mathbf{C_D}(s) + \overline{\mathbf{G}}_{\mathbf{oN}}(s)\mathbf{C_N}(s)\right]^{-1}\overline{\mathbf{G}}_{\mathbf{oD}}(s) \qquad (20.5.3)$$

$$\mathbf{T_o}(s) = \mathbf{G_{oN}}(s)\left[\overline{\mathbf{C}}_{\mathbf{D}}(s)\mathbf{G_{oD}}(s) + \overline{\mathbf{C}}_{\mathbf{N}}(s)\mathbf{G_{oN}}(s)\right]^{-1}\overline{\mathbf{C}}_{\mathbf{N}}(s) \qquad (20.5.4)$$

$$\mathbf{S_{uo}}(s) = \mathbf{C_N}(s)\left[\overline{\mathbf{G}}_{\mathbf{oD}}(s)\mathbf{C_D}(s) + \overline{\mathbf{G}}_{\mathbf{oN}}(s)\mathbf{C_N}(s)\right]^{-1}\overline{\mathbf{G}}_{\mathbf{oD}}(s) \qquad (20.5.5)$$

$$\mathbf{S_{io}}(s) = \mathbf{C_D}(s)\left[\overline{\mathbf{G}}_{\mathbf{oD}}(s)\mathbf{C_D}(s) + \overline{\mathbf{G}}_{\mathbf{oN}}(s)\mathbf{C_N}(s)\right]^{-1}\overline{\mathbf{G}}_{\mathbf{oN}}(s) \qquad (20.5.6)$$

$$\mathbf{S_{uo}}(s)\mathbf{G_o}(s) = \mathbf{C_N}(s)\left[\overline{\mathbf{G}}_{\mathbf{oD}}(s)\mathbf{C_D}(s) + \overline{\mathbf{G}}_{\mathbf{oN}}(s)\mathbf{C_N}(s)\right]^{-1}\overline{\mathbf{G}}_{\mathbf{oN}}(s) \qquad (20.5.7)$$

We recall that matrices $\overline{\mathbf{G}}_{\mathbf{oN}}(\mathbf{s})$, $\overline{\mathbf{G}}_{\mathbf{oD}}(\mathbf{s})$, $\mathbf{C_N}(\mathbf{s})$ and $\mathbf{C_D}(\mathbf{s})$ are matrices with stable rational entries. Then, Lemma 20.2 can be expressed in the following equivalent form.

Lemma 20.3. *Consider a one-d.o.f. MIMO feedback control loop, as shown in Figure 20.3 on page 625. Let the nominal plant model and the controller be expressed in MFD as in (20.5.1) and (20.5.2). Then the nominal loop is internally stable if and only if the* closed-loop characteristic matrix $\mathbf{A_{cl}}(s)$

$$\mathbf{A_{cl}}(s) \overset{\triangle}{=} \overline{\mathbf{G}}_{\mathbf{oD}}(s)\mathbf{C_D}(s) + \overline{\mathbf{G}}_{\mathbf{oN}}(s)\mathbf{C_N}(s) \qquad (20.5.8)$$

has all its zeros strictly in the LHP, where the zeros are defined to be the zeros of $\det\{\mathbf{A_{cl}}(s)\}$.

□□□

Remark 20.3. *Note that matrix* $\mathbf{A_{cl}}(s)$ *plays, for MIMO control loops, the same role as the one played by the polynomial* $A_{cl}(s)$ *for SISO control loops.*

Remark 20.4. *Say we define a controller by equations (20.3.33) to (20.3.36). Then, Lemma 20.1 on page 620 shows that* $\mathbf{A_{cl}}(s) = \mathbf{I}$, *and hence, the control loop is stable. Actually, this can also be seen from (20.3.38), if we set* $\Phi(s) = \mathbf{0}$. *Then (20.3.40) defines a relationship between* $U(s)$ *and* $Y(s)$ *(i.e., a feedback controller) of the form:*

$$U(s) = -\mathbf{C_N}(s)[\mathbf{C_D}(s)]^{-1}Y(s) \tag{20.5.9}$$

However, this controller really amounts to stabilizing feedback from state estimates generated by a stable observer. Hence, stability of the closed loop follows from the separation property, as in section §18.4.

Example 20.5. *A diagonal controller* $\mathbf{C}(s)$ *is proposed to control a MIMO plant with nominal model* $\mathbf{G_o}(s)$. *If* $\mathbf{C}(s)$ *and* $\mathbf{G_o}(s)$ *are given by*

$$\mathbf{G_o}(s) = \begin{bmatrix} \dfrac{2}{s+1} & \dfrac{1}{(s+1)(s+2)} \\[4mm] \dfrac{1}{(s+1)(s+2)} & \dfrac{2}{s+2} \end{bmatrix} ; \qquad \mathbf{C}(s) = \begin{bmatrix} \dfrac{2}{s} & 0 \\[3mm] 0 & \dfrac{1}{s} \end{bmatrix} \tag{20.5.10}$$

determine whether the closed loop is stable.

Solution

We use Lemma 20.3 on the page before to test stability. Thus, we need LMFD and RMFD for the plant model and the controller, respectively. A simple choice is

$$\overline{\mathbf{G}}_{\mathbf{oN}}(s) = \begin{bmatrix} 2(s+2) & 1 \\ 1 & 2(s+1) \end{bmatrix} ; \qquad \overline{\mathbf{G}}_{\mathbf{oD}}(s) = (s+1)(s+2)\mathbf{I} \tag{20.5.11}$$

and

$$\mathbf{C_N}(s) = \begin{bmatrix} 2 & 0 \\ 0 & 1 \end{bmatrix} ; \qquad \mathbf{C_D}(s) = s\mathbf{I} \tag{20.5.12}$$

Then

$$\begin{aligned} \mathbf{A_{cl}}(s) &= \overline{\mathbf{G}}_{\mathbf{oD}}(s)\mathbf{C_D}(s) + \overline{\mathbf{G}}_{\mathbf{oN}}(s)\mathbf{C_N}(s) \\[2mm] &= \begin{bmatrix} 2s^3 + 10s^2 + 18s + 8 & s^2 + 3s + 2 \\ s^2 + 3s + 2 & 2s^3 + 8s^2 + 11s + 4 \end{bmatrix} \end{aligned} \tag{20.5.13}$$

We now have to test whether the zeros of this polynomial matrix are all inside the open LHP. $\mathbf{A_{cl}}(s)$ is a polynomial matrix, so this can be simply tested, by computing the roots of its determinant. We have that

$$\det\left(\mathbf{A_{cl}}(s)\right) = 4s^6 + 36s^5 + 137s^4 + 272s^3 + 289s^2 + 148s + 28 \qquad (20.5.14)$$

All roots of $\det\left(\mathbf{A_{cl}}(s)\right)$ have negative real parts. Thus, the loop is stable.

20.5.2   Stability via Frequency Responses

The reader may well wonder whether tools from SISO analysis can be applied to test stability for MIMO systems. The answer is, in general, 'yes', but significant complications arise due to the multivariable nature of the problem. We will illustrate by showing how Nyquist theory (see section §5.7) might be extended to the MIMO case. If we assume that *only stable pole-zero cancellations occur* in a MIMO feedback loop, then the internal stability of the nominal loop is ensured by demanding that $S_o(s)$ be stable.

Consider now the function $F_o(s)$, defined as

$$F_o(s) = \det(\mathbf{I} + \mathbf{G_o}(s)\mathbf{C}(s)) = \prod_{i=1}^{m}(1 + \lambda_i(s)) \qquad (20.5.15)$$

where $\lambda_i(s)$, $i = 1, 2, \ldots, m$, are the eigenvalues of $\mathbf{G_o}(s)\mathbf{C}(s)$. The polar plots of $\lambda_i(j\omega)$, $i = 1, 2, \ldots, m$ on the complex plane are known as characteristic loci.

Comparing (20.5.15) and (20.4.5), we see that $S_o(s)$ is stable if and only if all zeros of $F_o(s)$ lie strictly inside the LHP.

If the Nyquist contour $\mathcal{C}_s = \mathcal{C}_i \cup \mathcal{C}_r$, shown in Figure 5.5 on page 141, is chosen, then we have the following theorem, which has been adapted from the Nyquist Theorem 5.1 on page 141.

Theorem 20.1. *If a proper open-loop transfer function $\mathbf{G_o}(s)\mathbf{C}(s)$ has P poles in the open RHP, then the closed loop has Z poles in the open RHP if and only if the polar plot that describes the combination of all characteristic loci (along the modified Nyquist path) encircles the point $(-1, 0)$ clockwise N=Z-P times.*

Proof

We previously observed that, because $F_o(s)$ is proper, $\mathcal{C}_r$ in $\boxed{s}$ maps into one point in $\boxed{F_o}$, which means that we are concerned only with the mapping of the imaginary axis $\mathcal{C}_i$. On this curve,

$$\arg\langle F_o(j\omega)\rangle = \sum_{i=1}^{m}\arg\langle(1 + \lambda_i(j\omega))\rangle \qquad (20.5.16)$$

Thus, any change in the angle of $F_o(j\omega)$ results from the combination of phase changes in the terms $1 + \lambda_i(j\omega)$. Hence, encirclements of the origin in the complex plane $\boxed{F_o}$ can be computed from the encirclements of $(-1; 0)$ by the combination of the characteristic-locus plots.

The relationship between the number of encirclements and the RHP poles and zeros of $F_o(s)$ follows the same construction as in Chapter 5.

□□□

Using the above result for design lies outside the scope of this text. The reader is directed to the references at the end of the chapter.

20.6   Steady-State Response for Step Inputs

Steady-state responses also share much in common with the SISO case. Here, however, we have vector inputs and outputs. Thus, we will consider step inputs coming from particular directions–i.e., applied to various combinations of inputs in the input vectors. This is achieved by defining

$$R(s) = K_r \frac{1}{s}; \qquad D_i(s) = K_{di} \frac{1}{s}; \qquad D_o(s) = K_{do} \frac{1}{s} \qquad (20.6.1)$$

where $K_r \in \mathbb{R}^m$, $K_{di} \in \mathbb{R}^m$, and $K_{do} \in \mathbb{R}^m$ are constant vectors.

It follows, provided that the loop is stable, that all signals in the loop will reach constant values at steady state. These values can be computed from (20.4.2)-(20.4.4), by using the final-value theorem, to yield

$$\lim_{t \to \infty} y(t) = \mathbf{T_o}(0)K_r + \mathbf{S_o}(0)K_{do} + \mathbf{S_{io}}(0)K_{di} \qquad (20.6.2)$$

$$\lim_{t \to \infty} u(t) = \mathbf{S_{uo}}(0)K_r - \mathbf{S_{uo}}(0)K_{do} - \mathbf{S_{uo}}(0)\mathbf{G_o}(0)K_{di} \qquad (20.6.3)$$

$$\lim_{t \to \infty} e(t) = \mathbf{S_o}(0)K_r - \mathbf{S_o}(0)K_{do} - \mathbf{S_{io}}(0)K_{di} \qquad (20.6.4)$$

It is also possible to examine the circumstances that lead to zero steady-state errors. By way of illustration, we have the following result for the case of step reference signals.

Lemma 20.4. *Consider a stable MIMO feedback loop, as in Figure 20.3 on page 625. Assume that the reference $R(s)$ is a vector of the form shown in equation (20.6.1). The steady-state error in the ith channel, $e_i(\infty)$, is zero if the ith row of $\mathbf{S_o}(0)$ is zero. Under these conditions, the ith row of $\mathbf{T_o}(0)$ is the elementary vector $e_i = [0 \ldots 0 \ 1 \ 0 \ldots 0]^T$*

Proof

Consider a reference vector, as in (20.6.1). Then the steady-state error $e_i(\infty)$ satisfies

$$e_i(\infty) = [\mathbf{S}(0)]_{i*}K_r = 0 \qquad (20.6.5)$$

where $[\mathbf{S_o}(0)]_{i*}$ is the ith row of $\mathbf{S_o}(0)$. However, $e_i(\infty) = 0$ for every constant vector K_r, as (20.6.5) is satisfied only if $[\mathbf{S}(0)]_{i*}$ has all its elements equal to zero– i.e., $\mathbf{S_o}$ loses rank at $s = 0$. This is exactly the definition of a zero in a MIMO system, as explained in section §B.5.

The property of the ith row of $\mathbf{T_o}(0)$ is a direct consequence of the identity $\mathbf{S_o}(s) + \mathbf{T_o}(s) = \mathbf{I}$ and the property of the ith row of $\mathbf{S_o}(0)$.

□□□

Analogous results hold for the disturbance inputs. We leave the reader to establish the corresponding conditions.

20.7 Frequency-Domain Analysis

We found, in the SISO case, that the frequency domain gave valuable insights into the response of a closed loop to various inputs. This is also true in the MIMO case. However, to apply these tools, we need to extend the notion of *frequency-domain gain* to the multivariable case. This is the subject of the next subsection.

20.7.1 Principal Gains and Principal Directions

Consider a MIMO system with m inputs and m outputs, having an $m \times m$ matrix transfer function $\mathbf{G}(s)$:

$$Y(s) = \mathbf{G}(s)U(s) \qquad (20.7.1)$$

We obtain the corresponding frequency response by setting $s = j\omega$. This leads to the question: *How can one define the* gain *of a MIMO system in the frequency domain?* To answer this question, we follow the same idea used to describe the gain in SISO systems, where we used the quotient between the absolute value of the (scalar) plant output $|Y(j\omega)|$ and the absolute value of the (scalar) plant input $|U(j\omega)|$. For the MIMO case, this definition of gain is inadequate, because we have to compare a vector of outputs with a vector of inputs. The resolution of this difficulty is to use vector norms instead of absolute values. Any suitable norm could be used. We will use $||r||$ to denote the norm of the vector v. For example, we could use the Euclidean norm, defined as follows:

$$||v|| = \sqrt{|v_1|^2 + |v_2|^2 + \ldots |v_n|^2} = \sqrt{v^H v} \qquad (20.7.2)$$

where v^H denotes conjugate transpose.

Furthermore, we recognize that, in the MIMO case, we have the problem of directionality: the magnitude of the output depends not only on the magnitude of the input (as in the scalar case) but also on the relative magnitude of the input components. For the moment, we fix the frequency, ω, and we note that, for fixed ω, the matrix transfer function is simply a complex matrix $\mathbf{G}$.

A possible way to define the MIMO system gain at frequency ω is then to choose a norm for the matrix $\mathbf{G}$ that considers the maximizing direction associated with the input for U. Thus, we define

$$||\mathbf{G}|| = \sup_{||U|| \neq 0} \frac{||\mathbf{G}U||}{||U||} \tag{20.7.3}$$

We call $||\mathbf{G}||$ the induced norm on $\mathbf{G}$ corresponding to the vector norm $||U||$. For example, when the vector norm is chosen to be the Euclidean norm,

$$||x|| = \sqrt{x^H x} \tag{20.7.4}$$

then we have the *induced spectral norm* for $\mathbf{G}$ defined by

$$||\mathbf{G}|| = \sup_{||U|| \neq 0} \frac{||\mathbf{G}U||}{||U||} = \sup_{||U|| \neq 0} \sqrt{\frac{U^H \mathbf{G}^H \mathbf{G} U}{U^H U}} \tag{20.7.5}$$

Actually, the above notion of induced norm is closely connected to the notion of "singular values." To show this connection, we recall the definition of singular values of an $m \times l$ complex matrix $\mathbf{\Gamma}$. The set of singular values of $\mathbf{\Gamma}$ is a set of cardinality $k = \min(l, m)$ defined by

$$(\sigma_1, \sigma_2, \ldots, \sigma_k) = \begin{cases} \sqrt{\text{eigenvalues of } \quad \mathbf{\Gamma}^H \mathbf{\Gamma}} & \text{if } m < l \\ \sqrt{\text{eigenvalues of } \quad \mathbf{\Gamma} \mathbf{\Gamma}^H} & \text{if } m \geq l \end{cases} \tag{20.7.6}$$

We note that the singular values are real positive values, because $\mathbf{\Gamma}^H \mathbf{\Gamma}$ and $\mathbf{\Gamma} \mathbf{\Gamma}^H$ are Hermitian matrices. We recall that $\mathbf{\Omega}(j\omega)$ is a Hermitian matrix if $\mathbf{\Omega}^H(j\omega) \triangleq \mathbf{\Omega}^T(-j\omega) = \mathbf{\Omega}(j\omega)$. It is customary to order the singular values, as follows:

$$\sigma_{max} = \sigma_1 \geq \sigma_2 \geq \sigma_3 \ldots \geq \sigma_k = \sigma_{min} \tag{20.7.7}$$

When we need to make explicit the connection between a matrix $\mathbf{\Gamma}$ and its singular values, we will denote them as $\sigma_i(\mathbf{\Gamma})$, for $i = 1, 2, \ldots, k$.

Another result of interest is that, if λ_{max} and λ_{min} are the maximum and minimum eigenvalues of a Hermitian matrix $\mathbf{\Lambda}$, respectively, then

$$\lambda_{max} = \sup_{||x|| \neq 0} \frac{x^H \mathbf{\Lambda} x}{||x||^2} ; \qquad \lambda_{min} = \inf_{||x|| \neq 0} \frac{x^H \mathbf{\Lambda} x}{||x||^2} \qquad (20.7.8)$$

We next apply these ideas to $||\mathbf{G}||$ as defined in (20.7.5). Taking $\mathbf{\Lambda} = \mathbf{G}^H \mathbf{G}$, we have that

$$||\mathbf{G}|| = \sup_{||U|| \neq 0} \sqrt{\frac{U^H \mathbf{G}^H \mathbf{G} U}{U^H U}} = \sqrt{\lambda_{max}\{\mathbf{G}^H \mathbf{G}\}} = \sigma_{max} \qquad (20.7.9)$$

where σ_{max}^2 is the maximum eigenvalue of $\mathbf{G}^H \mathbf{G}$. The quantity σ_{max} is the maximum singular value of $\mathbf{G}$. Thus, the induced spectral norm of $\mathbf{G}$ is the maximum singular value of $\mathbf{G}$. Some properties of singular values that are of interest in MIMO analysis and design, are summarized next. Consider two $m \times m$ matrices $\mathbf{\Omega}$ and $\mathbf{\Lambda}$ and a $m \times 1$ vector x. Denote by $\lambda_1(\mathbf{\Omega}), \lambda_2(\mathbf{\Omega}), \ldots, \lambda_m(\mathbf{\Omega})$ the eigenvalues of $\mathbf{\Omega}$. Then, the following properties apply to their singular values:

sv1 $\overline{\sigma}(\mathbf{\Omega}) = \sigma_{max}(\mathbf{\Omega}) = \sigma_1(\mathbf{\Omega}) = \max_{|x| \in \mathbb{R}} \dfrac{||\mathbf{\Omega} x||}{||x||}$ and

$\overline{\sigma}(\mathbf{\Lambda}) = \sigma_{max}(\mathbf{\Lambda}) = \sigma_1(\mathbf{\Lambda}) = \max_{|x| \in \mathbb{R}} \dfrac{||\mathbf{\Lambda} x||}{||x||}.$

sv2 $\underline{\sigma}(\mathbf{\Omega}) = \sigma_{min}(\mathbf{\Omega}) = \sigma_m(\mathbf{\Omega}) = \min_{|x| \in \mathbb{R}} \dfrac{||\mathbf{\Omega} x||}{||x||}$ and

$\underline{\sigma}(\mathbf{\Lambda}) = \sigma_{min}(\mathbf{\Lambda}) = \sigma_m(\mathbf{\Lambda}) = \min_{|x| \in \mathbb{R}} \dfrac{||\mathbf{\Lambda} x||}{||x||}$

sv3 $\underline{\sigma}(\mathbf{\Omega}) \leq |\lambda_i(\mathbf{\Omega})| \leq \overline{\sigma}(\mathbf{\Omega})$, for $i = 1, 2, \ldots, m$.

sv4 If $\mathbf{\Omega}$ is nonsingular, then $\underline{\sigma}(\mathbf{\Omega}) = [\overline{\sigma}(\mathbf{\Omega}^{-1})]^{-1}$ and $\overline{\sigma}(\mathbf{\Omega}) = [\underline{\sigma}(\mathbf{\Omega}^{-1})]^{-1}$

sv5 For any scalar α, $\sigma_i(\alpha \mathbf{\Omega}) = |\alpha| \sigma_i(\mathbf{\Omega})$, for $i = 1, 2, \ldots, m$.

sv6 $\overline{\sigma}(\mathbf{\Omega} + \mathbf{\Lambda}) \leq \overline{\sigma}(\mathbf{\Omega}) + \overline{\sigma}(\mathbf{\Lambda})$.

sv7 $\overline{\sigma}(\mathbf{\Omega} \mathbf{\Lambda}) \leq \overline{\sigma}(\mathbf{\Omega}) \overline{\sigma}(\mathbf{\Lambda})$.

sv8 $|\underline{\sigma}(\mathbf{\Omega}) - \overline{\sigma}(\mathbf{\Lambda})| \leq \overline{\sigma}(\mathbf{\Omega} + \mathbf{\Lambda}) \leq \underline{\sigma}(\mathbf{\Omega}) + \overline{\sigma}(\mathbf{\Lambda})$.

sv9 $\max_{i,k}[\mathbf{\Omega}]_{ik} \leq \overline{\sigma}(\mathbf{\Omega}) \leq m \times \max_{i,k}[\mathbf{\Omega}]_{ik}$

sv10 If $\mathbf{\Omega} = \mathbf{I}$, the identity matrix, then $\overline{\sigma}(\mathbf{\Omega}) = \underline{\sigma}(\mathbf{\Omega}) = 1$.

Some of the above properties result from the fact that the maximum singular value is a norm of the matrix; others originate from the fact that singular values are square roots of eigenvalues of a matrix.

Returning to the frequency-domain description of a system $\mathbf{G}$, the singular values of $\mathbf{G}$ become (real) functions of ω, and they are known as the *principal gains*. An important observation is that singular values and eigenvalues of matrix transfer functions *are not rational functions* of s (or of ω, in the frequency-response case).

An important consequence of the above definitions is that

$$\underline{\sigma}(\mathbf{G}(j\omega)) \leq \frac{||\mathbf{G}(j\omega)U(j\omega)||}{||U(j\omega)||} \leq \overline{\sigma}(\mathbf{G}(j\omega)) \qquad \forall \omega \in \mathbb{R} \qquad (20.7.10)$$

This implies that the minimum and maximum principal gains provide lower and upper bounds, respectively, for the magnitude of the gain for every input.

Assume now that, when $\omega = \omega_i$, we choose $U(s)$ so that $U_i \overset{\triangle}{=} U(j\omega_i)$ is an eigenvector of $\mathbf{G}^{\mathbf{H}}(j\omega_i)\mathbf{G}(j\omega_i)$–i.e.,

$$\mathbf{G}^{\mathbf{H}}(j\omega_i)\mathbf{G}(j\omega_i)U_i = \sigma_i^2 U_i \qquad (20.7.11)$$

We then have that

$$||Y|| = \sigma_i ||U_i|| \qquad (20.7.12)$$

and U_i is known as the *principal direction* associated with the singular value σ_i.

The norm defined in (20.7.5) is a point wise-in-frequency norm. In other words, it is a norm that depends on ω.

20.7.2   Tracking

We next consider the frequency-domain conditions necessary to give good tracking of reference signals. We recall that $E(s) = \mathbf{S_o}(s)R(s)$. We can thus obtain a combined measure of the magnitude of the errors in all channels by considering the Euclidean norm of $E(j\omega)$. Hence, consider

$$||E(j\omega)||_2 = \frac{||\mathbf{S_o}(j\omega)R(j\omega)||_2}{||R(j\omega)||_2}||R(j\omega)||_2 \leq \overline{\sigma}(\mathbf{S_o}(j\omega))||R(j\omega)||_2 \qquad (20.7.13)$$

We then see that errors are guaranteed small if $\overline{\sigma}(\mathbf{S_o}(j\omega))$ is small in the frequency band where $||R(j\omega)||_2$ is significant. Note that $\mathbf{S_o}(s) + \mathbf{T_o}(s) = \mathbf{I}$, and so

$$\overline{\sigma}(\mathbf{S_o}(j\omega)) \ll 1 \iff \overline{\sigma}(\mathbf{T_o}(j\omega)) \approx \underline{\sigma}(\mathbf{T_o}(j\omega)) \approx 1 \qquad (20.7.14)$$

We can go further, by appropriate use of the singular-value properties presented in the previous subsection. Consider, for example, property *sv4* on page 633; then

$$\overline{\sigma}(\mathbf{S_o}(j\omega)) = \overline{\sigma}\left([\mathbf{I} + \mathbf{G_o}(j\omega)\mathbf{C_o}(j\omega)]^{-1}\right) = (\underline{\sigma}(\mathbf{I} + \mathbf{G_o}(j\omega)\mathbf{C_o}(j\omega)))^{-1} \quad (20.7.15)$$

and, using properties *sv8* and *sv10*, we have that

$$(\underline{\sigma}(\mathbf{I} + \mathbf{G_o}(j\omega)\mathbf{C_o}(j\omega)))^{-1} \leq |\underline{\sigma}(\mathbf{G_o}(j\omega)\mathbf{C_o}(j\omega)) - 1|^{-1} \quad (20.7.16)$$

Thus, we see that errors in all channels are guaranteed small if $\underline{\sigma}(\mathbf{G_o}(j\omega)\mathbf{C_o}(j\omega))$ is made as large as possible over the frequency band where $||R(j\omega)||_2$ is significant.

Remark 20.5. *An interesting connection can be established with the steady-state analysis carried out in section §20.6. There, we stated that zero tracking errors at d.c. would be ensured if $\mathbf{S_o}(0) = 0$. This implies that*

$$\overline{\sigma}(\mathbf{S_o}(0)) = \underline{\sigma}(\mathbf{S_o}(0)) = 0; \qquad and \qquad \overline{\sigma}(\mathbf{G_o}(0)\mathbf{C}(0)) = \underline{\sigma}(\mathbf{G_o}(0)\mathbf{C}(0)) = \infty \quad (20.7.17)$$

□□□

We next turn our attention to the cost of good tracking in terms of the controller output energy.

We recall, from (20.4.8) and (20.4.3), that $U(s) = [\mathbf{G_o}(s)]^{-1}\mathbf{T_o}(s)R(s)$. Then

$$||U(j\omega)||_2 \leq \overline{\sigma}([\mathbf{G_o}(j\omega)]^{-1}\mathbf{T_o}(j\omega))||R(j\omega)||_2 \quad (20.7.18)$$

By using properties *sv4* and *sv7* (on page 633), the above equation can be expressed as

$$||U(j\omega)||_2 \leq \frac{\overline{\sigma}(\mathbf{T_o}(j\omega))}{\underline{\sigma}(\mathbf{G_o}(j\omega))}||R(j\omega)||_2 \quad (20.7.19)$$

Thus, good tracking (i.e., $\overline{\sigma}(\mathbf{T_o}(j\omega)) \approx 1$ in a frequency band where $\underline{\sigma}(\mathbf{G_o}(j\omega)) \ll 1$) might yield large control signals. Conversely, if a conservative design approach is followed, such that $\overline{\sigma}(\mathbf{T_o}(j\omega))$ and $\underline{\sigma}(\mathbf{G_o}(j\omega))$ do not differ significantly over the reference frequency band, then one can expect to avoid excessive control effort.

20.7.3   Disturbance Compensation

We next consider disturbance rejection. For the sake of illustration, we will consider only the input-disturbance case. If we examine equation (20.4.4), we conclude that compensation of output disturbances is similar to the tracking performance problem, because the reference frequency band and the output-disturbance frequency band are not disjoint. The same applies to the input-disturbance frequency band, where the relevant transfer function is $\mathbf{S_{io}}(s)$ instead of $\mathbf{S_o}(s)$.

For the input-disturbance case, we have that

$$||E||_2 \leq \overline{\sigma}(\mathbf{S_o}(j\omega)\mathbf{G_o}(j\omega))||D_i||_2 \qquad (20.7.20)$$

Furthermore, upon the applying of property $sv7$ on page 633, the above equation is rewritten as

$$||E||_2 \leq \overline{\sigma}(\mathbf{S_o}(j\omega))\overline{\sigma}(\mathbf{G_o}(j\omega))||D_i||_2 \qquad (20.7.21)$$

We can now use the results of our analysis for the tracking case, with a small twist.

Good input-disturbance compensation can be achieved if $\underline{\sigma}(\mathbf{G_o}(j\omega)\mathbf{C_o}(j\omega)) \gg 1$ over the frequency band where $\overline{\sigma}(\mathbf{G_o}(j\omega))||D_i||_2$ is significant. Comparison with the tracking case shows that input-disturbance rejection is normally less demanding, because $\overline{\sigma}(\mathbf{G_o}(j\omega))$ is usually low pass, and, hence, $\mathbf{G_o}$ itself will provide preliminary attenuation of the disturbance.

20.7.4   Measurement-Noise Rejection

The effect of measurement noise on MIMO loop performance can also be quantified by using singular values, as shown below. From (20.4.2), we have that, for measurement noise,

$$||Y||_2 \leq \overline{\sigma}(\mathbf{T_o}(j\omega))||D_m||_2 \qquad (20.7.22)$$

Thus, good noise rejection is achieved if $\overline{\sigma}(\mathbf{T_o}(j\omega)) \ll 1$ over the frequency band where the noise is significant. This is equivalent to demanding that $\overline{\sigma}(\mathbf{S_o}(j\omega)) \approx 1$. Measurement noise is usually significant in a frequency band above those of the reference and disturbances, so this requirement sets an effective limit to the closed-loop bandwidth (as measured by $\overline{\sigma}(\mathbf{T_o}(j\omega))$).

20.7.5   Directionality in Sensitivity Analysis

The preceding analysis produced upper and lower bounds that can be used as indicators of loop performance. However, the analysis presented so far has not emphasized one of the most significant features of MIMO systems, namely, directionality: the proportion in which every channel component appears in a signal vector (such as references and disturbances). To appreciate how this is connected to the previous analysis, we again consider the tracking problem.

Assume that the references for all loops have energy concentrated at a single frequency ω_r. Then, after the transient response has died out, the error vector (in phasor form) can be computed from

$$E(j\omega_r) = \mathbf{S_o}(j\omega_r)R(j\omega_r) \tag{20.7.23}$$

where $R(j\omega_r)$ is the reference vector (in phasor form). Assume, now, that $R(j\omega_r)$ coincides with the principal direction (see section §20.7.1) associated with the minimum singular value of $\mathbf{S_o}(j\omega_r)$; then

$$||E(j\omega_r)||_2 = \underline{\sigma}(\mathbf{S_o}(j\omega_r))||R(j\omega_r)||_2 \tag{20.7.24}$$

Comparing this with the result in (20.7.13), we see that, if the minimum and maximum singular values of $\mathbf{S_o}(j\omega_r)$ differ significantly, then the upper bound in (20.7.13) will be a loose bound for the actual result.

This brief discussion highlights the importance of directionality in MIMO problems. This will be a recurring theme in what follows.

To illustrate the above main issues, we consider the following example.

Example 20.6. *In a MIMO control loop, the complementary sensitivity is given by*

$$\mathbf{T_o}(s) = \begin{bmatrix} \dfrac{9}{s^2 + 5s + 9} & \dfrac{-s}{s^2 + 5s + 9} \\[3mm] \dfrac{s}{s^2 + 5s + 9} & \dfrac{3(s + 3)}{s^2 + 5s + 9} \end{bmatrix} \tag{20.7.25}$$

The loop has output disturbances given by

$$d_o(t) = \begin{bmatrix} K_1 \sin(\omega_d t + \alpha_1) & K_2 \sin(\omega_d t + \alpha_2) \end{bmatrix}^T \tag{20.7.26}$$

Determine the frequency ω_d, the ratio K_1/K_2, and the phase difference $\alpha_1 - \alpha_2$ that maximize the Euclidean norm of the stationary error, $||E||_2$.

Solution

In steady state, the error is a vector of sine waves with frequency ω_d. We then apply phasor analysis. The phasor representation of the output disturbance is

$$D_o = \begin{bmatrix} K_1 e^{j\alpha_1} & K_2 e^{j\alpha_2} \end{bmatrix}^T \tag{20.7.27}$$

We see from (20.4.4) that the error due to output disturbances is the negative of that for a reference signal. Then, from Equation (20.7.13), we have that, for every ratio K_1/K_2 such that $||D_o|| = 1$, the following holds:

$$||E(j\omega_d)||_2 = \max_{\omega \in \mathbb{R}} ||E(j\omega)||_2 \leq \max_{\omega \in \mathbb{R}} \overline{\sigma}(\mathbf{S_o}(j\omega)) = ||\mathbf{S_o}||_\infty \qquad (20.7.28)$$

The upper bound in (20.7.28) is reached precisely when the direction of the disturbance phasor coincides with that of the principal direction associated with the maximum singular value of $\mathbf{S_o}(j\omega_d)$. (See subsection §20.7.1.)

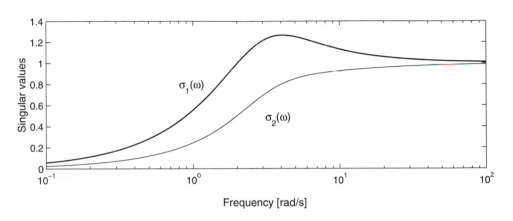

Figure 20.4. Singular values of the sensitivity function

We first obtain $\mathbf{S_o}(s)$, by applying the identity $\mathbf{T_o}(s) + \mathbf{S_o}(s) = \mathbf{I}$;

$$\mathbf{S_o}(s) = \begin{bmatrix} \dfrac{s(s+5)}{s^2+5s+9} & \dfrac{s}{s^2+5s+9} \\[3mm] \dfrac{-s}{s^2+5s+9} & \dfrac{s(s+2)}{s^2+5s+9} \end{bmatrix} \qquad (20.7.29)$$

We can now compute the value of ω at which $\overline{\sigma}(\mathbf{S_o}(j\omega))$ is maximal. The singular values of $\mathbf{S_o}(j\omega)$ are shown in Figure 20.4.

From here, we can see that $\overline{\sigma}(\mathbf{S_o}(j\omega))$ is maximum at $\omega \approx 4.1[rad/s]$. Thus, the maximizing disturbance frequency is $\omega_d = 4.1[rad/s]$.

*We then recall that the singular values of a matrix $\mathbf{\Omega}$ are the eigenvalues $\mathbf{\Omega}^H\mathbf{\Omega}$. Thus, to compute the principal directions of $\mathbf{S_o}(j\omega_d)$, we have to compute the eigenvalues of $\mathbf{S_o}(j\omega_d)\mathbf{S_o}^T(-j\omega_d)$. (MATLAB command **svd** can be used.)*

The principal directions associated with the two singular values, $\overline{\sigma}(\mathbf{S_o}(j\omega_d))$ and $\underline{\sigma}(\mathbf{S_o}(j\omega_d))$, are, respectively, given by

$$u_1 = \begin{bmatrix} 0.884 + j0.322 & -0.219 + j0.260 \end{bmatrix}^T \qquad (20.7.30)$$

$$u_2 = \begin{bmatrix} -0.340 + j0.003 & -0.315 + j0.886 \end{bmatrix}^T \qquad (20.7.31)$$

We are interested only in u_1, which can be expressed as

$$u_1 = \begin{bmatrix} 0.944\angle 0.35 & 0.340\angle 2.27 \end{bmatrix}^T \qquad (20.7.32)$$

Thus, the maximizing solution (i.e., the worst case) happens when

$$\frac{K_1}{K_2} = \frac{0.944}{0.34} = 2.774 \qquad and \qquad \alpha_1 - \alpha_2 = 0.35 - 2.27[rad] = -1.92[rad]$$

$$(20.7.33)$$

□□□

20.7.6 Directionality in Connection with Pole-Zero Cancellations

Directionality issues show up in connection also with pole-zero cancellations and loss of controllability or observability. Consider, for example, the set-up used in section §17.9, where two cascaded minimal systems $(\mathbf{A_1}, \mathbf{B_1}, \mathbf{C_1})$ and $(\mathbf{A_2}, \mathbf{B_2}, \mathbf{C_2})$ were studied. Here, we assume $u \in \mathbf{R}^m$, $u_2 = y_1 \in \mathbf{R}^p$, and $y \in \mathbf{R}^r$ with $\mathbf{B_2}$ of full column rank and $\mathbf{C_1}$ full row rank.

The composite system has realization $(\mathbf{A}, \mathbf{B}, \mathbf{C})$, where

$$\mathbf{A} = \begin{bmatrix} \mathbf{A_1} & \mathbf{0} \\ \mathbf{B_2 C_1} & \mathbf{A_2} \end{bmatrix} \qquad ; \mathbf{B} = \begin{bmatrix} \mathbf{B_1} \\ \mathbf{0} \end{bmatrix} \qquad (20.7.34)$$

$$\mathbf{C} = \begin{bmatrix} \mathbf{0} & \mathbf{C_2} \end{bmatrix} = 0 \qquad (20.7.35)$$

We know, from Chapter 3, that pole-zero cancellations play a role in loss of observability or controllability. However, in the MIMO case, directions are also important, as is shown in the following lemma.

Lemma 20.5. *The composite system loses observability if and only if β is a pole of system 1 and a zero of system 2 such that there exist an $x_1 \in$ the null space of $(\beta\mathbf{I} - \mathbf{A_1})$ and $\mathbf{C_1}x_1 \in$ the null space of $\mathbf{C_2} (\beta\mathbf{I} - \mathbf{A_2})^{-1} \mathbf{B_2}$.*

Proof

If: Immediate from the PBH test for (20.7.34), (20.7.35)
Only If: The only values of λ satisfying

$$(\mathbf{A} - \lambda\mathbf{I})x = 0 \qquad\qquad (20.7.36)$$

are eigenvalues of $\mathbf{A_1}$ *or* $\mathbf{A_2}$. *We consider the two cases separately.*

(a) *(λ is an eigenvalue of* $\mathbf{A}_1$*)* $\Rightarrow$ *(corresponding eigenvector is* $[x_1^T \quad x_2^T]^T$, *where* x_1 *is eigenvector of* $\mathbf{A}_1$ *and* $x_2 = (\lambda\mathbf{I} - \mathbf{A_2})^{-1}\mathbf{B_2}\mathbf{C_1}x_1$*). Now* $\mathbf{C_1}x_1 \neq 0$, *by virtue of the minimality of* $(\mathbf{A_1}, \mathbf{B_1}, \mathbf{C_1})$. *Hence,* $\mathbf{C_2}x_2$ *is zero (as required by loss of observability for the composite system) only if* $\mathbf{C_1}x_1 \neq 0$ *lies in the null space of* $\mathbf{C_2}(\lambda\mathbf{I} - \mathbf{A_2})^{-1}\mathbf{B_2}$. *However, this implies that* λ *is a zero of system 2 and that* $\mathbf{C_1}x_1$ *is a zero direction.*

(b) *(λ is an eigenvalue of* $\mathbf{A_2}$*)* $\Rightarrow$ *(corresponding eigenvector is* $[0 \quad x_2^T]^T$, *where* x_2 *is eigenvector of* $\mathbf{A_2}$*). Then, lack of observability for the composite system requires* $\mathbf{C_2}x_2 = 0$. *However, this violates the minimality of system 2. Hence, condition (a) is the only possibility.*

□□□

The corresponding result for controllability is as follows.

Lemma 20.6. *The composite system loses controllability if and only if* α *is a zero of system 1 and a pole of system 2 such that there exist* $x_2^T \in$ *the left null space of* $(\alpha\mathbf{I} - \mathbf{A_2})$ *and* $x_2^T\mathbf{B_2} \in$ *left null space of* $\mathbf{C_1}(\alpha\mathbf{I} - \mathbf{A_1})^{-1}\mathbf{B_1}$.

Proof

The steps to show this parallel those for the observability case presented above.

□□□

The pole-zero cancellation issues addressed by the two lemmas above are illustrated in the following example.

Example 20.7. *Consider two systems,* $\mathcal{S}_1$ *and* $\mathcal{S}_2$ *having, respectively, the transfer functions*

$$\mathbf{G_1}(s) = \begin{bmatrix} \dfrac{2s+4}{(s+1)(s+3)} & \dfrac{-2}{(s+1)(s+3)} \\ \dfrac{-3s-1}{(s+1)(s+3)} & \dfrac{-5s-7}{(s+1)(s+3)} \end{bmatrix} \qquad (20.7.37)$$

$$\mathbf{G_2}(s) = \begin{bmatrix} \dfrac{4}{(s+1)(s+2)} & \dfrac{-1}{s+1} \\ \dfrac{2}{s+1} & \dfrac{-1}{2(s+1)(s+2)} \end{bmatrix} \qquad (20.7.38)$$

We first build a state space representation for the system $\mathcal{S}_1$ by using the MAT-LAB command **ss**. We thus obtain the 4-tuple $(\mathbf{A_1}, \mathbf{B_1}, \mathbf{C_1}, \mathbf{0})$.

It is then straightforward (using MATLAB command **eig**) to compute the system eigenvalues, which are located at -1 and -3, with eigenvectors w_1 and w_2 given by

$$w_1^T = [0.8552 \quad 0.5184]; \qquad w_2^T = [-0.5184 \quad 0.8552] \qquad (20.7.39)$$

Also, this system has no zeros.

On the other hand, system $\mathcal{S}_2$ has three poles, located at -1, -2, and -2 and one zero, located at -3. This zero has a left direction μ and a right direction h, which are given by

$$\mu^T = [1 \quad 2]; \qquad h^T = [1 \quad -4] \qquad (20.7.40)$$

We observe that one pole of $\mathcal{S}_1$ coincides with one zero of $\mathcal{S}_2$. To examine the consequences of this coincidence we consider two cases of series connection.

$\mathcal{S}_1$ output is the input of $\mathcal{S}_2$:

To investigate a possible loss of observability, we have to compute $\mathbf{C_1} w_2$ and compare it with h. We first obtain $\mathbf{C_1} w_2 = [-1.414 \quad 5.657]^T$, from which we see that this vector is linearly dependent with h.

Thus, in this connection, there will be an unobservable mode, e^{-3t}.

$\mathcal{S}_2$ output is the input of $\mathcal{S}_1$:

To investigate a possible loss of controllability, we have to compute $w_2^T \mathbf{B_1}$ and compare it with μ. We have that $w_2^T \mathbf{B_1} = [-0.707 \quad -0.707]$. Thus, this vector is linearly independent of μ, and hence no loss of controllability occurs in this connection.

20.8 Robustness Issues

Finally, we extend the robustness results of section §5.9 for SISO to the MIMO case. As for the SISO case, MIMO models will usually be only approximate descriptions of any real system. Thus, the performance of the nominal control loop can significantly differ from the true or achieved performance. To gain some insight into this problem, we consider linear modeling errors, as we did for SISO systems. However, we now have the additional difficulty of dealing with matrices, for which the products do not commute. We will thus consider two equivalent forms for multiplicative modeling errors (MME):

$$\mathbf{G}(s) = (\mathbf{I} + \mathbf{G}_{\Delta\mathbf{l}}(s))\mathbf{G_o}(s) = \mathbf{G_o}(s)(\mathbf{I} + \mathbf{G}_{\Delta\mathbf{r}}(s)) \qquad (20.8.1)$$

where $\mathbf{G}_{\Delta\mathbf{l}}(s)$ and $\mathbf{G}_{\Delta\mathbf{r}}(s)$ are the left and right MME matrices, respectively. We observe from (20.8.1) that these matrices are related by

$$\mathbf{G}_{\Delta\mathrm{l}}(s) = \mathbf{G}_\mathbf{o}(s)\mathbf{G}_{\Delta\mathbf{r}}(s)[\mathbf{G}_\mathbf{o}(s)]^{-1}; \qquad \mathbf{G}_{\Delta\mathbf{r}}(s) = [\mathbf{G}_\mathbf{o}(s)]^{-1}\mathbf{G}_{\Delta\mathrm{l}}(s)\mathbf{G}_\mathbf{o}(s) \tag{20.8.2}$$

This equivalence allows us to derive expressions by using either one of the descriptions. For simplicity, we will choose the left MME matrix and will examine the two main sensitivities only: the sensitivity and the complementary sensitivity.

We can then derive expressions for the achievable sensitivities in (20.4.10)-(20.4.11) as functions of the modeling error. Using (20.8.1), we obtain

$$\mathbf{S}(s) = [\mathbf{I} + \mathbf{G}(s)\mathbf{C}(s)]^{-1} = [\mathbf{I} + \mathbf{G}_\mathbf{o}(s)\mathbf{C}(s) + \mathbf{G}_{\Delta\mathrm{l}}(s)\mathbf{G}_\mathbf{o}(s)\mathbf{C}(s)]^{-1}$$
$$= [\mathbf{I} + \mathbf{G}_\mathbf{o}(s)\mathbf{C}(s)]^{-1}[\mathbf{I} + \mathbf{G}_{\Delta\mathrm{l}}(s)\mathbf{T}_\mathbf{o}(s)]^{-1} = \mathbf{S}_\mathbf{o}(s)[\mathbf{I} + \mathbf{G}_{\Delta\mathrm{l}}(s)\mathbf{T}_\mathbf{o}(s)]^{-1} \tag{20.8.3}$$

$$\mathbf{T}(s) = \mathbf{G}(s)\mathbf{C}(s)[\mathbf{I} + \mathbf{G}(s)\mathbf{C}(s)]^{-1} = [\mathbf{I} + \mathbf{G}_{\Delta\mathrm{l}}(s)]\mathbf{T}_\mathbf{o}(s)[\mathbf{I} + \mathbf{G}_{\Delta\mathrm{l}}(s)\mathbf{T}_\mathbf{o}(s)]^{-1} \tag{20.8.4}$$

Note the similarity between the above expressions and those for the SISO case–see Equations (5.9.15)-(5.9.19). We can also use these expressions to obtain robustness results.

For example, the following robust stability theorem mirrors Theorem 5.3 on page 146.

Theorem 20.2. *Consider a plant with nominal and true transfer function* $\mathbf{G}_\mathbf{o}(s)$ *and* $\mathbf{G}(s)$, *respectively. Assume that they are related by (20.8.1). Also assume that a controller* $\mathbf{C}(s)$ *achieves nominal internal stability and that* $\mathbf{G}_\mathbf{o}(s)\mathbf{C}(s)$ *and* $\mathbf{G}(s)\mathbf{C}(s)$ *have the same number, P, of unstable poles. Then a sufficient condition for stability of the feedback loop obtained by applying the controller to the true plant is*

$$\overline{\sigma}(\mathbf{G}_{\Delta\mathrm{l}}(j\omega)\mathbf{T}_\mathbf{o}(j\omega)) < 1 \qquad \forall \omega \in \mathbb{R} \tag{20.8.5}$$

Proof

Define the function

$$F(s) = \det(\mathbf{I} + \mathbf{G}(s)\mathbf{C}(s)) \tag{20.8.6}$$

We then have, from Theorem 20.1 on page 629, that closed-loop stability for the true plant requires that $\arg\langle F(j\omega)\rangle$ *change by* $-2P\pi[rad]$ *as* ω *goes from* $-\infty$ *to* ∞.
Furthermore, upon using (20.8.1) and (20.5.15), we have that $F(s)$ *can also be expressed as*

$$F(s) = \det(\mathbf{I} + \mathbf{G_o}(s)\mathbf{C}(s)) \det(I + \mathbf{G_{\Delta l}}(s)\mathbf{T_o}(s)) = F_o(s) \det(I + \mathbf{G_{\Delta l}}(s)\mathbf{T_o}(s)) \tag{20.8.7}$$

Thus,

$$\arg\langle F(j\omega)\rangle = \arg\langle F_o(j\omega)\rangle + \arg\langle\det(I + \mathbf{G_{\Delta l}}(j\omega)\mathbf{T_o}(j\omega))\rangle \tag{20.8.8}$$

Let us denote the maximum (in magnitude) eigenvalue of $\mathbf{G_{\Delta l}}(j\omega)\mathbf{T_o}(j\omega)$ by $|\lambda|_{max}$. Then, from property sv3 (page 633), we have that

$$\overline{\sigma}(\mathbf{G_{\Delta l}}(j\omega)\mathbf{T_o}(j\omega)) < 1 \Longrightarrow |\lambda|_{max} < 1 \tag{20.8.9}$$

This implies that the change in $\arg\langle\det(I + \mathbf{G_{\Delta l}}(j\omega)\mathbf{T_o}(j\omega))\rangle$, as ω goes from $-\infty$ to ∞, is zero. The result follows.

□□□

An example that illustrates the quantification of these errors is shown next.

Example 20.8. *A MIMO plant has nominal and true models given by $\mathbf{G_o}(s)$ and $\mathbf{G}(s)$, respectively, where*

$$\mathbf{G_o}(s) = \begin{bmatrix} \dfrac{2}{s+1} & \dfrac{1}{(s+1)(s+2)} \\ \dfrac{1}{(s+1)(s+2)} & \dfrac{2}{s+2} \end{bmatrix} \tag{20.8.10}$$

$$\mathbf{G}(s) = \begin{bmatrix} \dfrac{20}{(s+1)(s+10)} & \dfrac{1}{(s+1)(s+2)} \\ \dfrac{1}{(s+1)(s+2)} & \dfrac{40}{(s+2)(s+20)} \end{bmatrix} \tag{20.8.11}$$

Find the left MME matrix $\mathbf{G_{\Delta l}}(s)$, and plot its singular values.

Solution

From Equation (20.8.1), we see that $\mathbf{G_{\Delta l}}(s)$ can be computed from

$$\mathbf{G_{\Delta l}}(s) = \mathbf{G}(s)[\mathbf{G_o}(s)]^{-1} - \mathbf{I} \tag{20.8.12}$$

This yields

$$\mathbf{G}_{\Delta 1}(s) = \begin{bmatrix} \dfrac{-4s^3 - 12s^2 - 8s}{4s^3 + 52s^2 + 127s + 70} & \dfrac{2s^2 + 4s}{4s^3 + 52s^2 + 127s + 70} \\[3mm] \dfrac{2s^2 + 2s}{4s^3 + 92s^2 + 247s + 140} & \dfrac{-4s^3 - 12s^2 + 8s}{4s^3 + 92s^2 + 247s + 140} \end{bmatrix} \qquad (20.8.13)$$

Note that, for $s = 0$, all matrix entries are zero. This is consistent with the nature of the error, because $\mathbf{G}(0) = \mathbf{G_o}(0)$.

The singular values of $\mathbf{G}_{\Delta 1}(s)$ are computed by using MATLAB commands, leading to the plots shown in Figure 20.5.

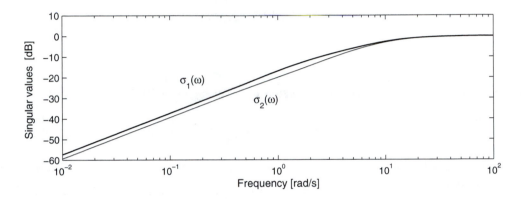

Figure 20.5. Singular values of MME matrix

20.9   Summary

- In previous chapters, we have considered the problem of controlling a single output by manipulating a single input (SISO).

- Many control problems, however, require that multiple outputs be controlled simultaneously; to do so, multiple inputs must be manipulated–usually, subtly orchestrated (MIMO).

  - Aircraft autopilot example: speed, altitude, pitch, roll, and yaw angles must be maintained; throttle, several rudders, and flaps are available as control variables.

  - Chemical process example: yield and throughput must be regulated; thermal energy, valve actuators, and various utilities are available as control variables.

- The key difficulty in achieving the necessary orchestration of inputs is the *multivariable interactions*, also known as *coupling*.

- From an input-output point of view, two fundamental phenomena arise from coupling. (See Figure 20.6.)

  a) a single input affects several outputs

  b) several inputs affect a single output

a)   b)

Figure 20.6. Two phenomena associated with multivariable interactions

- Multivariable interactions in the form, shown in Figure 20.6 add substantial complexity to MIMO control.

- Both state space and transfer-function models can be generalized to MIMO models.

- The MIMO transfer-function matrix can be obtained from a state space model by $\mathbf{G(s)} = \mathbf{C}(s\mathbf{I} - \mathbf{A})^{-1}\mathbf{B} + \mathbf{D}$.

- In general, if the model has m inputs, $u \in \mathbb{R}^m$, and l outputs, $y \in \mathbb{R}^l$, then

  ○ the transfer-function matrix consists of an $l \times m$ matrix of SISO transfer functions, and

  ○ for an n-dimensional state vector, $x \in \mathbb{R}^n$, the state space model matrices have dimensions $\mathbf{A} \in \mathbb{R}^{n \times n}$, $\mathbf{B} \in \mathbb{R}^{n \times m}$, $\mathbf{C} \in \mathbb{R}^{l \times n}$, $\mathbf{D} \in \mathbb{R}^{l \times m}$.

- Some MIMO model properties and analysis results generalize quite straightforwardly from SISO theory:

  ○ similarity transformations among state space realizations

  ○ observability and controllability

  ○ poles

- Other MIMO properties, usually due to interactions or the fact that matrices do not commute, are more subtle or complex than their SISO counterparts– e.g.,

  o zeros

  o left and right matrix fractions

20.10   Further Reading

Multivariable linear systems

Callier, F. and Desoer, C. (1982). *Multivariable Feedback Systems.* Springer-Verlag.

Kailath, T. (1980). *Linear Systems.* Prentice-Hall, Englewood Cliffs, N.J.

Maciejowski, J.M. (1989). *Multivariable Fedback Design.* Addison-Wesley, Wokingham, England.

Morari, M. and Zafiriou, E. (1989). *Robust Process Control.* Prentice-Hall, Englewood Cliffs, N.J.

Skogestad, S. and Postlethwaite, I. (1996). *Multivariable Feedback Control: Analysis and Design.* Wiley, New York.

Zhou, K., Doyle, J.C., and Glover, K. (1996).   *Robust and Optimal Control.* Prentice-Hall, Upper Saddle River, N.J.

Matrix fraction descriptions

Kailath, T. (1980). *Linear Systems.* Prentice-Hall, Englewood Cliffs, N.J.

Vidyasagar, M. (1985). *Control System Synthesis: A Factorization Approach.* MIT Press, Cambridge, Mass.

Frequency-domain analysis of MIMO systems

Hung, Y. and MacFarlane, A. (1982). Multivariable feedback. *Lecture notes in Control and Information Science*, 40. Springer-Verlag.

MacFarlane, A. (1979). *Frequency-Response Methods in control Systems.* IEEE Press.

Maciejowski, J.M. (1989). *Multivariable Fedback Design.* Addison-Wesley, Wokingham, England.

Mayne, D.Q. (1973). The design of linear multivariable systems. *Automatica*, 9(2):201-207.

Rosenbrock, H. (1970). *State Space and Multivariable Theory.* Wiley, New York.

Skogestad, S. and Postlethwaite, I. (1996). *Multivariable Feedback Control: Analysis and Design.* Wiley, New York.

Robustness in MIMO systems

Glover, K. (1986). Robust stabilization of linear multivariable systems. *International Journal of Control*, 43(3):741-766.

Green, M. and Limebeer, D. (1995). *Linear Robust Control.* Prentice-Hall, N.J.

McFarlane, D. and Glover, K. (1990). Robust controller design using normalized co-prime factor plant descriptions. *Lecture notes in Control and Information Science*, 138. Springer-Verlag.

Whittle, P. (1990). *Risk-sensitivity Optimal Control.* Wiley, New York.

Zhou, K. and Doyle, J.C. (1998). *Essentials of Robust Control.* Prentice-Hall, Upper Saddle River, N.J.

Zhou, K., Doyle, J.C., and Glover, K. (1996). *Robust and Optimal Control.* Prentice-Hall, Upper Saddle River, N.J.

Control of ammonia-synthesis converter

Bastiani, A. (2000). *Advanced control and optimization of an ammonia synthesis loop.* ME Thesis, Department of Electrical and Computer Engineering, The University of Newcastle, Australia.

20.11   Problems for the Reader

Problem 20.1. *Consider the following MIMO transfer functions:*

$$
\begin{bmatrix} \dfrac{-2s+1}{(s+1)^2} & \dfrac{2}{(s+4)(s+2)} \\[3mm] \dfrac{s+3}{(s+4)(s+2)} & \dfrac{2}{s+4} \end{bmatrix}
\quad ; \quad
\begin{bmatrix} \dfrac{2}{-s+1} & \dfrac{s+3}{(s+1)(s+2)} \\[3mm] \dfrac{s+3}{(s+1)(s+2)^2} & \dfrac{2}{s+2} \end{bmatrix}
$$

$$(20.11.1)$$

20.1.1 *For each case, build the Smith-McMillan form.*

20.1.2 *Compute the poles and zeros.*

20.1.3 *Compute the left and right directions associated with each zero.*

Problem 20.2. *For each of the systems in Problem 20.1, do the following.*

20.2.1 *Build a state space model.*

20.2.2 *Build a left and a right MFD.*

Problem 20.3. *Consider a strictly proper linear system having a state space model given by*

$$
\mathbf{A} = \begin{bmatrix} -3 & -1 \\ 1 & -2 \end{bmatrix} ; \quad
\mathbf{B} = \begin{bmatrix} 0 & 1 \\ 1 & 2 \end{bmatrix} ; \quad
\mathbf{C} = \begin{bmatrix} 0 & 1 \\ 0 & 1 \end{bmatrix}
\qquad (20.11.2)
$$

20.3.1 *Is this system of minimum phase?*

20.3.2 *Build a left and a right MFD.*

20.3.3 *Compute and plot the singular values of the transfer function.*

Problem 20.4. *A plant having a stable and minimum-phase nominal model $\mathbf{G_o}(s)$ is controlled using a one-d.o.f. feedback loop. The resulting complementary sensitivity, $\mathbf{T_o}(s)$, is given by*

$$\mathbf{T_o}(s) = \begin{bmatrix} \dfrac{4}{s^4 + 3s + 4} & \dfrac{\alpha s(s+5)}{(s^2 + 3s + 4)(s^2 + 4s + 9)} \\ 0 & \dfrac{9}{s^2 + 4s + 9} \end{bmatrix} \qquad (20.11.3)$$

Assume that $\alpha = 1$.

20.4.1 *Compute the closed-loop zeros, if any.*

20.4.2 *By inspection, we can see that* $\mathbf{T_o}(s)$ *is stable, but is the nominal closed loop internally stable?*

20.4.3 *Investigate the nature of the interaction by using step-output disturbances in both channels (at different times).*

Problem 20.5. *For the same plant as in Problem 20.4, do the following.*

20.5.1 *Compute its singular values for* $\alpha = 0.1$, $\alpha = 0.5$, *and* $\alpha = 5$.

20.5.2 *For each case, compare the results with the magnitude of the frequency response of the diagonal terms.*

Problem 20.6. *In a feedback control loop, we have that the open-loop transfer function is given by*

$$\mathbf{G_o}(s)\mathbf{C}(s) = \begin{bmatrix} \dfrac{s+2}{(s+4)s} & \dfrac{0.5}{s(s+2)} \\ \dfrac{0.5}{s(s+2)} & \dfrac{2(s+3)}{s(s+4)(s+2)} \end{bmatrix} \qquad (20.11.4)$$

20.6.1 *Is the closed-loop stable?*

20.6.2 *Compute and plot the singular values of* $\mathbf{T_o}(s)$.

Problem 20.7. *Consider a* 2×2 *MIMO system having its nominal model,* $\mathbf{G_o}(s)$, *given by*

$$\mathbf{G_o}(s) = \begin{bmatrix} \dfrac{120}{(s+10)(s+12)} & G_{12}(s) \\ G_{21}(s) & \dfrac{100}{s^2 + 13s + 100} \end{bmatrix} \qquad (20.11.5)$$

Determine the several sets of necessary conditions for $G_{12}(s)$ *and* $G_{21}(s)$ *such that the following occurs.*

20.7.1 *The system is significantly decoupled at all frequencies.*

20.7.2 *The system is completely decoupled at d.c. and at frequency $\omega = 3[rad/s]$.*

20.7.3 *The system is weakly coupled in the frequency band $[0\ ,\ 10]\,[rad/s]$.*

For each case, illustrate your answer with suitable values for $G_{12}(s)$ and $G_{21}(s)$.

Problem 20.8. *Consider the discrete-time system transfer function $\mathbf{G_q}(z)$ given by*

$$\mathbf{G_q}(z) = \frac{0.01}{z^3(z-0.4)^2(z-0.8)^2}\begin{bmatrix} 3z(z-0.4) & 20(z-0.4)^2(z-0.8) \\ 60z^2(z-0.4)(z-0.8)^2 & 12(z-0.4)(z-0.8) \end{bmatrix}$$

$$(20.11.6)$$

20.8.1 *Analyze the type of coupling in this system in different frequency bands.*

20.8.2 *Find the Smith–McMillan form, and compute the system poles and zeros.*

20.8.3 *Determine the system singular values.*

Problem 20.9. *Consider the one-d.o.f. control of a 2×2 MIMO stable plant. The controller transfer-function matrix $\mathbf{C}(s)$ is given by*

$$\mathbf{C}(s) = \begin{bmatrix} C_{11}(s) & C_{12}(s) \\ C_{21}(s) & C_{22}(s) \end{bmatrix}$$

$$(20.11.7)$$

It is required to achieve zero steady-state error for constant references and input disturbances in both channels. Determine which of the four controller entries should have integral action to meet this design specification.

Problem 20.10. *Consider a MIMO plant having a transfer function given by*

$$\mathbf{G_o}(s) = \frac{1}{(s+1)(s+2)^2}\begin{bmatrix} 1 & s+2 & s+1 \\ 0.25(s+1)^2 & -s+2 & 0.5(s+1)(s+2) \\ -0.5(s+2)^2 & 0 & 4(s+1) \end{bmatrix}$$

$$(20.11.8)$$

Assume that this plant has to be digitally controlled through zero-order sample-and-hold devices in all input channels. The sampling interval is chosen as $\Delta = 0.5[s]$.

*Find the discrete-time transfer function $\mathbf{H_q}(z) \overset{\triangle}{=} [\mathbf{G_o G_{h0}}]_q(z)$. (Hint: use MATLAB command **c2dm** on the state space model for the plant.)*

Problem 20.11. *Consider a MIMO plant having its transfer function given by*

$$\mathbf{G_o}(s) = \frac{1}{(s+1)(s+2)} \begin{bmatrix} 2 & 0.25e^{-s}(s+1) \\ 0.5(s+2) & 1 \end{bmatrix} \qquad (20.11.9)$$

Can this system be stabilized by using a diagonal controller with transfer function $\mathbf{C}(s) = k\mathbf{I_2}$, *for* $k \in \mathbb{R}$? *(Hint: use characteristic loci.)*

Problem 20.12. *Consider a discrete-time control loop where the sensitivity is given by*

$$\mathbf{S_{oq}}(z) = \frac{z-1}{(z-0.4)(z-0.5)} \begin{bmatrix} z & z-0.6 \\ z-0.4 & z+0.2 \end{bmatrix} \qquad (20.11.10)$$

Assume that both references are sinusoidal signals. Find the worst direction and frequency of the reference vector. (See solved problem 20.6.)

Chapter 21

EXPLOITING SISO TECHNIQUES IN MIMO CONTROL

21.1 Preview

In the case of SISO control, we found that one could use a wide variety of synthesis methods. Some of these carry over directly to the MIMO case. However, there are several complexities that arise in MIMO situations. For this reason, it is often desirable to use synthesis procedures that are in some sense automated. This will be the subject of the next few chapters. However, before we delve into the full complexity of MIMO design, it is appropriate that we pause to see when, if ever, SISO techniques can be applied to MIMO problems directly. In this chapter, then, we will see how far we can take SISO design into the MIMO case. In particular, we will study

- decentralized control as a mechanism for directly exploiting SISO methods in a MIMO setting

- robustness issues associated with decentralized control

21.2 Completely Decentralized Control

Before we consider a fully interacting multivariable design, it is often useful to check on whether a completely decentralized design can achieve the desired performance objectives. When applicable, the advantage of a completely decentralized controller, compared to a full MIMO controller, is that it is simpler to understand, is easier to maintain, and can be enhanced in a straightforward fashion (in the case of a plant upgrade).

The key simplifying assumption in completely decentralized control is that interactions can be treated as a form of disturbance. For example–if the first output, $y_1(t)$, of a $m \times m$ MIMO plant can be described (in Laplace-transform form) as

$$Y_1(s) = G_{11}(s)U_1(s) + \sum_{i=2}^{m} G_{1i}(s)U_i(s) \tag{21.2.1}$$

then the contribution of each input, other than u_1, could be considered as an output disturbance, $G_{1i}(s)U_i(s)$, for the first SISO loop. This approach is, of course, strictly incorrect; the essence of true disturbances is that they are independent inputs. However, under certain conditions (depending on magnitude and frequency content), this viewpoint can lead to acceptable results. We note that the idea of treating a MIMO plant as a set of SISO plants can be refined to structure the full multivariable system into sets of smaller multivariable systems that show little interaction with each other. However, for simplicity in the sequel, we will use *decentralized control* as meaning simply *complete decentralized control*.

Readers having previous exposure to practical control will realize that a substantial proportion of real-world systems will utilize decentralized architectures. Thus, one is led to ask the question, *is there ever a situation in which decentralized control will not yield a satisfactory solution?*. We will present several real-world examples later in Chapter 22 that require MIMO thinking to get a satisfactory solution. As a *textbook example* of where decentralized control can break down, consider the following MIMO example.

Example 21.1. *Consider a two-input, two-output plant having the transfer function*

$$\mathbf{G_o}(s) = \begin{bmatrix} G_{11}^o(s) & G_{12}^o(s) \\ G_{21}^o(s) & G_{22}^o(s) \end{bmatrix} \tag{21.2.2}$$

$$G_{11}^o(s) = \frac{2}{s^2 + 3s + 2} \qquad\qquad G_{12}^o(s) = \frac{k_{12}}{s+1} \tag{21.2.3}$$

$$G_{21}^o(s) = \frac{k_{21}}{s^2 + 2s + 1} \qquad\qquad G_{22}^o(s) = \frac{6}{s^2 + 5s + 6} \tag{21.2.4}$$

Let us say that k_{12} and k_{21} depend on the operating point (a common situation, in practice). We will consider four operating points.

Operating point 1 ($k_{12} = k_{21} = 0$)

Clearly, there is no interaction at this operating point.

Thus, we can safely design two SISO controllers. To be specific, say we aim for the following complementary sensitivities:

$$T_{o1}(s) = T_{o2}(s) = \frac{9}{s^2 + 4s + 9} \qquad (21.2.5)$$

The corresponding controller transfer functions are $C_1(s)$ and $C_2(s)$, where

$$C_1(s) = \frac{4.5(s^2 + 3s + 2)}{s(s+4)}; \qquad C_2(s) = \frac{1.5(s^2 + 5s + 6)}{s(s+4)} \qquad (21.2.6)$$

These independent loops perform as predicted by the choice of complementary sensitivities.

Operating point 2 ($k_{12} = k_{21} = 0.1$)

We leave the controller as it was for operating point 1. We apply a unit step in the reference for output 1 at $t = 1$ and a unit step in the reference for output 2 at $t = 10$. The closed-loop response is shown in Figure 21.1. These results would probably be considered very acceptable, even though the effects of coupling are now evident in the response.

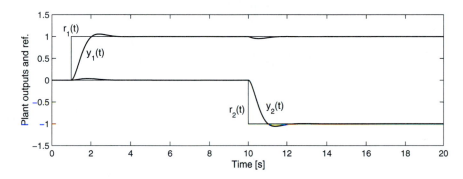

Figure 21.1. Effects of weak interaction in control loops with SISO design

Operating point 3 ($k_{12} = -1$, $k_{21} = 0.5$)

With the same controllers and from same test as for operating point 2, we obtain the results in Figure 21.2.

We see that a change in the reference in one loop now affects the output in the other loop significantly.

Operating point 4 ($k_{12} = -2$, $k_{21} = -1$)

Now a simulation with the same references indicates that the whole system becomes unstable. We see that the original SISO design has become unacceptable at this

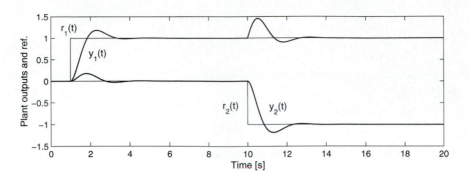

Figure 21.2. Effects of strong interaction in control loops with SISO design

final operating point. To gain further insight into the stability problem, we need to compute the four transfer functions that model the relationship between the two references, $R_1(s)$ and $R_2(s)$, and the two outputs, $Y_1(s)$ and $Y_2(s)$. These transfer functions can be organized in matrix form to yield

$$\begin{bmatrix} Y_1(s) \\ Y_2(s) \end{bmatrix} = (\mathbf{I} + \mathbf{G_o}(s)\mathbf{C}(s))^{-1}\mathbf{G_o}(s)\mathbf{C}(s) \begin{bmatrix} R_1(s) \\ R_2(s) \end{bmatrix} \tag{21.2.7}$$

where

$$\mathbf{G_o}(s) = \begin{bmatrix} G_{11}(s) & G_{12}(s) \\ G_{21}(s) & G_{22}(s) \end{bmatrix}; \qquad \mathbf{C}(s) = \begin{bmatrix} C_1(s) & 0 \\ 0 & C_2(s) \end{bmatrix} \tag{21.2.8}$$

If we consider the case in which $k_{12} = -2$ and $k_{21} = -1$, and we substitute (21.2.3), (21.2.4), and (21.2.6) into (21.2.7), we have that $(\mathbf{I}+\mathbf{G_o}(s)\mathbf{C}(s))^{-1}\mathbf{G_o}(s)\mathbf{C}(s)$ is given by

$$\frac{1}{d(s)} \begin{bmatrix} 9s^4 + 40.5s^3 + 67.5s^2 - 18s - 81 & -3s^5 - 30s^4 - 105s^3 - 150s^2 - 72s \\ -4.5s^4 - 31.5s^3 - 63s^2 - 36s & 9s^4 + 40.5s^3 + 67.5s^2 - 18s - 81 \end{bmatrix} \tag{21.2.9}$$

where

$$d(s) = s^6 + 10s^5 + 51s^4 + 134.5s^3 + 164.5s^2 + 18s - 81 \tag{21.2.10}$$

which is clearly unstable. (Actually, there is a closed-loop pole at 0.5322.)

We see that, although independent design of the controllers guarantees stability for the "noninteracting loops," when interaction appears in the plant model, instability or unacceptable performance may arise.

*The reader is encouraged to use the SIMULINK schematic in file **mimo1.mdl** to explore other interaction dynamics.*

□□□

We see from the above example that surprising issues can arise when MIMO interactions are ignored.

21.3   Pairing of Inputs and Outputs

If one is to use a decentralized architecture, then one needs to pair the inputs and outputs. In the case of an $m \times m$ plant transfer function, there are $m!$ possible pairings. However, physical insight can often be used to suggest sensible pairings or to eliminate pairings that have little hope of being useful.

One method that can be used to suggest pairings is a quantity known as the *Relative Gain Array* (RGA). For a MIMO system with matrix transfer function $\mathbf{G_o}(s)$, the RGA is defined as a matrix Λ with the ij^{th} element

$$\lambda_{ij} = [\mathbf{G_o}(0)]_{ij}[\mathbf{G_o}^{-1}(0)]_{ji} \tag{21.3.1}$$

where $[\mathbf{G_o}(0)]_{ij}$ and $[\mathbf{G_o}^{-1}(0)]_{ij}$ denote the ij^{th} element of the plant d.c. gain matrix and the ji^{th} element of the inverse of the d.c. gain matrix, respectively.

Note that $[\mathbf{G_o}(0)]_{ij}$ corresponds to the d.c. gain from the ith input, u_i, to the jth output, y_j, while *the rest of the inputs, u_l for $l \in \{1, 2, \ldots, i-1, i+1, \ldots, m\}$ are kept constant.* Also, $[\mathbf{G_o}^{-1}]_{ij}$ is the reciprocal of the d.c. gain from the ith input, u_i, to the jth output, y_j, while *the rest of the outputs, y_l for $l \in \{1, 2, \ldots, j-1, j+1, \ldots, m\}$ are kept constant.* Thus, the parameter λ_{ij} provides an indication of how sensible it is to pair the ith input with the jth output.

In particular, one usually aims to pick pairings such that the diagonal entries of Λ are large. One also tries to avoid pairings that result in negative diagonal entries in Λ.

Clearly this measure of interaction accounts for steady-state gain only. Nonetheless, it is an empirical observation that it works well in many cases of practical interest.

We illustrate by a simple example.

Example 21.2. *Consider again the system studied in Example 21.1 on page 654. It is a MIMO system having the transfer function*

$$\mathbf{G_o}(s) = \begin{bmatrix} \dfrac{2}{s^2 + 3s + 2} & \dfrac{k_{12}}{s+1} \\ \dfrac{k_{21}}{s^2 + 2s + 1} & \dfrac{6}{s^2 + 5s + 6} \end{bmatrix} \tag{21.3.2}$$

The RGA is then

$$\Lambda = \begin{bmatrix} \dfrac{1}{1 - k_{12}k_{21}} & \dfrac{-k_{12}k_{21}}{1 - k_{12}k_{21}} \\ \dfrac{-k_{12}k_{21}}{1 - k_{12}k_{21}} & \dfrac{1}{1 - k_{12}k_{21}} \end{bmatrix} \qquad (21.3.3)$$

For $1 > k_{12} > 0$, $1 > k_{21} > 0$, the RGA suggests the pairing (u_1, y_1), (u_2, y_2). We recall from section §20.2 that this pairing indeed worked very well for $k_{12} = k_{21} = 0.1$ and quite acceptably for $k_{12} = -1$, $k_{21} = 0.5$. In the latter case, the RGA is

$$\Lambda = \frac{1}{3}\begin{bmatrix} 2 & 1 \\ 1 & 2 \end{bmatrix} \qquad (21.3.4)$$

However, for $k_{12} = -2$, $k_{21} = -1$ we found in section §20.2 that the centralized controller based on the pairing (u_1, y_1), (u_2, y_2) was actually unstable. The corresponding RGA in this case is

$$\Lambda = \begin{bmatrix} -1 & 2 \\ 2 & -1 \end{bmatrix} \qquad (21.3.5)$$

which indicates that we probably should have changed to the pairing (u_1, y_2), (u_2, y_1).

□□□

Example 21.3 (Quadruple-tank apparatus, continued). *Consider again the quadruple-tank apparatus of Example 20.4 on page 623. We recall that this system had an approximate transfer function,*

$$\mathbf{G}(s) = \begin{bmatrix} \dfrac{3.7\gamma_1}{62s + 1} & \dfrac{3.7(1 - \gamma_2)}{(23s + 1)(62s + 1)} \\ \dfrac{4.7(1 - \gamma_1)}{(30s + 1)(90s + 1)} & \dfrac{4.7\gamma_2}{90s + 1} \end{bmatrix} \qquad (21.3.6)$$

We also recall that choice of γ_1 and γ_2 could change one of the system zeros from the left to the right-half plane. The RGA for this system is

$$\Lambda = \begin{bmatrix} \lambda & 1 - \lambda \\ 1 - \lambda & \lambda \end{bmatrix} \qquad where \quad \lambda = \frac{\gamma_1\gamma_2}{\gamma_1\gamma_2 - 1} \qquad (21.3.7)$$

For $1 < \gamma_1 + \gamma_2 < 2$, we recall from Example 20.4 on page 623 that the system is of minimum phase. If we take, for example, $\gamma_1 = 0.7$ and $\gamma_2 = 0.6$, then the RGA is

$$\mathbf{\Lambda} = \begin{bmatrix} 1.4 & -0.4 \\ -0.4 & 1.4 \end{bmatrix} \tag{21.3.8}$$

This suggests that we can pair (u_1, y_1) and (u_2, y_2).

Because the system is of minimum phase, the design of a decentralized controller is relatively easy in this case. For example, the following decentralized controller gives the results shown in Figure 21.3.

$$C_1(s) = 3\left(1 + \frac{1}{10s}\right); \qquad C_2(s) = 2.7\left(1 + \frac{1}{20s}\right) \tag{21.3.9}$$

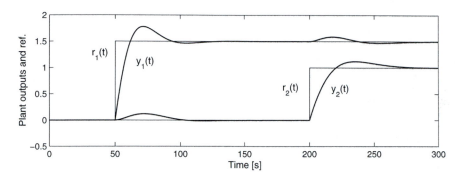

Figure 21.3. Decentralized control of a minimum-phase four-tank system

For $0 < \gamma_1 + \gamma_2 < 1$, we recall from Example 20.4 on page 623 that the system is of nonminimum phase. If we take, for example, $\gamma_1 = 0.43$ and $\gamma_2 = 0.34$, then the system has a NMP zero at $s = 0.0229$, and the relative gain array becomes

$$\mathbf{\Lambda} = \begin{bmatrix} -0.64 & 1.64 \\ 1.64 & -0.64 \end{bmatrix} \tag{21.3.10}$$

This suggests that (y_1, y_2) should be commuted for the purposes of decentralized control. This is physically reasonable, given the flow patterns produced in this case. This leads to a new RGA of

$$\mathbf{\Lambda} = \begin{bmatrix} 1.64 & -0.64 \\ -0.64 & 1.64 \end{bmatrix} \tag{21.3.11}$$

Note, however, that control will still be much harder than in the minimum-phase case. For example, the following decentralized controllers give the results[1] shown in Figure 21.4.

---

[1] Note the different time scale, in Figures 21.3 and 21.4.

$$C_1(s) = 0.5 \left(1 + \frac{1}{30s}\right); \qquad C_2(s) = 0.3 \left(1 + \frac{1}{50s}\right) \tag{21.3.12}$$

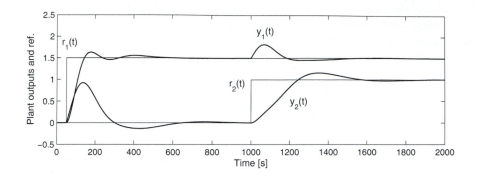

Figure 21.4. Decentralized control of a nonminimum-phase four-tank system

Comparing the results in Figures 21.3 and 21.4, we see evidence of greater interaction and slower responses in the nonminimum-phase case of Figure 21.4.

□□□

21.4   Robustness Issues in Decentralized Control

One way to carry out a decentralized control design is to use a diagonal *nominal* model. The off-diagonal terms then represent under-modeling, in the terminology of Chapter 3. Thus, say we have a model $\mathbf{G_o}(s)$, then the nominal model for decentralized control could be chosen as

$$\mathbf{G_o^d}(s) = diag\{g_{11}^o, \ldots, g_{mm}^o(s)\} \tag{21.4.1}$$

and the additive model error would be

$$\mathbf{G_\epsilon}(s) = \mathbf{G_o}(s) - \mathbf{G_o^d}(s); \qquad \mathbf{G_{\Delta l}}(s) = \mathbf{G_\epsilon}(s)[\mathbf{G_o^d}(s)]^{-1} \tag{21.4.2}$$

With this as a background, we can employ the robustness checks described in section §20.8. We recall that a sufficient condition for robust stability is

$$\overline{\sigma}\left(\mathbf{G_{\Delta l}}(j\omega)\mathbf{T_o}(j\omega)\right) < 1 \qquad \forall \omega \in \mathbb{R} \tag{21.4.3}$$

where $\overline{\sigma}\left(\mathbf{G_{\Delta l}}(j\omega)\mathbf{T_o}(j\omega)\right)$ is the maximum singular value of $\mathbf{G_{\Delta l}}(j\omega)\mathbf{T_o}(j\omega)$.
We illustrate by an example.

Example 21.4. *Consider again the system of Example 21.2 on page 657. In this case,*

$$
\mathbf{T_o}(s) = \frac{9}{s^2 + 4s + 9}\begin{bmatrix} 1 & 0 \\ 0 & 1 \end{bmatrix}
\qquad
\mathbf{G_o^d}(s) = \begin{bmatrix} \dfrac{2}{s^2 + 3s + 2} & 0 \\ 0 & \dfrac{6}{s^2 + 5s + 6} \end{bmatrix}
$$

$$(21.4.4)$$

$$
\mathbf{G_\epsilon}(s) = \begin{bmatrix} 0 & \dfrac{k_{12}}{s+1} \\ \dfrac{k_{21}}{s^2 + 2s + 1} & 0 \end{bmatrix}
\qquad
\mathbf{G_{\Delta l}}(s) = \begin{bmatrix} 0 & \dfrac{k_{12}(s^2 + 5s + 6)}{6(s+1)} \\ \dfrac{k_{21}(s+2)}{2(s+1)} & 0 \end{bmatrix}
$$

$$(21.4.5)$$

$$
\mathbf{G_{\Delta l}}(s)\mathbf{T_o}(s) = \begin{bmatrix} 0 & \dfrac{3k_{12}(s^2 + 5s + 6)}{2(s+1)(s^2 + 4s + 9)} \\ \dfrac{9k_{21}(s+2)}{2(s+1)(s^2 + 4s + 9)} & 0 \end{bmatrix}
$$

$$(21.4.6)$$

The singular values, in this case, are simply the magnitudes of the two off-diagonal elements. These are plotted in Figure 21.5 for normalized values $k_{12} = k_{21} = 1$.

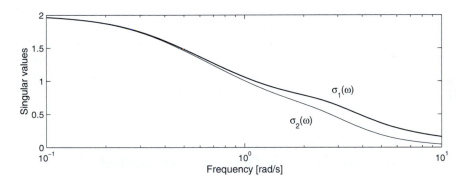

Figure 21.5. Singular values of $\mathbf{G_{\Delta l}}(j\omega)\mathbf{T_o}(j\omega)$

We see that a sufficient condition for robust stability of the decentralized control, with the pairing (u_1, y_1), (u_2, y_2), is that $|k_{12}| < 1$ and $|k_{21}| < 1$. Observe that this is conservative, but consistent with the performance results presented above.

□□□

An example that shows the interplay of RGA and robustness concepts follows.

Example 21.5. *Consider a MIMO system with*

$$\mathbf{G}(s) = \begin{bmatrix} \dfrac{1}{s+1} & 0.25\dfrac{10s+1}{(s+1)(s+2)} \\ 0.25\dfrac{10s+1}{(s+1)(s+2)} & \dfrac{2}{s+2} \end{bmatrix}; \qquad \mathbf{G_o}(s) = \begin{bmatrix} \dfrac{1}{s+1} & 0 \\ 0 & \dfrac{2}{s+2} \end{bmatrix}$$

$$(21.4.7)$$

We first observe that the RGA for the nominal model $\mathbf{G_o}(s)$ *is given by*

$$\Lambda = \begin{bmatrix} 1.0159 & -0.0159 \\ -0.0159 & 1.0159 \end{bmatrix} \qquad (21.4.8)$$

This value of the RGA might lead to the hypothesis that a correct pairing of inputs and outputs has been made and that the interaction is weak. Assume that we proceed to do a decentralized design leading to a diagonal controller $\mathbf{C}(s)$ *to achieve a complementary sensitivity* $\mathbf{T_o}(s)$, *where*

$$\mathbf{T_o}(s) = \frac{9}{s^2+4s+9}\begin{bmatrix} 1 & 0 \\ 0 & 1 \end{bmatrix}; \qquad \mathbf{C}(s) = \begin{bmatrix} \dfrac{9(s+1)}{s(s+4)} & 0 \\ 0 & \dfrac{9(s+2)}{2s(s+4)} \end{bmatrix} \qquad (21.4.9)$$

However, this controller, when applied to control $\mathbf{G}(s)$, *leads to closed-loop poles located at* -6.00, $-2.49 \pm j4.69$, $0.23 \pm j1.36$, *and* -0.50–*an "unstable closed loop"!*

The lack of robustness in this example can be traced to the fact that the required closed-loop bandwidth includes a frequency range where the off-diagonal frequency response is significant. Thus, in this example, closed-loop stability with decentralized control can be achieved only if we make the closed loop slower by significantly reducing the bandwidth of the diagonal terms in $\mathbf{T_o}(s)$.

□□□

21.5   Feedforward Action in Decentralized Control

Although it usually will not aid robust stability, the performance of decentralized controllers is often significantly enhanced by the judicious choice of feedforward action to reduce coupling. We refer to equation (21.2.1) and, for simplicity, we consider only the effect of the jth loop on the ith loop. We can then apply the feedforward ideas developed in Chapter 10 to obtain the architecture shown in Figure 21.6.

The feedforward gain $G_{ff}^{ji}(s)$ should be chosen in such a way that the coupling from the jth loop to the ith loop is compensated *in a particular, problem-dependent frequency band* $[0 \ \omega_{ff}]$–i.e., if

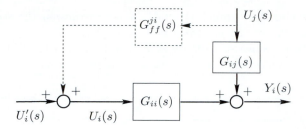

Figure 21.6. Feedforward action in decentralized control

$$G_{ff}^{ji}(j\omega)G_{ii}(j\omega) + G_{ij}(j\omega) \approx 0 \qquad \forall \omega \in [0 \ \ \omega_{ff}] \tag{21.5.1}$$

Equation (21.5.1) can also be written as

$$G_{ff}^{ji}(j\omega) \approx -[G_{ii}(j\omega)]^{-1}G_{ij}(j\omega) \qquad \forall \omega \in [0 \ \ \omega_{ff}] \tag{21.5.2}$$

from which we observe the necessity to build an inverse, and hence the associated issues arise again.

We illustrate by an example.

Example 21.6. *Consider again the system in Example 21.4 on page 661, with*
$k_{12} = -1$ *and* $k_{21} = 0.5$. *We recall the results presented in Figure 21.2 on page 656 for this case. We see that there is little coupling from the first to the second loop, but relatively strong coupling from the second to the first loop. This suggests feedforward from the second input to the first loop. We also choose* $G_{ff}^{ji}(s)$ *to completely compensate the coupling at d.c., i.e.,* $G_{ff}^{ji}(s)$ *is chosen to be a constant* $G_{ff}^{ji}(s) = \alpha$, *satisfying*

$$\alpha G_{11}(0) = -G_{12}(0) \implies \alpha = 1 \tag{21.5.3}$$

Then the modified MIMO system becomes modeled by

$$Y(s) = \mathbf{G_o}(s) \begin{bmatrix} U_1(s) \\ U_2(s) \end{bmatrix} = \mathbf{G_o}(s) \begin{bmatrix} 1 & 1 \\ 0 & 1 \end{bmatrix} \begin{bmatrix} U_1'(s) \\ U_2(s) \end{bmatrix} = \mathbf{G_o'}(s) \begin{bmatrix} U_1'(s) \\ U_2(s) \end{bmatrix} \tag{21.5.4}$$

where

$$\mathbf{G_o'}(s) = \begin{bmatrix} \dfrac{2}{s^2 + 3s + 2} & \dfrac{-s}{s^2 + 3s + 2} \\[3ex] \dfrac{0.5}{s^2 + 2s + 1} & \dfrac{6.5s^2 + 14.5s + 9}{(s^2 + 2s + 1)(s^2 + 5s + 6)} \end{bmatrix} \tag{21.5.5}$$

The RGA is now $\mathbf{\Lambda} = diag(1,1)$, and, when we redesign the decentralized controller, we obtain the results presented in Figure 21.7.

These results are superior to those in Figure 21.2 on page 656, where a decentralized controller performance was shown for the same case.

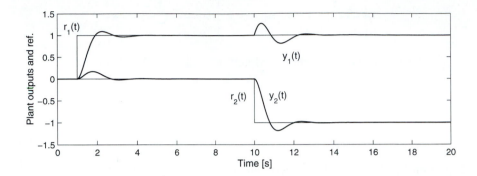

Figure 21.7. Performance of a MIMO decentralized control loop with interaction feedforward

□□□

Example 21.7 (Hold-up effect in reversing mill). *As an industrial application of feedforward control, we remind the reader of the example studied in subsection 10.6.1. There, we found that feedforward was crucial in overcoming the hold-up effect and hence achieving satisfactory closed-loop control.*

□□□

The above examples indicate that a little coupling introduced into the controller can be quite helpful. This, however, raised the question of how we can systematically design coupled controllers that rigorously take into account multivariable interaction. This motivates us to study the latter topic, which will be the topic of the next chapter. Before ending this chapter, we investigate whether there exist simple ways of converting an inherently MIMO problem to a set of SISO problems.

21.6   Converting MIMO Problems to SISO Problems

Many MIMO problems can be modified so that decentralized control becomes a more viable (or attractive) option. For example, one can sometimes use a precompensator to turn the resultant system into a more nearly diagonal transfer function.

To illustrate, say the nominal plant transfer function is $\mathbf{G_o}(s)$. If we introduce a precompensator $\mathbf{P}(s)$, then the control loop appears as in Figure 21.8.

The design of $\mathbf{C_p}(s)$ in Figure 21.8, can be based on the equivalent plant.

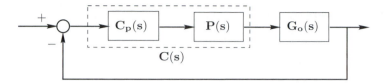

Figure 21.8. Feedback control with plant precompensation

$$\mathbf{H}(s) = \mathbf{G_o}(s)\mathbf{P}(s) \qquad (21.6.1)$$

Several comments are in order regarding this strategy:

(i) A first try at $\mathbf{P}(s)$ might be to approximate $\mathbf{G_o}(s)^{-1}$ in some way. For example, one might use the d.c. gain matrix $\mathbf{G_o}(0)^{-1}$ as a precompensator, assuming this exists. This was the strategy used in Example 21.6.

(ii) If dynamic precompensators are used, then one needs to check that no unstable pole-zero cancellations are introduced between the compensator and the original plant–see subsection §20.7.6.

(iii) Various measures of resultant interaction can be introduced. For example, the following terminology is frequently employed in this context.

Dynamically decoupled

Here, every output depends on one and only one input. The transfer-function matrix $\mathbf{H}(s)$ is diagonal for all s. In this case, the problem reduces to separate SISO control loops.

Band-decoupled and statically decoupled systems

When the transfer-function matrix $\mathbf{H}(j\omega)$ is diagonal only in a finite frequency band, we say that the system is decoupled in that band. In particular, we will say, when $\mathbf{H}(0)$ is diagonal, that the system is statically decoupled.

Triangularly coupled systems

A system is triangularly coupled when the inputs and outputs can be ordered in such a way that the transfer-function matrix $\mathbf{H}(s)$ is either upper or lower triangular, for all s. The coupling is then hierarchical. Consider the lower-triangular case. The first output depends only on the first input; the second output depends only on the first and second inputs; and in general, the kth output depends only on the first k inputs. Such systems are relatively easy to control by SISO controllers combined with feedforward action to compensate for the coupling.

Diagonally dominant systems

A MIMO system is said to be diagonally dominant if its $m \times m$ transfer function $\mathbf{H}(j\omega)$ is such that

$$|H_{kk}(j\omega)| > \sum_{\substack{i=1 \\ i \neq k}}^{m} |H_{ik}(j\omega)| \qquad \forall \omega \in \mathbb{R} \qquad (21.6.2)$$

Such systems are relatively decoupled, because condition (21.6.2) is similar to the situation in communications when the information in a given channel is stronger than the combined effect of the crosstalk interference from the other channels.

21.7   Industrial Case Study (Strip Flatness Control)

As an illustration of the use of simple precompensators to convert a MIMO problem into one in which SISO techniques can be employed, we consider the problem of strip flatness control in rolling mills. We have examined control problems in rolling mills earlier. See, in particular, Example 8.3 on page 208, Example 8.8 on page 226, and section §8.7. However, our earlier work focused on longitudinal (i.e., *along* the strip) control problems. That is, we have assumed that uniform reduction of the strip occurred across the width of the strip. This is not always the case in practice, so one needs to consider transversal (i.e., *across* the strip) control issues. This will be our concern here.

If rolling results in a nonuniform reduction of the strip thickness across the strip width, then a residual stress will be created, and buckling of the final product may occur. A practical difficulty is that flatness defects can be *pulled out* by the applied strip tensions, so that they are not visible to the mill operator. However, the buckling will become apparent as the coil is unwound or after it is slit or cut to length in subsequent processing operations.

There are several sources of flatness problems, including the following:

- roll thermal cambers

- incoming feed disturbances (profile, hardness, thickness)

- transverse temperature gradients

- roll stack deflections

- incorrect ground roll cambers

- roll wear

- inappropriate mill setup (reduction, tension, force, roll bending)

- lubrication effects

On the other hand, there are strong economic motives to control strip flatness, including the following:

- improved yield of prime-quality strip

- increased throughput, due to faster permissible acceleration, reduced threading delay, and higher rolling speed on shape-critical products

- more efficient recovery and operation on such downstream units as annealing and continuous-process lines

- reduced reprocessing of material on tension-leveling lines or temper-rolling mills

In this context, there are several control options to achieve improved flatness. These include roll tilt, roll bending, and cooling sprays. These typically can be separated by preprocessing the measured shape. Here, we will focus on the cooling spray option. Note that flatness defects can be measured across the strip by using a special instrument called a Shape Meter. A typical control configuration is shown in Figure 21.9.

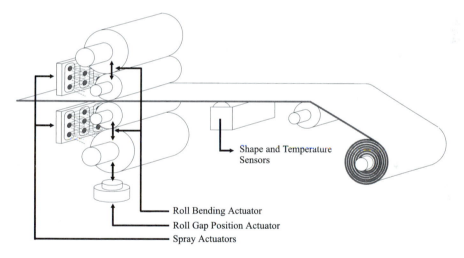

Figure 21.9. Typical flatness-control set-up for rolling mill

In this configuration, numerous cooling sprays are located across the roll, and the flow through each spray is controlled by a valve. The cool water sprayed onto the roll reduces the thermal expansion. The interesting thing is that each spray affects a large section of the roll, not just the section directly beneath it. This leads to an interactive MIMO system, rather than a series of decoupled SISO systems.

A simplified model for this system (ignoring nonlinear heat-transfer effects, etc.) is shown in the block diagram in Figure 21.10, where U denotes a vector of spray

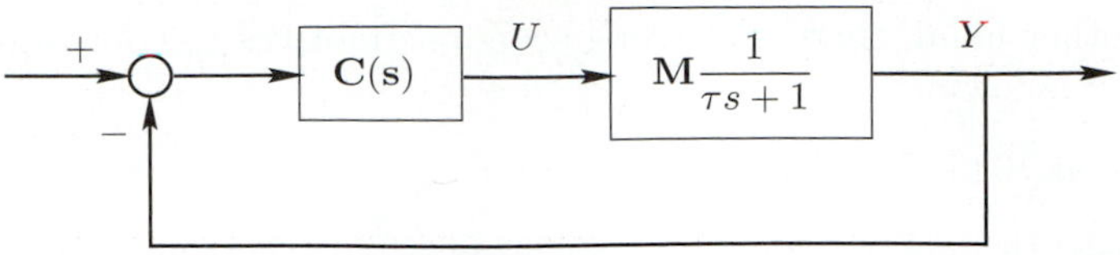

Figure 21.10. Simplified flatness-control feedback loop

valve positions and Y denotes the roll-thickness vector. (The lines indicate vectors rather than single signals).

The sprays affect the roll as described by this matrix $\mathbf{M}$:

$$\mathbf{M} = \begin{bmatrix} 1 & \alpha & \alpha^2 & \cdots \\ \alpha & 1 & & \\ \alpha^2 & & \ddots & \vdots \\ \vdots & & 1 & \alpha \\ & \cdots & \alpha & 1 \end{bmatrix} \tag{21.7.1}$$

The parameter α represents the level of interactivity in the system and is determined by the number of sprays present and how close together they are.

An interesting thing about this simplified model is that the interaction is captured totally by the d.c. gain matrix $\mathbf{M}$. This suggests that we could design an approximate precompensator by simply inverting this matrix. This leads to

$$\mathbf{M}^{-1} = \begin{bmatrix} \dfrac{1}{1-\alpha^2} & \dfrac{-\alpha}{1-\alpha^2} & 0 & \cdots & & 0 \\ \dfrac{-\alpha}{1-\alpha^2} & \dfrac{1+\alpha^2}{1-\alpha^2} & & & & \vdots \\ 0 & & \ddots & & & 0 \\ \vdots & & & \dfrac{1+\alpha^2}{1-\alpha^2} & \dfrac{-\alpha}{1-\alpha^2} \\ 0 & \cdots & 0 & \dfrac{-\alpha}{1-\alpha^2} & \dfrac{1}{1-\alpha^2} \end{bmatrix} \tag{21.7.2}$$

Using this matrix to decouple the system amounts to turning off surrounding sprays when a spray is turned on.

So we can (approximately) decouple the system simply by multiplying the control vector by this inverse. This set-up is shown in the block diagram below.

The nominal decoupled system then becomes simply $\mathbf{H}(s) = \text{diag} \frac{1}{(\tau s+1)}$. With this new model, the controller can be designed by using SISO methods. For example,

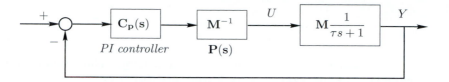

Figure 21.11. Flatness control with precompensation

a set of simple PI controllers linking each shape meter with the corresponding spray would seem to suffice. (We assume that the shape meters measure the shape of the rolls perfectly).

The reader is invited to interact with a model for this control system provided on the Web page.

Actually, control problems almost identical to the above can be found in many alternative industrial situations where there are longitudinal and traverse effects. To quote one other example, similar issues arise in paper making.

An interesting robustness issue, related to traverse frequency response, arises in all of these systems.

It might be felt that one can gain better and better accuracy by adding more and more sensors across the strip. However, this is not usually true in practice. The reason is that, if one thinks of the distance across the strip as a pseudo-time, then there is an associated frequency dimension. Hence, adding more actuators corresponds to attempting to increase the *pseudo-time* resolution–i.e., the *cross-machine bandwidth*. However, as we have seen in section §8.5, model errors tend to place an upper limit on achievable bandwidth. Thus, we might expect that model errors in the kind of two-dimensional control problem discussed above will similarly limit the degree of *pseudo-time* resolution. This can also be analyzed by using conventional MIMO robustness techniques. In particular, as we increase the number of *across-machine actuators* and *sensors*, then the extent of interaction increases, and the matrix $\mathbf{M}$ in (21.7.1) becomes ill-conditioned. We can illustrate the impact of this ill-conditioning by assuring that α has a small error. Thus, we can write the true $\mathbf{M}$ in terms of the nominal $\mathbf{M}_o$ as

$$\mathbf{M} = [\mathbf{M}_o + \Delta] \tag{21.7.3}$$

Hence, using the nominal inverse $\mathbf{M}_o^{-1}$ as a compensator leads to residual coupling, as can be seen from

$$\mathbf{M}_o^{-1}\mathbf{M} = [\mathbf{I} + \mathbf{M}_o^{-1}\Delta] \tag{21.7.4}$$

We note that the right MME in this case is given by

$$\mathbf{G}_{\Delta r}(s) = \mathbf{M}_o^{-1}\Delta \qquad\qquad (21.7.5)$$

$\mathbf{M}_o$ is near-singular, so $\mathbf{M}_o^{-1}\Delta$ can be extremely large. Thus, robust stability can easily be lost. (See section §20.8 for further discussion.)

There are several means that are typically used to avoid this problem. One can limit the number of actuator-sensor pairs, or one can limit the traverse bandwidth by restricting the control to a limited number of harmonics of the traverse dimension. The reader is again invited to investigate these robustness issues on the Web-site simulation.

□□□

21.8   Summary

- A fundamental decision in MIMO synthesis pertains to the choice of decentralized versus full MIMO control.

- Completely decentralized control

  o In completely decentralized control, the MIMO system is approximated as a set of independent SISO systems.

  o To do so, multivariable interactions are thought of as disturbances; this is an approximation, because the interactions involve feedback, whereas disturbance analysis actually presumes disturbances to be independent inputs.

  o When applicable, the advantage of completely decentralized control is that one can apply the simpler SISO theory.

  o Applicability of this approximation depends on the neglected interaction dynamics, which can be viewed as modeling errors; robustness analysis can be applied to determine their impact.

  o Chances of success are increased by judiciously pairing inputs and outputs (for example, by using the Relative Gain Array, RGA) and by using feedforward.

  o Feedforward is often a very effective tool in MIMO problems.

  o Some MIMO problems can be better treated as SISO problems if a pre-compensator is first used.

- There are several ways to quantify interactions in multivariable systems, including their structure and their strength.

  o Interactions can have a completely general structure (every input potentially affects every output) or display particular patterns, such as *triangular* or *dominant diagonal*; they can also display frequency-dependent patterns, such as being *statically decoupled* or *band-decoupled*.

- The lower the strength of interaction, the more nearly a system behaves like a set of independent systems that can be analyzed and controlled separately.

- Weak coupling can be due to the nature of the interacting dynamics or to a separation in frequency range or time scale.

- The stronger is the interaction, the more important it becomes to view the multi-input multi-output system and its interactions as a whole.

- Compared to the SISO techniques discussed so far, viewing the MIMO systems and its interactions as a whole requires generalized synthesis and design techniques and insight. These will be the topics of the following two chapters.

21.9   Further Reading

General

Skogestad, S. and Postlethwaite, I. (1996). *Multivariable Feedback Control: Analysis and Design*. Wiley, New York.

Decentralized control

Hovd, M. and Skogestad, S. (1994). Sequential design of decentralized controllers. *Automatica*, 30(10):1601-1607.

Sourlas, D. and Manousiouthakis, V. (1995). Best achievable decentralized performance. *IEEE Transactions on Automatic Control*, 40(11):1858-1871.

Zames, G. and Bensoussan, D. (1983). Multivariable feedback, sensitivity, and decentralized control. *IEEE Transactions on Automatic Control*, 28(11):1030-1035.

Relative gain array (RGA)

Bristol, E. (1966). On a new measure of interaction for multivariable process control. *IEEE Transactions on Automatic Control*, 11:133-134.

Stephanopoulos, G. (1984). *Chemical Process Control: an Introduction to Theory and Practice*. Prentice-Hall, Englewood Cliffs, N.J.

21.10   Problems for the Reader

Problem 21.1. *Consider again the system studied in Example 21.1 on page 654, with $k_{12} = -2$ and $k_{21} = -1$.*

21.1.1 *Use the feedforward strategy (see section §21.5) to make the equivalent plant transfer function $\mathbf{G}'_o(s)$ lower triangular in the frequency band $[0 \ \ \omega_{ff}] \ [rad/s]$, first for $\omega_{ff} = 1$ and then for $\omega_{ff} = 10$.*

21.1.2 *Repeat to achieve upper-triangular coupling.*

Problem 21.2. *Consider the upper-triangular coupling achieved in Problem 21.1. Test, using SIMULINK, the performance of the decentralized PI controller used in Example 21.1 on page 654, first with $\omega_{ff} = 1$ and then with $\omega_{ff} = 10$.*

Problem 21.3. *Assume a general 2×2 MIMO system.*
  Prove that if you apply the feedforward ideas of section §21.5 to obtain a diagonal equivalent plant (by first achieving upper-triangular coupling and then feedforward again to diagonalize the new system, or vice versa), this amounts to precompensating the plant.
  Verify this for the system used in 21.1.

Problem 21.4. *Assume that you have a plant having nominal model given by*

$$\mathbf{G_o}(s) = \begin{bmatrix} \dfrac{2e-0.5t}{s^2 + 3s + 2} & 0 \\ \dfrac{0.5}{(s+2)(\beta s + 1)} & \dfrac{6}{s^2 + 5s + 6} \end{bmatrix} \tag{21.10.1}$$

21.4.1 *Design, independently, SISO PI controllers for the two diagonal terms.*

21.4.2 *Evaluate the performance of your design, first for $\beta = 0.2$ and then for $\beta = -0.2$. Discuss.*

Problem 21.5. *Consider an upper-triangularly coupled $m \times m$ MIMO system.*
  Prove that an internally stable loop, based on SISO designs for the m diagonal terms, can be achieved, if and only if all off-diagonal transfer functions are stable.

Problem 21.6. *Consider a MIMO system having the transfer function*

$$\mathbf{G_o}(s) = \begin{bmatrix} \dfrac{8}{s^2 + 6s + 8} & \dfrac{0.5}{s + 8} \\[3mm] \dfrac{\beta}{s + 4} & \dfrac{6(s + 1)}{s^2 + 5s + 6} \end{bmatrix} \qquad (21.10.2)$$

21.6.1 *Use $\beta = -2$, and build a precompensator $\mathbf{P}(s)$ for this system, to achieve a diagonal precompensated plant $\mathbf{G_o}(s)\mathbf{P}(s)$. Design PID controllers for the diagonal transfer functions, and evaluate the performance by observing the plant outputs and inputs.*

21.6.2 *Repeat with $\beta = 2$. Discuss. (Hint: use the MATLAB command **zero** on $\mathbf{G_o}(s)$).*

Problem 21.7. *A MIMO system has the nominal model*

$$\mathbf{G_o}(s) = \begin{bmatrix} \dfrac{-s + 8}{s^2 + 3s + 2} & \dfrac{0.5(-s + 8)}{(s + 1)(s + 5)} \\[3mm] \dfrac{1}{s^2 + 2s + 1} & \dfrac{6}{s^2 + 5s + 6} \end{bmatrix} \qquad (21.10.3)$$

21.7.1 *Analyze the difficulties encountered in using precompensation to diagonalize the plant.*

21.7.2 *Factor the transfer function as*

$$\mathbf{G_o}(s) = \mathbf{G_{o1}}(s)\mathbf{G_{o2}}(s) = \begin{bmatrix} \dfrac{-s + 8}{s + 8} & 0 \\[3mm] 0 & 1 \end{bmatrix} \begin{bmatrix} \dfrac{s + 8}{s^2 + 3s + 2} & \dfrac{(s + 8)}{(s + 1)(s + 5)} \\[3mm] \dfrac{1}{s^2 + 2s + 1} & \dfrac{6}{s^2 + 5s + 6} \end{bmatrix}$$
$$(21.10.4)$$

and build an approximate (stable and proper) precompensator $\mathbf{P}(s)$ for $\mathbf{G_{o2}}(s)$.

21.7.3 *Design PI controllers for the diagonal terms of the precompensated plant, and evaluate your design.*

Part VII

MIMO CONTROL DESIGN

PREVIEW

We next examine techniques specifically aimed at MIMO control problems in which interaction cannot be ignored if one wants to achieve maximal performance. The first chapter in this part covers linear optimal control methods. These techniques are very frequently used in advanced control applications. Indeed, we will describe several real-world applications of these ideas. The next chapter covers Model Predictive Control. This is an extension of optimal control methods, to incorporate actuator and state constraints. These ideas have had a major impact on industrial control, especially in the petrochemical industries.

The final chapter covers fundamental design limitations in MIMO control. As in the SISO case, MIMO control-system design involves an intricate web of trade-offs. Indeed, many of the core issues that we have seen in the case of SISO systems have direct MIMO counterparts: e.g., open-loop poles, open-loop zeros, sensitivity functions, disturbances, and robustness. However, unlike SISO systems, these issues now have a distinctly directional flavor–it turns out that it matters which combination of inputs and outputs is involved in a particular property. These directional issues require us to be a little more careful in analysis, synthesis, and design of MIMO control loops. This thus sets the scene for the next five chapters.

Chapter 22

DESIGN VIA OPTIMAL
CONTROL TECHNIQUES

22.1 Preview

The previous chapter gave an introduction to MIMO control-system synthesis by showing how SISO methods could sometimes be used in MIMO problems. However, some MIMO problems require a fundamentally MIMO approach. These are the topic of the current chapter. We will emphasize methods based on optimal control theory. There are three reasons for this choice:

1 It is relatively easy to understand.

2 It has been used in a myriad of applications. (Indeed, the authors have used these methods on approximately 20 industrial applications.)

3 It is a valuable precursor to other advanced methods–e.g., Model Predictive Control, which is explained in the next chapter.

The analysis presented in this chapter builds on the results in Chapter 18, where state space design methods were briefly described in the SISO context. We recall, from that chapter, that the two key elements were

- state estimation by an observer

- state-estimate feedback

We will mirror these elements here for the MIMO case.

22.2 State-Estimate Feedback

Consider the following MIMO state space model having m inputs and p outputs.

$$\dot{x}(t) = \mathbf{A_o}x(t) + \mathbf{B_o}u(t) \qquad (22.2.1)$$
$$y(t) = \mathbf{C_o}x(t) \qquad (22.2.2)$$

where $x(t) \in \mathbb{R}^n$, $u(t) \in \mathbb{R}^m$, and $y(t) \in \mathbb{R}^p$.

By analogy with state-estimate feedback in the SISO case (as in Chapter 7), we seek a matrix $\mathbf{K} \in \mathbb{R}^{m \times n}$ such that $\mathbf{A_o} - \mathbf{B_o}\mathbf{K}$ has its eigenvalues in the LHP and an observer gain $\mathbf{J}$ such that $\mathbf{A_o} - \mathbf{J}\mathbf{C_o}$ has its eigenvalues in the LHP. Further, we will typically require that the closed-loop poles reside in some specified region in the left-half plane. Tools such as MATLAB provide solutions to these problems. We illustrate by a simple example.

Example 22.1. *Consider a MIMO plant having the nominal model*

$$\mathbf{G_o}(s) = \frac{1}{s(s+1)(s+2)} \begin{bmatrix} 2(s+1) & -0.5s(s+1) \\ s & 2s \end{bmatrix} \tag{22.2.3}$$

Say that the plant has step-type input disturbances in both channels.

Using state-estimate feedback ideas, design a multivariable controller which stabilizes the plant and, at the same time, ensures zero steady-state error for constant references and disturbances.

Solution

We first build state space models $(\mathbf{A_p}, \mathbf{B_p}, \mathbf{C_p}, \mathbf{0})$ and $(\mathbf{A_d}, \mathbf{B_d}, \mathbf{C_d}, \mathbf{0})$ for the plant and for the input disturbances, respectively. We follow the same basic idea as in the SISO case (in section §18.7)–we estimate not only the plant state $x_p(t)$ but also the disturbance vector $d_i(t)$. We then form the control law

$$u(t) = -K_p \hat{x}(t) - \hat{d}_i(t) + \overline{r}(t) \tag{22.2.4}$$

We recall that the disturbances are observable from the plant output, although they are unreachable from the plant input.

One pair of possible state space models is

$$\dot{x}_p(t) = \mathbf{A_p}x_p(t) + \mathbf{B_p}u(t) \qquad\qquad y(t) = \mathbf{C_p}x_p(t) \tag{22.2.5}$$
$$\dot{x}_d(t) = \mathbf{A_d}x_d(t) + \mathbf{B_d}u(t) \qquad\qquad d_i(t) = \mathbf{C_d}x_d(t) \tag{22.2.6}$$

where

$$\mathbf{A_p} = \begin{bmatrix} -3 & -2 & 0 & 0 \\ 1 & 0 & 0 & 0 \\ 0 & 0 & -2 & 2 \\ 0 & 0 & 0 & 0 \end{bmatrix}; \qquad \mathbf{B_p} = \begin{bmatrix} 1 & 2 \\ 0 & 0 \\ 0 & -0.5 \\ 1 & 0 \end{bmatrix}; \qquad \mathbf{C_p} = \begin{bmatrix} 0 & 0 & 1 & 0 \\ 0 & 1 & 0 & 0 \end{bmatrix}$$

$$\tag{22.2.7}$$

and

$$\mathbf{A_d} = \mathbf{0}; \qquad \mathbf{B_d} = \mathbf{0}; \qquad \mathbf{C_d} = \mathbf{I_2} \tag{22.2.8}$$

where $\mathbf{I_2}$ is the identity matrix in $\mathbb{R}^{2 \times 2}$.

The augmented state space model, $(\mathbf{A}, \mathbf{B}, \mathbf{C}, \mathbf{0})$, is then given by

$$\mathbf{A} = \begin{bmatrix} \mathbf{A_p} & \mathbf{B_p C_d} \\ \mathbf{0} & \mathbf{A_d} \end{bmatrix} = \begin{bmatrix} \mathbf{A_p} & \mathbf{B_p} \\ \mathbf{0} & \mathbf{0} \end{bmatrix} \qquad \mathbf{B} = \begin{bmatrix} \mathbf{B_p} \\ \mathbf{B_d} \end{bmatrix} = \begin{bmatrix} \mathbf{B_p} \\ \mathbf{0} \end{bmatrix} \qquad \mathbf{C} = \begin{bmatrix} \mathbf{C_p} & \mathbf{0} \end{bmatrix}$$

$$(22.2.9)$$

leading to a model with six states.

*We then compute the observer gain $\mathbf{J}$, choosing the six observer poles located at $-5, -6, -7, -8, -9, -10$. This is done using the MATLAB command **place** for the pair $(\mathbf{A}^T, \mathbf{C}^T)$.*

Next we compute the feedback gain K. We note from (22.2.4) that it is equivalent (with $\bar{r}(t) = 0$) to

$$u(t) = - \begin{bmatrix} \mathbf{K_p} & \mathbf{C_d} \end{bmatrix} \begin{bmatrix} \hat{x}_p(t) \\ \hat{x}_d(t) \end{bmatrix} \implies \mathbf{K} = \begin{bmatrix} \mathbf{K_p} & \mathbf{I_2} \end{bmatrix} \qquad (22.2.10)$$

*i.e., we need only compute $\mathbf{K_p}$. This is done by using the MATLAB command **place** for the pair $(\mathbf{A_p}, \mathbf{B_p})$. The poles in this case are chosen at $-1.5 \pm j1.32, -3$, and -5. Recall that the other two poles correspond to the uncontrollable disturbance states. They are located at the origin.*

The design is evaluated by applying step references and input disturbances in both channels, as follows:

$$r_1(t) = \mu(t-2); \qquad r_2(t) = -\mu(t-5); \qquad d_i^{(1)}(t) = \mu(t-10); \qquad d_i^{(2)}(t) = \mu(t-15)$$

$$(22.2.11)$$

where $d_i^{(1)}(t)$ and $d_i^{(2)}(t)$ are the first and second components of the input-disturbance vector respectively.

The results are shown in Figure 22.1

The design was based on an arbitrary choice of observer and controller poles (polynomials $E(s)$ and $F(s)$, respectively). Thus, the apparent triangular nature of the coupling in the design was not a design objective. However, the choice of the control law ensures that the disturbances are completely compensated in steady state.

*The file **mimo2.mdl** contains the SIMULINK schematic with the MIMO control loop for this example.*

□□□

We will not develop the idea of pole assignment further. Instead, we turn to an alternative procedure that deals with the MIMO case via optimization methods. A particularly nice approach for the design of $\mathbf{K}$ and $\mathbf{J}$ is to use quadratic optimization, because it leads to simple closed-form solutions. The details are provided in the following sections.

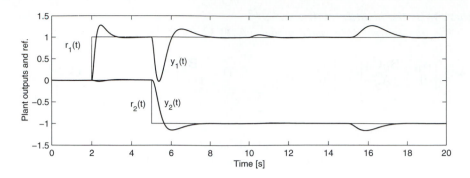

Figure 22.1. MIMO design based in state-estimate feedback

22.3 Dynamic Programming and Optimal Control

Consider a general nonlinear system with input $u(t) \in \mathbb{R}^m$, described in state space form by

$$\frac{dx(t)}{dt} = f(x(t), u(t), t) \tag{22.3.1}$$

where $x(t) \in \mathbb{R}^n$, together with the following optimization problem:

Problem (General optimal control problem). *Find an optimal input* $u^o(t)$*, for* $t \in [t_o, t_f]$*, such that*

$$u^o(t) = \arg\min_{u(t)} \left\{ \int_{t_o}^{t_f} \mathcal{V}(x, u, t) dt + g(x(t_f)) \right\} \tag{22.3.2}$$

where $\mathcal{V}(x, u, t)$ *and* $g(x(t_f))$ *are nonnegative functions.*

One possible way to solve this problem is to use Dynamic Programming ; this approach is based on the following result.

Theorem 22.1 (Optimality Principle (Bellman)). *If* $\{u(t) = u^o(t), t \in [t_o, t_f]\}$ *is the optimal solution for the above problem, then* $u^o(t)$ *is also the optimal solution over the (sub)interval* $[t_o + \Delta t, t_f]$*, where* $t_o < t_o + \Delta t < t_f$.

Proof

(By contradiction)

Denote by $x^o(t)$ the state evolution that results from applying $u^o(t)$ over the whole interval, i.e., for $t \in [t_o, t_f]$. We can then describe the optimal cost over the interval as

$$\int_{t_o}^{t_o+\Delta t} \mathcal{V}(x^o, u^o, t)dt + \int_{t_o+\Delta t}^{t_f} \mathcal{V}(x^o, u^o, t)dt + g(x^o(t_f)) \qquad (22.3.3)$$

Assume now that there exists an input $\tilde{u}(t)$ such that

$$\tilde{u}(t) = \arg\min_{u(t)} \left\{ \int_{t_o+\Delta t}^{t_f} \mathcal{V}(x, u, t)dt + g(x(t_f)) \right\} \qquad (22.3.4)$$

with corresponding state trajectory $\tilde{x}(t)$ such that $\tilde{x}(t_o + \Delta t) = x^o(t_o + \Delta t)$. Assume that

$$\int_{t_o+\Delta t}^{t_f} \mathcal{V}(\tilde{x}, \tilde{u}, t)dt + g(\tilde{x}(t_f)) < \int_{t_o+\Delta t}^{t_f} \mathcal{V}(x^o, u^o, t)dt + g(x^o(t_f)) \qquad (22.3.5)$$

Consider now the policy of applying $u(t) = u^o(t)$ in the interval $[t_o, t_o + \Delta t]$ and then $u(t) = \tilde{u}(t)$ over the interval $[t_o + \Delta t, t_f]$. The resulting cost over the complete interval would be

$$\int_{t_o}^{t_o+\Delta t} \mathcal{V}(x^o, u^o, t)dt + \int_{t_o+\Delta t}^{t_f} \mathcal{V}(\tilde{x}, \tilde{u}, t)dt + g(\tilde{x}(t_f)) \qquad (22.3.6)$$

By comparing (22.3.3) and (22.3.6), then, upon using (22.3.5), we conclude that the policy of applying $u(t) = u^o(t)$ for $t \in [t_o, t_o + \Delta t)$ and $\tilde{u}(t)$ for $t \in [t_o + \Delta t, t_f]$ would yield a smaller cost than applying $u(t) = u^o(t)$ over the entire interval. However this contradicts the assumption regarding the optimality of $u^o(t)$ over that interval.

□□□

We will next use Theorem 22.1 to derive necessary conditions for the optimal u. The idea is to consider a general time interval $[t, t_f]$, where $t \in [t_o, t_f]$, and then to use the Optimality Principle with an infinitesimal time interval $[t, t + \Delta t]$.

Denote by $J^o(x(t), t)$ the optimal (minimal) cost in the interval $[t, t_f]$, with initial state $x(t)$:

$$J^o(x(t), t) = \min_{\substack{u(\tau) \\ \tau \in [t, t_f]}} \left\{ \int_t^{t_f} \mathcal{V}(x, u, \tau)d\tau + g(x(t_f)) \right\} \qquad (22.3.7)$$

Then, upon applying Theorem 22.1, we have that

$$J^o(x(t), t) = \min_{\substack{u(\tau) \\ \tau \in [t, t+\Delta t]}} \left\{ \int_t^{t+\Delta t} \mathcal{V}(x, u, \tau)d\tau + J^o(x(t + \Delta t), t + \Delta t) \right\} \quad (22.3.8)$$

We shall now consider small perturbations around the point $(x(t), t)$, considering Δt to be a very small quantity. To achieve this, we expand $J^o(x(t + \Delta t), t + \Delta t)$ in a Taylor series. We thus obtain

$$J^o(x(t), t) = \min_{u(t)} \Big\{ \mathcal{V}(x(t), u(t), t)\Delta t +$$

$$J^o(x(t), t) + \frac{\partial J^o(x(t), t)}{\partial t}\Delta t + \left[\frac{\partial J^o(x(t), t)}{\partial x} \right]^T (x(t + \Delta t) - x(t)) + \mathcal{O}(x, t) \Big\}$$

$$(22.3.9)$$

where $\mathcal{O}(x, t)$ includes the high-order terms in the Taylor expansion.

If we now let $\Delta t \to dt$, then (22.3.9) yields

$$-\frac{\partial J^o(x(t), t)}{\partial t} = \min_{u(t)} \{ \mathcal{W}(x(t), u(t), t) \} \quad (22.3.10)$$

where

$$\mathcal{W}(x(t), u(t), t) = \mathcal{V}(x(t), u(t), t) + \left[\frac{\partial J^o(x(t), t)}{\partial x} \right]^T f(x(t), u(t), t) \quad (22.3.11)$$

Note that we have used the fact that

$$\lim_{\Delta t \to 0} \frac{x(t + \Delta t) - x(t)}{\Delta t} = \frac{dx(t)}{dt} \quad (22.3.12)$$

and Equation (22.3.1).

The optimal $u(t)$ satisfying the right-hand side of Equation (22.3.10) can be expressed symbolically as

$$u^o(t) = \mathcal{U}\left(\frac{\partial J^o(x(t), t)}{\partial x}, x(t), t \right) \quad (22.3.13)$$

which, when substituted into (22.3.10) leads to the following equations for the optimal cost:

$$-\frac{\partial J^o(x(t),t)}{\partial t} = \mathcal{V}(x(t),\mathcal{U},t) + \left[\frac{\partial J^o(x(t),t)}{\partial x}\right]^T f\big(x(t),\mathcal{U},t\big) \qquad (22.3.14)$$

The solution for this equation must satisfy the boundary condition

$$J^o(x(t_f),t_f) = g(x(t_f)) \qquad (22.3.15)$$

At this stage we cannot proceed further without being more specific about the nature of the original problem. We also note that we have implicitly assumed that the function $J^o(x(t),t)$ is well behaved, which means that it is continuous in its arguments and that it can be expanded in a Taylor series. It has been proved elsewhere (see references at the end of the chapter) that these conditions are sufficient to make (22.3.13) the optimal solution.

22.4   The Linear Quadratic Regulator (LQR)

We next apply the above general theory to the following problem.

Problem (The LQR problem). *Consider a linear time-invariant system having a state space model, as defined below:*

$$\frac{dx(t)}{dt} = \mathbf{A}x(t) + \mathbf{B}u(t) \qquad\qquad x(t_o) = x_o \qquad (22.4.1)$$

$$y(t) = \mathbf{C}x(t) + \mathbf{D}u(t) \qquad (22.4.2)$$

where $x \in \mathbb{R}^n$ is the state vector, $u \in \mathbb{R}^m$ is the input, $y \in \mathbb{R}^p$ is the output, $x_o \in \mathbb{R}^n$ is the state vector at time $t = t_o$, and $\mathbf{A}, \mathbf{B}, \mathbf{C},$ and $\mathbf{D}$ are matrices of appropriate dimensions.

Assume that we aim at driving the initial state x_o to the smallest possible value as soon as possible in the interval $[t_o, t_f]$, but without spending too much control effort (as measured by the magnitude of u) to achieve that goal. Then the optimal-regulator problem is defined as the problem of finding an optimal control $u(t)$ over the interval $[t_o, t_f]$ such that the following cost function is minimized:

$$J_u(x(t_o),t_o) = \int_{t_o}^{t_f} \left[x(t)^T \mathbf{\Psi} x(t) + u(t)^T \mathbf{\Phi} u(t)\right] dt + x(t_f)^T \mathbf{\Psi}_f x(t_f) \quad (22.4.3)$$

where $\mathbf{\Psi} \in \mathbb{R}^{n \times n}$ and $\mathbf{\Psi}_f \in \mathbb{R}^{n \times n}$ are symmetric nonnegative definite matrices and $\mathbf{\Phi} \in \mathbb{R}^{m \times m}$ is a symmetric positive definite matrix.

□□□

To solve this problem, the theory summarized in section §22.3 can be used. We first make the following connections between the general optimal problem and the LQR problem:

$$f(x(t), u(t), t) = \mathbf{A}x(t) + \mathbf{B}u(t) \tag{22.4.4}$$

$$\mathcal{V}(x, u, t) = x(t)^T \mathbf{\Psi} x(t) + u(t)^T \mathbf{\Phi} u(t) \tag{22.4.5}$$

$$g(x(t_f)) = x(t_f)^T \mathbf{\Psi}_f x(t_f) \tag{22.4.6}$$

Then

$$\mathcal{W}(x(t), u(t), t) \triangleq \mathcal{V}(x(t), u(t), t) + \left[\frac{\partial J^o(x(t), t)}{\partial x}\right]^T f\big(x(t), u(t), t\big)$$

$$= x(t)^T \mathbf{\Psi} x(t) + u(t)^T \mathbf{\Phi} u(t) + \left[\frac{\partial J^o(x(t), t)}{\partial x}\right]^T \big(\mathbf{A}x(t) + \mathbf{B}u(t)\big) \tag{22.4.7}$$

Thus, to obtain the optimal u as given by (22.3.13), we have to minimize $\mathcal{W}(x(t), u(t), t)$. This requires us to compute the gradient of $\mathcal{W}(x(t), u(t), t)$ with respect to u and then set this gradient equal to zero.

$$\frac{\partial \mathcal{W}}{\partial u} = 2\mathbf{\Phi}u(t) + \mathbf{B}^T \frac{\partial J^o(x(t), t)}{\partial x} \tag{22.4.8}$$

Now, setting $\frac{\partial \mathcal{W}}{\partial u} = 0$ shows that the optimal input, $u^o(t)$, is given by

$$u^o(t) = -\frac{1}{2}\mathbf{\Phi}^{-1}\mathbf{B}^T \frac{\partial J^o(x(t), t)}{\partial x} \tag{22.4.9}$$

We also observe that the Hessian of $\mathcal{W}$ with respect to u is equal to $\mathbf{\Phi}$, which, by initial assumption, is positive definite. This confirms that (22.4.9) gives a minimum for $\mathcal{W}$.

We next note that $J^o(x^o(t_f), t_f) = [x^o(t_f)]^T \mathbf{\Psi}_f x^o(t_f)$. We observe that this is a quadratic function of the state at time t_f. We show by induction that this is true at every time t. We proceed by assuming that the optimal cost has the form

$$J^o(x(t), t) = x^T(t)\mathbf{P}(t)x(t) \qquad \text{with} \qquad \mathbf{P}(t) = [\mathbf{P}(t)]^T \tag{22.4.10}$$

We then have that

$$\frac{\partial J^o(x(t), t)}{\partial x} = 2\mathbf{P}(t)x(t) \tag{22.4.11}$$

$$\frac{\partial J^o(x(t), t)}{\partial t} = x^T(t)\frac{d\mathbf{P}(t)}{dt}x(t) \tag{22.4.12}$$

If we now use (22.4.12) in (22.4.9),

---

the optimal control can be expressed as

$$u^o(t) = -\mathbf{K}_u(t)x(t) \qquad (22.4.13)$$

where $\mathbf{K}_u(t)$ is a time-varying gain, given by

$$\mathbf{K}_u(t) = \mathbf{\Phi}^{-1}\mathbf{B}^T\mathbf{P}(t) \qquad (22.4.14)$$

We also have from (22.4.7) that

$$\mathcal{W}(x(t), u^o(t), t) = x^T(t)\left(\mathbf{\Psi} - \mathbf{P}(t)\mathbf{B}\mathbf{\Phi}^{-1}\mathbf{B}^T\mathbf{P}(t) + 2\mathbf{P}(t)\mathbf{A}\right)x(t) \qquad (22.4.15)$$

To compute $\mathbf{K}_u(t)$, we first need to obtain $\mathbf{P}(t)$, which can be done by using (22.4.12) and (22.4.15) in (22.3.14), leading to

$$-x^T(t)\frac{d\mathbf{P}(t)}{dt}x(t) = x^T(t)\left(\mathbf{\Psi} - \mathbf{P}(t)\mathbf{B}\mathbf{\Phi}^{-1}\mathbf{B}^T\mathbf{P}(t) + 2\mathbf{P}(t)\mathbf{A}\right)x(t) \qquad (22.4.16)$$

We also note that $2x^T(t)\mathbf{P}(t)\mathbf{A}x(t) = x^T(t)\left(\mathbf{P}(t)\mathbf{A} + \mathbf{A}^T\mathbf{P}(t)\right)x(t)$. For (22.4.16) to hold for all $x(t)$, we require that

---

$$-\frac{d\mathbf{P}(t)}{dt} = \mathbf{\Psi} - \mathbf{P}(t)\mathbf{B}\mathbf{\Phi}^{-1}\mathbf{B}^T\mathbf{P}(t) + \mathbf{P}(t)\mathbf{A} + \mathbf{A}^T\mathbf{P}(t) \qquad (22.4.17)$$

Equation (22.4.17) is known as the **C**ontinuous **T**ime **D**ynamic **R**iccati **E**quation (CTDRE). This equation has to be solved backwards in time, to satisfy the boundary condition (22.4.6), (22.4.10):

$$\mathbf{P}(t_f) = \mathbf{\Psi}_f \qquad (22.4.18)$$

The above theory holds equally well for time-varying systems–i.e., when $\mathbf{A}, \mathbf{B}, \mathbf{\Phi}, \mathbf{\Psi}$ are all functions of time. However, in the time-invariant case, one can say much more about the properties of the solution. This is the subject of the next section.

22.5   Properties of the Linear Quadratic Optimal Regulator

Here we assume that $\mathbf{A}, \mathbf{B}, \mathbf{\Phi}, \mathbf{\Psi}$ are all time-invariant. We will be particularly interested in what happens at $t \to \infty$. We will summarize the key results here.

However, the reader is also referred to Appendix D, where a treatment of the results in greater depth is given. We will outline two versions of the results. In §22.5.1 we give a quick review of the basic properties as used in simple problems. In §22.5.2 we will outline more advanced results, such as might be needed in more sophisticated situations.

22.5.1   Quick Review of Properties

In many simple problems, two key assumptions hold:

(i) The system $(\mathbf{A}, \mathbf{B})$ is stabilizable from $u(t)$. (The reader may feel that this assumption is reasonable–after all if, the system isn't stabilizable, then clearly optimal control will not help recover the situation.)

(ii) The system states are all adequately *seen* by the cost function (22.4.3). Technically, this is stated as requiring that $(\mathbf{\Psi}^{\frac{1}{2}}, \mathbf{A})$ be detectable.

Under these conditions, the solution to the CTDRE, $\mathbf{P}(t)$, converges to a steady-state limit $\mathbf{P}_{\infty}^{\mathbf{s}}$ as $t_f \to \infty$. This limit has two key properties:

- $\mathbf{P}_{\infty}^{\mathbf{s}}$ is the only nonnegative solution of the matrix algebraic Riccati equation (22.5.1), in §22.5.2 obtained by setting $\frac{d\mathbf{P}(t)}{dt} = 0$ in (22.4.17).

- When this steady-state value is used to generate a feedback control law, as in (22.4.14), then the resulting closed-loop system is *stable*.

Clearly, the last property is of major importance.

22.5.2   More Detailed Review of Properties

This section refers frequently to Appendix D, so the reader is encouraged to view the appendix on his or her computer whilst reading this section.

In some applications, it is useful to know a bit more about the properties of the solution to the Optimal Control Problem than was given in subsection §22.5.1. Further results are outlined here.

Lemma 22.1. *If $\mathbf{P}(t)$ converges as $t_f \to \infty$, then the limiting value $\mathbf{P}_{\infty}$ satisfies the following Continuous-Time Algebraic Riccati Equation (CTARE):*

$$0 = \mathbf{\Psi} - \mathbf{P}_{\infty}\mathbf{B}\mathbf{\Phi}^{-1}\mathbf{B}^T\mathbf{P}_{\infty} + \mathbf{P}_{\infty}\mathbf{A} + \mathbf{A}^T\mathbf{P}_{\infty} \qquad (22.5.1)$$

Furthermore, provided $(\mathbf{A}, \mathbf{B})$ is stabilizable and $(\mathbf{A}, \mathbf{\Psi}^{\frac{1}{2}})$ has no unobservable modes on the imaginary axis, then there exists a unique positive semidefinite solution

$\mathbf{P}^s_\infty$ to the CTARE having the property that the system matrix of the closed-loop system, $\mathbf{A} - \mathbf{\Phi}^{-1}\mathbf{B}^T\mathbf{P}^s_\infty$, has all its eigenvalues in the OLHP. We call this solution the stabilizing solution of the CTARE. Other properties of the stabilizing solution are as follows:

a) If $(\mathbf{A}, \mathbf{\Psi}^{\frac{1}{2}})$ is detectable, the stabilizing solution is the only nonnegative solution of the CTARE.

b) If $(\mathbf{A}, \mathbf{\Psi}^{\frac{1}{2}})$ has unobservable modes in the OLHP, then the stabilizing solution is not positive definite.

c) If $(\mathbf{A}, \mathbf{\Psi}^{\frac{1}{2}})$ has an unobservable pole outside the OLHP, then, in addition to the stabilizing solution, there exists at least one other nonnegative solution to the CTARE. However, in this case, the stabilizing solution satisfies $\mathbf{P}^s_\infty - \overline{\mathbf{P}}_\infty \geq 0$, where $\overline{\mathbf{P}}_\infty$ is any other solution of the CTARE.

Proof

See Appendix D.

□□□

Convergence of the solution of the CTDRE to the stabilizing solution of the CTARE is addressed in the following lemma.

Lemma 22.2. *Provided that $(\mathbf{A}, \mathbf{B})$ is stabilizable, that $(\mathbf{A}, \mathbf{\Psi}^{\frac{1}{2}})$ has no unobservable poles on the imaginary axis, and that $\mathbf{\Psi} > \mathbf{P}^s_\infty$, then*

$$\lim_{t_f \to \infty} \mathbf{P}(t) = \mathbf{P}^s_\infty \tag{22.5.2}$$

(Provided that $(\mathbf{A}, \mathbf{\Psi}^{\frac{1}{2}})$ is detectable, then $\mathbf{\Psi} \geq \mathbf{0}$ suffices.)

Proof

See Appendix D.

□□□

We illustrate the above properties by a simple example.

Example 22.2. *Consider the scalar system*

$$\dot{x}(t) = ax(t) + u(t) \tag{22.5.3}$$

and the cost function

$$J = \psi_f x(t_f)^2 + \int_0^{t_f} \left(\psi x(t)^2 + u(t)^2 \right) dt \tag{22.5.4}$$

22.2.1 *Discuss this optimal control problem in the light of Lemma 22.1 and Lemma 22.2.*

22.2.2 *Discuss the convergence of the solutions of the CTDRE to* $\mathbf{P}^s_\infty$.

Solution

22.2.1 *The associated CTDRE is*

$$\dot{P}(t) = -2aP(t) + P(t)^2 - \psi; \qquad P(t_f) = \psi_f \qquad (22.5.5)$$

and the CTARE is

$$(P^s_\infty)^2 - 2aP^s_\infty - \psi = 0 \qquad (22.5.6)$$

Case 1 $\psi \neq 0$

> *Here,* $(\mathbf{A}, \mathbf{\Psi}^{\frac{1}{2}})$ *is completely observable (and thus detectable). From Lemma 22.1, part (a), there is only one nonnegative solution of the CTARE. This solution coincides with the stabilizing solution. Making the calculations, we find that the only nonnegative solution of the CTARE is*

$$P^s_\infty = \frac{2a + \sqrt{4a^2 + 4\psi}}{2} \qquad (22.5.7)$$

> *leading to the following feedback gain:*

$$K^s_\infty = a + \sqrt{a^2 + \psi} \qquad (22.5.8)$$

> *The corresponding closed-loop pole is at*

$$p_{cl} = -\sqrt{a^2 + \psi} \qquad (22.5.9)$$

> *This is clearly in the LHP, verifying that the solution is indeed the stabilizing solution.*

Case 2 $\psi = 0$

> *Here, we will consider three positive values for a: one positive, one negative, and one zero.*

> **(i)** $a > 0$. *In this case,* $(\mathbf{A}, \mathbf{\Psi}^{\frac{1}{2}})$ *has an observable pole outside the stability region. Then, from part (c) of Lemma 22.1, in addition to the stabilizing solution, there exists at least one other nonnegative solution to the CTARE. Making the calculations, we find that there*

are two nonnegative solutions of the CTARE: the stabilizing solution $\mathbf{P}_\infty^s = 2a$, and one other solution $\mathbf{P}_\infty'$.

We see that $\mathbf{P}_\infty^s - \mathbf{P}_\infty' > 0$ and that $\mathbf{P}_\infty^s$ gives a feedback-loop gain of $\mathbf{K}_\infty^s = 2a$, leading to closed-loop poles at $p_{cl} = -a$. This is clearly in the LHP, verifying that $\mathbf{P}_\infty^s$ is indeed the stabilizing solution.

(ii) $a < 0$. Here, $(\mathbf{A}, \mathbf{\Psi}^{\frac{1}{2}})$ is again detectable. Thus, from Lemma 22.1 on page 688, part (a), there is only one nonnegative solution of the CTARE. This solution coincides with the stabilizing solution.

Making the calculations, we find $\mathbf{P}_\infty^s = 0$, and the corresponding feedback gain is $\mathbf{K}_\infty^s = 0$, leading to closed-loop poles at $p_{cl} = a$, which is stable, because $a < 0$, by hypothesis.

(iii) $a = 0$. Here, $(\mathbf{A}, \mathbf{\Psi}^{\frac{1}{2}})$ has an unobservable pole on the stability boundary. This is, then, not covered by Lemma 22.1 on page 688. However, it can be verified that the CTARE has only one solution, namely $\mathbf{P}_\infty$, which is not stabilizing.

22.2.2 *To study the convergence of the solutions, we again consider two cases.*

Case 1 $\psi \neq 0$

Here $(\mathbf{A}, \mathbf{\Psi}^{\frac{1}{2}})$ is completely observable. Then $\mathbf{P}(t)$ converges to $\mathbf{P}_\infty^s$ given in (22.5.7) for any $\mathbf{P}(t_f) \geq 0$.

Case 2 $\psi = 0$

(i) $a > 0$. Here, $(\mathbf{A}, \mathbf{\Psi}^{\frac{1}{2}})$ has no unobservable poles on the $j\omega$ axis, but it is not detectable. Thus, $\mathbf{P}(t)$ converges to $\mathbf{P}_\infty^s$, provided that we choose $\mathbf{P}(t_f) > \mathbf{P}_\infty^s$.

(ii) $a < 0$. Here, $(\mathbf{A}, \mathbf{\Psi}^{\frac{1}{2}})$ is again detectable, and $\mathbf{P}(t)$ converges to $\mathbf{P}_\infty^s$ for any $\mathbf{P}(t_f) \geq 0$.

(iii) $a = 0$. Here $(\mathbf{A}, \mathbf{\Psi}^{\frac{1}{2}})$ has an unobservable pole on the $j\omega$ axis. Thus, Lemma 22.2 on page 689 does not apply. Actually, $\mathbf{P}(t)$ converges to **0** for any $\mathbf{P}(t_f) \geq 0$. However, in this case, zero is not a stabilizing solution.

□□□

Linear quadratic regulator theory is a powerful tool in control-system design. We illustrate its versatility in the next section by using it to solve the so-called Model Matching Problem (MMP).

22.6   Model Matching Based on Linear Quadratic Optimal Regulators

22.6.1   Problem Formulation

Many problems in control synthesis can be reduced to a problem of the following type:

> Given two stable transfer functions $\mathbf{M}(s)$ and $\mathbf{N}(s)$, find a stable transfer function $\Gamma(s)$ so that $\mathbf{N}(s)\Gamma(s)$ is *close* to $\mathbf{M}(s)$ in a quadratic norm sense.

We call this a model-matching problem. When $\mathbf{M}(s)$ and $\mathbf{N}(s)$ are matrix transfer functions, we need to define a suitable norm to measure *closeness*. By way of illustration, we consider a matrix $\mathbf{A} = [a_{ij}] \in \mathbb{C}^{p \times m}$ for which we define the *Fröbenius norm* as follows

$$||\mathbf{A}||_F = \sqrt{trace\,\mathbf{A}^H\mathbf{A}} = \sqrt{\sum_{i=1}^{p}\sum_{j=1}^{m} |a_{ij}|^2} \tag{22.6.1}$$

Under this norm, a suitable synthesis criterion might be

$$\Gamma^{\mathbf{o}} = \arg\min_{\Gamma \in \mathcal{S}} J_{\Gamma} \tag{22.6.2}$$

where

$$J_{\Gamma} = \frac{1}{2\pi} \int_{-\infty}^{\infty} \left\|\mathbf{M}(j\omega) - \mathbf{N}(j\omega)\Gamma(j\omega)\right\|_F^2 \, d\omega \tag{22.6.3}$$

and $\mathcal{S}$ is the class of stable transfer functions.

This problem can be converted into vector form by vectorizing $\mathbf{M}$ and Γ. For example, say that Γ is constrained to be lower triangular and that $\mathbf{M}$, $\mathbf{N}$, and Γ are 3×2, 3×2, and 2×2 matrices, respectively; then we can write

$$J_{\Theta} = \frac{1}{2\pi} \int_{-\infty}^{\infty} \left\|\mathbf{V}(j\omega) - \mathbf{W}(j\omega)\Theta(j\omega)\right\|_2^2 \, d\omega \tag{22.6.4}$$

where $\left\|\ \ \right\|_2$ denotes the usual Euclidean vector norm and where, in this special case,

$$
\mathbf{V}(s) = \begin{bmatrix} M_{11}(s) \\ M_{12}(s) \\ M_{21}(s) \\ M_{22}(s) \\ M_{31}(s) \\ M_{32}(s) \end{bmatrix}; \quad
\mathbf{W}(s) = \begin{bmatrix} N_{11}(s) & N_{12}(s) & 0 \\ 0 & 0 & N_{12}(s) \\ N_{21}(s) & N_{22}(s) & 0 \\ 0 & 0 & N_{22}(s) \\ N_{31}(s) & N_{32}(s) & 0 \\ 0 & 0 & N_{32}(s) \end{bmatrix}; \quad
\mathbf{\Theta}(s) = \begin{bmatrix} \Gamma_{11}(s) \\ \Gamma_{21}(s) \\ \Gamma_{22}(s) \end{bmatrix}
$$

$$(22.6.5)$$

22.6.2 Conversion to Time Domain

We next select a state space model for $\mathbf{V}(s)$ and $\mathbf{W}(s)$ of the form

$$\mathbf{V}(s) = \mathbf{C_1}[s\mathbf{I} - \mathbf{A_1}]^{-1}\mathbf{B_1} \qquad (22.6.6)$$

$$\mathbf{W}(s) = \mathbf{C_2}[s\mathbf{I} - \mathbf{A_2}]^{-1}\mathbf{B_2} \qquad (22.6.7)$$

where, in the special case of equation (22.6.5), $\mathbf{A_1}$, $\mathbf{B_1}$, $\mathbf{C_1}$, $\mathbf{A_2}$, $\mathbf{B_2}$, and $\mathbf{C_2}$ are matrices of dimensions $n_1 \times n_1$, $n_1 \times 1$, $6 \times n_1$, $n_2 \times n_2$, $n_2 \times 3$, and $6 \times n_2$, respectively.

Before proceeding to solve the model-matching problem, we make a slight generalization. In particular, it is sometimes desirable to restrict the size of $\mathbf{\Theta}$. We do this by generalizing (22.6.4) by introducing an extra term that weights $\mathbf{\Theta}$. This leads to

$$
J_{\mathbf{\Theta}} = \frac{1}{2\pi} \int_{-\infty}^{\infty} \left\{ \|\mathbf{V}(j\omega) - \mathbf{W}(j\omega)\mathbf{\Theta}(j\omega)\|_{\mathbf{\Gamma}}^2 + \|\mathbf{\Theta}(j\omega)\|_{\mathbf{R}}^2 \right\} d\omega \qquad (22.6.8)
$$

where $\mathbf{\Gamma}$ and $\mathbf{R}$ are nonnegative symmetrical matrices.

We can then apply Parseval's theorem (in subsection §4.10.2) to convert $J_{\mathbf{\Theta}}$ into the time domain. The transfer functions in (22.6.6) and (22.6.7) are stable and strictly proper, so this yields

$$
J_{\mathbf{\Theta}} = \int_0^{\infty} \left\{ \|y_1(t) - y_2(t)\|_{\mathbf{\Gamma}}^2 + \|u(t)\|_{\mathbf{R}}^2 \right\} dt \qquad (22.6.9)
$$

where

$$
\begin{bmatrix} \dot{x}_1(t) \\ \dot{x}_2(t) \end{bmatrix} = \begin{bmatrix} \mathbf{A_1} & 0 \\ 0 & \mathbf{A_2} \end{bmatrix} \begin{bmatrix} x_1(t) \\ x_2(t) \end{bmatrix} + \begin{bmatrix} 0 \\ \mathbf{B_2} \end{bmatrix} u(t); \qquad \begin{bmatrix} x_1(0) \\ x_2(0) \end{bmatrix} = \begin{bmatrix} \mathbf{B_1} \\ 0 \end{bmatrix} \qquad (22.6.10)
$$

$$
\begin{bmatrix} y_1(t) \\ y_2(t) \end{bmatrix} = \begin{bmatrix} \mathbf{C_1} & 0 \\ 0 & \mathbf{C_2} \end{bmatrix} \begin{bmatrix} x_1(t) \\ x_2(t) \end{bmatrix} \qquad (22.6.11)
$$

Using (22.6.10) and (22.6.11) in (22.6.9), we obtain

$$J_{\Theta} = \int_0^{\infty} \left\{ x(t)^T \Psi x(t) + u(t)^T \mathbf{R} u(t) \right\} dt \qquad (22.6.12)$$

where $x(t) = [x_1(t)^T \quad x_2(t)^T]$ and

$$\Psi = \begin{bmatrix} \mathbf{C_1}^T \\ -\mathbf{C_2}^T \end{bmatrix} \mathbf{\Gamma} \begin{bmatrix} \mathbf{C_1} & -\mathbf{C_2} \end{bmatrix} \qquad (22.6.13)$$

We recognize (22.6.10)-(22.6.12) as a standard LQR problem, as in section §22.4, where

$$\mathbf{A} = \begin{bmatrix} \mathbf{A_1} & 0 \\ 0 & \mathbf{A_2} \end{bmatrix}; \qquad \mathbf{B} = \begin{bmatrix} 0 \\ \mathbf{B_2} \end{bmatrix} \qquad (22.6.14)$$

Remark 22.1. *Note that, to achieve the transformation of the model-matching problem into a LQR problem, the key step is to link $\mathcal{L}^{-1}\left[\Theta(s)\right]$ to $u(t)$.*

22.6.3 Solution

We are interested in expressing $u(t)$ as a function of $x(t)$–i.e.,

$$u(t) = -\mathbf{K}x(t) = -[\mathbf{K_1} \quad \mathbf{K_2}] \begin{bmatrix} x_1(t) \\ x_2(t) \end{bmatrix} \qquad (22.6.15)$$

such that J_{Θ} in (22.6.12) is minimized. The optimal value of $\mathbf{K}$ is as in (22.4.14). We will also assume that the values of $\mathbf{A}$, $\mathbf{B}$, Φ, etc. are such that $\mathbf{K}$ corresponds to a stabilizing solution.

The final input $u(t)$ satisfies

$$\dot{x}(t) = \mathbf{A}x(t) + \mathbf{B}u(t) \qquad x(0) = \begin{bmatrix} \mathbf{B_1}^T & 0 \end{bmatrix}^T \qquad (22.6.16)$$
$$u(t) = -\mathbf{K}x(t) \qquad\qquad\qquad (22.6.17)$$

In transfer-function form, this is

$$U(s) = \Theta(s) = -\mathbf{K}\left(s\mathbf{I} - \mathbf{A} + \mathbf{B}\mathbf{K}\right)^{-1} \begin{bmatrix} \mathbf{B_1} \\ 0 \end{bmatrix} \qquad (22.6.18)$$

which, upon our using the special structure of $\mathbf{A}$, $\mathbf{B}$, and $\mathbf{K}$, yields

$$\Theta(s) = \left[-\mathbf{I} + \mathbf{K_2}\left(s\mathbf{I} - \mathbf{A_2} + \mathbf{B_2}\mathbf{K_2}\right)^{-1} \mathbf{B_2}\right] \mathbf{K_1}\left(s\mathbf{I} - \mathbf{A_1}\right)^{-1} \mathbf{B_1} \qquad (22.6.19)$$

22.7   Discrete-Time Optimal Regulators

The theory for optimal quadratic regulators for continuous-time systems can be extended in a straightforward way to provide similar tools for discrete-time systems. We will briefly summarize the main results, without providing proofs–they parallel those for the continuous-time case presented above.

Consider a discrete-time system having the following state space description:

$$x[k+1] = \mathbf{A}_q x[k] + \mathbf{B}_q u[k] \qquad (22.7.1)$$
$$y[k] = \mathbf{C}_q x[k] \qquad (22.7.2)$$

where $x \in \mathbb{R}^n$, $u \in \mathbb{R}^m$, and $y \in \mathbb{R}^p$.

For simplicity of notation, we will drop the subscript q from the matrices $\mathbf{A}_q, \mathbf{B}_q$, and $\mathbf{C}_q$ in the remainder of this section.

Consider now the cost function

$$J_u(x[k_o], k_o) = \sum_{k_o}^{k_f} \left(x[k]^T \mathbf{\Psi} x[k] + u[k]^T \mathbf{\Phi} u[k] \right) + x[k_f]^T \mathbf{\Psi}_f x[k_f] \qquad (22.7.3)$$

The optimal quadratic regulator is given by

$$u^o[k] = -\mathbf{K}_u[k] x[k] \qquad (22.7.4)$$

where $\mathbf{K}_u[k]$ is a time-varying gain, given by

$$\mathbf{K}_u[k] = \left(\mathbf{\Phi} + \mathbf{B}^T \mathbf{P}[k] \mathbf{B} \right)^{-1} \mathbf{B}^T \mathbf{P}[k] \mathbf{A} \qquad (22.7.5)$$

where $\mathbf{P}[k]$ satisfies the following **D**iscrete **T**ime **D**ynamic **R**iccati **E**quation (DTDRE).

$$\mathbf{P}[k] = \mathbf{A}^T \left(\mathbf{P}[k+1] - \mathbf{P}[k+1]\mathbf{B}\left(\mathbf{\Phi} + \mathbf{B}^T\mathbf{P}[k+1]\mathbf{B}\right)^{-1}\mathbf{B}^T\mathbf{P}[k+1] \right) \mathbf{A} + \mathbf{\Psi}$$

$$(22.7.6)$$

Note that this equation must also be solved backwards, subject to the boundary condition

$$\mathbf{P}[k_f] = \mathbf{\Psi}_f \qquad (22.7.7)$$

The steady-state $(k_f \rightarrow \infty)$ version of the control law (22.7.4) is given by

$$u^o[k] = -\mathbf{K}_\infty x[k] \qquad \text{where} \qquad \mathbf{K}_\infty = \left(\mathbf{\Phi} + \mathbf{B}^T\mathbf{P}_\infty\mathbf{B}\right)^{-1}\mathbf{B}^T\mathbf{P}_\infty\mathbf{A} \qquad (22.7.8)$$

where $\mathbf{K}_\infty$ and $\mathbf{P}_\infty$ satisfy the associated **D**iscrete **T**ime **A**lgebraic **R**iccati **E**quation (DTARE):

$$\mathbf{A}^T\left(\mathbf{P}_\infty - \mathbf{P}_\infty\mathbf{B}\left(\mathbf{\Phi} + \mathbf{B}^T\mathbf{P}_\infty\mathbf{B}\right)^{-1}\mathbf{B}^T\mathbf{P}_\infty\right)\mathbf{A} + \mathbf{\Psi} - \mathbf{P}_\infty = 0 \qquad (22.7.9)$$

with the property that $\mathbf{A} - \mathbf{B}\mathbf{K}_\infty$ has all its eigenvalues inside the stability boundary, provided that $(\mathbf{A}, \mathbf{B})$ is stabilizable and $(\mathbf{A}, \mathbf{\Psi}^{\frac{1}{2}})$ has no unobservable modes on the unit circle.

22.8   Connections to Pole Assignment

Note that, under reasonable conditions, the steady-state LQR ensures closed-loop stability. However, the connection to the precise closed-loop dynamics is rather indirect; it depends on the choice of $\mathbf{\Psi}$ and $\mathbf{\Phi}$. Thus, in practice, one usually needs to perform some trial-and-error procedure to obtain satisfactory closed-loop dynamics.

In some circumstances, it is possible to specify a region in which the closed-loop poles should reside and to enforce this in the solution. A simple example of this is when we require that the closed-loop poles have real part to the left of $s = -\alpha$, for $\alpha \in \mathbb{R}^+$. This can be achieved by first shifting the axis by the transformation

$$v = s + \alpha \qquad (22.8.1)$$

Then $\Re\{s\} = -\alpha \Longrightarrow \Re\{v\} = 0$.

To illustrate the application of this transformation, the Laplace form of (22.2.1) is

$$sX(s) = \mathbf{A_o}X(s) + \mathbf{B_o}U(s) \qquad (22.8.2)$$

whilst in the *v form* it is

$$vX(v) = (\mathbf{A_o} + \alpha\mathbf{I})X(v) + \mathbf{B_o}U(v) = \mathbf{A_o}'X(v) + \mathbf{B_o}U(v) \qquad (22.8.3)$$

Hence, if the closed-loop poles are in the left plane of $\boxed{v}$, they will be to the left of α in $\boxed{s}$. Note, however, that this procedure does not enforce any minimal damping on the poles. This is a nice example of a technique that ensures a particular constructed property (real part of closed-loop poles to the left of a given region) but says nothing about the consequential property of damping.

A slightly more interesting demand is to require that the closed-loop poles lie inside a circle with radius ρ and with center at $(-\alpha, 0)$, with $\alpha > \rho \geq 0$–i.e., the circle is entirely within the LHP.

This can be achieved by using a two-step procedure:

(i) We first transform the Laplace variable s to a new variable, ζ, defined as follows:

$$\zeta = \frac{s + \alpha}{\rho} \qquad (22.8.4)$$

This takes the original circle in $\boxed{s}$ to a unit circle in $\boxed{\zeta}$. The corresponding transformed state space model has the form

$$\zeta X(\zeta) = \frac{1}{\rho}(\alpha\mathbf{I} + \mathbf{A_o})X(\zeta) + \frac{1}{\rho}\mathbf{B_o}U(\zeta) \qquad (22.8.5)$$

(ii) One then treats (22.8.5) as the state space description of a *discrete-time system*. So, solving the corresponding discrete optimal control problem leads to a feedback gain $\mathbf{K}$ such that $\frac{1}{\rho}(\alpha\mathbf{I} + \mathbf{A_o} - \mathbf{B_o}\mathbf{K})$ has all its eigenvalues inside the unit disk. This in turn implies that, when the same control law is applied in continuous time, then the closed-loop poles reside in the original circle in $\boxed{s}$.

Example 22.3. *Consider a 2×2 multivariable system having the state space model*

$$\mathbf{A_o} = \begin{bmatrix} 1 & 1 & 1 \\ 2 & -1 & 0 \\ 3 & -2 & 2 \end{bmatrix}; \quad \mathbf{B_o} = \begin{bmatrix} 0 & 1 \\ 1 & 0 \\ 2 & -1 \end{bmatrix}; \quad \mathbf{C_o} = \begin{bmatrix} 1 & 0 & 0 \\ 0 & 1 & 0 \end{bmatrix}; \quad \mathbf{D_o} = \mathbf{0} \quad (22.8.6)$$

Find a state-feedback gain matrix $\mathbf{K}$ such that the closed-loop poles are all located in the disk with center at $(-\alpha; 0)$ and with radius ρ, where $\alpha = 6$ and $\rho = 2$.

Solution

We use the approach proposed above: we transform the complex variable s according to (22.8.4) and then solve a discrete-time optimal-regulator problem.

We first need the state space representation in the transformed space. This is evaluated by applying equation (22.8.5), which leads to

$$\mathbf{A}_\zeta = \frac{1}{\rho}(\alpha\mathbf{I} + \mathbf{A_o}) \qquad and \qquad \mathbf{B}_\zeta = \frac{1}{\rho}\mathbf{B_o} \qquad (22.8.7)$$

*The MATLAB command **dlqr**, with weighting matrices $\mathbf{\Psi} = \mathbf{I_3}$ and $\mathbf{\Phi} = \mathbf{I_2}$, is then used to obtain the optimal gain $\mathbf{K}_\zeta$, which is*

$$\mathbf{K}_\zeta = \begin{bmatrix} 7.00 & -4.58 & 7.73 \\ 3.18 & 7.02 & -4.10 \end{bmatrix} \qquad (22.8.8)$$

When this optimal gain is used in the original continuous-time system, the closed-loop poles, computed from $\det(s\mathbf{I} - \mathbf{A_o} + \mathbf{B_o}\mathbf{K}_\zeta) = 0$, are located at -5.13, -5.45, and -5.59. All these poles lie in the prescribed region, as expected.

□□□

Note that the above ideas can be extended to other cases in which the desired region can be transformed into the stability region for either the continuous- or discrete-time case by means of a suitable rational transformation. The reader is encouraged to explore other possibilities.

22.9   Observer Design

Next, we turn to the problem of state estimation. Here, we seek a matrix $\mathbf{J} \in \mathbb{R}^{n \times p}$ such that $\mathbf{A} - \mathbf{J}\mathbf{C}$ has its eigenvalues inside the stability region. Again, it is convenient to use quadratic optimization.

As a first step, we note that an observer can be designed for the pair $(\mathbf{C}, \mathbf{A})$ by simply considering an equivalent (called dual) control problem for the pair $(\mathbf{A}, \mathbf{B})$. To illustrate how this is done, consider the *dual* system with

$$\mathbf{A}' = \mathbf{A}^{\mathbf{T}} \qquad\qquad \mathbf{B}' = \mathbf{C}^{\mathbf{T}} \qquad (22.9.1)$$

Then, using any method for state-feedback design, we can find a matrix $\mathbf{K}' \in \mathbb{R}^{p \times n}$ such that $\mathbf{A}' - \mathbf{B}'\mathbf{K}'$ has its eigenvalues inside the stability region. Hence, if we choose $\mathbf{J} = (\mathbf{K}')^{\mathbf{T}}$, then we have ensured that $\mathbf{A} - \mathbf{J}\mathbf{C}$ has its eigenvalues inside the stability region. Thus, we have completed the observer design.

The procedure leads to a stable state estimation of the form

$$\dot{\hat{x}}(t) = \mathbf{A_o}\hat{x}(t) + \mathbf{B_o}u(t) + \mathbf{J}(y(t) - \mathbf{C}\hat{x}(t)) \qquad (22.9.2)$$

Of course, using the tricks outlined above for state-variable feedback, one can also use transformation techniques to ensure that the poles of the observer end up in any region that can be related to either the continuous- or the discrete-time case by a rational transformation.

We will show how the above procedure can be formalized by using Optimal Filtering theory. The resulting optimal filter is called a Kalman filter, to honor Kalman's seminal contribution to this problem.

22.10   Linear Optimal Filters

We will present two alternative derivations of the optimal filter–one based on stochastic modeling of the noise (subsection §22.10.1), and one based on deterministic assumptions (subsection §22.10.2). Readers can choose either derivations or both, depending upon the particular point of view they wish to adopt.

22.10.1   Derivation Based on a Stochastic Noise Model

In this section, we show how optimal-filter design can be set up as a quadratic optimization problem. This shows that the filter is *optimal* under certain assumptions regarding the signal-generating mechanism. In practice, this property is probably less important than the fact that the resultant filter has the right kind of *tuning knobs* so that it can be flexibly applied to a large range of problems of practical interest. The reader is encouraged to understand this core idea of the Kalman filter; it is arguably one of the most valuable tools in the control-system designer's *tool box*.

Consider a linear stochastic system of the form

$$dx(t) = \mathbf{A}x(t)dt + dw(t) \tag{22.10.1}$$
$$dy(t) = \mathbf{C}x(t)dt + dv(t) \tag{22.10.2}$$

where $dv(t)$ and $dw(t)$ are known as *orthogonal increment processes*. As a formal treatment of stochastic differential equations is beyond the scope of this book, it suffices here to think of the formal notation $\dot{w}(t)$, $\dot{v}(t)$ as white-noise processes with impulsive correlation:

$$\mathcal{E}\{\dot{w}(t)\dot{w}(\varsigma)^T\} = \mathbf{Q}\delta(t - \varsigma) \tag{22.10.3}$$
$$\mathcal{E}\{\dot{v}(t)\dot{v}(\varsigma)^T\} = \mathbf{R}\delta(t - \varsigma) \tag{22.10.4}$$

where $\mathcal{E}\{\circ\}$ denotes mathematical expectation and $\delta(\circ)$ is the Dirac-delta function. We can then *informally* write the model (22.10.1)-(22.10.2) as

$$\frac{dx(t)}{dt} = \mathbf{A}x(t) + \frac{dw(t)}{dt} \tag{22.10.5}$$

$$y'(t) = \frac{dy(t)}{dt} = \mathbf{C}x(t) + \frac{dv(t)}{dt} \tag{22.10.6}$$

We also assume that $\dot{w}(t)$ and $\dot{v}(t)$ are mutually uncorrelated. For readers familiar with the notation of spectral density for random processes, we are simply requiring that the spectral density for $\dot{w}(t)$ and $\dot{v}(t)$ be $\mathbf{Q}$ and $\mathbf{R}$, respectively.

Our objective will be to find a *linear* filter driven by $y'(t)$ that produces a state estimate $\hat{x}(t)$. We will optimize the filter by minimizing the quadratic function

$$J_t = \mathcal{E}\{\tilde{x}(t)\tilde{x}(t)^T\} \tag{22.10.7}$$

where

$$\tilde{x}(t) = \hat{x}(t) - x(t) \tag{22.10.8}$$

is the estimation error.

We will proceed to the solution of this problem in four steps.

Step 1

Consider a time-varying version of the model (22.10.5), given by

$$\frac{dx_z(t)}{dt} = \mathbf{A_z}(t)x(t) + \dot{w}_z(t) \tag{22.10.9}$$

$$y'_z(t) = \frac{dy_z(t)}{dt} = \mathbf{C_z}(t)x_z(t) + \dot{v}_z(t) \tag{22.10.10}$$

where $\dot{w}_z(t)$ and $\dot{v}_z(t)$ have zero mean and are uncorrelated, and

$$\mathcal{E}\{\dot{w}_z(t)\dot{w}_z(\zeta)^T\} = \mathbf{Q_z}(t)\delta(t - \zeta) \tag{22.10.11}$$

$$\mathcal{E}\{\dot{v}_z(t)\dot{v}_z(\zeta)^T\} = \mathbf{R_z}(t)\delta(t - \zeta) \tag{22.10.12}$$

For this model, we wish to compute $\overline{\mathbf{P}}(t) = \mathcal{E}\{x_z(t)x_z(t)^T\}$. We assume that $\mathcal{E}\{x_z(0)x_z(0)^T\} = \overline{\mathbf{P}_o}$, with $\dot{w}_z(t)$ uncorrelated with the initial state $x_z(0) = x_{oz}$.

Solution

The solution to (22.10.9) is

$$x_z(t) = \phi_z(t, 0)x_{oz} + \int_0^t \phi_z(t, \tau)\dot{w}_z(\tau)d\tau \qquad (22.10.13)$$

where $\phi_z(t_2, t_1) \in \mathbb{R}^{n \times n}$ is the state transition matrix for the system (22.10.9).
Then, squaring (22.10.13) and taking mathematical expectations, we have

$$\mathbf{P}(t) = \mathcal{E}\{x_z(t)x_z(t)^T\} = \phi_z(t, 0)\overline{\mathbf{P}}_o\phi_z(t, 0)^T + \int_0^t \phi_z(t, \tau)\mathbf{Q_z}(\tau)\phi_z(t, \tau)^T d\tau$$
$$(22.10.14)$$

where we have used (22.10.11).

Note that, by applying the Leibnitz rule (3.7.4) to (22.10.14), we obtain

$$\frac{d\overline{\mathbf{P}}(t)}{dt} = \mathbf{A_z}\overline{\mathbf{P}}(t) + \overline{\mathbf{P}}(t)\mathbf{A_z}^T + \mathbf{Q_z}(t) \qquad (22.10.15)$$

where we have also used the fact that $\dfrac{d}{dt}\phi(t, \tau) = \mathbf{A_z}(t)\phi(t, \tau)$.

Step 2

We now return to the original problem: to obtain an estimate, $\hat{x}(t)$, for the state, $x(t)$, in (22.10.5). We assume the following linear form for the filter:

$$\frac{d\hat{x}(t)}{dt} = \mathbf{A}\hat{x}(t) + \mathbf{J}(t)[y'(t) - \mathbf{C}x(t)] \qquad (22.10.16)$$

where $\mathbf{J}(t)$ is a time-varying gain yet to be determined. Note that (22.10.16) has the same structure as a standard linear observer, save that it has a time-varying observer gain.

Step 3

Assume that we are also given an initial state estimate $\hat{x}_o$ of statistical property

$$\mathcal{E}\{(x(0) - \hat{x}_o)(x(0) - \hat{x}_o)^T\} = \mathbf{P_o}, \qquad (22.10.17)$$

and assume, for the moment, that we are given some gain $\mathbf{J}(\tau)$ for $0 \le \tau \le t$. Derive an expression for

$$\begin{aligned} \mathbf{P}(t) &= \mathcal{E}\{(\hat{x}(t) - x(t))(\hat{x}(t) - x(t))^T\} \\ &= \mathcal{E}\{\tilde{x}(t)\tilde{x}(t)^T\} \end{aligned} \qquad (22.10.18)$$

Solution

Subtracting (22.10.16) from (22.10.5), we obtain

$$\frac{d\tilde{x}(t)}{dt} = (\mathbf{A} - \mathbf{J}(t)\mathbf{C})\tilde{x}(t) + \mathbf{J}(t)\dot{v}(t) - \dot{w}(t) \qquad (22.10.19)$$

We see that (22.10.19) is a time-varying system, and we can therefore immediately apply the solution to Step 1, after making the following connections:

$$x_z(t) \to \tilde{x}(t); \qquad \mathbf{A_z}(t) \to (\mathbf{A} - \mathbf{J}(t)\mathbf{C}); \qquad \dot{w}_z(t) \to \mathbf{J}(t)\dot{v}(t) - \dot{w}(t) \\ (22.10.20)$$

to conclude

$$\frac{d\mathbf{P}(t)}{dt} = (\mathbf{A} - \mathbf{J}(t)\mathbf{C})\mathbf{P}(t) + \mathbf{P}(t)(\mathbf{A} - \mathbf{J}(t)\mathbf{C})^T + \mathbf{J}(t)\mathbf{R}\mathbf{J}(t)^T + \mathbf{Q} \qquad (22.10.21)$$

subject to $\mathbf{P}(0) = \mathbf{P_o}$. Note that we have used the fact that $\mathbf{Q_z}(t) = \mathbf{J}(t)\mathbf{R}\mathbf{J}(t)^T + \mathbf{Q}$.

Step 4

We next choose $\mathbf{J}(t)$, at each time instant, so that $\dot{\mathbf{P}}$ is as small as possible.

Solution. We complete the square on the right-hand side of (22.10.21) by defining $\mathbf{J}(t) = \mathbf{J}^*(t) + \tilde{\mathbf{J}}(t)$ where $\mathbf{J}^*(t) = \mathbf{P}(t)\mathbf{C}^T\mathbf{R}^{-1}$. Substituting into (22.10.21) gives

$$\begin{aligned} \frac{d\mathbf{P}(t)}{dt} &= (\mathbf{A} - \mathbf{J}(t)\mathbf{C} - \tilde{\mathbf{J}}(t)\mathbf{C})\mathbf{P}(t) + \mathbf{P}(t)(\mathbf{A} - \mathbf{J}(t)\mathbf{C} - \tilde{\mathbf{J}}(t)\mathbf{C})^T \\ &\quad + (\mathbf{J}^*(t) + \tilde{\mathbf{J}}(t))\mathbf{R}(\mathbf{J}^*(t) + \tilde{\mathbf{J}}(t))^T + \mathbf{Q} \\ &= (\mathbf{A} - \mathbf{J}(t)\mathbf{C})\mathbf{P}(t) + \mathbf{P}(t)(\mathbf{A} - \mathbf{J}(t)\mathbf{C})^T \\ &\quad + \mathbf{J}^*(t)\mathbf{R}\mathbf{J}^*(t) + \mathbf{Q} + \tilde{\mathbf{J}}(t)\mathbf{R}(\tilde{\mathbf{J}}(t))^T \end{aligned} \qquad (22.10.22)$$

We clearly see that $\dot{\mathbf{P}}(t)$ is minimized at every time if we choose $\tilde{\mathbf{J}}(t) = 0$. Thus, $\mathbf{J}^*(t)$ is the optimal-filter gain, because it minimizes $\dot{\mathbf{P}}(t)$ (and hence $\mathbf{P}(t)$) for all t.

□□□

In summary, the optimal filter satisfies

$$\frac{d\hat{x}(t)}{dt} = \mathbf{A}\hat{x}(t) + \mathbf{J}^*(t)[y'(t) - \mathbf{C}\hat{x}(t)] \qquad (22.10.23)$$

where the optimal gain $\mathbf{J}^*(t)$ satisfies

$$\mathbf{J}^*(t) = \mathbf{P}(t)\mathbf{C}^T\mathbf{R}^{-1} \tag{22.10.24}$$

and $\mathbf{P}(t)$ is the solution to

$$\begin{aligned}\frac{d\mathbf{P}(t)}{dt} =&(\mathbf{A} - \mathbf{J}^*(t)\mathbf{C})\mathbf{P}(t) + \mathbf{P}(t)(\mathbf{A} - \mathbf{J}^*(t)\mathbf{C})^T \\ &+ \mathbf{J}^*(t)\mathbf{R}(\mathbf{J}^*(t))^T + \mathbf{Q}\end{aligned} \tag{22.10.25}$$

subject to $\mathbf{P}(0) = \mathbf{P_o}$.

Equation (22.10.25) can also be simplified to

$$\frac{d\mathbf{P}(t)}{dt} = \mathbf{Q} - \mathbf{P}(t)\mathbf{C}^T\mathbf{R}^{-1}\mathbf{C}\mathbf{P}(t) + \mathbf{P}(t)\mathbf{A}^T + \mathbf{A}\mathbf{P}(t) \tag{22.10.26}$$

The reader will recognize that the solution to the optimal linear filtering problem presented above has a very close connection to the LQR problem presented in section §22.4.

Remark 22.2. *It is important to note, in the above derivation, that it makes no difference whether the system is time varying (i.e., $\mathbf{A}, \mathbf{C}, \mathbf{Q}, \mathbf{R}$, etc. are all functions of time). This is often important in applications.*

Remark 22.3. *When we come to properties, these are usually restricted to the time-invariant case (or closely related cases–e.g., periodic systems). Thus, when discussing the* steady-state *filter, it is usual to restrict attention to the case in which $\mathbf{A}, \mathbf{C}, \mathbf{Q}, \mathbf{R}$, etc. are not explicit functions of time.*

The properties of the optimal filter therefore follow directly from the optimal LQR solutions, under the correspondences given in Table 22.10.1 below.

In particular, one is frequently interested in the steady-state optimal filter obtained when $\mathbf{A}$, $\mathbf{C}$, $\mathbf{Q}$ and $\mathbf{R}$ are time invariant and the filtering horizon tends to infinity. By duality with the optimal control problem, the steady-state filter takes the form

$$\frac{d\hat{x}(t)}{dt} = \mathbf{A}\hat{x} + \mathbf{J}_s^\infty(y' - \mathbf{C}\hat{x}) \tag{22.10.27}$$

Regulator	Filter
τ	$t-\tau$
t_f	0
$\mathbf{A}$	$-\mathbf{A}^T$
$\mathbf{B}$	$-\mathbf{C}^T$
$\boldsymbol{\Psi}$	$\mathbf{Q}$
$\boldsymbol{\Phi}$	$\mathbf{R}$
$\boldsymbol{\Psi}_f$	$\mathbf{P_o}$

Table 22.1. Duality between quadratic regulators and filters

where

$$\mathbf{J}_s^\infty = \mathbf{P}_\infty^s \mathbf{C}^T \mathbf{R}^{-1} \tag{22.10.28}$$

and $\mathbf{P}_\infty^s$ is the stabilizing solution of the following CTARE:

$$\mathbf{Q} - \mathbf{P}_\infty \mathbf{C}^T \mathbf{R}^{-1} \mathbf{C} \mathbf{P}_\infty + \mathbf{P}_\infty \mathbf{A}^T + \mathbf{A} \mathbf{P}_\infty = 0 \tag{22.10.29}$$

We will not labor the time-variant properties here; they are easily obtained from the corresponding properties of the optimal regulator. To illustrate, we state without proof the following facts that are the duals of those given in subsection §22.5.1 for the optimal regulator:

(i) Say that the system $(\mathbf{C}, \mathbf{A})$ is detectable from $y(t)$ (the reader may judge that this assumption is reasonable–after all, if the system isn't detectable, then there are *unstable states* whose response cannot be *seen* in the output, and it is heuristically reasonable that it would then be impossible to say anything sensible about these states by processing the output signal); and

(ii) say that the system states are all perturbed by noise, so that it is not possible to estimate them once and for all, and then let the open-loop model predict all future values. (Technically, this is stated as requiring that $(\mathbf{A}, \mathbf{Q}^{\frac{1}{2}})$ is stabilizable.)

Then, the optimal solution of the filtering Riccati equation (22.10.26) tends to a steady-state limit $\mathbf{P}_\infty^s$ as $t \to \infty$. This limit has two key properties:

- $\mathbf{P}_\infty^s$ is the only nonnegative solution of the matrix algebraic Riccati Equation (22.10.29) obtained by setting $\frac{d\mathbf{P}(t)}{dt}$ in (22.10.26).

- When this steady-state value is used to generate a steady-state observer, as in $\{(22.10.27), (22.10.28)\}$, then the observer has the property that $(\mathbf{A} - \mathbf{J}_s^\infty \mathbf{C})$ is a stability matrix.

This last property is important in applications. In particular, we recall that the error $\tilde{x}(t)$ between $\hat{x}(t)$ and $x(t)$ satisfies (22.10.19). The steady-state version of this equation is stable when $(\mathbf{A} - \mathbf{J}_s^\infty \mathbf{C})$ is a stability matrix.

22.10.2   Steady-State Filter as a Deterministic Model-Matching Problem

We next give an alternative, purely *deterministic* view of the *optimal* filter as a model-matching problem. Consider a state space model, as in $\{(22.10.1), (22.10.2)\}$ without noise:

$$\frac{dx(t)}{dt} = \mathbf{A}x(t); \qquad x(0) = x_o \tag{22.10.30}$$

$$y_o(t) = \mathbf{C}x(t) \tag{22.10.31}$$

Say that we are interested in estimating a scalar combination of the state, which we denote by

$$z_o(t) = \gamma^T x(t) \tag{22.10.32}$$

or, in transfer-function form

$$Z_o(s) = \gamma^T (s\mathbf{I} - \mathbf{A})^{-1} x_o \tag{22.10.33}$$

Our filter is required to be a linear time-invariant system driven by y_o. This we express in transfer-function form by requiring that

$$\hat{Z}_o(s) = \mathbf{F}(s) Y_o(s) \tag{22.10.34}$$

where $\mathbf{F}(s)$ is the $1 \times m$ row vector corresponding to the filter transfer function and

$$Y_o(s) = \mathbf{C}(s\mathbf{I} - \mathbf{A})^{-1}x_o \qquad (22.10.35)$$

Our remaining task is to find the optimal value for the transfer function $\mathbf{F}(s)$. Toward this goal, we want to consider all possible initial states x_o. We know that all such states can be formed as a linear combination of the basis vectors $e_1, \dots, e_n$, where $e_i = [0, \dots, 0, 1, 0 \dots, 0]^T$. Thus we vectorize (22.10.33), (22.10.34), (22.10.35) by changing x_o into each of the above vectors in turn–i.e., we introduce the row vectors

$$\mathbf{Z} = [z_1, \dots, z_n] \qquad (22.10.36)$$

$$\hat{\mathbf{Z}} = [\hat{z}_1, \dots, \hat{z}_n] \qquad (22.10.37)$$

and the $m \times n$ matrix

$$\mathbf{Y} = [y_1, \dots, y_n] \qquad (22.10.38)$$

where

$$Z_i(s) = \gamma^T(s\mathbf{I} - \mathbf{A})^{-1}e_i \qquad (22.10.39)$$

$$Y_i(s) = \mathbf{C}(s\mathbf{I} - \mathbf{A})^{-1}e_i \qquad (22.10.40)$$

$$\hat{Z}_i(s) = \mathbf{F}(s)\mathbf{Y_i}(s) \qquad (22.10.41)$$

We would like the row vector $\hat{\mathbf{Z}}(s)$ to be *near* $\mathbf{Z}(s)$ in some sense. However, we wish to place some weighting on the size of the filter to avoid large values of $\mathbf{F}$ that would magnify high-frequency components in measurement disturbances.

A criterion that trades off the closeness of $\mathbf{Z}$ to $\hat{\mathbf{Z}}$ with the size of $\mathbf{F}$ is the following quadratic criterion in the frequency domain:

$$J = \int_{-\infty}^{\infty} \left(\left\| \mathbf{Z}(j\omega)^T - \hat{\mathbf{Z}}(j\omega)^T \right\|_{\mathbf{Q}}^2 + \left\| \mathbf{F}(j\omega) \right\|_{\mathbf{R}}^2 \right) d\omega \qquad (22.10.42)$$

where we have somewhat arbitrarily introduced weightings $\mathbf{Q}$ and $\mathbf{R}$ to allow us to shift the balance between bringing $\hat{\mathbf{Z}}$ closer to $\mathbf{Z}$ versus having a large filter gain. Using (22.10.33), (22.10.34), (22.10.35), the criterion becomes

$$J = \int_{-\infty}^{\infty} \left(\left\| \mathbf{Z}^T - (\mathbf{FY})^T \right\|_{\mathbf{Q}}^2 + \left\| \mathbf{F}^T \right\|_{\mathbf{R}}^2 \right) d\omega \qquad (22.10.43)$$

It is interesting to transpose quantities in the above expression. This preserves the norm but shifts $\mathbf{F}$ to an intuitively interesting position. Doing this yields

$$J = \int_{-\infty}^{\infty} \left(\left\| \mathbf{Z}^T - \mathbf{Y}^T\mathbf{F}^T \right\|_{\mathbf{Q}}^2 + \left\| \mathbf{F}^T \right\|_{\mathbf{R}}^2 \right) d\omega \qquad (22.10.44)$$

We can now think of $\mathbf{F}^T$ as a control vector. In this form, we recognize (22.10.44) as being simply a standard model-matching problem, exactly as was discussed in section §22.6, with the correspondences

$$\mathbf{Y}^T = (s\mathbf{I} - \mathbf{A}^T)^{-1}\mathbf{C}^T \equiv \mathbf{W} = \mathbf{C_2}(s\mathbf{I} - \mathbf{A_2})^{-1}\mathbf{B_2} \qquad (22.10.45)$$

$$\mathbf{Z}^T = (s\mathbf{I} - \mathbf{A}^T)^{-1}\gamma \equiv \mathbf{V} = \mathbf{C_1}(s\mathbf{I} - \mathbf{A_1})^{-1}\mathbf{B_1} \qquad (22.10.46)$$

$$(22.10.47)$$

where $\mathbf{C_2} = \mathbf{C_1} = \mathbf{I}$, $\mathbf{A_1} = \mathbf{A_2} = \mathbf{A}^T$, $\mathbf{B_2} = \mathbf{C}^T$, and $\mathbf{B_1} = \gamma$.

From section §22.6, the optimal solution is

$$\mathbf{F}(s)^T = -[\mathbf{I} - \mathbf{K_2}(s\mathbf{I} - \mathbf{A}^T + \mathbf{C}^T\mathbf{K_2})^{-1}\mathbf{C}^T][\mathbf{K_1}(s\mathbf{I} - \mathbf{A}^T)^{-1}\gamma] \qquad (22.10.48)$$

or

$$\mathbf{F}(s) = \gamma^T[s\mathbf{I} - \mathbf{A}]^{-1}\mathbf{K_1}^T[-\mathbf{I} + \mathbf{C}(s\mathbf{I} - \mathbf{A} + \mathbf{K_2}^T\mathbf{C})^{-1}\mathbf{K_2}^T] \qquad (22.10.49)$$

Now, because of the special structure of the problem, the Riccati equation and optimal gains have a special form. Specifically, we have that the Riccati equation is

$$0 = \tilde{\boldsymbol{\Psi}} - \mathbf{P}\tilde{\mathbf{B}}\tilde{\boldsymbol{\Phi}}^{-1}\tilde{\mathbf{B}}^T\mathbf{P} + \mathbf{P}\tilde{\mathbf{A}} + \tilde{\mathbf{A}}^T\mathbf{P} \qquad (22.10.50)$$

which, upon using the special structure of $\tilde{\mathbf{A}}$, $\tilde{\mathbf{B}}$, $\tilde{\boldsymbol{\Psi}}$, and $\tilde{\boldsymbol{\Phi}}$, gives

$$0 = \begin{bmatrix} \mathbf{Q} & -\mathbf{Q} \\ -\mathbf{Q} & \mathbf{Q} \end{bmatrix} - \begin{bmatrix} \mathbf{P_3}\mathbf{C}^T\mathbf{R}^{-1}\mathbf{C}\mathbf{P_3} & \mathbf{P_3}\mathbf{C}\mathbf{R}^{-1}\mathbf{C}\mathbf{P_2} \\ \mathbf{P_2}\mathbf{C}^T\mathbf{R}^{-1}\mathbf{C}\mathbf{P_3}^T & \mathbf{P_2}\mathbf{C}^T\mathbf{R}^{-1}\mathbf{C}\mathbf{P_2} \end{bmatrix} \qquad (22.10.51)$$

$$+ \begin{bmatrix} \mathbf{P_1}\mathbf{A}^T & \mathbf{P_3}\mathbf{A}^T \\ \mathbf{P_3}^T\mathbf{A}^T & \mathbf{P_2}\mathbf{A}^T \end{bmatrix} + \begin{bmatrix} \mathbf{A}\mathbf{P_1} & \mathbf{A}\mathbf{P_3} \\ \mathbf{A}\mathbf{P_3}^T & \mathbf{A}\mathbf{P_2} \end{bmatrix}$$

where we have used

$$\tilde{\mathbf{A}} = \begin{bmatrix} \mathbf{A_1}^T & 0 \\ 0 & \mathbf{A_2}^T \end{bmatrix}; \quad \tilde{\mathbf{B}} = \begin{bmatrix} 0 \\ \mathbf{C}^T \end{bmatrix}; \quad \mathbf{P} = \begin{bmatrix} \mathbf{P_1} & \mathbf{P_3} \\ \mathbf{P_3}^T & \mathbf{P_2} \end{bmatrix}; \quad \tilde{\boldsymbol{\Phi}} = \mathbf{R} \quad (22.10.52)$$

We note that, in our problem, $\mathbf{A_1} = \mathbf{A_2} = \mathbf{A}^T$ and $\mathbf{B_2} = \mathbf{C}^T$ and that $\bar{\mathbf{P}} = \mathbf{P_2}$ satisfies the following separate equation:

$$0 = \mathbf{Q} - \bar{\mathbf{P}}\mathbf{C}^T\mathbf{R}^{-1}\mathbf{C}\bar{\mathbf{P}} + \bar{\mathbf{P}}\mathbf{A}^T + \mathbf{A}\bar{\mathbf{P}} \qquad (22.10.53)$$

Also, we have that $\mathbf{P_3}$ satisfies

$$0 = -\mathbf{Q} - \bar{\mathbf{P}}\mathbf{C}^T\mathbf{R}^{-1}\mathbf{C}\mathbf{P_3}^T + \mathbf{P_3}^T\mathbf{A}^T + \mathbf{A}\mathbf{P_3}^T \tag{22.10.54}$$

Adding (22.10.53) and (22.10.54) gives

$$-\bar{\mathbf{P}}\mathbf{C}^T\mathbf{R}^{-1}\mathbf{C}(\mathbf{P_3}^T + \bar{\mathbf{P}}) + (\mathbf{P_3}^T + \bar{\mathbf{P}})\mathbf{A}^T + \mathbf{A}(\mathbf{P_3}^T + \bar{\mathbf{P}}) = 0 \tag{22.10.55}$$

which leads to $\mathbf{P_3}^T = -\bar{\mathbf{P}} = \mathbf{P_3}$
   Finally,

$$[\mathbf{K_1} \quad \mathbf{K_2}] = \mathbf{\Phi}^{-1}[\mathbf{0} \quad \mathbf{B_2}^T]\begin{bmatrix} \mathbf{P_1} & \mathbf{P_3} \\ \mathbf{P_3} & \mathbf{P_2} \end{bmatrix} \tag{22.10.56}$$

$$= \mathbf{\Phi}^{-1}[\mathbf{B_2}^T\mathbf{P_3} \quad \mathbf{B_2}^T\mathbf{P_2}]$$

$$= [-\mathbf{R}^{-1}\mathbf{C}\bar{\mathbf{P}} \quad \mathbf{R}^{-1}\mathbf{C}\bar{\mathbf{P}}]$$

Setting $\mathbf{J}^T = -\mathbf{K_1} = \mathbf{K_2}$, then,

$$\mathbf{J} = \bar{\mathbf{P}}\mathbf{C}^T\mathbf{R}^{-1} \tag{22.10.57}$$

Finally, substituting (22.10.57) into (22.10.49), we have

$$F(s) = \gamma^T[s\mathbf{I} - \mathbf{A}]^{-1}\mathbf{J}[\mathbf{I} - \mathbf{C}(s\mathbf{I} - \mathbf{A} + \mathbf{J}\mathbf{C})^{-1}\mathbf{J}] \tag{22.10.58}$$

$$= \gamma^T[s\mathbf{I} - \mathbf{A}]^{-1}[\mathbf{I} - \mathbf{J}\mathbf{C}(s\mathbf{I} - \mathbf{A} + \mathbf{J}\mathbf{C})^{-1}]\mathbf{J}$$

$$= \gamma^T(s\mathbf{I} - \mathbf{A})^{-1}[s\mathbf{I} - \mathbf{A} + \mathbf{J}\mathbf{C} - \mathbf{J}\mathbf{C}](s\mathbf{I} - \mathbf{A} + \mathbf{J}\mathbf{C})^{-1}\mathbf{J}$$

$$= \gamma^T[s\mathbf{I} - \mathbf{A} + \mathbf{J}\mathbf{C}]^{-1}\mathbf{J}$$

We notice that γ^T appears as a factor outside of the filter. Thus, we have actually found a result that can be translated to one which is independent of the choice of γ. In particular, we note from (22.10.32) and (22.10.34), that the optimal filter for estimating x is simply $(s\mathbf{I} - \mathbf{A} + \mathbf{J}\mathbf{C})^{-1}\mathbf{J}$. This transfer function can be realized in state space form as

$$\dot{\hat{x}}(t) = (\mathbf{A} - \mathbf{J}\mathbf{C})\hat{x}(t) + \mathbf{J}y(t) \tag{22.10.59}$$

which we recognize as being identical to the steady-state filter given in (22.10.27).

Remark 22.4. *The above derivation gives a purely deterministic view of the optimal filtering and allows us to trade off the size of the error versus the size of the filter.*

Other connections with deterministic optimal control theory are explored in §D.5 of the Appendix.

□□□

22.10.3   Discrete-Time Optimal Quadratic Filter

As we did in section §22.7 for the control problem, we can readily develop discrete forms for the optimal filter. As in section §22.7, we will simply quote the result; the theory parallels the continuous case.

Consider a discrete-time system having the following state space description:

$$x[k + 1] = \mathbf{A}x[k] + \mathbf{B}u[k] + w[k] \tag{22.10.60}$$

$$y[k] = \mathbf{C}[k] + v[k] \tag{22.10.61}$$

where $w[k] \in \mathbb{R}^n$ and $v[k] \in \mathbb{R}^p$ are uncorrelated stationary stochastic processes, with covariances given by

$$\mathcal{E}\{w[k]w^T[\varsigma]\} = \mathbf{Q}\delta_K[k - \varsigma] \tag{22.10.62}$$

$$\mathcal{E}\{v[k]v^T[\varsigma]\} = \mathbf{R}\delta_K[k - \varsigma] \tag{22.10.63}$$

where $\mathbf{Q} \in \mathbb{R}^{n \times n}$ is a symmetric nonnegative definite matrix and $\mathbf{R} \in \mathbb{R}^{p \times p}$ is a symmetric positive definite matrix.

Consider now the following observer to estimate the system state:

$$\hat{x}[k + 1] = \mathbf{A}\hat{x}[k] + \mathbf{B}u[k] + \mathbf{J}_o[k]\big(y[k] - \mathbf{C}\hat{x}[k]\big) \tag{22.10.64}$$

Furthermore, assume that the initial state $x[0]$ satisfies

$$\mathcal{E}\{(x[0] - \hat{x}[0])(x[0] - \hat{x}[0])^T\} = \mathbf{P_o} \tag{22.10.65}$$

Then the optimal choice (in a quadratic sense) for $\mathbf{J}_o[k]$ is given by

$$\mathbf{J}_o[k] = \mathbf{A}\mathbf{P}[k]\mathbf{C}^T\big(\mathbf{R} + \mathbf{C}\mathbf{P}[k]\mathbf{C}^T\big)^{-1} \tag{22.10.66}$$

where $\mathbf{P}[k]$ satisfies the following discrete-time dynamic Riccati equation (DTDRE).

$$\mathbf{P}[k+1] = \mathbf{A}\left(\mathbf{P}[k] - \mathbf{P}[k]\mathbf{C}^T\left(\mathbf{R} + \mathbf{C}\mathbf{P}[k]\mathbf{C}^T\right)^{-1}\mathbf{C}\mathbf{P}[k]\right)\mathbf{A}^T + \mathbf{Q} \quad (22.10.67)$$

which can be solved forward in time, subject to

$$\mathbf{P}[0] = \mathbf{P_o} \qquad (22.10.68)$$

Remark 22.5. *Note again that* $\mathbf{A}, \mathbf{C}, \mathbf{Q}, \mathbf{R}$, *etc. can all be time varying for the general filter. However, when we come to discuss the steady-state filter, it is usually required that the problem be time invariant (or, at least, transferable to this form, as is the case for periodic systems).*

The steady-state $(k \to \infty)$ filter gain satisfies the DTARE given by

$$\mathbf{A}\left[\mathbf{P}_\infty - \mathbf{P}_\infty\mathbf{C}^T\left(\mathbf{R} + \mathbf{C}\mathbf{P}_\infty\mathbf{C}^T\right)^{-1}\mathbf{C}\mathbf{P}_\infty\right]\mathbf{A}^T + \mathbf{Q} = \mathbf{P}_\infty \qquad (22.10.69)$$

The above results show that the duality between optimal regulators and optimal filters is also present in discrete-time systems.

22.10.4 Stochastic Noise Models

In the above development, we have simply represented the noise as a white-noise sequence $(\{\omega(k)\})$ and a white measurement-noise sequence $(\{v(k)\})$. Actually, this is much more general than it may seem at first sight. For example, it can include *colored* noise having an arbitrary rational noise spectrum. The essential idea is to model this noise as the output of a linear system (i.e., a filter) driven by white noise. Thus, say that a system is described by

$$x(k+1) = \mathbf{A}x(k) + \mathbf{B}u(k) + \omega_c(k) \qquad (22.10.70)$$
$$y(k) = \mathbf{C}x(k) + v(k) \qquad (22.10.71)$$

where $\{\omega_c(k)\}$ represents *colored noise*–noise that is white noise passed through a filter. Then we can add the additional noise model $\{(22.10.70), (22.10.71)\}$ to the description. For example, let the noise filter be

$$x'(k+1) = \mathbf{A}'x(k) + \omega(k) \qquad (22.10.72)$$
$$\omega_c(k) = \mathbf{C}'x'(k) \qquad (22.10.73)$$

where $\{\omega(k)\}$ is a white-noise sequence.

Combining (22.10.70) to (22.10.73) gives a composite system *driven by white noise*, of the form

$$\bar{x}(k+1) = \bar{A}\bar{x}(k) + \bar{B}u(k) + \bar{\omega}(k) \qquad (22.10.74)$$
$$y(k) = \bar{C}\bar{x}(k) + v(k) \qquad (22.10.75)$$

where

$$\bar{x}(k) = [x(k)^T, x'(k)^T]^T \qquad (22.10.76)$$
$$\bar{\omega}(k) = [0, \omega(k)^T] \qquad (22.10.77)$$

$$\bar{A} = \begin{bmatrix} A & C' \\ 0 & A' \end{bmatrix}; \qquad \bar{B} = \begin{bmatrix} B \\ 0 \end{bmatrix}; \qquad \bar{C} = \begin{bmatrix} C & 0 \end{bmatrix} \qquad (22.10.78)$$

22.10.5   Optimal Prediction

In the case of gaussian noise, the Kalman filter actually produces the mean of the conditional distribution for $x(k)$ given the past data: $\{y(\ell), u(\ell); \; \ell = 0, \dots, k-2, k-1\}$. This can be written as

$$\hat{x}(k) = E\{x(k)/y(\ell), u(\ell); \quad \ell = 0, \dots, k-1\}$$
$$\triangleq \hat{x}(k|k-1) \qquad (22.10.79)$$

The unpredictability of white noise causes the optimal prediction of future states to be obtained by simply iterating the model with the future white noise set equal to its expected value, which is zero. Thus, consider the composite model $\{(22.10.74), (22.10.75)\}$, then for a known input $\{u(k)\}$, the predicted future states satisfy

$$\tilde{x}(j+1) = \bar{A}\tilde{x}(j) + \bar{B}u(j); \quad j = k, k+1, k+2, \dots \qquad (22.10.80)$$

with initial condition $\tilde{x}(k) = \hat{x}(k|k-1)$. Note that $\tilde{x}(j)$ can also be written as

$$\tilde{x}(j) = E\{x(j)|y(\ell), u(\ell); \quad \ell = 0, \dots, k-1\}; \quad j \geq k$$
$$\triangleq \hat{x}(j|k-1); \quad j \geq k \qquad (22.10.81)$$

22.10.6   State Estimation with Continuous Models and Sampled Observations

A problem that is frequently met in practice arises from an underlying continuous-time system whose data is collected at discrete sample points. Indeed, the sample points can, in some applications, be nonuniformly spaced. This still falls under the general discrete time-varying filtering problem.

What we need to do is to propagate the continuous system between samples and then update at the sample points by using an appropriate discrete model. To illustrate, say that the system is described (formally) by (22.10.5) and (22.10.6):

$$\dot{x} = \mathbf{A}x + \dot{\omega} \tag{22.10.82}$$

$$y' = \mathbf{C}x + \dot{\upsilon} \tag{22.10.83}$$

where $\dot{\omega}$, $\dot{\upsilon}$ have spectral density $\mathbf{Q}$ and $\mathbf{R}$ respectively.

Before we can sensibly sample y', we need first to pass it through an appropriate anti-aliasing filter. One such filter is the so-called *integrate and dump* filter. In this case, the kth sample is given by

$$\bar{y}[k] = \frac{1}{t_k - t_{k-1}} \int_{t_{k-1}}^{t_k} y'(t) dt \tag{22.10.84}$$

where t_k denotes the time of the kth sample.

The discrete observation $\bar{y}[k]$ can be related to a discrete state space model of the form

$$\bar{x}[k+1] = A_q[k]\bar{x}[k] + \bar{\omega}[k] \tag{22.10.85}$$

$$\bar{y}[k] = C_q[k]\bar{x}[k] + \bar{\upsilon}[k] \tag{22.10.86}$$

By integrating the model $\{(22.10.82), (22.10.83)\}$, it can be seen that

$$A_q[k] = e^{A(t_{k+1} - t_k)} \simeq I + A(t_{k+1} - t_k) \tag{22.10.87}$$

$$C_q[k] = \frac{1}{(t_{k+1} - t_k)} \int_{t_k}^{t_{k+1}} Ce^{A\tau} d\tau \simeq C \tag{22.10.88}$$

$$E\{\bar{\omega}[k]\bar{\omega}[k]^T\} = \bar{Q} \simeq Q(t_{k+1} - t_k) \tag{22.10.89}$$

$$E\{\bar{\upsilon}[k]\bar{\upsilon}[k]^T\} = \bar{R} \simeq \frac{R}{(t_{k+1} - t_k)} \tag{22.10.90}$$

The discrete model can then be used to obtain the state estimates *at the sample points*. The continuous state can then be interpolated between samples by using the open-loop model

$$\dot{\hat{x}}(t) = A\hat{x}(t); \qquad \text{for} \qquad t\epsilon[t_k, t_{k+1}] \qquad (22.10.91)$$

with initial condition

$$\hat{x}(t_k) = \hat{\bar{x}}[k] \qquad (22.10.92)$$

22.11   State-Estimate Feedback

Finally, we can combine the state estimation provided by (22.9.2) with the state-variable feedback determined earlier to yield the following state-estimate feedback-control law:

$$u(t) = -\mathbf{K}\hat{x}(t) + \bar{r}(t) \qquad (22.11.1)$$

Note that Theorem 18.1 also applies here; the closed-loop poles resulting from the use of (22.11.1) are the union of the eigenvalues that result from the use of the state feedback together with the eigenvalues associated with the observer. Note, in particular, that the proof of Theorem 18.1 does not depend on the scalar nature of the problem.

22.12   Transfer-Function Interpretation

As in section §18.5, the state-estimation feedback law described above can be given a polynomial interpretation.

In particular, Lemma 18.5 on page 539 still holds for the steady-state observer; i.e., $\hat{X}(s)$ can be expressed as

$$\hat{X}(s) = (s\mathbf{I} - \mathbf{A_o} + \mathbf{JC_o})^{-1}(\mathbf{B_o}U(s) + \mathbf{J}Y(s)) = \mathbf{T_1}(s)U(s) + \mathbf{T_2}(s)Y(s) \qquad (22.12.1)$$

where $\mathbf{T_1}(s)$ and $\mathbf{T_2}(s)$ are two stable transfer-function matrices given by

$$\mathbf{T_1}(s) \triangleq (s\mathbf{I} - \mathbf{A_o} + \mathbf{JC_o})^{-1}\mathbf{B_o} \qquad (22.12.2)$$

$$\mathbf{T_2}(s) \triangleq (s\mathbf{I} - \mathbf{A_o} + \mathbf{JC_o})^{-1}\mathbf{J} \qquad (22.12.3)$$

Thus, the feedback control law based on the state estimate is given by

$$\begin{aligned} U(s) &= -\mathbf{K}\hat{X}(s) - \overline{R}(s) \\ &= -\mathbf{K}\mathbf{T_1}(s)U(s) - \mathbf{K}\mathbf{T_2}(s)Y(s) + \overline{R}(s) \end{aligned} \qquad (22.12.4)$$

This can be written in an equivalent form as

$$\overline{\mathbf{C}}_{\mathbf{D}}(s)U(s) = -\overline{\mathbf{C}}_{\mathbf{N}}(s)Y(s) + \overline{R}(s) \qquad (22.12.5)$$

where $\overline{\mathbf{C}}_{\mathbf{D}}(s)$ and $\overline{\mathbf{C}}_{\mathbf{N}}(s)$ are stable matrices given by

$$\overline{\mathbf{C}}_{\mathbf{D}}(s) = \mathbf{I} + \mathbf{K}[s\mathbf{I} - \mathbf{A_o} + \mathbf{JC_o}]^{-1}\mathbf{B_o} = \frac{\overline{\mathbf{N}}_{\mathbf{CD}}(s)}{E(s)} \qquad (22.12.6)$$

$$\overline{\mathbf{C}}_{\mathbf{N}}(s) = \mathbf{K}[s\mathbf{I} - \mathbf{A_o} + \mathbf{JC_o}]^{-1}\mathbf{J} = \frac{\overline{\mathbf{N}}_{\mathbf{CN}}(s)}{E(s)} \qquad (22.12.7)$$

where $\overline{\mathbf{N}}_{\mathbf{CD}}(s)$ and $\overline{\mathbf{N}}_{\mathbf{CN}}(s)$ are polynomial matrices, and where $E(s)$ is the observer polynomial given by

$$E(s) = \det(s\mathbf{I} - \mathbf{A_o} + \mathbf{JC_o}) \qquad (22.12.8)$$

From (22.12.5), we see that $\mathbf{C}(s) = [\overline{\mathbf{C}}_{\mathbf{D}}(s)]^{-1}\overline{\mathbf{C}}_{\mathbf{N}}(s)$ is a LMFD for the controller. Furthermore, this controller satisfies the following lemma.

Lemma 22.3. *Consider the controller* $\mathbf{C}(s) = [\overline{\mathbf{C}}_{\mathbf{D}}(s)]^{-1}\overline{\mathbf{C}}_{\mathbf{N}}(s)$, *with* $\overline{\mathbf{C}}_{\mathbf{D}}(s)$ *and* $\overline{\mathbf{C}}_{\mathbf{N}}(s)$ *defined in (22.12.6) and (22.12.6), respectively. Then a state space realization for this controller is given by the $4-$tuple* $(\mathbf{A_o} - \mathbf{JC_o} - \mathbf{B_oK}, \mathbf{J}, \mathbf{K}, \mathbf{0})$:

$$\mathbf{C}(s) = \mathbf{K}[s\mathbf{I} - \mathbf{A_o} + \mathbf{JC_o} + \mathbf{B_oK}]^{-1}\mathbf{J} \qquad (22.12.9)$$

Proof

We first apply the matrix inversion lemma to obtain

$$[\overline{\mathbf{C}}_{\mathbf{D}}(s)]^{-1} = \mathbf{I} - \mathbf{K}[s\mathbf{I} - \mathbf{A_o} + \mathbf{JC_o} + \mathbf{B_oK}]^{-1}\mathbf{B_o} \qquad (22.12.10)$$

We next form the product $[\overline{\mathbf{C}}_{\mathbf{D}}(s)]^{-1}\overline{\mathbf{C}}_{\mathbf{N}}(s)$ *and use (22.12.10) and (22.12.7), from which, after some straightforward matrix manipulations, the result follows.*
□□□

The state space implementation of the controller allows the direct use of the results of the observer and state-feedback synthesis stages. This is useful not only in simulation but also when implementing the controller in practice.

It is also possible to obtain a RMFD for the controller. In particular, from Lemma 20.1 on page 620, it follows that

$$[\overline{\mathbf{C}}_{\mathbf{D}}(s)]^{-1}\overline{\mathbf{C}}_{\mathbf{N}}(s) = \mathbf{C}_{\mathbf{N}}(s)[\mathbf{C}_{\mathbf{D}}(s)]^{-1} \qquad (22.12.11)$$

where

$$\mathbf{C_D}(s) = \mathbf{I} - \mathbf{C_o}(s\mathbf{I} - \mathbf{A_o} + \mathbf{B_o}\mathbf{K})^{-1}\mathbf{J} = \frac{\mathbf{N_{CD}}(s)}{F(s)} \qquad (22.12.12)$$

$$\mathbf{C_N}(s) = \mathbf{K}(s\mathbf{I} - \mathbf{A_o} + \mathbf{B_o}\mathbf{K})^{-1}\mathbf{J} = \frac{\mathbf{N_{CN}}(s)}{F(s)} \qquad (22.12.13)$$

where $\mathbf{N_{CD}}(s)$ and $\mathbf{N_{CN}}(s)$ are polynomial matrices, and where $F(s)$ is the state-feedback polynomial given by

$$F(s) = \det(s\mathbf{I} - \mathbf{A_o} + \mathbf{B_o}\mathbf{K}) \qquad (22.12.14)$$

Similarly, by using (20.3.24) and (20.3.25) and Lemma 20.1 on page 620, the plant transfer function can be written as

$$\begin{aligned} \mathbf{G_o}(s) &= \mathbf{C_o}[s\mathbf{I} - \mathbf{A_o}]^{-1}\mathbf{B_o} \\ &= [\overline{\mathbf{G}}_{\mathbf{oD}}(s)]^{-1}\overline{\mathbf{G}}_{\mathbf{oN}}(s) = \mathbf{G_{oN}}(s)[\mathbf{G_{oD}}(s)]^{-1} \end{aligned} \qquad (22.12.15)$$

where

$$\overline{\mathbf{G}}_{\mathbf{oN}}(s) = \mathbf{C_o}(s\mathbf{I} - \mathbf{A_o} + \mathbf{J}\mathbf{C_o})^{-1}\mathbf{B_o} \qquad (22.12.16)$$

$$\overline{\mathbf{G}}_{\mathbf{oD}}(s) = \mathbf{I} - \mathbf{C_o}(s\mathbf{I} - \mathbf{A_o} + \mathbf{J}\mathbf{C_o})^{-1}\mathbf{J} \qquad (22.12.17)$$

$$\mathbf{G_{oN}}(s) = \mathbf{C_o}(s\mathbf{I} - \mathbf{A_o} + \mathbf{B_o}\mathbf{K})^{-1}\mathbf{B_o} \qquad (22.12.18)$$

$$\mathbf{G_{oD}}(s) = \mathbf{I} - \mathbf{K}(s\mathbf{I} - \mathbf{A_o} + \mathbf{B_o}\mathbf{K})^{-1}\mathbf{B_o} \qquad (22.12.19)$$

We then have the following result.

Lemma 22.4. *The sensitivity and complementary sensitivity resulting from the control law 22.12.5 are, respectively,*

$$\mathbf{S_o}(s) = \mathbf{C_D}(s)\overline{\mathbf{G}}_{\mathbf{oD}}(s) = \mathbf{I} - \mathbf{G_{oN}}(s)\overline{\mathbf{C}}_{\mathbf{N}}(s) \qquad (22.12.20)$$

$$\mathbf{T_o}(s) = \mathbf{G_{oN}}(s)\overline{\mathbf{C}}_{\mathbf{N}}(s) = \mathbf{I} - \mathbf{C_D}(s)\overline{\mathbf{G}}_{\mathbf{oD}}(s) \qquad (22.12.21)$$

Proof

From Lemma 20.1 on page 620, we have that

$$\overline{\mathbf{G}}_{\mathbf{oD}}(s)\mathbf{C_D}(s) + \overline{\mathbf{G}}_{\mathbf{oN}}(s)\mathbf{C_N}(s) = \overline{\mathbf{C}}_{\mathbf{D}}(s)\mathbf{G_{oD}}(s) + \overline{\mathbf{C}}_{\mathbf{N}}(s)\mathbf{G_{oN}}(s) = \mathbf{I} \quad (22.12.22)$$

By using (22.12.22) in (20.5.3) and (20.5.4), the result follows.

□□□

From Lemma 22.4, we see again that the closed-loop poles are located at the zeros of $E(s)$ and $F(s)$, as in the SISO case.

22.13 Achieving Integral Action in LQR Synthesis

An important aspect not addressed so far is that optimal control and optimal state-estimate feedback do not automatically introduce integral action. The latter property is an architectural issue that has to be *forced* onto the solution.

We refer the reader to section §18.7, where various ways of introducing integral action (or general IMP) in state-estimate feedback were discussed. For example, from part (b) of section §18.7, we have seen that one way of forcing integral action is to put a set of integrators at the output of the plant.

This can be described in state space form as

$$\dot{x}(t) = \mathbf{A}x(t) + \mathbf{B}u(t) \qquad (22.13.1)$$
$$y(t) = \mathbf{C}x(t)$$
$$\dot{z}(t) = -y(t)$$

As before, we can use an observer (or Kalman filter) to estimate x from u and y. Hence, in the sequel we will assume (without further comment) that x and z are directly measured. The composite system can be written in state space form as

$$\dot{x}'(t) = \mathbf{A}'x'(t) + \mathbf{B}'u(t) \qquad (22.13.2)$$

where

$$x' = \begin{bmatrix} x(t) \\ z(t) \end{bmatrix}; \qquad \mathbf{A}' = \begin{bmatrix} \mathbf{A} & \mathbf{0} \\ -\mathbf{C} & \mathbf{0} \end{bmatrix}; \qquad \mathbf{B}' = \begin{bmatrix} \mathbf{B} \\ \mathbf{0} \end{bmatrix} \qquad (22.13.3)$$

We then determine state feedback (from $x'(t)$) to stabilize the composite system. Of course, a key issue is that the composite system be controllable. A sufficient condition for this is provided in the following.

Lemma 22.5. *The cascaded system (22.13.2)-(22.13.3) is completely controllable, provided that the original system is controllable and that there is no zero of the original system at the origin.*[1]

Proof

The result follows immediately from Lemma 20.6 on page 640.

□□□

The final architecture of the control system would then appear as in Figure 22.2.

It is readily seen that the above idea can be extended to any disturbance or reference model, by incorporating the appropriate disturbance-generating polynomials

---

[1] Recall that, if a plant has one or more zeros at the origin, its output cannot be steered to a constant value by a bounded control signal.

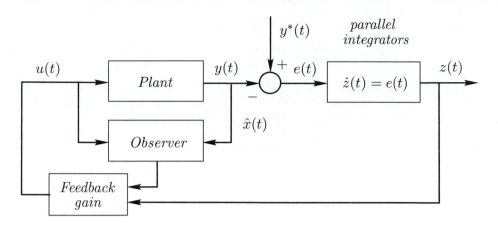

Figure 22.2. Integral action in MIMO control

(Internal Model Principle) in place of the parallel integrators in Figure 22.2. The corresponding version of Lemma 22.5 simply requires that the poles of the disturbance/reference models not be zeros of the original system, exactly as discussed in subsection §20.7.6.

We illustrate the above idea with the following example.

Example 22.4. *Consider a system with state space description given by* $(\mathbf{A}, \mathbf{B}, \mathbf{C}, \mathbf{0})$, *where*

$$
\mathbf{A} = \begin{bmatrix} -1 & 2 \\ -5 & -3 \end{bmatrix} ; \qquad \mathbf{B} = \begin{bmatrix} -1 & 0 \\ -1 & 2 \end{bmatrix} ; \qquad \mathbf{C} = \begin{bmatrix} -1 & 0 \\ 1 & 1 \end{bmatrix} \qquad (22.13.4)
$$

Synthesize a linear quadratic optimal regulator that forces integral action and where the cost function is built with $\mathbf{\Psi}$ *and* $\mathbf{\Phi}$, *where* $\mathbf{\Psi} = \mathbf{I}$ *and* $\mathbf{\Phi} = 0.01 * diag(0.1, 1)$.

Solution

We first choose the poles for the plant-state observer to be located at -5 *and* -10. *This leads to*

$$
\mathbf{J} = \begin{bmatrix} -4 & 6 \\ 5 & 2 \end{bmatrix} \qquad (22.13.5)
$$

Then the regulator will feed the plant input with a linear combination of the plant-state estimates and the integrator states. We choose to build an optimal regulator yielding closed-loop poles located in a disk with center at $s = -3$ *and unit radius. (See section §22.8). We thus obtain an optimal feedback gain, given by*

$$\mathbf{K} = \begin{bmatrix} -4.7065 & -2.0200 & -8.0631 & 0.0619 \\ -4.8306 & 0.3421 & -8.0217 & -4.0244 \end{bmatrix} \tag{22.13.6}$$

*This design is simulated in a loop with the same structure as in Figure 22.2 on the page before. (See SIMULINK file **cint.mdl**.) The tracking performance is illustrated in Figure 22.3, where step references have been applied.*

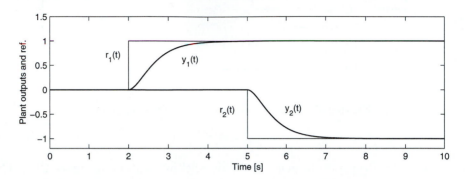

Figure 22.3. Tracking performance of a LQR design with integration

22.14   Industrial Applications

Multivariable design based on LQR theory and the Kalman filter accounts for thousands of real-world applications. Indeed, it was recently pointed out that there are approximately 400 patents per year based around the Kalman filter alone. The authors of this book have also applied these methods in numerous real-world applications. Our belief is that the Kalman filter is one of the most powerful tools available to the control-system designer.

The key issue in using these techniques in practice lies in the problem formulation; once the problem has been properly posed, the solution is usually rather straightforward. Much of the success in applications of this theory depends on the formulation, so we will conclude this chapter with brief descriptions of four real-world applications that the authors have been personally involved with. Our purpose in presenting these applications is not to give full details–that would take us well beyond the scope of this text. Instead, we hope to give sufficient information so that the reader can see the methodology used in applying these methods to different types of problems.

The particular applications that we will review are

- Geostationary satellite tracking

- Zinc Coating-Mass estimation in continuous galvanizing lines

- Roll-Eccentricity estimation in thickness control for rolling mills

- Vibration control in flexible structures

22.14.1   Geostationary Satellite Tracking

It is known that so-called geostationary satellites actually appear to *wobble* in the sky. The period of this wobble is one sidereal day. If one wishes to point a receiving antenna exactly at a satellite so as to maximize the received signal, then it is necessary to track this perceived motion. The required pointing accuracy is typically to within a few hundredths of a degree. The physical set-up is as shown in Figure 22.4.

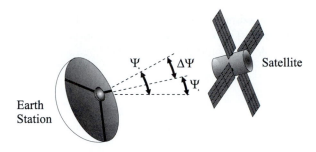

Figure 22.4. Satellite and antenna angle definitions

One could use an open-loop solution to this problem, as follows: Given a model (e.g., a list of pointing angles versus time), the antenna could be pointed in the correct orientation as indicated by position encoders. This technique is used in practice, but it suffers from the following practical issues:

- It requires high absolute accuracy in the position encoders, antenna, and reflector structure.

- It also requires regular *maintenance* to put in new model parameters.

- It cannot compensate for wind, thermal, and other time-varying effects on the antenna and reflector.

This motivates the use of a closed-loop solution. In such a solution, the idea is to move the antenna periodically so as to find the direction of maximum signal strength. However, the data so received are noisy for several reasons, including the following:

- noise in the received signal, p;

- variations in the signal intensity transmitted from the satellite;

- imprecise knowledge of the beam pattern for the antenna; and

- the effect of wind gusts on the structure and the reflector.

It is a reasonable hypothesis that we can smooth this data by using a Kalman filter. Toward this end, we need first to build a model for the orbit. Now, as seen from the earth, the satellite executes a periodic motion in the two axes of the antenna (azimuth and elevation). Several harmonics are present but the dominant harmonic is the fundamental. This leads to a model of the form

$$y(t) = \Psi_s(t) = x_1 + x_2 \sin \omega t + x_3 \cos \omega t \qquad (22.14.1)$$

where $\Psi_s(t)$ is, say, the azimuth angle as a function of time. The frequency ω in this application is known. There are several ways of describing this model in state space form. For example, we could use

$$\frac{d}{dt} \begin{pmatrix} x_1 \\ x_2 \\ x_3 \end{pmatrix} = 0 \qquad (22.14.2)$$

$$y(t) = C(t)x(t) \qquad (22.14.3)$$

$$\text{where} \qquad C(t) = [1, \sin \omega t, \cos \omega t] \qquad (22.14.4)$$

This system is time-varying (actually periodic). We can then immediately apply the Kalman filter to estimate $x_1, x_2 and x_3$ from noisy measurements of $y(t)$.

A difficulty with the above formulation is that equation (22.14.2) has zeros on the stability boundary. Hence, inspection of Lemma 22.2 on page 689 indicates that the steady-state filter could experience difficulties. This is resolved by adding a small amount of fictitious noise to the right-hand side of the model (22.14.2) to account for orbit imperfections. Also, note that this is an example of the kind of problem mentioned in section §22.10.6, because the data on satellite location can arrive at discrete times that are nonuniformly spaced.

To finalize the filter design, the following information must be supplied:

- a guess at the current satellite orientation;

- a guess at the covariance of the initial state error ($P(0)$);

- a guess at the measurement-noise intensity (R); and

- a rough value for the added process noise intensity (Q).

A commercial system built around the above principles has been designed and built at the University of Newcastle, Australia. This system is marketed under the trade name ORBTRACK® and has been used in many real-world applications ranging from Australia to Indonesia and Antarctica.

□□□

22.14.2   Zinc Coating-Mass Estimation in Continuous Galvanizing Lines

A diagram of a continuous galvanizing line is shown in Figure 22.5. An interesting feature of this application is that the sheet being galvanized is a meter or so wide and many hundreds of meters long.

The strip passes through a zinc pot (as in the figure). Subsequently, excess zinc is removed by air knives. The strip then moves through a cooling section, and finally the coating mass is measured by a traversing X-ray gauge–see Figure 22.6.

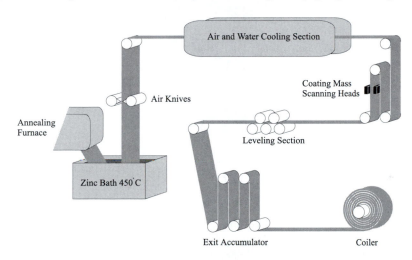

Figure 22.5. Schematic diagram of continuous galvanizing line

If one combines the lateral motion of the X-ray gauge with the longitudinal motion of the strip, then one obtains the zig-zag measurement pattern shown in Figure 22.7.

Because of the sparse measurement pattern, it is highly desirable to smooth and interpolate the coating-mass measurements. The Kalman filter is a possible tool to carry out this data-smoothing function. However, before we can apply this tool, we need a model for the relevant components in the coating-mass distribution. The relevant components include the following.

- *Shape Disturbances* (arising from shape errors in the rolling process)

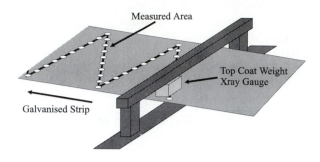

Figure 22.6. Traversing X-ray gauge

Figure 22.7. Zig-zag measurement pattern

These can be described by band-pass-filtered noise components, by using a model of the form

$$\dot{x}_1 = -\omega_1 x_1 - \left[\frac{\omega_2\omega_1}{\omega_2 - \omega_1}\right] n \tag{22.14.5}$$

$$\dot{x}_2 = -\omega_2 x_2 - \left[\frac{\omega_2^2}{\omega_1 - \omega_2}\right] n \tag{22.14.6}$$

$$y_{sd} = (1,1) \begin{pmatrix} x_1 \\ x_2 \end{pmatrix} \tag{22.14.7}$$

- *Cross Bow* (a quadratic term arising from nonuniform coating effects)

  This is a quadratic function of distance across the strip and is modeled by

$$\dot{x}_3 = 0 \qquad \text{(22.14.8)}$$

$$y_{cb} = \{d(t)[d(t) - W]\}x_3 \qquad \text{(22.14.9)}$$

where $d(t)$ denotes the distance from the left edge of the strip and W denotes the total strip width

- *Skew* (due to misalignment of the knife jet)

  This is a term that increases linearly with distance from the edge. It can thus be modeled by

$$\dot{x}_4 = 0 \qquad \text{(22.14.10)}$$

$$y_{sc} = \{d(t)\}x_4 \qquad \text{(22.14.11)}$$

- *Eccentricity* (due to out-of-round in the rolls)

  Say that the strip velocity is v_s and that the roll radius is r.   Then this component can be modeled as

$$\dot{x}_5 = 0 \qquad \text{(22.14.12)}$$

$$\dot{x}_6 = 0 \qquad \text{(22.14.13)}$$

$$y_e = \{\sin\left(\frac{v_s}{r}\right)t, \cos\left(\frac{v_s}{r}\right)t\}\begin{pmatrix} x_5 \\ x_6 \end{pmatrix} \qquad \text{(22.14.14)}$$

- *Strip Flap* (due to lateral movement of the strip in the vertical section of the galvanizing line)

  Let $f(t)$ denote the model for the flap; then this component is modeled by

$$\dot{x}_7 = 0 \qquad \text{(22.14.15)}$$

$$y_f = \{f(t)\}x_7 \qquad \text{(22.14.16)}$$

- *Mean Coating Mass* (the mean value of the zinc layer)

This can be simply modeled by

$$\dot{x}_8 = 0 \tag{22.14.17}$$

$$y_m = x_8 \tag{22.14.18}$$

Putting all of the equations together gives us an 8th-order model of the form

$$\dot{x} = \mathbf{A}x + \mathbf{B}n \tag{22.14.19}$$

$$z = y = \mathbf{C}(t)x + v \tag{22.14.20}$$

$$\mathbf{A} = \begin{bmatrix} -\omega_1 & 0 & 0 & 0 & 0 & 0 & 0 & 0 \\ 0 & -\omega_2 & 0 & 0 & 0 & 0 & 0 & 0 \\ 0 & 0 & 0 & 0 & 0 & 0 & 0 & 0 \\ 0 & 0 & 0 & 0 & 0 & 0 & 0 & 0 \\ 0 & 0 & 0 & 0 & 0 & 0 & 0 & 0 \\ 0 & 0 & 0 & 0 & 0 & 0 & 0 & 0 \\ 0 & 0 & 0 & 0 & 0 & 0 & 0 & 0 \\ 0 & 0 & 0 & 0 & 0 & 0 & 0 & 0 \end{bmatrix} \; ; \quad \mathbf{B} = \begin{bmatrix} -\left(\frac{\omega_2\omega_1}{\omega_2-\omega_1}\right) \\ -\left(\frac{\omega_2^2}{\omega_1-\omega_2}\right) \\ 0 \\ 0 \\ 0 \\ 0 \\ 0 \\ 0 \end{bmatrix} \tag{22.14.21}$$

$$\mathbf{C} = [1, 1, d(t)[d(t) - W], d(t), \sin\left(\frac{v_s}{d}t\right), \cos\left(\frac{v_s}{d}t\right), f(t), 1] \tag{22.14.22}$$

Given the above model, one can apply the Kalman filter to estimate the coating-thickness model. The resultant model can then be used to interpolate the thickness measurement. Note that here the Kalman filter is actually periodic, reflecting the periodic nature of the X-ray traversing system.

A practical form of this algorithm is part of a commercial system for Coating-Mass Control developed in collaboration with the authors of this book by a company (Industrial Automation Services Pty. Ltd.). The system has proven to be very beneficial in reducing times to accurately estimate coating mass.

22.14.3   Roll-Eccentricity Compensation in Rolling Mills

The reader will recall that rolling-mill thickness-control problems were described in Chapter 8. A schematic of the set-up is shown in the Figure 22.8.

In Figure 22.8, we have used the following symbols:

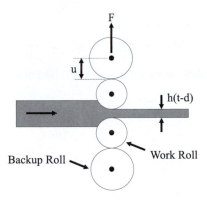

Figure 22.8. Rolling-mill thickness control

$F(t):$ Force
$h(t):$ Exit-thickness Measurement
$u(t):$ Unloaded Roll Gap (the control variable)

In Chapter 8, it was argued that the following virtual sensor (called a BISRA gauge) could be used to estimate the exit thickness and thus eliminate the transport delay from mill to measurement.

$$\hat{h}(t) = \frac{F(t)}{M} + u(t) \qquad (22.14.23)$$

However, one difficulty that we have not previously mentioned with this virtual sensor is that the presence of eccentricity in the rolls (see Figure 22.9) significantly affects the results.

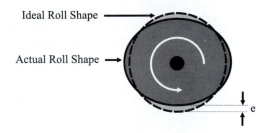

Figure 22.9. Roll eccentricity

To illustrate why this is so, let e denote the roll eccentricity. Then the true roll force is given by

$$F(t) = M(h(t) - u(t) + e(t)) \qquad (22.14.24)$$

In this case, the previous estimate of the thickness obtained from the force actually gives

$$\hat{h}(t) = h(t) + e(t) \qquad (22.14.25)$$

Thus, $e(t)$ represents an error, or disturbance term, in the virtual sensor output, one due to the effects of eccentricity.

The error $e(t)$ in the virtual sensor output turns out to be very important in this particular application. There is thus strong motivation to try to remove the effect of eccentricity from the estimate $\hat{h}(t)$.

A key property that allows us to make progress on the problem is that $e(t)$ is actually (almost) periodic, because it arises from eccentricity in the four rolls of the mill (two work rolls and two back-up rolls). Also, the roll angular velocities are easily measured in this application by using position encoders. From this data, one can determine a multi-harmonic model for the eccentricity, of the form

$$e(t) = \sum_{k=1}^{N} \alpha_k \sin \omega_k t + \beta_k \cos \omega_k t \qquad (22.14.26)$$

where α_i and β_i are unknown but $\omega_1 \ldots \omega_N$ are known.

Moreover, each sinusoidal input can be modeled by a state space model of the form

$$\dot{x}_1^k(t) = \omega_k x_2^k(t) \qquad (22.14.27)$$
$$\dot{x}_2^k(t) = -\omega_k x_1^k(t) \qquad (22.14.28)$$

Finally, consider any given measurement, say the force $F(t)$. We can think of $F(t)$ as comparing the above eccentricity components buried in noise:

$$y(t) = F(t) = \sum_{k=1}^{N} x_1^k(t) + x_2^k(t) + n(t) \qquad (22.14.29)$$

We can then, in principle, apply the Kalman filter to estimate $\{x_1^k(t), x_2^k(t); k = 1, \ldots, N\}$ and hence to correct the measured force measurements for eccentricity.

The final control system is as shown in the Figure 22.10.

An interesting feature of this problem is that there is some practical benefit in using the general time-varying form of the Kalman filter rather than the steady-state filter. The reason is that, in steady state, the filter acts as a narrow band-pass

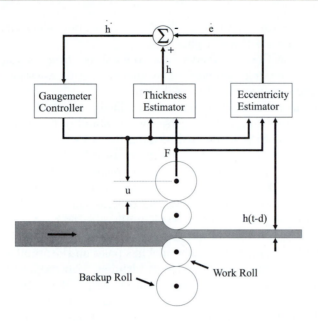

Figure 22.10. Final roll eccentricity compensated control system

filter bank centered on the harmonic frequencies. This is, heuristically, the correct steady-state solution. However, an interesting fact that the reader can readily verify is that the transient response time of a narrow band-pass filter is inversely proportional to the filter bandwidth. This means that, in steady state, one has the following fundamental design trade-off:

- On the one hand, one would like to have a narrow band-pass, to obtain good frequency selectivity and hence good noise rejection.

- On the other hand, one would like to have a wide band-pass, to minimize the initial transient period.

This is an inescapable dichotomy for any time-invariant filter.

Perhaps the astute reader might say: *Well, why not start with a wide-band filter, to minimize the transient, but then narrow the filter band down as the signal is acquired.* If you thought of this, then you are exactly correct. Indeed, this is precisely what the time-varying Kalman filter does. In particular, if one starts the solution of the Riccati equation by setting the initial condition $P(t_o)$ to some large value, then the resultant filter will initially have a wide bandwidth. As the Riccati equation solution proceeds, then $P(t)$ will decrease, along with the associated filter bandwidth. Indeed, the time-varying Kalman filter, gives an *optimal* time-varying trade-off between minimizing the transient time and accurately locking onto the harmonic components.

Thus, the trick to get the best trade-off between filter selectivity and transient response time is to use a time-varying filter.

Another question that the reader may well ask is: *How narrow does the final filter bandwidth ultimately become?* The answer to this question raises another fundamental trade-off in optimal-filter design.

The reader can verify that the model described in $\{(22.14.27), (22.14.28)\}$ has uncontrollable modes (from process noise) on the stability boundary. Careful reading of the properties of the associated filtering Riccati equation given in Appendix D indicates that $P(t)$ converges to a steady-state value in this case but that this happens to be a *zero* matrix. Moreover, the corresponding filter is then not stable. You might well ask: *What has gone wrong?* Well, nothing, really. We told the filter in this ideal formulation that there was no process noise. In this case, the filter could (asymptotically) estimate the states with zero error. Also, once the states were *found*, then they could be predicted into the future by using the resonant model $\{(22.14.27), (22.14.28)\}$. This model has poles on the stability boundary and is thus not stable! (Indeed, the filter bandwidth has been reduced to zero).

The difficulty is that we have over-idealized the problem. In practice, the frequencies in the eccentricity signal will not be exactly known (on account of, e.g., slip). Also, the harmonic components will not be time-invariant (on account of, e.g., roll-heating effects). To model these effects, we can simply add a little process noise to the right-hand side of the model $\{(22.14.27), (22.14.28)\}$. In this case, the model becomes stabilizable *from the noise*. The steady-state solution of the Riccati equation then turns out to be positive definite and the final filter will be stable and of finite bandwidth.

This leads us to another design trade-off. In particular, if one says that the model $\{(22.14.27), (22.14.28)\}$ is poorly known (by setting the process noise intensity Q to a large value), then the final filter bandwidth will be large. On the other hand, if one says that the model $\{(22.14.27), (22.14.28)\}$ has a high degree of validity (by setting the process noise intensity to a small value), then the final filter bandwidth will be small.

In practice, once one has been through such a design as the one above, then it is often possible to capture the key ideas in a simpler fashion. Indeed, this is the case here. In the final implementation, on-line solution of a Riccati equation is typically not used. Instead, a time-varying filter bank is employed that captures the essence of the lessons learnt from the formal *optimal* design. This kind of final *wrap-up* of the problem is typical in applications.

The final system, as described above, has been patented under the name AUSREC® and is available as a commercial product from Industrial Automation Services Pty. Ltd.

□□□

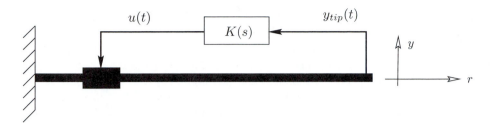

Figure 22.11. Vibration control by using a piezoelectric actuator

22.14.4   Vibration Control in Flexible Structures

As was pointed out in the introduction to this book, new actuators often open up exciting new opportunities for enhanced control. For example, there is substantial interest in new types of electromagnetic actuators and devices that take advantage of the high energy density of smart-material transducer elements. Smart-material-based transducer elements include piezoelectrics, electrostrictions, magnitorestrictions, and shape memory alloys.

Many application areas exist for these technologies, especially in the control of lightly decoupled systems.

To illustrate the principles involved in this kind of application, we will consider the problem of controller design for the piezoelectric laminate beam of Figure 22.11. In this system, the measurements are taken by a displacement sensor that is attached to the tip of the beam, and a piezoelectric patch is used as the actuator. The purpose of the controller is to minimize beam vibrations. It is easy to see that this is a regulator problem; hence, a LQG controller can be designed to reduce the unwanted vibrations.

To find the dynamics of structures such as the beam in Figure 22.11, one has to solve a particular partial differential equation that is known as the Bernoulli-Euler beam equation. By using modal analysis techniques, it is possible to show that a transfer function of the beam would consist of an infinite number of very lightly damped second-order resonant terms–that is, the transfer function from the voltage that is applied to the actuator to the displacement of the tip of the beam can be described by

$$G(s) = \sum_{i=1}^{\infty} \frac{\alpha_i}{s^2 + 2\zeta_i \omega_i s + \omega_i^2}. \qquad (22.14.30)$$

In general, one is interested in designing a controller only for a particular bandwidth. As a result, it is common practice to truncate (22.14.30) by keeping the first N modes that lie within the bandwidth of interest.

In this example, we consider a particular system described in the references given at the end of the chapter. We consider only the first six modes of this system.

i	α_i	$\omega_i \ (rad/sec)$
1	9.72×10^{-4}	18.95
2	0.0122	118.76
3	0.0012	332.54
4	-0.0583	651.660
5	-0.0013	1077.2
6	0.1199	1609.2

The transfer function is then

$$G(s) = \sum_{i=1}^{6} \frac{\alpha_i}{s^2 + 2\zeta_i \omega_i s + \omega_i^2}. \tag{22.14.31}$$

Here, ζ_i's are assumed to be 0.002 and α_i's as are shown in the Table here.

We first design a Linear Quadratic Regulator. Here, the $\boldsymbol{\Psi}$ matrix is chosen to be

$$\boldsymbol{\Psi} = 0.1483 diag(\omega_1^2, 1, \dots, \omega_6^2, 1).$$

The reason for this choice of state-weighting matrix is that, with this choice of $\boldsymbol{\Psi}$, the quadratic form $x(t)^T \boldsymbol{\Psi} x(t)$ represents the total energy of the beam at each time corresponding to its first six resonant modes. The control-weighting matrix is also, somewhat arbitrarily, chosen as $\boldsymbol{\Phi} = 10^{-8}$. Next, a Kalman-filter state estimator is designed with $\mathbf{Q} = 0.08I$ and $\mathbf{R} = 0.005$.

To show the closed-loop performance of the controller, we have plotted in Figure 22.12 the open-loop and closed-loop impulse responses of the system. It can be observed that the LQG controller can considerably reduce structural vibrations. In Figure 22.13, we have plotted the open-loop and closed-loop frequency responses of the beam. It can be observed that the LQG controller has significantly damped the first three resonant modes of the structure.

22.15   Summary

- Full multivariable control incorporates the interaction dynamics rigorously and explicitly.

- The fundamental SISO synthesis result that, under mild conditions, the nominal closed-loop poles can be assigned arbitrarily carries over to the MIMO case.

- Equivalence of state-feedback and frequency-domain pole placement by solving the (multivariable) Diophantine Equation carries over as well.

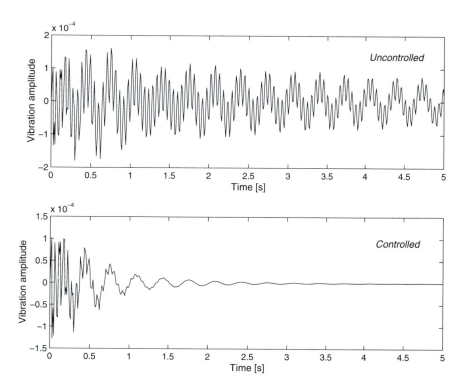

Figure 22.12. Open-loop and closed-loop impulse responses of the beam

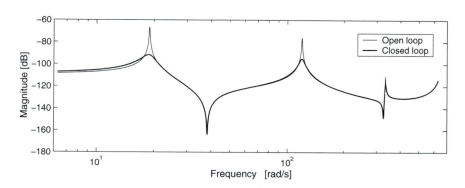

Figure 22.13. Open-loop and closed-loop frequency responses of the beam

- The complexities of multivariable systems, cause criterion-based synthesis (briefly alluded to in the SISO case) to gain additional motivation.

- A popular family of criteria are functionals involving quadratic forms of control error and control effort.

- For a general nonlinear formulation, the optimal solution is characterized by a two-point boundary-value problem.

- In the linear case (the so-called linear quadratic regulator, LQR), the general problem reduces to the solution of the continuous-time dynamic Riccati equation, which can be feasibly solved, leading to time-variable state feedback.

- After initial conditions decay, the optimal time-varying solution converges to a constant state feedback, the so-called steady-state LQR solution.

- It is frequently sufficient to neglect the initial transient of the strict LQR and only implement the steady-state LQR.

- The steady-state LQR is equivalent

    ○ to a model-matching approach, where a desired complementary sensitivity is specified and a controller is computed that matches it as closely as possible according to some selected measure, and

    ○ to pole placement, where a closed-loop polynomial is specified and a controller is computed to achieve it.

- Thus, LQR, model matching, and pole placement are mathematically equivalent, although they do offer different tuning parameters.

Equivalent synthesis techniques	*Tuning parameters*
LQR	relative penalties on control error versus control effort
model matching	closed-loop complementary sensitivity reference model and weighted penalty on the difference to the control loop
pole placement	closed-loop polynomial

- These techniques can be extended to discrete-time systems.

- There is a very close connection to the dual problem of filtering: inferring a state from a related (but not exactly invertible) set of measurements.

- Optimal-filter design based on quadratic criteria leads again to a Riccati equation.

- The filters can be synthesized and interpreted equivalently in a

   - linear quadratic,

   - model-matching, or

   - pole-placement

  framework.

- The arguably most famous optimal-filter formulation, the Kalman filter, can be given a stochastic or a deterministic interpretation, depending on taste.

- The LQR does not automatically include integral action; thus, rejection of constant or other polynomial disturbances must be enforced via the Internal Model Principle.

22.16   Further Reading

General optimal control

Bellman, R. (1957). *Dynamic Programming*. Princeton University Press.

Bryson, A. and Ho, YC.. (1969). *Applied Optimal Control*. Blaisdell.

Linear quadratic regulator

Anderson, B. and Moore, J. (1971).   *Linear Optimal Control*.   Prentice-Hall, Englewood Cliffs, N.J.

Athans, M. and Falb, P. (1966). *Optimal control*. McGraw-Hill.

Kalman, R.E. (1960b). When is a linear control systems optimal. *ASME J.Basic Eng.*, 86:51-60.

Kwakernaak, H. and Sivan, R. (1972). *Linear Optimal Control Systems*. Wiley-Interscience, New York.

Zhou, K., Doyle, J.C., and Glover, K. (1996).   *Robust and Optimal Control*. Prentice-Hall, Upper Saddle River, N.J.

Linear optimal filter

Anderson, B. and Moore, J. (1979). *Optimal Filtering*. Prentice-Hall, Englewood Cliffs, N.J.

Bucy, R. and Joseph, P. (1968). *Filtering for Stochastic Process with Applications to Guidance*. Wiley-Interscience.

Goodwin, G.C. and Sin, K. (1984). *Adaptive Filtering Prediction and Control*. Prentice-Hall, Englewood Cliffs, N.J.

Kailath, T. (1968). An innovations aproach to least-square estimation. Part I: Linear filtering in additive white noise. *IEEE Transactions on Automatic Control*, 13(6):647-655.

Kailath, T. (1974). A view of three decades of linear filtering theory. *IEEE Trans. on information theory*, 20(6):145-181.

Kalman, R.E. and Bucy, R. (1961). New results in linear filtering and prediction theory. *Trans ASME J.Basic Eng.*, 83:95-107.

Feuer, A. and Goodwin, G.C. (1996). *Sampling in Digital Signal Processing and Control*. Birkhöusser, Boston.

Riccati equation properties

Bittanti, S., Laub, A.J., and Willems, J.C.(1996). *The Riccati Equation*. Springer-Verlag, Berlin.

Chan, S., Goodwin, G., and Sin, K. (1984). Convergence properties of the Riccati difference equation in optimal filtering of nonstabilizable systems. *IEEE Transactions on Automatic Control*, 29(2):110-118.

De Souza, C., Gevers, M., and Goodwin, G.C. (1986). Riccati equation in optimal filtering of nonstabilizable systems having singular state transition matrices. *IEEE Transactions on Automatic Control*, 31:831-839.

Kucera, V. (1972). The discrete Riccati equation of optimal control. *Kibernetika*, 8(5):430-447.

Lancaster, P. and Rodman, L. (1995). *Algebraic Riccati equations*. Oxford University Press, Oxford.

Willems, J.C. (1970). *Stability Theory of Dynamical Systems*. T. Nelson, London.

Wimmer, H.K. (1985). Monotonicity of maximal solution of algebraic Riccati equations. *Systems and Control Letters*, 5(5):317-319.

Zhou, K., Doyle, J.C., and Glover, K. (1996). *Robust and Optimal Control*. Prentice-Hall, Upper Saddle River, N.J.

Stochastic separation theorem

Simon, H. (1956). Dynamic programing under uncertainty with a quadratic crite-rion function. *Econometrica*, 24:74.

Control of piezoelectric laminate beam

Moheimani, S.O.R., Pota, H.R., and Petersen, I.R. (1998). Active control of vibra-tions in a piezoelectric laminate cantilevered beam–A spatial LQG approach. *Proc. IEEE CDC*, Tampa, Florida.

22.17   Problems for the Reader

Problem 22.1. *A MIMO plant has a nominal transfer function $\mathbf{G_o}(s)$, and it is being controlled in closed loop. Synthesize a MIMO controller such that the dominant poles of the closed loop are the roots of $s^2 + 13s + 100$. Assume that*

$$\mathbf{G_o}(s) = \frac{1}{(s+4)(s+5)} \begin{bmatrix} 2 & -1 \\ 0.5 & 3 \end{bmatrix} \tag{22.17.1}$$

Problem 22.2. *For the same plant as in Problem 22.1, synthesize an optimal controller such that all the closed-loop poles are inside a circle centered at $(-8,0)$ and with radius equal to 2.*

Problem 22.3. *Consider a plant having a model given by*

$$\mathbf{G_o}(s) = \frac{1}{s^3 - 2s^2 - 6s + 7} \begin{bmatrix} 3s - 2 & s^2 - 2s + 3 \\ s^2 - 3s + 3 & 2s - 6 \end{bmatrix} \tag{22.17.2}$$

Determine a controller, in transfer-function form, that provides zero steady-state errors for constant references and disturbances and, at the same time, leads to closed-loop natural modes that decay at least as fast as e^{-2t}.

Problem 22.4. *Repeat the previous problem, but assume now that a digital controller has to be designed. The sampling interval is chosen to be $\Delta = 0.1[s]$.*

Problem 22.5. *Design a modification for the discrete-time LQR synthesis such that the resulting closed-loop poles are inside a circle of given radius ρ ($\rho < 1$).*

Problem 22.6. *Consider a plant having the model*

$$\mathbf{G_o}(s) = \frac{1}{s+2} \begin{bmatrix} e^{-0.5s} & 0.5 \\ -0.2 & e^{-0.25s} \end{bmatrix} \tag{22.17.3}$$

22.6.1 *Using the LQR approach based on state-estimate feedback, design a digital regulator, with $\Delta = 0.25$ [s], such that all the closed-loop poles are located inside the circle with radius 0.5. (Use $\mathbf{\Phi} = \mathbf{I}$ and $\mathbf{\Psi} = \mathbf{I}$.)*

22.6.2 *Compute the resulting closed-loop poles.*

22.6.3 *Repeat the design, using pole assignment in such a way that the closed-loop poles are the same as in the LQR design approach.*

Problem 22.7. *Consider a plant described, in state space form, by the $4 - tuple$ $(\mathbf{A_o}, \mathbf{B_o}, \mathbf{C_o}, 0)$. Assume that this plant is completely controllable and completely observable. Further assume that the closed-loop poles for the control of this plant are allocated via LQR design. The cost function is given in (22.4.3) (page 685), where*

$$t_f = \infty; \quad \mathbf{\Psi}_f = 0; \quad \mathbf{\Psi} = \mathbf{C_o^T C_o}; \quad \mathbf{\Phi} = \rho^2 \mathbf{I} \tag{22.17.4}$$

with ρ a positive scalar constant.

22.7.1 *Prove that the closed-loop poles are the LHP roots of*

$$\det \left(\mathbf{I} + \frac{1}{\rho^2} \mathbf{G_o}(s) \mathbf{G_o}(-s)^T \right) \tag{22.17.5}$$

where $\mathbf{G_o}(s)$ is the system transfer-function matrix–i.e., $\mathbf{G_o}(s) = \mathbf{C_o}[s\mathbf{I} - \mathbf{A_o}]^{-1}\mathbf{B_o}$. (Hint: recall that the closed-loop poles for the LQR are the stable eigenvalues of the Hamiltonian matrix given in (D.1.8).)

22.7.2 *Use the result from 22.7.1 to analyze the location of the closed-loop poles for small (cheap-control) and large (expensive-control) values of ρ. Look, in particular, to the cases of unstable and nonminimum-phase plants.*

Problem 22.8. *Apply the result proved in problem 22.7 to the control of a plant having a model given by*

$$\mathbf{G_o}(s) = \frac{1}{(s+3)(s+5)} \begin{bmatrix} 3(-s+5) & 2 \\ 1.5 & 1 \end{bmatrix} \tag{22.17.6}$$

Problem 22.9. *A discrete-time noise measurement, $y[k]$, of an unknown constant α satisfies the model*

$$y[k] = \alpha + v[k] \tag{22.17.7}$$

where $v[k]$ is an uncorrelated stationary process with variance σ^2.

22.9.1 *Find an optimal filter to estimate α.*

22.9.2 *Build a SIMULINK scheme to evaluate your filter in steady state.*

Problem 22.10. *Consider the case of a discrete-time noise measurement, $y[k]$, of a sine wave of known frequency but unknown amplitude and unknown phase:*

$$y[k] = A \sin(\omega_o k + \alpha) + v[k] \tag{22.17.8}$$

where $v[k]$ is an uncorrelated stationary process with variance σ^2.

22.10.1 *Build a state model to generate the sine wave.*

22.10.2 *Find an optimal filter to estimate A and α.*

22.10.2 *Build a SIMULINK scheme to evaluate your filter in steady state.*

Problem 22.11. *Consider the following problem of model matching:*

$$J = \frac{1}{2\pi} \int_{-\infty}^{\infty} \left| M(j\omega) - N(j\omega)Q(j\omega) \right|^2 d\omega \tag{22.17.9}$$

where

$$M(s) = \frac{9}{s^2 + 4s + 9}; \qquad N(s) = \frac{-s + 8}{(s+8)(s+2)} \tag{22.17.10}$$

22.11.1 *Translate the problem into an LQR problem that uses $R = 10^{-3}$.*

22.11.2 *Solve the LQR problem, and find the optimal value for $Q(s)$.*

Chapter 23

MODEL PREDICTIVE
CONTROL

23.1 Preview

As was mentioned in Chapter 11, all real-world control problems are subject to constraints of various types. The most common constraints are actuator constraints (amplitude and slew-rate limits). In addition, many problems also have constraints on state variables (e.g., maximal pressures that cannot be exceeded, minimum tank levels).

In many design problems, these constraints can be ignored, at least in the initial design phase. However, in other problems, these constraints are an inescapable part of the problem formulation, because the system operates near a constraint boundary. Indeed, in many process-control problems, the optimal steady-state operating point is frequently *at* a constraint boundary. In these cases, it is desirable to be able to carry out the design so as to include the constraints from the beginning.

Chapter 11 described methods (for dealing with constraints) based on anti-wind-up strategies. These are probably perfectly adequate for simple problems–especially SISO problems. Also, these methods can be extended to certain MIMO problems, as we shall see in Chapter 26. However, in more complex MIMO problems–especially those having both input and state constraints–it is frequently desirable to have a more formal mechanism for dealing with constraints in MIMO control-system design.

We describe here one such mechanism based on Model Predictive Control. This has actually been a major success story in the application of modern control. More than 2,000 applications of this method have been reported in the literature, predominantly in the petrochemical area. Also, the method is being increasingly used in electromechanical control problems, as control systems push constraint boundaries. Its main advantages are the following:

- It provides a *one-stop-shop* for MIMO control in the presence of constraints.

- It is one of the few methods that allow one to treat state constraints.

739

- Several commercial packages are available that give industrially robust versions of the algorithms aimed at chemical process control.

There are many alternative ways of describing MPC, including polynomial and state space methods. Here we will give a brief introduction to the method, using a state space description. This is intended to acquaint the reader with the basic ideas of MPC (e.g., the notion of receding-horizon optimization). It also serves as an illustration of some of the optimal control ideas introduced in Chapter 22, so as to provide motivation for MPC.

One of us (Graebe) has considerable practical experience with MPC in his industry, where it is credited with enabling very substantial annual financial savings through improved control performance. Another of us (Goodwin) has applied the method to several industrial control problems (including *added water* control in a sugar mill, rudder roll stabilization of ships, and control of a food extruder) where constraint handling was a key consideration.

Before delving into Model Predictive Control in detail, we pause to revisit the anti-wind-up strategies introduced in Chapter 11.

23.2   Anti-Wind-Up Revisited

We assume, for the moment, that the complete state of a system is directly measured. Then, if one has a time-invariant model for a system and if the objectives and constraints are time-invariant, it follows that the control policy should be expressible as a fixed mapping from the state to the control. That is, the optimal control policy will be expressible as

$$u_x^o(t) = h(x(t)) \tag{23.2.1}$$

for some static mapping $h(\circ)$. What remains is to give a characterization of the mapping $h(\circ)$. For general constrained problems, it will be difficult to give a simple parameterization to $h(\circ)$. However, we can think of the anti-wind-up strategies of Chapter 11 as giving a particular simple (ad-hoc) parameterization of $h(\circ)$. Specifically, if the control problem is formulated, in the absence of constraints, as a linear quadratic regulator, then we know from Chapter 22 that the unconstrained infinite horizon policy is of the form of Equation (23.2.1), where $h(\circ)$ takes the simple linear form

$$h^o(x(t)) = -K_\infty x(t) \tag{23.2.2}$$

for some constant-gain matrix K_∞. (See section §22.5.) If we then consider a SISO problem in which the control only is constrained, then the anti-wind-up form of (23.2.2) is (per section §18.9) given simply by

$$h(x(t)) = sat\{h^o(x(t))\} \qquad (23.2.3)$$

Now, we found in Chapter 11 that the above scheme actually works remarkably well, at least for simpler problems.

We illustrate by a further simple example.

Example 23.1. *Consider a continuous-time double-integrator plant that is sampled at an interval $\Delta = 1$ second. The corresponding discrete-time model is of the form*

$$x(k+1) = \mathbf{A}x(k) + \mathbf{B}u(k) \qquad (23.2.4)$$
$$y(k) = \mathbf{C}x(k) \qquad (23.2.5)$$

where

$$\mathbf{A} = \begin{bmatrix} 1 & 1 \\ 0 & 1 \end{bmatrix}; \quad \mathbf{B} = \begin{bmatrix} 0.5 \\ 1 \end{bmatrix}; \quad \mathbf{C} = \begin{bmatrix} 1 & 0 \end{bmatrix} \qquad (23.2.6)$$

We choose to use infinite horizon LQR theory to develop the control law. Within this framework, we use the following weighting matrices:

$$\Psi = \mathbf{C}^T\mathbf{C}, \qquad \Phi = 0.1 \qquad (23.2.7)$$

The unconstrained optimal control policy, as in (23.2.2), is then found via the methods described in Chapter 22.

We next investigate the performance of this control law under different scenarios. We first consider the case when the control is unconstrained.

Then, for an initial condition of $x(0) = (-10,0)^T$, the output response and input signal are as shown in Figures 23.1 and 23.2 on the next page. Figure 23.3 on the following page shows the optimal response on a phase-plane plot.

We next assume that the input must satisfy the mild constraint $|u(k)| \leq 5$. Applying the anti-wind-up policy (23.2.3) leads to the response shown in Figure 23.4 on page 743, with the corresponding input signal as in Figure 23.5 on page 743. The response is again shown on a phase-plane plot in Figure 23.6 on page 743.

Inspection of Figures 23.4, 23.5, and 23.6 on page 743 indicates that the simple anti-wind-up strategy of Equation (23.2.3) has produced a perfectly acceptable response in this case.

□□□

The above example would seem to indicate that one need never worry about fancy methods. Indeed, it can be shown that the control law of Equation (23.2.3) is actually *optimal* under certain circumstances–see problem 23.10 at the end of the chapter. However, if we make the constraints more stringent, the situation changes, as we see in the next example.

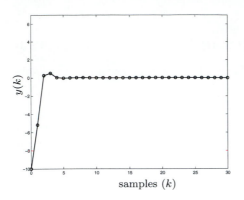

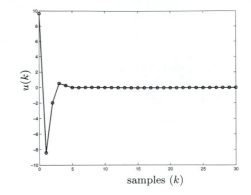

Figure 23.1. Output response without constraints

Figure 23.2. Input response without constraints

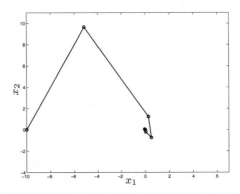

Figure 23.3. Phase-plane plot without constraints

Example 23.2. *Consider the same set-up as in Example 23.1, save that the input is now required to satisfy the more severe constraint $|u(k)| \leq 1$. Notice that this constraint is 10% of the initial unconstrained input shown in Figure 23.2. Thus, this is a relatively severe constraint. The control law of Equation (23.2.3) now leads to the response shown in Figure 23.7 on page 744, with corresponding input as in Figure 23.8 on page 744. The corresponding phase-plane plot is shown in Figure 23.9 on page 744.*

Inspection of Figure 23.7 on page 744 indicates that the simple policy of Equation (23.2.3) is not performing well and, indeed, has resulted in large overshoot in this case.

□□□

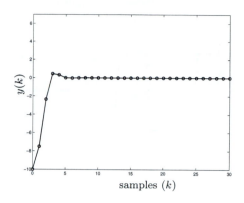

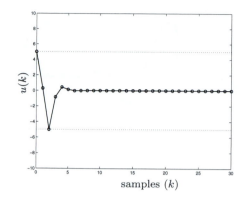

Figure 23.4. Output response with input constraint $|u(k)| \leq 5$

Figure 23.5. Input response with constraint $|u(k)| \leq 5$

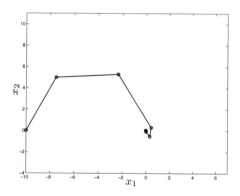

Figure 23.6. Phase-plane plot with constraint $|u(k)| \leq 5$

The reader may wonder what has gone wrong in the above example. A clue is given in Figure 23.9 on the following page. We see that the initial input steps have caused the velocity to build up to a large value. If the control were unconstrained, this large velocity would help us get to the origin quickly. However, because of the limited control authority, the system *braking capacity* is restricted, and hence large overshoot occurs. In conclusion, it seems that the control policy of Equation (23.2.3) has been too *shortsighted* and has not been able to account for the fact that *future* control inputs would be constrained as well as the current control input. The solution would seem to be to try to *look ahead* (i.e., predict the future response) and to take account of *current* and *future* constraints in deriving the control policy. This leads us to the idea of Model Predictive Control.

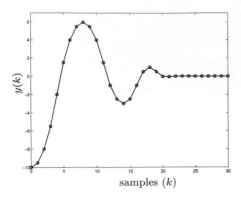

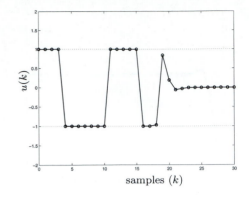

samples (k) samples (k)

Figure 23.7. Output response with constraint $|u(k)| \leq 1$

Figure 23.8. Input response with constraint $|u(k)| \leq 1$

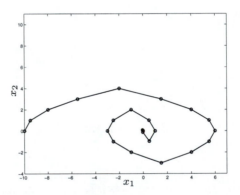

Figure 23.9. Phase-plane plot with constraint $|u(k)| \leq 1$

23.3   What is Model Predictive Control?

Model Predictive Control is a control algorithm based on solving an on-line *optimal* control problem. A *receding horizon* approach is used, which can be summarized in the following steps:

(i) At time k and for the current state $x(k)$, solve, on-line, an open-loop optimal control problem over some future interval, taking into account the *current* and *future* constraints.

(ii) Apply the first step in the optimal control sequence.

(iii) Repeat the procedure at time $(k+1)$, using the current state $x(k+1)$.

The solution is converted into a closed-loop strategy by using the measured value of $x(k)$ as the current state. When $x(k)$ is not directly measured, then one can obtain a closed-loop policy by replacing $x(k)$ by an estimate provided by some form of observer. The latter topic is taken up in section §23.6. For the moment, we assume that $x(k)$ is measured. Then, in a general nonlinear setting, the method is as follows:

Given a model

$$x(\ell + 1) = f(x(\ell), u(\ell)), \qquad x(k) = x \qquad (23.3.1)$$

the MPC at event (x, k) is computed by solving a constrained optimal control problem:

$$\mathcal{P}_N(x): \qquad V_N^o(x) = \min_{U \in \mathcal{U}_N} V_N(x, U) \qquad (23.3.2)$$

where

$$U = \{u(k), u(k+1), \dots, u(k+N-1)\} \qquad (23.3.3)$$

$$V_N(x, U) = \sum_{\ell=k}^{k+N-1} L(x(\ell), u(\ell)) + F(x(k+N)) \qquad (23.3.4)$$

and $\mathcal{U}_N$ is the set of U that satisfy the constraints over the entire interval $[k, k + N - 1]$:

$$\begin{aligned} u(\ell) &\in \mathbb{U} & \ell = k, k+1, \dots, k+N-1 & \qquad (23.3.5) \\ x(\ell) &\in \mathbb{X} & \ell = k, k+1, \dots, k+N & \qquad (23.3.6) \end{aligned}$$

together with the terminal constraint

$$x(k+N) \in W \qquad (23.3.7)$$

Usually, $\mathbb{U} \subset \mathbb{R}^m$ is convex and compact, $\mathbb{X} \subset \mathbb{R}^n$ is convex and closed, and W is a set that can be selected appropriately to achieve stability.

In the above formulation, the model and cost function are time invariant. Hence, one obtains a time-invariant feedback control law. In particular, we can set $k = 0$ in the open-loop control problem without loss of generality. Then, at event (x, k), we solve

$$\mathcal{P}_N(x): \qquad V_N^o(x) = \min_{U \in \mathcal{U}_N} V_N(x, U) \qquad (23.3.8)$$

where

$$U = \{u(0), u(1), \ldots, u(N-1)\} \tag{23.3.9}$$

$$V_N(x, U) = \sum_{\ell=0}^{N-1} L(x(\ell), u(\ell)) + F(x(N)) \tag{23.3.10}$$

subject to the appropriate constraints. Standard optimization methods are used to solve the above problem.

Let the minimizing control sequence be

$$U_x^o = \{u_x^o(0), u_x^o(1), \ldots, u_x^o(N-1)\} \tag{23.3.11}$$

Then the actual control applied at time k is the first element of this sequence, i.e.

$$u = u_x^o(0) \tag{23.3.12}$$

Time is then stepped forward one instant, and the above procedure is repeated for another N-step-ahead optimization horizon. The first input of the new N-step-ahead input sequence is then applied. The above procedure is repeated endlessly. The idea is illustrated in Figure 23.10 on the next page. Note that only the shaded inputs are actually applied to the plant.

The above MPC strategy *implicitly* defines a time-invariant control policy $h(\circ)$ as in (23.2.1)–i.e., a static mapping $h : \mathbb{X} \to \mathbb{U}$ of the form

$$h(x) = u_x^o(0) \tag{23.3.13}$$

The method originally arose in industry as a response to the need to deal with constraints. The associated literature can be divided into four generations, as follows:

- First Generation (1970's)–impulse or step-response linear models, quadratic cost function, and ad-hoc treatment of constraints.

- Second Generation (1980's)–linear state space models, quadratic cost function, input and output constraints expressed as linear inequalities, and quadratic programming used to solve the constrained optimal control problem.

- Third Generation (1990's)–several levels of constraints (soft, hard, ranked), mechanisms to recover from infeasible solutions.

- Fourth Generation (late 1990's)–nonlinear problems, guaranteed stability, and robust modifications.

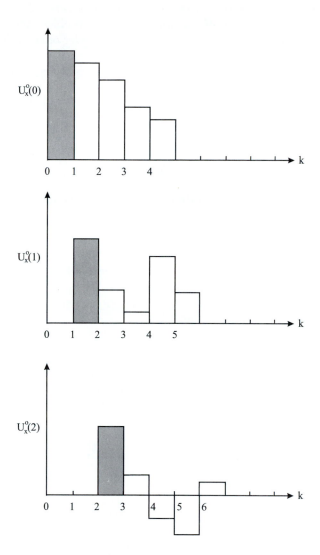

Figure 23.10. Receding-horizon control principle

23.4   Stability

A remarkable property of MPC is that one can establish stability of the resultant feedback system (at least with full state information). This is made possible by the fact that the value function of the optimal control problem acts as a Lyapunov function for the closed-loop system. We remind the reader of the brief introduction to Lyapunov stability given in Chapter 19.

For clarity of exposition, we make the following simplifying assumptions.

A1 : An additional constraint is placed on the final state of the receding-horizon optimization problem–namely, that

$$x(N) = 0 \qquad\qquad (23.4.1)$$

where $x(N)$ is the terminal state resulting from the control sequence U_x^o as in (23.3.11).

A2 : $L(x, u)$ is positive definite in both arguments.

Theorem 23.1. *Consider the system* (23.3.1) *controlled by the receding-horizon MPC algorithm* (23.3.8) *to* (23.3.11) *and subject to the terminal constraint* (23.4.1). *This control law renders the resultant closed-loop system globally asymptotically stable.*

Proof

Under regularity assumptions, the value function $V_N^o(\cdot)$ is positive definite and proper: $V(x) \to \infty$ as $||x|| \to \infty$. It can therefore be used as a Lyapunov function for the problem. We recall that, at event (x, k), MPC solves

$$\mathcal{P}_N(x): \qquad V_N^o(x) = \min_{U \in \mathcal{U}_N} V_N(x, U) \qquad\qquad (23.4.2)$$

$$V_N(x, U) = \sum_{\ell=0}^{N-1} L(x(\ell), u(\ell)) + F(x(N)) \qquad\qquad (23.4.3)$$

subject to constraints.

We denote the optimal open-loop control sequence solving $\mathcal{P}_N(x)$ as

$$U_x^o = \{u_x^o(0), u_x^o(1), \dots, u_x^o(N-1)\} \qquad\qquad (23.4.4)$$

We recall that inherent in the MPC strategy is the fact that the actual control applied at time k is the first value of this sequence,

$$u = h(x) = u_x^o(0) \qquad (23.4.5)$$

Let $x(1) = f(x, h(x))$, and let $x(N)$ be the terminal state resulting from the application of U_x^o. Note that we are assuming $x(N) = 0$.

A *feasible* solution (but not the optimal one) for the second step in the receding-horizon computation $\mathcal{P}_N(x_1)$, is then:

$$\tilde{U}_x = \{u_x^o(1), u_x^o(2), \dots, u_x^o(N-1), 0\} \qquad (23.4.6)$$

Then the increment of the Lyapunov function, upon using the true MPC optimal input and when moving from x to $x(1) = f(x, h(x))$, satisfies

$$\begin{aligned} \Delta_h V_N^o(x) &\triangleq V_N^o(x(1)) - V_N^o(x) \\ &= V_N(x(1), U_{x_1}^o) - V_N(x, U_x^o) \end{aligned} \qquad (23.4.7)$$

However, $U_{x_1}^o$ is optimal, so we know that

$$V_N(x(1), U_{x_1}^o) \le V_N(x(1), \tilde{U}_x) \qquad (23.4.8)$$

where $\tilde{U}_x$ is the suboptimal sequence defined in (23.4.6). Hence, we have

$$\Delta_h V_N^o(x) \le V_N(x(1), \tilde{U}_x) - V_N(x, U_x^o) \qquad (23.4.9)$$

Using the fact the $\tilde{U}_x$ shares $(N-1)$ terms in common with U_x^o, we can see that the right-hand side of (23.4.9) satisfies

$$V_N(x(1), \tilde{U}_x) - V_N(x, U_x^o) = -L(x, h(x)) \qquad (23.4.10)$$

where we have used the facts that U_x^o leads to $x(N) = 0$ (by assumption) and hence that $\tilde{U}_x$ leads to $x(N+1) = 0$.

Finally, using (23.4.9) and (23.4.10), we have

$$\Delta_h V_N^o(x) \le -L(x, h(x)) \qquad (23.4.11)$$

When $L(x, u)$ is positive definite in both arguments, then stability follows immediately from Theorem 19.1 on page 586.

□□□

Remark 23.1. *For clarity, we have used rather restrictive assumptions in the above proof, so as to keep it simple. Actually, both assumptions, (A1) and (A2), can be significantly relaxed. For example, assumption (A1) can be replaced by the assumption that $x(N)$ enters a terminal set in which "nice properties" hold. Similarly, assumption (A2) can be relaxed to requiring that the system be detectable in the cost function. We leave the interested reader to pursue these embellishments in the references given at the end of the chapter.*

□□□

Remark 23.2. *Beyond the issue of stability, a user of MPC would clearly also be interested in what, if any, performance advantages are associated with the use of this algorithm. In an effort to (partially) answer this question, we pause to revisit Example 23.2 on page 742.*

Example 23.3. *Consider again the problem described in Example 23.2, with input constraint $|u(k)| \leq 1$. We recall that the shortsighted policy used in Example 23.2 led to large overshoot.*

Here, we consider the cost function given in (23.3.10), with $N = 2$ and such that $F(x(N))$ is the optimal unconstrained *infinite horizon cost and $L(x(\ell), u(\ell))$ is the incremental cost associated with the underlying LQR problem as described in Example 23.1.*

The essential difference between Example 23.2 and the current example is that, whereas in Example 23.2 we considered only the input constraint on the present control in the optimization, here we consider the constraint on the present and the next step. Thus, the derivation of the control policy is not quite as "short sighted" as was previously the case.

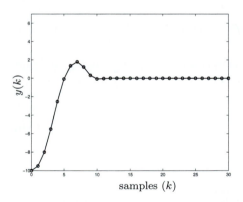

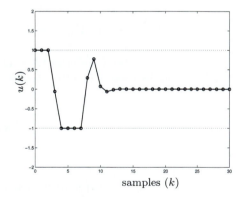

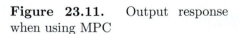

Figure 23.11. Output response when using MPC

Figure 23.12. Input response when using MPC

Figure 23.11 shows the resultant response, and Figure 23.12 shows the corresponding input. The response is shown in phase-plane form in Figure 23.13. Com-

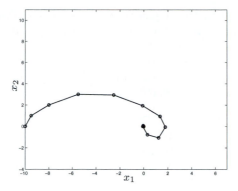

Figure 23.13. Phase-plane plot when using MPC

paring Figures 23.11 and 23.13 with Figures 23.7 and 23.9 on page 744 confirms the advantages of taking into account future constraints in deriving the control policy.

□□□

23.5   Linear Models with Quadratic Cost Function

So far, we have described the MPC algorithm in a rather general nonlinear setting. However, it might reasonably be expected that further insights can be obtained if one specializes the algorithm to cover the linear case with quadratic cost function. This will be the topic of the current section. Again, we will use a state space set-up. Let us therefore assume that the system is described by the following linear time-invariant model:

$$x(\ell + 1) = \mathbf{A}x(\ell) + \mathbf{B}u(\ell) \qquad (23.5.1)$$

$$y(\ell) = \mathbf{C}x(\ell) \qquad (23.5.2)$$

where $x(\ell) \in \mathbb{R}^n$, $u(\ell) \in \mathbb{R}^m$, and $y(\ell) \in \mathbb{R}^m$.

We assume that $(\mathbf{A}, \mathbf{B}, \mathbf{C})$ is stabilizable and detectable and that 1 is not an eigenvalue of $\mathbf{A}$. So as to illustrate the principles involved, we go beyond the set-up described in section §23.2 to include reference tracking and disturbance rejection. Thus, consider the problem of tracking a constant set-point y_s and rejecting a time-varying output disturbance $\{d(\ell)\}$: we wish to regulate, to zero, the error

$$e(\ell) = y(\ell) + (d(\ell) - y_s) \qquad (23.5.3)$$

It will be convenient in the sequel to make no special distinction between the output disturbance and the set-point, so we define an *equivalent* output disturbance d_e as the external signal:

$$d_e(\ell) = d(\ell) - y_s \tag{23.5.4}$$

Without loss of generality, we take the current time as 0.

Given knowledge of the external signal d_e and the current state measurement $x(0)$ (signals that will be later replaced by on-line estimates), our aim is to find the M-move control sequence $\{u(0), u(1), \ldots, u(M-1)\}$ that minimizes the finite-horizon performance index:

$$
\begin{aligned}
J_o =& [x(N) - x_s]^T \mathbf{\Psi}_f [x(N) - x_s] \\
&+ \sum_{\ell=0}^{N-1} e^T(\ell) \mathbf{\Psi} e(\ell) \\
&+ \sum_{\ell=0}^{M-1} [u(\ell) - u_s]^T \mathbf{\Phi} [u(\ell) - u_s]
\end{aligned}
\tag{23.5.5}
$$

where $\mathbf{\Psi} \geq 0$, $\mathbf{\Phi} > 0$, $\mathbf{\Psi}_f \geq 0$. Note that this cost function is a slight generalization of the one given in Equation (22.7.3) of Chapter 22.

In (23.5.5), N is the prediction horizon, $M \leq N$ is the control horizon.

We assume that $d_e(\ell)$ contains time-varying components as well as a constant component. Let $\bar{d}_e$ denote the constant (steady-state) component.

In equation (23.5.5), we then let u_s and x_s denote the corresponding steady-state values of u and x:

$$u_s = -[\mathbf{C}(\mathbf{I} - \mathbf{A})^{-1}\mathbf{B}]^{-1}\bar{d}_e \tag{23.5.6}$$

$$x_s = (\mathbf{I} - \mathbf{A})^{-1}\mathbf{B}u_s \tag{23.5.7}$$

The minimization of (23.5.5) is performed on the assumption that the control reaches its steady-state value after M steps–that is, $u(\ell) = u_s$, $\forall \ell \geq M$.

In the presence of constraints on the input and output, this dynamic optimization problem simply amounts to finding the solution of a nondynamic constrained quadratic program. Indeed, from (23.5.1), and using $\mathbf{B}u_s = (I - \mathbf{A})x_s$, we can write

$$X - X_s = \mathbf{\Gamma} U + \mathbf{\Lambda} x(0) - \overline{X}_s \tag{23.5.8}$$

where

$$
X = \begin{bmatrix} x(1) \\ x(2) \\ \vdots \\ x(N) \end{bmatrix} ; \quad X_s = \begin{bmatrix} x_s \\ x_s \\ \vdots \\ x_s \end{bmatrix} ; \quad U = \begin{bmatrix} u(0) \\ u(1) \\ \vdots \\ u(M-1) \end{bmatrix} ; \quad \Lambda = \begin{bmatrix} \mathbf{A} \\ \mathbf{A}^2 \\ \vdots \\ \mathbf{A}^N \end{bmatrix} ; \qquad (23.5.9)
$$

$$
\Gamma = \begin{bmatrix} \mathbf{B} & 0 & \cdots & 0 & 0 \\ \mathbf{AB} & \mathbf{B} & \cdots & 0 & 0 \\ \vdots & \vdots & \ddots & \vdots & \vdots \\ \mathbf{A}^{M-1}\mathbf{B} & \mathbf{A}^{M-2}\mathbf{B} & \cdots & \mathbf{AB} & \mathbf{B} \\ \mathbf{A}^M\mathbf{B} & \mathbf{A}^{M-1}\mathbf{B} & \cdots & \mathbf{A}^2\mathbf{B} & \mathbf{AB} \\ \vdots & \vdots & \ddots & \vdots & \vdots \\ \mathbf{A}^{N-1}\mathbf{B} & \mathbf{A}^{N-2}\mathbf{B} & \cdots & \cdots & \mathbf{A}^{N-M}\mathbf{B} \end{bmatrix} ; \quad \overline{X}_s = \begin{bmatrix} x_s \\ x_s \\ \vdots \\ x_s \\ \mathbf{A}x_s \\ \vdots \\ \mathbf{A}^{N-M}x_s \end{bmatrix}
$$

Then, using (23.5.8), (23.5.9), and

$$
\overline{\mathbf{\Psi}} = \mathrm{diag}[\mathbf{C}^T\mathbf{\Psi}\mathbf{C}, \dots, \mathbf{C}^T\mathbf{\Psi}\mathbf{C}, \mathbf{\Psi}_f] \tag{23.5.10}
$$

$$
\overline{\mathbf{\Phi}} = \mathrm{diag}[\mathbf{\Phi}, \dots, \mathbf{\Phi}] \tag{23.5.11}
$$

$$
U_s = \begin{bmatrix} u_s^T & u_s^T & \cdots & u_s^T \end{bmatrix}^T \tag{23.5.12}
$$

$$
\tilde{D} = [d_e(0)^T - \bar{d}_e^T, d_e(1)^T - \bar{d}_e^T, \dots, d_e(N-1)^T - \bar{d}_e, 0]^T \tag{23.5.13}
$$

$$
Z = \mathrm{diag}[\mathbf{C}^T\mathbf{\Psi}, \mathbf{C}^T\mathbf{\Psi}, \dots, \mathbf{C}^T\mathbf{\Psi}] \tag{23.5.14}
$$

we can express (23.5.5) as

$$
\begin{aligned}
J_o &= e^T(0)\mathbf{\Psi}e(0) + (X - X_s)^T\overline{\mathbf{\Psi}}(X - X_s) + (U - U_s)^T\overline{\mathbf{\Phi}}(U - U_s) \\
&\quad + 2(X - X_s)^T Z\tilde{D} + \tilde{D}^T \mathrm{diag}[\mathbf{\Psi}, \dots, \mathbf{\Psi}]\tilde{D} \\
&= \overline{J}_o + U^T\mathbf{W}U + 2U^T\mathbf{V} \tag{23.5.15}
\end{aligned}
$$

In (23.5.15), $\overline{J}_o$ is independent of U, and

$$
\mathbf{W} = \mathbf{\Gamma}^T\overline{\mathbf{\Psi}}\mathbf{\Gamma} + \overline{\mathbf{\Phi}}, \qquad \mathbf{V} = \mathbf{\Gamma}^T\overline{\mathbf{\Psi}}\left[\Lambda x(0) - \overline{X}_s\right] - \overline{\mathbf{\Phi}}U_s + \mathbf{\Gamma}^T Z\tilde{d}. \tag{23.5.16}
$$

From (23.5.15), it is clear that, if the design is *unconstrained*, J_o is minimized by taking

$$
\boxed{U = -\mathbf{W}^{-1}\mathbf{V}} \tag{23.5.17}
$$

Next, we introduce constraints into the problem formulation.

Magnitude and rate constraints on both the *input* and the *output* of the plant can be expressed as follows:

$$u_{\min} \le u(k) \le u_{\max}; \quad k = 0, 1, \ldots, M - 1 \qquad (23.5.18)$$
$$y_{\min} \le y(k) \le y_{\max}; \quad k = 1, 2, \ldots, N - 1 \qquad (23.5.19)$$
$$\Delta u_{\min} \le u(k) - u(k - 1) \le \Delta u_{\max}; \quad k = 0, 1, \ldots, M - 1 \qquad (23.5.20)$$

These constraints can be expressed as *linear* constraints on U of the form

$$LU \le K \qquad (23.5.21)$$

Remark 23.3. *As an illustration for the single-input single-output case, L and K take the following form:*

$$L = \begin{bmatrix} I \\ -I \\ D \\ -D \\ W \\ -W \end{bmatrix}; \quad K = \begin{bmatrix} \overline{U}_{\max} \\ \overline{U}_{\min} \\ \overline{Y}_{\max} \\ \overline{Y}_{\min} \\ \overline{V}_{\max} \\ \overline{V}_{\min} \end{bmatrix} \qquad (23.5.22)$$

where
I is the $M \times M$ identity matrix (M is the control horizon).
W is the following $M \times M$ matrix

$$W = \begin{bmatrix} 1 & & & \\ -1 & \ddots & & 0 \\ & \ddots & \ddots & \\ 0 & & -1 & 1 \end{bmatrix} \qquad (23.5.23)$$

D is the following $(N - 1) \times M$ matrix

$$
D = \begin{bmatrix}
CB & & & & \\
CAB & & \ddots & & \\
\vdots & & \ddots & & \ddots \\
\vdots & & & \ddots & & \ddots \\
CA^{M-1}B & \cdots & & CAB & & CB \\
CA^{M}B & \cdots & & \cdots & & CAB \\
\vdots & & & & & \\
CA^{N-2}B & \cdots & & CA^{N-M}B & CA^{N-M-1}B
\end{bmatrix} \tag{23.5.24}
$$

$$
\overline{U}_{\max} = \begin{bmatrix} u_{\max} \\ \vdots \\ u_{\max} \end{bmatrix}; \qquad \overline{U}_{\min} = \begin{bmatrix} -u_{\min} \\ \vdots \\ -u_{\min} \end{bmatrix} \tag{23.5.25}
$$

$$
\overline{V}_{\max} = \begin{bmatrix} u(0) + \Delta u_{\max} \\ \Delta u_{\max} \\ \vdots \\ \Delta u_{\max} \end{bmatrix}; \qquad \overline{V}_{\min} = \begin{bmatrix} -u(0) - \Delta u_{\min} \\ -\Delta u_{\min} \\ \vdots \\ -\Delta u_{\min} \end{bmatrix} \tag{23.5.26}
$$

$$
\overline{Y}_{\max} = \begin{bmatrix} y_{\max} - CAx_o - d_1 \\ \vdots \\ \vdots \\ y_{\max} - CA^{N-1}x_o - d_{N-1} \\ -\sum_{i=0}^{N-M-2} CA^i Bu_s \end{bmatrix}; \quad \overline{Y}_{\min} = \begin{bmatrix} -y_{\min} + CAx_o + d_1 \\ \vdots \\ \vdots \\ -y_{\min} + CA^{N-1}x_o + d_{N-1} \\ +\sum_{i=0}^{N-M-2} CA^i Bu_s \end{bmatrix} \tag{23.5.27}
$$

and $u_{\max}$, $u_{\min}$, $\Delta u_{\max}$, $\Delta u_{\min}$, $y_{\max}$, $y_{\min}$ *are the upper and lower limits on* $u(k)$, $(u(k) - u(k-1))$, *and* $y(k)$ *respectively.*

□□□

The optimal solution

$$
U^{OPT} = \left[(\bar{u}^{OPT}(0))^T, (\bar{u}^{OPT}(1))^T, \ldots, (\bar{u}^{OPT}(M-1))^T \right]^T \tag{23.5.28}
$$

can be numerically computed via the following static-optimization problem:

$$
U^{OPT} = \arg \min_{\substack{U \\ LU \le K}} U^T \mathbf{W} U + 2U^T \mathbf{V} \tag{23.5.29}
$$

That this optimization is a convex problem is due to the quadratic cost and linear constraints. Also, standard numerical procedures (called Quadratic Programming algorithms or QP for short) are available to solve this subproblem. Many software packages (e.g., MATLAB) provide standard tools to solve QP problems. Thus, we will not dwell on this aspect here.

Note that we apply only the first control, $u^{OPT}(0)$. Then the whole procedure is repeated at the next time instant, with the optimization horizon kept constant.

Remark 23.4. *In the above development, we have assumed that 1 is not an eigenvalue of A. There are several ways to treat the case when 1 is an eigenvalue of A. For example, in the single-input single-output case, when A has a single eigenvalue at 1, then $u_s = 0$. Also, we can write the state space model so that the integrator is shifted to the output, i.e.,*

$$\bar{x}(k+1) = \overline{A}\bar{x}(k) + \overline{B}u(k) \tag{23.5.30}$$
$$x'(k+1) = \overline{C}\bar{x}(k) + x'(k)$$
$$y(k) = x'(k)$$

where $\overline{A}$, $\overline{B}$, and $\overline{C}$ correspond to the state space model of the reduced-order plant, i.e., the plant without the integrator.

With this transformation, $[\overline{A} - I]$ is nonsingular, and hence

$$\begin{bmatrix} \bar{x} \\ x' \end{bmatrix}_s = \begin{bmatrix} 0 \\ \vdots \\ 0 \\ -\bar{d}_e \end{bmatrix} \tag{23.5.31}$$

The MPC problem is then solved in terms of these transformed state variables. The reader is encouraged to consider more general cases.

□□□

23.6   State Estimation and Disturbance Prediction

In the above description of the MPC algorithm, we have assumed, for simplicity of initial exposition, that the state of the system (including all disturbances) was available. However, it is relatively easy to extend the idea to the case when the states are not directly measured. The essential idea involved in producing a version of MPC that requires only output data is to replace the unmeasured state by an on-line estimate provided by a suitable observer. For example, one could use a Kalman filter (see Chapter 22) to estimate the current state $x(k)$ from observations of the output and input $\{y(\ell), u(\ell); \ell = \dots, k-2, k-1\}$. Disturbances can also be included in this strategy by including the appropriate noise-shaping filters in a composite model, as described in subsection §22.10.4.

The MPC algorithm also requires that future states and disturbances be predicted, so that the optimization over the future horizon can be carried out. To do this we simply propagate the deterministic model obtained from the composite system and noise model as described in subsection §22.10.5–see, in particular, Equation (22.10.80).

23.6.1   One-Degree-of-Freedom Controller

A one-degree-of-freedom output-feedback controller is obtained by including the set-point in the quantities to be estimated. The resultant output-feedback MPC strategy is schematically depicted in Figure 23.14, where $G_o(z) = \mathbf{C}(z\mathbf{I} - \mathbf{A})^{-1}\mathbf{B}$ is the plant transfer function and the observer has the structure discussed in subsections §22.10.4 and §22.10.5. In Figure 23.14, QP denotes the quadratic program used to solve the constrained optimal control problem.

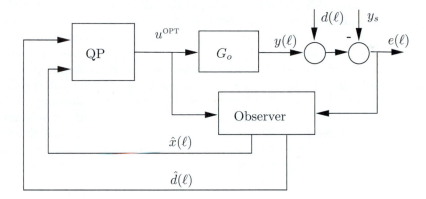

Figure 23.14. One-degree-of-freedom MPC architecture

23.6.2   Integral Action

An important observation is that the architecture described in Figure 23.14 gives a form of integral action. In particular, y is taken to the set-point y_s, irrespective of the true plant description (provided that a steady state is reached and that, in this steady state, u is not constrained). To show this, we argue as follows.

Say that $\{d(\ell)\}$ contains only constant values: then the observer needed to estimate these unknown constants will be of the form

$$\hat{d}_e(\ell + 1) = \hat{d}_e(\ell) + J_d(e(\ell) - \mathbf{C}\hat{x}(\ell) - \hat{d}_e(\ell)) \qquad (23.6.1)$$

where $\{\hat{x}(\ell)\}$ denotes the estimate of the other plant states.

We see that, if a steady state is reached, then it must be true that $\hat{d}_e(\ell+1) = \hat{d}_e(\ell)$, and hence with $J_d \neq 0$ we have

$$e(\ell) = \mathbf{C}\hat{x}(\ell) + \hat{d}_e(\ell) \tag{23.6.2}$$

Now the control law has the property that the model output, $\mathbf{C}\hat{x}(\ell) + \hat{d}_e(\ell)$, is taken to zero in steady state. Thus, from (23.6.2), it must also be true that the actual quantity $\{e(\ell)\}$ as measured at the plant output must also be taken to zero in steady state. However, inspection of Figure 23.14 then indicates that, in this steady-state condition, the tracking error is zero.

23.7   Rudder Roll Stabilization of Ships

Here we present a realistic application of Model Predictive Control to rudder roll stabilization of ships. We will adapt the algorithm described in sections §23.5 and §23.6. The motivation for this particular case study follows.

It is desirable to reduce the rolling motion of ships (produced by wave action) so as to prevent cargo damage and improve crew efficiency and passenger comfort. Conventional methods for ship roll stabilization include water tanks, stabilization fins, and bilge keels. Another alternative is to use the rudder for roll stabilization as well as for course keeping. However, using the rudder simultaneously for course keeping and for roll reduction is nontrivial, because only one actuator is available to deal with two objectives. Therefore, frequency separation of the two models is required, and design trade-offs will limit the achievable performance. However, rudder roll stabilization is attractive because no extra equipment needs to be added to the ship.

An important issue is that the rudder mechanism is usually limited in amplitude and in slew rate. Hence, this is a suitable problem for Model Predictive Control.

In order to describe the motion of a ship, six independent coordinates are necessary. The first three coordinates and their time derivatives correspond to the position and translational motion; the other three coordinates and their time derivatives correspond to orientation and rotational motion. For marine vehicles, the six different motion components are named *surge, sway, heave, roll, pitch,* and *yaw.* The most generally used notation for these quantities are x, y, z, ϕ, θ, and ψ, respectively. Figure 23.15 shows the six coordinate definitions and the reference frame most generally adopted.

Figure 23.15 also shows the *rudder angle*, which is usually noted as δ.

The roll motion due to the rudder presents a faster dynamic response than does the rudder-produced yaw motion, because of the different moments of inertia of the hull associated with each direction. This property is exploited in rudder roll stabilization.

The common method of modeling the ship motion is based on the use of Newton's

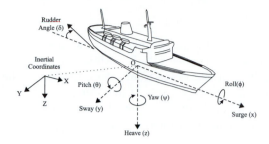

Figure 23.15. Magnitudes and conventions for ship motion description

equation in surge, sway, and roll. By using this approach, a nonlinear model is obtained, from which a linearized state space model can be derived that depends upon five states: $[v, r, p, \phi, \psi]^T$.

The model is a function of the ship speed. However, this is not of much concern; the model is quasi-constant over a reasonable speed range. This allows the use of gain scheduling to update the model according to different speeds of operation.

Undesired variations of the roll and yaw angles are generated mainly by waves. These wave disturbances are usually considered at the output, and the complete model of the ship dynamics, including the output equation, is of the form

$$\dot{x} = \mathbf{A}x + \mathbf{B}\delta \tag{23.7.1}$$

$$y = \mathbf{C}x + d_{wave} \tag{23.7.2}$$

Typically, only the roll and yaw are directly measured: $y := [\phi, \psi]^T$. In Equation (23.7.2), d_{wave} is the wave-induced disturbance on the output variables.

The wave disturbances can be characterized in terms of their frequency spectrum. This frequency spectrum can be simulated by using filtered white noise. The filter used to approximate the spectrum is usually a second-order one of the form

$$H(s) = \frac{K_w s}{s^2 + 2\xi\omega_o s + \omega_o^2} \tag{23.7.3}$$

where K_w is a constant describing the wave intensity, ξ is a damping coefficient, and ω_o is the dominant wave frequency. This model produces decaying *sine-wave-like* disturbances. The resultant colored noise representing the effect of wave disturbances is added to the two components of y, with appropriate scaling. (Actually, the wave motion affects mainly the roll motion.)

The design objectives for rudder roll stabilization are:

- Increase the damping and reduce the roll amplitude;

- Control the heading of the ship.

The rudder is constrained in amplitude and slew rate.

For merchant ships, the rudder rate is within the range from 3 to 4 deg/sec; for military vessels, it is in the range from 3 to 8 deg/sec. The maximum rudder excursion is typically in the range from 20 to 35 deg.

We will adopt the Model Predictive Control algorithm described in section §23.5.

The following model was used for the ship–see the references given at the end of the chapter.

$$\mathbf{A} = \begin{bmatrix} -0.1795 & -0.8404 & 0.2115 & 0.9665 & 0 \\ -0.0159 & -0.4492 & 0.0053 & 0.0151 & 0 \\ 0.0354 & -1.5594 & -0.1714 & -0.7883 & 0 \\ 0 & 0 & 1 & 0 & 0 \\ 0 & 1 & 0 & 0 & 0 \end{bmatrix} \tag{23.7.4}$$

$$\mathbf{B} = \begin{bmatrix} 0.2784 \\ -0.0334 \\ -0.0894 \\ 0 \\ 0 \end{bmatrix} \tag{23.7.5}$$

This system was sampled with a zero-order hold and sampling interval equal to 0.5 sec.

Details of the Model Predictive Control optimization criterion of Equation (23.4.6) were

$$\mathbf{\Psi} = \begin{bmatrix} 90 & 0 \\ 0 & 3 \end{bmatrix} ; \qquad \mathbf{\Phi} = 0.1 \tag{23.7.6}$$

Optimization horizon, $N = 20$; Control horizon, $M = 18$.

Remark 23.5. *Note that the continuous time system (23.7.4) and (23.7.5) contains a pure integrator. Hence, the discretized system has an eigenvalue at 1; however, for this problem, we have taken the set-point on the output to be zero and the disturbance has zero mean. Thus, the problem reduces to a regulation problem where $x_s = 0$, $u_s = 0$.*

□□□

Only roll, ϕ, and yaw, ψ, were assumed to be directly measured, so a Kalman filter was employed to estimate the 5 system states and 2 noise states in the composite model. The output disturbance was predicted by using the methodology described in section §23.6.

The set-points for the two system outputs were taken to be zero, and there were no constant off-sets in the disturbance model. Thus, the problem reduces to a regulation problem.

In the following plots we show the roll angle (with and without rudder roll stabilization), the yaw angle (with and without rudder roll stabilization), and the rudder angle with roll stabilization.

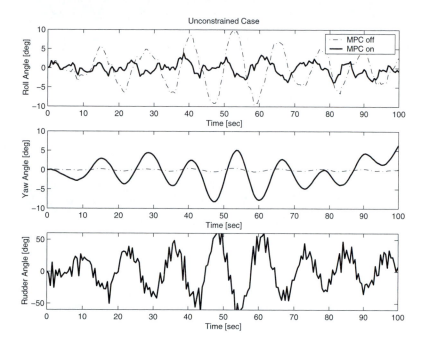

Figure 23.16. Ship motion with no constraints on rudder motion

Figure 23.16 shows the results for the case in which no constraints are applied to the rudder amplitude or slew rate. In this case, a 66% reduction in the standard deviation of the roll angle was achieved relative to no stabilization. The price paid for this is a modest increase in the standard deviation in the yaw angle. Notice, also, that the rudder angle reaches 60 degrees and has a very fast slew rate.

Figure 23.17 shows the results for a maximum rudder angle of 30 degrees and a maximum slew rate of 15 deg/sec. In this case, a 59% reduction in the standard deviation of the roll angle was achieved relative to no stabilization. Notice that the rudder angle satisfies the given constraints via the application of the MPC algorithm.

Figure 23.18 shows the results for a maximum rudder angle of 20 degrees and maximum slew rate of 8 deg/sec. In this case, a 42% reduction in the standard deviation of the roll angle was achieved relative to no stabilization. Notice that the more severe constraints on the rudder are again satisfied via the application of the MPC algorithms.

□□□

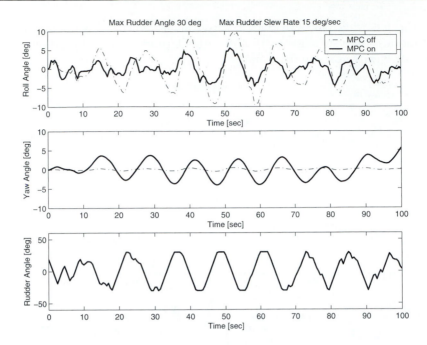

Figure 23.17. Ship motion when rudder is constrained to maximum angle of 30 degrees and maximum slew rate of 15 degrees/sec

23.8 Summary

- MPC provides a systematic procedure for dealing with constraints (both input and state) in MIMO control problems.

- It has been widely used in industry.

- Remarkable properties of the method can be established–e.g., global asymptotic stability provided that certain conditions are satisfied (e.g., appropriate weighting on the final state).

- The key elements of MPC for linear systems are

  ○ state space (or equivalent) model,

  ○ on-line state estimation (including disturbances),

  ○ prediction of future states (including disturbances),

  ○ on-line optimization of future trajectory subject to constraints by using Quadratic Programming, and

  ○ implementation of the first step of the control sequence.

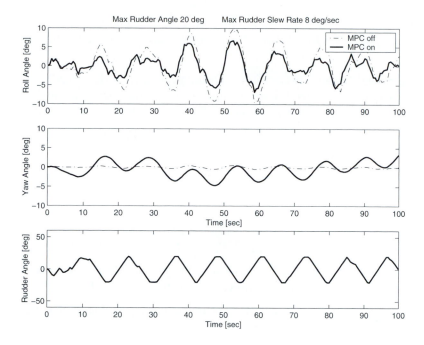

Figure 23.18. Ship motion when rudder is constrained to maximum angle of 20 degrees and maximum slew rate of 8 degrees/sec

23.9   Further Reading

Optimal control background

Bellman, R. (1957). *Dynamic Programming.* Princeton University Press.

Kalman, R.E. (1960). Contributions to the theory of optimal control. *Boletín Sociedad Matemática Mexicana*, 5:102-119.

Lee, E. and Markus, L. (1967). *Foundations of Optimal Control Theory.* Wiley, New York.

Receding-horizon Model Predictive Control

Cutler, C. and Ramaker, B. (1989). Dynamic matrix control. A computer control algorithm. *Proceedings Joint Automatic Control Conference*, San Francisco, California.

Garcia, C.E. and Morshedi, A. (1986). Quadratic programming solution of dynamic matrix control (QDMC). *Chemical Engineering Communications*, 46:73-87.

Grosdidier, P., Froisy, B. and Hammann, M. (1988). The IDCOM-M controller. *Proceedings of the 1988 IFAC Workshop on Model Based Process Control*, pp. 31-36.

Keerthi, S.S. and Gilbert, E.G. (1988). Optimal, infinite horizon feedback laws for a general class of constrained discrete time systems: Stability and moving-horizon approximations. *Journal of Optimization Theory and Applications*, 57:265-293.

Mayne, D.Q., Rawlings, J.B., Rao, C.V., and Scokaert, P. (2000). Model Predictive Control: A review. *Automatica*, to appear.

Michalska, H. and Mayne, D.Q. (1993). Robust receding horizon control of constrained nonlinear systems. *IEEE Trans. on Automatic Control*, 38:1623-1632.

Qin, S. and Badgwell, T. (1997) An overview of industrial Model Predictive Control technology. *Chemical Process Control–V, CACHE, AIChE*, 232-256.

Richalet, J., Rault, A., Testud, J., and Papon, J. (1978). Model predictive heuristic control: Applications to industrial processes. *Automatica*, 14:413-428.

Thomas, Y. (1975). Linear quadratic optimal estimation and control with receding horizon. *Electronics Letters*, 11:19-21.

Stability of MPC

Chen, C.C. and Shaw, L. (1982). On receding horizon feedback control. *Automatica*, 18:349-352.

De Doná, J.A. and Goodwin, G.C. (1999). Characterisation of regions in which Model Predictive Control polices have a finite dimensional parameterisation. *SIAM Journal of Control and Optimization*, Submitted.

Gilbert, E.G. and Tan, K.T. (1991). Linear systems with state and control constraints: The theory and application of maximal output admissible sets. *IEEE Trans. on Automatic Control*, AC-36:1008-1020.

Jadbabaie, A., Yu, J., and Hauser, J. (1999). Unconstrained receding horizon control of nonlinear systems. *IEEE Trans. on Automatic Control*, Submitted.

Keerthi, S.S. and Gilbert, E.G. (1988). Optimal, infinite horizon feedback laws for a general class of constrained discrete time systems: Stability and moving-horizon approximations. *Journal of Optimization Theory and Applications*, 57:265-293.

Kleinman, B. (1970). An easy way to stabilize a linear constant system. *IEEE Trans. on Automatic Control*, 15(12):693.

Mayne, D.Q., Rawlings, J.B., Rao, C.V., and Scokaert, P.O.M. (1999). Constrained Model Predictive Control: Stability and optimality. *Automatica*, 36: Accepted.

Polak, E. (1997). *Optimization: Algorithms and Consistent Approximations*. Springer-

Verlag, New York.

Rawlings, J.B. and Muske, K.R. (1993). Stability of constrained receding horizon control. *IEEE Trans. on Automatic Control*, 38(10):1512-1516.

Scokaert, P.O.M. and Rawlings, J.B. (1998). Constrained linear quadratic regulation. *IEEE Trans. on Automatic Control*, 43(8):1163-1169.

Rudder roll stabilization

Fossen, T. (1994). *Guidance and Control of Ocean Vehicles.* John Wiley and Sons Ltd.

Perez, T., Tzeng, C-Y., and Goodwin, G.C. (1999). *Model predictive rudder roll stabilization control for ships.* Technical Report No. 9946, Department of Electrical and Computer Engineering, The University of Newcastle, Australia (submitted for publication).

Blanke, M. and Christensen, A. (1993). *Rudder-roll damping autopilot robustness to sway-yaw-roll couplings.* SCSS, Proceedings, 10th SCSS, Ottawa, Canada.

23.10   Problems for the Reader

Problem 23.1. *Consider a discrete-time single-input linear regulator problem for which the control is to be designed via a linear quadratic cost function. Say that the input is required to satisfy a constraint of the form $|u(k)| \leq \Delta$. Use a receding-horizon MPC set-up for this problem. Show that if, in the solution to the optimization problem, only the first control is saturated, then*

$$u(k) = -\text{sat}(Kx(k)) \qquad\qquad (23.10.1)$$

is the optimal constrained control, where $u(k) = -Kx(k)$ is the optimal unconstrained control. (Hint: use Dynamic Programming, and examine the last step of the argument.)

Problem 23.2. *Why does the result given in Problem 23.1 not, in general, hold if future controls in the optimization problem also reach saturation levels?*

Problem 23.3. *Extend the result in Problem 23.1 to the 2-input case.*

Problem 23.4. *Consider the following scalar state space model:*

$$x(k+1) = ax(k) + bu(k); \qquad x(0) \text{ given} \qquad (23.10.2)$$

Let us consider the following fixed-horizon cost function:

$$J = \sum_{k=o}^{N} x(k)^2 \qquad\qquad (23.10.3)$$

Use Dynamic Programming to show that the optimal control law is given by

$$u(k) = -\frac{a}{b}x(k) \qquad\qquad (23.10.4)$$

Note that this brings the state to zero in one step–i.e., $x(1) = 0$.

Problem 23.5. *(Harder) Consider the same set-up as in Problem 23.4, save that we now require that the input satisfy the constraint $|u(k)| \leq \Delta$. Show that the optimal receding-horizon control is given by*

$$u(k) = -\text{sat}(\frac{a}{b}x(k))$$ (23.10.5)

(Hint: use Dynamic Programming.)

Problem 23.6. *Compare the results in Problems 23.4 and 23.5 with the result in 23.1. Comment on the connections and differences.*

Problem 23.7. *Consider the quadratic cost function*

$$J = (\mathbf{\Gamma} - \mathbf{\Omega}U)^T(\mathbf{\Gamma} - \mathbf{\Omega}U)$$ (23.10.6)

where $\mathbf{\Gamma}$ and U are vectors and $\mathbf{\Omega}$ is an $N \times M$ matrix.

23.7.1 *Show that the minimum of J is achieved (when U is unconstrained) by*

$$U = U_o = (\mathbf{\Omega}^T\mathbf{\Omega})^{-1}\mathbf{\Omega}^T\mathbf{\Gamma}$$ (23.10.7)

23.7.2 *Say that U is required to satisfy the following set of linear equality constraints:*

$$\mathbf{L}U = \mathbf{C}$$ (23.10.8)

where $\mathbf{C}$ is a vector of dimension S ($S \leq M$) and $\mathbf{L}$ is an $S \times M$ matrix. Show that the cost function J is minimized, subject to this constraint, by

$$U = U_o' = U_o - (\mathbf{\Omega}^T\mathbf{\Omega})^{-1}\mathbf{L}^T(\mathbf{L}(\mathbf{\Omega}^T\mathbf{\Omega})^{-1}\mathbf{L}^T)^{-1}(\mathbf{L}U_o - \mathbf{C})$$ (23.10.9)

where U_o is the unconstrained solution found in part (23.7.1).

Problem 23.8. *Using the results in Problem 23.7, show that the optimal solution to the receding-horizon linear quadratic regulator with linear inequality constraints always has the following affine form:*

$$u(k) = -\overline{K}x(k) + \bar{b}$$ (23.10.10)

where $\overline{K}$ and $\bar{b}$ depend upon which constraints are active (i.e., upon which inequalities are actually equalities for this particular initial state $x(k)$).

Problem 23.9. *The models presented below are related to real-world case studies. The authors have found these models useful in testing various MIMO design strategies. It is suggested that the reader use these examples to test his or her understanding of MIMO control. Things that might be tried include*

(i) decentralized SISO designs

(ii) centralized LQR design

(iii) appling various constraints to the inputs of the systems and using anti-wind-up or MPC methods to design controllers that take account of these constraints

23.9.1 *This example relates to a ship's steering system. $Y1$ is the heading angle of the ship, $Y2$ the roll angle of the ship, $U1$ the rudder angle, and $U2$ the fin angle. In theory, the rudder is used to control the heading of the ship and the fin is used to control the roll angle of the ship. Of course, there is coupling between these two control signals. A model covering this situation is:*

$$Y(s) = \begin{bmatrix} \dfrac{0.034(21.6s+1)}{s(-59.1s+1)(8.1s+1)} & \dfrac{0.00662(14.1s+1)}{s(-59.1s+1)(8.1s+1)} \\[4mm] \dfrac{-0.416(11.4s+1)(-6.8s+1)}{(-51.9s+1)(8.1s+1)(45.118s^2+0.903s+1)} & \dfrac{-0.245}{(45.118s^2+0.903s+1)} \end{bmatrix} U$$

$$(23.10.11)$$

23.9.2 *This problem relates to a distillation column. A model frequently used to describe this kind of system is*

$$Y(s) = \begin{bmatrix} \dfrac{0.66e^{-2.6s}}{6.7s+1} & \dfrac{-0.61e^{-3.5s}}{8.64s+1} & \dfrac{-0.49e^{-s}}{9.06s+1} \\[4mm] \dfrac{1.11e^{-6.5s}}{3.25s+1} & \dfrac{-2.36e^{-3s}}{5s+1} & \dfrac{-1.2e^{-1.2s}}{7.09s+1} \\[4mm] \dfrac{-0.3468e^{-9.2s}}{8.15s+1} & \dfrac{0.462e^{-9.4s}}{10.9s+1} & \dfrac{(10.1007s+0.87)e^{-s}}{73.132s^2+22.69s+1} \end{bmatrix} U(s)$$

$$(23.10.12)$$

23.9.3 *This example is related to a mining system called an AG mill. The purpose of this system is to take large rocks and to cause them to grind against each other so as to produce rocks of a small size. The model used to describe this is*

$$Y(s) = \begin{bmatrix} \dfrac{0.105e^{-65s}}{83s+1} & \dfrac{-0.082e^{-80s}}{1766s+1} & \dfrac{-0.0575e^{-460s}}{167s+1} \\[3ex] \dfrac{-0.0468e^{-140s}}{1864s+1} & \dfrac{0.00122}{s} & \dfrac{0.115e^{-120s}}{1984s+1} \\[3ex] \dfrac{0.00253}{s} & 0 & \dfrac{-0.00299}{s} \end{bmatrix} U(s) \quad (23.10.13)$$

Chapter 24

FUNDAMENTAL LIMITATIONS IN MIMO CONTROL

24.1 Preview

Arguably, the best way to learn about real design issues is to become involved in practical applications. We hope that the reader gained some feeling for the lateral thinking that is typically needed in most real-world problems, from reading the various case studies that we have presented. In particular, we point to the four MIMO case studies described in Chapter 22.

In this chapter, we will adopt a more abstract stance and extend the design insights of Chapters 8 and 9 to the MIMO case. As a prelude to this, we recall that, in Chapter 17, we saw that, by a combination of an observer and state-estimate feedback, the closed-loop poles of a MIMO system can be exactly (or approximately) assigned, depending on the synthesis method used. However, as in the SISO case, this leaves open two key questions: where the poles should be placed, and what the associated sensitivity trade-off issues are. This raises fundamental design issues that are the MIMO versions of the topics discussed in Chapters 8 and 9.

It was shown in Chapters 8 and 9 that the open-loop properties of a SISO plant impose fundamental and unavoidable constraints on the closed-loop characteristics that are achievable. For example, we have seen that, for a one-degree-of-freedom loop, a double integrator in the open-loop transfer function implies that the integral of the error due to a step reference change must be zero. We have also seen that real RHP zeros necessarily imply undershoot in the response to a step reference change.

As might be expected, similar concepts apply to multivariable systems. However, whereas in SISO systems one has only the frequency (or time) axis along which to deal with the constraints, in MIMO systems there is also a spatial dimension: one can trade-off limitations between different outputs as well as on a frequency-by-frequency basis. This means that it is also necessary to account for the interactions between outputs, rather than simply being able to focus on one output at a time.

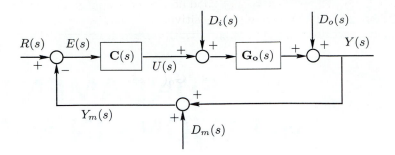

Figure 24.1. MIMO feedback loop

These issues will be explored below.

24.2   Closed-Loop Transfer Function

We consider the MIMO loop of the form shown in Figure 24.1.

We describe the plant model $\mathbf{G_o}(s)$ and the controller $\mathbf{C}(s)$ in LMFD and RMFD form as

$$\mathbf{G_o}(s) = \mathbf{G_{oN}}(s)[\mathbf{G_{oD}}(s)]^{-1} = \left[\, \overline{\mathbf{G}}_{\mathbf{oD}}(s)\right]^{-1}\overline{\mathbf{G}}_{\mathbf{oN}}(s) \qquad (24.2.1)$$

$$\mathbf{C}(s) = \mathbf{C_N}(s)[\mathbf{C_D}(s)]^{-1} = \left[\, \overline{\mathbf{C}}_{\mathbf{D}}(s)\right]^{-1}\overline{\mathbf{C}}_{\mathbf{N}}(s) \qquad (24.2.2)$$

We recall from Lemma 20.3 on page 627 that, for closed-loop stability, it is necessary and sufficient that matrix $\mathbf{A_{cl}}(s)$ be stably invertible, where

$$\mathbf{A_{cl}}(s) \triangleq \overline{\mathbf{G}}_{\mathbf{oD}}(s)\mathbf{C_D}(s) + \overline{\mathbf{G}}_{\mathbf{oN}}(s)\mathbf{C_N}(s) \qquad (24.2.3)$$

For the purpose of the analysis in this chapter, we will continue working under the assumption that the MIMO plant is square, i.e., its model is an $m \times m$ transfer-function matrix. We also assume that $\mathbf{G_o}(s)$ is nonsingular for almost all s and, in particular, that $\det \mathbf{G_o}(0) \neq 0$.

For future use, we denote the ith column of $\mathbf{S_o}(s)$ as $[\mathbf{S_o}(s)]_{*i}$ and the kth row of $\mathbf{T_o}(s)$ as $[\mathbf{T_o}(s)]_{k*}$; so

$$\mathbf{S_o}(s) = \left[[\mathbf{S_o}(s)]_{*1} \quad [\mathbf{S_o}(s)]_{*2} \quad \cdots \quad [\mathbf{S_o}(s)]_{*m}\right]; \qquad \mathbf{T_o}(s) = \begin{bmatrix} [\mathbf{T_o}(s)]_{1*} \\ [\mathbf{T_o}(s)]_{2*} \\ \cdots \\ [\mathbf{T_o}(s)]_{m*} \end{bmatrix}$$

$$(24.2.4)$$

From Chapter 20, we observe that good nominal tracking is, as in the SISO case, connected to the issue of having low sensitivity in certain frequency bands. Upon examining this requirement, we see that it can be met if we can make

$$[\mathbf{I} + \mathbf{G_o}(j\omega)\mathbf{C_o}(j\omega)]^{-1}\mathbf{G_o}(j\omega)\mathbf{C_o}(j\omega) \approx \mathbf{I} \qquad (24.2.5)$$

for all ω in the frequency bands of interest.

24.3   MIMO Internal Model Principle

In SISO control design, a key design objective is usually to achieve zero steady-state errors for certain classes of references and disturbances. However, we have also seen that this requirement can produce secondary effects on the transient behavior of these errors. In MIMO control design, similar features appear, as we next demonstrate.

In Chapter 20, we showed that, to achieve zero steady-state errors to step reference inputs on each channel, we require that

$$\mathbf{T_o}(0) = \mathbf{I} \iff \mathbf{S_o}(0) = \mathbf{0} \qquad (24.3.1)$$

We have seen earlier in the book that a sufficient condition to obtain this result is that we can write the controller as

$$\mathbf{C}(s) = \frac{1}{s}\overline{\mathbf{C}}(s) \qquad \text{where} \qquad \det(\overline{\mathbf{C}}(0)) \neq 0 \qquad (24.3.2)$$

This is usually achieved in practice by placing one integrator in each error channel immediately after we form $e_i(t) = r_i(t) - y_i(t)$, $i = 1, 2, \ldots, m$.

The above idea can be generalized to cover more complex reference inputs (ramps, sinusoids, etc.) and disturbances. All that is needed is to adjoin the appropriate reference or disturbance-generating function. This leads to a MIMO version of the Internal Model Principle.

24.4   The Cost of the Internal Model Principle

As in the SISO case, the Internal Model Principle comes at a cost. As an illustration, the following result extends Lemma 8.1 on page 211 to the multivariable case.

Lemma 24.1. *If zero steady-state errors are required to a ramp reference input on the rth channel, then it is necessary that*

$$\lim_{s \to 0} \frac{1}{s}[\mathbf{S_o}(s)]_{*r} = 0 \qquad (24.4.1)$$

and, as a consequence, in a one-d.o.f. loop,

$$\int_0^\infty e_i^r(t)dt = 0 \qquad i = 1, 2, \ldots, m \qquad\qquad (24.4.2)$$

where $e_i^r(t)$ denotes the error in the ith channel resulting from a step reference input on the rth channel.

Proof

As for Lemma 8.1.

$\square\square\square$

It is interesting to note the essentially multivariable nature of the result in Lemma 24.1: the integral of all channel errors is zero, in response to a step reference in only one channel. We will observe similar situations in the case of RHP poles and zeros.

Furthermore, Lemma 24.1 shows that *all* components of the MIMO plant output will overshoot their stationary values when a step reference change occurs on the rth channel. We emphasize that this is due to the presence of a double integrator in that channel. The only exception to this effect will be for the outputs of those channels for which there is no coupling with channel r.

24.5 RHP Poles and Zeros

In the case of SISO plants, we found that performance limitations are intimately connected to the presence of open-loop RHP poles and zeros. We shall find that this is also true in the MIMO case. As a prelude to developing these results, we first review the appropriate definitions of poles and zeros from Chapter 20.

Consider the plant model $\mathbf{G_o}(s)$ as in (24.2.1). We recall from Chapter 20 that z_o is a zero of $\mathbf{G_o}(s)$, with corresponding left directions $h_1^T, h_2^T, \ldots, h_{\mu_z}^T$, if

$$\det(\mathbf{G_{oN}}(z_o)) = 0 \quad \text{and} \quad h_i^T(\mathbf{G_{oN}}(z_o)) = 0 \quad i = 1, 2, \ldots, \mu_z \qquad (24.5.1)$$

Similarly, we say that η_o is a pole of $\mathbf{G_o}(s)$, with corresponding right directions $g_1, g_2, \ldots, g_{\mu_p}$, if

$$\det(\overline{\mathbf{G}}_{\mathbf{oD}}(\eta_o)) = 0 \quad \text{and} \quad (\overline{\mathbf{G}}_{\mathbf{oD}}(\eta_o))g_i = 0 \quad i = 1, 2, \ldots, \mu_p \qquad (24.5.2)$$

If we now assume that z_o and η_o are not canceled by the controller, then the following lemma holds.

Lemma 24.2. *With z_o and η_o defined as above,*

$$
\begin{array}{lll}
\mathbf{S_o}(\eta_o)g_i = 0 & i = 1, 2, \ldots \mu_p & (24.5.3) \\
\mathbf{T_o}(\eta_o)g_i = g_i & i = 1, 2, \ldots \mu_p & (24.5.4) \\
h_i^T \mathbf{T_o}(z_o) = 0 & i = 1, 2, \ldots \mu_z & (24.5.5) \\
h_i^T \mathbf{S_o}(z_o) = h_i^T & i = 1, 2, \ldots \mu_z & (24.5.6)
\end{array}
$$

Proof

Immediate from using (20.5.3), (20.5.4), (24.5.1), (24.5.2), and the identity $\mathbf{S_o}(s) + \mathbf{T_o}(s) = \mathbf{I}$.

□□□

We see that, as in the SISO case, open-loop poles (i.e., the poles of $\mathbf{G_o}(s)\mathbf{C}(s)$) become zeros of $\mathbf{S_o}(s)$, and open-loop zeros (i.e., the zeros of $\mathbf{G_o}(s)\mathbf{C}(s)$) become zeros of $\mathbf{T_o}(s)$. However, we must now consider the associated directional properties to have a better understanding of MIMO performance issues.

24.6   Time-Domain Constraints

We saw in Lemma 8.3 that the presence of RHP poles and zeros had certain implications for the time responses of closed-loop systems. We have the following MIMO version of Lemma 8.3.

Lemma 24.3. *Consider a MIMO feedback control loop having stable closed-loop poles located to the left of* $-\alpha$ *for some* $\alpha > 0$. *Also, assume that zero steady-state error occurs for reference step inputs in all channels. Then, for a plant zero* z_o *with left directions* $h_1^T, h_2^T, \ldots, h_{\mu_z}^T$ *and a plant pole* η_o *with right directions* $g_1, g_2, \ldots, g_{\mu_p}$ *satisfying* $\Re(z_o) > -\alpha$ *and* $\Re(\eta_o) > -\alpha$, *we have the following:*

(i) *For a positive unit reference step on the* r^{th} *channel,*

$$
\int_0^\infty h_i^T e(t) e^{-z_o t} dt = \frac{h_{ir}}{z_o}; \qquad i = 1, 2, \ldots, \mu_z \qquad (24.6.1)
$$

where h_{ir} *is the* r^{th} *component of* h_i.

(ii) *For a (positive or negative) unit-step output disturbance in direction* g_i, $i = 1, 2, \ldots, \mu_p$, *the resulting error,* $e(t)$, *satisfies*

$$
\int_0^\infty e(t) e^{-\eta_o t} dt = 0 \qquad (24.6.2)
$$

(iii) *For a (positive or negative) unit reference step in the rth channel, and provided that z_o is in the RHP,*

$$\int_0^\infty h_i^T y(t) e^{-z_o t} dt = 0; \qquad i = 1, 2, \ldots, \mu_z \qquad (24.6.3)$$

Proof

(i) We know that

$$E(s) = \mathbf{S_o}(s) R(s) = \int_0^\infty e(t) e^{-st} dt \qquad (24.6.4)$$

If we left-multiply both sides by h_i^T, we obtain

$$h_i^T E(s) = h_i^T \mathbf{S_o}(s) R(s) = h_i^T \mathbf{S_o}(s) \frac{R_o}{s} = \int_0^\infty h_i^T e(t) e^{-st} dt \qquad (24.6.5)$$

where R_o is a constant vector. Evaluating equation (24.6.5) at $s = z_o$ (which lies in the region of convergence of the Laplace transform), and using (24.5.6), we have that

$$h_i^T E(z_o) = h_i^T \mathbf{S_o}(z_o) \frac{R_o}{z_o} = h_i^T \frac{R_o}{z_o} = \int_0^\infty h_i^T e(t) e^{-z_o t} dt \qquad (24.6.6)$$

from which the result follows upon taking all references equal to zero save for the one on the rth channel.

(ii) We know that, for the given output disturbance $D_o(s) = \frac{1}{s} g_i$, the error is given by

$$E(s) = -\mathbf{S_o}(s) D_o(s) = -\mathbf{S_o}(s) \frac{1}{s} g_i = \int_0^\infty e(t) e^{-st} dt \qquad (24.6.7)$$

from which the result follows from evaluating (24.6.7) at $s = \eta_o$ and using (24.5.3).

(iii) The proof relies on the fact that the region of convergence of the Laplace Transform for $y(t)$ and $r(t)$ includes the open RHP, where z_o lies.

Then

$$Y(s) = \mathbf{T_o}(s) R(s) = \int_0^\infty y(t) e^{-st} dt \qquad (24.6.8)$$

If we left-multiply both sides by h_i^T, we obtain

$$h_i^T Y(s) = h_i^T \mathbf{T_o}(s) R(s) = \int_0^\infty h_i^T y(t) e^{-st} dt \qquad (24.6.9)$$

The result follows upon evaluating (24.6.9) at $s = z_o$ and using (24.5.5).

□□□

Remark 24.1. *When z_o and η_o are in the RHP, parts (ii) and (iii) of Lemma 24.3 hold for any input whose Laplace Transform converges for $\Re(s) > 0$.*

□□□

Comparing Lemma 24.3 with Lemma 8.3 clearly shows the multivariable nature of these constraints. For example, (24.6.2) holds for disturbances coming from a particular direction. Also, (24.6.1) applies to particular combinations of the errors. Thus, the undershoot property can (sometimes) be shared amongst different error channels, depending on the directionality of the zeros.

Example 24.1 (Quadruple-tank apparatus continued). *Consider again the quadruple-tank apparatus in Example 20.4 on page 623 and Example 21.3 on page 658. For the case $\gamma_1 = 0.43$, $\gamma_2 = 0.34$, there is a nonminimum-phase zero at $z_o = 0.0229$. The associated left zero direction is approximately $\begin{bmatrix} 1 & -1 \end{bmatrix}$.*

Hence, from Lemma 24.3 on page 775, we have

$$\int_0^\infty (y_1(t) - y_2(t)) e^{-z_o t} dt = 0 \qquad (24.6.10)$$

$$\int_0^\infty (e_1(t) - e_2(t)) e^{-z_o t} dt = \frac{(-1)^{i-1}}{z_o} \qquad (24.6.11)$$

for a unit step in the ith channel reference.

The zero in this case is an interaction zero; *hence, we do not necessarily get undershoot in the response. However, there are constraints on the extent of interaction that must occur. This explains the high level of interaction observed in Figure 21.4 on page 660.*

We actually see from (24.6.10) and (24.6.11) that there are two ways one can deal with this constraint.

(i) *If we allow coupling in the final response, then we can spread the constraint between outputs; i.e., we can satisfy (24.6.10) and (24.6.11) by having $y_2(t)$, and hence $e_2(t)$, respond when a step is applied to channel 1 reference, and vice-versa. This might allow us to avoid undershoot, at the expense of having interaction. The amount of interaction needed grows as the bandwidth increases beyond z_o.*

(ii) *If we design and achieve (near) decoupling, then only one of the outputs can be nonzero after each individual reference changes.*

This implies that undershoot must occur in this case. Also, we see that under-shoot will occur in both channels (i.e., the effect of the single RHP zero now influences both channels). This is an example of spreading *resulting from dynamic decoupling.*

As usual, the amount of undershoot required to satisfy (24.6.10) and (24.6.11) grows as the bandwidth increases beyond z_o.

□□□

It is interesting to see the impact of dynamic decoupling on (24.6.1). If we can achieve a design with this property (a subject to be analyzed in greater depth in Chapter 26), then it necessarily follows, that for a reference step in the rth channel, there will be no effect on the other channels:

$$e_k(t) = 0 \qquad \text{for} \quad \forall k \neq r, \quad \forall t > 0 \tag{24.6.12}$$

Then (24.6.1) gives

$$\int_0^\infty h_{ir} e_r(t) e^{-z_o t} dt = \frac{h_{ir}}{z_o}; \qquad i = 1, 2, \ldots, \mu_z \tag{24.6.13}$$

or, for $h_{ir} \neq 0$,

$$\int_0^\infty e_r(t) e^{-z_o t} dt = \frac{1}{z_o} \tag{24.6.14}$$

which is exactly the constraint applicable to the SISO case.

We thus conclude that dynamic decoupling removes the possibility of sharing the *zero constraint* amongst different error channels. Another interesting observation is that (24.6.14) holds for every z_o having a direction with *any* nonzero component in the rth position. Thus, whereas (24.6.1) holds for a combination of errors, (24.6.14) applies for a scalar error, and there is a separate constraint for every zero having a direction with any nonzero component in the rth position. Thus, (24.6.14) implies that one zero might appear in more than one error constraint. This can be thought of as a cost of decoupling. The only time that a zero does not *spread its influence* over many channels is when the corresponding zero direction has only one nonzero component. We then say that the corresponding zero direction is *canonical*.

Example 24.2. *Consider the following transfer function:*

$$\mathbf{G_o}(s) = \begin{bmatrix} \dfrac{s-1}{(s+1)^2} & \dfrac{2(s-1)}{(s+1)^2} \\[2ex] \dfrac{1}{(s+1)^2} & \dfrac{\epsilon}{(s+1)^2} \end{bmatrix} \tag{24.6.15}$$

We see that $z_o = 1$ is a zero with direction $h_1^T = \begin{bmatrix} 1 & 0 \end{bmatrix}$ that we see is canonical. In this case, (24.6.1) gives

$$\int_0^\infty \begin{bmatrix} 1 & 0 \end{bmatrix}^T e(t) e^{-t} dt = \int_0^\infty e_1(t) e^{-t} dt = \frac{1}{z_o} = 1 \qquad (24.6.16)$$

for a step input on the first channel. Note that, in this case, this is the same as (24.6.14), i.e., there is no additional cost to decoupling. However, if we instead consider the plant

$$\mathbf{G_o}(s) = \begin{bmatrix} \dfrac{s-1}{(s+1)^2} & \dfrac{1}{(s+1)^2} \\[3mm] \dfrac{2(s-1)}{(s+1)^2} & \dfrac{\epsilon}{(s+1)^2} \end{bmatrix} \qquad (24.6.17)$$

then the situation changes significantly.

In this case, $z_o = 1$ is a zero with direction $h_1^T = \begin{bmatrix} \epsilon & -1 \end{bmatrix}$ that we see is non-canonical. Thus, (24.6.1) gives for a step reference in the first channel that

$$\int_0^\infty \begin{bmatrix} \epsilon & -1 \end{bmatrix}^T e(t) e^{-t} dt = \int_0^\infty (\epsilon e_1(t) - e_2(t)) e^{-t} dt = \frac{\epsilon}{z_o} = \epsilon \qquad (24.6.18)$$

and for a step reference in the second channel that

$$\int_0^\infty (\epsilon e_1(t) - e_2(t)) e^{-t} dt = \frac{-1}{z_o} = -1 \qquad (24.6.19)$$

If, on the other hand, we insist on dynamic decoupling, we obtain for a unit step reference in the first channel that

$$\int_0^\infty e_1(t) e^{-t} dt = \frac{1}{z_o} = 1 \qquad (24.6.20)$$

and for a step reference in the second channel that

$$\int_0^\infty e_2(t) e^{-t} dt = \frac{1}{z_o} = 1 \qquad (24.6.21)$$

Clearly, in this example, a small amount of coupling from channel 1 into channel 2 can be very helpful in satisfying (24.6.18), when $\epsilon \neq 0$, whereas (24.6.20) and (24.6.21) requires the zero to be dealt with, in full, in both scalar channels.

□□□

The time-domain constraints explored above are also matched by frequency-domain constraints that are the MIMO extensions of the SISO results presented in Chapter 9. These are explored next.

24.7 Poisson Integral Constraints on MIMO Complementary Sensitivity

We will develop the MIMO versions of results presented in section §9.5.

Note that the vector $\mathbf{T_o}(s)g_i$ can be premultiplied by a matrix $\mathbf{B_i}(s)$ to yield a vector $\tau_i(s)$:

$$\tau_i(s) = \mathbf{B_i}(s)\mathbf{T_o}(s)g_i = \begin{bmatrix} \tau_{i1}(s) \\ \tau_{i2}(s) \\ \vdots \\ \tau_{im}(s) \end{bmatrix} ; \qquad i = 1, 2, \ldots, \mu_p \qquad (24.7.1)$$

where $\mathbf{B_i}(s)$ is a diagonal matrix in which each diagonal entry, $[\mathbf{B_i}(s)]_{jj}$, is a scalar inverse Blaschke product, constructed so that $\ln(\tau_{ij}(s))$ is an analytic function in the open RHP. This means that

$$[\mathbf{B_i}(s)]_{jj} = \prod_{k=1}^{r_j} \frac{s + z_{jk}^*}{s - z_{jk}^*} \qquad (24.7.2)$$

where z_{jk}, $k = 1, 2, \ldots, r_j$, are the NMP zeros of the jth element of vector $\mathbf{T_o}(s)g_i$.

We also define a column vector $\bar{g}_i(s)$ as follows:

$$\bar{g}_i(s) = \mathbf{B_i}(s)g_i = \begin{bmatrix} [\mathbf{B_i}(s)]_{11}g_{i1}(s) \\ \vdots \\ [\mathbf{B_i}(s)]_{mm}g_{im}(s) \end{bmatrix} = \begin{bmatrix} \bar{g}_{i1}(s) \\ \vdots \\ \bar{g}_{im}(s) \end{bmatrix} \qquad (24.7.3)$$

We next define a set of integers, ∇_i, corresponding to the indices of the nonzero elements of g_i:

$$\nabla_i = \{r | g_{ir} \neq 0\}; \qquad i = 1, 2, \ldots, \mu_p \qquad (24.7.4)$$

We can now state a theorem that describes the constraints imposed on the complementary sensitivity by the presence of unstable open-loop poles.

Theorem 24.1 (Complementary sensitivity and unstable poles). *Consider a MIMO system with an unstable pole located at $s = \eta_o = \alpha + j\beta$ and having associated directions $g_1, g_2, \ldots, g_{\mu_p}$; then*

(i)

$$\frac{1}{\pi} \int_{-\infty}^{\infty} \ln|[\mathbf{T_o}(j\omega)]_{r*}g_i| d\Omega(\eta_o, \omega) = \ln|B_{ir}(\eta_o)g_{ir}|; \qquad r \in \nabla_i; \qquad i = 1, 2, \ldots, \mu_p$$

$$(24.7.5)$$

(ii)

$$\frac{1}{\pi} \int_{-\infty}^{\infty} \ln|[\mathbf{T_o}(j\omega)]_{r*}g_i| \, d\Omega(\eta_o, \omega) \geq \ln|g_{ir}|; \qquad r \in \nabla_i; \qquad i = 1, 2, \dots, \mu_p$$

(24.7.6)

where

$$d\Omega(\eta_o, \omega) = \frac{\alpha}{\alpha^2 + (\omega - \beta)^2} \, d\omega \implies \int_{-\infty}^{\infty} d\Omega(\eta_o, \omega) = \pi \qquad (24.7.7)$$

Proof

(i) *We first note that* $\ln(\tau_{ir}(s)) = \ln([\mathbf{B_i}(s)]_{rr}[\mathbf{T_o}(s)]_{r*}g_i)$ *is a scalar function, analytic in the open RHP.*

If we evaluate $\ln(\tau_{ir}(s))$ *at* $s = \eta_o$, *then, upon using (24.5.4) and (24.7.3), we obtain*

$$\ln(\tau_{ir}(\eta_o)) = \ln([\mathbf{B_i}(\eta_o)]_{rr}[\mathbf{T_o}(\eta_o)]_{r*}g_i) = \overline{g}_{ir}(\eta_o) \qquad (24.7.8)$$

We can next apply the Poisson-Jensen formula given in Lemma C.1. The result then follows.

(ii) *We note that, because* z_{ij} *and* η_o *are in the RHP, then* $|[\mathbf{B_i}(\eta_o)]_{rr}| \geq 1$; *thus,*

$$\ln|[\mathbf{B_i}(\eta_o)]_{rr}g_{ir}| = \ln|[\mathbf{B_i}(\eta_o)]_{rr}|\ln|g_{ir}| \geq \ln|g_{ir}| \qquad (24.7.9)$$

Remark 24.2. *Although (24.7.5) is a more precise conclusion than (24.7.6), it is a constraint that depends on the controller. The result presented in the following corollary is independent of the controller.*

$\square\square\square$

Corollary 24.1. *Consider theorem 24.1; then (24.7.6) can also be written as*

$$\int_{-\infty}^{\infty} \ln|[\mathbf{T_o}(j\omega)]_{rr}| \, d\Omega(\eta_o, \omega) \geq \int_{-\infty}^{\infty} \ln\left|\frac{[\mathbf{T_o}(j\omega)]_{rr}g_{ir}}{\sum_{k \in \nabla}[\mathbf{T_o}(j\omega)]_{rk}g_{ik}}\right| d\Omega(\eta_o, \omega) \quad (24.7.10)$$

Proof

We first note that

$$\ln|[\mathbf{To}(j\omega)]_{r*}g_i| = \ln\left|\sum_{k \in \nabla} g_{ik}[\mathbf{T_o}(j\omega)]_{kr}\right| = \ln|[\mathbf{T_o}(j\omega)]_{rr}| + \ln(g_{ir})$$

$$- \ln\left|\frac{g_{ir}[\mathbf{T_o}(j\omega)]_{rr}}{\sum_{k \in \nabla} g_{ik}[\mathbf{T_o}(j\omega)]_{kr}}\right| \qquad (24.7.11)$$

The result follows upon using (24.7.11) in (24.7.6) and upon using the property of the Poisson kernel in (24.7.7).

□□□

24.8   Poisson Integral Constraints on MIMO Sensitivity

When the plant has NMP zeros, a result similar to that in Theorem 24.1 on page 780 can be established for the sensitivity function, $\mathbf{S_o}(s)$.

We first note that the vector $h_i^T \mathbf{S_o}(s)$ can be postmultiplied by a matrix $\mathbf{B_i'}(s)$ to yield a vector $v_i(s)$:

$$v_i(s) = h_i^T \mathbf{S_o}(s)\mathbf{B_i'}(s) = \begin{bmatrix} v_{i1}(s) & v_{i2}(s) & \cdots & v_{im}(s) \end{bmatrix}; \qquad i = 1, 2, \ldots, \mu_z \tag{24.8.1}$$

where $\mathbf{B_i'}(s)$ is a diagonal matrix in which each diagonal entry, $[\mathbf{B_i'}(s)]_{jj}$, is a scalar inverse Blaschke product, constructed so that $\ln(v_{ij}(s))$ is an analytic function in the open RHP. This means that

$$[\mathbf{B_i'}(s)]_{jj} = \prod_{k=1}^{r_j} \frac{s + p_{jk}^*}{s - p_{jk}^*} \tag{24.8.2}$$

where p_{jk}, $k = 1, 2, \ldots, r_j'$, are the unstable poles of the jth element of the row vector $h_i^T \mathbf{S_o}(s)$.

We also define a row vector, $\overline{h}_i(s)$, where

$$\overline{h}_i(s) = h_i^T(s)\mathbf{B_i'}(s) = \begin{bmatrix} h_{i1}^T(s)[\mathbf{B_i'}(s)]_{11} \\ \vdots \\ h_{im}^T(s)[\mathbf{B_i'}(s)]_{mm} \end{bmatrix} = \begin{bmatrix} \overline{h}_{i1}(s) & \cdots & \overline{h}_{im}(s) \end{bmatrix} \tag{24.8.3}$$

We next define a set of integers ∇_i' corresponding to the indices of the nonzero elements of h_i:

$$\nabla_i' = \{r | h_{ir} \neq 0\}; \qquad i = 1, 2, \ldots, \mu_z \tag{24.8.4}$$

We then have the following result.

Theorem 24.2 (Sensitivity and NMP zeros). *Consider a MIMO plant having a NMP zero at $s = z_o = \gamma + j\delta$, with associated directions $h_1^T, h_2^T, \ldots, h_{\mu_z}^T$; then the sensitivity in any control loop for that plant satisfies*

(i)

$$\frac{1}{\pi} \int_{-\infty}^{\infty} \ln|h_i^T[\mathbf{S_o}(j\omega)]_{*r}|d\Omega(z_o, \omega) = \ln|h_{ir}[\mathbf{B_i'}(\mathbf{z_o})]_{rr}|; \qquad r \in \nabla_i'; \qquad i = 1, 2, \ldots, \mu_p$$

$$\tag{24.8.5}$$

and

(ii)

$$\frac{1}{\pi}\int_{-\infty}^{\infty}\ln|h_i^T[\mathbf{S_o}(j\omega)]_{*r}|d\Omega(z_o,\omega) \geq \ln|h_{ir}|; \quad r \in \nabla_i'; \quad i = 1, 2, \ldots, \mu_p$$

$$(24.8.6)$$

where

$$d\Omega(z_o,\omega) = \frac{\gamma}{\gamma^2 + (\omega - \delta)^2}d\omega \implies \int_{-\infty}^{\infty}d\Omega(z_o,\omega) = \pi \qquad (24.8.7)$$

Proof

The proof follows along the same lines as that of theorem 24.1.

□□□

Corollary 24.2. *Consider the set-up in Theorem 24.2 on the facing page. Then (24.8.6) can also be written as*

$$\int_{-\infty}^{\infty}\ln|[\mathbf{S_o}(j\omega)]_{rr}|\,d\Omega(z_o,\omega) \geq \int_{-\infty}^{\infty}\ln\left|\frac{h_{ir}[\mathbf{S_o}(j\omega)]_{rr}}{\sum_{k\in\nabla'}h_{ik}[\mathbf{S_o}(j\omega)]_{kr}}\right|d\Omega(z_o,\omega) \quad (24.8.8)$$

Proof

We first note that

$$\ln\left|h_i^T[\mathbf{S_o}(j\omega)]_{*r}\right| = \ln\left|\sum_{k\in\nabla'}h_{ik}[\mathbf{S_o}(j\omega)]_{kr}\right| = \ln|[\mathbf{S_o}(j\omega)]_{rr}| + \ln(h_{ir})$$

$$- \ln\left|\frac{h_{ir}[\mathbf{S_o}(j\omega)]_{rr}}{\sum_{k\in\nabla'}h_{ik}[\mathbf{S_o}(j\omega)]_{kr}}\right| \qquad (24.8.9)$$

The result follows upon using (24.8.9) in (24.8.6) and using the property of the Poisson kernel in (24.8.7).

□□□

24.9   Interpretation

Theorem 24.2 on the preceding page shows that in MIMO systems, as is the case in SISO systems, there is a sensitivity trade-off along a frequency-weighted axis. To explore the issue further, we consider the following lemma.

Lemma 24.4. *Consider the lth column $(l \in \nabla'_i)$ in the expression (24.8.6), i.e., the case when the lth sensitivity column, $[\mathbf{S_o}]_{*l}$, is considered. Furthermore, assume that some design specifications require that*

$$|\mathbf{S_o}(j\omega)|_{kl} \le \epsilon_{kl} \ll 1 \qquad \forall \omega \in [0, \omega_c]; k = 1, 2, \ldots, m \qquad (24.9.1)$$

Then the following inequality must be satisfied:

$$\|[\mathbf{S_o}]_{ll}\|_\infty + \sum_{\substack{k=1 \\ k \ne l}}^{m} \left| \frac{h_{ik}}{h_{il}} \right| \|[\mathbf{S_o}]_{kl}\|_\infty \ge \left(\epsilon_{ll} + \sum_{\substack{k=1 \\ k \ne l}}^{m} \left| \frac{h_{ik}}{h_{il}} \right| \epsilon_{kl} \right)^{-\dfrac{\psi(\omega_c)}{\pi - \psi(\omega_c)}} \qquad (24.9.2)$$

where

$$\psi(\omega_c) = \int_{o}^{\omega_c} \left[\frac{\gamma}{\gamma^2 + (\omega - \delta)^2} + \frac{\gamma}{\gamma^2 + (\omega + \delta)^2} \right] d\omega \qquad (24.9.3)$$

Proof

We first note that

$$|h_i^T[\mathbf{S_o}(j\omega)]_{*l}| \le \sum_{k=1}^{m} |h_{ik}|[\mathbf{S_o}(j\omega)]_{kl}$$

$$\le \begin{cases} L_c \triangleq \sum_{k=1}^{m} |h_{ik}|\epsilon_{kl} & for\ -\omega_c \le \omega \le \omega_c \\ L_\infty \triangleq \sum_{k=1}^{m} |h_{ik}|\|[\mathbf{S_o}]_{kl}\|_\infty & for\ |\omega| > \omega_c \end{cases} \qquad (24.9.4)$$

Consider now equation (24.8.6). We first split the integration interval, $(-\infty, \infty)$, into $(-\infty, -\omega_c) \cup [-\omega_c, 0) \cup [0, \omega_c) \cup [\omega_c, \infty)$, and then we apply (24.8.6), replacing $|h_i^T[\mathbf{S_o}(j\omega)]_{*r}|$ by its upper bound in each interval:

$$\frac{1}{\pi} \int_{-\infty}^{\infty} \ln|h_i^T[\mathbf{S_o}(j\omega)]_{*r}|d\Omega(z_o, \omega) \le \frac{\ln(L_\infty)}{\pi} \int_{-\infty}^{-\omega_c} d\Omega(z_o, \omega)$$

$$+ \frac{\ln(L_c)}{\pi} \int_{-\omega_c}^{\omega_c} d\Omega(z_o, \omega) + \frac{\ln(L_\infty)}{\pi} \int_{\omega_c}^{\infty} d\Omega(z_o, \omega)$$

$$= \frac{\ln(L_\infty)}{\pi} \int_{-\infty}^{\infty} d\Omega(z_o, \omega) + \frac{\ln(L_c) - \ln(L_\infty)}{\pi} \int_{-\omega_c}^{\omega_c} d\Omega(z_o, \omega) \qquad (24.9.5)$$

We then note, upon using (24.8.7) and (24.9.3), that

$$\frac{\ln(L_c) - \ln(L_\infty)}{\pi} \int_{-\omega_c}^{\omega_c} d\Omega(z_o, \omega) = \frac{\ln(L_c) - \ln(L_\infty)}{\pi} \psi(\omega_c) \qquad (24.9.6)$$

and, upon using (24.8.7)

$$\frac{\ln(L_\infty)}{\pi} \int_{-\infty}^{\infty} d\Omega(z_o, \omega) = \ln(L_\infty) \qquad (24.9.7)$$

If we now use (24.9.4)-(24.9.7) *in* (24.8.6), *we obtain*

$$\pi \ln(|h_{il}|) \leq (\pi - \psi(\omega_c)) \ln(L_\infty) + \psi(\omega_c) \ln(L_c) \qquad (24.9.8)$$

And the result follows upon using the definition of L_c and L_∞ from (24.9.4).

□□□

These results are similar to those derived for SISO control loops, because we also obtain lower bounds for sensitivity peaks. Furthermore, these bounds grow with bandwidth requirements, i.e., when we want ϵ_{kl} smaller and/or when we specify a larger ω_c.

However, a major difference is that in the MIMO case the bound in (24.9.2) refers to a linear combination of sensitivity peaks. This combination is determined by the directions associated to the NMP zero under consideration. We will further embellish this result in section §26.6, where we will add the additional constraint that the closed-loop system should be fully diagonally decoupled.

24.10   An Industrial Application: Sugar Mill

24.10.1   Model Description

In this section, we consider the design of a controller for a typical industrial process. It has been chosen because it includes significant multivariable interaction, a nonself regulating nature, and nonminimum-phase behavior. A detailed description of the process modeling and control can be found in the reference given at the end of the chapter.

The sugar mill unit under consideration constitutes one of multiple stages in the overall process. A schematic diagram of the Milling Train is shown in Figure 24.2. A single stage of this Milling Train is shown in Figure 24.3 on page 787.

In this mill, bagasse (in intermediate form), which is milled sugar cane, is delivered via the intercarrier from the previous mill stage. This is then fed into the buffer chute. The material is fed into the pressure-feed chute via a flap mechanism and pressure-feed rolls. The flap mechanism can be manipulated via a hydraulic actuator to modify the exit geometry and volume of the buffer chute. The pressure-feed rolls grip the bagasse and force it into the pressure-feed chute.

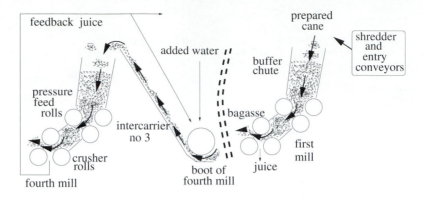

Figure 24.2. A Sugar Milling Train

Cane is then fed to the pressure-feed chute through the mill rolls. The pressure-feed chute increases the compression ratio of the bagasse, because the pressure-feed roll surface speed is slightly higher than the mill roll speed. The mill rolls are grooved so that no expressed juice can escape. The grooves are given a rough surface so that the mill rolls can grip the cane.

The mill rolls and pressure-feed rolls are driven by a steam turbine through a gear system. The steam turbine speed is controlled via a governor feedback system with an adjustable set-point.

For the purpose of maximal juice extraction, the process requires the control of two quantities: the buffer chute height, $h(t)$, and the mill torque, $\tau(t)$. For the control of these variables, the flap position, $f(t)$, and the turbine speed set-point, $\omega(t)$, may be used. For control purposes, this plant can thus be modeled as a MIMO system with 2 inputs and 2 outputs. In this system, the main disturbance, $d(t)$, originates in the variable feed to the buffer chute.

In this example, regulation of the height in the buffer chute is less important for the process than regulation of the torque. Indeed, the purpose of this chute is typically to filter out sudden volume changes in the feed.

After the applying of phenomenological considerations and the performing of different experiments with incremental step inputs, a linearized plant model was obtained. The outcome of the modeling stage is shown in Figure 24.4.

From Figure 24.4, we can compute the nominal plant model in RMFD form, linking the inputs $f(t)$ and $\omega(t)$ to the outputs $\tau(t)$ and $h(t)$ as

$$\mathbf{G_o}(s) = \mathbf{G_{oN}}(s)[\mathbf{G_{oD}}(s)]^{-1} \qquad (24.10.1)$$

where

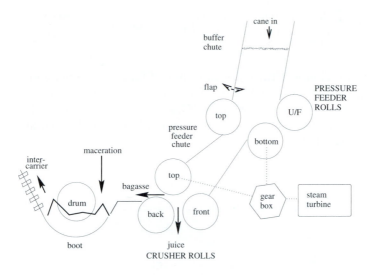

Figure 24.3. Single crushing mill

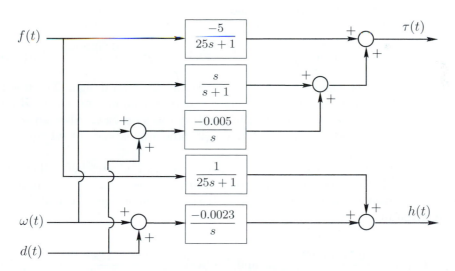

Figure 24.4. Sugar mill linearized block model

$$\mathbf{G_{oN}}(s) = \begin{bmatrix} -5 & s^2 - 0.005s - 0.005 \\ 1 & -0.0023(s+1) \end{bmatrix}; \qquad \mathbf{G_{oD}}(s) = \begin{bmatrix} 25s+1 & 0 \\ 0 & s(s+1) \end{bmatrix}$$

$$(24.10.2)$$

We can now compute the poles and zeros of $\mathbf{G_o}(s)$. The poles of $\mathbf{G_o}(s)$ are the zeros of $\mathbf{G_{oD}}(s)$, i.e., $(-1, -0.04, 0)$. The zeros of $\mathbf{G_o}(s)$ are the zeros of $\mathbf{G_{oN}}(s)$, i.e., the values of s that are roots of $\det(\mathbf{G_{oN}}(s)) = 0$; this leads to $(-0.121, 0.137)$. Note that the plant model has a *nonminimum-phase zero*, located at $s = 0.137$.

We also have that

$$\mathbf{G_o}(0.137) = \begin{bmatrix} -1.13 & 0.084 \\ 0.226 & -0.0168 \end{bmatrix} \qquad (24.10.3)$$

From (24.10.3) we have that the direction associated with the NMP zero is given by

$$h^T = \begin{bmatrix} 1 & 5 \end{bmatrix} \qquad (24.10.4)$$

24.10.2 SISO Design

Before attempting any MIMO design, we start by examining a SISO design using two separate PID controllers. In this approach, we initially ignore the cross-coupling terms in the model transfer function $\mathbf{G_o}(s)$, and we carry out independent PID designs for the resulting two SISO models

$$G_{11}(s) = \frac{-5}{25s+1}; \qquad \text{and} \qquad G_{22}(s)\frac{-0.0023}{s} \qquad (24.10.5)$$

An iterative procedure was used to determine the final design. The iterations were initially guided by a desire to obtain fast responses in both loops. However, the design was subsequently refined by limiting the loop bandwidths so as not to exceed the magnitude of the NMP zero. (Bandwidths of around $0.1[rad/s]$ were achieved.)

Note that for the second loop we decided to use a PID instead of a PI controller, because $G_{22}(s)$ has one pole at the origin, which suggests that two controller zeros are more convenient to partially compensate the additional lag. The final controllers were given by

$$C_1(s) = -\frac{0.5s + 0.02}{s}; \qquad \text{and} \qquad C_2(s) = -\frac{20s^2 + 10s + 0.2}{s^2 + s} \qquad (24.10.6)$$

To illustrate the limitations of this approach and the associated trade-offs, Figure 24.5 shows the performance of the loop under the resultant SISO-designed PID controllers.

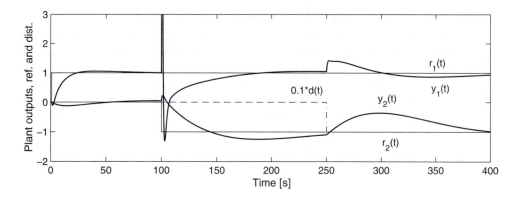

Figure 24.5. Loop performance with SISO design

In this simulation, the (step) references and disturbance were set as follows:

$$r_1(t) = \mu(t-1); \qquad r_2(t) = \mu(t-100); \qquad d(t) = -10\mu(t-250) \quad (24.10.7)$$

The following observations follow from the results shown in Figure 24.5.

(i) Interaction between the loops is strong. In particular, we observe that a reference change in channel 2 (height) will induce strong perturbations of the output in channel 1 (torque).

(ii) Both outputs exhibit nonminimum-phase behavior. However, due to the design-imposed limitation on the bandwidth, this is not very strong in either of the outputs in response to a change in its own reference. Notice, however, that the transient in y_1 in response to a reference change in r_2 is–because of the interaction neglected in the design–clearly of nonminimum phase.

(iii) The effects of the disturbance on the outputs show mainly low-frequency components. This is due to the fact that abrupt changes in the feed rate are filtered out by the buffer.

The above design suffers from a problem: that equal importance has been given to the control of both variables. This leads to the undesirable feature that the performance of channel 1 is severely perturbed by changes in the second channel. For that reason, we will next explore a design approach that accounts for the interacting structure of this plant.

Remark 24.3. *The reader is encouraged to try different PID settings for this problem, using SIMULINK file* **sugpid.mdl**. *It is also interesting to verify that, by introducing input saturation in both channels, the dynamics of both loops is significantly slowed down.*

24.10.3   MIMO Design. Preliminary Considerations

We now consider a full MIMO design. We begin by analyzing the main issues that will affect the MIMO design. They can be summarized in the following considerations.

(i) The compensation of the *input* disturbance requires that integration be included in the controller to be designed.

(ii) To ensure internal stability, the NMP zero must not be canceled by the controller. Thus, $\mathbf{C}(s)$ should not have poles at $s = 0.137$.

(iii) In order to avoid the possibility of input saturation, the bandwidth should be limited. We will work in the range of 0.1-$0.2[rad/s]$.

(iii) The location of the NMP zero suggests that the dominant mode in the channel(s) affected by that zero should not be faster than $e^{-0.137t}$. Otherwise, responses to step reference and step input disturbances will exhibit significant undershoot. (See section §24.6.)

(iv) From (24.10.4), the left direction, $h^T = \begin{bmatrix} 1 & 5 \end{bmatrix}$, associated with the NMP zero is not a canonical direction. Hence, if dynamic decoupling is attempted, the NMP zero will affect both channels. (See section §26.5.) Thus, a controller having triangular structure is desirable to restrict the NMP zero to only one channel.

24.10.4   MIMO Design. Dynamic Decoupling

The analysis in the previous section leads to the conclusion that a triangular design is the best choice. However, we first consider a fully dynamic decoupled design, to provide a benchmark against which the triangular decoupled design can be compared.

We choose

$$\mathbf{M}(s) = \mathbf{G_o}(s)\mathbf{C}(s) = \begin{bmatrix} M_{11}(s) & 0 \\ 0 & M_{22}(s) \end{bmatrix} \tag{24.10.8}$$

which leads to the complementary sensitivity

$$\mathbf{T_o}(s) = \begin{bmatrix} T_{11}(s) & 0 \\ 0 & T_{22}(s) \end{bmatrix} = \begin{bmatrix} \dfrac{M_{11}(s)}{1 + M_{11}(s)} & 0 \\ 0 & \dfrac{M_{22}(s)}{1 + M_{22}(s)} \end{bmatrix} \tag{24.10.9}$$

Then

$$\mathbf{C}(s) = \begin{bmatrix} C_{11}(s) & C_{12}(s) \\ C_{21}(s) & C_{22}(s) \end{bmatrix} = [\mathbf{G_o}(s)]^{-1}\mathbf{M}(s) \qquad (24.10.10)$$

Also,

$$[\mathbf{G_o}(s)]^{-1} = \begin{bmatrix} 25s+1 & 0 \\ 0 & s \end{bmatrix} \begin{bmatrix} \dfrac{n_{11}(s)}{d^-(s)d^+(s)} & \dfrac{n_{12}(s)}{d^-(s)d^+(s)} \\[2ex] \dfrac{n_{21}(s)}{d^-(s)d^+(s)} & \dfrac{n_{22}(s)}{d^-(s)d^+(s)} \end{bmatrix} \qquad (24.10.11)$$

where

$$n_{11}(s) = 0.0023(s+1) \qquad n_{12}(s) = s^2 - 0.005s - 0.005 \qquad (24.10.12)$$
$$n_{21}(s) = s+1 \qquad\qquad n_{22}(s) = 5(s+1) \qquad\qquad\qquad (24.10.13)$$
$$d^-(s) = s+0.121 \qquad\quad d^+(s) = s - 0.137 \qquad\qquad\qquad (24.10.14)$$

Then, from (24.10.8) and (24.10.10), the controller is given by

$$\mathbf{C}(s) = \begin{bmatrix} \dfrac{(25s+1)n_{11}(s)M_{11}(s)}{d^-(s)d^+(s)} & \dfrac{(25s+1)n_{12}(s)M_{22}(s)}{d^-(s)d^+(s)} \\[3ex] \dfrac{sn_{21}M_{11}(s)}{d^-(s)d^+(s)} & \dfrac{sn_{22}(s)M_{22}(s)}{d^-(s)d^+(s)} \end{bmatrix} \qquad (24.10.15)$$

$\mathbf{C}(s)$ should not have poles at $s = 0.137$, so the polynomial $d^+(s)$ should be canceled in the *four fraction matrix entries* in (24.10.15). This implies that

$$M_{11}(0.137) = M_{22}(0.137) = 0 \qquad (24.10.16)$$

Furthermore, we need to completely compensate the input disturbance, so we require integral action in the controller (in addition to the integral action in the plant). We thus make the following choices:

$$M_{11}(s) = \frac{(s-0.137)p_{11}(s)}{s^2 l_{11}(s)}; \qquad \text{and} \qquad M_{22}(s) = \frac{(s-0.137)p_{22}(s)}{s^2 l_{22}(s)} \qquad (24.10.17)$$

where $p_{11}(s), l_{11}(s), l_{22}(s),$ and $p_{22}(s)$ are chosen by using polynomial pole-placement techniques. (See Chapter 7.) For simplicity, we choose the same denominator polynomial for $T_{11}(s)$ and $T_{22}(s)$, $A_{cl}(s) = (s+0.1)^2(s+0.2)$. Note that, with this choice, we respect the constraint of having a dominant mode with decay speed bounded (above) by the NMP zero. This leads, after solving the pole assignment equation, to

$$p_{11}(s) = p_{22}(s) = -(0.472s + 0.0015); \qquad l_{11}(s) = l_{22}(s) = s + 0.872$$

$$(24.10.18)$$

With these values, the controller is calculated from (24.10.15). A simulation was run with this design and with the same conditions as for the decentralized PID case, i.e.,

$$r_1(t) = \mu(t - 1); \qquad r_2(t) = \mu(t - 100); \qquad d(t) = -10\mu(t - 250) \quad (24.10.19)$$

The results are shown in Figure 24.6.

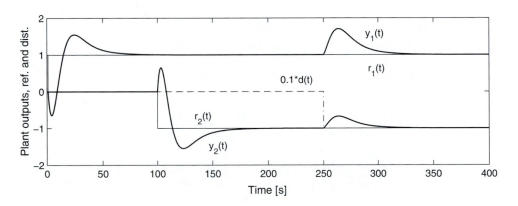

Figure 24.6. Loop performance with dynamic decoupling design

The results shown in Figure 24.6 confirm the two key issues underlying this design strategy: the channels are dynamically decoupled, and the NMP zero affects both channels.

24.10.5   MIMO Design. Triangular Decoupling

The results in the previous section verify that dynamic decoupling is not an ideal solution for this particular problem, because the nonminimum-phase behavior of the plant is dispersed over both outputs, rather than being isolated in the less important output, height $(y_2(t))$. We therefore aim for a triangular structure where the more important torque channel will be chosen as the decoupled loop.

The resultant triangular structure will have the form

$$\mathbf{M}(s) = \mathbf{G_o}(s)\mathbf{C}(s) = \begin{bmatrix} M_{11}(s) & 0 \\ M_{21}(s) & M_{22}(s) \end{bmatrix} \tag{24.10.20}$$

which leads to the complementary sensitivity

$$\mathbf{T_o}(s) = \begin{bmatrix} T_{11}(s) & T_{12}(s) \\ T_{21}(s) & T_{22}(s) \end{bmatrix} = \begin{bmatrix} \dfrac{M_{11}(s)}{1 + M_{11}(s)} & 0 \\ \dfrac{M_{21}(s)}{(1 + M_{11}(s))(1 + M_{22}(s))} & \dfrac{M_{22}(s)}{1 + M_{22}(s)} \end{bmatrix} \tag{24.10.21}$$

Then

$$\mathbf{C}(s) = \begin{bmatrix} C_{11}(s) & C_{12}(s) \\ C_{21}(s) & C_{22}(s) \end{bmatrix} = [\mathbf{G_o}(s)]^{-1}\mathbf{M}(s) \tag{24.10.22}$$

Also,

$$[\mathbf{G_o}(s)]^{-1} = \begin{bmatrix} 25s + 1 & 0 \\ 0 & s \end{bmatrix} \begin{bmatrix} \dfrac{n_{11}(s)}{d^-(s)d^+(s)} & \dfrac{n_{12}(s)}{d^-(s)d^+(s)} \\ \dfrac{n_{21}(s)}{d^-(s)d^+(s)} & \dfrac{n_{22}(s)}{d^-(s)d^+(s)} \end{bmatrix} \tag{24.10.23}$$

where

$$n_{11}(s) = 0.0023(s+1) \qquad n_{12}(s) = s^2 - 0.005s - 0.005 \tag{24.10.24}$$
$$n_{21}(s) = s+1 \qquad\qquad n_{22}(s) = 5(s+1) \tag{24.10.25}$$
$$d^-(s) = s + 0.121 \qquad\quad d^+(s) = s + 0.137 \tag{24.10.26}$$

Then, from (24.10.20) and (24.10.22), the controller is given by

$$\mathbf{C}(s) = \begin{bmatrix} \dfrac{(25s+1)(n_{11}(s)M_{11}(s) + n_{12}(s)M_{21}(s))}{d^-(s)d^+(s)} & \dfrac{sn_{12}(s)M_{22}(s)}{d^-(s)d^+(s)} \\ \dfrac{(25s+1)(n_{21}M_{11}(s) + n_{22}(s)M_{21}(s))}{d^-(s)d^+(s)} & \dfrac{sn_{22}(s)M_{22}(s)}{d^-(s)d^+(s)} \end{bmatrix} \tag{24.10.27}$$

$\mathbf{C}(s)$ should not have poles at $s = 0.137$, so the polynomial $d^+(s)$ should be canceled in the *four fraction matrix entries* in (24.10.27). This implies that

$$M_{22}(0.137) = 0 \qquad (24.10.28)$$

$$n_{11}(s)M_{11}(s) + n_{12}(s)M_{21}(s)\Big|_{s=0.137} = 0 \qquad (24.10.29)$$

$$n_{21}(s)M_{11}(s) + n_{22}(s)M_{21}(s)\Big|_{s=0.137} = 0 \qquad (24.10.30)$$

Equations (24.10.29) and (24.10.30) lead to

$$\frac{M_{21}(s)}{M_{11}(s)}\bigg|_{s=0.137} = -\frac{n_{11}(s)}{n_{12}(s)}\bigg|_{s=0.137} \qquad (24.10.31)$$

$$\frac{M_{21}(s)}{M_{11}(s)}\bigg|_{s=0.137} = -\frac{n_{21}(s)}{n_{22}(s)}\bigg|_{s=0.137} \qquad (24.10.32)$$

respectively. Note that (24.10.31) and (24.10.32) are simultaneously satisfied: for every zero, z_o, of $\mathbf{G_o}(s)$, we have that

$$n_{11}(s)n_{22}(s) - n_{12}(s)n_{21}(s)\Big|_{s=z_o} = 0 \qquad (24.10.33)$$

The next step is to make some choices.

(i) $M_{11}(s)$, $M_{21}(s)$ and $M_{22}(s)$ are chosen to have two poles at the origin. One of these comes from the plant model and the other is added to ensure steady-state compensation of step input disturbances.

(ii) $M_{11}(s)$ is chosen to achieve a bandwidth of around $0.15[rad/s]$ in channel 1. A possible choice is

$$M_{11}(s) = \frac{0.15s + 0.01}{s^2} \iff T_{11}(s) = \frac{0.15s + 0.01}{s^2 + 0.15s + 0.01} \qquad (24.10.34)$$

(iii) A simple choice for $M_{21}(s)$ is to assign two poles to the origin, i.e.,

$$M_{21}(s) = \frac{\alpha}{s^2} \qquad (24.10.35)$$

where α is a parameter to be determined.

(iv) $M_{22}(s)$ is chosen to have two poles at the origin and to satisfy (24.10.28). We choose

$$M_{22}(s) = \frac{(s - 0.137)p_{22}(s)}{s^2 l_{22}(s)} \tag{24.10.36}$$

where $p_{22}(s)$ and $l_{22}(s)$ are polynomials in s to be determined.

With the above choices and the design considerations, we can proceed to compute the controller.

First, α in (24.10.35) is computed to satisfy (24.10.30); this leads to $\alpha = -0.0061$, which, in turn, yields

$$C_{21}(s) = \frac{0.15(s + 1)}{s(s + 0.121)} \tag{24.10.37}$$

$$C_{11}(s) = \frac{-0.0058(25s + 1)(s + 0.0678)}{s^2(s + 0.121)} \tag{24.10.38}$$

We also have, from (24.10.36), that the complementary sensitivity in channel 2 is given by

$$T_{22}(s) = \frac{M_{22}(s)}{1 + M_{22}(s)} = \frac{(s - 0.137)p_{22}(s)}{s^2 l_{22}(s) + (s - 0.137)p_{22}(s)} \tag{24.10.39}$$

Then, $p_{22}(s)$ and $l_{22}(s)$ can be computed by using polynomial pole-assignment techniques, as we did for the decoupled design. We choose the denominator polynomial for $T_{22}(s)$ to be equal to $(s + 0.1)^2(s + 0.2)$. With this choice, we obtain

$$M_{22}(s) = \frac{(-s + 0.137)(0.4715s + 0.0146)}{s^2(s + 0.8715)} \tag{24.10.40}$$

This leads to

$$C_{12}(s) = \frac{-(25s + 1)(s^2 - 0.005s - 0.005)(0.4715s + 0.0146)}{s^2(s + 0.121)(s + 0.8715)} \tag{24.10.41}$$

$$C_{22}(s) = \frac{-5(s + 1)(0.4715s + 0.0146)}{s(s + 0.121)(s + 0.8715)} \tag{24.10.42}$$

With the controller designed above, simulations were performed to assess the performance of the MIMO control loop concerning tracking and disturbance compensation.

Unit step references and a unit step disturbance were applied, as follows:

$$r_1(t) = \mu(t-1); \qquad r_2(t) = -\mu(t-100); \qquad d(t) = -10\mu(t-250)$$
$$(24.10.43)$$

The results of the simulation, shown in Figure 24.7, motivate the following observations.

(i) The output of channel 1 is now unaffected by changes in the reference for channel 2. However, the output of channel 2 is affected by changes in the reference for channel 1. The asymmetry is consistent with the choice of a lower-triangular complementary sensitivity, $\mathbf{T_o}(s)$.

(ii) The nonminimum-phase behavior is evident in channel 2 but does not show up in the output of channel 1. This has also been achieved by choosing a lower-triangular $\mathbf{T_o}(s)$; that is, the open-loop NMP zero is a canonical zero of the closed-loop.

(iii) The transient compensation of the disturbance in channel 1 has also been improved with respect to the fully decoupled loop. Compare the results shown in Figure 24.6.

(iv) The step disturbance is completely compensated in steady state. This is due to the integral effect in the control for both channels.

(v) The output of channel one exhibits a significant overshoot (around 20%). This was predicted in section §24.6 for any loop having a double integrator.

□□□

The reader can evaluate the loop performance on his (her) own design by using the SIMULINK file **sugmill.mdl**.

24.11   Nonsquare Systems

In most of the above treatment, we have assumed equal number of inputs and outputs. However, in practice, there are either excess inputs (*fat systems*) or extra measurements (*tall systems*). We briefly discuss these two scenarios below.

Excess inputs

Say we have m inputs and p outputs, where $m > p$. In broad terms, the design alternatives can be characterized under four headings.

a) **Squaring up**

Because we have extra degrees of freedom in the input, it is possible to control extra variables (even though they need not be measured). One possible strategy is to use an observer to estimate the missing variables. Indeed, when the

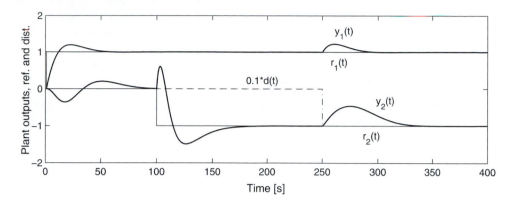

Figure 24.7. Loop performance with triangular design

observer estimates a further $m - p$ variables, then we can design a controller for an $m \times m$ system. Note that, the observer will normally be unbiased and so the transfer function from the input to the estimated variables is nominally identical to the transfer function to the true, but unmeasured, variables. Hence, no additional complexity arises.

b) **Coordinated control**

Another, and very common, situation, is where p inputs are chosen as the primary control variables, but other variables from the remaining $m - p$ inputs are used in some fixed, or possibly dynamic, relationships to the primary controls.

A particularly common instance of this approach is where one uses fuel flow as a primary control variable (say to control temperature), but then ties air-flow to the primary variable in a fixed ratio so as to achieve stoichiometric burning.

c) **Soft load sharing**

If one decides to simply control the available measurements, then one can share the load of achieving this control between the excess inputs. This can be achieved via various optimization approaches (e.g., quadratic).

d) **Hard load sharing**

It is often the case that one has a subset of the inputs (say of dimension p) that is a preferable choice from the point of view of precision or economics, but that these have limited amplitude or authority. In this case, other inputs can be called upon to assist. A nice example of this kind of situation is where one wishes to accurately control the flow of a material over a wide range. In this application, the appropriate strategy is to use a *big* valve to regulate the flow for large errors and then allow a *small* valve to trim the flow to the desired value. This is a switching control strategy. More will be said about this in the next chapter.

Excess outputs

Here we assume that $p > m$. In this case, we cannot hope to control each of the measured outputs independently at all times. We investigate three alternative strategies.

- **Squaring down**

  Although all the measurements should be used in obtaining state estimates, only m quantities can be independently controlled. Thus, any part of the controller that depends on state-estimate feedback should use the full set of measurements; however, set-point injection should be carried out only for a subset of m variables (be they estimated or measured).

- **Soft sharing control**

  If one really wants to control more variables than there exist inputs, then it is possible to define their relative importance by using a suitable performance index. For example, one might use a quadratic performance index with different weightings on different outputs. Note, of course, that zero steady-state error cannot, in general, be forced in all loops. Thus integral action (if desired) cannot be applied to more than m outputs.

- **Switching strategies**

  It is also possible to take care of m variables at any one time by use of a switching law. This law might include time-division multiplexing or some more sophisticated decision structure. An example of this has been presented in section §11.4, where a single control is used to impose both state constraints and output tracking by switching between two linear controllers, one designed for each purpose.

The availability of extra inputs or outputs can also be very beneficial in allowing one to achieve a satisfactory design in the face of fundamental performance limitations. We illustrate by an example.

Example 24.3 (Inverted pendulum). *We recall the inverted-pendulum problem discussed in Example 9.4 on page 256. We saw that this system, when considered as a single-input (force applied to the cart), single-output (cart position) problem, has a real RHP pole that has a larger magnitude than a real RHP zero. Although this problem is, formally, controllable, it was argued that this set-up, when viewed in the light of fundamental performance limitations, is practically impossible to control, on account of severe and unavoidable sensitivity peaks.*

However, the situation changes dramatically if we also measure the angle of the pendulum. This leads to a single input (force) and two outputs (cart position, $y(t)$, and angle, $\theta(t)$). This system can be represented in block-diagram form as in Figure 24.8.

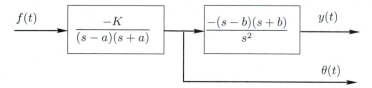

Figure 24.8. One-input, two-output inverted-pendulum model

Note that this nonsquare system has poles at $(0, 0, a, -a)$ but no finite (MIMO) zeros. Thus, one might reasonably expect that the very severe limitations which existed for the SISO system no longer apply to this nonsquare system.

We use the same numerical values as in Example 9.4: $K = 2$, $a = \sqrt{20}$, and $b = \sqrt{10}$. Then a suitable nonsquare controller turns out to be

$$U(s) = \begin{bmatrix} C_y(s) & C_\theta(s) \end{bmatrix} \begin{bmatrix} R(s) - Y(s) \\ -\Theta(s) \end{bmatrix} \tag{24.11.1}$$

where $R(s) = \mathcal{L}\left[r(t)\right]$ is the reference for the cart position, and

$$C_y(s) = -\frac{4(s + 0.2)}{s + 5}; \qquad C_\theta(s) = -\frac{150(s + 4)}{s + 30} \tag{24.11.2}$$

Figure 24.9 on the following page shows the response of the closed-loop system for $r(t) = \mu(t - 1)$, i.e., a unit step reference applied at $t = 1$.

An interesting observation is that the nonminimum-phase zero lies between the input and $y(t)$. Thus, irrespective of how the input is chosen, the performance limitations due to that zero remain. For example, we have for a unit reference step that

$$\int_0^\infty \left[r(t) - y(t)\right] e^{-bt} dt = \frac{1}{b} \tag{24.11.3}$$

It is readily verified that this holds for the upper plot in Figure 24.9. In particular, the presence of the nonminimum-phase zero places an upper limit on the closed-loop bandwidth irrespective of the availability of the measurement of the angle.

The key issue that explains the advantage of using nonsquare control in this case is that the second controller effectively shifts the unstable pole to the stability region. Thus there is no longer a conflict between a small NMP zero and a large unstable pole, and we need only to pay attention to the bandwidth limitations introduced by the NMP zero.

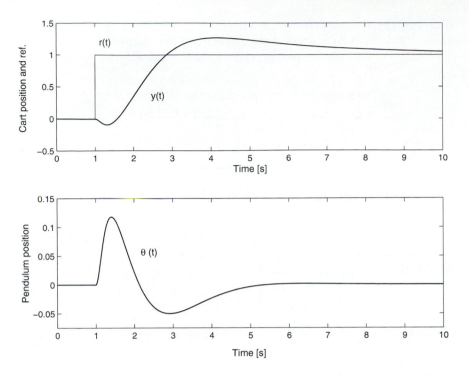

Figure 24.9. Step response in a nonsquare control for the inverted pendulum

24.12   Discrete-Time Systems

The results presented in the previous sections also apply, *mutatis mutandis*, to discrete-time systems. The main differences are related to these issues:

(i) different stability region; and

(ii) the fact, that in discrete time, pure delays appear simply as poles at the origin (in the z-plane), whereas, in continuous time, they lead to transcendental functions of s.

Point (ii) above is particularly important, because it allows finite-dimensional techniques to be applied to time-delay systems.

24.13   Summary

- Analogously to the SISO case, MIMO performance specifications can generally not be addressed independently from another, because they are linked by a web of trade-offs.

- A number of the SISO fundamental algebraic laws of trade-off generalize rather directly to the MIMO case:

  - $\mathbf{S_o}(s) = \mathbf{I} - \mathbf{T_o}(s)$, implying a trade-off between speed of response to a change in reference or rejecting disturbances ($\mathbf{S_o}(s)$ small) versus necessary control effort, sensitivity to measurement noise, or modeling errors ($\mathbf{T_o}(s)$ small);

  - $Y_m(s) = -\mathbf{T_o}(s)D_m(s)$, implying a trade-off between the bandwidth of the complementary sensitivity and sensitivity to measurement noise;

  - $\mathbf{S_{uo}}(s) = [\mathbf{G_o}(s)]^{-1}\mathbf{T_o}(s)$, implying that a complementary sensitivity with bandwidth significantly higher than the open loop will generate large control signals;

  - $\mathbf{S_{io}}(s) = \mathbf{S_o}(s)\mathbf{G_o}(s)$, implying a trade-off between input and output disturbances; and

  - $\mathbf{S}(s) = \mathbf{S_o}(s)\mathbf{S_\Delta}(s)$, where $\mathbf{S_\Delta}(s) = [\mathbf{I} + \mathbf{G_{\Delta l}}(s)\mathbf{T_o}(s)]^{-1}$, implying a trade-off between the complementary sensitivity and robustness to modeling errors.

- There also exist frequency- and time-domain trade-offs due to unstable poles and zeros.

  - *Qualitatively*, they parallel the SISO results in that (in a MIMO measure) low bandwidth in conjunction with unstable poles is associated with increasing overshoot, whereas high bandwidth in conjunction with unstable zeros is associated with increasing undershoot.

  - *Quantitatively*, the measure in which the above is true is more complex than in the SISO case: the effects of under- and overshoot, as well as of integral constraints, pertain to linear combinations of the MIMO channels.

- MIMO systems are subject to the additional design specification of desired decoupling.

- Decoupling is related to the time- and frequency-domain constraints via directionality.

  - The constraints due to open-loop NMP zeros with noncanonical directions can be isolated in a subset of outputs, if triangular decoupling is acceptable.

  - Alternatively, if dynamic decoupling is enforced, the constraint is dispersed over several channels.

- Advantages and disadvantages of completely decentralized control, dynamical and triangular decoupling designs were illustrated with an industrial case study.

Sugar-Mill Case Study		
Design	*Advantage*	*Disadvantage*
Decentralized	Simpler SISO theory can be used	Interactions are ignored; poor performance is achieved.
Dynamic Decoupling	Outputs can be controlled separately	Both outputs must obey the lower bandwidth constraint due to the one NMP zero.
Triangular Decoupling	Most important output decoupled and without NMP constraint	The second (albeit less important) output is affected by the first output and NMP constraint.

24.14 Further Reading

MIMO frequency-domain constraints

Chen, J. (1995). Sensitivity integral relation and design trade-offs in linear multivariable feedback systems. *IEEE Transactions on Automatic Control*, 40(10):1700-1716.

Chen, J. and Nett, C. (1995). Sensitivity integrals for multivariable discrete-time systems. *Automatica*, 31(8):113-124.

Freudenberg, J.S. and Looze, D.P. (1988). *Frequency-Domain Properties of Scalar and Multivariable Feedback Systems*. Springer-Verlag, New York.

Gómez, G. and Goodwin, G.C. (1995). Vectorial sensitivity constraints for linear multivariable systems. *Proceedings of the 34th CDC, New Orleans, LA*, 4:4097-4102.

Gómez, G. and Goodwin, G.C. (1996). Integral constraints on sensitivity vectors for multivariable linear systems. *Automatica*, 32(4):499-518.

Skogestad, S. and Postlethwaite, I. (1996). *Multivariable Feedback Control: Analysis and Design*. Wiley, New York.

Sule, V. and Athani, V. (1991). Directional sensitivity trade-offs in multivariable feedback systems. *Automatica*, 27(5):869-872.

Sugar mill application

West, M. (1997). *Modelling and control of a sugar crushing station.* ME Thesis, Department of Electrical and Computer Engineering, The University of Newcastle, Australia.

Sensitivity issues

Boyd, S.P. and Barratt, C.H. (1991). *Linear Controller Design–Limits of Performance.* Prentice-Hall, Englewood Cliffs, N.J.

Doyle, J.C. and Stein, G. (1981). Multivariable feedback design: Concepts for a classic/modern synthesis. *IEEE Transactions on Automatic Control*, 26(1):4-16.

Postlethwaite, I., Edwards, W.J., and MacFarlane, A. (1981). Principal gains and principal phases in the analysis of linear multivariable feedback systems. *IEEE Transactions on Automatic Control*, 26(1):32-46.

Postlethwaite, I. and MacFarlane, A. (1979). *A Complex Variable Approach to the analysis of linear Multivariable Feedback Systems.* Lecture notes in control and information sciences, Vol. 12, Springer-Verlag.

Robustness issues

Dahleh, M. and Diaz-Bobillo, I. (1995). *Control of Uncertain Systems.* Prentice-Hall, Englewood Cliffs, N.J.

Stoorvogel, A. (1992). *The H_∞ Control Problem: a state space approach.* Prentice-Hall, Englewood Cliffs, N.J.

Zhou, K., Doyle, J.C., and Glover, K. (1996). *Robust and Optimal Control.* Prentice-Hall, Upper Saddle River, N.J.

24.15   Problems for the Reader

Problem 24.1. *Consider a 2×2 MIMO plant being the transfer function*

$$\mathbf{G_o}(s) = \frac{20}{(s+2)(s+5)^2} \begin{bmatrix} s-2 & -s-1 \\ s-2 & -5 \end{bmatrix} \qquad (24.15.1)$$

24.1.1 *Compute the system NMP zeros and the associated left and right directions.*

24.1.2 *Determine the sensitivity constraints due to those zeros.*

Problem 24.2. *Consider a 2×2 MIMO plant having the nominal model*

$$\mathbf{G_o}(s) = \frac{1}{(s+1)^2} \begin{bmatrix} s-2 & 1 \\ -2 & -2 \end{bmatrix} \qquad (24.15.2)$$

24.2.1 *Show that this system has an interacting NMP zero, and compute its associated left direction(s).*

24.2.2 *Is this case analogous to the quadruple-tank case analyzed in Example 24.1 on page 777? Discuss.*

Problem 24.3. *Consider the same plant as in Problem 24.2. Assume that a feedback control loop has to be designed to achieve zero steady-state errors for constant references and a bandwidth of 0.5 [rad/s] in both channels.*

24.3.1 *Design a controller to achieve a diagonal complementary sensitivity.*

24.3.2 *Design a controller to achieve a lower-triangular complementary sensitivity.*

24.3.3 *Compare the performance of the two designs.*

Problem 24.4. *Consider a MIMO plant having a nominal model satisfying*

$$\mathbf{G_o}(s) = \mathbf{G_{oN}}(s)[\mathbf{G_{oD}}(s)]^{-1} \qquad where \quad \mathbf{G_{oD}}(3) = \begin{bmatrix} 2 & 0 \\ -1 & 0 \end{bmatrix} \qquad (24.15.3)$$

24.4.1 *Determine the directions associated to the unstable pole.*

24.4.2 *Determine the constraints that will affect the controller output $u(t)$ for a step-output disturbance in the ith channel $(i = 1, 2)$.*

Problem 24.5. *Derive the time-domain constraints for discrete-time systems in the presence of NMP zeros and unstable poles.*

Problem 24.6. *Consider a 2×2 MIMO plant having the transfer function*

$$\mathbf{G_o}(s) = \frac{20}{(s-2)(s+5)^2} \begin{bmatrix} 1 & 0.2(s+5) \\ -0.125(s+5)^2 & 2 \end{bmatrix} \qquad (24.15.4)$$

This plant has to be digitally controlled, with sampling interval $\Delta = 0.1$ [s] and a zero-order hold.

24.6.1 *Compute the model for the sampled-data system, $[\mathbf{G_{ho}G_o}]_q (z)$.*

24.6.2 *Determine the time-domain constraints arising from unstable poles and NMP zeros (if any).*

Problem 24.7. *Consider a feedback control loop for a 2×2 MIMO plant having a NMP zero at $s = 2$, with (unique) associated left direction $h^T = [1 \quad -1]$; then, equation (24.6.3) says that, for a step input in either of the channels, we have that*

$$\int_0^\infty h^T y(t) e^{-z_o t} dt = \int_0^\infty (y_1(t) - y_2(t)) e^{-2t} dt = 0 \qquad (24.15.5)$$

where $y_1(t)$ and $y_2(t)$ are the plant outputs in channels 1 and 2, respectively.

Equation (24.15.5) might suggest that we can avoid undershoot if the controller is designed to achieve a complementary sensitivity where $T_{11}(s) = T_{21}(s)$ and $T_{12}(s) = T_{22}(s)$, because then we can have $y_1(t) - y_2(t) = 0$ for all t. Why is this not a sensible idea?

Problem 24.8. *Consider a MIMO system having the model*

$$\mathbf{G_o}(s) = \frac{2}{(s+1)^2(s+2)} \begin{bmatrix} 2(-s+3) & 0.5(-s+3) & (-s+\alpha) \\ 0.5(s+1) & -\beta & -(s+2) \\ -1 & 0.5 & 2.5 \end{bmatrix} \qquad (24.15.6)$$

If $\alpha = 3$ and $\beta = 1$, find all time-domain constraints for the feedback control of this system. (The reader is reminded that there can be more than one left direction associated with a NMP zero.)

Part VIII

ADVANCED MIMO CONTROL

PREVIEW

This final part of the book covers some advanced ideas in MIMO control. We begin, in Chapter 25, with the MIMO extension of the controller parameterizations described in Chapter 15 for the SISO case. Finally, in Chapter 26, we bring together many ideas from the book. Our aim here is to describe some advanced design ideas. However, we also want to illustrate to the reader that, having reached this point, he or she can understand quite sophisticated design issues. In particular, we show how full dynamic and partial decoupling can be achieved via one- and two-degree-of-freedom controllers. We also show how dynamic decoupling can be retained in the presence of actuator amplitude and slew-rate limitations.

Chapter 25

MIMO CONTROLLER PARAMETERIZATIONS

25.1 Preview

In this chapter, we will extend the SISO design methods of Chapter 15 to the MIMO case. We will find that many issues are common between the SISO and MIMO cases. However, there are distinctive issues in the MIMO case that warrant separate treatment. The key factor leading to these differences is once again the fact that MIMO systems have spatial coupling, i.e., each input can affect more than one output and each output can be affected by more than one input. The consequences of this are far-reaching. Examples of the difficulties that arise from these interactions include stability, nonminimum-phase zeros with their directionality properties, and tracking performance.

Notwithstanding these differences, the central issue in MIMO control-system design still turns out to be that of (approximate) inversion. Again, because of interactions, inversion is more intricate than in the SISO case, and we will thus need to develop more sophisticated tools for achieving this objective.

25.2 Affine Parameterization: Stable MIMO Plants

We refer the reader to Chapter 15, and in particular to section §15.3, where we presented the parameterization of all stabilizing controllers for a stable linear system. The generalization to the multivariable case is straightforward. Indeed, all controllers that yield a stable closed-loop for a given open-loop stable plant having nominal transfer function $\mathbf{G_o}(s)$ can be expressed as

$$\mathbf{C}(s) = [\mathbf{I} - \mathbf{Q}(s)\mathbf{G_o}(s)]^{-1}\mathbf{Q}(s) = \mathbf{Q}(s)[\mathbf{I} - \mathbf{G_o}(s)\mathbf{Q}(s)]^{-1} \qquad (25.2.1)$$

where $\mathbf{Q}(s)$ is any stable proper transfer-function matrix.

The resulting nominal sensitivity functions are

$$\mathbf{T_o}(s) = \mathbf{G_o}(s)\mathbf{Q}(s) \tag{25.2.2}$$
$$\mathbf{S_o}(s) = \mathbf{I} - \mathbf{G_o}(s)\mathbf{Q}(s) \tag{25.2.3}$$
$$\mathbf{S_{io}}(s) = (\mathbf{I} - \mathbf{G_o}(s)\mathbf{Q}(s))\mathbf{G_o}(s) \tag{25.2.4}$$
$$\mathbf{S_{uo}}(s) = \mathbf{Q}(s) \tag{25.2.5}$$

These transfer-function matrices are simultaneously stable if and only if $\mathbf{Q}(s)$ is stable. A key property of (25.2.2) to (25.2.5) is that they are affine in the matrix $\mathbf{Q}(s)$. We will exploit this property below when we discuss various design issues.

Remark 25.1. *Note that the following are a valid LMFD and RMFD for the nominal plant and controller.*

$$\overline{\mathbf{C}}_{\mathbf{D}}(s) = \mathbf{I} - \mathbf{Q}(s)\mathbf{G_o}(s) \qquad \overline{\mathbf{C}}_{\mathbf{N}}(s) = \mathbf{Q}(s) \tag{25.2.6}$$
$$\mathbf{C}_{\mathbf{D}}(s) = \mathbf{I} - \mathbf{G_o}(s)\mathbf{Q}(s) \qquad \mathbf{C}_{\mathbf{N}}(s) = \mathbf{Q}(s) \tag{25.2.7}$$
$$\overline{\mathbf{G}}_{\mathbf{oD}}(s) = \mathbf{I} \qquad \overline{\mathbf{G}}_{\mathbf{oN}}(s) = \mathbf{G_o}(s) \tag{25.2.8}$$
$$\mathbf{G}_{\mathbf{oD}}(s) = \mathbf{I} \qquad \mathbf{G}_{\mathbf{oN}}(s) = \mathbf{G_o}(s) \tag{25.2.9}$$

Actually, the above choices show that (22.12.22) is satisfied and hence that closed-loop stability is guaranteed by having $\mathbf{Q}(s)$ stable. We will find (25.2.6) and (25.2.7) a convenient MFD for the controller that will facilitate subsequent design procedures.

□□□

An idealized target sensitivity function is $\mathbf{T}(s) = \mathbf{I}$. We then see from (25.2.2) and (25.2.3) that the design of $\mathbf{Q}(s)$ reduces to the problem of finding an (approximate) right inverse for $\mathbf{G_o}(s)$, such that the trade-offs are approximately met in the different frequency regions.

We refer the reader to section §15.3.2, where the following issues arose in the SISO problem of finding approximate inverses:

- nonminimum-phase zeros

- model relative degree

- disturbance trade-offs

- control effort

- robustness

- uncontrollable modes

These same issues appear in the MIMO case, but they are compounded by directionality issues. In the sequel, we will explore the issues of MIMO relative degree, robustness, and nonminimum-phase zeros.

25.3   Achieved Sensitivities

As in the SISO case, we need to distinguish between the nominal sensitivities and the achieved sensitivities. The results of section 20.8 apply here as well. For example, the achieved sensitivity is given, as in (20.8.3), by

$$\mathbf{S}(s) \; = \; \mathbf{S_o}(s)[\mathbf{I} + \mathbf{G_{\Delta l}}(s)\mathbf{T_o}(s)]^{-1} \tag{25.3.1}$$
$$= \; [\mathbf{I} - \mathbf{G_o}(s)\mathbf{Q}(s)][\mathbf{I} + \mathbf{G_\epsilon}(s)\mathbf{Q}(s)]^{-1} \tag{25.3.2}$$

where $\mathbf{G_\epsilon}(s)$ is the additive model error, defined in analogy to (4.12.1) as

$$\mathbf{G}(s) = \mathbf{G_o}(s) + \mathbf{G_\epsilon}(s) \tag{25.3.3}$$

25.4   Dealing with Model Relative Degree

We recall from Chapter 15 that, in the SISO case, we dealt with model relative-degree issues by simply introducing extra filtering to render the appropriate transfer-function biproper. This same principle applies to the MIMO case, save that the filter needed to achieve a biproper matrix transfer function is a good deal more interesting than in the SISO case. To understand this, we need to pause to study the issue of MIMO relative degree.

25.4.1   MIMO Relative Degree

The issue of MIMO model degree is more subtle than in the SISO case. In the following sections, we define MIMO relative degree and explore some of its properties.

(a) Interactor matrices

We recall that the relative degree of a SISO model, amongst other things, sets a lower limit to the relative degree of the complementary sensitivity. In the SISO case, we say that the relative degree of a (scalar) transfer function $G(s)$ is the degree of a polynomial $p(s)$ such that

$$\lim_{s \to \infty} p(s)G(s) = K \qquad \text{where } 0 < |K| < \infty \tag{25.4.1}$$

This means that $p(s)G(s)$ is biproper, i.e., $(p(s)G(s))^{-1}$ is also proper.

We can actually make this polynomial $p(s)$ unique if we require that it should belong to the class of polynomials $\mathcal{P} = \{s^k | k \in \mathbb{N}\}$.

In the MIMO case, every entry in the transfer-function matrix $\mathbf{G}(s)$ can have a different relative degree. Thus, to generate a multivariable version of the scalar polynomial $p(s)$ in (25.4.1), we will need to consider the individual entries and their interactions. To see how this can be done, consider an $m \times m$ matrix $\mathbf{G}(s)$.

We will show that there exist matrices $\boldsymbol{\xi}_{\mathbf{L}}(s)$ and $\boldsymbol{\xi}_{\mathbf{R}}(s)$ such that the following properties (which are the multivariable analogues of (25.4.1)) hold.

$$\lim_{s \to \infty} \boldsymbol{\xi}_{\mathbf{L}}(s)\mathbf{G}(s) = \mathbf{K}_{\mathbf{L}} \qquad\qquad 0 < |\det(\mathbf{K}_{\mathbf{L}})| < \infty \qquad\qquad (25.4.2)$$

$$\lim_{s \to \infty} \mathbf{G}(s)\boldsymbol{\xi}_{\mathbf{R}}(s) = \mathbf{K}_{\mathbf{R}} \qquad\qquad 0 < |\det(\mathbf{K}_{\mathbf{R}})| < \infty \qquad\qquad (25.4.3)$$

This result is established in the following theorem.

Theorem 25.1. *Consider a square transfer-function $m \times m$ matrix $\mathbf{G}(s)$, nonsingular almost everywhere in s. Then there exist unique transfer matrices $\boldsymbol{\xi}_{\mathbf{L}}(s)$ and $\boldsymbol{\xi}_{\mathbf{R}}(s)$ (known as the left and right interactor matrices, respectively) such that (25.4.2) and (25.4.3) are satisfied, such that*

$$\boldsymbol{\xi}_{\mathbf{L}}(s) = \mathbf{H}_{\mathbf{L}}(s)\mathbf{D}_{\mathbf{L}}(s) \qquad\qquad (25.4.4)$$

$$\mathbf{D}_{\mathbf{L}}(s) = \operatorname{diag}\left(s^{p_1}, \ldots, s^{p_m}\right) \qquad\qquad (25.4.5)$$

$$\mathbf{H_L}(s) = \begin{bmatrix} 1 & 0 & \cdots & \cdots & 0 \\ h_{21}^L(s) & 1 & \cdots & \cdots & 0 \\ h_{31}^L(s) & h_{32}^L(s) & \ddots & & \vdots \\ \vdots & \vdots & & \ddots & \vdots \\ h_{m1}^L(s) & h_{m2}^L(s) & \cdots & \cdots & 1 \end{bmatrix} \tag{25.4.6}$$

$$\boldsymbol{\xi_R}(s) = \mathbf{D_R}(s)\mathbf{H_R}(s) \tag{25.4.7}$$

$$\mathbf{D_R}(s) = \operatorname{diag}\left(s^{q_1}, \ldots, s^{q_m}\right) \tag{25.4.8}$$

$$\mathbf{H_R}(s) = \begin{bmatrix} 1 & h_{12}^R(s) & h_{13}^R(s) & \cdots & h_{1m}^R(s) \\ 0 & 1 & h_{23}^R(s) & \cdots & h_{2m}^R(s) \\ \vdots & \vdots & \ddots & & \vdots \\ \vdots & \vdots & & \ddots & \vdots \\ 0 & 0 & \cdots & \cdots & 1 \end{bmatrix} \tag{25.4.9}$$

and such that $h_{ij}^L(s)$ and $h_{ij}^R(s)$ are polynomials in s, satisfying $h_{ij}^L(0) = 0$ and $h_{ij}^R(0) = 0$.

Proof

[By construction]

We first recall that $\mathbf{G}(s)$ can always be expressed in RMFD in such a way that $\mathbf{G_{oN}}(s)$ and $\mathbf{G_{oD}}(s)$ are right coprime polynomial matrices and that $\mathbf{G_{oD}}(s)$ is column proper. Let n_d be the sum of the degrees of the columns of $\mathbf{G_{oD}}(s)$, and let n_n be the degree of $\det(\mathbf{G_{oN}}(s))$; then the relative degree of $\mathbf{G}(s)$ is $n_d - n_n$, and properness of $\mathbf{G}(s)$ implies that $n_n \leq n_d$.

We will prove the theorem for the left interactor. The case of the right interactor can be proved by mirroring the arguments with regard to rows and columns.

Consider first the ith row of $\mathbf{G}(s)$, $[\mathbf{G}(s)]_{i*}$. Then there exists a minimum nonnegative integer n_i such that

$$\lim_{s \to \infty} s^{n_i} [\mathbf{G}(s)]_{i*} = f_i^T \tag{25.4.10}$$

where f_i^T is a row vector, not identically zero and having finite entries.
Then one can proceed to construct $\boldsymbol{\xi_L}(s)$ as follows:

(i) Choose the first row of $\boldsymbol{\xi_L}(s)$, $[\boldsymbol{\xi_L}(s)]_{1*}$, as

$$[\boldsymbol{\xi_L}(s)]_{1*} = [s^{n_1} \ 0 \ 0 \ldots 0] \tag{25.4.11}$$

Then

$$[\mathbf{K_L}]_{1*} = \lim_{s \to \infty} [\boldsymbol{\xi_L}(s)]_{1*}\mathbf{G}(s) = r_1^T \tag{25.4.12}$$

where r_1^T is a row vector such that $r_1^T = f_1^T$

(ii) Consider the second-row vector, f_2^T. If f_2^T is linearly independent of r_1^T, then choose the second row of $\boldsymbol{\xi_L}(s)$, $[\boldsymbol{\xi_L}(s)]_{2*}$, as

$$[\boldsymbol{\xi_L}(s)]_{2*} = [0 \ s^{n_2} \ 0 \ 0 \ldots 0] \tag{25.4.13}$$

This will lead to

$$[\mathbf{K_L}]_{2*} = \lim_{s \to \infty} [\boldsymbol{\xi_L}(s)]_{2*}\mathbf{G}(s) = r_2^T \tag{25.4.14}$$

(iii) If f_2 is linearly dependent on r_1, i.e., if there exists a nonzero β_2^1 such that $f_2^T = \beta_2^1 r_1^T$, then we cannot choose $[\boldsymbol{\xi_L}(s)]_{2*}$ as in (25.4.13), because then the matrix $\mathbf{K_L}$ in (25.4.2) would be singular. Instead, we form the row vector

$$[\boldsymbol{\xi_L}(s)]_{2*}^1 = s^{n_2^1}([0 \ s^{n_2} \ 0 \ 0 \ldots 0] - \beta_2^1[\boldsymbol{\xi_L}(s)]_{1*}) \tag{25.4.15}$$

where n_2^1 is the unique integer such that

$$\lim_{s \to \infty} [\boldsymbol{\xi_L}(s)]_{2*}^1 \mathbf{G}(s) = (r_2^1)^T \tag{25.4.16}$$

and r_2^1 is a vector not identically zero and with finite entries.

If r_2^1 is linearly independent of r_1, then choose the second row of $\boldsymbol{\xi_L}(s)$, $[\boldsymbol{\xi_L}(s)]_{2*}$, as

$$[\boldsymbol{\xi_L}(s)]_{2*} = [\boldsymbol{\xi_L}(s)]_{2*}^1 \tag{25.4.17}$$

If r_2^1 is linearly dependent on r_1, then there exists a nonzero β_2^2 such that $r_2 = \beta_2^2 r_1$. Next, form the row vector

$$[\boldsymbol{\xi}_{\mathbf{L}}(s)]_{2*}^2 = (s^{n_2^2}([\boldsymbol{\xi}_{\mathbf{L}}(s)]_{2*}^1 - \beta_2^1 [\boldsymbol{\xi}_{\mathbf{L}}(s)]_{1*}) \tag{25.4.18}$$

where n_2^2 is the unique integer such that

$$\lim_{s \to \infty} [\boldsymbol{\xi}_{\mathbf{L}}(s)]_{2*}^2 \mathbf{G}(s) = (r_2^2)^T \tag{25.4.19}$$

where r_2^2 is a vector not identically zero and with finite entries.

If r_2^2 is linearly independent of r_1, then choose the second row of $\boldsymbol{\xi}_{\mathbf{L}}(s)$, $[\boldsymbol{\xi}_{\mathbf{L}}(s)]_{2*}$, as

$$[\boldsymbol{\xi}_{\mathbf{L}}(s)]_{2*} = [\boldsymbol{\xi}_{\mathbf{L}}(s)]_{2*}^2 \tag{25.4.20}$$

If $r_2^{(2)}$ is linearly dependent on r_1, then the process is repeated until either linear independence is achieved or the kth attempt yields $n_1 + n_2^k = n_d - n_n$, in which case make $p_2 = 0$ and the corresponding off-diagonal terms $h_{21}, \ldots, h_{2(m-1)}$ equal zero. Note that $n_1 + n_2^k$ can never be larger than $n_d - n_n$, because this latter value is the relative degree of the matrix.

(iv) Proceed with the other rows in a similar fashion.

□□□

We illustrate the procedure with the following example.

Example 25.1. *Consider the transfer-function matrix* $\mathbf{G}(s)$ *given by*

$$\mathbf{G}(s) = \begin{bmatrix} (s+1)^2 & (s+1) \\ 2(s+1) & 1 \end{bmatrix} [(s+1)^3 \mathbf{I}]^{-1} \tag{25.4.21}$$

Then $n_d = 6$ *and* $n_n = 2$. *We also have that* $n_1 = 1$, *with* $f_1 = [1 \; 0]^T$, *and* $n_2 = 2$, *with* $f_2 = [2 \; 0]^T$.

(i) *We first form*

$$[\boldsymbol{\xi}_{\mathbf{L}}(s)]_{1*} = [s^{n_1} \; 0 \; 0 \ldots 0] = [s \; 0] \tag{25.4.22}$$

Then,

$$\lim_{s \to \infty} [\boldsymbol{\xi}_{\mathbf{L}}(s)]_{1*} [\mathbf{G}(s)]_{1*} = r_1^T = [1 \; 0] \tag{25.4.23}$$

(ii) *Consider the row vector* f_2^T. *Because* f_2 *is linearly dependent on* r_1, *with* $\beta_2^1 = 2$, *i.e.,* $f_2 = 2r_1$, *we then choose the second row of* $\boldsymbol{\xi}_{\mathbf{L}}(s)$, $[\boldsymbol{\xi}_{\mathbf{L}}(s)]_{2*}$ *as*

$$[\boldsymbol{\xi}_{\mathbf{L}}(s)]_{2*}^1 = s^x([0 \ s^2] - \beta_2^1[\boldsymbol{\xi}_{\mathbf{L}}(s)]_{1*}) = [-2s^{1+x} \ s^{2+x}] \qquad (25.4.24)$$

where x *is found by noting that*

$$\lim_{s \to \infty} [\boldsymbol{\xi}_{\mathbf{L}}(s)]_{2*}^1 \mathbf{G}(s) = (r_2^1)^T = \lim_{s \to \infty} \begin{bmatrix} \dfrac{-2s^{1+x}}{(s+1)^2} & \dfrac{-s^{2+x} - 2s^{1+x}}{(s+1)^3} \end{bmatrix} \qquad (25.4.25)$$

from which we obtain $x = 1$. *This leads to*

$$r_2^1 = [-2 \ -1]^T \qquad (25.4.26)$$

which is linearly independent of f_1^T. *Thus, the choice* (25.4.24), *with* $x = 1$, *is a valid choice as the second row of the interactor matrix. Thus,*

$$\boldsymbol{\xi}_{\mathbf{L}}(s) = \begin{bmatrix} s & 0 \\ -2s^2 & s^3 \end{bmatrix} \qquad (25.4.27)$$

□□□

Remark 25.2. *It is straightforward to see that the interactors can be defined by using diagonal matrices* $\mathbf{D}_{\mathbf{L}}(s)$ *and* $\mathbf{D}_{\mathbf{R}}(s)$ *in* (25.4.4) *and* (25.4.7), *with arbitrary polynomial diagonal entries with degrees* $p_1, p_2, \ldots, p_m$, *which are invariants of the interactor representation of a given matrix* $\mathbf{G}(s)$. *This flexibility is important, because we can always choose stable polynomials, implying that the inverses of* $\boldsymbol{\xi}_{\mathbf{L}}(s)$ *and* $\boldsymbol{\xi}_{\mathbf{R}}(s)$ *are also stable.*

(b) Interpretation

Interactor matrices play a central role in control, because they define the (multivariable) relative degree. For example, they define the minimum achievable relative degree of any complementary sensitivity obtainable by using a proper controller.

The left interactor matrix has the interpretation that $\boldsymbol{\xi}_{\mathbf{L}}(s)Y(s)$ is a particular combination of *predicted outputs* having the special property that the transfer function connecting $\boldsymbol{\xi}_{\mathbf{L}}(s)Y(s)$ to $U(s)$ has nonzero high-frequency gain. This property can be used to develop a prototype control law by setting the *predicted output* $\boldsymbol{\xi}_{\mathbf{L}}(s)Y(s)$ equal to some desired value. This interpretation can be best seen from a simple discrete-time example:

Example 25.2. *Consider the following transfer function, expressed in terms of the Z-transform variable:*

$$\mathbf{G_{oq}}(z) = \frac{1}{z^2} \begin{bmatrix} z & 2 \\ 3z & 4 \end{bmatrix} \tag{25.4.28}$$

In agreement with the definitions above, the left interactor $\boldsymbol{\xi_L}(z)$ turns out to be

$$\boldsymbol{\xi_L}(z) = \begin{bmatrix} z & 0 \\ -3z^2 & z^2 \end{bmatrix} \tag{25.4.29}$$

By using $\boldsymbol{\xi_L}(z)$, we can define a new variable $Y_{qL}(z)$ as follows:

$$Y_{qL}(z) \triangleq \boldsymbol{\xi_L}(z)Y_q(z) = \boldsymbol{\xi_L}(z)\mathbf{G_{oq}}(z)U_q(z)$$

$$= \begin{bmatrix} zY_q(z)^{(1)} \\ -3z^2Y_q(z)^{(1)} + z^2Y_q(z)^{(2)} \end{bmatrix} = \begin{bmatrix} 1 & 2z^{-1} \\ 0 & -2 \end{bmatrix} U_q(z) \tag{25.4.30}$$

where $Y_q(z)^{(1)}$ and $Y_q(z)^{(2)}$ denote the first and second elements in the output vector $Y_q(z)$, respectively.

We see from the rightmost expression in (25.4.30) that we have built a predictor for a combination of the future outputs that depends on present and past values of the inputs. Moreover, the dependence on the present inputs is via an invertible matrix. Thus, there exists a choice for the present inputs that brings $y_L[k] = \mathcal{Z}^{-1}[Y_{qL}(z)]$ to any desired value, $y_L^*[k]$.

We define $Y_{qL}^*(z)$ as $\boldsymbol{\xi_L}(1)Y_q^*(z)$. We then define the control law by setting $Y_{qL}(z)$ equal to $Y_{qL}^*(z)$. This leads to the proper control law

$$\begin{bmatrix} 1 & 2z^{-1} \\ 0 & -2 \end{bmatrix} U_q(z) = Y_{qL}^*(z) \tag{25.4.31}$$

or

$$u[k]^{(1)} = -2u[k-1]^{(2)} + y_{qL}^*[k]^{(1)} \qquad and \qquad u[k]^{(2)} = -0.5y_{qL}^*[k]^{(2)} \tag{25.4.32}$$

This leads to the result

$$Y_{qL}(z) = \boldsymbol{\xi_L}(z)Y_q(z) = Y_{qL}^*(z) = \boldsymbol{\xi_L}(1)Y_q^*(z) \qquad (25.4.33)$$

Finally,

$$Y_q(z) = [\boldsymbol{\xi_L}(z)]^{-1}\boldsymbol{\xi_L}(1)Y_q^*(z)$$

$$= \begin{bmatrix} z^{-1} & 0 \\ 3z^{-1} & z^{-2} \end{bmatrix} \begin{bmatrix} 1 & 0 \\ -3 & 1 \end{bmatrix} Y_q^*(z) \qquad (25.4.34)$$

$$= \begin{bmatrix} z^{-1} & 0 \\ 3z^{-1}(1-z^{-1}) & z^{-2} \end{bmatrix} Y_q^*(z)$$

We can evaluate the resultant time response of the system as follows.

A unit step in $y_1^*[k]$ (the first component of $y^*[k]$) produces the following output sequences:

$$\{y_1[k]\} = [0\ 1\ 1\ 1\cdots] \qquad (25.4.35)$$
$$\{y_2[k]\} = [0\ 3\ 3\ 0\cdots] \qquad (25.4.36)$$

Similarly, a unit step in $y_2^*[k]$ (the second component of $y^*[k]$) produces

$$y_1[k] = [0\ 0\ 0\ 0\cdots] \qquad (25.4.37)$$
$$y_2[k] = [0\ 0\ 1\ 1\cdots] \qquad (25.4.38)$$

Note that $y^{(1)}$ responds in one sample but produces significant coupling into $y^{(2)}$.

We can remove this coupling by accepting a longer delay in the prediction. For example, instead of (25.4.33), we might use

$$Y_{qL}^*(z) = \boldsymbol{\xi_L}(z) \begin{bmatrix} z^{-2} & 0 \\ 0 & z^{-2} \end{bmatrix} Y_q^*(z) \qquad (25.4.39)$$

If we then set $Y_{qL}(z) = Y_{qL}^*(z)$, we obtain the causal control law

$$\xi_{\mathbf{L}}(z) \begin{bmatrix} z^{-2} & 0 \\ 0 & z^{-2} \end{bmatrix} Y^*(z) = \begin{bmatrix} 1 & 2z^{-1} \\ 0 & -2 \end{bmatrix} U_q(z) \tag{25.4.40}$$

which gives

$$Y_q(z) = \begin{bmatrix} z^{-2} & 0 \\ 0 & z^{-2} \end{bmatrix} Y_q^*(z) \tag{25.4.41}$$

So we see that we now have a dynamically decoupled system but that an extra delay (i.e., a zero at $z = \infty$) has been introduced in the response to a unit step in channel 1.

In summary, we see the following:

(i) *The interactor captures the minimum delay structure in the system, i.e., the structure of the zeros at ∞.*

(ii) *Exploiting this minimum delay structure in control design can lead to dynamic coupling.*

(iii) *We can get dynamic decoupling, but it could come at the expense of extra delay, i.e., extra zeros at ∞.*

Actually these kind of properties will be mirrored in more general MIMO designs, which we discuss below.

□□□

We can develop an analogous interpretation for the right interactor. Indeed, if we write

$$\mathbf{G_o}(s) = \mathbf{G_o}(s)\xi_{\mathbf{R}}(s)[\xi_{\mathbf{R}}(s)]^{-1} \tag{25.4.42}$$

then we see that $[\xi_{\mathbf{R}}(s)]^{-1}U(s)$ is a particular combination of *past* inputs that has the property that the high-frequency gain of the transfer function relating $[\xi_{\mathbf{R}}(s)]^{-1}U(s)$ to the output $Y(s)$ is nonsingular. This can also be used to develop a prototype control law by attempting to find a particular combination of *past* controls that bring the *current* output to some desired value. We again illustrate by a simple discrete-time example.

Example 25.3. *Consider again the transfer function given in (25.4.28). For this example, the right interactor is*

$$\boldsymbol{\xi_R}(z) = \begin{bmatrix} z & 0 \\ 0 & z^2 \end{bmatrix} \quad and \quad \mathbf{G_{oq}}(z)\boldsymbol{\xi_R}(z) = \begin{bmatrix} 1 & 2 \\ 3 & 4 \end{bmatrix} \tag{25.4.43}$$

We can use this to build a predictor for the current output *in terms of* combinations of past inputs. *In particular, we have*

$$Y_q(z) = \mathbf{G_{oq}}(z)U_q(z) = (\mathbf{G_{oq}}(z)\boldsymbol{\xi_R}(z))([\boldsymbol{\xi_R}(z)]^{-1}U_q(z))$$

$$= \begin{bmatrix} 1 & 2 \\ 3 & 4 \end{bmatrix} \begin{bmatrix} z^{-1}U_q^{(1)}(z) \\ z^{-2}U_q^{(2)}(z) \end{bmatrix} \tag{25.4.44}$$

where $U_q^{(1)}$ and $U_q^{(2)}$ are the first and second components of the input vector, respectively.

We can now attempt to determine that particular combination of past inputs that brings the current output to suitable past values of $y^*[k]$.

We define $U_q(z)$ by setting

$$\mathbf{G_o}(s)\boldsymbol{\xi_R}(s)[\boldsymbol{\xi_R}(s)]^{-1}U_q(z) = \begin{bmatrix} z^{-n_1} & 0 \\ 0 & z^{-n_2} \end{bmatrix} Y_q^*(z) \tag{25.4.45}$$

This leads to

$$U_q(z) = \boldsymbol{\xi_R}(z)[\mathbf{G_{oq}}(z)\boldsymbol{\xi_R}(z)]^{-1} \begin{bmatrix} z^{-n_1} & 0 \\ 0 & z^{-n_2} \end{bmatrix} Y_q^*(z) \tag{25.4.46}$$

where n_1 and n_2 are chosen to make the controller causal. For our example, we end up with

$$U_q(z) = \begin{bmatrix} z & 0 \\ 0 & z^2 \end{bmatrix} \begin{bmatrix} 1 & 2 \\ 3 & 4 \end{bmatrix}^{-1} \begin{bmatrix} z^{-2} & 0 \\ 0 & z^{-2} \end{bmatrix} Y_q^*(z) \tag{25.4.47}$$

This leads to

$$Y_q(z) = \begin{bmatrix} z^{-2} & 0 \\ 0 & z^{-2} \end{bmatrix} Y_q^*(z) \tag{25.4.48}$$

□□□

25.4.2   Approximate Inverses

We next show how interactors can be used to construct approximate inverses accounting for relative degree.

A crucial property of $\boldsymbol{\xi_L}(s)$ and $\boldsymbol{\xi_R}(s)$ is that

$$\boldsymbol{\Lambda_R}(s) \triangleq \mathbf{G_o}(s)\boldsymbol{\xi_R}(s) \qquad \text{and} \qquad \boldsymbol{\Lambda_L}(s) \triangleq \boldsymbol{\xi_L}(s)\mathbf{G_o}(s) \tag{25.4.49}$$

are both biproper transfer functions having nonsingular high-frequency gain. This simplifies the problem of inversion. Note that $\boldsymbol{\Lambda_R}(s)$ and $\boldsymbol{\Lambda_L}(s)$ both have a state space representation of the form

$$\dot{x}(t) = \mathbf{A}x(t) + \overline{\mathbf{B}}u(t) \tag{25.4.50}$$
$$y(t) = \mathbf{C}x(t) + \overline{\mathbf{D}}u(t) \tag{25.4.51}$$

where $\det \overline{\mathbf{D}} \neq 0$. Note also that $\mathbf{A}$ and $\mathbf{C}$ are the same as in the plant description (22.2.1)-(22.2.2).

The key point about (25.4.50) and (25.4.51) is that the exact inverse of $\boldsymbol{\Lambda}(s)$ can be obtained by simply reversing the roles of input and output, to yield the following state space realization of $[\boldsymbol{\Lambda}(s)]^{-1}$:

$$\dot{x}(t) = \mathbf{A}x(t) + \overline{\mathbf{B}}\,\overline{\mathbf{D}}^{-1}(y(t) - \mathbf{C}x(t)) \tag{25.4.52}$$
$$= \mathbf{A_\lambda}x(t) + \mathbf{B_\lambda}\tilde{u}(t) \tag{25.4.53}$$
$$\overline{u}(t) = \overline{\mathbf{D}}^{-1}(y(t) - \mathbf{C}x(t)) \tag{25.4.54}$$
$$= \mathbf{C_\lambda}x(t) + \mathbf{D_\lambda}\tilde{u}(t) \tag{25.4.55}$$

where $\tilde{u}(t)$ denotes the input to the inverse, $\tilde{u}(t) = y(t)$, and $\overline{u}(t)$ denotes the output of the inverse. Also, in (25.4.52) to (25.4.54),

$$\mathbf{A_\lambda} = \mathbf{A} - \overline{\mathbf{B}}\,\overline{\mathbf{D}}^{-1}\mathbf{C} \qquad\qquad \mathbf{B_\lambda} = \overline{\mathbf{B}}\,\overline{\mathbf{D}}^{-1} \tag{25.4.56}$$
$$\mathbf{C_\lambda} = -\overline{\mathbf{D}}^{-1}\mathbf{C} \qquad\qquad \mathbf{D_\lambda} = \overline{\mathbf{D}}^{-1} \tag{25.4.57}$$

We can use $[\mathbf{\Lambda_L}(s)]^{-1}$ or $[\mathbf{\Lambda_R}(s)]^{-1}$ to construct various approximations to the inverse of $\mathbf{G_o}(s)$. For example,

$$\mathbf{G_R^{inv}}(s) \triangleq [\boldsymbol{\xi_L}(s)\mathbf{G_o}(s)]^{-1}\boldsymbol{\xi_L}(0) \qquad (25.4.58)$$

is an approximate *right inverse* with the property

$$\mathbf{G_o}(s)\mathbf{G_R^{inv}}(s) = [\boldsymbol{\xi_L}(s)]^{-1}\boldsymbol{\xi_L}(0) \qquad (25.4.59)$$

which is lower triangular, and equal to the identity matrix at d.c.
    Similarly

$$\mathbf{G_L^{inv}}(s) \triangleq \boldsymbol{\xi_R}(0)[\mathbf{G_o}(s)\boldsymbol{\xi_R}(s)]^{-1} \qquad (25.4.60)$$

is an approximate *left inverse* with the property

$$\mathbf{G_L^{inv}}(s)\mathbf{G_o}(s) = \boldsymbol{\xi_R}(0)[\boldsymbol{\xi_R}(s)]^{-1} \qquad (25.4.61)$$

which is also lower triangular, and equal to the identity matrix at d.c.
    With the above tools in hand, we return to the original problem of constructing $\mathbf{Q}(s)$ as an (approximate) inverse for $\mathbf{G_o}(s)$. For example, we could choose $\mathbf{Q}(s)$ as

$$\mathbf{Q}(s) = [\mathbf{\Lambda_L}(s)]^{-1}\boldsymbol{\xi_L}(0) = [\boldsymbol{\xi_L}(s)\mathbf{G_o}(s)]^{-1}\boldsymbol{\xi_L}(0) \qquad (25.4.62)$$

With this choice, we find that

$$\begin{aligned}
\mathbf{T_o}(s) &= \mathbf{G_o}(s)\mathbf{Q}(s) \\
&= \mathbf{G_o}(s)[\mathbf{\Lambda_L}(s)]^{-1}\boldsymbol{\xi_L}(0) \qquad (25.4.63)\\
&= [\boldsymbol{\xi_L}(s)]^{-1}\boldsymbol{\xi_L}(0)
\end{aligned}$$

    Thus, by choice of the relative-degree-modifying factors $(s + \alpha)$ we can make $\mathbf{T_o}(s)$ equal to $\mathbf{I}$ at d.c. and triangular at other frequencies, with bandwidth determined by the factors $(s + \alpha)$ used in forming $\boldsymbol{\xi_L}(s)$.

25.5   Dealing with NMP Zeros

25.5.1   Z-Interactors

We saw in subsection §25.4.1 that interactor matrices are a convenient way of describing the relative degree or zeros at ∞ of a plant. Also, we saw that the interactor matrix can be used to precompensate the plant so as to isolate the zeros at

∞, thus allowing a *proper* inverse to be computed for the remainder. The same basic idea can be used to describe the structure of finite zeros. The appropriate transformations are known as *z-interactors*. They allow a precompensator to be computed that isolates particular finite zeros. In particular, when applied to isolating the nonminimum-phase zeros and combined with interactors for the zeros at ∞, *z-interactors* allow a *stable* and *proper* inverse to be computed.

The extension of Theorem 25.1 on page 814 to a finite zero is the following.

Lemma 25.1. *Consider a nonsingular $m \times m$ matrix $\mathbf{G}(s)$. Assume that this matrix has a real NMP zero located at $s = z_o$. Then there exist upper- and lower-triangular matrices, $\boldsymbol{\psi}_{\mathbf{R}}(s)$ and $\boldsymbol{\psi}_{\mathbf{L}}(s)$, respectively, such that*

$$\lim_{s \to z_o} \boldsymbol{\psi}_{\mathbf{L}}(s)\mathbf{G}(s) = \mathbf{K}_{\mathbf{Lz}} \qquad 0 < |\det(\mathbf{K}_{\mathbf{Lz}})| < \infty \qquad (25.5.1)$$

$$\lim_{s \to z_o} \mathbf{G}(s)\boldsymbol{\psi}_{\mathbf{R}}(s) = \mathbf{K}_{\mathbf{Rz}} \qquad 0 < |\det(\mathbf{K}_{\mathbf{Rz}})| < \infty \qquad (25.5.2)$$

with

$$\boldsymbol{\psi}_{\mathbf{L}}(s) = \mathbf{H}_{\mathbf{L}}(v)\mathbf{D}_{\mathbf{L}}(v) \qquad (25.5.3)$$

$$\mathbf{D}_{\mathbf{L}}(v) = \text{diag}\,(v^{p_1}, \dots, v^{p_m}) \qquad (25.5.4)$$

$$\mathbf{H}_{\mathbf{L}}(v) = \begin{bmatrix} 1 & 0 & \cdots & \cdots & 0 \\ h_{21}^L(v) & 1 & \cdots & \cdots & 0 \\ h_{31}^L(v) & h_{32}^L(v) & \ddots & & \vdots \\ \vdots & \vdots & & \ddots & \vdots \\ h_{m1}^L(v) & h_{m2}^L(v) & \cdots & \cdots & 1 \end{bmatrix} \qquad (25.5.5)$$

$$\boldsymbol{\psi}_{\mathbf{R}}(s) = \mathbf{D}_{\mathbf{R}}(v)\mathbf{H}_{\mathbf{R}}(v) \qquad (25.5.6)$$

$$\mathbf{D}_{\mathbf{R}}(v) = \text{diag}\,(v^{q_1}, \dots, v^{q_m}) \qquad (25.5.7)$$

$$\mathbf{H}_{\mathbf{R}}(v) = \begin{bmatrix} 1 & h_{12}^R(v) & h_{13}^R(v) & \cdots & h_{1m}^R(v) \\ 0 & 1 & h_{23}^R(v) & \cdots & h_{2m}^R(v) \\ \vdots & \vdots & \ddots & & \vdots \\ \vdots & \vdots & & \ddots & \vdots \\ 0 & 0 & \cdots & \cdots & 1 \end{bmatrix} \qquad (25.5.8)$$

where $h_{ij}^L(v)$ and $h_{ij}^R(v)$ are polynomials in v satisfying $h_{ij}^L(0) = 0$ and $h_{ij}^R(0) = 0$ and where

$$v = -\frac{sz_o}{s - z_o} \qquad (25.5.9)$$

Proof

We observe that the mapping (25.5.9) converts the zero at $s = z_o$ to a zero at $v = \infty$. We can then apply Theorem 25.1.

□□□

Remark 25.3. *As was the case in Theorem 25.1, z-interactors can also be defined by using diagonal matrices $\mathbf{D_L}(v)$ and $\mathbf{D_R}(v)$ with arbitrary polynomial diagonal entries with degrees $p_1, p_2, \ldots, p_m$, which are invariants of the interactor representation for a given matrix $\mathbf{G}(s)$. In particular, we can replace v by $(v + \alpha)$, where $\alpha \in \mathbb{R}^+$. Then (25.5.9) has to be transformed, accordingly, to*

$$v = -\frac{s(\alpha + z_o)}{s - z_o} \tag{25.5.10}$$

Then $h_{ij}^L(v)$ and $h_{ij}^R(v)$ become polynomials in $(v + \alpha)$, satisfying $h_{ij}^L(-\alpha) = 0$ and $h_{ij}^R(-\alpha) = 0$.

Remark 25.4. *Note that these interactors remove the NMP zero, in the sense that the products $\boldsymbol{\psi_L}(s)\mathbf{G}(s)$ and $\mathbf{G}(s)\boldsymbol{\psi_R}(s)$ are nonsingular at $s = z_o$. This also means that $[\boldsymbol{\psi_L}(s)]^{-1}$ and $[\boldsymbol{\psi_R}(s)]^{-1}$ are singular for $s = z_o$.*

We illustrate the above ideas with the following example.

Example 25.4. *Consider a plant having the nominal model*

$$\mathbf{G_o}(s) = \frac{1}{(s+1)(s+2)} \begin{bmatrix} s+1 & 1 \\ 2 & 1 \end{bmatrix} \tag{25.5.11}$$

We observe that this system has a NMP zero located at $s = z_o = 1$. We define the transformation (25.5.9), which yields

$$v = -\frac{s z_o}{(s - z_o)} = -\frac{s}{s-1} \iff s = \frac{v}{v+1} \tag{25.5.12}$$

This transformation, when applied to $\mathbf{G_o}(s)$ leads to

$$\mathbf{V_o}(v) \overset{\triangle}{=} \mathbf{G_o}(s)\Big|_{s=\frac{v}{v+1}} = \frac{1}{(2v+1)(3v+2)} \begin{bmatrix} (v+1)(2v+1) & (v+1)^2 \\ 2(v+1)^2 & (v+1)^2 \end{bmatrix} \tag{25.5.13}$$

We will next compute the left interactor (for relative degree) for this transformed plant. Note that, although every entry in $\mathbf{V_o}(v)$ is biproper, the matrix itself is strictly proper, because its determinant vanishes for $v = \infty$. We use the construction procedure outlined in the proof of Theorem 25.1.

Note that $n_1 = 0$ with $f_1 = \frac{1}{6}[2 \quad 1]$ and $n_2 = 0$ with $f_2 = \frac{1}{6}[2 \quad 1]$.

(i) We first form

$$[\boldsymbol{\xi_L}]_{1*}(v) = [v^{n_1} \ 0 \ 0 \dots 0] = [1 \ 0] \tag{25.5.14}$$

Then

$$\lim_{v \to \infty} [\boldsymbol{\xi_L}]_{1*}(v)\mathbf{V_o}(v) = r_1^T = [1 \ 0] \tag{25.5.15}$$

(ii) Consider the row vector f_2^T. f_2 is linearly dependent of r_1, with $\beta_2^1 = 1$, i.e., $f_2 = r_1$. We then choose the second row of $\boldsymbol{\xi_L}(v)$, $[\boldsymbol{\xi_L}(v)]_{2*}$, as

$$[\boldsymbol{\xi_L}(v)]_{2*}^1 = v^x([0 \ 1] - \beta_2^1[\boldsymbol{\xi_L}(v)]_{1*}) = [-v^x \ v^x] \tag{25.5.16}$$

where x is found by noting that

$$\lim_{v \to \infty} [\boldsymbol{\xi_L}(v)]_{2*}^1 \mathbf{V_o}(v) = (r_2^1)^T = \lim_{v \to \infty} \left[\frac{v^x(v+1)}{(3v+2)(2v+1)} \quad 0 \right] \tag{25.5.17}$$

From this, we obtain $x = 1$. This leads to

$$r_2^1 = [1 \ 0]^T \tag{25.5.18}$$

which is linearly independent of f_1^T. Thus, the choice (25.5.16), with $x = 1$, is a valid choice as the second row of the interactor matrix. Thus,

$$\boldsymbol{\xi_L}(v) = \begin{bmatrix} 1 & 0 \\ -v & v \end{bmatrix} \tag{25.5.19}$$

(iii) We want to have $[\boldsymbol{\xi_L}(v)]^{-1}$ stable, without losing the essential nature of the interactor, so we can replace v by $v + \alpha$ with, say, $\alpha = 1$, leading to

$$\boldsymbol{\xi_L}(v) = \begin{bmatrix} 1 & 0 \\ -(v+1) & (v+1) \end{bmatrix} \tag{25.5.20}$$

We can now transform back the interactor matrix $\boldsymbol{\xi_L}(v)$, to obtain $\boldsymbol{\psi_L}(s)$ as

$$\boldsymbol{\psi_L}(s) = \begin{bmatrix} 1 & 0 \\ \dfrac{s+1}{s-1} & -\dfrac{s+1}{s-1} \end{bmatrix} \quad and \quad [\boldsymbol{\psi_L}(s)]^{-1} = \begin{bmatrix} 1 & 0 \\ 1 & -\dfrac{s-1}{s+1} \end{bmatrix} \tag{25.5.21}$$

Note that $\det(\boldsymbol{\psi_L}(s)) = -\frac{s+1}{s-1}$. This ensures that $\det(\boldsymbol{\psi_L}(s)\mathbf{G_o}(s))$ is non-singular for $s = z_o = 1$.

25.5.2   Q Synthesis by using Interactors and Z-Interactors

We will next show how z-interactors can be used to design the controller in the Q-parameterized structure for NMP plants.

For illustration purposes, say that $\mathbf{G_o}$ is a stable transfer-function matrix having one NMP zero located at $s = z_o$. We recall from subsection §25.5.1 that, if $\boldsymbol{\psi_L}(s)$ is a left z-interactor matrix, and if $\boldsymbol{\psi_R}(s)$ is a right z-interactor matrix, then

$$\lim_{s \to z_o} \boldsymbol{\psi_L}(s)\mathbf{G_o}(s) = \mathbf{K_{Lz}} \qquad 0 < |\det(\mathbf{K_{Lz}})| < \infty \qquad (25.5.22)$$

$$\lim_{s \to z_o} \mathbf{G_o}(s)\boldsymbol{\psi_R}(s) = \mathbf{K_{Rz}} \qquad 0 < |\det(\mathbf{K_{Rz}})| < \infty \qquad (25.5.23)$$

We also recall that, by construction, matrices $\boldsymbol{\psi_L}(s)$ and $\boldsymbol{\psi_R}(s)$ have, amongst others, the following properties:

- $\boldsymbol{\psi_L}(s)$ and $\boldsymbol{\psi_R}(s)$ are unstable matrices;

- $\boldsymbol{\psi_L}(s)$ and $\boldsymbol{\psi_R}(s)$ are lower- and upper-triangular matrices, respectively, having entries of the form

$$\left[-\frac{z_o(\alpha + s)}{s - z_o}\right]^k \qquad \text{where} \quad k \in \mathbb{N} \quad \text{and} \quad \alpha \in \mathbb{R}^+ \qquad (25.5.24)$$

- $[\boldsymbol{\psi_L}(s)]^{-1}$ and $[\boldsymbol{\psi_R}(s)]^{-1}$ are stable triangular matrices having diagonal elements of the form

$$\left[-\frac{s - z_o}{z_o(\alpha + s)}\right]^k \qquad (25.5.25)$$

- $\lim_{s \to \infty} \boldsymbol{\psi_L}(s) = \mathbf{Z_L}$ and $\lim_{s \to \infty} \boldsymbol{\psi_R}(s) = \mathbf{Z_R}$ are matrices with finite nonzero determinant.

- $\det\big(\boldsymbol{\psi_L}(0)\big) = \det\big(\boldsymbol{\psi_R}(0)\big) = 1$. (This property is a consequence of (25.5.24).)

Consider the following choice for $\mathbf{Q}(s)$:

$$\mathbf{Q}(s) = [\boldsymbol{\xi_L}(s)\mathbf{H_o}(s)]^{-1}\boldsymbol{\xi_L}(s)\mathbf{D_Q}(s) = [\mathbf{H_o}(s)]^{-1}\mathbf{D_Q}(s) \qquad (25.5.26)$$

where $\mathbf{H_o}(s)$ is the plant precompensated by the left z-interactor $\boldsymbol{\psi_L}(s)$:

$$\mathbf{H_o}(s) \triangleq \boldsymbol{\psi_L}(s)\mathbf{G_o}(s) \qquad (25.5.27)$$

and $\boldsymbol{\xi_L}(s)$ is the left interactor for the zero at ∞. $\mathbf{D_Q}(s)$ in (25.5.26) is a stable matrix with other properties to be defined below.

Note that, by virtue of the precompensation, $\mathbf{H_o}(s)$ is a stable and minimum-phase matrix.

Using (25.2.2), (25.5.26), and (25.5.27), we see that the complementary sensitivity is given by

$$\mathbf{T_o}(s) = \mathbf{G_o}(s)\mathbf{Q}(s) = [\boldsymbol{\psi_L}(s)]^{-1}\mathbf{H_o}(s)\mathbf{Q}(s) = [\boldsymbol{\psi_L}(s)]^{-1}\mathbf{D_Q}(s) \qquad (25.5.28)$$

We can now see that $\mathbf{D_Q}(s)$ can be used to shape the structure and frequency response of the MIMO loop (as constrained by the unavoidable presence of the plant nonminimum-phase zero). Thus, $\mathbf{D_Q}(s)$ should be chosen to achieve the desired dynamics in the different channels. For example, we can always achieve triangular coupling by choosing $\mathbf{D_Q}(s)$ as a lower-triangular matrix. The relative degree of the entries in $\mathbf{D_Q}(s)$ must be chosen to make $\mathbf{Q}(s)$ proper (usually, biproper).

Note that the presence of $[\boldsymbol{\psi_L}(s)]^{-1}$ in (25.5.28) ensures internal stability, by forcing the NMP zero of $\mathbf{G_o}$ in $\mathbf{T_o}(s)$. Also, $[\boldsymbol{\psi_L}(s)]^{-1}$ contributes to the loop dynamics with poles located at $s = -\alpha$ (see (25.5.25)), which are usually chosen much larger than the required bandwidth.

We illustrate by a simple example.

Example 25.5. *Consider a plant having the nominal model*

$$\mathbf{G}(s) = \frac{1}{(s+1)(s+3)} \begin{bmatrix} s+1 & 1 \\ 2 & 1 \end{bmatrix} \qquad (25.5.29)$$

Design a MIMO control loop with a bandwidth of, approximately, $0.5[rad/s]$ in each channel.

Solution

This is a stable and strictly proper plant. However, it has a NMP zero located at $s = z_o = 1$. We can thus use z-interactors to synthesize $\mathbf{Q}(s)$. (See subsections §25.5.1 and §25.5.2.)

We first need to find the left z-interactor $\boldsymbol{\psi_L}(s)$ for the matrix $\mathbf{G_o}(s)$ and the matrix $\mathbf{H_o}(s)$.

To compute $\boldsymbol{\psi_L}(s)$, we note that this plant is similar to that in example 25.4, where $\alpha = 1$ was chosen. It is then straightforward to prove that

$$\psi_{\mathbf{L}}(s) = \begin{bmatrix} 1 & 0 \\ \dfrac{s+1}{s-1} & -\dfrac{s+1}{s-1} \end{bmatrix} \quad and \quad [\psi_{\mathbf{L}}(s)]^{-1} = \begin{bmatrix} 1 & 0 \\ 1 & -\dfrac{s-1}{s+1} \end{bmatrix} \quad (25.5.30)$$

We then compute $\mathbf{H_o}(s)$, which is given by

$$\mathbf{H_o}(s) = \psi_{\mathbf{L}}(s)\mathbf{G_o}(s) = \frac{1}{(s+1)(s+3)} \begin{bmatrix} s+1 & 1 \\ s+1 & 0 \end{bmatrix} \quad (25.5.31)$$

We also need to compute the exact model inverse $[\mathbf{G_o}(s)]^{-1}$. This is given by

$$[\mathbf{G_o}(s)]^{-1} = \frac{(s+1)(s+3)}{(s-1)} \begin{bmatrix} 1 & -1 \\ -2 & s+1 \end{bmatrix} \quad (25.5.32)$$

From (25.5.28), and the above expressions we have that

$$\mathbf{T_o}(s) = [\psi_{\mathbf{L}}(s)]^{-1}\mathbf{D_Q}(s) = \begin{bmatrix} 1 & 0 \\ 1 & -\dfrac{s-1}{s+1} \end{bmatrix} \mathbf{D_Q}(s) \quad (25.5.33)$$

We first consider a choice of $\mathbf{D_Q}(s)$ to make $\mathbf{T_o}(s)$ diagonal. This can be achieved with a lower triangular $\mathbf{D_Q}(s)$:

$$\mathbf{D_Q}(s) = \begin{bmatrix} D_{11}(s) & 0 \\ D_{21}(s) & D_{22}(s) \end{bmatrix} \quad (25.5.34)$$

Then,

$$\mathbf{T_o}(s) = \begin{bmatrix} 1 & 0 \\ 1 & -\dfrac{s-1}{s+1} \end{bmatrix} \begin{bmatrix} D_{11}(s) & 0 \\ D_{21}(s) & D_{22}(s) \end{bmatrix} = \begin{bmatrix} D_{11}(s) & 0 \\ D_{11} - \dfrac{s-1}{s+1}D_{21}(s) & -\dfrac{s-1}{s+1}D_{22}(s) \end{bmatrix}$$
$$(25.5.35)$$

and

$$\mathbf{Q}(s) = [\mathbf{H_o}(s)]^{-1}\mathbf{D_Q}(s) = (s+1)(s+3) \begin{bmatrix} 0 & \dfrac{1}{s+1} \\ 1 & -1 \end{bmatrix} \begin{bmatrix} D_{11}(s) & 0 \\ D_{21}(s) & D_{22}(s) \end{bmatrix}$$

$$= \begin{bmatrix} (s+3)D_{21}(s) & (s+3)D_{22}(s) \\ (s+1)(s+3)(D_{11}(s) - D_{21}(s)) & -(s+1)(s+3)D_{22}(s) \end{bmatrix}$$

$$(25.5.36)$$

We immediately observe from (25.5.35) that, to make $\mathbf{T_o}(s)$ *diagonal, it is necessary that* $(s+1)D_{11}(s) = (s-1)D_{21}(s)$. *This means that the NMP zero will appear in channel 1 as well as in channel 2 of the closed loop. As we will see in Chapter 26, the phenomenon that decoupling forces NMP zeros into multiple channels is not peculiar to this example but is more generally a trade-off associated with full dynamic decoupling.*

If, instead, we decide to achieve only triangular decoupling, we can avoid having the NMP zero in channel 1. Before we choose $\mathbf{D_Q}(s)$, *we need to determine, from (25.5.36), the necessary constraints on the degrees of* $D_{11}(s)$, $D_{21}(s)$, *and* $D_{22}(s)$, *so as to achieve a biproper* $\mathbf{Q}(s)$.

From (25.5.36), we see that $\mathbf{Q}(s)$ *is biproper if the following conditions are simultaneously satisfied:*

(c1) relative degree of $D_{22}(s)$ *equal to 2;*

(c2) relative degree of $D_{21}(s)$ *equal to 1;*

(c3) relative degree of $D_{11}(s) - D_{21}(s)$ *equal to 2.*

Conditions c2 and c3 are simultaneously satisfied if $D_{11}(s)$ *has relative degree 1, while at the same time* $D_{11}(s)$ *and* $D_{21}(s)$ *have the same high-frequency gain.*

Furthermore, we assume that it is required that the MIMO control loop be decoupled at low frequencies. (Otherwise, steady-state errors appear for constant references and disturbances.) From equation (25.5.35), we see that this goal is attained if $D_{11}(s)$ *and* $D_{21}(s)$ *have low-frequency gains of equal magnitudes but opposite signs.*

A suitable choice that simultaneously achieves both biproperness of $\mathbf{Q}(s)$ *and decoupling at low frequencies is*

$$D_{21}(s) = \frac{s - \beta}{s + \beta} D_{11}(s) \qquad (25.5.37)$$

where $\beta \in \mathbb{R}^+$ *is much larger than the desired bandwidth, say* $\beta = 5$.

We can now proceed to choose $\mathbf{D_Q}(s)$ in such a way that (25.5.37) is satisfied, together with the bandwidth constraints.

A possible choice is

$$
\mathbf{D_Q}(s) = \begin{bmatrix} \dfrac{0.5(s+0.5)}{s^2 + 0.75s + 0.25} & 0 \\[4mm] \dfrac{0.5(s+0.5)(s-5)}{(s^2+0.75s+0.25)(s+5)} & \dfrac{0.25}{s^2+0.75s+0.25} \end{bmatrix} \tag{25.5.38}
$$

The performance of this design can be evaluated via simulation with the SIMULINK file **mimo3.mdl**. Note that you must first run the MATLAB program in file **pmimo3.m**.

□□□

Remark 25.5. When $\mathbf{G_o}(s)$ has $n_z > 1$ (different) NMP zeros, say at $z_{o1}, \ldots, z_{on_z}$, then the above approach can still be applied if we choose

$$
\mathbf{Q}(s) = [\boldsymbol{\xi}_\mathbf{L}(s)\mathbf{H_o}(s)]^{-1}\,\boldsymbol{\xi}_\mathbf{L}(s)\mathbf{D_Q}(s) \tag{25.5.39}
$$

where $\mathbf{H_o}(s)$ is now defined as

$$
\mathbf{H_o}(s) \triangleq \prod_{k=1}^{n_z} \boldsymbol{\psi}_\mathbf{L}(s)^{(k)}\mathbf{G_o}(s) \tag{25.5.40}
$$

and where $\boldsymbol{\psi}_\mathbf{L}(s)^{(k)}$ is the z-interactor for the pole located at $s = z_{ok}$.

25.5.3   Q Synthesis as a Model-Matching Problem

Although the use of interactors and z-interactors gives insight into the principal possibilities and fundamental structure of Q-synthesis for MIMO design, the procedure described in subsection §25.5.2 is usually not appropriate for numerically carrying out the synthesis. In addition to being difficult to automate as a numerical algorithm, the procedure would require the analytical removal of unstable pole-zero cancellation, which is awkward. The difficulty lies entirely with the z-interactor part of the computation; robust state space methods are readily available for computing the normal interactor.

In this section, we therefore investigate an alternative method for computing a stable approximate inverse by using the model-matching procedures of section §22.6. This circumvents the need for using z-interactors.

We first turn the plant into biproper form by using a normal interactor, $\boldsymbol{\xi}_\mathbf{L}(s)$.

Let us assume that the target complementary sensitivity is $\mathbf{T}^*(\mathbf{s})$. We know from (25.2.2) that, under the MIMO affine parameterization for the controller, the nominal sensitivity is $\mathbf{G_o}(s)\mathbf{Q}(s)$. Hence we can convert the Q-synthesis problem into a model-matching problem by choosing $\mathbf{Q}(s)$ to minimize

$$J = \frac{1}{2\pi}\int_{-\infty}^{\infty}\left\|\mathbf{M}(j\omega) - \mathbf{N}(j\omega)\mathbf{\Gamma}(j\omega)\right\|_F^2\,d\omega \qquad (25.5.41)$$

where $\mathbf{M}$, $\mathbf{N}$, and $\mathbf{\Gamma}$ correspond to $\mathbf{T}^*$, $\mathbf{\xi_L}\mathbf{G_o}$, and $\mathbf{Q}$, respectively. We can then, at least in principle, apply the methods of subsection §25.5.4 to synthesize $\mathbf{Q}$.

We begin by examining $\mathbf{Q}$ one column at a time. For the ith column, we have

$$J_i = \frac{1}{2\pi}\int_{-\infty}^{\infty}\left\|[\mathbf{M}(j\omega)]_{*i} - \mathbf{N}(j\omega)[\mathbf{\Gamma}(j\omega)]_{*i}\right\|_F^2\,d\omega \qquad (25.5.42)$$

As in subsection §25.5.4, we convert to the time domain and use

$$\dot{x}_1(t) = \mathbf{A_1}x_1(t); \qquad x_1(0) = \mathbf{B_1} \qquad (25.5.43)$$
$$y_1(t) = \mathbf{C_1}x_1(t) \qquad (25.5.44)$$

to represent the system with transfer function $[\mathbf{M}(s)]_{*i}$, and we use

$$\dot{\tilde{x}}_2(t) = \mathbf{A_2}\tilde{x}_2(t) + \tilde{\mathbf{B}}_2 u(t) \qquad (25.5.45)$$
$$z_2(t) = \mathbf{C_2}\tilde{x}_2(t) + \tilde{\mathbf{D}}u(t) \qquad (25.5.46)$$

to represent the system with biproper transfer function $\mathbf{\xi_L}(s)\mathbf{G_o}(s)$. Also, for square plants, we know from the properties of $\mathbf{\xi_L}(s)$ that $\det\{\mathbf{D}\} \neq 0$.

This seems to fit the theory given in section §22.6 for Model Matching. However, an important difference is that here we have no weighting on the control effort $u(t)$. This was not explicitly allowed in the earlier work. Thus, we pause to extend the earlier results to cover the case where no control weighting is used.

25.5.4   Model Matching with Unweighted Solution

Here we consider a special case of the general MMP set-up. Specifically, in the basic MMP, as described in (22.6.3), we seek an answer to the problem in which the *size* of $\mathbf{\Theta}$ is of no concern. In this case, we take $\mathbf{R} = 0$ and $\mathbf{\Gamma} = \mathbf{I}$.

However, a difficulty is that this setting leads to an equivalent LQR problem where the weighting penalizing the control effort is zero. This was not allowed in the original LQR set-up. It can be rectified in a number of ways. We will show how one can get around the problem by using a left interactor $\mathbf{\xi_L}(s)$. (See subsection §25.4.1.) We choose the elements of the interactors as $(\tau s + 1)^n$ (for τ small) to change the state space model for y_2 into one for $Z_2(s) = \mathbf{\xi_L}(s)Y_2(s)$ having a direct path from u to z_2. Then $z_2(t)$ satisfies a state space model of the form

$$\dot{\tilde{x}}_2(t) = \mathbf{A_2}\tilde{x}_2(t) + \mathbf{\tilde{B}_2}u(t) \tag{25.5.47}$$

$$z_2(t) = \mathbf{C_2}\tilde{x}_2(t) + \mathbf{\tilde{D}}u(t) \tag{25.5.48}$$

where $\mathbf{\tilde{D}}^T\mathbf{\tilde{D}}$ is nonsingular.

The optimal control form of the MMP then changes to

$$J = \int_0^\infty \|y_1(t) - z_2(t)\|_2^2 \, dt = \int_0^\infty \begin{bmatrix} \tilde{x}^T(t) & u^T(t) \end{bmatrix} \begin{bmatrix} \mathbf{\tilde{\Psi}} & \mathbf{S}^T \\ \mathbf{S} & \mathbf{\Phi} \end{bmatrix} \begin{bmatrix} \tilde{x}(t) \\ u(t) \end{bmatrix} \, dt \tag{25.5.49}$$

where

$$\frac{d\tilde{x}(t)}{dt} = \mathbf{\tilde{A}}\tilde{x}(t) + \mathbf{\tilde{B}}u(t); \qquad \tilde{x}(0) = \begin{bmatrix} \mathbf{B_1} \\ \mathbf{0} \end{bmatrix} \tag{25.5.50}$$

$$\mathbf{\Phi} = \mathbf{\tilde{D}}^T\mathbf{\tilde{D}} \qquad \mathbf{\tilde{C}} = \begin{bmatrix} \mathbf{C_1} & -\mathbf{C_2} \end{bmatrix} \qquad \mathbf{\tilde{\Psi}} = \mathbf{\tilde{C}}^T\mathbf{\tilde{C}} \tag{25.5.51}$$

$$\mathbf{S}^T = -\mathbf{\tilde{C}}^T\mathbf{\tilde{D}} \qquad \mathbf{\tilde{A}} = \begin{bmatrix} \mathbf{A_1} & \mathbf{0} \\ \mathbf{0} & \mathbf{A_2} \end{bmatrix} \qquad \mathbf{\tilde{B}} = \begin{bmatrix} \mathbf{0} \\ \mathbf{\tilde{B}_2} \end{bmatrix} \tag{25.5.52}$$

The cost function has cross coupling between $\tilde{x}(t)$ and $u(t)$, but this can be removed by redefining the control law as

$$u(t) = \overline{u}(t) - \mathbf{\Phi}^{-1}\mathbf{S}\tilde{x}(t) \tag{25.5.53}$$

This transforms the problem into

$$\frac{\tilde{x}(t)}{dt} = \overline{\mathbf{A}}\tilde{x}(t) + \overline{\mathbf{B}}\overline{u}(t) \tag{25.5.54}$$

$$J_t = \int_0^\infty \left\{ \tilde{x}(t)^T \, \overline{\mathbf{\Psi}} \, \tilde{x}(t) + \overline{u}(t)^T \, \overline{\mathbf{\Phi}} \, \overline{u}(t) \right\} dt \tag{25.5.55}$$

where

$$\overline{\mathbf{A}} = \tilde{\mathbf{A}} - \tilde{\mathbf{B}}\mathbf{\Phi}^{-1}\mathbf{S} \tag{25.5.56}$$

$$\overline{\mathbf{B}} = \tilde{\mathbf{B}} \tag{25.5.57}$$

$$\overline{\mathbf{\Psi}} = \tilde{\mathbf{C}}^T[\mathbf{I} - \tilde{\mathbf{D}}\mathbf{\Phi}^{-1}\tilde{\mathbf{D}}^T]\tilde{\mathbf{C}} \tag{25.5.58}$$

$$\overline{\mathbf{\Phi}} = \mathbf{\Phi} = \tilde{\mathbf{D}}^T\tilde{\mathbf{D}} \tag{25.5.59}$$

Now we see from section §22.5 that a stabilizing solution to the problem exists, provided that $(\overline{\mathbf{A}}, \overline{\mathbf{\Psi}}^{\frac{1}{2}})$ has no unobservable zeros on the imaginary axis. Under these conditions, we can express the optimal $\overline{u}(t)$ as

$$\overline{u}(t) = -\overline{\mathbf{K}}_s\tilde{x}(t) = - \begin{bmatrix} \mathbf{K}_s^1 & \mathbf{K}_s^2 \end{bmatrix} \tilde{x}(t) \tag{25.5.60}$$

or, using (25.5.53), as

$$u(t) = -\tilde{\mathbf{K}}\tilde{x}(t) = - \begin{bmatrix} \tilde{\mathbf{K}}_1 & \tilde{\mathbf{K}}_2 \end{bmatrix} \tilde{x}(t) \tag{25.5.61}$$

where

$$\tilde{\mathbf{K}}_1 = \mathbf{K}_s^1 - \tilde{\mathbf{D}}^{-1}\mathbf{C}_1$$

$$\tilde{\mathbf{K}}_2 = \mathbf{K}_s^2 + \tilde{\mathbf{D}}^{-1}\mathbf{C}_2$$

As in subsection §22.6.3, we seek that particular state feedback for the original system that solves the optimization problem. Thus, $u(t)$ can be generated from

$$\frac{d\tilde{x}(t)}{dt} = \tilde{\mathbf{A}}\tilde{x}(t) + \tilde{\mathbf{B}}u(t); \qquad \tilde{x}(0) = \begin{bmatrix} \mathbf{B}_1 \\ \mathbf{0} \end{bmatrix} \tag{25.5.62}$$

$$u(t) = -\tilde{\mathbf{K}}\tilde{x}(t) \tag{25.5.63}$$

Taking the Laplace transform gives the final realization of the optimal $\mathbf{\Theta}(s)$ as

$$\mathbf{\Theta}(s) = [-\mathbf{I} + \tilde{\mathbf{K}}_2 \left(s\mathbf{I} - \mathbf{A}_2 + \tilde{\mathbf{B}}_2\tilde{\mathbf{K}}_2\right)^{-1} \tilde{\mathbf{B}}_2] \tilde{\mathbf{K}}_1 \left(s\mathbf{I} - \mathbf{A}_1\right)^{-1} \tilde{\mathbf{B}}_1 \tag{25.5.64}$$

Remark 25.6. *The above model-matching procedure constrains $\mathbf{\Theta}(s)$ to be stable. However, as we have argued in Chapters 7 and 15, it is usually necessary (e.g., in the presence of input disturbances) to distinguish between stable and desirable closed-loop pole locations. The method described above can be modified so that $\mathbf{\Theta}(s)$ is constrained to have poles in various desirable regions, as in section §22.8.*

Remark 25.7. *In the case when $\tilde{\mathbf{D}}$ is square and nonsingular, then we see from (25.5.58) that $\overline{\mathbf{\Psi}}$ will be identically zero. This will be the situation when we use this algorithm for Q synthesis–see below.*

25.5.5 Application to Q Synthesis for Nonminimum-Phase Zeros

We can now apply the results of subsection §25.5.4 to the Q Synthesis problem. Applying (25.5.64) to the problem posed in (25.5.45) and (25.5.46) leads to the result

$$[\mathbf{\Theta}(s)]_{*i} = [-\mathbf{I} + \tilde{\mathbf{K}}_2 \left(s\mathbf{I} - \mathbf{A}_2 + \tilde{\mathbf{B}}_2\tilde{\mathbf{K}}_2\right)^{-1} \tilde{\mathbf{B}}_2] \, \tilde{\mathbf{K}}_1 \left(s\mathbf{I} - \mathbf{A}_1\right)^{-1} \tilde{\mathbf{B}}_1 \qquad (25.5.65)$$

where

$$\tilde{\mathbf{K}}_1 = \mathbf{K}_1^{\mathrm{s}}(s) - \tilde{\mathbf{D}}^{-1}\mathbf{C}_1 \qquad (25.5.66)$$

$$\tilde{\mathbf{K}}_2 = \mathbf{K}_2^{\mathrm{s}}(s) - \tilde{\mathbf{D}}^{-1}\mathbf{C}_2 \qquad (25.5.67)$$

Actually, the associated LQR problem has some special features because, as was pointed out in Remark 25.7, $\Psi = \mathbf{0}$–i.e., there is no weighting on the states in the cost function.

We therefore need to apply the results in Appendix D carefully to understand the fine detail of the solution. We refer to the transformed problem given in (25.5.54) to (25.5.59). For this set-up, we have the following:

(i)

When the system is of minimum phase, then $\mathbf{A}_\lambda$ is stable. It then follows from Appendix D that the only positive semi-definite solution to the CTARE is $\mathbf{P}_\infty = 0$, giving the optimal feedback gain as $\mathbf{K}_1^{\mathrm{s}}(s) = \mathbf{0}$, $\mathbf{K}_2^{\mathrm{s}}(s) = \mathbf{0}$. Note that $\mathbf{P}(t)$ converges to $\mathbf{P}_\infty$ for any positive definite initial condition. Substituting into (25.5.65), (25.5.66), (25.5.67) gives

$$[\mathbf{\Theta}(s)]_{*i} = [-\mathbf{I} + \tilde{\mathbf{D}}^{-1}\mathbf{C}_2 \left(s\mathbf{I} - \mathbf{A}_2 + \tilde{\mathbf{B}}_2\tilde{\mathbf{D}}^{-1}\mathbf{C}_2\right)^{-1} \tilde{\mathbf{B}}_2] \, [-\tilde{\mathbf{D}}^{-1}\mathbf{C}_1 \left(s\mathbf{I} - \mathbf{A}_1\right)^{-1} \tilde{\mathbf{B}}_1]$$
$$(25.5.68)$$

$$= [\tilde{\mathbf{D}}^{-1} - \tilde{\mathbf{D}}^{-1}\mathbf{C}_2 \left(s\mathbf{I} - \mathbf{A}_2 + \tilde{\mathbf{B}}_2\tilde{\mathbf{D}}^{-1}\mathbf{C}_2\right)^{-1} \tilde{\mathbf{B}}_2\tilde{\mathbf{D}}^{-1}] \, [\mathbf{C}_1 \left(s\mathbf{I} - \mathbf{A}_1\right)^{-1} \tilde{\mathbf{B}}_1]$$
$$(25.5.69)$$

The reader is invited to check that this is exactly equal to $[\mathbf{\Lambda}(s)]^{-1}[\mathbf{T}^*]_{*i}(s)$ where $[\mathbf{\Lambda}(s)]^{-1}$ has the state space realization $\{(25.4.56), (25.4.57)\}$. Thus, in the minimum-phase case, the model-matching algorithm has essentially resulted in the exact inverse of $\boldsymbol{\xi_L}(s)\mathbf{G_o}(s)$ multiplied by the target complementary sensitivity $\mathbf{T}^*(s)$. This is heuristically reasonable.

(ii)

When the system is of nonminimum phase, then $\mathbf{K_1^s}(s)$ and $\mathbf{K_2^s}(s)$ will, in general, be nonzero, ensuring that the eigenvalues of $\mathbf{A_2} - \tilde{\mathbf{B}}_2\tilde{\mathbf{K}}_2$ lie in the stabilizing region. Indeed, it can be shown that the eigenvalues of $\mathbf{A_2} - \tilde{\mathbf{B}}_2\tilde{\mathbf{K}}_2$ correspond to the stable eigenvalues of $\mathbf{A_2}$ together with the unstable eigenvalues of $\mathbf{A_2}$ reflected through the stability boundary. Furthermore $\mathbf{K_2^s}$ satisfies the following CTARE:

$$0 = \mathbf{P_{22}}\tilde{\mathbf{B}}_2\overline{\mathbf{\Phi}}^{-1}\tilde{\mathbf{B}}_2^T\mathbf{P_{22}} + \mathbf{P_{22}}\mathbf{A_2} + \mathbf{A_2}^T\mathbf{P_{22}} \qquad (25.5.70)$$

$$\mathbf{K_2^s} = \overline{\mathbf{\Phi}}^{-1}\tilde{\mathbf{B}}_2\mathbf{P_{22}} \qquad (25.5.71)$$

Finally, the solution has the form

$$[\mathbf{Q}(s)]_{*i}(s) = \mathbf{L}(s)[\mathbf{\Gamma}(s)]_{*i} \qquad (25.5.72)$$

where

$$\mathbf{L}(s) = -\mathbf{I} + \tilde{\mathbf{K}}_2\left(s\mathbf{I} - \mathbf{A_2} + \tilde{\mathbf{B}}_2\tilde{\mathbf{K}}_2\right)^{-1}\tilde{\mathbf{B}}_2 \qquad (25.5.73)$$

and

$$[\mathbf{\Gamma}(s)]_{*i} = \tilde{\mathbf{K}}_1\left(s\mathbf{I} - \mathbf{A_1}\right)^{-1}\tilde{\mathbf{B}}_1 \qquad (25.5.74)$$

Furthermore, putting the columns of $\mathbf{Q}(s)$ side by side, we have

$$\mathbf{Q}(s) = \mathbf{L}(s)\mathbf{\Gamma}(s); \qquad \mathbf{\Gamma}(s) = \left[[\mathbf{\Gamma}(s)]_{*1}, \ [\mathbf{\Gamma}(s)]_{*2}\ldots[\mathbf{\Gamma}(s)]_{*m}\right] \qquad (25.5.75)$$

where $\mathbf{\Gamma}(s)$ is affine in $\mathbf{T}^*(s)$.

Remark 25.8. *The above design (although optimal in an $\mathcal{L}_2$ sense) does not guarantee that $\mathbf{Q}(0) = [\mathbf{G_o}(0)]^{-1}$, i.e., there is no guarantee of integral action in the controller. To achieve integral action we can do one of two things:*

(i) *Multiply $\mathbf{Q}(s)$ on the left by a constant matrix $\mathbf{M}$ to yield $\mathbf{M}\mathbf{Q}(0) = [\mathbf{G_o}(0)]^{-1}$.*

(ii) *Follow the procedure described in section §16.3.4–i.e., express*

$$\mathbf{Q}(s) = [\mathbf{G_o}(0)]^{-1} + s\overline{\mathbf{Q}}(s) \qquad (25.5.76)$$

and optimize $\overline{\mathbf{Q}}(s)$.

We also modify the cost function to include a weighting function $\mathbf{W}(s) = \mathbf{I}/s$.

Remark 25.9. *The above procedure finds an exact inverse for minimum-phase stable systems and an approximate inverse for nonminimum-phase systems. For the nonminimum-phase case, the plant unstable zeros are reflected in the stability boundary, to then appear as poles in $\mathbf{Q}(s)$. Additional poles in $\mathbf{Q}(s)$ are introduced by $[\xi_{\mathbf{L}}(s)]^{-1}$, which appears as part of the inverse system.*
Three interesting questions arise concerning this design procedure.

(i) *How do we assign the poles of $\mathbf{Q}(s)$ to some region, rather than simply ensure that $\mathbf{Q}(s)$ is stable? This can be achieved with the transformations introduced in section §22.8.*

(ii) *How do we choose the factors $(s + \alpha)$ in the interactor? This is related, amongst other things, to the NMP zeros that are present. As we have seen in Chapter 24, the NMP zeros place an upper limit on the bandwidth that is desirable to use. Other factors that determine $(s + \alpha)$ are the allowable input range and the presence of measurement noise.*

(iii) *How do we impose additional design constraints such as diagonal decoupling? This topic requires additional work and is taken up in Chapter 26.*

Example 25.6. *Consider a 2×2 MIMO plant having the nominal model*

$$\mathbf{G_o}(s) = \begin{bmatrix} \dfrac{-1}{s+2} & \dfrac{2}{s+1} \\[3mm] \dfrac{2}{s+2} & \dfrac{7(-s+1)}{(s+1)(s+2)} \end{bmatrix} \tag{25.5.77}$$

This is a stable but nonminimum-phase system, with poles at -1, -2, and -2 and a zero at $s = 5$.
The target sensitivity function is chosen as

$$\mathbf{T}^*(s) = \frac{9}{s^2 + 4s + 9} \begin{bmatrix} 1 & 0 \\ 0 & 1 \end{bmatrix} \tag{25.5.78}$$

To cast this into the problem formulation outlined above, we next reparameterize $\mathbf{Q}(s)$ to force integration in the feedback loop. We thus use

$$\mathbf{Q}(s) = \mathbf{G_o}(0)^{-1} + s\overline{\mathbf{Q}}(s) = \frac{1}{15} \begin{bmatrix} -14 & 8 \\ 4 & 2 \end{bmatrix} + s\overline{\mathbf{Q}}(s) \tag{25.5.79}$$

and introduce a weighting function $\mathbf{W_s}(s) = \mathbf{I}/s$. Then, with reference to (22.6.3), we have that

$$\mathbf{M}(s) = \mathbf{W_s}(s)\left(\mathbf{T}^*(s) - \mathbf{G_o}(s)\mathbf{G_o}(0)^{-1}\right) \quad and \quad \mathbf{N}(s) = \mathbf{G_o}(s) \tag{25.5.80}$$

Thus,

$$\mathbf{M}(s) = \begin{bmatrix} \dfrac{-1.46s^2 + 1.13s + 5.8}{(s^2 + 4s + 9)(s+1)(s+2)} & \dfrac{0.267}{(s+1)(s+2)} \\[4mm] \dfrac{3.73}{(s+1)(s+2)} & \dfrac{-0.13s^2 + 6.466s + 17.8}{(s^2 + 4s + 9)(s+1)(s+2)} \end{bmatrix} \quad (25.5.81)$$

To solve the problem by following the approach presented above, we need to build the left interactor, $\boldsymbol{\xi}_{\mathbf{L}}(s)$, for $\mathbf{N}(s)$. This interactor is given by $\boldsymbol{\xi}_{\mathbf{L}}(s) = s\mathbf{I}$, leading to

$$\boldsymbol{\xi}_L(\alpha)^{-1}\boldsymbol{\xi}_L(s+\alpha)\mathbf{N}(s) = \frac{\tau s + 1}{(s+1)(s+2)}\begin{bmatrix} -(s+1) & 2(s+2) \\ 2(s+1) & 7(-s+1) \end{bmatrix} \quad (25.5.82)$$

where $\tau = \alpha^{-1} = 0.1$.
Upon solving the quadratic problem, we get

$$\overline{\mathbf{Q}}(s) = \frac{\begin{bmatrix} \overline{Q}_{11}(s) & \overline{Q}_{12}(s) \\ \overline{Q}_{21}(s) & \overline{Q}_{22}(s) \end{bmatrix}}{s^4 + 19s^3 + 119s^2 + 335s + 450} \quad (25.5.83)$$

$$\overline{Q}_{11}(s) = 7.11s^3 + 83.11s^2 + 337.11s + 328.67$$
$$\overline{Q}_{12}(s) = 3.55s^3 + 31.56s^2 + 123.56s + 189.33$$
$$\overline{Q}_{21}(s) = 2.22s^3 + 24.00s^2 + 100.44s + 142.67$$
$$\overline{Q}_{22}(s) = 1.11s^3 + 12.0s^2 + 50.22s + 71.33$$

Finally, by using (25.5.79), we recover $\mathbf{Q}(s)$ as

$$Q(s) = \frac{\begin{bmatrix} Q_{11}(s) & Q_{12}(s) \\ Q_{21}(s) & Q_{22}(s) \end{bmatrix}}{s^4 + 19s^3 + 119s^2 + 335s + 450} \quad (25.5.84)$$

$$Q_{11}(s) = 6.18s^4 + 65.38s^3 + 226.04s^2 + 16.00s - 420.00$$
$$Q_{12}(s) = 4.09s^4 + 41.69s^3 + 187.02s^2 + 368.00s + 240.00$$
$$Q_{21}(s) = 2.49s^4 + 29.07s^3 + 132.18s^2 + 232.00s + 120.00$$
$$Q_{22}(s) = 1.24s^4 + 14.53s^3 + 66.09s^2 + 116.00s + 60$$

The design is tried with unit step references. The results are shown in Figure 25.1.

25.5.6 Q Synthesis of Decentralized Controllers

As a potential application of the MIMO Q parameterization, we consider the design of a decentralized controller for a plant G_o. Say that we are given a full multivariable solution with complementary sensitivity $\mathbf{H_o}$. Say that we choose a particular

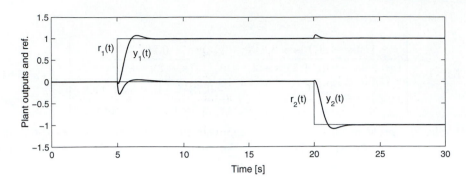

Figure 25.1. Optimal quadratic design–step reference tracking

pairing of the inputs and outputs, which without loss of generality we take to be $(u_1, y_1) \ldots (u_m, y_m)$.

Then, define the following diagonal nominal model with associated additive error $\mathbf{G}_\epsilon(s)$.

$$\mathbf{G_o^d}(s) = \text{diag}[g_{11}^o(s), \ldots, g_{mm}^o(s)], \quad \mathbf{G}_\epsilon(s) = \mathbf{G_o}(s) - \mathbf{G_o^d}(s) \qquad (25.5.85)$$

Say that the plant is stable; then we can use the stable $\mathbf{Q}$ parameterization of all decentralized controllers that stabilize $\mathbf{G_o^d}(s)$:

$$\mathbf{C_o^d}(s) = \mathbf{Q_o^d}[\mathbf{I} - \mathbf{G_o^d}(s)\mathbf{Q_o^d}(s)]^{-1} \qquad (25.5.86)$$

where $\mathbf{Q_o^d}(s)$ and $\mathbf{C_o^d}(s)$ are diagonal.

The achieved sensitivity on the full plant is (per equation (25.3.1))

$$\mathbf{S_A}(s) = [\mathbf{I} - \mathbf{G_o^d}(s)\mathbf{Q_o^d}(s)][\mathbf{I} + \mathbf{G}_\epsilon(s)\mathbf{Q_o^d}(s)]^{-1} \qquad (25.5.87)$$

We can then design $\mathbf{Q_o^d}(s)$ by minimizing a measure of the weighted sensitivity error,

$$J = \int_0^\infty \left\| (\mathbf{I} - \mathbf{H_o}(s) - \mathbf{S_A}(s))\mathbf{W}(s) \right\|_F^2 d\omega \qquad (25.5.88)$$

where $\mathbf{W}(s) = [\mathbf{I} + \mathbf{G}_\epsilon(s)\mathbf{Q_o^d}(s)]$ and $\| \quad \|_F$ denotes the Fröbenius norm.

We see that J takes the form

$$J = \int_0^\infty \left\| \mathbf{M} - \mathbf{N}\mathbf{Q_o^d} \right\|_F^2 \, d\omega \qquad (25.5.89)$$

where $\mathbf{M}(s) = \mathbf{H_o}$ and $\mathbf{N}(s) = (-\mathbf{H_o}\mathbf{G}_\epsilon(s) + \mathbf{G_o}(s))$.

However, this is exactly the form of the model-matching problem discussed in section §22.6. Hence, we can design $\mathbf{Q_o^d}(s)$ to bring the achieved sensitivity, $\mathbf{S_A}(s)$, *close to* the target sensitivity, $\mathbf{S_T}(s) - \mathbf{H_o}(s)$, where *closeness* is measured via (25.5.88).

A normalized measure of the cost of using decentralized control would then be

$$\nu = \frac{J^o}{\int_0^\infty \left\| \mathbf{S_T}(s)\mathbf{W}(s) \right\|_F^2 \, d\omega} \qquad (25.5.90)$$

where J^o is the minimum value of (25.5.88).

Remark 25.10. *In the above example, we have implicitly taken stable to mean desirable. If one wishes to restrict the region in which the closed-loop poles lie, then one needs to constrain $\mathbf{Q}$ as was done in the SISO case.*

We next study the Q synthesis problem for open-loop unstable systems. Note that, as in Chapter 15, we take the term *unstable open-loop poles* to include *undesirable open-loop poles*, such as resonant modes.

25.6 Affine Parameterization: Unstable MIMO Plants

We consider a LMFD and RMFD for the plant of the form

$$\mathbf{G_o}(s) = [\overline{\mathbf{G}}_{\mathbf{oD}}(s)]^{-1}\overline{\mathbf{G}}_{\mathbf{oN}}(s) = \mathbf{G_{oN}}(s)[\mathbf{G_{oD}}(s)]^{-1} \qquad (25.6.1)$$

Note that one such representation can be obtained as in subsection §20.3.4. In this case, we have that $\overline{\mathbf{G}}_{\mathbf{oN}}(s), \overline{\mathbf{G}}_{\mathbf{oD}}(s), \mathbf{G_{oN}}(s)$ and $\mathbf{G_{oD}}(s)$ are given by the expressions (22.12.16) to (22.12.19).

Observe also that, if $\mathbf{G_o}(s)$ is unstable, then $\overline{\mathbf{G}}_{\mathbf{oD}}(s)$ and $\mathbf{G_{oD}}(s)$ will be of nonminimum phase. Similarly, if $\mathbf{G_o}(s)$ is of nonminimum phase, then so are $\overline{\mathbf{G}}_{\mathbf{oN}}(s)$ and $\mathbf{G_{oN}}(s)$.

The following result is the MIMO version of the result described in section §15.7.

Lemma 25.2 (Affine parameterization for unstable MIMO plants). *Consider a plant described in MFD as in (25.6.1), where $\overline{\mathbf{G}}_{\mathbf{oN}}(s), \overline{\mathbf{G}}_{\mathbf{oD}}(s), \mathbf{G_{oN}}(s)$, and $\mathbf{G_{oD}}(s)$ are a coprime factorization as in Lemma 20.1–i.e., one satisfying*

$$\begin{bmatrix} \overline{\mathbf{C}}_{\mathbf{D}}(s) & \overline{\mathbf{C}}_{\mathbf{N}}(s) \\ -\overline{\mathbf{G}}_{\mathbf{N}}(s) & \overline{\mathbf{G}}_{\mathbf{D}}(s) \end{bmatrix} \begin{bmatrix} \mathbf{G_D}(s) & -\mathbf{C_N}(s) \\ \mathbf{G_N}(s) & \mathbf{C_D}(s) \end{bmatrix} = \mathbf{I} \qquad (25.6.2)$$

Then the class of all stabilizing controllers for the nominal plant can be expressed as

$$\mathbf{C}(s) = \mathbf{C}_{\mathbf{N}\Omega}(s)[\mathbf{C}_{\mathbf{D}\Omega}(s)]^{-1} = [\overline{\mathbf{C}}_{\mathbf{D}\Omega}(s)]^{-1}[\overline{\mathbf{C}}_{\mathbf{N}\Omega}(s)] \qquad (25.6.3)$$

where

$$\overline{\mathbf{C}}_{\mathbf{D}\Omega}(s) = \overline{\mathbf{C}}_{\mathbf{D}}(s) - \Omega(s)\overline{\mathbf{G}}_{\mathbf{oN}}(s) \qquad (25.6.4)$$

$$\overline{\mathbf{C}}_{\mathbf{N}\Omega}(s) = \overline{\mathbf{C}}_{\mathbf{N}}(s) + \Omega(s)\overline{\mathbf{G}}_{\mathbf{oD}}(s) \qquad (25.6.5)$$

$$\mathbf{C}_{\mathbf{D}\Omega}(s) = \mathbf{C}_{\mathbf{D}}(s) - \mathbf{G}_{\mathbf{oN}}(s)\Omega(s) \qquad (25.6.6)$$

$$\mathbf{C}_{\mathbf{N}\Omega}(s) = \mathbf{C}_{\mathbf{N}}(s) + \mathbf{G}_{\mathbf{oD}}(s)\Omega(s) \qquad (25.6.7)$$

where $\Omega(s)$ is any stable $m \times m$ proper transfer matrix.

Proof

(i) **Sufficiency**

The resulting four sensitivity functions in this case are

$$\mathbf{S}_{\mathbf{o}}(s) = (\mathbf{C}_{\mathbf{D}}(s) - \mathbf{G}_{\mathbf{oN}}(s)\Omega(s))\overline{\mathbf{G}}_{\mathbf{oD}}(s) \qquad (25.6.8)$$

$$\mathbf{T}_{\mathbf{o}}(s) = \mathbf{G}_{\mathbf{oN}}(s)(\overline{\mathbf{C}}_{\mathbf{N}}(s) + \Omega(s)\overline{\mathbf{G}}_{\mathbf{oD}}(s)) \qquad (25.6.9)$$

$$\mathbf{S}_{\mathbf{io}}(s) = (\mathbf{C}_{\mathbf{D}}(s) - \mathbf{G}_{\mathbf{oN}}(s)\Omega(s))\overline{\mathbf{G}}_{\mathbf{oN}}(s) \qquad (25.6.10)$$

$$\mathbf{S}_{\mathbf{u}}(s) = \mathbf{G}_{\mathbf{oD}}(s)(\overline{\mathbf{C}}_{\mathbf{N}}(s) + \Omega(s)\overline{\mathbf{G}}_{\mathbf{oD}}(s)) \qquad (25.6.11)$$

These transfer functions are clearly stable if $\Omega(s)$ is stable.

(ii) **Necessity**

As in Chapter 15, this result follows from the algebraic structure of the problem; see the references for details.

The controller described above is depicted in Figure 25.2.
From Figure 25.2, if we make the special choice

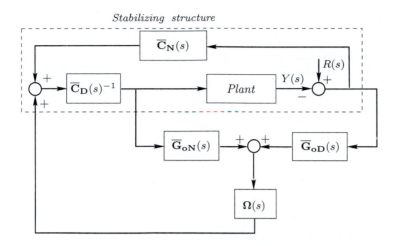

Figure 25.2. Q parameterisation for MIMO unstable plants

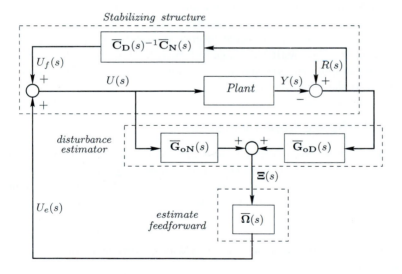

Figure 25.3. Alternative Q parameterization for MIMO unstable plants with restricted $\mathbf{\Omega}(s)$

$$\mathbf{\Omega}(s) = \overline{\mathbf{C}}_{\mathbf{D}}(s)\overline{\mathbf{\Omega}}(s) \tag{25.6.12}$$

we can represent the system as in Figure 25.3.

In this case, we have

$$
\begin{aligned}
\mathbf{S_o}(s) &= (\mathbf{C_D}(s) - \mathbf{G_{oN}}(s)\overline{\mathbf{C}}_{\mathbf{D}}(s)\overline{\mathbf{\Omega}}(s))\overline{\mathbf{G}}_{\mathbf{oD}}(s) \\
&= (\mathbf{I} - [\mathbf{C_D}(s)]^{-1}\mathbf{G_{oN}}(s)\overline{\mathbf{C}}_{\mathbf{D}}(s)\overline{\mathbf{\Omega}}(s))\mathbf{C_D}(s)\overline{\mathbf{G}}_{\mathbf{oD}}(s) \\
&= (\mathbf{I} - [\mathbf{C_D}(s)]^{-1}(\mathbf{C_D}(s)\overline{\mathbf{G}}_{\mathbf{oN}}(s))\overline{\mathbf{\Omega}}(s))\mathbf{C_D}(s)\overline{\mathbf{G}}_{\mathbf{oD}}(s) \\
&= (\mathbf{I} - \overline{\mathbf{G}}_{\mathbf{oN}}(s)\overline{\mathbf{\Omega}}(s))\mathbf{C_D}(s)\overline{\mathbf{G}}_{\mathbf{oD}}(s) \\
&= \mathbf{S}_{\overline{\mathbf{\Omega}}}(s)\mathbf{S_C}(s) \tag{25.6.13}
\end{aligned}
$$

where $\mathbf{S_C}(s)$ is the sensitivity achieved with the prestabilizing controller (as in Figure 25.3 on the page before) and $\mathbf{S}_{\overline{\mathbf{\Omega}}}(s)$ is the sensitivity function

$$\mathbf{S}_{\overline{\mathbf{\Omega}}}(s) = \mathbf{I}(s) - \overline{\mathbf{G}}_{\mathbf{oN}}(s)\overline{\mathbf{\Omega}}(s) \tag{25.6.14}$$

We recognize $\mathbf{S}_{\overline{\mathbf{\Omega}}}(s)$ as having the form of (25.2.3) for the stable *equivalent plant* $\overline{\mathbf{G}}_{\mathbf{oN}}(s)$. Thus, the techniques developed earlier in this chapter for Q design in the stable open-loop case can be used to design $\overline{\mathbf{\Omega}}(s)$.

Note, however, that it is desirable to ensure that $\mathbf{S_C}(s)$ is sensible, or else this will negatively interact with the choice of $\mathbf{S}_{\overline{\mathbf{\Omega}}}(s)$. For example, if $\mathbf{S_C}(s)$ is not diagonal, then making $\mathbf{S}_{\overline{\mathbf{\Omega}}}(s)$ diagonal does not give dynamic decoupling. We take this topic up in the next chapter.

25.7   State Space Implementation

We have seen in Chapter 15 that there exists a nice state space interpretation of the class of all stabilizing controllers for the open-loop unstable case. A similar interpretation applies to the MIMO case. This interpretation is particular useful in the MIMO case, where a state space format greatly facilitates design and implementation.

We begin with the transfer-function form of the controller shown in Figure 25.3 on the preceding page. We give a state space form for each of the blocks in the controller, as follows.

(i)   Stabilizing structure

This is readily achieved by state-estimate feedback. All that is required is an observer (possibly designed via Kalman-filter theory) and stabilizing state-estimate feedback (possibly designed via LQR theory). Thus, we write

$$e(t) = y(t) - r(t) \tag{25.7.1}$$

$$\dot{\hat{x}}_1(t) = \mathbf{A}\hat{x}_1(t) + \mathbf{B}u(t) + \mathbf{J_1}(e(t) - \mathbf{C}\hat{x}_1) - \mathbf{B}u_e(t) \tag{25.7.2}$$

$$u(t) = -\mathbf{K}\hat{x}_1(t) + u_e(t) \tag{25.7.3}$$

In the above expressions, $u_e(t)$ corresponds to the inverse Laplace transform of the quantity $U_e(s)$ in Figure 25.3 on page 843. We have injected $u_e(t)$ into the above equations so as to correctly position $\overline{\mathbf{C}}_\mathbf{D}(s)^{-1}$ in the loop. This is verified by noting that (25.7.2) can be written in transfer-function form as

$$\hat{X}_1(s) = \mathbf{T_1}(s)U(s) + \mathbf{T_2}(s)E(s) - \mathbf{T_1}(s)U_e(s) \tag{25.7.4}$$

where

$$\mathbf{T_1}(s) = (s\mathbf{I} - \mathbf{A} + \mathbf{J_1}\mathbf{C})^{-1}\mathbf{B} \tag{25.7.5}$$
$$\mathbf{T_2}(s) = (s\mathbf{I} - \mathbf{A} + \mathbf{J_1}\mathbf{C})^{-1}\mathbf{J_1} \tag{25.7.6}$$

Then, equation (25.7.3) becomes

$$U(s) = -\mathbf{K}\mathbf{T_1}(s)U(s) - \mathbf{K}\mathbf{T_2}(s)E(s) + \mathbf{K}\mathbf{T_1}(s)U_e(s) + U_e(s) \tag{25.7.7}$$

This equation has the form

$$U(s) = -\overline{\mathbf{C}}_\mathbf{D}(s)^{-1}\overline{\mathbf{C}}_\mathbf{N}(s)E(s) + \overline{\mathbf{C}}_\mathbf{D}(s)^{-1}\overline{\mathbf{C}}_\mathbf{D}(s)U_e(s) \tag{25.7.8}$$
$$= -\overline{\mathbf{C}}_\mathbf{D}(s)^{-1}\overline{\mathbf{C}}_\mathbf{N}(s)E(s) + U_e(s) \tag{25.7.9}$$

where

$$\overline{\mathbf{C}}_\mathbf{D}(s) = \mathbf{I} + \mathbf{K}\mathbf{T_1}(s) \qquad \text{and} \qquad \overline{\mathbf{C}}_\mathbf{N}(s) = \mathbf{K}\mathbf{T_2}(s) \tag{25.7.10}$$

We see that (25.7.9) corresponds to Figure 25.2 on page 843.

(ii) Disturbance estimator

We also know from Chapter 20 that the LMFD in the disturbance block can be realized by an observer:

$$\dot{\hat{x}}_2(t) = \mathbf{A}\hat{x}_2(t) + \mathbf{B}u(t) + \mathbf{J_2}(e(t) - \mathbf{C}\hat{x}_2(t)) \tag{25.7.11}$$

$$\xi(t) = -(e(t) - \mathbf{C}\hat{x}_2(t)) \tag{25.7.12}$$

These equations immediately have the form[1]

$$\Xi(s) = -\overline{\mathbf{G}}_{\mathbf{oD}}(s)E(s) + \overline{\mathbf{G}}_{\mathbf{oN}}(s)U(s) \tag{25.7.13}$$

where

$$\overline{\mathbf{G}}_{\mathbf{oD}}(s) = \mathbf{I} - \mathbf{C}(s\mathbf{I} - \mathbf{A} + \mathbf{J_2C})^{-1}\mathbf{J_2} \tag{25.7.14}$$

$$\overline{\mathbf{G}}_{\mathbf{oN}}(s) = \mathbf{C}(s\mathbf{I} - \mathbf{A} + \mathbf{J_2C})^{-1}\mathbf{B} \tag{25.7.15}$$

We see that (25.7.13) corresponds to the estimate feedforward block in Figure 25.3.

(iii) Design of estimate feedforward block

The block $\overline{\Omega}(s)$ can be designed by inverting $\overline{\mathbf{G}}_{\mathbf{oN}}(s)$ where, from (25.7.15), we see that $\overline{\mathbf{G}}_{\mathbf{oN}}(s)$ has the state space model

$$\dot{\hat{x}}'(t) = \mathbf{A}\hat{x}'(t) + \mathbf{B}u(t) - \mathbf{J_2C}\hat{x}'(t) \tag{25.7.16}$$

$$\xi'(t) = \mathbf{C}\hat{x}'(t) \tag{25.7.17}$$

By introducing a left interactor, this model can be converted to a biproper form:

---

[1] Note that $\Xi(s) = \mathcal{L}[\xi(t)]$.

$$\dot{\hat{x}}''(t) = \mathbf{A}\hat{x}''(t) + \mathbf{B}u(t) - \mathbf{J}\mathbf{C_2}\hat{x}''(t) \qquad (25.7.18)$$
$$\xi''(t) = \mathbf{C}''\hat{x}''(t) + \mathbf{D}''u(t) \qquad (25.7.19)$$

where $\mathbf{D}''$ is square and nonsingular. Provided that this system is of minimum phase, we can immediately invert this system by reversing the roles of input and output.

Note that the controller will be biproper if $\overline{\Omega}(s)$ is biproper. This follows, because $\overline{\mathbf{G}_{\mathbf{oD}}}(s)$ is biproper.

25.8   Summary

- The generalization of the affine parameterization for a stable multivariable model $\mathbf{G_o}(s)$ is given by the controller representation
  $\mathbf{C}(s) = [\mathbf{I} - \mathbf{Q}(s)\mathbf{G_o}(s)]^{-1}\mathbf{Q}(s) = \mathbf{Q}(s)[\mathbf{I} - \mathbf{G_o}(s)\mathbf{Q}(s)]^{-1}$
  yielding the nominal sensitivities
  $\mathbf{T_o}(s) = \mathbf{G_o}(s)\mathbf{Q}(s)$
  $\mathbf{S_o}(s) = \mathbf{I} - \mathbf{G_o}(s)\mathbf{Q}(s)$
  $\mathbf{S_{io}}(s) = [\mathbf{I} - \mathbf{G_o}(s)\mathbf{Q}(s)]\mathbf{G_o}(s)$
  $\mathbf{S_{uo}}(s) = \mathbf{Q}(s)$

- The associated achieved sensitivity, when the controller is applied to $\mathbf{G}(s)$, is given by
  $\mathbf{S}(s) = \mathbf{S_o}(s)[\mathbf{I} + \mathbf{G_\epsilon}(s)\mathbf{Q}(s)]^{-1}$
  where $\mathbf{G_\epsilon}(s) = \mathbf{G}(s) - \mathbf{G_o}(s)$ is the additive modeling error.

- In analogy to the SISO case, key advantages of the affine parameterization include the following:

  - explicit stability of the nominal closed loop if and only if $\mathbf{Q}(s)$ is stable;
  - highlighting the fundamental importance of invertibility, i.e., the achievable and achieved properties of $\mathbf{G_o}(s)\mathbf{Q}(s)$ and $\mathbf{G}(s)\mathbf{Q}(s)$; and
  - sensitivities that are affine in $\mathbf{Q}(s)$–this facilitates criterion-based synthesis, which is particularly attractive for MIMO systems.

- Again in analogy to the SISO case, inversion of stable MIMO systems involves two key issues:

  - relative degree–i.e., the structure of zeros at infinity; and
  - inverse stability–i.e., the structure of NMP zeros.

- Because of directionality, both of these attributes exhibit additional complexity in the MIMO case.

- The structure of zeros at infinity is captured by the left or right interactor ($\boldsymbol{\xi}_{\mathbf{L}}(s)$ or $\boldsymbol{\xi}_{\mathbf{R}}(s)$, respectively).

- Thus, $\boldsymbol{\xi}_{\mathbf{L}}(s)\mathbf{G_o}(s)$ is biproper, i.e., its determinant is a nonzero bounded quantity for $s \to \infty$.

- The structure of NMP zeros is captured by the left or right z-interactor ($\boldsymbol{\psi}_{\mathbf{L}}(s)$ or $\boldsymbol{\psi}_{\mathbf{R}}(s)$, respectively).

- Thus, *analytically*, $\boldsymbol{\psi}_{\mathbf{L}}(s)\mathbf{G_o}(s)$ is a realization of the inversely stable portion of the model–i.e., the equivalent to the minimum-phase factors in the SISO case.

- However, the realization $\boldsymbol{\psi}_{\mathbf{L}}(s)\mathbf{G_o}(s)$

    ○ is nonminimal, and

    ○ generally involves cancellations of unstable pole-zero dynamics (the NMP zero dynamics of $\mathbf{G_o}(s)$).

- Thus, the realization $\boldsymbol{\psi}_{\mathbf{L}}(s)\mathbf{G_o}(s)$

    ○ is useful for analyzing the fundamentally achievable properties of the key quantity $\mathbf{G_o}(s)\mathbf{Q}(s)$, subject to the stability of $\mathbf{Q}(s)$, and

    ○ is generally not suitable for either implementation or inverse implementation, because it involves unstable pole-zero cancellation.

- A stable inverse suitable for implementation is generated by model matching, which leads to a particular linear quadratic regulator (LQR) problem which is solvable via Riccati equations.

- If the plant model is unstable, controller design can be carried out in two steps:

    (i) prestabilization, for example via LQR; then

    (ii) detailed design, by applying the theory for stable models to the prestabilized system.

- All of the above results can be interpreted equivalently in either a transfer-function or a state space framework; for MIMO systems, the state space framework is particularly attractive for numerical implementation.

25.9   Further Reading

Affine parameterization and synthesis

Desoer, C., Liu, R., Murray, J., and Saeks, R. (1980). Feedback systems design: The fractional representation approach to analysis and synthesis. *IEEE Transactions on Automatic Control*, 25(3):399-412.

Morari, M. and Zafiriou, E. (1989). *Robust Process Control*. Prentice-Hall, Englewood Cliffs, N.J.

Zhou, K., Doyle, J.C., and Glover, K. (1996). *Robust and Optimal Control*. Prentice-Hall, Upper Saddle River, N.J.

Interactors

Wolowich, W. and Falb, P. (1976). Invariants and canonical forms under dynamic compensation. *SIAM Journal on Control and Optimization*, 14(6):996-1008.

MIMO relative degree and interactors

Goodwin, G.C., Feuer, A., and Gómez, G. (1997). A state space tehcnique for the evaluation of diagonalizing compensator. *Systems and Control Letters*, 32(3):173-177.

Z-Interactors

Weller, S.R. and Goodwin, G.C. (1995). Partial decoupling of unstable linear multivariable systems. In *Proceedings of the 3rd European Control Conference, Rome, Italy*, pages 2539-2544.

Weller, S.R. and Goodwin, G.C. (1996). Controller design for partial decoupling of linear multivariable systems. *International Journal of Control*, 63(3):535-556.

Decentralized control

Goodwin, G.C., Seron, M.M., and Salgado, M.E. (1998). H_2 design of decentralized controllers. In *Proceedings of the 1999 American Control Conference*.

Güçlü, A. and Özgüler, B. (1986). Diagonal stabilization of linear multivariable systems. *International Journal of Control*, 43(3):965-980.

Hovd, M. and Skogestad, S. (1994). Sequential design of decentralized controllers. *Automatica*, 30(10):1601-1607.

Wang, S.H. and Davison, E.J. (1973). On the stabilization of decentralized control systems. *IEEE Transactions on Automatic Control*, 18(5):473-478.

25.10   Problems for the Reader

Problem 25.1. *Consider a stable MIMO plant having the state space model* $(\mathbf{A_o}, \mathbf{B_o}, \mathbf{C_o}, 0)$. *Assume that state-estimate feedback is used to design the one-d.o.f. controller in IMC form.*

Show that the controller $\mathbf{Q}(s)$ *has a state space realization given by the* $4 - tuple$ $(\mathbf{A_Q}, \mathbf{B_Q}, \mathbf{C_Q}, 0)$, *where*

$$
\mathbf{A_Q} = \begin{bmatrix} \mathbf{A_o} - \mathbf{B_o K} & -\mathbf{J C_o} \\ 0 & \mathbf{A_o} - \mathbf{J C_o} \end{bmatrix} ; \quad \mathbf{B_Q} = \begin{bmatrix} \mathbf{J} \\ \mathbf{J} \end{bmatrix} ; \quad \mathbf{C_Q} = \begin{bmatrix} \mathbf{K} & 0 \end{bmatrix} \quad (25.10.1)
$$

where $\mathbf{J}$ *and* $\mathbf{K}$ *are the observer gain and the state-feedback gain respectively. (Hint: Use Lemma 22.3 to obtain a state space model for* $\mathbf{Q}(s) = \mathbf{C}(s)[\mathbf{I} + \mathbf{G_o}(s)\mathbf{C}(s)]^{-1}$. *Then, perform a suitable similarity transformation on the state.)*

Problem 25.2. *Consider a linear plant having the model*

$$
\mathbf{G_o}(s) = \frac{1}{(s+1)^3} \begin{bmatrix} (s+1)^2 & (s+1) \\ -2(s+1) & 1 \end{bmatrix} \quad (25.10.2)
$$

25.2.1 *Find the left interactor matrix,* $\boldsymbol{\xi}_\mathbf{L}(s)$, *assuming that the loop bandwidth must be at least* $3[rad/s]$.

25.2.2 *Obtain* $\mathbf{\Lambda_L}(s) \triangleq \boldsymbol{\xi}_\mathbf{L}(s)\mathbf{G_o}(s)$ *and its state space representation.*

25.2.3 *Obtain* $\mathbf{Q}(s)$, *in the IMC architecture, as defined in* (25.4.62) *on page 824. (Use MATLAB routine **minv.m**.)*

25.2.4 *Evaluate your design by using reference inputs in the frequency band* $[0,3][rad/s]$.

Problem 25.3. *A discrete-time system is formed from the following continuous transfer function by using a zero-order hold and sampling interval of* 0.1 *seconds.*

$$
\mathbf{G_o}(s) = \frac{1}{(s+1)(s+2)} \begin{bmatrix} 1 & 3 \\ -2 & 1 \end{bmatrix} \quad (25.10.3)
$$

25.3.1 *Find a simple controller that stabilizes this system.*

25.3.2 *Parameterize all stabilizing controllers.*

Problem 25.4. *Consider a continuous-time system having the nominal model*

$$\mathbf{G_o}(s) = \frac{1}{(s+1)(s+2)} \begin{bmatrix} s+4 & 3 \\ 2 & 1 \end{bmatrix} \qquad (25.10.4)$$

Find two matrices $\mathbf{V}(s)$ and $\mathbf{W}(s)$ such that $\mathbf{V}(s)\mathbf{G_o}(s)\mathbf{W}(s)$ is stable, biproper, and of minimum phase.

Problem 25.5. *Consider the same plant model as in the previous example, and define $\overline{\mathbf{\Lambda}}(s)$ as*

$$\overline{\mathbf{\Lambda}}(s) = \frac{1}{(s+1)^3} \begin{bmatrix} (\alpha')^{-1}(s+\alpha')(s+1)^2 & (s+1) \\ -2(s+1) & (\alpha')^{-3}(s+\alpha')^3 \end{bmatrix} \qquad (25.10.5)$$

where α' is chosen equal to the value of α in the building of the left interactor. Note that $\overline{\mathbf{\Lambda}}(s)$ is a biproper matrix, because $\det(\overline{\mathbf{\Lambda}}(\infty))$ has a finite nonzero value.

25.5.1 Obtain a state space realization for $\overline{\mathbf{\Lambda}}(s)$.

25.5.2 Compare $\overline{\mathbf{\Lambda}}(s)$ and $\mathbf{\Lambda_L}(s)$ used in problem 25.2 (use singular values).

25.5.3 Build a control loop with the IMC architecture, choosing $\mathbf{Q}(s) = [\overline{\mathbf{\Lambda}}(s)]^{-1}$. Evaluate under the same conditions used in problem 25.2. Compare and discuss.

Problem 25.6. *Consider an unstable MIMO plant having the model*

$$\mathbf{G_o}(s) = \frac{1}{(s+1)(s+2)(s-1)} \begin{bmatrix} 4(s+1)(s+2) & (s+1) \\ (s+2)(s-1) & 2(s-1) \end{bmatrix} \qquad (25.10.6)$$

25.6.1 *Find a state space model for this plant and determine the gain matrices* **K** *and* **J** *to implement a state-estimate feedback scheme. The loop must exhibit zero steady-state errors for constant references and disturbances. The closed transient should have at least the same decay speed as* e^{-4t}. *Evaluate your design.*

25.6.2 *Using the results presented in section* §25.6, *synthesize* **Q**(s) *to implement a one-d.o.f. control loop with performance similar to that obtained from the design above.*

Problem 25.7. *Show how you would incorporate integration into the solution for decentralized control proposed in subsection* §25.5.6.

Chapter 26

DECOUPLING

26.1 Preview

An idealized requirement in MIMO control-system design is that of decoupling. As was discussed in section §21.6, decoupling can take various forms, ranging from *static* (where decoupling is demanded only for constant reference signals) up to *full dynamic* (where decoupling is demanded at all frequencies). Clearly, full dynamic decoupling is a stringent demand. Thus, in practice, it is more usual to seek dynamic decoupling over some desired bandwidth. If a plant is dynamically decoupled, then changes in the set-point of one process variable lead to a response in that process variable but *all other* process variables remain constant. The advantages of such a design are intuitively clear; e.g., a temperature may be required to be changed, but it may be undesirable for other variables (e.g., pressure) to suffer any associated transient.

This chapter describes the design procedures necessary to achieve dynamic decoupling. In particular, we discuss

- dynamic decoupling for stable minimum-phase systems

- dynamic decoupling for stable nonminimum-phase systems

- dynamic decoupling for open-loop unstable systems.

As might be expected, full dynamic decoupling is a strong requirement and is generally not *cost-free*. We will thus also quantify the performance cost of decoupling by using frequency-domain procedures. These allow a designer to assess *a-priori* whether the cost associated with decoupling is acceptable in a given application.

Of course, some form of decoupling is a very common requirement. For example, static decoupling is almost always desirable. The question then becomes, over what bandwidth will decoupling (approximately) be asked for? It will turn out that the additional cost of decoupling is a function of open-loop poles and zeros in the right-half plane. Thus, if one is restricting decoupling in some bandwidth, then by focusing attention on those open-loop poles and zeros that fall within this bandwidth, one can get a feel for the cost of decoupling over that bandwidth. In this sense, the results presented in this chapter are applicable to almost all MIMO design problems, because some form of decoupling over a limited bandwidth (usually around d.c.) is almost always required.

We will also examine the impact of actuator saturation on decoupling. In the case of static decoupling, it is necessary to avoid integrator wind-up. This can be achieved by using methods that are analogous to the SISO case treated in Chapter 11. In the case of full dynamic decoupling, special precautions are necessary to maintain decoupling in the face of actuator limits. We show that this is indeed possible by appropriate use of MIMO anti-wind-up mechanisms.

26.2   Stable Systems

We first consider the situation in which the open-loop poles of the plant are located in *desirable* locations. We will employ the affine-parameterization technique described in Chapter 25 to design a controller that achieves full dynamic decoupling.

26.2.1   Minimum-Phase Case

We refer to the general Q-design procedure outlined in Chapter 25.

To achieve dynamic decoupling, we make the following choice for $\mathbf{Q}(s)$:

$$\mathbf{Q}(s) = \boldsymbol{\xi}_{\mathbf{R}}(s)[\boldsymbol{\Lambda}_{\mathbf{R}}(s)]^{-1}\mathbf{D}_{\mathbf{Q}}(s) \tag{26.2.1}$$

$$\boldsymbol{\Lambda}_{\mathbf{R}}(s) = \mathbf{G}_{\mathbf{o}}(s)\boldsymbol{\xi}_{\mathbf{R}}(s) \tag{26.2.2}$$

$$\mathbf{D}_{\mathbf{Q}}(s) = \text{diag}\left(\frac{1}{p_1(s)}, \frac{1}{p_2(s)}, \cdots, \frac{1}{p_m(s)}\right) \tag{26.2.3}$$

where $\boldsymbol{\xi}_{\mathbf{R}}(s)$ is the right interactor defined in (25.4.3), and $p_1(s), p_2(s), \ldots p_m(s)$ are stable polynomials chosen to make $\mathbf{Q}(s)$ proper. Actually, these polynomials can be shown to have the corresponding column degrees of the left interactor for $\mathbf{G}_{\mathbf{o}}(s)$. The polynomials $p_1(s), p_2(s), \ldots p_m(s)$ should be chosen to have unit d.c. gain.

We observe that, with the above choice, we achieve the following nominal complementary sensitivity:

$$\mathbf{T}_{\mathbf{o}}(s) = \mathbf{G}_{\mathbf{o}}(s)\mathbf{Q}(s) \tag{26.2.4}$$

$$= \mathbf{G}_{\mathbf{o}}(s)\boldsymbol{\xi}_{\mathbf{R}}(s)[\boldsymbol{\Lambda}_{\mathbf{R}}(s)]^{-1}\mathbf{D}_{\mathbf{Q}}(s) \tag{26.2.5}$$

$$= \mathbf{G}_{\mathbf{o}}(s)\boldsymbol{\xi}_{\mathbf{R}}(s)[\mathbf{G}_{\mathbf{o}}(s)\boldsymbol{\xi}_{\mathbf{R}}(s)]^{-1}\mathbf{D}_{\mathbf{Q}}(s) \tag{26.2.6}$$

$$= \text{diag}\left(\frac{1}{p_1(s)}, \frac{1}{p_2(s)}, \cdots, \frac{1}{p_m(s)}\right) \tag{26.2.7}$$

We see that this is diagonal, as required. The associated control-system structure would then be as shown in Figure 26.1.

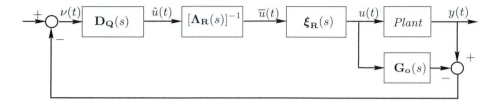

Figure 26.1. IMC decoupled control of stable MIMO plants

Actually, the above design is not unique. For example, an alternative choice for $\mathbf{Q}(s)$ is

$$\mathbf{Q}(s) = [\mathbf{\Lambda_L}(s)]^{-1}\boldsymbol{\xi_L}(s)\mathbf{D_Q}(s) \qquad (26.2.8)$$

where $\mathbf{D_Q}(s)$ is given by (26.2.3).

Note also that $\mathbf{D_Q}(s)$ can have the more general structure

$$\mathbf{D_Q}(s) = \mathrm{diag}\,(t_1(s), t_2(s), \cdots, t_m(s)) \qquad (26.2.9)$$

where $t_1(s), t_2(s), \ldots t_m(s)$ are proper stable transfer functions having relative degrees equal to the corresponding column degrees of the left interactor for $\mathbf{G_o}(s)$. The transfer functions $t_1(s), t_2(s), \ldots, t_m(s)$ should be chosen to have unit d.c. gain.

We illustrate by a simple example.

Example 26.1. *Consider a stable 2×2 MIMO system having the nominal model*

$$\mathbf{G_o}(s) = \frac{1}{(s+1)^2(s+2)} \begin{bmatrix} 2(s+1) & -1 \\ (s+1)^2 & (s+1)(s+2) \end{bmatrix} \qquad (26.2.10)$$

Choose a suitable matrix $\mathbf{Q}(s)$ to control this plant, using the affine parameterization, in such a way that the MIMO control loop is able to track references of bandwidths less than or equal to $2[rad/s]$ and $4[rad/s]$ in channels 1 and 2, respectively.

Solution

We will attempt to obtain a decoupled design, i.e., to obtain a complementary sensitivity matrix given by

$$\mathbf{T_o}(s) = \mathrm{diag}(T_{11}(s), T_{22}(s)) \qquad (26.2.11)$$

where $T_{11}(s)$ and $T_{22}(s)$ will be chosen to have bandwidths $2[rad/s]$ and $4[rad/s]$ in channels 1 and 2, respectively.

Then, $\mathbf{Q}(s)$ must ideally satisfy

$$\mathbf{Q}(s) = [\mathbf{G_o}(s)]^{-1}\mathbf{T_o}(s). \qquad (26.2.12)$$

We also note that the matrix $\mathbf{G_o}(s)$ is stable and has its zeros strictly inside the LHP. Then the only difficulty in obtaining an inverse of $\mathbf{G_o}(s)$ arises from the need to achieve a proper (biproper) $\mathbf{Q}$. This may look simple to achieve; it would suffice to add a large number of fast, stables poles to $T_{11}(s)$ and $T_{22}(s)$. However, having the relative degrees larger than strictly necessary usually adds unwanted lag at high frequencies. We might also be interested in obtaining a biproper $\mathbf{Q}(s)$ to implement anti-wind-up architectures (later in the chapter). Here is where interactors play a useful role. We choose the structure (26.2.8), from which we see that $\mathbf{T_o}(s) = \mathbf{D_Q}(s)$. Hence, the relative degrees of $T_{11}(s)$ and $T_{22}(s)$ will be chosen equal to the degrees of the first and second column of the left interactor for $\mathbf{G_o}(s)$, respectively.

We then follow subsection §25.4.1 to compute the left interactor $\boldsymbol{\xi}_{\mathbf{L}}(s)$. This leads to

$$\boldsymbol{\xi}_{\mathbf{L}}(s) = \mathrm{diag}\big((s+\alpha)^2, (s+\alpha)\big); \qquad \alpha \in \mathbb{R}^+ \qquad (26.2.13)$$

Then (26.2.12) can also be written as

$$\mathbf{Q}(s) = [\boldsymbol{\xi}_{\mathbf{L}}(s)\mathbf{G_o}(s)]^{-1}\boldsymbol{\xi}_{\mathbf{L}}(s)\mathbf{T_o}(s). \qquad (26.2.14)$$

Hence, $\mathbf{Q}(s)$ is proper if and only if $\mathbf{T_o}(s)$ is chosen so as to make $\boldsymbol{\xi}_{\mathbf{L}}(s)\mathbf{T_o}(s)$ proper.

Thus, possible choices for $T_{11}(s)$ and $T_{22}(s)$ are

$$T_{11}(s) = \frac{4}{s^2 + 3s + 4} \quad and \quad T_{22}(s) = \frac{4(s+4)}{s^2 + 6s + 16} \qquad (26.2.15)$$

The reader is invited to check that these transfer functions have the required bandwidths.

To obtain the final expression for $\mathbf{Q}(s)$, *we next need to compute* $[\mathbf{G_o}(s)]^{-1}$, *which is given by*

$$[\mathbf{G_o}(s)]^{-1} = \frac{s+2}{2s+5} \begin{bmatrix} (s+1)(s+2) & 1 \\ \\ -(s+1)^2 & 2(s+1) \end{bmatrix} \tag{26.2.16}$$

from which we finally obtain

$$\mathbf{Q}(s) = [\mathbf{G_o}(s)]^{-1}\mathbf{T_o}(s) = \frac{s+2}{2s+5} \begin{bmatrix} \dfrac{4(s+1)(s+2)}{s^2+3s+4} & \dfrac{4(s+4)}{s^2+6s+16} \\ \\ \dfrac{-4(s+1)^2}{s^2+3s+4} & \dfrac{8(s+4)(s+1)}{s^2+6s+16} \end{bmatrix} \tag{26.2.17}$$

*The above design can be tested by using SIMULINK file **mimo4.mdl**.*

□□□

The above design procedure is limited to minimum-phase systems. In particular, it is clear that $\mathbf{Q}(s)$, chosen as in (26.2.1) or as in (26.2.8), is stable if and only if $\mathbf{G_o}(s)$ is of minimum phase, because $[\mathbf{\Lambda_R}(s)]^{-1}$ and $[\mathbf{\Lambda_L}(s)]^{-1}$ involve an inverse of $\mathbf{G_o}(s)$. We therefore need to modify $\mathbf{Q}(s)$ so as to ensure stability when $\mathbf{G_o}(s)$ is of nonminimum phase. A way of doing this is described in the next subsection.

26.2.2   Nonminimum-Phase Case

We will begin with the state space realization for $[\mathbf{\Lambda_R}(s)]^{-1}$, defined by a 4-tuple $(\mathbf{A_\lambda}, \mathbf{B_\lambda}, \mathbf{C_\lambda}, \mathbf{D_\lambda})$[1]. We will denote by $\tilde{u}(t)$ the input to this system. Our aim is to modify $[\mathbf{\Lambda_R}(s)]^{-1}$ so as to achieve two objectives: (i) render the transfer function stable, whilst (ii) retaining its diagonalizing properties.

To this end, we define the following subsystem, which is driven by the ith component of $\tilde{u}(t)$

$$\dot{x}_i(t) = \mathbf{A_i}x_i(t) + \mathbf{B_i}\tilde{u}_i(t) \tag{26.2.18}$$
$$v_i(t) = \mathbf{C_i}x_i(t) + \mathbf{D_i}\tilde{u}_i(t) \tag{26.2.19}$$

where $v_i(t) \in \mathbb{R}^m$, $\tilde{u}_i(t) \in \mathbb{R}$, and $(\mathbf{A_i}, \mathbf{B_i}, \mathbf{C_i}, \mathbf{D_i})$ is a minimal realization of the transfer function from the ith component of $\tilde{u}(t)$ to the complete vector output $\overline{u}(t)$. Thus $(\mathbf{A_i}, \mathbf{B_i}, \mathbf{C_i}, \mathbf{D_i})$ is a minimal realization of $(\mathbf{A_\lambda}, \mathbf{B_\lambda}e_i, \mathbf{C_\lambda}, \mathbf{D_\lambda}e_i)$, where e_i is the ith column of the $m \times m$ identity matrix.

---

[1]Note that the NMP zeros of $\mathbf{G_o}(s)$ are eigenvalues of $\mathbf{A_\lambda}$–i.e., $\mathbf{A_\lambda}$ is assumed unstable here.

We next apply stabilizing state feedback to each of these subsystems–i.e., we form

$$\tilde{u}_i(t) = -\mathbf{K_i}x_i(t) + \bar{r}_i(t); \qquad i = 1, 2, \ldots, m \qquad (26.2.20)$$

where $\bar{r}_i(t) \in \mathbb{R}$. The design of $\mathbf{K_i}$ can be done in any convenient fashion–e.g., by linear quadratic optimization.

Finally we add together the m vectors $v_1(t), v_2(t), \ldots v_m(t)$ to produce an output, which can be renamed $\bar{u}(t)$:

$$\bar{u}(t) = \sum_{i=1}^{m} v_i(t) \qquad (26.2.21)$$

With the above definitions, we are in position to establish the following result.

Lemma 26.1.   *(a) The transfer function from $\bar{r}(t) = [\bar{r}_1(t)\ \bar{r}_2(t)\ \ldots \bar{r}_m(t)]^T$ to $\bar{u}(t)$ is given by*

$$\mathbf{W}(s) = [\mathbf{\Lambda_R}(s)]^{-1}\mathbf{D_z}(s) \qquad (26.2.22)$$

where

$$[\mathbf{\Lambda_R}(s)]^{-1} = (\mathbf{C_\lambda}[s\mathbf{I} - \mathbf{A_\lambda}]^{-1}\mathbf{B_\lambda} + \mathbf{D_\lambda}) \qquad (26.2.23)$$

$$\mathbf{D_z}(s) = \text{diag}\left\{[1 + \mathbf{K_i}[s\mathbf{I} - \mathbf{A_i}]^{-1}\mathbf{B_i}]^{-1}\right\} \qquad (26.2.24)$$

(b) $\mathbf{\Lambda_R}(s)\mathbf{W}(s)$ is a diagonal matrix.

(c) $\mathbf{W}(s)$ has a state space realization as in (26.2.18) to (26.2.21).

Proof

(a) From equations (26.2.18)-(26.2.20), the transfer function from $\tilde{r}_i(t)$ to $v_i(t)$ is given by

$$\mathbf{W_i}(s) = (\mathbf{C_i} - \mathbf{D_i}\mathbf{K_i})[s\mathbf{I} - \mathbf{A_i} + \mathbf{B_i}\mathbf{K_i}]^{-1}\mathbf{B_i} + \mathbf{D_i} \qquad (26.2.25)$$

Using the matrix inversion lemma (Lemma 18.2 on page 530) and (26.2.18)-(26.2.20), we obtain

$$\mathbf{W_i}(s) = (\mathbf{C_\lambda}[s\mathbf{I} - \mathbf{A_\lambda}]^{-1}\mathbf{B_\lambda} + \mathbf{D_\lambda})e_i[1 + \mathbf{K_i}[s\mathbf{I} - \mathbf{A_i}]^{-1}\mathbf{B_i}]^{-1} \qquad (26.2.26)$$

Then the transfer function from $\bar{r}(t)$ to $\bar{u}(t)$, defined in (26.2.21), is given by

$$\mathbf{W}(s) = \sum_{i=1}^{m} \mathbf{W_i}(s)e_i^T \tag{26.2.27}$$

$$= (\mathbf{C_\lambda}[s\mathbf{I} - \mathbf{A_\lambda}]^{-1}\mathbf{B_\lambda} + \mathbf{D_\lambda}) \operatorname{diag}\left\{[1 + \mathbf{K_i}[s\mathbf{I} - \mathbf{A_i}]^{-1}\mathbf{B_i}]^{-1}\right\} \tag{26.2.28}$$

(b) Immediate from the definition of $[\mathbf{\Lambda_R}(s)]^{-1}$.

(c) By construction.

□□□

Returning now to the problem of determining $\mathbf{Q}(s)$, we choose

$$\mathbf{Q}(s) = \mathbf{\xi_R}(s)\mathbf{W}(s)\mathbf{D_Q}(s) \tag{26.2.29}$$

This is equivalent to

$$\mathbf{Q}(s) = \mathbf{\xi_R}(s)[\mathbf{\xi_R}(s)]^{-1}[\mathbf{G_o}(s)]^{-1}\mathbf{D_z}(s)\mathbf{D_Q}(s) \tag{26.2.30}$$

where $\mathbf{D_Q}(s)$ is as in (26.2.9) (or (26.2.3)) and $\mathbf{D_z}(s)$ is as in (26.2.24) and we have used (26.2.20). Then

$$\mathbf{Q}(s) = [\mathbf{G_o}(s)]^{-1}\mathbf{D_z}(s) \operatorname{diag}\left\{t_1(s), t_2(s), \cdots, t_m(s)\right\} \tag{26.2.31}$$

where $t_1(s), t_2(s), \ldots, t_m(s)$ are as in (26.2.9). Finally, we see that the resulting nominal complementary sensitivity is

$$\mathbf{T_o}(s) = \operatorname{diag}\left\{[1 + \mathbf{K_i}[s\mathbf{I} - \mathbf{A_i}]^{-1}\mathbf{B_i}]^{-1}t_i(s)\right\} \tag{26.2.32}$$

Note that any nonminimum-phase zeros in the plant are retained in $\mathbf{T_o}(s)$, which is a requirement for internal stability.

The reader will notice that it is implicit in (26.2.32) that some of the NMP zeros have been duplicated and appear in multiple diagonal elements. At worst, each NMP zero will appear in every diagonal term. Precisely how many zeros will appear and in which channel, depends on the degree of each minimum realization involved in the determination of the models in (26.2.18)-(26.2.20). We will not prove

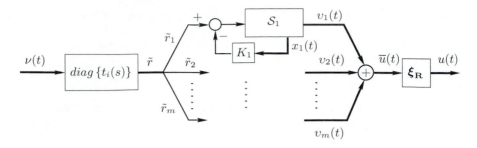

Figure 26.2. Diagonal decoupling MIMO controller ($\mathbf{Q}(s)$)

it here, but the NMP zeros appearing in (26.2.32) are diagonalizing invariants and appear in all possible diagonalized closed loops. Thus, spreading NMP dynamics into other channels is a trade-off inherently associated with decoupling.

The final implementation of $\mathbf{Q}(s)$ is as in Figure 26.2.

In Figure 26.2, $\mathcal{S}_1$ denotes the system described in state space form by the 4-tuple $(\mathbf{A_1}, \mathbf{B_1}, \mathbf{C_1}, \mathbf{D_1})$ as in (26.2.18) and (26.2.19). Vector signals have been depicted with thick lines.

The previous analysis has had the goal of finding a stable inverse that ensures dynamic decoupling. Of course, other structures are possible.

Example 26.2. *Consider a plant having the nominal model*

$$\mathbf{G_o}(s) = \frac{1}{(s+1)^2} \begin{bmatrix} s+2 & -3 \\ -2 & 1 \end{bmatrix} \tag{26.2.33}$$

This model has a NMP zero at $s = 4$. To synthesize a controller following the ideas presented above, we first compute a right interactor matrix, which turns out to have the general form $\boldsymbol{\xi_R}(s) = \operatorname{diag}\{s+\alpha,\ (s+\alpha)^2\}$. For numerical simplicity, we choose $\alpha = 1$. Then

$$[\boldsymbol{\Lambda_R}(s)]^{-1} = [\mathbf{G_o}(s)\boldsymbol{\xi_R}(s)]^{-1} = \frac{1}{s-4} \begin{bmatrix} s+1 & 3(s+1) \\ 2 & s+2 \end{bmatrix} \tag{26.2.34}$$

and a state space realization for $[\boldsymbol{\Lambda_R}(s)]^{-1}$ is

$$\mathbf{A_\lambda} = \begin{bmatrix} 4 & 0 \\ 0 & 4 \end{bmatrix}; \quad \mathbf{B_\lambda} = \mathbf{I}; \quad \mathbf{C_\lambda} = \begin{bmatrix} 5 & 15 \\ 2 & 6 \end{bmatrix}; \quad \mathbf{D_\lambda} = \begin{bmatrix} 1 & 3 \\ 0 & 1 \end{bmatrix} \tag{26.2.35}$$

We next compute $(\mathbf{A_i}, \mathbf{B_i}, \mathbf{C_i}, \mathbf{D_i})$ as a minimal realization of $(\mathbf{A_\lambda}, \mathbf{B_\lambda} e_i, \mathbf{C_\lambda}, \mathbf{D_\lambda} e_i)$, for $i = 1$ and $i = 2$. This computation yields

$$\mathbf{A_1} = 4 \qquad \mathbf{B_1} = 1 \qquad \mathbf{C_1} = [5 \quad 2]^T \qquad \mathbf{D_1} = [1 \quad 0]^T \quad (26.2.36)$$

$$\mathbf{A_2} = 4 \qquad \mathbf{B_2} = 1 \qquad \mathbf{C_2} = [15 \quad 6]^T \qquad \mathbf{D_2} = [3 \quad 1]^T \quad (26.2.37)$$

These subsystems can be stabilized by state feedback with gains $\mathbf{K_1}$ *and* $\mathbf{K_2}$, *respectively. For this case, each gain is chosen to shift the unstable pole at* $s = 4$ *to a stable location, say* $s = -10$, *which leads to* $\mathbf{K_1} = \mathbf{K_2} = 14$. *Thus,* $\mathbf{D_z}(s)$ *in* (26.2.24) *is a* 2×2 *diagonal matrix given by*

$$\mathbf{D_z}(s) = \frac{s-4}{s+10}\,\mathbf{I} \qquad (26.2.38)$$

We finally choose $\mathbf{D_Q}(s)$ *in* (26.2.29) *to achieve a bandwidth approximately equal to* 3 [rad/s], *say*

$$\mathbf{D_Q}(s) = \text{diag}\left\{ \frac{-9(s+10)}{4(s^2+4s+9)} \quad \frac{-90}{4(s^2+4s+9)} \right\} \qquad (26.2.39)$$

Note that the elements $t_1(s)$ *and* $t_2(s)$ *in* $\mathbf{D_Q}(s)$ *have been chosen of relative degree equal to the corresponding column degrees of the interactor* $\boldsymbol{\xi}_\mathbf{R}(s)$. *Also, their d.c. gains have been chosen to yield unit d.c. gain in the complementary sensitivity* $\mathbf{T_o}(s)$, *leading to*

$$\mathbf{T_o}(s) = \text{diag}\left\{ \frac{-9(s-4)}{4(s^2+4s+9)} \quad \frac{-90(s-4)}{4(s+10)(s^2+4s+9)} \right\} \qquad (26.2.40)$$

26.3 Pre- and PostDiagonalization

The transfer-function matrix $\mathbf{Q}(s)$ presented in (26.2.29) is actually a right-diagonalizing compensator for a stable (but not necessarily minimum-phase) plant. This can be seen by noting that (as in (26.2.32))

$$\mathbf{G_o}(s)\boldsymbol{\Pi}_\mathbf{R}(s) = \text{diag}\left\{ [1 + \mathbf{K_i}[s\mathbf{I} - \mathbf{A_i}]^{-1}\mathbf{B_i}]^{-1}t_i(s) \right\} \qquad (26.3.1)$$

where

$$\begin{aligned} \boldsymbol{\Pi}_\mathbf{R}(s) &= \mathbf{Q}(s) \\ &= \boldsymbol{\xi}_\mathbf{R}(s)\mathbf{W}(s)\mathbf{D_Q}(s) \end{aligned} \qquad (26.3.2)$$

We will find later that it is sometimes also desirable to have a left-diagonalizing compensator. We could derive such a compensator from first principles. However, a simple way is to first form

$$\overline{\mathbf{G}}_\mathbf{o}(s) = \mathbf{G_o^T}(s) \qquad (26.3.3)$$

We find a right-diagonalizing compensator $\overline{\boldsymbol{\Pi}}_\mathbf{R}(s)$ for $\overline{\mathbf{G}}_\mathbf{o}(s)$ by using the method outlined above. We then let $\boldsymbol{\Pi}_\mathbf{L}(s) = \overline{\boldsymbol{\Pi}}_\mathbf{R}(s)^T$, which has the following property.

$$\Pi_{\mathbf{L}}(s)\mathbf{G_o}(s) = \overline{\Pi}_{\mathbf{R}}^{\mathbf{T}}(s)\overline{\mathbf{G}}_{\mathbf{o}}(s)^T = [\overline{\mathbf{G}}_{\mathbf{o}}(s)\Pi_{\mathbf{R}}(s)]^T \qquad (26.3.4)$$

which is a diagonal matrix by construction.

Example 26.3. *Consider the same plant as in Example 26.2 on page 860. Then*

$$\Pi_{\mathbf{R}}(s) = \xi_{\mathbf{R}}(s)\mathbf{W}(s)\mathbf{D_Q}(s) = \frac{-9(s+1)^2}{4(s^2+4s+9)(s+10)}\begin{bmatrix} s+10 & 120 \\ s+10 & 40(s+2) \end{bmatrix} \qquad (26.3.5)$$

If we repeat the procedure in Example 26.2, but this time for $\mathbf{G}_{\mathbf{o}}^{\mathbf{T}}(s)$, we have that

$$[\mathbf{\Lambda_R}(s)]^{-1} = [\mathbf{G}_{\mathbf{o}}^{\mathbf{T}}(s)\xi_{\mathbf{R}}(s)]^{-1} = \frac{1}{s-4}\begin{bmatrix} s+1 & 2(s+1) \\ 3 & s+2 \end{bmatrix} \qquad (26.3.6)$$

In this case, a state space realization for $[\mathbf{\Lambda_R}(s)]^{-1}$ is

$$\mathbf{A}_\lambda = \begin{bmatrix} 4 & 0 \\ 0 & 4 \end{bmatrix}; \quad \mathbf{B}_\lambda = \begin{bmatrix} 2 & 0 \\ 0 & 4 \end{bmatrix}; \quad \mathbf{C}_\lambda = \begin{bmatrix} 2.5 & 2.5 \\ 1.5 & 1.5 \end{bmatrix}; \quad \mathbf{D}_\lambda = \begin{bmatrix} 1 & 2 \\ 0 & 1 \end{bmatrix} \quad (26.3.7)$$

Then the minimal realizations $(\mathbf{A_i}, \mathbf{B_i}, \mathbf{C_i}, \mathbf{D_i})$ for $(\mathbf{A}_\lambda, \mathbf{B}_\lambda e_i, \mathbf{C}_\lambda, \mathbf{D}_\lambda e_i)$, for $i = 1$ and $i = 2$, are

$$\mathbf{A_1} = 4 \qquad \mathbf{B_1} = 2 \qquad \mathbf{C_1} = [2.5 \quad 1.5]^T \qquad \mathbf{D_1} = [1 \quad 0]^T \quad (26.3.8)$$
$$\mathbf{A_2} = 4 \qquad \mathbf{B_2} = 4 \qquad \mathbf{C_2} = [2.5 \quad 1.5]^T \qquad \mathbf{D_2} = [2 \quad 1]^T \quad (26.3.9)$$

These subsystems can be stabilized by state feedback with gains $\mathbf{K_1}$ and $\mathbf{K_2}$, respectively. We make the same choice as in Example 26.2: we shift the unstable pole at $s = 4$ to a stable location, say $s = -10$, as is achieved with $\mathbf{K_1} = 7$ and $\mathbf{K_2} = 3.5$. Thus, $\mathbf{D_z}(s)$ is a 2×2 diagonal matrix given by

$$\mathbf{D_z}(s) = \frac{s-4}{s+10}\mathbf{I} \qquad (26.3.10)$$

$\mathbf{D_Q}(s)$ is chosen as (26.2.39), to achieve the same bandwidth of 3 [rad/s]. This finally yields

$$\Pi_{\mathbf{L}}(s) = \Pi_{\mathbf{R}}^{\mathbf{T}}(s) = \frac{-9(s+1)^2}{4(s^2+4s+9)(s+10)}\begin{bmatrix} s+10 & 2(s+10) \\ 30 & 10(s+2) \end{bmatrix} \qquad (26.3.11)$$

26.4   Unstable Systems

We next turn to the problem of designing a decoupling controller for an unstable MIMO plant. Here we have an additional complexity: some minimal feedback is necessary to ensure stability. To gain insight into this problem, we will present four alternative design choices:

(i) a two-degree-of-freedom design based on prefiltering the reference;

(ii) a two-degree-of-freedom design using the affine parameterization;

(iii) a design based on one-degree-of-freedom state feedback; and

(iv) a design integrating both state feedback and the affine parameterization.

26.4.1   Two-Degree-of-Freedom Design Based on PreFiltering the Reference

If one requires full dynamic decoupling for reference-signal changes only, then this can be readily achieved by first stabilizing the system by using some suitable controller $C(s)$ and then using prefiltering of the reference signal. The essential idea is illustrated in Figure 26.3.

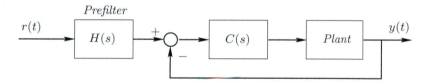

Figure 26.3. Prefilter design for full dynamic decoupling

Say that the plant has transfer function $\mathbf{G_o}(s)$; then the closed-loop transfer function linking $R(s)$ to $Y(s)$ in Figure 26.3 is

$$\mathbf{G_{cl}}(s) = \left[\mathbf{I} + \mathbf{G_o}(s)\mathbf{C}(s)\right]^{-1}\mathbf{G_o}(s)\mathbf{C}(s)\mathbf{H}(s) \qquad (26.4.1)$$

To achieve decoupling, one then need only choose $\mathbf{H}(s)$ as a right-diagonalizing precompensator for the stable transfer function $\left[\mathbf{I} + \mathbf{G_o}(s)\mathbf{C}(s)\right]^{-1}\mathbf{G_o}(s)\mathbf{C}(s)$. We illustrate by an example.

Example 26.4. *Consider the plant*

$$\mathbf{G_o}(s) = \mathbf{G_{oN}}(s)[\mathbf{G_{oD}}(s)]^{-1} \qquad (26.4.2)$$

where

$$\mathbf{G_{oN}}(s) = \begin{bmatrix} -5 & s^2 \\ 1 & -0.0023 \end{bmatrix}; \qquad \mathbf{G_{oD}}(s) = \begin{bmatrix} 25s+1 & 0 \\ 0 & s(s+1)^2 \end{bmatrix} \qquad (26.4.3)$$

(i) *Convert to state space form, and evaluate the zeros.*

(ii) *Design a prestabilizing controller to give static decoupling for reference signals.*

(iii) *Design a prefilter to give full dynamic decoupling for reference signals.*

Solution

(i)

If we compute $\det(\mathbf{G_{oN}}(s))$ *we find that this is a nonminimum-phase system having zeros at* $s = \pm 0.1072$*. We wish to design a controller that achieves dynamic decoupling. We proceed in a number of steps.*

Step 1. State space model

The state space model for the system is

$$\dot{x}_p(t) = \mathbf{A_o}x_p(t) + \mathbf{B_o}u(t) \qquad (26.4.4)$$
$$y(t) = \mathbf{C_o}x_p(t) + \mathbf{D_o}u(t) \qquad (26.4.5)$$

where

$$\mathbf{A_o} = \begin{bmatrix} -0.04 & 0 & 0 & 0 \\ 0 & -2 & -1 & 0 \\ 0 & 1 & 0 & 0 \\ 0 & 0 & 1 & 0 \end{bmatrix} \qquad \mathbf{B_o} = \begin{bmatrix} 1 & 0 \\ 0 & 1 \\ 0 & 0 \\ 0 & 0 \end{bmatrix} \qquad (26.4.6)$$

$$\mathbf{C_o} = \begin{bmatrix} -0.2 & 1 & 0 & 0 \\ 0.04 & 0 & 0 & -0.0023 \end{bmatrix} \qquad \mathbf{D_o} = 0 \qquad (26.4.7)$$

(ii)

We will design a stabilizing controller under the architecture shown in Figure 26.4
We design an observer for the state $x_p(t)$*, given the output* $y(t)$*. This design uses Kalman-filter theory with* $\mathbf{Q} = \mathbf{B_o}\mathbf{B_o}^T$ *and* $\mathbf{R} = 0.05\mathbf{I}_{2\times 2}$*.*
The optimal observer gains turn out to be

$$\mathbf{J} = \begin{bmatrix} -3.9272 & 1.3644 \\ 2.6120 & 0.1221 \\ -0.6379 & 0.1368 \\ -2.7266 & -4.6461 \end{bmatrix} \qquad (26.4.8)$$

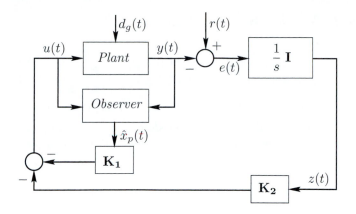

Figure 26.4. Optimal quadratic design with integral action

We wish to have zero steady-state errors in the face of step input disturbances. We therefore follow the procedure of section §22.13 and introduce an integrator with transfer function $\mathbf{I}/s$ at the output of the system (after the comparator). That is, we add [2]

$$\dot{z}(t) = -y(t) = -\mathbf{C}_o x_p(t) \qquad (26.4.9)$$

We can now define a composite state vector $\overline{x}(t) = [x_p^T(t) \quad z^T(t)]^T$ leading to the composite model

$$\dot{\overline{x}}(t) = \overline{\mathbf{A}}\,\overline{x}(t) + \overline{\mathbf{B}}\,u(t) \qquad (26.4.10)$$

where

$$\overline{\mathbf{A}} = \begin{bmatrix} \mathbf{A}_o & \mathbf{0} \\ -\mathbf{C}_o & \mathbf{0} \end{bmatrix}; \qquad \overline{\mathbf{B}} = \begin{bmatrix} \mathbf{B}_o \\ \mathbf{0} \end{bmatrix} \qquad (26.4.11)$$

We next consider the composite system and design a state controller via LQR theory.
We choose

$$\mathbf{\Psi} = \begin{bmatrix} \mathbf{C}_o{}^T\mathbf{C}_o & 0 & 0 \\ 0 & 0.005 & 0 \\ 0 & 0 & 0.1 \end{bmatrix}; \qquad \mathbf{\Phi} = 2\mathbf{I}_{2\times 2} \qquad (26.4.12)$$

---

[2]For simplicity we consider the reference $r(t)$ to be equal to 0 here.

leading to the feedback gain $\mathbf{K} = [\mathbf{K_1} \quad \mathbf{K_2}]$, *where*

$$\mathbf{K_1} = \begin{bmatrix} 0.1807 & -0.0177 & 0.1011 & -0.0016 \\ -0.0177 & 0.1496 & 0.0877 & 0.0294 \end{bmatrix}; \qquad \mathbf{K_2} = \begin{bmatrix} 0.0412 & -0.1264 \\ 0.0283 & 0.1844 \end{bmatrix}$$

$$(26.4.13)$$

   This leads to the equivalent closed loop shown in Figure 26.5 (where we have ignored the observer dynamics, because these disappear in steady state).

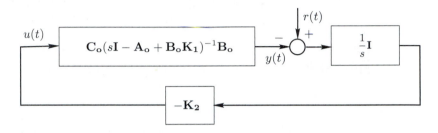

Figure 26.5. Equivalent closed loop (ignoring the observer dynamics)

   The resulting closed-loop responses for unit step references are shown in Figure 26.6, where $r_1(t) = \mu(t-1)$ *and* $r_2(t) = -\mu(t-501)$. *Note that, as expected, the system is "statically decoupled", but significant coupling occurs during transients, especially following the step in the second reference.*

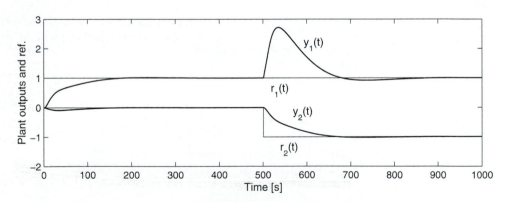

Figure 26.6. Statically decoupled control

(iii)

The closed loop has the transfer function

$$\mathbf{T_o}(s) = (\mathbf{I} + \tilde{\mathbf{G}}(s))^{-1}\tilde{\mathbf{G}}(s) \qquad (26.4.14)$$

where

$$\tilde{\mathbf{G}}(s) = \mathbf{C_o}(s\mathbf{I} - \mathbf{A_o} + \mathbf{B_o}\mathbf{K_1})^{-1}\mathbf{B_o}\mathbf{K_2}\frac{1}{s} \qquad (26.4.15)$$

This is a stable proper transfer function. Note, however, that this is of nonminimum phase, because the original plant was of nonminimum phase.

We use the techniques of subsection §26.2.2 to design an inverse that retains dynamic decoupling in the presence of nonminimum-phase zeros. To use those techniques, the equivalent plant is the closed-loop system with transfer function (26.4.14) and with state space model given by the 4-tuple $(\mathbf{A_e}, \mathbf{B_e}, \mathbf{C_e}, \mathbf{0})$, *where*

$$\mathbf{A_e} = \begin{bmatrix} \mathbf{A_o} - \mathbf{B_o}\mathbf{K_1} & \mathbf{B_o}\mathbf{K_2} \\ -\mathbf{C_o} & \mathbf{0} \end{bmatrix}; \quad \mathbf{B_e} = \begin{bmatrix} \mathbf{0} & \mathbf{I} \end{bmatrix}^T; \quad \mathbf{C_e} = \begin{bmatrix} \mathbf{C_o} & \mathbf{0} \end{bmatrix} \qquad (26.4.16)$$

An interactor for this closed-loop system is

$$\boldsymbol{\xi_L}(s) = \begin{bmatrix} (s+\alpha)^2 & 0 \\ 0 & (s+\alpha)^2 \end{bmatrix}; \qquad \alpha = 0.03$$

This leads to an augmented system having the state space model $(\mathbf{A'_e}, \mathbf{B'_e}, \mathbf{C'_e}, \mathbf{D'_e})$ *with*

$$\mathbf{A'_e} = \mathbf{A_e}$$
$$\mathbf{B'_e} = \mathbf{B_e}$$
$$\mathbf{C'_e} = \alpha^2\mathbf{C_e} + 2\alpha\mathbf{C_e}\mathbf{A_e} + \mathbf{C_e}\mathbf{A_e}^2$$
$$\mathbf{D'_e} = \mathbf{C_e}\mathbf{A_e}\mathbf{B_e}$$

The exact inverse then has the state space model $(\mathbf{A_\lambda}, \mathbf{B_\lambda}, \mathbf{C_\lambda}, \mathbf{D_\lambda})$, *where*

$$\mathbf{A_\lambda} = \mathbf{A'_e} - \mathbf{B'_e}[\mathbf{D'_e}]^{-1}\mathbf{C'_e}$$
$$\mathbf{B_\lambda} = \mathbf{B'_e}[\mathbf{D'_e}]^{-1}\mathbf{C'_e}$$
$$\mathbf{C_\lambda} = -[\mathbf{D'_e}]^{-1}\mathbf{C'_e}$$
$$\mathbf{D_\lambda} = [\mathbf{D'_e}]^{-1}$$

We now form the two subsystems as in subsection §26.2.2. We form minimal realizations of these two systems, which we denote by $(\mathbf{A_1}, \mathbf{B_1}, \mathbf{C_1}, \mathbf{D_1})$ *and*

$(\mathbf{A_2}, \mathbf{B_2}, \mathbf{C_2}, \mathbf{D_2})$. *We determine stabilizing feedback for these two systems by using LQR theory with*

$$\mathbf{\Psi}_1 = \mathbf{C_1}^T \mathbf{C_1} \qquad\qquad\qquad \mathbf{\Phi}_1 = 10^6$$
$$\mathbf{\Psi}_2 = \mathbf{C_2}^T \mathbf{C_2} \qquad\qquad\qquad \mathbf{\Phi}_2 = 10^7$$

We then implement the precompensator as in Figure 26.2, where we choose

$$t_1(s) = \alpha^2 \frac{1 - \mathbf{K_1}[\mathbf{A_1}]^{-1}\mathbf{B_1}}{(s + \alpha)^2}$$
$$t_2(s) = \alpha^2 \frac{1 - \mathbf{K_2}[\mathbf{A_2}]^{-1}\mathbf{B_2}}{(s + \alpha)^2}$$

where $\mathbf{K_1}$, $\mathbf{K_2}$ *now represent the stabilizing gains for the two subsystems, as in subsection* §26.2.2.

The resulting closed-loop responses for step references are shown in Figure 26.7, where $r_1(t) = \mu(t - 1)$ *and* $r_2(t) = -\mu(t - 501)$. *Note that, as expected, the system is now "fully decoupled" from the reference to the output response.*

The reader is invited to simulate and study the input/output behavior of the prefilter, $H(s)$. *Note the subtly coordinated interaction in the reference signals as seen by the plant (output of* $H(s)$). *It would be virtually impossible for a human operator to manipulate the references, by hand, so that one plant output changed without inducing a transient in the other output.*

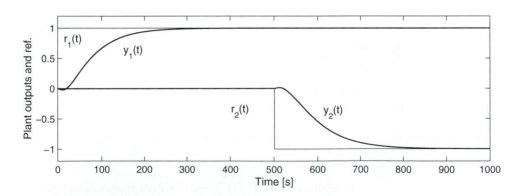

Figure 26.7. Dynamically decoupled control

26.4.2   Two-Degree-of-Freedom Design Based on the Affine Parameterization

Here, we present an alternative two-degree-of-freedom decoupling design using the ideas of Chapter 25.

We first recall the affine parameterization of all stabilizing controllers for a not necessarily stable system, given in Lemma 25.2 on page 841. This representation is redrawn in Figure 26.8.

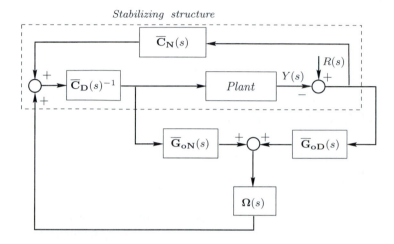

Figure 26.8. Q parameterization for MIMO unstable plants

Next consider the arrangement in Figure 26.9 on the following page, which incorporates two degrees of freedom.

It can readily be shown that the nominal transfer function from $r(t)$ to $y(t)$ is given by

$$\mathbf{H}_{cl}(s) = \mathbf{G}_{oN}(s)\mathbf{\Gamma}(s) \tag{26.4.17}$$

Hence, we simply need to choose $\mathbf{\Gamma}(s)$ as a stable right-diagonalizing compensator for the *special stable plant* $\mathbf{G}_{oN}(s)$. This can be done as in section §26.2 and section §26.3. Note that this will give independent design of the reference-to-output transfer function (determined by $\mathbf{\Gamma}(s)$) from the disturbance sensitivities (determined by $\overline{\mathbf{\Omega}}(s)$).

26.4.3   One-Degree-of-Freedom Design using State Feedback

The design methods presented in subsections §26.4.1 and §26.4.2 achieve diagonal decoupling for reference signals only, because they are based on two-d.o.f. architectures.

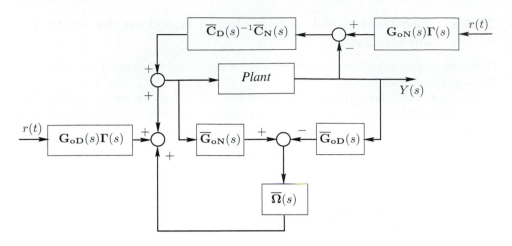

Figure 26.9. Q parameterization with two-d.o.f. for unstable MIMO plants

In this subsection, we will show how the prestabilizing loop can be designed (when this is possible!) to achieve a diagonal nominal sensitivity. We aim to achieve dynamic decoupling for both reference signals and output disturbances.

We first express the plant by a LMFD-RMFD pair, as follows:

$$\mathbf{G}(s) = [\overline{\mathbf{G}}_{\mathbf{oD}}(s)]^{-1}\overline{\mathbf{G}}_{\mathbf{oN}}(s) = \mathbf{G}_{\mathbf{oN}}(s)[\mathbf{G}_{\mathbf{oD}}(s)]^{-1} \qquad (26.4.18)$$

We then use the following steps.

Step 1. We first use the methods of section §26.3 to find stable pre- and post-compensators $\mathbf{\Pi}_{\mathbf{L}}(s)$ and $\mathbf{\Pi}_{\mathbf{R}}(s)$ such that

$$\mathbf{G}_{\mathbf{oN}}(s)\mathbf{\Pi}_{\mathbf{R}}(s) = \mathbf{D}_1(s) \qquad \text{and} \qquad \mathbf{\Pi}_{\mathbf{L}}(s)\overline{\mathbf{G}}_{\mathbf{oD}}(s) = \mathbf{D}_2(s) \qquad (26.4.19)$$

where $\mathbf{D}_1(s)$ and $\mathbf{D}_2(s)$ are diagonal.

By design $\mathbf{G}_{\mathbf{oN}}(s)$ and $\overline{\mathbf{G}}_{\mathbf{oD}}(s)$ are stable, but not necessarily of minimum phase. We thus recognize this subproblem as the problem of finding stable pre- and postdiagonalizing compensators for stable transfer matrices. This is precisely the problem solved in section §26.3.

Now if it were possible to choose

$$\overline{\mathbf{C}}_{\mathbf{N}}(s) = \mathbf{\Pi}_{\mathbf{R}}(s) \qquad \text{and} \qquad \mathbf{\Pi}_{\mathbf{L}}(s) = \mathbf{C}_{\mathbf{D}}(s) \qquad (26.4.20)$$

and it were true that Lemma 20.1 on page 620 held, then we would have achieved diagonal $\mathbf{T_o}(s)$ and $\mathbf{S_o}(s)$ as required. The difficulty is that solutions to (26.4.19), in general, will not satisfy Lemma 20.1 on page 620. We thus cannot use (26.4.20) directly but instead proceed as follows.

Step 2.   We solve the following equation (if possible), for stable diagonal $\tilde{\mathbf{D}}_1(s)$ and $\tilde{\mathbf{D}}_2(s)$ such that

$$\mathbf{D_1}(s)\tilde{\mathbf{D}}_1(s) + \tilde{\mathbf{D}}_2(s)\mathbf{D_2}(s) = \mathbf{I} \qquad (26.4.21)$$

where $\mathbf{D_1}(s)$ and $\mathbf{D_2}(s)$ are as in (26.4.19).

Notice that (26.4.21) is a set of m independent scalar Bezout identities. We know that this is solvable if and only if the numerators of the corresponding diagonal terms in $\mathbf{D_1}(s)$ and $\mathbf{D_2}(s)$ have no common unstable factors.

It has been shown (as in references at the end of the chapter) that this is actually a necessary and sufficient condition for the existence of a one-d.o.f. diagonalizing controller. Hence, if there is no stable diagonal solution to (26.4.21), then decoupling in a one-d.o.f. architecture is not possible by any method! Actually, $\mathbf{D_1}(s)$ and $\mathbf{D_2}(s)$ are commonly called the diagonal structures for the numerator $\mathbf{G_{oN}}(s)$ and denominator $\overline{\mathbf{G}}_{\mathbf{oD}}(s)$, respectively. Having unstable common modes in these structures implies that one-d.o.f. decoupling would necessarily lead to unstable pole-zero cancellations, which is not permissible. We thus assume the contrary and proceed to step 3.

Step 3.   We build the controller $\mathbf{C}(s)$ in a RMFD form as

$$\mathbf{C}(s) = \mathbf{C_N}(s)[\mathbf{C_D}(s)]^{-1}$$

where

$$\mathbf{C_N}(s) = \mathbf{G_{oD}}(s)\overline{\mathbf{C}}_{\mathbf{N}}(s)[\overline{\mathbf{G}}_{\mathbf{oD}}(s)]^{-1}$$
$$\mathbf{C_D}(s) = \tilde{\mathbf{D}}_2(s)\mathbf{\Pi_L}(s)$$

together with

$$\overline{\mathbf{C}}_{\mathbf{N}}(s) = \mathbf{\Pi_R}(s)\tilde{\mathbf{D}}_1(s) \qquad (26.4.22)$$

We then have the following result.

Lemma 26.2.   *The above controller, if it can be constructed, ensures the following:*

(i) *The feedback loop is stable–i.e.,*

$$\overline{\mathbf{G}}_{\mathbf{oD}}(s)\mathbf{C}_{\mathbf{D}}(s) + \overline{\mathbf{G}}_{\mathbf{oN}}(s)\mathbf{C}_{\mathbf{N}}(s) = \mathbf{I}$$

(ii) *The nominal sensitivity is diagonal–i.e.,*

$$\mathbf{S}_{\mathbf{o}}(s) = \mathbf{C}_{\mathbf{D}}(s)\overline{\mathbf{G}}_{\mathbf{oD}}(s) \quad \text{is diagonal}$$

(iii) *An alternative LMFD for* $\mathbf{C}(s)$ *is*

$$\mathbf{C}(s) = [\overline{\mathbf{C}}_{\mathbf{D}}(s)]^{-1}\overline{\mathbf{C}}_{\mathbf{N}}(s)$$

where $\overline{\mathbf{C}}_{\mathbf{N}}(s)$ *is as in (26.4.22) and*

$$\overline{\mathbf{C}}_{\mathbf{D}}(s) = [\mathbf{I} - \overline{\mathbf{C}}_{\mathbf{N}}(s)\mathbf{G}_{\mathbf{oN}}(s)][\mathbf{G}_{\mathbf{oD}}(s)]^{-1}$$

Proof

(i)

$$\overline{\mathbf{G}}_{\mathbf{oD}}(s)\mathbf{C}_{\mathbf{D}}(s) + \overline{\mathbf{G}}_{\mathbf{oN}}(s)\mathbf{C}_{\mathbf{N}}(s) = \overline{\mathbf{G}}_{\mathbf{oD}}(s)\tilde{\mathbf{D}}_{\mathbf{2}}(s)\mathbf{\Pi}_{\mathbf{L}}(s)$$
$$+ \overline{\mathbf{G}}_{\mathbf{oN}}(s)\mathbf{G}_{\mathbf{oD}}(s)\overline{\mathbf{C}}_{\mathbf{N}}(s)[\overline{\mathbf{G}}_{\mathbf{oD}}(s)]^{-1} = (\overline{\mathbf{G}}_{\mathbf{oD}}(s)\tilde{\mathbf{D}}_{\mathbf{2}}(s)\mathbf{\Pi}_{\mathbf{L}}(s)\overline{\mathbf{G}}_{\mathbf{oD}}(s) +$$
$$\overline{\mathbf{G}}_{\mathbf{oN}}(s)\mathbf{G}_{\mathbf{oD}}(s)\mathbf{\Pi}_{\mathbf{R}}(s)\tilde{\mathbf{D}}_{\mathbf{1}}(s))[\overline{\mathbf{G}}_{\mathbf{oD}}(s)]^{-1}$$
$$= (\overline{\mathbf{G}}_{\mathbf{oD}}(s)\tilde{\mathbf{D}}_{\mathbf{2}}(s)\mathbf{D}_{\mathbf{2}}(s) + \overline{\mathbf{G}}_{\mathbf{oD}}(s)\mathbf{G}_{\mathbf{oN}}(s)\mathbf{\Pi}_{\mathbf{R}}(s)\tilde{\mathbf{D}}_{\mathbf{1}}(s))[\overline{\mathbf{G}}_{\mathbf{oD}}(s)]^{-1}$$
$$= (\overline{\mathbf{G}}_{\mathbf{oD}}(s)\tilde{\mathbf{D}}_{\mathbf{2}}(s)\mathbf{D}_{\mathbf{2}}(s) + \overline{\mathbf{G}}_{\mathbf{oD}}(s)\mathbf{D}_{\mathbf{1}}(s)\tilde{\mathbf{D}}_{\mathbf{1}}(s))[\overline{\mathbf{G}}_{\mathbf{oD}}(s)]^{-1}$$
$$= \overline{\mathbf{G}}_{\mathbf{oD}}(s)(\tilde{\mathbf{D}}_{\mathbf{2}}(s)\mathbf{D}_{\mathbf{2}}(s) + \mathbf{D}_{\mathbf{1}}(s)\tilde{\mathbf{D}}_{\mathbf{1}}(s))[\overline{\mathbf{G}}_{\mathbf{oD}}(s)]^{-1} = \mathbf{I} \quad (26.4.23)$$

(ii)

$$\mathbf{S}_{\mathbf{o}}(s) = \mathbf{C}_{\mathbf{D}}(s)[\overline{\mathbf{G}}_{\mathbf{oD}}(s)\mathbf{C}_{\mathbf{D}}(s) + \overline{\mathbf{G}}_{\mathbf{oN}}(s)\mathbf{C}_{\mathbf{N}}(s)]^{-1}\overline{\mathbf{G}}_{\mathbf{oD}}(s)$$
$$= \mathbf{C}_{\mathbf{D}}(s)\overline{\mathbf{G}}_{\mathbf{oD}}(s) \qquad\qquad [\textit{by using part (i)}]$$
$$= \tilde{\mathbf{D}}_{\mathbf{2}}(s)\mathbf{\Pi}_{\mathbf{L}}(s)\overline{\mathbf{G}}_{\mathbf{oD}}(s) = \tilde{\mathbf{D}}_{\mathbf{2}}(s)\mathbf{D}_{\mathbf{2}}(s) \qquad\qquad (26.4.24)$$

which is diagonal.

□□□

The reader is warned that, although the above procedure is analytically correct, it is not suitable for implementation, because (26.4.24) is not a minimal realization and contains implicit unstable pole-zero cancellations that need to be performed algebraically prior to implementation.

26.5   Zeros of Decoupled and Partially Decoupled Systems

We have seen above that NMP zeros and unstable poles significantly affect the ease with which decoupling can be achieved. Indeed, the analysis above suggests that a single RHP zero or pole might need to be dealt with in multiple loops if decoupling is a design requirement. We illustrate this below by a different argument. Consider a plant having the nominal model $\mathbf{G_o}(s)$. Assume that this model has a NMP zero located at $s = z_o$, with direction $h^T = [h_1 \ h_2 \ \dots h_m]$, i.e.,

$$h^T \mathbf{G_o}(z_o) = 0 \qquad (26.5.1)$$

Let us assume that a controller $\mathbf{C}(s)$ is designed to achieve dynamic decoupling, i.e., to obtain diagonal sensitivity matrices. This means that the open-loop transfer-function matrix $\mathbf{M}(s) \triangleq \mathbf{G_o}(s)\mathbf{C}(s)$ must also be diagonal–i.e.,

$$\mathbf{M}(s) = \mathrm{diag}(M_{11}(s), M_{22}(s), \dots, M_{mm}(s)) \qquad (26.5.2)$$

then,

$$h^T \mathbf{M}(z_o) = \begin{bmatrix} h_1 M_{11}(z_o) & h_2 M_{11}(z_o) & \dots & h_m M_{11}(z_o) \end{bmatrix} = 0 \qquad (26.5.3)$$

Thus, we must have that $h_i M_{ii}(z_o) = 0$, for $i = 1, 2, ..., m$. This implies that $M_{ii}(z_o) = 0$ for all i such that the corresponding component h_i is nonzero.

A similar situation arises with unstable poles. In that case, we consider the direction g associated with an unstable pole η_o. We recall that η_o is an unstable pole if there exists a nonzero vector $g = [g_1 \ g_2 \ \dots g_m]^T$ such that, when $\mathbf{G_o}(s)$ is expressed in LMFD, $\mathbf{G_o}(s) = [\overline{\mathbf{G}_{oD}}(s)]^{-1}\overline{\mathbf{G}_{oN}}(s)$; then

$$\overline{\mathbf{G}_{oD}}(\eta_o)g = 0 \Longrightarrow [\mathbf{G_o}(\eta_o)]^{-1}g = 0 \qquad (26.5.4)$$

Dynamic decoupling implies that $[\mathbf{M}(s)]^{-1}$ must also be diagonal. Thus

$$[\mathbf{M}(z_o)]^{-1}g = [(M_{11}(z_o))^{-1}g_1 \ (M_{11}(z_o))^{-1}g_2 \ \dots (M_{mm}(z_o))^{-1}g_m]^T = 0 \quad (26.5.5)$$

Thus, we must have that $(M_{ii}(z_o))^{-1}g_i = 0$, for $i = 1, 2, ..., m$. This implies that $M_{ii}(z_o) = 0$ for every i such that the corresponding component g_i is nonzero.

To obtain a more complete understanding of this phenomenon, we have the following example, where, for simplicity, we address only the issue of NMP zeros.

Example 26.5. *Consider a plant having the nominal model*

$$\mathbf{G_o}(s) = \frac{1}{(s+1)(s+2)} \begin{bmatrix} s+1 & 2 \\ 1 & 1 \end{bmatrix} \qquad (26.5.6)$$

This plant has a zero at $s = z_o = 1$, with geometric multiplicity $\mu_z = 1$ and correspondingly left direction $h^T = [1 \; -2]$.

Assume that we aim at having a decoupled MIMO control loop–i.e, one with a diagonal complementary sensitivity matrix. Then the open-loop transfer-function matrix $\mathbf{M}(s) = \mathbf{G_o}(s)\mathbf{C}(s)$ must also be diagonal–i.e.,

$$\mathbf{G_o}(s)\mathbf{C}(s) = \begin{bmatrix} M_{11}(s) & 0 \\ 0 & M_{22}(s) \end{bmatrix} \tag{26.5.7}$$

giving

$$\mathbf{T_o}(s) = \begin{bmatrix} \dfrac{M_{11}(s)}{1 + M_{11}(s)} & 0 \\ 0 & \dfrac{M_{22}(s)}{1 + M_{22}(s)} \end{bmatrix} \tag{26.5.8}$$

From (26.5.7), the controller must satisfy

$$\mathbf{C}(s) = [\mathbf{G_o}(s)]^{-1} \begin{bmatrix} M_{11}(s) & 0 \\ 0 & M_{22}(s) \end{bmatrix} \tag{26.5.9}$$

The inverse of the plant model is given by

$$[\mathbf{G_o}(s)]^{-1} = \frac{(s+1)(s+2)}{(s-1)} \begin{bmatrix} 1 & -2 \\ -1 & s+1 \end{bmatrix} \tag{26.5.10}$$

This confirms the presence of a NMP zero located at $s = 1$.
Then, using (26.5.10) in (26.5.9), we obtain

$$\mathbf{C}(s) = \frac{(s+1)(s+2)}{(s-1)} \begin{bmatrix} M_{11}(s) & -2M_{22}(s) \\ -M_{11}(s) & (s+1)M_{22}(s) \end{bmatrix} \tag{26.5.11}$$

It is straightforward to show, by using (20.4.8), that the control sensitivity matrix, $\mathbf{S_{uo}}(s)$, is given by

$$\mathbf{S_{uo}}(s) = [\mathbf{G_o}(s)]^{-1} \begin{bmatrix} \dfrac{M_{11}(s)}{1 + M_{11}(s)} & 0 \\ 0 & \dfrac{M_{22}(s)}{1 + M_{22}(s)} \end{bmatrix} \tag{26.5.12}$$

from which it follows that the loop will not be internally stable unless both $M_{11}(s)$ **and** $M_{22}(s)$ vanish at $s = z_o$.

We then see that the NMP zero must appear in both channels to avoid instability. Consider next a modified model given by

$$\mathbf{G_o}(s) = \frac{1}{(s+1)(s+2)} \begin{bmatrix} -s+1 & -s+1 \\ 2 & 1 \end{bmatrix} \qquad (26.5.13)$$

We note that this plant also has a NMP zero, located at $s = z_o = 1$. To get the same open-loop transfer function as in (26.5.7), the controller should now be chosen as

$$\mathbf{C}(s) = \frac{(s+1)(s+2)}{(s-1)} \begin{bmatrix} M_{11}(s) & (-s+1)M_{22}(s) \\ 2M_{11}(s) & (-s+1)M_{22}(s) \end{bmatrix} \qquad (26.5.14)$$

Then the stability of the loop can be achieved without requiring that $M_{22}(s)$ vanish at $s = z_o$. However, it is necessary that $M_{11}(s)$ vanish at $s = z_o$.

We observe that the direction associated with the NMP zero in this case is $h^T = [1\ 0]$.

We next investigate whether there is something to be gained by not forcing dynamic decoupling. We will consider the original plant having the model given in (26.5.6). We choose to have an upper-triangular sensitivity matrix–i.e., the open-loop transfer-function matrix is required to have the form

$$\mathbf{G_o}(s)\mathbf{C}(s) = \begin{bmatrix} M_{11}(s) & M_{12}(s) \\ 0 & M_{22}(s) \end{bmatrix} \qquad (26.5.15)$$

and

$$\mathbf{T_o}(s) = \begin{bmatrix} \dfrac{M_{11}(s)}{1 + M_{11}(s)} & \dfrac{M_{12}(s)}{(1 + M_{11}(s))(1 + M_{22}(s))} \\ 0 & \dfrac{M_{22}(s)}{1 + M_{22}(s)} \end{bmatrix} \qquad (26.5.16)$$

Then the required controller has the form

$$\mathbf{C}(s) = \frac{(s+1)(s+2)}{(s-1)} \begin{bmatrix} M_{11}(s) & M_{12}(s) - 2M_{22}(s) \\ -M_{11}(s) & -M_{12}(s) + (s+1)M_{22}(s) \end{bmatrix} \qquad (26.5.17)$$

and

$$\mathbf{S_{uo}}(s) = \frac{(s+1)(s+2)}{s-1} \begin{bmatrix} \dfrac{M_{11}(s)}{1 + M_{11}(s)} & \dfrac{M_{12}(s) - 2M_{22}(s)(1 + M_{11}(s))}{(1 + M_{11}(s))(1 + M_{22}(s))} \\ -\dfrac{M_{11}(s)}{1 + M_{11}(s)} & \dfrac{-M_{12}(s) + (s+1)M_{22}(s)(1 + M_{11}(s))}{(1 + M_{11}(s))(1 + M_{22}(s))} \end{bmatrix}$$
$$(26.5.18)$$

We then conclude that, to achieve closed-loop stability, we must have that $M_{11}(s)$ vanishes at $s = 1$–i.e., the NMP zero must appear in channel 1. We also need that

$$[M_{12}(s) - 2M_{22}(s)]_{s=1} = 0 \qquad (26.5.19)$$

$$[-M_{12}(s) + (s+1)M_{22}(s)]_{s=1} = 0 \qquad (26.5.20)$$

i.e., it is not necessary that the NMP zero appear in channel 2 also. Instead, we can choose the coupling term $M_{12}(s)$ in such a way that (26.5.19) and (26.5.20) are satisfied. The reader is encouraged to investigate why both conditions are always simultaneously satisfied.

It is evident that, had we chosen a lower-triangular coupling structure, then the NMP zero would have appeared in channel 2, but not necessarily in channel 1.

This example confirms that a single NMP zero can be transferred to more than one channel when full dynamic decoupling is enforced. The specific way in which this happens depends on the direction associated with that zero.

This situation, which can also occur with unstable poles, yields additional performance trade-offs, given the unavoidable trade-offs that arise from the presence of NMP zeros and unstable poles.

The example also confirms the conclusion that triangular decoupling can improve the overall design trade-offs compared to those which apply with full dynamic decoupling.

□□□

26.6 Frequency-Domain Constraints for Dynamically Decoupled Systems

Further insight into the multivariable nature of frequency-domain constraints can be obtained by examining the impact of decoupling on sensitivity trade-offs, as in Chapter 24. In particular, we suggest that the reader review Theorem 24.2 on page 782.

Consider a MIMO control loop where $\mathbf{S_o}(s)$ and, consequently, $\mathbf{T_o}(s)$ are diagonal stable matrices.

Lemma 26.3. *Consider a MIMO plant with a NMP zero at $s = z_o = \gamma + j\delta$, with associated directions $h_1^T, h_2^T \ldots h_{\mu_z}^T$.*

Assume, in addition, that $\mathbf{S_o}(s)$ is diagonal; then, for any value of r such that $h_{ir} \neq 0$,

$$\int_{-\infty}^{\infty} \ln|[\mathbf{S_o}(j\omega)]_{rr}|d\Omega(z_o, \omega) = 0; \qquad for \qquad r \in \nabla_i' \qquad (26.6.1)$$

Proof

The loop is decoupled, so we have that $[\mathbf{S_o}(s)]_{kr} = 0$ for all $k \in \{1, 2, \ldots, m\}$, except for $k = r$. This implies that

$$h_i^T[\mathbf{S_o}(j\omega)]_{*r} = h_{ir}[\mathbf{S_o}(j\omega)]_{rr} \qquad (26.6.2)$$

Upon using this property in equation (24.8.6) of Theorem 24.2 on page 782, we obtain

$$\frac{1}{\pi}\int_{-\infty}^{\infty} \ln|h_{ir}[\mathbf{S_o}(j\omega)]_{rr}|d\Omega(z_o, \omega) = \frac{\ln(|h_{ir}|)}{\pi}\int_{-\infty}^{\infty} d\Omega(z_o, \omega)$$

$$+ \frac{1}{\pi}\int_{-\infty}^{\infty} \ln|[\mathbf{S_o}(j\omega)]_{rr}|d\Omega(z_o, \omega) \geq \ln(|h_{ir}|) \quad (26.6.3)$$

Then the result follows on using (24.8.7).

□□□

Corollary 26.1. *Under the same hypothesis of Lemma 26.3, if the MIMO loop is decoupled (diagonal sensitivity matrix) and the design specification is $|[\mathbf{S_o}(j\omega)]_{rr}| \leq \epsilon_{rr} \ll 1$ for $\omega \in [0, \omega_r]$, then*

$$||[\mathbf{S_o}(j\omega)]_{rr}||_\infty \geq \left(\frac{1}{\epsilon_{rr}}\right)^{\frac{\psi(\omega_r)}{\pi - \psi(\omega_r)}} \qquad (26.6.4)$$

Proof

As in equation (24.9.2) of section §24.9 and using (26.6.1) where $\psi(\omega_r)$ is as in (24.9.3).

□□□

We also have the following corresponding result for the complementary sensitivity function.

Lemma 26.4. *Consider a MIMO system with an unstable pole located at $s = \eta_o = \alpha + j\beta$ and having associated directions $g_1, \ldots, g_{\mu_p}$. Assume, in addition, that $\mathbf{T_o}(s)$ is diagonal; then, for any value of r such that $g_{ir} \neq 0$,*

$$\int_{-\infty}^{\infty} \ln|[\mathbf{T_o}(j\omega)]_{rr}|d\Omega(\eta_o, \omega) = 0; \qquad for \qquad r \in \nabla_i \qquad (26.6.5)$$

Proof

The loop is decoupled, so we have that $[\mathbf{T_o}(s)]_{rk} = 0$ *for all* $k \in \{1, 2, \ldots, m\}$, *except for* $k = r$. *This implies that*

$$[\mathbf{T_o}(j\omega)]_{r*}g_i = g_{ir}[\mathbf{T_o}(j\omega)]_{rr} \tag{26.6.6}$$

Upon using this property in equation (24.7.6) of Theorem 24.1 on page 780, we obtain

$$\frac{1}{\pi} \int_{-\infty}^{\infty} \ln|g_{ir}[\mathbf{T_o}(j\omega)]_{rr}|d\Omega(\eta_o, \omega) = \frac{\ln(|g_{ir}|)}{\pi} \int_{-\infty}^{\infty} d\Omega(\eta_o, \omega)$$

$$+ \frac{1}{\pi} \int_{-\infty}^{\infty} \ln|[\mathbf{T_o}(j\omega)]_{rr}|d\Omega(\eta_o, \omega) \geq \ln(|g_{ir}|) \tag{26.6.7}$$

Then the result follows upon using (24.7.7).

□□□

26.7   The Cost of Decoupling

We can now investigate the cost of dynamic decoupling, by comparing the results in Chapter 24 (namely Lemma 24.4 on page 784) with those in corollary 26.1. To make the analysis more insightful, we assume that the geometric multiplicity μ_z of the zero is 1–i.e., there is only one left direction, h_1, associated with the particular zero.

We first assume that h_1 has more than one element different from zero–i.e., that the cardinality of ∇'_1 is larger than one. We then compare equations (24.8.8) (applicable to a loop with interaction) and (26.6.1) (applicable to a dynamically decoupled MIMO loop). In the first equation, we see that the right-hand side of the inequality can be negative for certain combinations of nonzero off-diagonal sensitivities. Thus, it is feasible to use off-diagonal sensitivities to reduce the lower bound on the diagonal sensitivity peak. This can be interpreted as a two-dimensional sensitivity trade-off, because it involves a spatial as well as a frequency dimension.

A similar analysis can be carried out regarding equations (24.9.2) and (26.6.4). We then observe that in (24.9.2) the lower bound on $||[\mathbf{S_o}(j\omega)]_{rr}||_\infty$ can be made smaller by proper selection of the off-diagonal sensitivity specifications. Thus the sensitivity constraint arising from (24.9.2) can be made softer than that in (26.6.4), which corresponds to the decoupled case.

The conclusion from the above analysis is that it is less restrictive, from the point of view of design trade-offs and constraints, to have an interacting MIMO control loop, compared to a dynamically decoupled one. However, it is a significant fact that to draw these conclusions we relied on the fact that h_1 had more than one nonzero element. If that is not the case, i.e., if only $h_{1r} \neq 0$ (the corresponding direction is canonical), then there is no additional trade-off imposed by requiring a decoupled closed loop.

When h_1 has only one nonzero element, $h_{1r} \neq 0$, it means that the rth sensitivity column is zero, i.e. $[\mathbf{S_o}]_{*r}(z_o) = 0$, and that the remaining columns of $\mathbf{S_o}(z_o)$ are linearly independent. The zero is then associated only with the rth input, $u_r(t)$. On the contrary, if h_1 has more than one nonzero element, then the zero is associated with a combination of inputs.

The other simplification in the above analysis was that $\mathbf{G_o}(s)$ is a full-rank $m \times m$ matrix and $\mathbf{G_o}(z_o)$ is of rank $m - 1$–i.e., the null space associated with the zero at $s = z_o$ is of dimension $\mu_z = 1$. When this constraint is lifted, we might have the favorable situation that the directions $h_1^T, h_2^T, \ldots, h_{\mu_z}^T$ can be chosen in such a way that every one of these directions has only one nonzero element. This suggests that there would be no cost to dynamic decoupling.

We illustrate the above analysis with the following example.

Example 26.6. *Consider the following MIMO system:*

$$\mathbf{G_o}(s) = \begin{bmatrix} \dfrac{1-s}{(s+1)^2} & \dfrac{s+3}{(s+1)(s+2)} \\ \dfrac{1-s}{(s+1)(s+2)} & \dfrac{s+4}{(s+2)^2} \end{bmatrix} = \mathbf{G_{oN}}(s)[\mathbf{G_{oD}}(s)]^{-1}\mathbf{I} \quad (26.7.1)$$

where

$$\mathbf{G_{oN}}(s) = \begin{bmatrix} (1-s)(s+2)^2 & (s+1)(s+2)(s+3) \\ (1-s)(s+2)(s+3) & (s+1)^2(s+4) \end{bmatrix} \quad (26.7.2)$$

$$\mathbf{G_{oD}}(s) = (s+1)^2(s+2)^2 \quad (26.7.3)$$

(i) *Determine the location of RHP zeros and their directions.*

(ii) *Evaluate the integral constraints on sensitivity that apply without enforcing dynamic decoupling, and obtain bounds on the sensitivity peak.*

(iii) *Evaluate the integral constraints on sensitivity that apply if dynamic decoupling is required, and obtain bounds on the sensitivity peak.*

(iv) *Compare the bounds obtained in parts (ii) and (iii).*

Solution

(i) *The zeros of the plant are the roots of* $\det(\mathbf{G_{oN}}(s))$*–i.e., the roots of* $-s^6 - 11s^5 - 43s^4 - 63s^3 + 74s + 44$*. Only one of these roots, namely the one located at* $s = 1$*, lies in the RHP. Thus,* $z_o = 1$*, and*

$$d\Omega(z_o, \omega) = \frac{1}{1 + \omega^2} d\omega \tag{26.7.4}$$

We then compute $\mathbf{G_o}(1)$ as

$$\mathbf{G_o}(1) = \begin{bmatrix} 0 & \frac{2}{3} \\ 0 & \frac{5}{9} \end{bmatrix} \tag{26.7.5}$$

from which it can be seen that the dimension of the null space is $\mu_z = 1$ and the (only) associated (left) direction is $h^T = [5 \ -6]$. Clearly, this vector has two nonzero elements, so we could expect that there will be additional design trade-offs arising from decoupling.

(ii) Applying Theorem 24.2 on page 782 part (ii) for $r = 1$ and $r = 2$, we obtain, respectively

$$\frac{1}{\pi} \int_\infty^\infty \ln |5[\mathbf{S_o}(j\omega)]_{11} - 6[\mathbf{S_o}(j\omega)]_{21}| \frac{1}{1 + \omega^2} d\omega \geq \ln(5) \tag{26.7.6}$$

$$\frac{1}{\pi} \int_\infty^\infty \ln |5[\mathbf{S_o}(j\omega)]_{12} - 6[\mathbf{S_o}(j\omega)]_{22}| \frac{1}{1 + \omega^2} d\omega \geq \ln(6) \tag{26.7.7}$$

If we impose design requirements, as in Lemma 24.4 on page 784, we have, for the interacting MIMO loop, that

$$\|[\mathbf{S_o}]_{11}\|_\infty + \frac{6}{5}\|[\mathbf{S_o}]_{21}\|_\infty \geq \left(\frac{1}{\epsilon_{11} + \frac{6}{5}\epsilon_{21}}\right)^{\frac{\psi(\omega_c)}{\pi - \psi(\omega_c)}} \tag{26.7.8}$$

$$\|[\mathbf{S_o}]_{22}\|_\infty + \frac{5}{6}\|[\mathbf{S_o}]_{12}\|_\infty \geq \left(\frac{1}{\epsilon_{22} + \frac{5}{6}\epsilon_{21}}\right)^{\frac{\psi(\omega_c)}{\pi - \psi(\omega_c)}} \tag{26.7.9}$$

(iii) If we require dynamic decoupling, expressions (26.7.6) and (26.7.7) simplify, respectively, to

$$\frac{1}{\pi} \int_\infty^\infty \ln |[\mathbf{S_o}(j\omega)]_{11}| \frac{1}{1 + \omega^2} d\omega \geq 0 \tag{26.7.10}$$

$$\frac{1}{\pi} \int_\infty^\infty \ln |[\mathbf{S_o}(j\omega)]_{22}| \frac{1}{1 + \omega^2} d\omega \geq 0 \tag{26.7.11}$$

With dynamic decoupling, (26.7.8) and (26.7.9) simplify to

$$\|[\mathbf{S_o}]_{11}\|_\infty \geq \left(\frac{1}{\epsilon_{11}}\right)^{\frac{\psi(\omega_c)}{\pi - \psi(\omega_c)}} \qquad\qquad (26.7.12)$$

$$\|[\mathbf{S_o}]_{22}\|_\infty \geq \left(\frac{1}{\epsilon_{22}}\right)^{\frac{\psi(\omega_c)}{\pi - \psi(\omega_c)}} \qquad\qquad (26.7.13)$$

(iv) *To quantify the relationship between the magnitude of the bounds in the coupled and the decoupled situations, we use an indicator κ_{1d}, formed as the quotient between the right-hand sides of inequalities (26.7.8) and (26.7.12):*

$$\kappa_{1d} \triangleq \left(1 + \frac{6}{5}\lambda_{1\epsilon}\right)^{-\frac{\psi(\omega_c)}{\pi - \psi(\omega_c)}} \qquad where \qquad \lambda_{1\epsilon} \triangleq \frac{\epsilon_{21}}{\epsilon_{11}} \qquad (26.7.14)$$

Thus, $\lambda_{1\epsilon}$ is a relative measure of interaction in the direction from channel 1 to channel 2.

The issues discussed above are captured in graphical form in Figure 26.10.

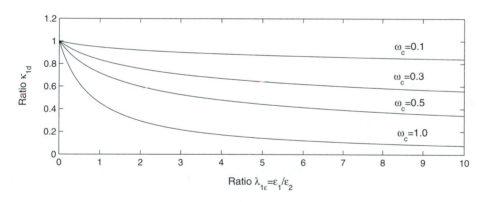

Figure 26.10. Cost of decoupling in terms of sensitivity-peak lower bounds

In Figure 26.10, we show a family of curves, each corresponding to a different bandwidth ω_c. Each curve represents, for the specified bandwidth, the ratio between the bounds for the sensitivity peaks as a function of the decoupling indicator, $\lambda_{1\epsilon}$. We can summarize our main observations as follows.

a) *When $\lambda_{1\epsilon}$ is very small, there is virtually no effect of channel 1 into channel 2 (at least in the frequency band $[0, \omega_c]$); then, the bounds are very close ($\kappa_{1d} \approx 1$).*

b) As $\lambda_{1\epsilon}$ increases, we are allowing the off-diagonal sensitivity to become larger than the diagonal sensitivity in $[0, \omega_c]$. The effect of this manifests itself in $\kappa_{1d} < 1$, i.e. in bounds for the sensitivity peak that are smaller than for the decoupled situation.

c) If we keep $\lambda_{1\epsilon}$ fixed and we increase the bandwidth, then the advantages of using a coupled system also grow.

Also note that the left-hand sides of (26.7.8) and (26.7.12) are different. In particular, (26.7.8) can be written as

$$
\begin{aligned}
\|[\mathbf{S_o}]_{11}\|_\infty &\geq \left(\frac{1}{\epsilon_{11} + \frac{6}{5}\epsilon_{21}} \right)^{\frac{\psi(\omega_c)}{\pi - \psi(\omega_c)}} - \frac{6}{5}\|[\mathbf{S_o}]_{21}\|_\infty \\[2mm]
&\geq \left(\frac{1}{\epsilon_{11} + \frac{6}{5}\epsilon_{21}} \right)^{\frac{\psi(\omega_c)}{\pi - \psi(\omega_c)}} - \frac{6}{5}\epsilon_{21}
\end{aligned}
\tag{26.7.15}
$$

□□□

We see, from the above example, that decoupling can be relatively cost-free, depending upon the bandwidth over which one requires that the closed-loop system operate. This is in accord with intuition, because zeros become significant only when one pushes the bandwidth beyond their locations.

26.8   Input Saturation

Finally, we explore the impact that input saturation has on linear controllers that enforce decoupling. We will also develop anti-wind-up mechanisms that preserve decoupling in the face of saturation, using methods that are the MIMO equivalent of the SISO anti-wind-up methods of Chapter 11.

We assume that our plant is modeled as a square system, with input $u(t) \in \mathbb{R}^m$ and output $y(t) \in \mathbb{R}^m$. We also assume that the plant input is subject to saturation.[3] Then, if $u^{(i)}(t)$ corresponds to the plant input in the ith channel, $i = 1, 2, \ldots, m$, the saturation is described by

$$
u^{(i)}(t) = Sat\langle \hat{u}^{(i)}(t) \rangle \triangleq
\begin{cases}
u^{(i)}_{max} & \text{if } \hat{u}^{(i)}(t) > u^{(i)}_{max}, \\[2mm]
\hat{u}^{(i)}(t) & \text{if } u^{(i)}_{min} \leq \hat{u}^{(i)}(t) \leq u^{(i)}_{max}, \\[2mm]
u^{(i)}_{min} & \text{if } \hat{u}^{(i)}(t) < u^{(i)}_{min}.
\end{cases}
\tag{26.8.1}
$$

For simplicity of notation, we further assume that the linear region is symmetrical with respect to the origin, i.e. $|u^{(i)}_{min}| = |u^{(i)}_{max}| = u^{(i)}_{sat}$, for $i = 1, 2, \ldots, m$. We will describe the saturation levels by $u_{sat} \in \mathbb{R}^m$, where

---

[3]A similar analysis can be developed for input slew limitations.

$$u_{sat} \triangleq \begin{bmatrix} u_{sat}^{(1)} & u_{sat}^{(2)} & \cdots & u_{sat}^{(m)} \end{bmatrix}^T \tag{26.8.2}$$

The essential problem with input constraints, as in Chapter 11, is that the control signal can wind up during periods of saturation.

26.9   MIMO Anti-Wind-Up Mechanism

In Chapter 11, the wind-up problems were dealt with by using a particular implementation of the controller. This idea can be easily extended to the MIMO case, as follows.

Assume that the controller transfer-function matrix, $\mathbf{C}(s)$, is biproper–i.e.,

$$\lim_{s \to \infty} \mathbf{C}(s) = \mathbf{C}_\infty \tag{26.9.1}$$

where $\mathbf{C}_\infty$ is nonsingular. The multivariable version of the anti-wind-up scheme of Figure 11.6 on page 298 is as shown in Figure 26.11.

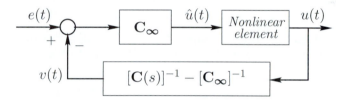

Figure 26.11. Anti-wind-up controller implementation. MIMO case

In the scalar case, we found that the nonlinear element in Figure 26.11 could be thought of in many different ways–e.g., as a simple saturation or as a reference governor. However, for SISO problems, all these procedures turn out to be equivalent. In the MIMO case, subtle issues arise from the way that the desired control, $\hat{u}(t)$, is projected into the allowable region. We will explore three possibilities:

   (i)   simple saturation
   (ii)   input scaling
   (iii) error scaling

(i) Simple saturation. Input saturation is the direct analog of the scalar case and simply requires that (26.8.1) be inserted into as the nonlinear element in Figure 26.11.

(ii) Input scaling. Here, compensation is achieved by scaling down the *controller output vector* $\hat{u}(t)$ to a new vector, $\beta\hat{u}(t)$, every time that one (or more) component of $\hat{u}(t)$ exceeds its corresponding saturation level. The scaling factor, β, is chosen in such a way that $u(t) = \beta\hat{u}(t)$–i.e., the controller is forced to come back just to the linear operation zone. This idea is shown schematically in Figure 26.12.

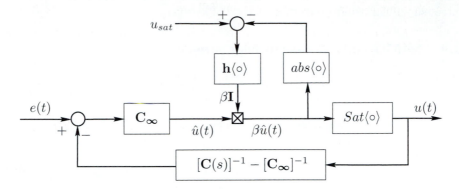

Figure 26.12. Scheme to implement the scaling of controller outputs

(iii) Error scaling. The third scheme is built by scaling the *error vector* down to bring the loop just into the linear region. We refer to Figure 26.13.

Note that $\hat{u}$ can be changed only instantaneously by modifying w_2, because $w_1(t)$ is generated through a strictly proper transfer function. Hence, the scaling of the error is equivalent to bringing w_2 to a value such that $\hat{u}$ is just inside the linear region.

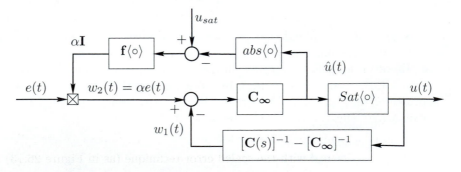

Figure 26.13. Implementation of anti-wind-up via error scaling

In Figure 26.13, the block $f\langle \circ \rangle$ denotes a function that generates the scaling factor $0 < \alpha < 1$. We observe that the block with transfer-function matrix

$[\mathbf{C}(s)]^{-1} - \mathbf{C}_\infty$ is strictly proper, so that any change in the error vector $e(t)$ will translate immediately into a change in the vector $\hat{u}(t)$. Instead of introducing abrupt changes in $e(t)$ (and thus in $\hat{u}(t)$), a gentler strategy can be used. An example of this strategy is to generate α as the output of a first-order dynamic system with unit d.c. gain, time constant τ, and initial condition $\alpha(0) = 1$. The input to this $v(t)$ is generated as follows:

(i) At the time instant that one (or more) of the components of $\hat{u}(t)$ hits a saturation limit, set $v(t) = 0$, where $\alpha(t)$ decreases towards 0 at a speed controlled by τ.

(ii) As soon as all components of $\hat{u}(t)$ are in the linear region, set $v(t) = 1$. Then $\alpha(t)$ grows towards 1 at a speed also determined by τ.

The three techniques discussed above can be summarized in the following graph, which illustrates the case $m = 2$.

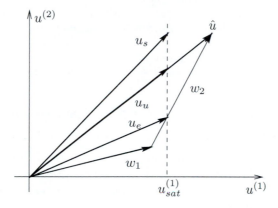

Figure 26.14. Effects of different techniques for dealing with saturation in MIMO systems

In Figure 26.14, we have

$\hat{u}$: raw control signal

u_s : control signal which results from directly saturating $u^{(1)}$

u_u : control signal obtained with the scaled control technique (as in Figure 26.12)

u_e : control signal obtained with the scaled error technique (as in Figure 26.13)

$$(26.9.2)$$

Remark 26.1. *From Figure 26.14, it is evident that, when w_1 lies outside the linear region, then there might be no scaling of the error that brings the control back to the linear region and, at the same time, preserves directionality. In this case, one can revert to one of the other strategies–e.g., simple saturation.*

Remark 26.2. *Note that error scaling is equivalent to introducing a nonlinear controller gain, and none of the above schemes guarantees that the control loop will be stable. In particular, when the plant is open-loop unstable, saturation of the plant input can drive the plant states to values from which it may be impossible to recapture them with the limited control magnitude available.*

We will illustrate the above ideas by a simple example.

Example 26.7. *Consider a MIMO process having the nominal model*

$$\mathbf{G_o}(s) = \frac{1}{(s^2 + 2s + 4)} \begin{bmatrix} -s + 2 & 2s + 1 \\ -3 & -s + 2 \end{bmatrix} \qquad with \qquad \det(\mathbf{G_o}(s)) = \frac{s^2 + 2s + 7}{(s^2 + 2s + 4)^2}$$

$$(26.9.3)$$

For this plant, carry out the following:

(a) *Design a dynamically decoupling controller to achieve a closed-loop bandwidth of approximately 3 [rad/s].*

(b) *Examine what happens if the controller output in the first channel saturates at ±2.5.*

(c) *Explore the effectiveness of the three anti-wind-up procedures outlined above.*

Solution

(a)

Note that this model is stable and of minimum phase. Therefore, dynamic decoupling is possible without significant difficulties.

We are required to design a controller to achieve a closed-loop bandwidth of around 3[rad/s]. Denote by $\mathbf{M}(s)$ the (diagonal) open-loop transfer function, $\mathbf{G_o}(s)\mathbf{C}(s) = \mathbf{M}(s)$. We need to choose $\mathbf{M}(s)$ of minimum relative degree, in such a way that the controller is biproper. We choose

$$\mathbf{C}(s) = [\boldsymbol{\xi}_{\mathbf{L}}(s)\mathbf{G_o}(s)]^{-1}\boldsymbol{\xi}_{\mathbf{L}}(s)\mathbf{M}(s) = [\boldsymbol{\xi}_{\mathbf{L}}(s)\mathbf{G_o}(s)]^{-1}\boldsymbol{\xi}_{\mathbf{L}}(s) \begin{bmatrix} M_1(s) & 0 \\ 0 & M_2(s) \end{bmatrix}$$

$$(26.9.4)$$

where $\boldsymbol{\xi}_{\mathbf{L}}(s)$ is the left interactor for $\mathbf{G_o}(s)$–see section §25.4.1. We then observe that $\mathbf{M}(s)$ must be chosen in such a way that

$$\lim_{s \to \infty} \boldsymbol{\xi}_{\mathbf{L}}(s)\mathbf{M}(s) = \mathbf{K_M} \qquad (26.9.5)$$

with $\mathbf{K_M}$ a bounded nonsingular matrix.

Using the procedure explained in subsection §25.4.1, we find that

$$\boldsymbol{\xi}_{\mathbf{L}}(s) = \begin{bmatrix} s & 0 \\ 0 & s \end{bmatrix} \tag{26.9.6}$$

Thus, the relative degree of $M_1(s)$ and $M_2(s)$ must be 1 to obtain a biproper controller, to satisfy (26.9.1). We also require integration in both channels to ensure zero steady-state error for constant references and disturbances. Say that we choose

$$\mathbf{M}(s) = \frac{2(s+2)}{s(s+1)}\mathbf{I_2} \quad \Longleftrightarrow \quad \mathbf{T_o}(s) = \frac{2(s+2)}{s^2 + 3s + 4}\mathbf{I_2} \tag{26.9.7}$$

This leads to the following controller:

$$\mathbf{C}(s) = \frac{2(s+2)(s^2 + 2s + 4)}{s(s+1)(s^2 + 2s + 7)}\begin{bmatrix} -s+2 & -2s-1 \\ 3 & -s+2 \end{bmatrix} \tag{26.9.8}$$

The design is evaluated by assuming that step references are applied at both channels but at different times. We choose

$$r(t) = \begin{bmatrix} \mu(t-1) & 1.5\mu(t-10) \end{bmatrix}^T \tag{26.9.9}$$

We first run a simulation that assumes that there is no input saturation. The results are shown in Figure 26.15. Observe that full dynamic decoupling has indeed been achieved.

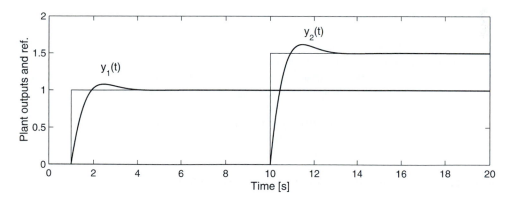

Figure 26.15. Decoupled design in the absence of saturation

(b)

We run a second simulation including saturation for the controller output in the first channel, at symmetrical levels ± 2.5. The results are shown in Figure 26.16.

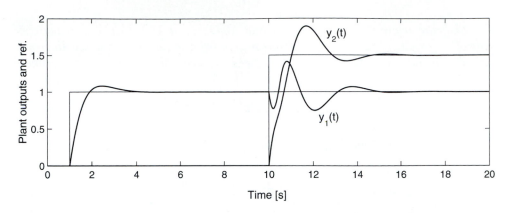

Figure 26.16. Linear decoupled design–saturation in channel 1, at ± 2.5

Clearly, the results are very poor. This is due to wind-up effects in the controller that have not been compensated. We therefore next explore anti-wind-up procedures.

(c)

We examine the three anti-wind-up procedures described above.

(i) Simple saturation *The results of simply putting a saturation element into the nonlinear element of Figure 26.11 on page 883 are shown in Figure 26.17. It can be seen that this is unsatisfactory–indeed, the results are similar to those seen in part (b), where no anti-wind-up mechanism was used.*

(ii) Input scaling *When the idea shown in Figure 26.12 is applied to our example, the control output $u_1(t)$ is automatically adjusted, as shown in Figure 26.18. In this figure, $u_1(t)$ and $\boldsymbol{\eta}(t)$ are depicted. Note that $0 < \beta < 1$ over the time interval when the control signal $\hat{u}_1(t)$ is beyond the saturation bound (in this example, -2.5).*

*However, a rather disappointing result is observed regarding the plant outputs. They are shown in Figure 26.19. The SIMULINK schematic can be found in file **mmawu.mdl**.*

The results shown in Figure 26.19 show a slight improvement over those obtained by using the pure anti-wind-up mechanism (i).

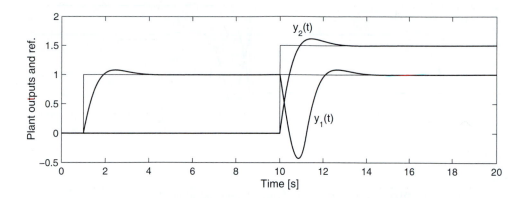

Figure 26.17. Decoupled linear design with saturation in channel 1 and anti-wind-up scheme (i)

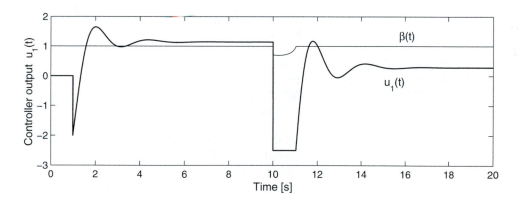

Figure 26.18. Controller output (channel 1) when using control scaling

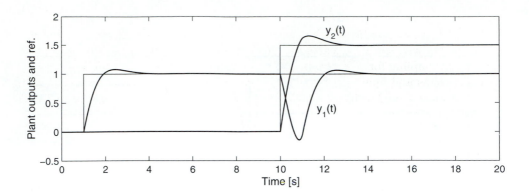

Figure 26.19. Plant outputs when using control scaling

(iii) Error scaling *When the error-scaling strategy is applied to our example, we obtain the results shown in Figure 26.20.*

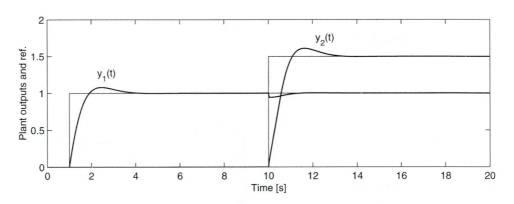

Figure 26.20. Plant outputs when using scaled errors

*In the case shown in Figure 26.20, a value of 0.1 was chosen for τ. The results are remarkably better than those produced by the rest of the strategies treated so far. The SIMULINK schematic can be found in file **mmawe.mdl**. Actually, full dynamic decoupling is essentially retained here–the small coupling evident in Figure 26.20 is due to the implementation of the error scaling via a (fast) dynamical system.*

□□□

In this example, we found that the anti-wind-up mechanism incorporating error scaling proved the most effective. Indeed, it was shown that it preserved full dynamic decoupling for reference-signal changes. We have tested the idea on many other problems and found it to give excellent results in almost all cases.

Further ideas on MIMO saturating actuators are explored in the references at the end of the chapter.

26.10 Summary

- Recall these key closed-loop specifications shared by SISO and MIMO design:

  - continued compensation of disturbances
  - continued compensation of model uncertainty
  - stabilization of open-loop unstable systems

  whilst not

  - becoming too sensitive to measurement noise
  - generating excessive control signals

  and accepting inherent limitations due to

  - unstable zeros
  - unstable poles
  - modeling error
  - frequency- and time-domain integral constraints

- Generally, MIMO systems also exhibit additional complexities due to

  - directionality (several inputs acting on one output)
  - dispersion (one input acting on several outputs)
  - and the resulting phenomenon of coupling

- Designing a controller for closed-loop compensation of this MIMO coupling phenomenon is called *decoupling*.

- Recall that there are different degrees of decoupling, including the following:

  - static (i.e., $\mathbf{T_o}(0)$ is diagonal);
  - triangular (i.e., $\mathbf{T_o}(s)$ is triangular); and
  - dynamic (i.e., $\mathbf{T_o}(s)$ is diagonal).

- Due to the fundamental law that $\mathbf{S_o}(s) + \mathbf{T_o}(s) = \mathbf{I}$, if $\mathbf{T_o}$ exhibits any of these decoupling properties, so does $\mathbf{S_o}$.

- The severity and types of the trade-offs associated with decoupling depend on

    ○ whether the system is of minimum phase;

    ○ the directionality and cardinality of nonminimum-phase zeros;

    ○ unstable poles.

- If all of a system's unstable zeros are canonical (their directionality affects one output only), then their adverse effect is not spread to other channels by decoupling, provided that the direction of decoupling is congruent with the direction of the unstable zeros.

- The price for dynamically decoupling a system having noncanonical nonminimum-phase zeros of simple multiplicity is that

    ○ the effect of the nonminimum-phase zeros is potentially spread across several loops; and,

    ○ therefore, although the loops are decoupled, each of the affected loops needs to observe the bandwidth and sensitivity limitations imposed by the unstable zero dynamics.

- If one accepts the less stringent triangular decoupling, the effect of dispersing limitations due to nonminimum-phase zeros can be minimized.

- Depending on the case, a higher cardinality of nonminimum-phase zeros can either enforce or mitigate the adverse effects.

- If a system is also open-loop unstable, there may not be any way at all to achieve full dynamic decoupling with a one-d.o.f. controller, although it is always possible with a two-d.o.f. architecture for reference-signal changes.

- If a system is essentially linear but exhibits such actuator nonlinearities as input or slew-rate saturations, then the controller design must reflect this appropriately.

- Otherwise, the MIMO generalization of the SISO wind-up phenomenon can occur.

- MIMO wind-up manifests itself in two aspects of performance degradation:

    ○ transients due to growing controller states; and

    ○ transients due to the nonlinearity impacting on directionality.

- The first of these two phenomena . . .

    . . . is in analogy to the SISO case.

> ... is due to the saturated control signals not being able to annihilate the control errors sufficiently fast compared to the controller dynamics; therefore the control states continue to grow in response to the nondecreasing control. These *wound up* states produce the transients when the loop emerges from saturation.

> ... can be compensated by a direct generalization of the SISO anti-wind-up implementation.

- The second phenomenon ...

  > ... is specific to MIMO systems.

  > ... is due to uncompensated interactions arising from the input vectors losing its original design direction.

- In analogy to the SISO case, there can be regions in state space from which an open-loop unstable MIMO system with input saturation cannot be stabilized by any control.

- More severely than in the SISO case, MIMO systems are difficult to control in the presence of input saturation, even if the linear loop is stable and the controller is implemented with anti-wind-up. This is due to saturation changing the directionality of the input vector.

- This problem of preserving decoupling in the presence of input saturation can be addressed by anti-wind-up schemes that scale the control *error* rather than the control *signal*.

26.11   Further Reading

Dynamic decoupling

Desoer, C. and Gündes, A. (1986). Decoupling linear multi-input multi-output plants by dynamic output feedback. An algebraic theory. *IEEE Transactions on Automatic Control*, 31(8):744-750.

Falb, P. and Wolowich, W. (1967). Decoupling in the design and synthesis of multivariable control system. *Automatica*, 12:651-669.

Gilbert, E.G. (1969). The decoupling of multivariable systems by state feedback. *SIAM Journal of Control and Optimization*, 7(1):50-63.

Goodwin, G.C., Feuer, A., and Gómez, G. (1997). A state space technique for the evaluation of diagonalizing compensator. *Systems and Control Letters*, 32(3):173-177.

Hammer, J. and Khargonekar, P.P. (1984). Decoupling of linear delay equations. *Journal Mathematical System Theory*, pages 135-137.

Hautus, M. and Heymann, M. (1983). Linear feedback decoupling–transfer function analysis. *IEEE Transactions on Automatic Control*, 28(8):823-832.

Lin, C-A. and Hsie, T.F. (1991). Decoupling controller design for linear multivariable plants. *Automatica*, 36:485-489.

Morse, A. and Wonham, W. (1973). Status of noninteracting control. *IEEE Transactions on Automatic Control*, 16:568-581.

Williams, T. and Antsaklis, P. (1986). A unifying approach to the decoupling of linear multivariable systems. *International Journal of Control*, 44(1):181-201.

Wonham, W. (1985). *Linear Multivariable Control: A Geometric Approach*. Springer-Verlag, 3rd edition.

Decoupling invariants

Commault, C., Descusse, J., Dion, J.M., Lafay, J., and Malabre, M. (1986). New decoupling invariants: the essential orders. *International Journal of Control*, 44(3):689-700.

Dion, J. M. and Commault, C. (1988). The minimal delay decoupling problem: feedback implementation with stability. *SIAM Journal of Control and Optimization*, 26(1):66-81.

Gündes, A. (1990). Parameterization of all decoupling compensators and all achievable diagonal maps for the unity-feedback system. In *Proceedings of the 29th CDC, Hawaii*, pages 2492-2493.

Lin, C-A. (1995). Necessary and sufficient conditions for existence of decoupling controllers. *IEEE Transactions on Automatic Control*, 42(8):1157-1161.

Cost of decoupling

Gómez, G. and Goodwin, G.C. (1996). Integral constraints on sensitivity vectors for multivariable linear systems. *Automatica*, 32(4):499-518.

26.12   Problems for the Reader

Problem 26.1. *A discrete-time MIMO system has a nominal model with transfer function* $\mathbf{G_q}(z)$, *where*

$$\mathbf{G_q}(z) = \frac{1}{(z - 0.7)(z - 0.9)} \begin{bmatrix} z - a & -0.5 \\ 0.5 & z \end{bmatrix} \qquad (26.12.1)$$

26.1.1 *For* $a = 1$, *build a decoupled loop such that the closed-loop poles are located inside the circle with radius 0.4.*

26.1.2 *Repeat for* $a = 2$.

Problem 26.2. *Consider a* **stable** *MIMO plant having nominal transfer function* $\mathbf{G_o}(s)$.

26.2.1 *If* $\mathbf{G_o}(s)$ *is of minimum phase, discuss the feasibility of designing a controller such that the input sensitivity* $\mathbf{S_{io}}(s)$ *is diagonal.*

26.2.2 *Repeat your analysis for the case when* $\mathbf{G_o}(s)$ *is of nonminimum phase.*

Problem 26.3. *Consider a MIMO system having the model*

$$\mathbf{G_o}(s) = \frac{2}{(s+1)^2(s+2)} \begin{bmatrix} 2(-s+3) & 0.5(-s+3) & (-s+\alpha) \\ 0.5(s+1) & -\beta & -(s+2) \\ -1 & 0.5 & 2.5 \end{bmatrix} \qquad (26.12.2)$$

with $\alpha = -2$ *and* $\beta = 1$. *It is desired to achieve dynamic decoupling.*

26.3.1 *Is it possible without spreading the nonminimum-phase zeros in to the three channels?*

26.3.2 *If* $\alpha = 3$ *and* $\beta = -1$, *are your conclusions still valid?*

Problem 26.4. *Consider again the plant of problem 26.3.*

26.4.1 *Repeat problem 26.3, but aiming for a triangular (lower and upper) design.*

26.4.2 *Discuss the issue of dynamic decoupling cost when* $\beta = -1$ *and* α *has an uncertain value around* $\alpha = 3$. *(This is a hard problem.)*

Problem 26.5. *Consider a process with the nominal model*

$$\mathbf{G_o}(s) = \frac{1}{(s+1)(s+2)} \begin{bmatrix} 2 & -1 \\ 0.5 & s+2 \end{bmatrix} \qquad (26.12.3)$$

Design a Q − controller to achieve dynamic decoupling and zero steady-state error for constant references and step disturbances.

Problem 26.6. *Discuss the difficulties of achieving dynamic decoupling for input disturbances for stable (although not necessarily minimum-phase) plants.*

Problem 26.7. *Consider a plant with the same nominal model as in problem 26.5. Design a digital controller, assuming that the sampling rate is $\Delta = 0.1[s]$.*

Name Index

Ahlen, A., 604, 605
Anderson, B., 382, 733
Antsaklis, P., 891
Araki, M., 397
Åström, K., 39, 175, 192, 348, 381, 382, 452
Athani, V., 802
Athans, M., 733

Badgwell, T., 764
Balakrishnan, V., 484
Banach, 593
Barratt, C.H., 484, 803
Bastiani, A., 647
Bellman, R., 733, 763
Bensoussan, D., 671
Bernstein, D., 307
Bittanti, S., 734
Black, H.W., 19
Blanke, M., 765
Bode, H., 19, 260
Bohlin, T., 60, 484
Bongiorno, J., 452, 483
Boyd, S.P., 484, 803
Braslavsky, J.H., 260, 397, 484, 485, 555, 605
Bristol, E., 671
Brockett, R., 152, 522
Brown, J.W., 260
Bryant, G., 235
Bryson, A., 733
Bucy, R., 734

Cadzow, J., 348
Callier, F., 646
Campbell, D.P., 60
Campo, P., 307
Cannon, R., 60
Chan, S., 734
Chen, C-T., 521, 555
Chen, C.C., 764
Chen, J., 802
Chen, T., 397
Christensen, A., 765
Churchill, R.V., 260
Clark, M., 235
Commault, C., 892
Crisafulli, S., 175, 605
Cutler, C., 763

D'Azzo, J., 39
Dahleh, M., 803
Daoutidis, P., 308
Davison, E.J., 484, 847
De Doná, J.A., 308, 764
De Souza, C., 734
Dendle, D., 235
Descusse, J., 892
Desoer, C., 452, 605, 646, 846, 891
Diaz-Bobillo, I., 803
Dion, J.M, 892
Dion, J.M., 892
Distefano, J., 109
Doeblin, E.O., 39
Doetsch, G., 108
Dorato, P.E., 483

897

Subject Index